ADVANCES IN AI FOR BIOMEDICAL INSTRUMENTATION, ELECTRONICS AND COMPUTING

This book contains the proceedings of 5th International Conference on Advances in AI for Biomedical Instrumentation, Electronics and Computing (ICABEC - 2023), which provided an international forum for the exchange of ideas among researchers, students, academicians, and practitioners.

It presents original research papers on subjects of AI, Biomedical, Communications & Computing Systems. Some interesting topics it covers are enhancing air quality prediction using machine learning, optimization of leakage power consumption using hybrid techniques, multi-robot path planning in complex industrial dynamic environment, enhancing prediction accuracy of earthquake using machine learning algorithms and advanced machine learning models for accurate cancer diagnostics.

Containing work presented by a diverse range of researchers, this book will be of interest to students and researchers in the fields of Electronics and Communication Engineering, Computer Science Engineering, Information Technology, Electrical Engineering, Electronics and Instrumentation Engineering, Computer applications and all inter-disciplinary streams of Engineering Sciences.

PROCEEDINGS OF THE 5TH INTERNATIONAL CONFERENCE ON ADVANCES IN AI FOR BIOMEDICAL INSTRUMENTATION, ELECTRONICS AND COMPUTING (ICABEC 2023), GHAZIABAD, INDIA, 22–23 DECEMBER 2023

Advances in AI for Biomedical Instrumentation, Electronics and Computing

Edited by

Vibhav Sachan
KIET Group of Institutions, Delhi-NCR, Ghaziabad, Uttar Pradesh, India

Shahid Malik
Indian Institute of Technology, Delhi

Ruchita Gautam and Parvin Kumar
KIET Group of Institutions, Delhi-NCR, Ghaziabad, Uttar Pradesh, India

CRC Press
Taylor & Francis Group
Boca Raton London New York Leiden

CRC Press is an imprint of the
Taylor & Francis Group, an **informa** business

A BALKEMA BOOK

First published 2024
by CRC Press/Balkema
4 Park Square, Milton Park, Abingdon, Oxon, OX14 4RN

and by CRC Press/Balkema
2385 NW Executive Center Drive, Suite 320, Boca Raton FL 33431

CRC Press/Balkema is an imprint of the Taylor & Francis Group, an informa business

British Library Cataloguing-in-Publication Data
A catalogue record for this book is available from the British Library

Library of Congress Cataloging-in-Publication Data
A catalog record has been requested for this book

ISBN: 978-1-032-64298-7 (hbk)
ISBN: 978-1-032-64474-5 (pbk)
ISBN: 978-1-032-64475-2 (ebk)

DOI: 10.1201/9781032644752

Typeset in Times New Roman
by MPS Limited, Chennai, India

Table of Contents

Preface xiii

About the Conference xv

Ensemble of diverse pre-trained convolutional neural networks for enhanced
diabetic retinopathy image classification 1
S.K. Feroz Ahmed, Y. Yaswanth Chowdary, Jyostna Devi Bodapati &
Lakshmana Kumar Yenduri

Enhancing air quality prediction using machine learning algorithms and
boosting techniques 7
Rushvi Sajja, S.K. Feroz Ahmed, Jyostna Devi Bodapati &
Lakshmana Kumar Yenduri

Predicting missed appointments in dental clinic: Enhanced risk assessment
with multiple learning techniques 13
Jeffin Joseph, S. Senith, A. Alfred Kirubaraj & S.R. Jino Ramson

Optimization of leakage power consumption in D Flip-Flop using hybrid
technique 19
Amiya Prakash & Priyanka Jain

YOLO based road anomaly detection for Indian roads 25
Jyoti Tripathi, Bijendra Kumar, Karan Ahlawat, Sanskriti Jain &
Ashwani Chauhan

Job recommendation system based on content-based filtering 30
Anu Saini, Jyoti Tripathi, Mohd Faraz & Mohd Kaif

Wondervision: A smart stick for visually impaired people 36
Sunita Tiwari, Manjeet Singh Pangtey, Golu Kumar, Raghav K. Jha &
Sambhav Anand

C band application-oriented H-shaped patch antenna design 42
B. Jeyapoornima & J. Joselin Jeya Sheela

Microcavity based 2D Photonics crystal biosensor for detection of
COVID-19 47
Vijay Laxmi Kalyani & Virendra Kumar Sharma

Prognosis of heart disease by utilizing data mining techniques 52
Sartaj Ahmad, Vaishnavi Awasthi, Srishti Chaurasia & Ajay Agarwal

LSTM model for stock price prediction 57
Nissankara Lakshmi Prasanna, Rajesh Babu Yallamanda,
Rama devi Gunnam & T. Kameswara Rao

A systematic study of image/data sharing security methods using Modified
Advanced Encryption standards (MAES) 63
B. Satyaramamanohar A., T. Bernatin & P.S.D. Anvesh

Compact-size frequency reconfigurable antenna for sub-6 GHz 5G applications 68
Sachin Kumar, Himanshu Nagpal, Ghanshyam Singh & L. Harlan

ASIC acceleration of floating-point arithmetic: Empowering fast and
reliable computations 73
Kapileswarreddy Siddireddy & Ganapathi Hegde

Design and comparative analysis of inverter based OTA using FinFET
and CNTFET 79
Mohammad Aleem Farshori & M. Nizamuddin

Automatic room light controller and visitor counter using microcontroller 84
*Aman Kumar Yadav, Amrit Raj Singh, Amit Kumar, Amit Kumar,
Amritanshu Singh & Niraj Singh Mehta*

Investigation of DDMZM and LiNb-MZM based radio over fiber link
for a future-oriented communication system 89
*Balram Tamrakar, Shubham Shukla, Gauri Brijaria, Manjari Singh,
Manu Gupta, Shweta Sharma & Mayank Kashyap*

Medical assistance bot: DOBO 95
*Sulekha Saxena, Shobhit Singh Rawat, Yashdeep Tyagi, Yash Rajput &
Suyash Kumar Srivastava*

Performance analysis of WDM-FSO system using RAMAN-EDFA and
MIMO for light rain 99
Abhinav Pratap Shahi, Abhishek Tripathi, Ankit Kumar & R.L. Yadava

Optimization of 60 GHz MIMO OFDM radio over fiber system for next
generation networks2 106
*Balram Tamrakar, Mohd Sharique Alam, Shubham Verma, Ujjwal Tyagi,
Mohd Khalid & Mohd Salman Zaidi*

Review on multi-robot path planning in complex industrial dynamic
environment using improved nature inspired techniques 111
Subhash Yadav, Shubham Shukla & Ritu Tiwari

Network pharmacology and molecular docking approach to unveil the
mechanism of amantadine for Parkinson's disease 116
*Nimesh Bhardwaj, Himani Tyagi, Siddhi Gupta, Pooja Gangwar,
Shubham Chaudhary, Mansi Singh, Vinay Kumar & Praveen Kr. Dixit*

Development of smart and secure billing mechanism for stores 121
*Himanshu Chaudhary, Hitesh Gupta, Kunal Gupta, Naman Srivastava,
Niranjan Kalra, Raghunandan Bajaj, Praveen Kumar & Abhishek Sharma*

Food restaurant management system 127
*Vinay, Shivam Tyagi, Shivam Bhardwaj, Shantanu, Sachin Tyagi,
Mohit Tyagi & Satya Prakash Singh*

Tobacco use among the youth: Prediction, prevalence and prevention 133
*Pravesh Singh, Rakshit Srivastava, Adarsh Singh, Pratishtha Srivastava &
Harsh Kansal*

Automated car parking indicator system 137
Hunny Pahuja, Rohit Verma, Abhishek Pandey, Akash Singh & Satyansh Kumar

Deep learning solution for real-time violence detection in video streams   142
Sachin Tyagi, Shivansh Tyagi, Varun Bagga, Sourish Bansal,
Siddharth Goswami, Yash Verma, Mohit Tyagi & Satya Prakash Singh

Disease diagnosis with instant first aid and medicine recommender   147
Akanksha Shukla, Shivani Agarwal, Nandita Goyal &
Kartikeya Patel

Versatile 4-bit signed binary multiplier for complex digital circuits   153
Deepti Maurya & Uma Sharma

Path finding visualizer   158
Ankit, Anushka, Akash, Shahid, S. Ranjan & Richa Srivastava

Exploring smartphone-driven indoor localization systems: A systematic
literature review   163
Prashant Kumar, Chandan Kumar, Gaurav Jain & Aakanshi Gupta

Enhancing sensing and prediction accuracy of earthquake using machine
learning algorithms   171
Surbhi Bharti, Priya Pahwa, Sakshi Gupta & Ashwni Kumar

Intelligent smart-watch for enhanced cardiovascular health management   177
S.T. Rama, S. Heeravathi, R. Ragul & N. Shanmugasundaram

Advancements in robotics and prosthetics: A comprehensive study of
voice-controlled hand prosthetic gloves   182
Ruchita Gautam, Himanshu Chaudhary, Apoorv Sharma, Gautam Matta,
Ayush Jain, Aryan Sharma & Adhyan Kaushik

A review on analysis and prediction methodologies of crop
yield pattern   187
Neha, Anil Ahlawat & Himanshu Chaudhary

A review paper on comparative analysis of various reconfiguration
techniques under partial shading environments   192
Pushpender, S. Khatoon & M.F. Jalil

Novel memristor-CMOS based domino self-resetting half-adder design
for fast and low-power biomedical applications   197
Monica Gupta, Kirti Gupta & M. Bhutani

Prediction of pulmonary arterial hypertension from HRCT chest in
post-COVID patients using double attention U-NeT model   204
Azra Nazir, Roohie Naaz, Shaima Qureshi & Nidha Nazir

Comprehensive review on impact of nanosatellites on human medical
infrastructure   209
Abhishek Sharma, Suryansh Dev, Shambhavi Kumari, Riya Jain,
Satyam Kumar Jha, Mohit Kumar, Praveen Kumar & Ruchita Gautam

Onboard compression and preprocessing methods for LEO
satellite imagery: A review   213
Abhishek Sharma, Nidhi Singh, Yashvardhan Srivastava, Shubhi Sharma,
Krishna Pratap Singh, Harsh Jaiswal, Praveen Kumar & Parvin Kumar

Satellite security: Navigating threats and implementing safeguards in
the modern space age   218
Abhishek Sharma, Divyanshi Srivastava, Suryanshi Singh, Ira Nafees,
Satvik Aggarwal, Himanshu Chaudhary & Praveen Kumar

UAV meets VANET: A hybrid model for performance enhancement
in smart cities   222
Hunny Pahuja, Manoj Sindhwani & Shippu Sachdeva

Vending machine for medicine using automated arm mechanism   228
Pankaj Kumar, Kaushal Saraf, Prakhar Nigam, Rishi Yadav,
Kartik Verma & Divya Sharma

Implementation and design of traffic light controller using Verilog   232
Kavya Kumar Tayal, Ajay Suri & Rohit Vikram Singh Bhadauria

Design and implementation of Vedic multiplier   237
Khushi Mishra & Ajay Suri

Smart lifesaving helmet for coal miners   242
Vipin Kumar Verma, Daksh Singhal, Deepak Kr Bari,
Anuj Kumar Chaudhary, Dev Gupta & Anurag Dubey

Beyond the ballot box: Crafting a resilient voting system with
MySQL and PHP   248
Abhas Kanungo, Ananya Pandey, Asthana Somya Subhash,
Anoushka Bharti & Alok Kumar

Enhancing decision-making for breast cancer through advanced machine
learning and data analytics   253
Sahil Aggarwal, Priyanshu Aggarwal, Satyam Chauhan, R. Dheivanai &
Ritu Pahwa

Beyond borders: Redefining dynamics of nativeness and migration on
cyberspace   259
Leanora Pereira Madeira

PlaceTech: An AI-enabled solution for smart placement   264
Khalid Alfatmi, Sakshi Pande, Makarand Shahade, Vaibhavi Suryawanshi,
Kalpesh Badgujar, Vishwajit Patil & Mayuri Kulkarni

Credit card fraud detection with formula-based authentication   270
P. Asha, Bhavya Babu, S. Prince Mary, B. Ankayarkanni,
M.D. Anto Praveena & A. Christy

Analysis of various topic modeling algorithms using internal-quality metrics 276
Astha Goyal & Indu Kashyap

Cancer prediction: A comparative analysis of advanced machine learning
models for accurate diagnostics   283
Greeshma Arya, Maithili Singh & Abha Bhardwaj

Comprehensive study of reviews from social media using machine learning
techniques   288
Rohini G. Khalkar, M.S. Bewoor, Sampat P. Medhane & Gauri Rao

ArthritisCare: Empowering wellness through personalized arthritis detection
and physiotherapy exercise recommendation 294
Vijaylaxmi Bittal, Makarand Shahade, Isha Wagh, Pratiksha Yeshi,
Pritam Lokhande, Hindraj Patil & Khalid Alfatmi

Rights reach: AI-powered legal assistance for the physically challenged 300
Vijaylaxmi Bittal, Sakshi Pingale, Makarand Shahade, Manish Patil,
Ketaki Patil, Manashri Patil & Khalid Alfatmi

Simplifying legal language: An AI-powered approach to enhance
document accessibility 306
Kiran Somwanshi, Khalid Alfatmi, Ashwini Vibhandik, Darshana Karbhari,
Gagan Jarsodiwala, Rahul Relan & Makarand Shahade

Fuel theft detection system with SMS alert using microcontroller 311
Vipin Kumar Verma, Nilesh Yadav, Pankaj Gupta, Prateek Dhar Dubey,
Purushottam Mani Tripathi & Vikram Vishwakarma

Unmasking attacker identity behind the VPN 316
A.S. Awate, B.N. Nandwalkar, M.R. Shahade, D.B. Mali, H.V. Patil,
H.R. Waghare & H.R. Patil

e-Nidan: Autism spectrum disorder detection using machine learning 322
Ashish Awate, Krutika Yeola, Makarand Shahade, Vaishnavee Patil,
Mayuri Vispute, Hemshri Amrutkar & Bhushan Nandwalkar

Real-time automated fabric defect detection system 327
Umakant Mandawkar, Makarand Shahade, Samruddhi Wadekar,
Chetan Kachhava, Yash Patil & Sakshi Mandwekar

Comprehensive analysis of communication quality: Signal to noise ratio
with respect to bit error rate for nanosatellite beacon 333
Vidushi Pandey, Ayush Singh, Abhishek Sharma, Gati Saraswat, Amrita Singh,
Ayush Yadav & Ruchita Gautam

Single page optimization techniques using react 338
Abhishek Pokhriyal, Saurav Pratihasta, Shubh Kansal, Vasu Goyal,
Shourya Singh & Shruti Mishra

DAM: Drone Automation and Mapping 343
Pranjal Agarwal, Prajesh Pratap Singh & Sharad Gupta

Development of an advanced, cost-effective prosthetic limb with an
EMG sensor 348
Shivam Kesarwani, Shrey Shekhar, Shruti Mishra, Abhinav Singh, Somya Tyagi,
S.P. Singh, Neelesh Ranjan Srivastava, Sachin Tyagi & Mohit Tyagi

Detection of disease and appointment to the Doctor 353
Yashasvi Singh, Sanskar Gupta, Rituz Gupta, Riya Tyagi & Amit Kumar

Image encryption with switching effects 356
Sunil Sriharsha Gudimella, Umesh Ghanekar & Kundan Kumar

Design of fuzzy based PID controller for effective delivery of
syringe pump 362
Saba Parveen, Munna Khan & Kashif Sherwani

An innovative hybrid full adder design for low-power VLSI circuit applications 368
A. Sharma, N.S. Singha, R. Yadav & A. Kumar

Performance evaluation of full adder cells implemented in CMOS technology 374
Yash Pathak & Dharmendra Kumar Jhariya

Code converter realization in IoT by TCP/IP network layer through node MCU and LabVIEW 380
G. Dhanabalan, H.B. Michael Rajan & R. Ashok

Prediction of spam reviews using feature-driven opinion mining deep learning model 385
Surya Prakash Sharma, Laxman Singh, Nagesh Sharma, Abdul Khalid & Rajdev Tiwari

Design and implementation of automatic street light systems 391
Mohit Tyagi, Kanishka Chauhan, Hardik Mitra, Raman Pundhir, Naman Gupta, Sachin Tyagi & Satya Prakash Singh

Deep learning based computerized diagnosis of breast cancer using digital mammograms 396
Laxman Singh, Rekha Kashyap, Sovers Singh Bisht, Nagesh Sharma & Surya Prakash Sharma

Shoe extension using ultrasonic sensor and gyroscope for blinds 402
Shikha Agarwal, Aarti Chaudhary, Veena Bharti & Shivam Umrao

Dynamic power allocation technique for IOT applications in mobile edge computing 407
Ashvini Joshi, Anjulata Yadav & Amit Naik

Review on smart landmine and landmine detection 411
Hans Kumar, Abhas Kanungo, Kartik Chaudhary, Jatin Tomar, Priyanshu & Harshit Yadav

Design of multiplexer in 90nm technology using energy recovery logic circuit 416
Sitaram Kumar, Amit Kumar & Dharmendra K. Jhariya

Blockchain in education: A revolutionary paradigm for enhanced security and transparency 422
Shikha Agarwal, Aarti Chaudhary, Komal Shivhare & Ashish Bajpai

Analysis of different categories of prediction methods in intelligent transport VANET system 429
R. Gracelin Sheeba & N. Edna Elizabeth

CNN sight: Precision detection in gangrene diagnostics 436
Priyanshu Aggarwal, Sahil Aggarwal, Harshita & Ritu Pahwa

Implementation of crowbar protection in DFIG 443
Farhat Nasim, Shahida Khatoon, Ibraheem & Mohammad Shahid

A simplified low-cost portable ventilator design 450
Mohd Shadaab, Shahida Khatoon & Mohammad Shahid

Generating maximum power in photovoltaic systems using HHO-based
embedded controllers   455
T.P. Sujithkumar, Shanmugasundaram, V. Rajendran & Debarchita Mishra

5G Network – Deployment, status and roadmap in Indian telecom
ecosystem   461
Manu Srivastava

Iteration-based reduction in cell population for biomedical applications   466
Ashutosh Mishra & Piyush Kumar Tripathi

Unraveling the power of AI assistants   473
*Abhinav Karn, Prashant Kumar Singh, Chirag Agarwal, Ayush Verma,
Deepak Singh & Mupnesh Kumari*

Retinal diseases analysis and detection – A comprehensive review   480
P. Renuka, V. Sumitra, P. Latha & K. Swaminathan

MediSafe – Enhancing secure medical data management and doctor–patient
communication   486
*Neha Rajas, Hrishikesh Potnis, Chinmayee Prabhu, Dnyaneshwari Pote,
Payal Powar & Sujal Powar*

Face detection attendance system in Artificial Intelligence   491
*Simran Kaur Arora, Priyanka Behki, Gourav Batar, Vivek Tiwari &
Siya Jindal*

Smart helmet for bike riders   497
*Devansh Gupta, Sanchit Jain, Agrim Chauhan, Tanay Srivastava,
Abhinav Saini, Praveen Kumar, Himanshu Chaudhary & Abhishek Sharma*

Performance investigation of an improved high speed WDM RoFSO link in
foggy and rainy weather   502
Kamaldeep Kaur & Abhimanyu Nain

Automated plant monitoring system with WebCAM and shadow shelter   507
*Sulekha Saxena, Ritik Kumar, Harsh Singh, Rishabh Kumar,
Mohd. Harmain & Meet Choudhary*

Social media news verifier   511
Ritik Rana, Rohit Jha, Saksham Singh, Sarthak Choudhary & Chirag Arora

Review of performance parameters of PV array based on different
configurations operating in mismatch scenarios   516
Aisha Naaz, Mohd Faisal Jalil, Shahida Khatoon & Pushpender

Correlates of integrated marketing communication with respect to banking
industry in India   522
P. Jain, A. Saihjpal, N. Aggarwal & A. Kaur

Non-contact temperature detection system   527
*Vikas Nandeshwar, Devang Bissa, Sarthak Biyani, Darshan Biradar,
Bilal Khan, Pratham Bisen & Saif Bichu*

A review of humidifier for healthcare   534
*Garima Bhargava, Jassi Sandhu, Lakshmi Tiwari, Lokender Singh,
Shreya Dubey, Parvin Kumar & Vipin Kumar*

Design of ECG monitoring system using NI LabVIEW 538
Shivansh Sinha, Shraddha Tripathi, Sumit Srivastava, Shubhi Sharma,
Krishna Pratap Singh, Parvin Kumar, Abhishek Sharma & Vipin Kumar

High-performance dual band graphene slotted antenna for terahertz
applications 543
R. Yadav, S. Sood & V.S. Pandey

Australian wildfire visualization 549
Manjinder Kaur, Roop Lal, Ankur Pandey, Anurag Singh & Dipak Rajbhar

Live code sync 554
Aditya Kumar & Er. Manjinder Kaur

Farm automation using NodeMCU 560
Vikas Nandeshwar, Ishawar Borade, Atharva Borade, Atharva Bonde,
Tanmay Bora, Om Bobade & Vishal Bokare

Multiple disease prediction using machine learning algorithms 567
Parth Dayal, Deepansh Sharma, Aman Agarwal, Himanshu Chaudhary,
Ruchita Gautam, Praveen Kumar & Abhishek Sharma

Helping hand for handicap (triple h) 573
Sumit Sharma, Shagun Kumar, Shri Bihari Singh, Satyam Singh &
Divya Sharma

Single-input voltage mode differentiator using DDCCTA and grounded
passive elements 577
Priyanka Jain & Chandra Shekhar

Discrete hartley transform using recursive algorithm 583
Vivek Singh, Dhwani Kaushal & Priyanka Jain

Experimental study on variants of Gaussian mixture model for segmentation 589
Sanjeev Kumar Katti, Shrinivas D. Desai, Vishwanath P. Baligar &
Gururaj N. Bhadri

Real-time-abuse detection model 595
Ayush Kumar, Aryan Nigam, Aradhana Tripathi, Aftab Khan,
Nigam Kumar Mishra & Rochak Bajpai

Comparative analysis of machine learning models for sentiment analysis on
X (twitter) dataset 600
Akshat Singh, Akanksha Singh, Anisha Kumari, Aryan Chauhan &
Richa Srivastava

Design of a 1-bit full adder in hybrid logic for high end computing in
biomedical instrumentation 607
A. Tomar, V.K. Sachan, J. Kandpal, N. Singh, P. Chauhan & S. Bhandari

Author index 613

Preface

Dear Delegates,

It is with great pleasure that we extend a warm welcome to all of you on behalf of the editorial team as the Chief Editor of this prestigious 5th International Conference on **Advances in AI for Biomedical Instrumentation, Electronics and Computing (ICABEC 2023)**. This will provide a forum for researchers around the globe to explore and discuss various aspects of Biomedical, Electronics & Computing.

In the fast-paced landscape of biomedical research, electronics development, and computing innovations, staying abreast of the latest breakthroughs is crucial. The papers and contributions contained in these proceedings reflect the tireless efforts of researchers, academics, and industry professionals who are at the forefront of driving progress in these fields. As per the market estimates the global market involving biomedical instrumentation was around USD 10.56 billion and estimated to reach 16.78 billion by year 2026, at CAGR of approximately 6.3%. This growth trajectory requires innovative designs and applications to satiate the large demand from the industry.

This volume encompasses a diverse range of topics, from groundbreaking biomedical discoveries that have the potential to revolutionize healthcare to the latest developments in electronics and computing that promise to reshape the way we interact with technology. Our contributors have delved into the intricacies of their respective disciplines, presenting novel approaches, methodologies, and solutions that will undoubtedly inspire and inform the wider scientific community.

The conference comprises of various parallel sessions including keynote sessions. Each session will be addressed by outstanding experts who will highlight the recent advances in various facets of Biomedical, Instrumentation and Electronics Engineering This conference will offer the rich resource for budding engineers to develop their own individual perspective to understand the challenges and the proposed solutions during the discussions.

The organizing committee is extremely gratified by the tremendous response to the call for papers. More than 250 papers were submitted by researchers, academicians, and students on a wide range of topics. The conference is approved by CRC Press (Taylor & Francis Group). The proceedings will be published online by CRC Press (Taylor & Francis Group). The conference received more than 250 papers. The acceptance rate of the conference is about 45%. Speakers and participants across the globe are participating in the conference.

We hope that the conference provides a forum for researchers to collaborate and work on novel ideas.

The Editors

About the Conference

International Conference on Advances in AI for Biomedical Instrumentation, Electronics and Computing (ICABEC 2023) is going to be held in the Department of Electronics and Communication Engineering, KIET Group of Institutions, Delhi-NCR, Ghaziabad, Uttar Pradesh, India on Dec 22-23, 2023.

The introduction of new technologies drastically transforms the future of the biomedical field as well as Electronics and Communication Engineering, therefore, the aim of the conference is to provide a forum for researchers around the globe to explore and discuss on various aspects of the biomedical field. The conference consists of various sessions and includes keynote and parallel sessions. Each session will be addressed by outstanding experts who will highlight the recent advances in various facets of Biomedical Instrumentation, sensors and Electronics and Communication Engineering. It will also offer the budding researchers an opportunity to present their work in front of eminent experts. Discussion on the latest innovations, trends and practical concerns and challenges faced in these fields are also encouraged.

Innovation will bring added value and widen the employment base. The present conference attempt to establish some robust and mutual interactions and cooperation with scientific people, educationalists, doctors, governmental and non-governmental organizations, and industries across local, national and international levels to meet the objectives of creating contacts as possible to provide a means for scientific and industrial exchanges, boosting the scientific and research collaborations with other scientific centres in the country and around the world.

The scope of this conference is to provide a platform for researchers, engineers, academicians as well as industrial professionals from all over the world to present their research results and development activities in the field of Electronics and Communication Engineering, Computer Science Engineering, Information Technology, Electrical Engineering, Electronics and Instrumentation Engineering, Computer applications and all interdisciplinary streams of Engineering Sciences, having central focus on AI for Biomedical Instrumentation, Electronics and Computing. The conference is being organized specifically to help the industries to derive benefits from the advances of next Technologies.

ICABEC - 2023 is soliciting original, previously unpublished, and high-quality research papers addressing research challenges and advances in the tracks mentioned below.

The conference will highlight the related focused areas based on the following themes:

Category-1: AI-Based Biomedical Instrumentation

Signal Processing and Image Analysis for Biomedical Instrumentation
Topics related to processing and analyzing medical signals and images using AI, such as ECG, EEG, MRI, and CT scans.

Wearable and Implantable Medical Devices and Systems
Focus on the design, development, and application of AI in wearable and implantable medical devices, such as smart sensors, pacemakers, and artificial limbs.

Intelligent Medical Robotics and Automation
This track explore the use of AI and machine learning in medical robotics and automation, including robotic surgery, drug delivery, and patient monitoring.

Human Machine Interfaces and Brain Computer Interfaces

This track cover the development and application of AI-based human machine and brain-computer interfaces in healthcare.

Category-2: Applications of AI and ML for Electronics and Signal Processing

Machine Learning and Data Analytics for Biomedical Applications

This track cover the application of AI and machine learning in biomedical data analysis and decision-making, such as disease diagnosis, prognosis, and treatment planning.

Healthcare Information Systems and Electronic Health Records

Explore the use of AI and machine learning in managing and analyzing electronic health records and healthcare information systems, including clinical decision support systems and disease registries.

Big Data and Cloud Computing for Biomedical Applications

This track could focus on the use of AI and cloud computing in handling and analyzing large-scale biomedical data, such as genomics, proteomics, and electronic health records.

Category-3: Generalized Tracks for Space Technologies, Electronics Engineering, and Image Processing

Space based medical Technologies and Satellite Program

This track covers the use of AI and machine learning in designing and developing medical technologies for space applications, such as remote diagnosis and Health Monitoring, Space based Medical Technologies

Image Processing for Biomedical Applications

This track focus on the use of image processing techniques in biomedical research and healthcare, including medical image analysis and computer-aided diagnosis.

Electronic Circuits and devices for Biomedical Applications

This track explore the applications of electronic engineering principles, circuits and techniques in biomedical research and healthcare.

Drug Discovery and Development

This track includes the development of various drug discoveries for medical applications.

Advances in AI for Biomedical Instrumentation, Electronics and Computing – Sachan et al. (eds)
© 2024 The Author(s), ISBN 978-1-032-64298-7

Ensemble of diverse pre-trained convolutional neural networks for enhanced diabetic retinopathy image classification

S.K. Feroz Ahmed* & Y. Yaswanth Chowdary*
Department of Advanced Computer Science and Engineering, VFSTR Deemed to be University, Andhra Pradesh, India
ORCID ID: 0009-0008-9921-4370, 0009-0002-0495-3135

Jyostna Devi Bodapati*
Department of Advanced Computer Science and Engineering, Vignan University, Andhra Pradesh, India
ORCID ID: 0009-0006-3337-2398

Lakshmana Kumar Yenduri*
Senior Member, IEEE, San Francisco Bay Area

ABSTRACT: Diabetic retinopathy (DR) presents a significant threat to individuals with diabetes, affecting the retinal blood vessels and potentially leading to vision impairment or loss. Early detection plays a pivotal role in mitigating vision-related complications, and recent advancements in deep learning offer promising avenues for enhancing diagnostic accuracy. In this study, we introduce an ensemble model comprising six distinct pre-trained Convolutional Neural Network (CNN) architectures for early detection of diabetic retinopathy. The proposed model encompasses various stages, including data collection, pre-processing, augmentation, and modeling. Leveraging the benchmark APTOS19 dataset, our ensemble approach achieves an impressive accuracy of 85.18%, surpassing the performance of individual models involved.

1 INTRODUCTION

One of the significant risks faced by individuals with diabetes is the susceptibility to a diabetic-related eye condition known as Diabetic Retinopathy. Early detection of this condition is paramount in preventing vision impairment or loss. However, recent advancements in deep learning have demonstrated promising outcomes in enhancing the accuracy and efficiency of diagnosis (De Calleja et al. 2014). Deep learning, a subset of machine learning, focuses on developing algorithms capable of learning from extensive datasets to make informed decisions. In the realm of medical diagnosis, deep learning algorithms can analyze large repositories of retinal images and identify specific features associated with diabetic retinopathy . This approach holds considerable potential for automating the detection process, facilitating early diagnosis, and enabling timely intervention (Acharya et al. 2009).

In contrast to traditional methods reliant on manual interpretation by ophthalmologists, deep learning algorithms can directly learn from labeled training data and automatically categorize images based on predefined criteria (Anant et al. 2017). This not only reduces subjectivity but also enhances consistency across various diagnoses. An additional advantage is early diagnosis, a crucial factor in mitigating the progression of diabetic retinopathy. Within

*Corresponding Authors: ahmedshaik.0862@gmail.com, yarrayashwanth991@gmail.com, jyostna.bodapati82@gmail.com and lakshmanyenduri@gmail.com

DOI: 10.1201/9781032644752-1

the domain of deep learning, various techniques have been devised for diagnosing diabetic retinopathy, with convolutional neural networks (CNNs) being particularly effective in image classification tasks due to their capability to capture spatial patterns within images.

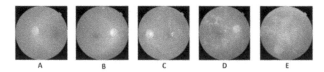

Figure 1.   Stages of diabetic retinopathy from No DR to Severe DR.

Several studies in the literature focuses on the Detection of Diabetic Retinopathy (DDR), employing Deep Learning models such as Densenet-169, GoogleNet, and general CNN models. Despite these efforts, the highest accuracy reported in previous papers remains at 81%. Some studies have explored modifications like changing kernel sizes and resizing images, resulting in negligible improvements.

For our studies, we adopted a strategy of utilizing multiple Convolutional Neural Network (CNN) models with the aim of achieving the highest possible accuracy. APTOS 2019 dataset, the benchmark dataset for retinopathy is used for experimental validation of the proposed work. Considering the computational resources available, we trained these models with the highest feasible configurations. As a result, the ensemble of all employed methods yields an accuracy of 85.32%. This demonstrates a significant improvement over the accuracy reported in previous studies and also significantly better than any of the single model involved in the study.

2   LITERATURE SURVEY

The following are the observations from the mentioned papers on Diabetic Retinopathy.

Few significant contributions to the field of diabetic retinopathy (DR) detection and analysis are summarized in this section.

Computer-based detection of diabetic retinopathy was introduced and achieves an accuracy rate exceeding 85% (Acharya et al. 2009). This study focused on analyzing various features such as blood vessels, microaneurysms, exudates, and hemorrhages using 331 fundus images to enhance patient care and clinical decision-making.

Building upon this foundation, Anant proposed a model for detecting type 2 diabetes through image mining (Anant et al. 2017). Leveraging Image Processing and Data Mining techniques on the DIARETDB1 database, their research demonstrated an impressive accuracy rate of 97.95% in identifying diabetic retinopathy by utilizing texture and wavelet characteristics from the dataset.

In a similar vein, an approach for automated DR detection using a Support Vector Machine (SVM) classifier was presented (Dhanasekaran et al. 2013). Their machine learning approach, particularly the SVM Classifier, was employed for decision-making on the severity level of Diabetic Retinopathy, showcasing the potential of machine learning in this domain.

Further advancements were made by employing Convolutional Neural Networks (CNNs) for hand-crafted feature extraction to detect red lesions within the eye's retina (Orlando et al. 2018). This integration of manual and deep learning feature extraction methods aimed to improve diabetic retinopathy analysis by harnessing the power of CNNs.

Expanding beyond diabetic retinopathy, machine learning and data mining techniques were utilized to predict heart disease while considering both benefits and drawbacks (Preetha et al. 2020). Their work proposed a different approach using machine learning and data mining techniques compared to traditional heart disease prediction methods, showcasing the versatility of these methodologies in medical research.

A quantitative methodology was developed to establish new parameters for proliferative diabetic retinopathy detection (Sadda *et al.* 2020). By considering the position, volume, and size of lesions, their study aimed to enhance retinal disease prediction, utilizing techniques such as Ultrawide Field Image Lesion Segmentation, Quantification Lesion Parameters, and Statistical Analysis. These collective efforts underscore the interdisciplinary nature of medical research and highlight the potential for innovative solutions to complex healthcare challenges.

3 MODELING

The proposed approach involved in modeling diabetic retinopathy detection involves several key steps. Each of those steps is discussed in this section.

Firstly, the input data is divided into training, validation, and test sets. The model then loads and extracts images and labels from fundus data files, followed by pre-processing and augmentation operations. Hyper-parameter setup, regularization techniques, and optimization algorithms are explained as part of the method's structure. Finally, network training and performance computations are conducted.

3.1 *Dataset summary*

In this study, two separate datasets are utilized: the APTOS19 dataset and the eyePACS dataset. The eyePACS dataset comprises 35,126 pre-processed retinal images, while the APTOS19 dataset consists of 3,593 files with both training and test images. The images are of size 255 × 255, and the datasets are divided into training and test sets.

3.2 *Pre-processing*

Since the chosen dataset is already pre-processed, the images are resized to 244 × 244, and the data is split into training and validation sets, with 80% of the data allocated for training and 20% for validation.

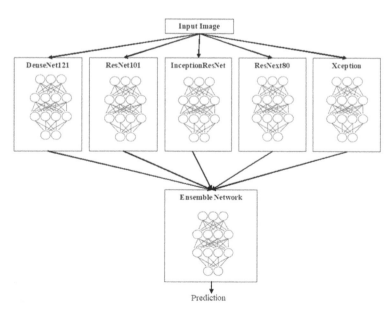

Figure 2. Architecture of the proposed ensemble of diverse pre-trained convolutional neural network models.

3.3 *Custom CNN models*

Custom Convolutional Neural Network (CNN) models play a crucial role in pixel-based image recognition tasks. These models offer flexibility in terms of architecture, allowing researchers to experiment with different configurations to optimize accuracy. CNN architectures can vary significantly in terms of their structure, including the number of hidden layers, convolutional layers, dense layers, normalization techniques, and dropout layers. This diversity enables researchers to explore a wide range of possibilities to achieve the best performance for a given task. Each layer within a CNN communicates information with one another, facilitating the learning process and ultimately enhancing the accuracy of image categorization. Through the hierarchical arrangement of layers, CNNs can effectively capture and represent complex patterns and features present in images.

Training a CNN involves various techniques, including feature extraction and data feeding. Feature extraction entails dividing the layers of the model into blocks of different sizes and extracting relevant features from these blocks. These features can then be aggregated using methods such as addition, averaging, concatenation, or maximum pooling. Once the features are extracted, they are fed into the model, which predicts the output class for the provided image. By iteratively adjusting the model parameters through processes like backpropagation, the CNN learns to accurately classify images based on the features extracted from them. Overall, custom CNN models offer a powerful framework for image recognition tasks, providing researchers with the flexibility and tools necessary to optimize accuracy and performance for specific applications.

4 PROPOSED ENSEMBLE APPROACH

The proposed ensemble approach involves combining seven distinct neural network models: DensNet121, ResNet101, InceptionResNetV3, RegNetX080, and Xception. The input data undergoes training with each individual model and subsequently with ensembles comprising four, five, and all seven models. Notably, the resulting accuracies vary with each combination, highlighting the dynamic nature of ensemble learning. Finally, the accuracies of each model configuration are evaluated and tabulated for comparison.

Below is the brief overview of the architectures of each model:

4.1 *ResNet101*

Renowned for its exceptional performance in image classification tasks, ResNet-101 has demonstrated state-of-the-art results across various benchmarks, including the extensive ImageNet dataset. Despite its remarkable accuracy, ResNet-101 computational demands are substantial due to its complex architecture, which includes a significant number of layers and parameters (Bodapati *et al.* 2021).

4.2 *EffnetV2B3*

EfficientNet is a group of convolutional neural networks designed to achieve high accuracy in image classification while prioritizing computational efficiency. It employs a unique scaling technique to adjust depth, width, and resolution dimensions uniformly, thereby striking a balance between performance and efficiency (Bodapati *et al.* 2023).

4.3 *RegNetX-080*

RegNetX-080 is tailored for image classification tasks and is part of the RegNet family of models, which aims to provide efficient and scalable network architectures. RegNetX models strike a balance between accuracy and efficiency by leveraging techniques such as linear bottleneck blocks, group convolutions, and network width multiplier adjustments (Bodapati *et al.* 2021).

4.4 *Densenet-121*

DenseNet-121 belongs to the DenseNet family of models, known for their densely connected layers. DenseNet addresses the vanishing gradient problem by establishing direct connections between every layer in a feed-forward manner. This dense connectivity facilitates feature reuse and gradient flow throughout the network, enabling effective learning with fewer parameters (Bodapati *et al.* 2023).

4.5 *Xception*

Xception, or Extreme Inception, is a deep convolutional neural network architecture introduced by François Chollet. Based on the Inception architecture, Xception incorporates depth-wise separable convolutions, enhancing its efficiency and effectiveness in various computer vision tasks (Qummar *et al.* 2019).

5   EXPERIMENT RESULTS

The data is processed through every model that has been proposed and also the data is trained with the ensembled model of all the proposed models. The accuracies for both datasets for different network models have been recorded as tabulated below.

196	**2**	**1**	**0**	**0**
3	19	8	0	0
2	8	69	2	6
0	0	7	6	4
0	2	14	6	11

Model = DenseNet121
Accuracy = 83.8

194	**3**	**2**	**0**	**0**
3	14	12	0	1
1	4	78	4	0
0	0	10	6	1
0	4	13	7	9

Model = ResNet101
Accuracy = 81.6

194	**3**	**2**	**0**	**0**
3	14	12	0	1
1	4	78	4	0
0	0	10	6	1
0	4	13	7	9

Model = InceptionResNet
Accuracy = 84.42

196	**2**	**1**	**0**	**0**
5	14	10	0	1
2	5	77	3	0
0	0	10	7	0
0	1	15	5	12

Model = RegNetX80
Accuracy = 83.87

195	**2**	**2**	**0**	**0**
4	14	12	0	0
2	3	76	4	2
0	0	10	4	3
0	2	11	3	17

Model = Xception
Accuracy = 83

196	**2**	**1**	**0**	**0**
3	16	10	0	1
1	5	77	2	2
0	0	11	5	1
0	2	12	6	13

Model = Ensemble
Accuracy = 85.7

In a depth-wise separable convolution, the convolutional operation undergoes two distinct stages: depth-wise convolutions and pointwise convolutions. Initially, the depth-wise convolution applies a single filter to each input channel independently, followed by pointwise convolutions that amalgamate the outputs of the depth-wise convolutions through 1×1 convolutions to produce the final feature maps.

6   CONCLUSION

Diabetes mellitus affects various bodily systems, including the nervous system, kidneys, and retina, leading to organ damage. Diabetic Retinopathy (DR) specifically manifests as a consequence of diabetes affecting the eyes. In this study, an ensemble model comprising Convolutional Neural Networks (CNNs) is proposed for the classification of Diabetic Retinopathy (DDR). The ensemble incorporates CNN architectures such as VGG16, VGG19, InceptionV3, EffNetV2B3, Xception, DenseNet101, and DenseNet169. Individually, each model achieved accuracies of 83.8, 81.6, 82.7, 84.6, 83.87, and 85, respectively, while the ensemble model yielded the highest accuracy of 85.7%. This culmination underscores the efficacy of ensemble learning in enhancing diagnostic accuracy for Diabetic Retinopathy detection.

REFERENCES

Acharya U. R., Lim C. M., Ng E. Y. K., Chee C. and Tamura T., "Computer-based detection of diabetes retinopathy stages using digital fundus images," *Proc. Inst. Mech. Eng. Part H J. Eng. Med.*, vol. 223, no. 5, pp. 545–553, 2009, doi: 10.1243/09544119JEIM486.

Anant K. A., Ghorpade T. and Jethani V., "Diabetic retinopathy detection through image mining for type 2 diabetes," *in 2017 International Conference on Computer Communication and Informatics*, ICCCI 2017, 2017, doi: 10.1109/ICCCI.2017.8117738.

Bodapati, J. D. "Enhancing brain tumor diagnosis using a multi-architecture deep convolutional neural network on MRI scans." *Inf. Dyn. Appl* 2.1 (2023): 42–50.

Bodapati, Jyostna Devi, and Bharadwaj Bagepalli Balaji. "Self-adaptive stacking ensemble approach with attention based deep neural network models for diabetic retinopathy severity prediction." *Multimedia Tools and Applications* (2023): 1–20.

Bodapati, Jyostna Devi, and Bharadwaj Bagepalli Balaji. "TumorAwareNet: Deep representation learning with attention based sparse convolutional denoising autoencoder for brain tumor recognition." *Multimedia Tools and Applications* (2023): 1–19.

Bodapati, Jyostna Devi, RamaKrishna Sajja, and Veeranjaneyulu Naralasetti. "An Efficient Approach for Semantic Segmentation of Salt Domes in Seismic Images Using Improved UNET Architecture." *Journal of The Institution of Engineers (India): Series B* 104.3 (2023): 569–578.

Bodapati, Jyostna Devi. "Modified self-training based statistical models for image classification and speaker identification." *International Journal of Speech Technology* 24.4 (2021): 1007–1015.

Bodapati, Jyostna Devi. "SAE-PD-Seq: sequence autoencoder-based pre-training of decoder for sequence learning tasks." *Signal, Image and Video Processing* 15.7 (2021): 1453–1459.

De Calleja J., Tecuapetla L., and Medina M. A., *"LBP and Machine Learning for Diabetic Retinopathy Detection,"* pp. 110–117, 2014.

Gandhi M. and Dhanasekaran R., "Diagnosis of diabetic retinopathy using morphological process and SVM classifier," *Int. Conf. Commun. Signal Process.* ICCSP 2013 – Proc., pp. 873–877, 2013, doi: 10.1109/iccsp.2013.6577181.

Kakani, Vamsi, *et al.* "Post-COVID Chest Disease Monitoring using self adaptive Convolutional Neural Network." *2023 IEEE 8th International Conference for Convergence in Technology (I2CT)*. IEEE, 2023.

Orlando J. I., Prokofyeva E., del Fresno M. and Blaschko M. B., "An ensemble deep learning based approach for red lesion detection in fundus images," *Comput. Methods Programs Biomed.*, vol. 153, pp. 115–127, 2018, doi: 10.1016/j.cmpb.2017.10.017.

Preetha S., Chandan N., Darshan N.K. and Gowrav P. B., "Diabetes Disease Prediction Using Machine Learning," *Int. J. Recent trends Eng. Res.*, vol. 6, no. 5, 2020, doi: 10.23883/IJRTER.2020.6029.65Q5H.

Qummar, Sehrish, *et al.* "A deep learning ensemble approach for diabetic retinopathy detection." *Ieee Access* 7 (2019): 150530–150539.

Sadda S. R. *et al.*, "Quantitative assessment of the severity of diabetic retinopathy," *Am. J. Ophthalmol.*, 2020, doi: 10.1016/j.ajo.2020.05.021.

Advances in AI for Biomedical Instrumentation, Electronics and Computing – Sachan et al. (eds)
© 2024 The Author(s), ISBN 978-1-032-64298-7

Enhancing air quality prediction using machine learning algorithms and boosting techniques

Rushvi Sajja*
Department of Advanced Computer Science and Engineering, VFSTR Deemed to be University, Andhra Pradesh, India

S.K. Feroz Ahmed*
Department of Advanced Computer Science and Engineering, VFSTR Deemed to be University, Andhra Pradesh, India
ORCID ID: 0009-0008-9921-4370

Jyostna Devi Bodapati*
Department of Advanced Computer Science and Engineering, Vignan University, Andhra Pradesh, India
ORCID ID: 0009-0006-3337-2398

Lakshmana Kumar Yenduri*
Senior Software Engineer, IEEE Senior Member, San Francisco Bay Area
ORCID ID: 0009-0002-8345-2979

ABSTRACT: Air pollution, exacerbated by industrialization and urbanization, poses a significant environmental and public health challenge worldwide. Particular concern is directed towards PM2.5, which poses greater health risks than other airborne impurities combined. Accurately predicting PM2.5 levels is crucial for effective pollution management. This study focuses on predicting the Air Quality Index (AQI) as a measure of air pollution, employing machine learning algorithms to enhance predictive accuracy. Specifically, we utilize the Random Forest Classifier and AdaBoost boosting algorithm to develop a predictive model. Our approach involves data categorization and sensitivity analysis to ensure robust air quality predictions.

Keywords: Air Quality, Random Forest Classifier, SVM, Air Quality Index (AQI), Classification, Normalization, Boosting

1 INTRODUCTION

Air pollution is a pressing global concern with profound implications for public health and the environment. Exposure to poor air quality can result in a myriad of respiratory and cardiovascular ailments, allergies, and premature mortality. Of particular concern is the air quality in the vicinity of educational institutions, such as schools, colleges, and universities, which is susceptible to pollution due to the high volume of vehicular traffic (Xiaojun, C. *et al.* 2015). It is imperative to regulate air quality near these institutions to safeguard the health of students and faculty members. Accurate prediction of air quality enables governments, public health authorities, and individuals to implement proactive measures to mitigate the adverse effects of pollution (Raipure

---

*Corresponding Authors: rushvisajja@gmail.com, ahmedshaik.0862@gmail.com, jyostna.bodapati82@gmail.com and lakshmanyenduri@gmail.com

DOI: 10.1201/9781032644752-2

et al. 2015). The application of air quality prediction systems spans various sectors, each benefiting from its predictive capabilities. In the realm of public health, these systems play a vital role in issuing early warnings and disseminating information about potential air pollution events. Environmental monitoring relies on air quality prediction systems to assess the impact of diverse pollution sources on the ecosystem. Additionally, industries emitting pollutants, such as power plants, factories, and refineries, can optimize their operations by leveraging the insights provided by these systems (Xi *et al.* 2015). This study proposes a predictive model that utilizes the Random Forest Algorithm and the AdaBoost Boosting Algorithm to calculate air quality. The model is trained on a dataset comprising 16 attributes, with the 'AQI_Bucket' attribute transformed into 'AQI_quality' to enhance predictive accuracy. Data for model training is sourced from City_data, and the resultant model is characterized by its ability to process large datasets with minimal time complexity and high accuracy. The Air Quality Index (AQI) serves as a crucial metric for assessing air quality, and our proposed system leverages this measure to achieve superior performance (Gómez *et al.* 2017).

2 LITERATURE SURVEY

In a study by Shaban, Khaled Bashir, Abdullah Kadri, and Eman Rezk, published in the IEEE Sensors Journal in 2016, the authors delved into urban air pollution monitoring systems with forecasting models. They emphasized the importance of integrating real-time monitoring data with forecasting models to facilitate proactive decision-making and effective pollution control measures (Shaban *et al.* 2016). Another notable contribution comes from Xiaojun, Chen, Liu Xianpeng, and Xu Peng, who explored IoT-based air pollution monitoring and forecasting systems in their paper published in the ICCCS. They focused on the integration of IoT devices, data collection, and analysis techniques to enhance air pollution monitoring capabilities (Xiaojun *et al.* 2015). Gómez, Jorge E., *et al.* conducted research on the use of IoT technology for monitoring environmental variables in urban areas. Their study, published in Procedia Computer Science in 2017, provided insights into the advancements in IoT-based environmental monitoring systems, aiming to improve urban environmental quality (Gómez *et al.* 2017). Xi, Xia, *et al.* presented a comprehensive evaluation of air pollution prediction improvement using machine learning methods at the IEEE International Conference on Service Operations Logistics, and Informatics (SOLI) in 2015. Their work highlighted the effectiveness of machine learning algorithms in enhancing air pollution prediction accuracy (Xi *et al.* 2015). Raipure, Shwetal, and Deepak Mehetre investigated the use of wireless sensor networks (WSNs) for pollution monitoring in metropolitan cities. Their paper, presented at the International Conference on Communications and Signal Processing (ICCSP) in 2015, emphasized the integration of WSNs, data collection, and analysis techniques for efficient pollution monitoring and management (Raipure *et al.* 2015). In a study by Fuertes, Walter, *et al.* presented at the IEEE/ACM 19th International Symposium on Distributed Simulation and Real-Time Applications (DS-RT) in 2015, the authors explored the potential of distributed systems and the Internet of Things (IoT) for low-cost, real-time air pollution wireless monitoring. Their research aimed to leverage distributed systems and IoT technologies to enable real-time monitoring of air pollution levels (Fuertes *et al.* 2015).

3 PROPOSED MODEL

Machine learning encompasses computations that learn from complex data. The initial step in modeling involves data pre-processing, which includes handling null attributes, removing unnecessary data, normalization, data splitting, dimensionality reduction, and deploying a variety of prediction, classification, and scoring models. The model can effectively detect patterns and engage in self-learning. The classification methods employed in this model are Support Vector Machine (SVM) and Random Forest classifiers.

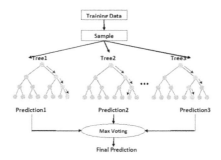

Figure 1.   Proposed air quality prediction model.

3.1   *Handling null values and removing useless attributes*

To address null values, several strategies are employed, such as replacing them with the mean, median, or mode of the data within that attribute, or by using a global value. Additionally, unnecessary data, including attributes not utilized in the model, are removed to streamline the dataset and improve computational efficiency.

3.2   *Normalization*

Normalization is a crucial pre-processing technique utilized to convert the values of numerical columns in the dataset to a common scale, particularly when the dataset contains values from diverse ranges. By ensuring that all elements lie within the range of 0 to 1, normalization facilitates fair comparison and prevents features with larger scales from dominating the learning process. Various normalization techniques, including Standard Scaler, z-score, and Minmax Normalization, can be applied based on the specific requirements of the model. In this case, Minmax Normalization is employed, which performs a linear transformation on the data, preserving the relationship among the original data while ensuring that all data falls within the specified range (0, 1).

3.3   *Support vector machine and random forest for classification*

SVM is a powerful supervised learning algorithm utilized for classification tasks. It constructs a hyperplane to effectively separate data belonging to different classes, leveraging support vectors to define the boundary between classes. SVM employs various types of kernels, such as linear, polynomial, and radial basis function (RBF), to facilitate classification in both linearly separable and non-linearly separable datasets. Random Forest, on the other hand, is an ensemble learning method renowned for its versatility and high predictive accuracy. By aggregating predictions from multiple decision trees built on different subsets of the input dataset, Random Forest mitigates overfitting and enhances generalization performance. This makes it suitable for handling both classification and regression tasks across various domains.

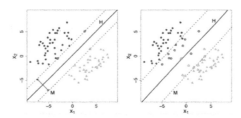

Figure 2.   Overview of the hyper plane produced by SVM algorithm. Hyper plane generated by SVM for Separating the classes for classification **2a)** without Error **2b)** With Error.

Figure 2 depicts an overview of the hyperplane produced by the Support Vector Machine (SVM) algorithm. The hyperplane generated by SVM serves to separate classes in a classification task, facilitating the classification of data points into distinct categories. Two scenarios are presented: Figure 2a illustrates the hyperplane generated by SVM without error, showcasing its effectiveness in accurately separating the classes in the dataset. In this scenario, the hyperplane effectively delineates the boundaries between different classes, enabling reliable classification of data points. Figure 2b depicts the hyperplane produced by SVM with error, indicating instances where the algorithm may struggle to perfectly separate the classes. Despite potential errors or misclassifications, the hyperplane still serves as a valuable decision boundary, albeit with some imperfections. Support Vector Machine (SVM) is a supervised learning algorithm that utilizes support vectors to delineate the boundary between different classes in a dataset. This boundary, known as the hyperplane, serves to separate the classes effectively. The hyperplane can be linear or nonlinear, depending on the nature of the data, and SVM employs various types of kernels such as linear, polynomial, and quadratic to achieve optimal classification accuracy. On the other hand, Random Forest is a widely used machine learning algorithm capable of handling both classification and regression tasks. It employs an ensemble learning approach wherein multiple decision trees are trained on different subsets of the input dataset. By averaging the predictions of these trees, Random Forest enhances predictive accuracy while mitigating the risk of overfitting.

3.5 *Quadratic discriminant analysis*

Quadratic Discriminant Analysis (QDA) is a variant of Linear Discriminant Analysis (LDA) where each class is associated with its own covariance matrix. This approach is particularly useful when it is known that different classes have distinct covariance structures. However, QDA does not offer dimensionality reduction capabilities, unlike LDA.

3.6 *Adaboost*

Adaboost, short for Adaptive Boosting, is a powerful machine learning algorithm that has gained significant traction in recent years. Its key strength lies in its ability to iteratively improve the performance of weak learners, subsequently enhancing the overall predictive accuracy of the model. This essay aims to provide a comprehensive overview of Adaboost, delving into its fundamental components, operational mechanisms, and applications across various domains. Additionally, it seeks to elucidate the advantages of Adaboost while also addressing any limitations or challenges associated with its implementation.

4 EXPERIMENTAL ANALYSIS

The experimental analysis involves the application of classification models such as Random Forest and SVM, followed by dimensionality reduction using Quadratic Discriminant Analysis (QDA), and subsequently boosting the model's accuracy using the Adaboost algorithm. The dataset utilized for this analysis is "city_data," sourced from Kaggle, encompassing 16 attributes including "City," "Date," "PM2.5," "PM10," "NO," "NO2," "NOX," "NH3," among others. The target attribute considered is "AQI_Bucket," which comprises six distinct classes ranging from "Poor" to "Good," indicating air quality levels. The dataset consists of 29,532 instances of data. Data preprocessing involves handling null attributes and removing unnecessary data, followed by normalization using appropriate techniques. The dataset is then split into training and testing sets, with 99.5% of the data allocated for training and 0.5% for testing, a configuration chosen for its superior accuracy performance. The classification models are trained both with and without dimensionality reduction, and their performance is evaluated. It is observed that the Random Forest Classification model outperforms the SVM Classification model in terms of accuracy.

4.1 *Dataset summary*

The dataset utilized in this experimental analysis, known as "city_data" and sourced from Kaggle, encompasses a total of 16 attributes. These attributes include various parameters related to air quality such as "PM2.5," "PM10," "NO," "NO2," "NOX," "NH3," alongside others. The dataset also includes categorical variables like "City" and "Date." The target attribute of interest is "AQI_Bucket," which categorizes air quality levels into six distinct classes ranging from "Poor" to "Good." In total, the dataset comprises 29,532 instances, each providing a snapshot of air quality measurements and corresponding attributes for specific cities and dates.

5 RESULTS

The experimental analysis using classification and boosting algorithms provided insightful results. Without dimensionality reduction but with boosting, the model achieved an accuracy of 81.44888. On the other hand, when dimensionality reduction was applied but boosting was not, the accuracy slightly increased to 81.45.

	SVM Classifier			Random Forest Classifier		
Class	Precision	Recall	F-score	Precision	Recall	F-score
1	82	61	70	85	78	81
2	67	71	69	70	82	76
3	58	35	43	69	62	65
4	74	88	81	86	86	86
5	0	0	0	84	69	75
6	69	70	70	80	83	82

The findings underscore the critical importance of monitoring air quality, particularly in the context of industrial development that often results in pollutant emissions. The AdaBoost algorithm emerged as particularly effective, demonstrating superior accuracy compared to other methods. Specifically, the combination of Random Forest and the boosting capability of AdaBoost yielded the highest accuracy of 83%.

	SVM Classifier			Random Forest Classifier		
Class	Precision	Recall	F-score	Precision	Recall	F-score
1	83	61	71	80	76	78
2	67	73	64	72	78	75
3	60	35	43	67	64	66
4	74	90	80	86	87	87
5	0	0	0	75	64	69
6	65	71	67	81	82	82

This research highlights the significance of employing advanced algorithms such as AdaBoost for air quality prediction. While this study focused on a specific set of algorithms, it also suggests the potential for exploring additional algorithms to further enhance accuracy in air quality prediction tasks.

6 CONCLUSION

In conclusion, the pursuit of predicting air quality emerges as a critical endeavor in today's industrial landscape, where burgeoning pollution levels necessitate vigilant monitoring and proactive measures. Through this project, AdaBoost emerges as a standout algorithm, particularly potent when coupled with Random Forest, showcasing an impressive accuracy rate of 83%. This underscores the pivotal role of advanced machine learning techniques in addressing complex environmental challenges. By harnessing algorithms like AdaBoost, we not only deepen our comprehension of air quality dynamics but also fortify our ability to formulate effective strategies for pollution control and mitigation. Moving forward, sustained exploration and experimentation with diverse algorithms will be imperative in refining air quality prediction models and augmenting their efficacy. Through the power of machine learning, we can forge a path towards more informed decision-making and proactive interventions to safeguard our environment and public health from the pervasive impact of air pollution.

REFERENCES

Bodapati, Jyostna Devi. "Modified self-training based statistical models for image classification and speaker identification." *International Journal of Speech Technology* 24.4 (2021): 1007–1015.

Bodapati, Jyostna Devi. "SAE-PD-Seq: sequence autoencoder-based pre-training of decoder for sequence learning tasks." *Signal, Image and Video Processing* 15.7 (2021): 1453–1459.

Bodapati, J. D. "Enhancing brain tumor diagnosis using a multi-architecture deep convolutional neural network on MRI scans." *Inf. Dyn. Appl* 2.1 (2023): 42–50.

Bodapati, Jyostna Devi, and Bharadwaj Bagepalli Balaji. "Self-adaptive stacking ensemble approach with attention based deep neural network models for diabetic retinopathy severity prediction." *Multimedia Tools and Applications* (2023): 1–20.

Bodapati, Jyostna Devi, and Bharadwaj Bagepalli Balaji. "TumorAwareNet: Deep representation learning with attention based sparse convolutional denoising autoencoder for brain tumor recognition." *Multimedia Tools and Applications* (2023): 1–19.

Bodapati, Jyostna Devi, RamaKrishna Sajja, and Veeranjaneyulu Naralasetti. "An Efficient Approach for Semantic Segmentation of Salt Domes in Seismic Images Using Improved UNET Architecture." *Journal of The Institution of Engineers (India): Series B* 104.3 (2023): 569–578.

Fuertes, Walter, *et al.* "Distributed system as internet of things for a new low-cost, air pollution wireless monitoring on real time." *2015 IEEE/ACM 19th International Symposium on Distributed Simulation and Real Time Applications (DS-RT).* IEEE, 2015.

Gómez, Jorge E., *et al.* "IoT for environmental variables in urban areas." *Procedia computer science* 109 (2017): 67–74.

Kakani, Vamsi, *et al.* "Post-COVID Chest Disease Monitoring using self adaptive convolutional neural Network." *2023 IEEE 8th International Conference for Convergence in Technology (I2CT).* IEEE, 2023.

Raipure, Shwetal, and Deepak Mehetre. "Wireless sensor network based pollution monitoring system in metropolitan cities." *2015 International Conference on Communications and Signal Processing (ICCSP).* IEEE, 2015.

Shaban, Khaled Bashir, Abdullah Kadri, and Eman Rezk. "Urban air pollution monitoring system with forecasting models." *IEEE Sensors Journal* 16.8 (2016): 2598–2606.

Xi, Xia, *et al.* "A comprehensive evaluation of air pollution prediction improvement by a machine learning method." *2015 IEEE international conference on service operations and logistics, and informatics (SOLI).* IEEE, 2015.

Xiaojun, C., L. Xianpeng, and X. Peng. "IOT-based air pollution monitoring and forecasting system-IEEE Conference Publication." *Ieeexplore. ieee. org* (2015).

Advances in AI for Biomedical Instrumentation, Electronics and Computing – Sachan et al. (eds)
© 2024 The Author(s), ISBN 978-1-032-64298-7

Predicting missed appointments in dental clinic: Enhanced risk assessment with multiple learning techniques

Jeffin Joseph, S. Senith & A. Alfred Kirubaraj
Karunya Institute of Technology and Sciences, Coimbatore, India

S.R. Jino Ramson
Global Foundries US LL2, Vermont, USA

ABSTRACT: The prediction of missed appointments in dental clinics is a critical challenge that impacts the efficient utilization of resources and the delivery of timely care. In the study, the researchers predicted the dental missed appointments from categorical Electronic Health Records data using Logistic Regression, Random Forest, LightGBM, Extra Trees, and Artificial Neural Network (ANN). The study attempted to address the class imbalance issue through the implementation of Adaptive Synthetic Sampling (ADASYN). The Performance is measured in terms of Accuracy, Precision, Recall and Area under the Receiver Operating Characteristic Curve (AUC). The Logistic Regression, and LightGBM approaches, outperformed other models in predicting missed appointments, offering insights for researchers to adopt the best predictive model, and highlighting the significance of enhancing dental clinics' operational efficiency and effectiveness.

1   INTRODUCTION

Dental clinics hold an indispensable role in promoting oral health and overall patient well-being. However, the persistent challenge of missed appointments significantly undermines the efficiency and effectiveness of healthcare delivery. This issue disrupts daily clinic operations and leads to the underutilization of resources, causing delays in providing essential services to those in need. Consequently, both researchers and dental clinics face the crucial task of developing effective strategies to predict and mitigate the occurrence of missed appointments, with the goal of improving patient engagement and optimizing the scheduling process. In a study conducted by (Alabdulkarim *et al.* 2022), machine learning techniques were employed to predict dental missed appointments, resulting in a Receiver Operating Characteristic Curve (ROC) of 0.72 and an F measure of 66.5%. Similarly, a distinct study established a substantial big data framework, in which Gradient Boosting emerged as the most successful approach, achieving an accuracy of 79% and an 81% ROC score in predicting missed appointments (Daghistani *et al.* 2020). Another comprehensive study undertook a systematic review encompassing 727 articles, revealing significant patient attributes and prevailing determinants of missed appointments (Dantas *et al.* 2018). In parallel, (Fan *et al.* 2021) predicted online outpatient missed appointments, resulting in an 11.1% patient no-show rate, with Bagging emerging as the algorithm yielding the highest AUC. In another study in the Saudi Arabian context, AlMuhaideb *et al.* (2019) implemented diverse machine learning algorithms, achieving accuracy rates of 76.44% and 77.13%. Notably, another study introduced an interpretable deep learning-based methodology, successfully identifying 83% of no-shows during appointment scheduling (Liu *et al.* 2022). Another study employing various predictive models, Logistic Regression, Artificial Neural Networks, and Naïve Bayes Classifier, showcased commendable AUC values (Mohammadi *et al.* 2018). This literature highlights the growing body of research dedicated to predicting and addressing missed dental appointments, with various studies employing diverse machine learning techniques and big data frameworks. While previous studies have demonstrated promising results in terms of prediction accuracy and identification of relevant factors, the current study advances this field further by utilizing a comprehensive set of models Logistic Regression, Random Forest, Extra Trees, Artificial Neural Network (ANN), and LightGBM for predicting dental missed appointments. The study also addressed the class imbalance issue through the implementation of Adaptive Synthetic Sampling (ADASYN). The findings of the study indicate that Logistic Regression and LightGBM outperform other models, offering valuable insights for improving the effectiveness of appointment scheduling and patient engagement in dental clinics.

2 MODEL DEVELOPMENT

2.1 *Analysis of variables used in prediction*

In this study, a publicly available dataset is utilized (10.7717/peerj-cs.1147/supp-1). To ensure data quality, duplicates are removed, and new variables are created from the existing ones to facilitate prediction. The final prediction dataset comprises 156,599 unique records. Continuous variables in the dataset are transformed into categorical ones to enhance prediction accuracy. Additionally, missing values in the Nationality field are indicated as such to maintain information integrity and minimize bias. The Lead Time, calculated from the booking and appointment dates, is a notable derived variable. The variables used in the study for prediction of missed appointments are Age, Gender, Marital Status, Nationality, Appointment Week Day, Appointment Month, Time of Appointment, Lead Time, Appointment Duration, Doctor ID, SMS Send, Temperature of the Day, and Weather Condition of the day. The output variable, "Missed Appointment" consists of two categories: "Show" and "No-show" For an in-depth understanding of the variables utilized for predicting No-show, refer to Figures 1 and 2.

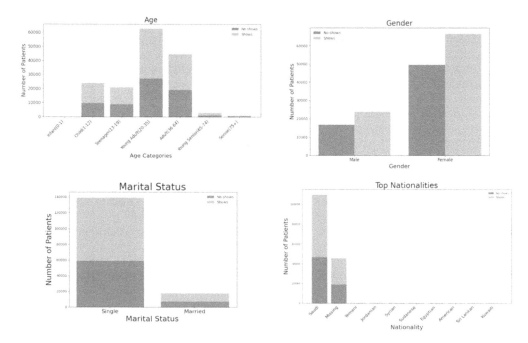

Figure 1.   Analysis of variables Age, Gender, Marital Status and Top Nationalities, with Shows and No-Shows.

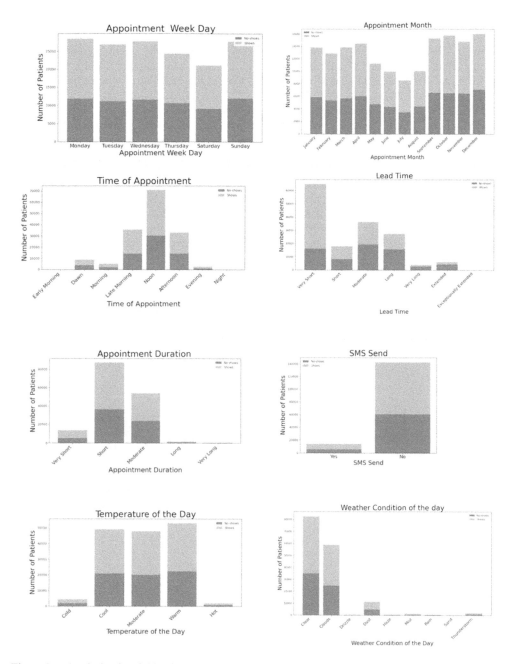

Figure 2. Analysis of variables Appointment Week Day, Appointment Month, Time of Appointment, Lead Time, Appointment Duration, SMS Send, Temperature of the Day and Weather Condition of the day, with Shows and No-Shows.

The variable "Age" comprised seven categories: Infant (0-1), Child (1-12), Teenager (13-19), Young Adult (20-35), Adult (36-64), Young Senior (65-74), and Senior (75+). The variable "Gender" had two categories: Male and Female. "Marital Status" consisted of Single and Married. The variable "Nationality" encompassed 28 categories, including 27 nationalities and a 'missing' category.

Another variable, "Appointment Weekday," included categories Monday through Sunday except Friday. The variable "Appointment Month" contained categories from January through December. Another variable, "Time of Appointment," was divided into eight categories: Early Morning: 3:00 AM – 5:59 AM, Dawn: 6:00 AM – 7:59 AM, Morning: 8:00 AM – 9:59 AM, Late Morning: 10:00 AM – 11:59 AM, Noon: 12:00 PM – 12:59 PM, Afternoon: 1:00 PM – 4:59 PM, Evening: 5:00 PM – 7:59 PM, Night: 8:00 PM – 11:59 PM, Midnight: 12:00 AM – 2.59 AM. The variable "Lead Time" had categories: Very Short (0 to 7 days), Short (8 to 14 days), Moderate (15 to 30 days), Long (31 to 90 days), Very Long (91 to 180 days), Extended (181 to 400 days), Exceptionally Extended (Above 400 days). Another variable, "Appointment Duration," was categorized as Very Short (5 to 15), Short (16 to 30), Moderate (31 to 60), Long (61 to 120), and Very Long (121 to 300). "Doctor ID" is a variable that included 105 unique IDs of different doctors with whom appointments were made. Another variable, "SMS Send," had two categories: Yes and No. The variable "Temperature of the Day" had categories: Cold (4°C – 10°C), Cool (11°C – 20°C), Moderate (21°C – 30°C), Warm (31°C – 40°C), Hot (41°C – 43°C). Additionally, the variable "Weather Condition of the day" encompassed nine categories: Clear, Clouds, Drizzle, Dust, Haze, Mist, Rain, Sand, and Thunderstorm.

2.2 Implementation of the prediction models

Step 1: Data Cleaning and Variable Extraction: The original dataset is cleaned; inconsistencies or missing values are removed and the relevant variables or features from the dataset that are essential for the predictive model are identified and extracted. Figure 3 shows the Implementation Process of the Prediction Models.

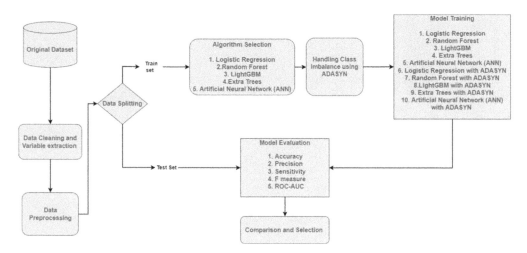

Figure 3.   Implementation process of prediction models.

Step 2: Data Preprocessing: Continuous variables were transformed into categorical variables, and categorical variables were managed using one-hot encoding. Subsequently, the data was divided into predictor variables (X) and the target variable (Y).

Step 3: Data Splitting: The preprocessed data was split into a training set and a testing set. 80% of the data was used for training, and 20% for testing. This helped assess model performance.

Step 4: Algorithm Selection: The machine learning and deep learning techniques chosen for prediction included Logistic Regression, Random Forest, LightGBM, Extra Trees, and Artificial Neural Networks (ANN).

Step 5: Handling Class Imbalance: Addressed the class imbalance issue by using Adaptive Synthetic Sampling (ADASYN) as mentioned in the study (He *et al.* 2008). ADASYN generated synthetic samples for the minority class to balance the dataset.

Step 6: Model Training: Each selected model was trained on the training dataset. Logistic Regression, Random Forest, LightGBM, Extra Trees, and ANN were trained separately using the training dataset. Additionally, Logistic Regression with ADASYN, Random Forest with ADASYN, LightGBM with ADASYN, Extra Trees with ADASYN, and ANN with ADASYN were trained separately using the training dataset.

Step 7: Model Evaluation: The performance of each trained model was evaluated using the test dataset. Performance metrics, including Accuracy, Precision, Sensitivity (recall), F-measure, and Area under the Receiver Operating Characteristic curve (ROC-AUC), were calculated for each model to assess their performance.

Step 8: Comparison and Selection: The performance metrics of all models were compared to determine the best-performed model.

3 PERFORMANCE ANALYSIS AND DISCUSSION

The performance measures of each model are shown in Table 1.

Table 1. Performance measures.

	Accuracy	Precision	Sensitivity	F-measure	ROC-AUC
LR	0.66	0.60	0.59	0.59	0.65
LR with ADASYN	0.66	0.60	0.58	0.59	0.65
Random Forest	0.64	0.59	0.53	0.56	0.63
Random Forest with ADASYN	0.64	0.59	0.53	0.56	0.63
LightGBM	0.66	0.61	0.55	0.58	0.65
LightGBM with ADASYN	0.66	0.60	0.60	0.60	0.65
Extra Trees	0.63	0.57	0.54	0.56	0.62
Extra Trees with ADASYN	0.63	0.57	0.55	0.56	0.62
ANN	0.62	0.56	0.51	0.53	0.61
ANN with ADASYN	0.64	0.59	0.51	0.55	0.62

The study encompasses the exploration of prediction using traditional, ensemble learning, and deep learning techniques. Both Logistic Regression (LR) and its implementations with Adaptive Synthetic Sampling (ADASYN) exhibit comparatively better performance, achieving an Accuracy of 0.66 and ROC-AUC of 0.65. LightGBM also demonstrates an improved accuracy of 0.66 and displays better AUC values, suggesting its suitability for the task. Random Forest and Extra Trees show comparable accuracies of 0.64, both before and after implementing ADASYN to address class imbalance. Notably, the use of ADASYN does not consistently enhance model performance across all metrics. The Artificial Neural Network (ANN) exhibits relatively lower performance. Both Logistic Regression and LightGBM showcase notable performance, achieving an Accuracy of 0.66 and an ROC-AUC of 0.65, respectively, demonstrating competitive outcomes. These findings also underline the significance of considering various metrics like Precision, Recall, F-measure, and ROC-AUC in addition to Accuracy when evaluating model performance in prediction tasks. The variations in results highlight the influence of the algorithm choice on predictive performance.

4 CONCLUSION

The study underscores the diversity in model performances, with both the Logistic Regression and LightGBM models emerging as particularly promising models. The effectiveness of applying ADASYN becomes evident in its uneven augmentation of predictive capabilities across various models and highlights the importance of selecting the right algorithm and preprocessing techniques based on the specific situation. Thus, the study holds significance, by providing valuable insights into the effective prediction of missed dental appointments through the utilization of diverse machine learning algorithms and addressing class imbalance, thereby contributing to enhanced resource utilization and timely patient care in dental clinics.

5 FUTURE RESEARCH

In the realm of future research, there is an opportunity to explore hybrid models that combine various algorithms in order to enhance the accuracy of predicting missed appointments at dental clinics. Additionally, the inclusion of historical missed appointment data could provide valuable insights into recurring trends and patient behaviors, potentially enhancing predictive models to better understand the complexities of patient decision-making.

REFERENCES

Alabdulkarim, Y., Almukaynizi, M., Alameer, A., Makanati, B., Al-thumairy, R., & Almaslukh, A. 2022. Predicting no-shows for dental appointments. *PeerJ Computer Science*, 8, e1147.

AlMuhaideb, S., Alswailem, O., Alsubaie, N., Ferwana, I., & Alnajem, A. 2019. Prediction of hospital no-Show appointments through artificial intelligence algorithms. *Annals of Saudi Medicine*, 39(6), 373–381

Daghistani, T., AlGhamdi, H., Alshammari, R., & AlHazme, R. H. 2020. Predictors of outpatients' no-show: Big data analytics using apache spark. *Journal of Big Data*, 7(1).

Dantas, L. F., Fleck, J. L., Cyrino Oliveira, F. L., & Hamacher, S. 2018. no-shows in appointment scheduling – a systematic literature review. *Health Policy*, 122(4), 412–421.

Fan, G., Deng, Z., Ye, Q., & Wang, B. 2021. Machine learning-based prediction models for patients no-show in online outpatient appointments. *Data Science and Management*, 2, 45–52.

Liu, D., Shin, W.-Y., Sprecher, E., Conroy, K., Santiago, O., Wachtel, G., & Santillana, M. 2022. Machine learning approaches to predicting no-shows in pediatric medical appointment. *Npj Digital Medicine*, 5(1).

He, H., Bai, Y., Garcia, E. A., & Li, S. 2008. ADASYN: Adaptive synthetic sampling approach for imbalanced learning. In *2008 IEEE International Joint Conference on Neural Networks (IEEE World Congress on Computational Intelligence)*, Hong Kong, 2008 (pp. 1322–1328).

Mohammadi, I., Wu, H., Turkcan, A., Toscos, T., & Doebbeling, B. N. 2018. Data analytics and modeling for appointment no-show in community health centers. *Journal of Primary Care & Community Health*, 9, 215013271881169.

Advances in AI for Biomedical Instrumentation, Electronics and Computing – Sachan et al. (eds)
© 2024 The Author(s), ISBN 978-1-032-64298-7

Optimization of leakage power consumption in D Flip-Flop using hybrid technique

Amiya Prakash*
Delhi Technological University Delhi
G. L. B. I. T. M. Greater Noida

Priyanka Jain*
Delhi Technological University Delhi

ABSTRACT: In deep submicron regime high performance digital circuits, leakage power accounts for a sizable fraction of the total power consumption. An idle circuit only experiences leakage as a source of power consumption. Thus, it's crucial to minimize power leakage in portable systems. In this paper we propose the reduction of leakage power in sequential circuits using two strategies, including transistor stacking and altering the threshold voltage level of PMOS and NMOS through body biased in the circuits. In this paper the simulation show that the proposed Transmission gate based D flip-flop using adjusts substrate biased voltage level circuit has the least leakage power dissipation of 6.086 μW with a delay of 2.978 nS. This is because the adjust substrate biased voltage level circuit reduces the leakage current by adjusting the substrate bias voltage. The substrate bias voltage is the voltage applied to the substrate of a transistor. The substrate bias voltage affects the threshold voltage of the transistor, which in turn affects the leakage current. By adjusting the substrate bias voltage, the leakage current can be reduced.

Keywords: Leakage current, Leakage power, Transmission gate, Substrate biased, stacking effect

1 INTRODUCTION

1.1 *D-Flip Flop*

A flip-flop is a digital circuit that has two stable states. It can be used to store a single bit of data. Flip-flops are used in many digital circuits, such as counters, registers, and memories. A D flip-flop is a type of flip-flop that stores the data that is present on its input when the clock signal is high presented in (Rabaey 2003). The data is then held in the flip-flop until the next clock signal. D flip-flops can be constructed using two SR latches. The SR latch is a simple flip-flop that can be made using two NAND gates. The SR latch has two inputs, S and R, and two outputs, Q and Q'. Author presented in (Rabaey 2003). To construct a D flip-flop, we connect an inverter between the S and R inputs of the SR latch. This creates a single data input, D. When the clock signal is high, the data on the D input is transferred to the output, Q. When the clock signal is low, the output remains unchanged. The D flip-flop can be used to store a single bit of data. The data can be stored in the flip-flop for as long as the clock signal is high. When the clock signal is low, the data is lost and it is used in many digital circuits such as counters, registers, and memories. Here are some of the applications of D flip-flops:

---

*Corresponding Authors: amiyprakash@gmail.com and priyankajain@dtu.ac.in

DOI: 10.1201/9781032644752-4

1.1.1 *Counters*

A counter is a circuit that can count the number of clock pulses that have occurred. A D flip-flop can be used to create a simple counter.

1.1.2 *Registers*

A register is a circuit that can store a sequence of data. A D flip-flop can be used to create a simple register.

1.1.3 *Memories*

A memory is a circuit that can store a large amount of data. A D flip-flop can be used to create a simple memory.

Author presented in (Wang 2006). The D flip-flop is frequently used in the following situations and it is used in registers, RAM (random access memory), counters, frequency dividers, transparent data latches, and other electronic devices. Author presented in (Soeleman and Roy 1999) High performance portable systems with improved dependability have been developed as a result of the semi- conductor device industry's rapid growth. Due to the finite amount of battery power available in such portable applications, minimizing current usage is crucial. As a result, power dissipation is now acknowledged as a high priority issue for VLSI circuit design. In today's high performance Microprocessors, leakage power can account for up to 50% of overall power usage. So the power lost by a circuit while it is in standby or sleep state is known as leakage power dissipation. Leakage power is determined by equation 1.

(Calhoun *et al.* 2005), while a circuit's propagation delay (Tpd) is determined by equation 2 (Calhoun *et al.* 2005).

$$P_{leak} = I_{leak} * V_{dd} \qquad (1)$$

Where P_{leak} is the leakage power, I_{leak} is the leakage current when the transistor is in the off-state and V_{dd} is the supply voltage.

$$T_{pd} \propto V_{dd}/(V_{dd} - V_{th})^2 \qquad (2)$$

Where the propagation delay is denoted by T_{pd} and threshold voltage of the transistor is denoted by V_{th}.

1.2 *Basic principles of transmission gate*

The basic concept of Transmission gate logic (PTL) in electronics refers to a number of logic families utilized in the integrated circuits. It reduces the number of transistor by removing redundant transistors Author presented in (Chung *et al.* 1997). In Paper (Kao *et al.* 2002) Author presented that however transistors are used as a switch to pass the logic level amongst different nodes of the circuit. We can connect directly voltage supply instead of switches, which reduces the number of active devices at the cost of the decreases voltage level of high and low voltage levels of each stages. Author presented in (Weste and Harris 2004). In series each transistor is less saturated at output than input. If we use several devices in chain of the series connection in a logic path, conventionally designed logic gate may be needed to restore the signal voltage to the original value.

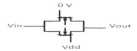

Figure 1.   Transmission gate schematic.

If multiple devices are coupled in series in large circuits, contrarily, traditional CMOS logic switches transistor to connect the output to one of the power supply rails (similar to an open drain design), maintaining the logic voltage levels in a sequential chain presented in (Kai *et al.* 2017). To ensure acceptable performance, circuit simulation may be necessary. A periodic clock signal apply '0' at PMOS and $V_{dd} = 1$ at NMOS drives the Transmission gate in Figure 1 presented in (Jaehyun *et al.* 2009), which functions as an access switch to charge or discharge the parasitic capacitance C_x based on the input signal Vin. The logic "1" transfer (charging the capacitance C_x to a logic-high level) and the logic "0" transfer (charging the parasitic capacitance C_x to a logic-low level) presented in (Dokic 2013).

2 RELATED WORK

Flip-flops are essential timing components in digital circuits and have a significant impact on their speed and power usage. Research on high speed and low power dissipation is actively being conducted. In (Chun and Chen 2016) author presented mixed threshold voltage based flip flop with reduced leakage power at the cost of slightly increase in either setup time or propagation delay. In (Kumre *et al.* 2019) Author presented leakage power reduction techniques for adiabatic sequential circuits based on two phase Transmission gate adiabatic logic using power gating techniques. In (Gangele 2015). Author present ultra-low power flip-flops using two leakage diode and analysis of master slave latches and flip flops. In (Upadhyay *et al.* 2012) author describe a new techniques based on transmission gate for leakage power reduction using dual threshold CMOS scheme.

Figure 2. D-Flip Flop using transmission gate schematic.

Figure 3. Transient response of conventional transmission gate D-Flip Flop.

2.1 *D-Flip Flop using transmission gate*

Author presented in (Saini and Mehra 2012) Modern electronics extensively employ sequential logic and memory storage systems, which leads to the creation of basic memory elements with low power and high speed. The D flip-flop (D-FF) is among the most significant basic memory components. In this section, three distinct CMOS D flip-flop architectures are presented. The D flip-flop combines two D latches, one master and one slave. For the master-slave latches in Design-I, presented in (A. Chandra kasan and Broder sen 1995) Transmission gates (PT) and inverters are used, as shown in Figure 2. When the PMOS loop transistor is active, that is when clk = 0, the two chained inverters are in the memory state and right hand side chain inverter acts in the opposite way. Figure 2. Shows the schematic of positive edge triggered flip flop that means the state of flip flop changes its state during rising edge of the clock. We can see the transient response of D-flip flop using Transmission gate in Figure 3 where output follows the inputs with delay and controlled by clock input of the D-flip flop presented in (Ye *et al.* 1998). Total power consumed by the conventional D Flip-flop by using Transmission gate is 32.257 μW shows in Figure 4.

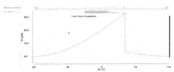

Figure 4. Total power dissipation of the conventional transmission gate D-Flip Fop.

Figure 5. D-Flip Flop using Stacking Techniques schematic.

3 PROPOSED POWER REDUCTION TECHNIQUES

3.1 *D-Flip Flop using transmission gate with stacking techniques*

Stacking is the techniques by which the leakage power reduces, because more than one transistor is connected in series. Figure 5 shows the design-II that uses the stacking techniques in the same schematic design-I. When more than one transistor in the stack is turned OFF, the leakage current flowing through the stack of series connected transistors decreases. When two or more transistors switched off are stacked on top of each other as in Figure.5, they dissipate less leakage power as compared to a single transistor used in the circuit. This is so because every transistor in the stack creates a tiny reverse bias between the source and gate of the transistor immediately below it, raising the threshold voltage of the bottom transistor and making it more leakage-resistant. As a result, as the current flows from V_{dd} to GND, the overall leakage current and related leakage power through the Transmission gate drop. If there is no natural stacking of transistors in a circuit, then two transistors, each of width W/2, are used in place of one transistor of width W to achieve the stacking effect. As we can see simulated result in Figure 6 that the total power dissipation reduces from 32.257 μW (conventional Transmission gate based D-Flip flop) to 12.629 μW (Stacking techniques based D-flip flop).

3.2 *D-Flip Flop using transmission gate with combination of stacking and substrate biased techniques*

We present D-Flip flop using Transmission gate with additional techniques as a combination of stacking as well as Substrate (Body) biased techniques simultaneously for reducing leakage power. Figure 6 shows the schematic of D-flip flop using stacking and substrate biased techniques, where four stages of conventional CMOS inverter have been used. As we know that, in conventional CMOS inverter generally the substrate (Body) of PMOS is connected to V_{dd} and substrate (Body) of NMOS is connected to GND.

Instead of the substrate (Body) of PMOS is connected to V_{dd}, we connect more positive Bias voltage at the body of PMOS and similarly we connect more negative substrate bias voltage at the substrate (Body) of NMOS. Now we can describe the effect of Substrate (Body) biased by the relationship between threshold voltage and substrate biased is given by

$$V_T(V_{SB}) = V_{T0} + \gamma \left(\sqrt{|1 - 2\varphi \quad V_{SB}|} - \sqrt{|-2\varphi \quad V_F|} \right) \tag{3}$$

Where $'\gamma'$ is substrate bias or body effect coefficient, Φ_F is the Fermi potential, (V_{SB}) is the Substrate bias voltage and V_{TO} is the threshold voltage for zero substrate voltage.

$$\gamma = \sqrt{\frac{2qN_{A\epsilon_{Si}}}{C_{ox}}} \tag{4}$$

Where C_{ox} is the gate oxide capacitance per unit area, ϵ_{si} is dielectric constant and N_A is the substrate doping density.

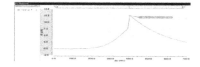

Figure 6.   Total Power dissipation of Stacking Technique.

3.3 *Drain current under substrate bias voltage*

Author presented, the drain current in linear and saturation region with substrate bias voltage is given by

$$I_D(\text{Linear}) = \frac{\mu_{n}.C_{ox}}{2} \quad \frac{W}{L} 2 \left(V_{gs} - V_T(V_{SB}) \right) V_{DS} - V_{DS}^2 \tag{5}$$

$$I_D(\text{Sat}) = \frac{\mu_{n}.C_{ox}}{2} \quad \frac{W}{L} \left(V_{gs} - V_T(V_{SB}) \right)^2 (1 + \lambda.V_{DS}) \tag{6}$$

In this paper we have used only the term V_T instead of V_T (V_{SB}) to express the general (Substrate bias dependent) threshold voltage, the substrate-bias effect can significantly change the value of the threshold voltage and, hence, the current capability of the MOSFET.

With this modification, we finally arrive at a complete first-order characterization of the drain (channel) current a nonlinear function of the terminal voltages.

$$I_D = f(V_{GS}, V_{DS}, V_{SB}) \tag{7}$$

Note that the threshold voltage V_T and the terminal voltages V_{GS}, V_{DS}, are all negative and V_{SB} is connected to the VDD for the PMOS transistor, similarly the threshold voltage V_T and the terminal voltages V_{GS}, V_{DS}, are all positive and V_{SB} is connected to zero for the NMOS transistor. Due to the inverse relationship between leakage current and threshold voltage, leakage current decreases as substrate bias voltage of PMOS increases. By using these techniques, we can reduce the overall leakage power of a CMOS inverter and corresponding leakage power of D-Flip flop, which is similar to how leakage current decreases as negative substrate bias voltage increases for NMOS. So that we proposed the designing of D-flip-flops using combination of Stacking and Substrate (body) biased schematic shown in Figure 7.

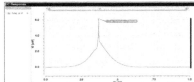

Figure 7.   D-Flip Flop using Stacking and Substrate Biasing schematic.

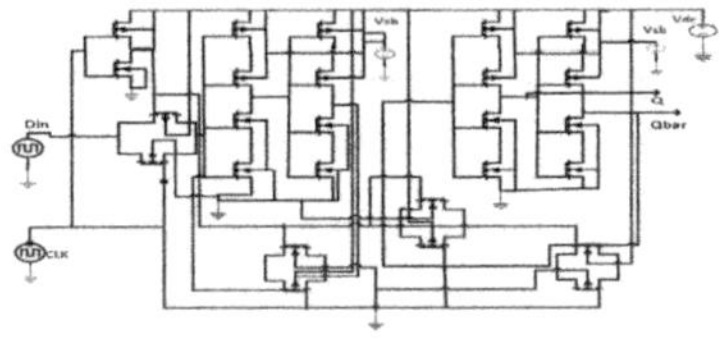

Figure 8.   Total Power dissipation of combined Stacking and substrate biased Technique.

Table 1.   Total power and delay comparison of D-FFs with reduction techniques.

Reduction Techniques	Total power dissipation	Delay
D Flip-Flop based on Transmission gate	32.5276 μW	2.357 nS
D Flip-Flop based on Transmission gate with Stacking techniques	12.6229 μW	2.59 nS
D Flip-Flop based on Transmission gate with Stacking techniques and Substrate biasing.	6.86 μW	

4   SIMULATION RESULTS

This paper proposes three CMOS implementations of DFFs employing transmission gates, transmission gates with stacking techniques, and Stacking with substrate biasing techniques. Average power and delay for all three configurations at 1 GHz clock frequency are shown in Table 1. Stacking with substrate biasing techniques has the lowest power dissipation at the cost of slightly increasing in propagation delay. As we can see in respective table which shows the different power reduction techniques result in terms of total Power dissipation and delay are shown. The total power dissipation and delay of conventional D-Flip Flop based on Transmission gate are 32.5276 μW and 2.357 nS respectively and the proposed both techniques D Flip-Flop based on Transmission gate with stacking and D-Flip Flop based on Transmission gate with simultaneously stacking and biased techniques reduces total power by 61.2 % and 81.28 % respectively with slightly increases delay of the circuit. All the design simulated in 90 nm technology using Cadence virtuoso tools for low power applications.

5   CONCLUSION

In this paper three CMOS implementations of DFFs using Transmission gate, transmission gates with stacking and transmission gates with stacking and substrate biased of all the PMOS are proposed. The average power and delay of designs at 1 GHz clock frequency are presented .The DFF design using stacking and stacking with substrate biased gates has the maximum delay (2.978nS) and maximum total power dissipation (6.0865 μW). The total power of all the designs decreases when reduction techniques are applied. The percentage reduction of total power is more with the proposed stacking and substrate biased technique. The design of D-FF using transmission gate with stacking technique gives the minimum power dissipation of 12.6229 μW. The design using stacking and substrate biased give least power dissipation of 6.0865 μW. Therefore for low power applications stacking and substrate biased design can be used in 90 nm technology.

REFERENCES

Calhoun B.H., Wang A., Chandra kasan A. Modeling and sizing for minimum energy operation in sub threshold circuits. *IEEE Journal of Solid State Circuits.* 2005 Sep; 40(1):1778–86.

Chandra kasan P. and Broder sen R. W., *"Low Power Digital CMOS Design"* Kluwer Academic, Boston, MA, 1995, pp. 105.

Chun J. W. and Chen C.Y.R., "Leakage power reduction using the body bias and pinner ordering technique," *IEICE Electronics Express*, vol. 13, no. 3, 2016, Art. no.13.2015-1052, https://doi.org/10.1587/elex.13. 20151052.

Chung I.Y., Park Y.J., Shick Min H. A new SOI inverter using dynamic threshold for low power applications. *IEEE Electron Device Letters.* 1997 Jun; 18(6): 248–50.Vol 9

Deepak subramanyan S., Nunez A., *"Analysis of Sub Threshold Leakage Reduction in CMOS Digital Circuits,"* in Proc. 13th NASA VLSI Symp. June 2000.

Dokic B. L., "A review on energy efficient CMOS digital logic," *Engineering, Technology & Applied Science Research*, vol. 3, no. 6, pp. 552–561, Dec. 2013, https://doi.org/10.48084/etasr-389.

February 2016, www.indjst.org Indian Journal of Science and Technology.

Gangele M. and Patra K. P., "Comparative analysis of lector and stack technique to reduce the leakage current in CMOS circuits," *International Journal of Research in Engineering and Technology*, vol. 4, no. 7, pp. 92–100, Jul. 2015, https://doi.org/10.15623/ijret.2015.0407015.

Jaehyun Kim, Chungki Oh and Youngsoo Shin, (2009) "Minimizing leakage power of sequential circuits through mixed-Vt flip-flops and multi-Vt combinational gates", *Journal ACM Transactions on Design Automation of Electronic Systems*, Volume 15 Issue 1, December 2009.

Johnson M. C., Soma Sekhar D., Chiou L. Y., and Roy K., "Leakage control with efficient use of transistor stacks in single threshold CMOS," *IEEE Trans. VLSI Syst.*, vol. 10, pp. 1–5, Feb. 2002.

Johnson M. C., Roy K., and Soma sekhar D. "A model for leakage control by transistor stacking", *Technical Report* TR-ECE 97–12, Purdue University, Department of ECE. Jiren Yuan and Christer Svensson, "High-Speed CMOS Circuit Technique" *IEEE Journal of Solid-State Circuits*, vol. 24, no.1, February 1989.

Kai W. W., binti Ahmad N. and bin Jabber M. H., "Variable Body Biasing (VBB) based VLSI design approach to reduce static power," *International Journal of Electrical and Computer Engineering*, vol. 7, no. 6, pp. 3010–3019, Dec. 2017, https://doi.org/10.11591/ijece.v7i6.pp3010-3019.

Kao J.T., Miyazaki M, Chandra kasan A. "Multiply accumulate unit using an adaptive supply voltage and body bias architecture", *IEEE Journal of Solid State Circuits.* 2002 Nov; 37(11):1545–54.

Kumre L., Shrivastava B. P. and Rai N., "Comparative analysis Of CMOS inverter for low leakage power," *International Journal Of Scientific & Technology Research*, vol. 8, no. 9, pp. 1598–1601, Sep. 2019.

Rabaey J.M., Chandrakasan A, Nikolic B. *Digital Integrated Circuits: A Design Perspective.* 2nd ed. UK; Pearson Education: 2003.

Research and Applications (IJERA), Vol. 2, Issue 3, May-Jun 2012, pp. 640–646.

Saini P., Mehra R. "Leakage power reduction in CMOS VLSI Circuits", *International Journal of Computer Applications (0975 – 8887)* Volume 55– No.8, October 2012.

Soeleman H., Roy K. Ultra low power sub-threshold digital logic circuits. *Proceedings of IEEE Conference on Low Power Electronics and Design*; San Diego, CA, USA. 1999 Aug 17. pp. 94–6.

Upadhyay H., Choubey A., Nigam K. "Comparison among different CMOS inverter with stack keeper approach in VLSI design" International Journal of EngineeringWang A, Calhoun BH, Chandra kasan A. *Sub Threshold Design for Ultra-low Power Systems.* 1st ed. US; Springer 2006.

Weste NH, Harris D. CMOS VLSI design – A circuits and systems perspective. 3rd ed. US; *Pearson Education*: 2004.

Ye Y., Borkar S. and De V., "Standby Leakage Reduction in High-Performance Circuits Using Transistor Stack Effects," 1998 Symposium on VLSI Circuit, to appear.

Advances in AI for Biomedical Instrumentation, Electronics and Computing – Sachan et al. (eds)
© 2024 The Author(s), ISBN 978-1-032-64298-7

YOLO based road anomaly detection for Indian roads

Jyoti Tripathi
Research Scholar, Netaji Subhas University of Technology, Dwarka, New Delhi, India
Assistant Professor, G B Pant DSEU Okhla-1 Campus, New Delhi, India

Bijendra Kumar
Professor, Department of CSE, Netaji Subhas University of Technology, Dwarka, New Delhi, India

Karan Ahlawat, Sanskriti Jain & Ashwani Chauhan
G B Pant DSEU Okhla-1 Campus, New Delhi, India

ABSTRACT: Road anomalies are found to be one of the major concerns behind road accidents. Early detection of anomalies such as potholes, obstacles, and weather conditions can reduce incident risks such as traffic jams and accidents. For Indian roads, the study related to anomaly detection was missing. This is the initial step toward the creation of a model to address the anomaly detection problem. The dataset used in this project was the Dataset for Indian Road Scenarios (DIRS 21) collected from IEEE, which was unlabeled and hence needed to be annotated. Computer vision annotation was used to annotate the image, and a YOLO model was developed for the evaluation and detection of anomalies.

1 INTRODUCTION

India has the second-largest road network in the world, spanning 5.89 million kilometers, making it the most popular mode of transportation in the country. The road network contributes around 3–4% of the total GDP and is one of the government's most valuable assets (SIRU 2022). However, with more than 1 lakh fatalities annually due to road anomalies, poor quality of roads, and abnormal driving patterns, it is a challenging proposition.

Abrupt lane changing is another prominent reason for accidents in India, accounting for roughly 7000–8000 accidents a year on national highways and contributing to 10%–12% of the total casualties. The primary cause of abrupt lane changes on roads is the road environment or overtaking. A traffic infraction due to lane changes or driving in the wrong lanes accounts for 5%–7% of road infractions (Farooq *et al.* 2021).

In urban areas, poor-quality roads increase commuters' overall travel time and city pollution level. Roads in rural areas are nightmares due to their unstructured nature and poor construction quality. Since the country has diverse terrain and population, it is challenging to construct and maintain quality roads (Atlas and Boots 2022). Since the road network is an indirect contributor to the country's economy and prosperity, we therefore need to analyse road anomalies and help commuters travel safely on the road.

This work aims to classify the anomalies as positive and negative to cater to drivers' needs. It is also helpful in predicting accident-prone and disaster-struck areas, which will help the vehicles they are supposed to reach later. This work is also beneficial to civic bodies concerning the maintenance of the roads. This work aims to annotate the image dataset for Indian Road Scenarios (DIRS21), which consists of high-resolution images. It is a paid dataset available on IEEE. It has over 5000 images captured by a 12-megapixel camera with an aperture of f/1.8 and an image pixel resolution of 1920 × 1080. These images were captured while driving on

DOI: 10.1201/9781032644752-5

25

the highways and in the city in normal traffic conditions, illumination, and clear weather. Since the dataset's images were raw with noise or corrupted data, feature extraction analysis can be done. It is the process of converting the raw data to features that are helpful to conclude crucial information from the original dataset for various purposes (Kumar *et al.* 2021).

As of now, most of the existing work included a dataset of foreign roads; the differentiator of the project was this dataset. Indian roads are diversified and are constructed according to the different terrains present in the country, which was also one of the major reasons for choosing the dataset. The dataset was unlabeled as per the model's requirements. Computer vision annotation tools were used to annotate the data. The annotated data was exported into YOLO format for further analysis. The rest of this paper is organised as follows: Section 2 illustrates the data preprocessing. Section 3 describes the YOLO detection, and results and conclusions are discussed in Section 4.

2 DATA PREPROCESSING

An image annotation tool, cvat.ai, was used to annotate the data. CVAT is an open-source image annotation tool developed by Intel. It is useful for annotating and labelling the dataset online or by installing the software. It has high applications in the fields of object detection, tracking, image classification, and more. The annotated dataset can be stored in various formats, making it easier to use for predictions. Job tasks were created by uploading the zip file of the dataset, and then each image was manually annotated and classified into the prediction labels.

We use rectangular-shaped annotations. After annotations were complete, they were converted into a Yolo annotation format using CVAT, giving us the image and a.txt file containing the coordinates of each annotation. The annotations were done by taking two classes: positive and negative anomalies. These were the two main classes that characterised the various anomalies on the road. Positive anomalies were depicted on the positive y-axis, such as the objects piled up on the road. Whereas, negative anomalies depicted the anomalies present on the negative y-axis of the road, such as potholes and cracks. Table 1 exhibits the examples of the image and the

Table 1. Various image annotations using CVAT annotation tool.

Figure 1. Unannotated image of the road.

Figure 2. Annotated negative anomaly image.

Figure 3. Unannotated image with objects on the road.

Figure 4. Annotated positive anomaly image.

Figure 5. Unannotated image.

Figure 6. Annotated image with obstruction on road identified as positive anomaly.

26

annotated image. Figures 1 and 2 show the annotation of the negative image anomaly, i.e., a pothole. Figures 3 and 4 represent the annotation of a positive image anomaly on the road surface, including stray dogs, sudden diversion boards, and other articles on the road.

The objective of dividing the dataset into two classes was to simplify the analysis and problem-solving. The images were then exported into YOLO 1.1 format for further use.

3 YOLO DETECTION

YOLO is an object detection model that uses deep neural networks. YOLO stands for "You Only Look Once." The algorithm requires single-forward propagation through a neural network to detect objects. The model processes the entire image at once, goes through the network once, and can see things in real-time.

The algorithm works using the following techniques:

- Residual Blocks-These divide the image into various grids with a dimension of A × A.
- Intersection Over Union-It describes how boxes overlap, and YOLO uses IOU to provide the output box that surrounds the object perfectly.
- Bounding Box Regression – The outline highlights the object in the image with the following attributes, which are bounding box coordinates such as Width,Height, Bounding Box Centre and Classes.

The most popular implementation is YOLOv5 by Ultralytics. The latest performance is YOLOv7. The new YOLOv7 shows the best speed-to-accuracy balance compared to state-of-the-art object detectors. YOLOv7 surpasses all previous object detectors in speed and accuracy, ranging from 5 FPS to 160 FPS (Shivaprasad 2020 and Maj 2023).

Here, the focus is on processing the custom dataset to tune the hyperparameters. We trained YOLOv7 on 4 sets of hyperparameters and YOLOv5 on 2 sets of hyperparameters. They are adjusted to fit the problem at hand, such as the learning rate for the optimizer. The various operations on the images, such as sheer, rotation, flip, etc., are done by the YOLO network for data augmentation. It helps create new data for training the existing model and avoids overfitting. When starting the network training, some command-line parameters must be specified, which greatly impacts the results.

These are as follows -

- IMAGE SIZE – The images are resized by shrinking down, enlarging, padding, etc., to the size specified here. Usually, the larger size means much more memory consumption and better recognition of smaller details. For our purposes, the measures used were 1280, 640, and 320 pixels, respectively.
- EPOCHS – The number of times the model cycles through the full training dataset. Usually, the number of training epochs is always high. The sizes utilized in our runs were between 100 and 300.
- CONFIG – The configuration file specified in this parameter dictates the network architecture used in the model. Based on the network's depth (number of hidden layers and number of units in layers), there are 5 different neural networks, from nano to extra-large, each consuming increasing amounts of memory and time to train.
- WEIGHTS – This is one of the most important parameters of the network. It lets us choose a set of weights we would like to use for the model. This implies that we can use pre-trained weights (corresponding to the selected configuration) for the training, yielding better results or custom weights based on the dataset. On the other hand, if our data set is large enough (>10k images, with equal distribution of the classes), we can choose to randomly initialize the weights to train weights highly customized to the data set.

A single round of training takes anywhere from 40 minutes to 3 hours, depending on the different parameters, and the hardware of the machine used to train the data (Boesch 2023).

The flowchart depicting the process of the YOLO model is shown in Figure 7. The dataset used is IEEE DIRS21. The dataset was unannotated and did not include any predefined classes.

The data preparation was required by manually annotating the images of the dataset using the cvat.ai tool. Image annotation has been achieved by placing the bounding boxes on the object of interest. Further, the annotations were exported in YOLO format, with each image having an associated text file containing information about each bounding box on the image.

The model training could be done either by using the pre-trained weights of the YOLO model or by training the weights from scratch. Here, the training of the weights had been done from scratch with the following steps:

- Select the architecture of the network to use (can be specified as a YAML file).
- Select the size images will be resized to in-memory for training, according to compute restrictions.
- Train all the layers of the network from scratch, determining the weight for all connections.
- Train till stable loss is reached.
- Evaluate model performance and predict results.

Lastly, the predictions of the result were made by the trained Yolo model. The input image is processed and converted into an appropriate format for input into the YOLO network. The image is passed through the trained network. It produces a set of bounding boxes with class confidence and other parameters.

4 RESULTS AND CONCLUSIONS

Intersection over Union (IOU) is an important metric that is calculated for evaluation purposes. It helps us to define other terms related to it more quantifiable way. IOU metric is defined as the degree of overlap between the ground truth (gt) and prediction (pd). The ground truth and prediction can be of any shape and are calculated by the formula in Equation 1:

$$IOU = area(gt \cap pd)/area(gt \cup pd) \tag{1}$$

IOU ranges from 0 to 1, with 0 signifying no overlap and 1 signifying a perfect overlap between the predicted and ground truth bounding boxes.

In YOLO, IOU is used as a threshold to determine whether a predicted bounding box is a true positive or a false positive. If the IOU value between a predicted bounding box and a ground truth bounding box is above a certain threshold (typically 0.5 or 0.75), the predicted bounding box is considered a true positive. If the IOU value is below the threshold, the predicted bounding box is considered a false positive.

The important parameter used to measure the performance of a YOLO model and their equations illustrated below and furthermore these parameters are depicted in the graph between the number of epochs and the parameter for various test cases:

Figure 7.   Flowchart of the process for image annotation.

1. *Box = Center of the object + Coverage of Area*   (2)

2. Objectivenss
3. Classification

4. *Precision = TP/(TP + FP)*   (3)

5. *Recall = TP/(TP + FN)*   (4)

6. Average Precision $AP@\alpha = \int_{0}^{1} p(r)dr$ (5)

7. Mean Average Precision $AP@\alpha = \dfrac{1}{n}\sum_{i=1}^{n} APi$ (6)

Here, n is the number of classes to classify the image intovalBox, valObjectiveness, and valClassification represents the result of objectives and classification over the validation dataset. Here, the results have been generated by varying the input image size and runtime parameters.

The results of **YOLO** with an image size of 320 pixels are presented in Figure 8. The box coverage area is high consistently at the various epochs, indicating that the model covers a wide area around the object in the input image. However, the objectiveness of the result is low, suggesting that the model failed to identify the correct objects. The precision of the model was initially high for a small count of epochs but became zero for the latter epochs. This resulted in poor classification and precision. The recall was initially at 0.1 for a small number of epochs but then dropped to zero as the number of epochs increased. The mean average precision (mAP) was initially high for a small number of epochs but decreased significantly as the number of epochs increased. The recall also followed a similar pattern.

Figure 8.   Yolo5 Model results on the image of 320 pixels.

Figure 9.   YOLO7 Model results on image of 640 pixels.

Figure 9 exhibit the Yolo results when the image size was set to 640 pixels and it shows the similar parametric values as we obtained in Figure 8.

REFERENCES

Atlas & Boots (2022) *World's Worst Countries to Drive in – Ranked.*
Boesch, G. (2023) Yolov7: *The Most Powerful Object Detection Algorithm (2023 Guide).* viso.ai.
Farooq, D., Moslem, S., Jamal, A., Butt, F. M., Almarhabi, Y., Faisal Tufail, R., & Almoshaogeh, M. (2021). Assessment of significant factors affecting frequent lane-changing related to road safety: An integrated approach of the AHP–BWM model. *International Journal of Environmental Research and Public Health*, 18(20), 10628.
Kumar. S., Naginneni S., and Parasuraman S. (2021). *DIRS21-Dataset for Indian Road Scenarios.*
Maj M. (2023). *Object Detection and Image Classification with Yolo. KDnuggets.* (n.d.).
SIRU, (2022) Development in highway and road networks to boost economic gro... *Investment Promotion and Facilitation Agency.*
Shivaprasad P. (2020) *A Comprehensive Guide to Object Detection Using YOLO Framework-Part II.*

Advances in AI for Biomedical Instrumentation, Electronics and Computing – Sachan et al. (eds)
© 2024 The Author(s), ISBN 978-1-032-64298-7

Job recommendation system based on content-based filtering

Anu Saini, Jyoti Tripathi, Mohd Faraz & Mohd Kaif
G.B. Pant DSEU Okhla I Campus

ABSTRACT: Searching for jobs among the numerous listings on the internet can be quite time-intensive. Therefore, it is crucial for job recommender systems to offer highly accurate predictions that align with the user's preferences. Our initiative employs a content-based recommendation strategy to propose suitable job opportunities to job seekers. The system takes user inputs, including skills, experience, and qualifications, and employs the content-based recommendation method. This involves using techniques like cosine similarity and Term Frequency-Inverse Document Frequency. Term Frequency-Inverse Document Frequency, a well-established Natural Language Processing technique, is harnessed to convert text into a diverse set of vectors. Subsequently, the cosine similarity metric is used to gauge the resemblance between two vectors, both for determining vector similarity and for assessing word similarity. Adopting this approach brings about enhancements in accuracy, quality, and scalability, all while keeping memory usage minimal.

Keywords: Recommendation System, Jobs, Content-Based Filtering

1 INTRODUCTION

According to a recent study, choosing a career field can be challenging for most college graduates. Currently, numerous websites offer extensive information about diverse job opportunities. However, for students, this process is excessively time-consuming, as they are required to sift through substantial volumes of data across various platforms to identify suitable employment prospects. It can be confusing for them because different job opportunities are available on different websites at different time intervals. Therefore, there is a need for a system that recommends jobs based on an analysis of an individual's skills.

Most recommendation systems fall into two categories: collaborative filtering and content-based recommendation. Content-based recommendation algorithms group users by comparing representations of items and user profiles, with a user profile reflecting the user's interests. On the other hand, collaborative filtering relies on peer judgment to predict the interests of others. In collaborative filtering, a target user is compared to every other user to identify those who are similar to the target user and share their interests.

Recommendation systems are employed by numerous businesses to enhance user engagement and improve job seekers' experiences. One advantage of recommendation systems is an increase in client satisfaction. The job recommendation system is robust and essential. However, job recommendation systems also face challenges such as poor suggestion quality and scalability issues when relying solely on a collaborative approach.

At some point in their lives, everyone seeks employment, leading to the development of numerous software applications designed to assist individuals in this endeavor. One such tool takes the form of an online recruitment platform, enabling individuals to effortlessly discover jobs aligned with their interests with just a single click. Many people are looking to transition from their current field to IT, enrolling in online courses and conducting random job searches. Nowadays, IT fields attract many students, but they often struggle to determine which domain is the right fit for them. To address this situation, candidates need a job recommendation

DOI: 10.1201/9781032644752-6

system that analyzes their skills, experience, and qualifications to recommend a suitable job. Content-based recommendation algorithms are well-known for their effectiveness; for example, according to Business Insider Australia, Netflix estimates that 20% of video views come from searches, while 80% come from their recommendation system.

Our system will operate in a machine learning environment where it analyzes user data and matches the user's skill set with the jobs currently available on various job portals using web scraping and content-based filtering, which makes it easier to scale to a large number of users. Subsequently, the system will provide the user with recommendations for the most optimal job opportunities, ensuring that the user receives results that meet their expectations. Users will also have the option to read classified overviews about organizations, which will assist them in choosing the best job offered.

2 RELATED WORK

This section discusses the various studies spanning various types of research carried out on the aforementioned topic of Job and related works. The advantages and setbacks faced by the researchers/studies are discussed. (Jorge Valverde-Rebaza *et al.* 2018), here introduces a framework for job recommendation. This framework simplifies the understanding of how job recommendations work and allows for the utilization of various text processing and recommendation techniques tailored to the preferences of the job recommender system's designer. They also released a new dataset containing profiles of job seekers and open positions. (Ioannis Paparrizos *et al.* 2011) proposed a supervised learning model for anticipating career moves and offering employment recommendations. Their model achieved significantly higher accuracy compared to a basic baseline predictor, trained on a large dataset of job transitions found on the web. (Islam A. Heggo *et al.* 2018) the objective is to establish a successful online recruitment platform that satisfies the high standards of both recruiters and job seekers. This is achieved through the development of a hybrid recommendation algorithm that considers content, behavior, demographics, and other significant recommendation data. The paper also explores optimal ranking criteria and how these attributes can be utilized as input for the hybrid algorithm to create superior job recommendation models. (Poonam Manjare *et al.* 2017) presents a job portal offering an efficient way for job seekers to access information about available positions. The primary objective of this project is to match graduates with market demands and enhance online job portals by addressing existing system issues. It also possesses the capability for flexible high-level management and standardized content services. (Dheeraj Bahl *et al.* 2016) allows users to express their interests using its well-organized resources. The system utilizes user profiles and employment data to generate personalized job and candidate recommendations. It employs various types of graphs, including directed, weighted, and multi-relational graphs, to model the collected data. (Ravita Mishraet *et al.* 2016) emphasizing two specific challenges: the lack of appropriate evaluation metrics and scalability issues. The discussion also covers strategies to enhance data collection and conversion to strengthen observation and product analysis. Addressing privacy concerns while meeting modeling requirements and employing multiple methodologies, including MF, KNN, and GPA, we focus on privacy and security issues. (Manal Alghieth *et al.* 2019) proposes and implements a general map-based job recommender system within the Saudi Arabian job-seeking context. This system utilizes content-based techniques, including cosine similarity, to recommend jobs and helps Saudi job seekers efficiently locate their desired positions using interactive maps. This initiative aligns with Saudi Arabia's goal of achieving digital transformation, a key objective of the Saudi Vision 2030. (V. M. Deshmukh *et al.* 2018) this recommendation system employs various recommendation approaches, including simple matching, collaborative filtering, and content-based matching, to determine the similarity between items in the matrix using similarity metrics. The method was successfully implemented, and various recommendation strategies were compared. (Wenxing Hong *et al.* 2013) offers an extensive analysis of four online job recommender systems (JRSs) from four different perspectives: user profiling, recommendation strategies, recommendation output, and user feedback. It specifically highlights the distinctions between various online JRSs and underscores their pros and cons. Designing recommendation systems poses a significant challenge due to the diverse features of job seekers. (Juhi Dhameliya *et al.* 2019) combined multiple approaches to enhance job recommendations. Unfortunately, there is a shortage of benchmark datasets for testing. Content-based filtering (CBF) necessitates extensive data preprocessing due to the varied job description formats. Experiments revealed that using Word2Vec in CBF could lead to false recommendations by including irrelevant skills. However, Word2Vec in collaborative filtering (CF) improved accuracy.

3 PROPOSED APPROACH

Many job portals provide extensive information on a wide range of job opportunities. However, for students, this process becomes exceedingly time-consuming as they must sift through vast amounts of data across various websites to find suitable job options. This confusion arises because different job opportunities are scattered across different websites. Therefore, the need for a Job Recommendation System arises, designed to assist individuals in finding jobs based on their skills, qualifications, and experience.

3.1 *Flowchart*

The user searches by entering qualifications, skills, experience, etc. The system preprocesses the scraped data, analyzes it, and applies content-based filtering based on the user's input to provide similar job recommendations. See Figure 1 for the system workflow.

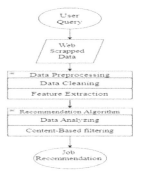

Figure 1.   Flowchart of the proposed system.

1. The user inputs essential information such as skills, qualifications, years of experience, etc.
2. The scraped dataset is cleaned, and essential features are extracted for further processing.
3. After preprocessing the dataset, feature vectorization is performed using the TfidfVectorizer from the sklearn library to convert textual data into feature vectors. This process generates the required sparse matrix.
4. To find similar jobs, we compute Cosine Similarity among the vectors of the sparse matrix using the cosine similarity function from the sklearn library.
5. After these processes, the algorithm seeks the closest match to the user's input within the dataset. It generates a list of comparable jobs based on similarity scores and arranges them accordingly.
6. A list of recommended similar jobs based on user information will be provided.

4 RESULT AND IMPLEMENTATION

4.1 *Creating the dataset*

To create the real-time dataset, we scraped a job portal using Python modules like Selenium and BeautifulSoup. Figure 2. illustrates the web scraping process of a job portal and converting the scraped data into a data frame for further processing.

```
In [79]: import requests
         from selenium import webdriver
         from bs4 import BeautifulSoup
         import time
         import pandas as pd

In [80]: df = pd.DataFrame(columns=['Title','Company','Skills','Experience','Qualification','Salary','Location',

In [81]: List = ["web-developer-jobs","data-engineer-jobs", "android-developer-jobs","computer-support-jobs","c)
         for i in List:
             url="https://www.naukri.com/"+i
             print(url)
             for param in range(1,2):
                 driver = webdriver.Chrome("D:\\Selenium\\chromedriver.exe")
                 driver.get(url)
                 time.sleep(10)
                 soup = BeautifulSoup(driver.page_source,'html5lib')
                 driver.close()
                 results = soup.find(class_='list')
                 job_elems = results.find_all('article',class_='jobTuple')
```

Figure 2.   Creating the dataset.

4.2 *Preprocessing the data*

Data cleaning has been applied to the created data frame, and 'stopwords' have been removed from the 'Skills' and 'Qualification' labels. See Figure 3 for data preprocessing.

Figure 3.   Pre-processing the data.

4.3 *Vectorization of text*

Now, textual data is vectorized using the TF-IDF technique to create similarity matrices for 'Skills' and 'Qualification' labels, as shown in Figure 4.

```
In [11]: df_skills[['Exp','last']] = df_skills['experience'].str.split('-', n=1, expand=True)
         df_skills.drop(df_skills.columns[[3,10]], axis=1, inplace=True)
         df_skills.rename(columns={"Exp": "Experience"}, inplace=True)
         df_skills
         df_skills.to_csv('file1.csv')

In [12]: tf = TfidfVectorizer(analyzer='word',ngram_range=(1, 2),min_df=0, stop_words='english')
         tfidf_matrix = tf.fit_transform(df_skills['Skills'])
         tfidf_matrix_qual = tf.fit_transform(df_skills['Qualification'])
         print(tfidf_matrix.shape)
         print(tfidf_matrix_qual.shape)

         (220, 2208)
         (220, 201)

In [13]: tfidf_matrix
Out[13]: <220x2208 sparse matrix of type '<class 'numpy.float64'>'
             with 4195 stored elements in Compressed Sparse Row format>

In [14]: import pickle
         file="tfidf_vectorizer_skills.pickle"
         fileobj=open(file,'wb')
         pickle.dump(tfidf_matrix,fileobj)
         fileobj.close()
         fileobj=open(file,'rb')
         mm=pickle.load(fileobj)
         print(mm)
         (0, 650)         0.26926301315180773
```

Figure 4.   Vectorization of text.

33

4.4 Finding cosine similarity

Using TF-IDF vectorization, Cosine Similarity is computed for each record in the sparse matrix to determine similar vectors, as shown in Figure 5.

```
In [16]:
        cosine_sim = cosine_similarity(tfidf_matrix, tfidf_matrix)
        cosine_sim.shape
Out[16]: (220, 220)

In [17]:
        cosine_sim_qual = cosine_similarity(tfidf_matrix_qual, tfidf_matrix_qual)
        cosine_sim_qual.shape
Out[17]: (220, 220)

In [18]: cosine_sim[1]

In [19]: cosine_sim_qual[1]
```

Figure 5. Finding cosine similarity.

4.5 Getting recommendations index

Here, index values for recommended jobs are computed by enumerating the Cosine Similarity matrix, ensuring that the most similar jobs are listed at the top. Refer to Figure 6. for illustration.

```
from sklearn.metrics.pairwise import linear_kernel, cosine_similarity
from fuzzywuzzy import process, fuzz
import pickle

def get_recommandation(input_skills,input_exp,input_qual,tfidf_matrix,tfidf_matrix_qual,dfa_skills,dfa_qual,indices,indices_qual):
    cosine_sim = linear_kernel(tfidf_matrix, tfidf_matrix)
    cosine_sim_qual = linear_kernel(tfidf_matrix_qual, tfidf_matrix_qual)

    idx = indices[process.extractOne(input_skills.lower(), indices.index, scorer=fuzz.token_sort_ratio)[0]]
    idx_qual = indices_qual[process.extractOne(input_qual.lower(), indices_qual.index, scorer=fuzz.token_sort_ratio)[0]]

    if isinstance(idx, pd.Series):
        idx = idx[0]

    if isinstance(idx_qual, pd.Series):
        idx_qual = idx_qual[0]

    sim_scores = list(enumerate(cosine_sim[idx]))
    sim_scores = sorted(sim_scores, key=lambda x: x[1], reverse=True)

    sim_scores_qual = list(enumerate(cosine_sim_qual[idx_qual]))
    sim_scores_qual = sorted(sim_scores_qual, key=lambda x: x[1], reverse=True)

    job_indices = [i[0] for i in sim_scores]
    job_indices_qual = [i[0] for i in sim_scores_qual]

    return indices.iloc[job_indices], job_indices, indices_qual.iloc[job_indices_qual], job_indices_qual
```

Figure 6. Getting recommendations index.

4.6 Taking user input

Here, user input includes skillset, qualifications, experience, etc., as shown in Figure 7.

```
if __name__ == "__main__":
    df_skills = pd.read_csv('skills.csv', usecols=['Skills'])
    df_qual = pd.read_csv('quals.csv', usecols=['Qualification'])

    main_df = pd.read_csv('jobs_recommends.csv', index_col='Index')
    indices = pd.read_csv('skills_indices.csv')
    indices = pd.Series(indices.index, index=indices['Skills'])
    with open("tfidf_vectorizer_skills.pickle", "rb") as f:
        tfidf_matrix = pickle.load(f)
    indices_qual = pd.read_csv('quals_indices.csv')
    indices_qual = pd.Series(indices_qual.index, index=indices_qual['Qualification'])
    with open("tfidf_vectorizer_skills_qual.pickle", "rb") as f:
        tfidf_matrix_qual = pickle.load(f)
    input_skill = input("Enter the skills: ")
    input_exp=input("Enter the years of experience: ")
    input_qua = input("Enter your qualification: ")
    input_sk=input_skill.lower()
    input_ex=input_exp.lower()
    input_q=input_qua.lower()
    recommandations,job_indices,recommandations_q,job_indices_qual = get_recommandation(input_sk,input_ex,input_qua,tfidf_matrix,

Enter the skills: Android Development
Enter the years of Experience: 2
Enter your qualification: b tech
```

Figure 7. Taking user input.

4.7 Output

Jobs are recommended to the user based on their entered details, as shown in Figure 8.

```
dt=datas.loc[datas.index[index_1s]]
dt.iloc[:12]
```

Index	Title	Company	Skills	Experience	Qualification	Salary	Location
42	Need Android Developer For Gurgaon (Off roll) 30K	TeamLease	Mobile Application Development,Android	2-5 Yrs	B.Tech/B.E. in Any Specialization	3,00,000 4,25,000 P.A.	Gurgaon/Gurugram
39	Android Developer	Bar Code India	Android Application Development,Java,android d...	2-5 Yrs	B.Tech/B.E. in Electronics/Telecommunication.	Not disclosed	Gurgaon/Gurugram
41	Flutter Developer - IOS/Android Apps	Appscrip	IOS mobile software development,Android,automa	0-1 Yrs	B.Sc in Any Specialization, BCA in Any Special.	Not disclosed	Kolkata, Mumbai, Hyderabad/Secunderabad Luckn
53	Android Developer	Ather Energy	java,kotlin,debugging,Android SDK,Git,performa	2-5 Yrs	Any Graduate	Not disclosed	Pune
46	Android Developer	Kashyaps Hr Solutions	Mobile development,Programming,JSON,HR,Android	2-4 Yrs	BCA in Computers,MCA in Computers	Not disclosed	Noida, Ghaziabad, New Delhi, Gurgaon/Gurugram
55	Android Developer - Mobile App Designing	Hyrezy Talent Solution	Mobile App Designing,Android,Java,Core Java,An	0-5 Yrs	BCA in Any Specialization, B.Tech/B.E. in Any	Not disclosed	Hyderabad/Secunderabad
51	Android Application Developer	Tecxpert Software Pvt Ltd	Mobile Applications,web application development	2-7 Yrs	BCA in Any Specialization, B.E.A/ B.M.S in Any.	Not disclosed	Delhi / NCR

Figure 8. Output.

5 CONCLUSION

Our primary objective is to suggest job opportunities from a dataset based on user-provided information. This project offers a solution for job recommendation using content-based filtering. By considering user inputs such as skills, experience, and qualifications, we can recommend suitable jobs. Our recommendation system relies on content-based filtering rather than collaborative filtering to provide personalized job suggestions, ensuring scalability for a large user base. Additionally, the content-based approach offers superior accuracy and quality compared to the collaborative approach, thanks to its utilization of domain knowledge for tailored recommendations.

REFERENCES

Deshmukh V.M. and Roshan G. Belsare (2018) "Employment recommendation system using matching, collaborative filtering and content based recommendation". *In International Journal of Computer Applications Technology and Research*, Vol. 7, Issue 6.

Dheeraj Bahl, Vinay Desai, Shree kumar Vibhandik, Isra Fatma (2016) "A profile based job recommender system". In *International Research Journal of Engineering and Technology*, Vol. 3, Issue 11.

Ioannis Paparrizos, Barla Cambazoglu B., Aristides Gionis (2011) "Machine learned job recommendation". *In Proceedings of the Fifth ACM Conference on Recommender Systems*, Chicago, IL, USA, pp. 325–328.

Islam A. Heggo and Nashwa Abdelbaki (2018) "Hybrid information filtering engine for personalized job recommender system". In *Advances in Intelligent Systems and Computing*, Vol. 723, pp. 553–563.

Jorge Valverde-Rebaza, Ricardo Puma-Alvarez, Paul Bustios, Nathalia C. Silva (2018) "Job recommendation Based on job seeker skills: An empirical study". In *Proceedings of the Text2StoryIR'18 Workshop*, Grenoble, France.

Juhi Dhameliya, Nikita Desai (2019) "Job recommendation system using content and collaborative filtering based techniques". In *International Journal of Soft Computing and Engineering (IJSCE)*, Vol. 9, Issue 3.

Manal Alghieth and Amal A. Shargabi (2019) "A map-based job recommender model". *In International Journal of Advanced Computer Science and Applications*, Vol. 10, No. 9.

Poonam Manjare, Jyoti kumbhar, Sayali ovhal, Rajnandini Munde (2017) "An effective job recruitment system using content-based filtering". In *International Research Journal of Engineering and Technology*, Vol. 4, Issue 3.

Ravita Mishra and Sheetal Rathi (2019) "Efficient and scalable job recommender system using collaborative filtering". In *Lecture Notes in Electrical Engineering* 2019, Vol. 601, pp. 842–856.

Wenxing Hong, Siting Zheng, Huan Wang (2013) "A job recommender system based on user clustering". In *Journal of computers*, Vol. 8, No. 8, pp. 1960–1967.

Advances in AI for Biomedical Instrumentation, Electronics and Computing – Sachan et al. (eds)
© 2024 The Author(s), ISBN 978-1-032-64298-7

Wondervision: A smart stick for visually impaired people

Sunita Tiwari, Manjeet Singh Pangtey, Golu Kumar, Raghav K. Jha & Sambhav Anand
G B Pant DSEU Okhla-1 Campus, Delhi, India

ABSTRACT: Sight is a precious gift for living beings, allowing us to capture and admire the beauty of nature through our eyes. Eyes serve as fundamental organs for the survival of any living being. Unfortunately, a significant portion of the global population lacks this gift. This challenge is even more pronounced in developing and impoverished countries where essential infrastructure is lacking, thereby complicating the mobility of the visually impaired. This work aims to enhance the daily lives of the visually impaired by providing them with increased confidence and independence both indoors and outdoors. By incorporating a buzzer, the proposed device will alert users, ensuring safer transitions from one point to another. This device can be conveniently attached to the visually impaired individual's cane, effectively transforming it into a smart cane. Notably, one of the device's strengths lies in its affordability, which sets it apart from other assistive devices for the visually impaired. This affordability enables individuals from economically disadvantaged backgrounds to easily acquire the device.

Keywords: Assistive technology, IOT, Visually impaired, Arduino UNO, Ultrasonic Sensors

1   INTRODUCTION

The assistive technology for the visually impaired is a developing research area. These man-made assistive devices benefit the visually impaired by making them more independent. The condition is called visual impairment when the person faces a loss of sight and it is beyond the full recovery even after using glasses or contact lenses. Assistive technology can help such people by enabling them to perform routine activities like indoor and outdoor movements, locating objects and obstacles, etc. There are various assistive technologies available in the market for visually impaired people; most of them are either challenging to carry or they are very expensive for people of underdeveloped or developing countries. Over the years, different commercial applications were developed, and among them, the most popular applications are the GPS powered applications like Mobile Geo, Braille Note GPS (Sterr 1998), MoBIC (Petrie 1996) etc., computer vision-based application like The vOICe (Auvray 2007), NAVI (Ponchillia 2007), ENVS (Benjamin 1973), TVS (Sivan and Darsan 2016) etc.

According to the National Blindness and Visual Impairment survey (Prasad 2020), 0.32% of the total population of India suffers from complete blindness, and our work has been pursued considering making their life easier. The proposed product is designed in such a way that it can be easily attachable to the cane of the visually impaired, making it a smart cane and helping users detect obstacles in their path using ultrasonic waves and sensors. If there is an obstacle in the path and there is a chance of collision, the device will start beeping with the help of a buzzer, signaling the user to change their path in order to avoid the collision. The device also has an emergency alert system, which can be used to send SOS signals to family members or any registered mobile number.

The technologies being used are Arduino, jumping wires, ultrasonic sensors, and buzzers. The device is built taking special consideration to cost as we want to make the device affordable to everyone who needs it. The device is easy to use, and the user will need little adaption time so that the user can feel comfortable using it in a very short time, and it doesn't need much training time and cost.

DOI: 10.1201/9781032644752-7

2 RELATED WORK

Considerable efforts and scientific endeavors have been dedicated to facilitating the mobility of the visually impaired through the development of new technologies and innovative solutions. However, numerous limitations persist within these systems, which we aim to address and enhance through this project. The subsequent section outlines the limitations observed in some of these existing systems.

Shove *et al.* (Prasad 2003) presented a method for individuals with visual impairments, wherein they introduced two distinct categories of sounds. The primary focus of their research is on noise recognition. In a similar vein, Yuan and Manduchi (2005) introduced the notion of a functional triangle within his devised apparatus. This triangle aids in object identification. However, a notable drawback of this functionality lies in its limitation to detecting objects at a rate of only 15 per second. Pereira (Pereira 2015) presents the "Blind Guide," a sensory-driven ultrasound network designed to assist visually impaired individuals. The paper offers a comprehensive approach to detecting blind individuals who typically rely on a white cane or guide dogs, as well as a visual detection method for those who don't use such aids. This work showcases remarkable innovation, although it encountered the challenge of affordability for the middle-class and economically disadvantaged populations in India. This issue is particularly relevant due to India's higher temperatures compared to Western regions, as the elevated heat can lead to sweating, potentially affecting nerves and other electronic components.

Sabarish *et al.* (Sabarish 2013) put forth a system proposal similar to our project, albeit with the inclusion of vibrators in their device. Despite facing setbacks, they persisted in their efforts. Espinosa *et al.* (Espinosa 1998) introduced an idea that came with a higher-than-usual cost. Unfortunately, this pricing consideration excluded economically disadvantaged individuals from accessing such technology, thereby impeding our shared goal. Pooja Sharma (Sharma 2015) designed a device for the visually impaired. However, this device's object detection is constrained to a specific width within its range. JM. Benjamin (Benjamin 1973) introduced a concept involving a laser beam functioning in three directions: one at a 45-degree elevation, one parallel to the ground, and one providing sharp depth perception. Notably, his system's primary limitation is its unsuitability for outdoor activities.

Compared to the current research, the proposed work asserts its affordability and user-friendliness, employing a cane as a tool for visually impaired individuals. Additionally, it offers the capability to send emergency alerts promptly when needed.

3 SYSTEM DESIGN

Assistive technology for the visually impaired is a continually evolving field. These devices offer vital support to those with visual impairments, enhancing independence. Visual impairment, when sight loss remains despite corrective measures, drives the need for such technology. Assistive tools aid daily tasks, including navigation and object detection. Despite market options, accessibility and cost barriers persist, particularly in underdeveloped regions.

Notably, GPS-powered and computer vision-based applications, such as Mobile Geo, The vOICe, and NAVI, have gained popularity. The paper presents the solution in the form of a smart cane.

The necessary hardware components for the system consist of an Arduino UNO, ultrasonic sensors, a buzzer, a 5 mm red LED, jumper wires, and an ESP32 wifi module. The software employed is Arduino, an open-source editing platform widely utilized for designing such systems. Arduino, acting as a microcontroller, facilitates real-time sensing and control of objects within an environment. This microcontroller is both cost-effective and extensively used across various Internet of Things (IoT) applications. The device is capable of receiving inputs from sources such as light on a sensor, a physical button, or a message and converting these inputs into corresponding outputs like LED light blinking online content publication, or message transmission. The operational voltage requirement stands at 5V, and the device possesses 32KB of flash memory. In Figure 1, an image depicts an Arduino UNO board, ultrasonic sensor, jumper wires, and the sensor itself.

The device is wired according to the following configuration. The ground of the buzzer is linked to Arduino GND. The positive terminal of the buzzer connects to Arduino pin 12. The ultrasonic sensor is appropriately connected as well. The VCC pin of the ultrasonic sensor is connected to Arduino pin VCC, the GND pin to Arduino pin GND, the Trig pin to Arduino pin 3, and the Echo pin to Arduino PIN 2. This ensures the correct setup of the device. In Figure 2, the circuit design of the proposed system is depicted, including the ESP32 module. The ESP32 is a potent WiFi + Bluetooth/Bluetooth LE module extensively employed across a broad spectrum of IoT applications, ranging from low-power sensor networks to highly demanding tasks.

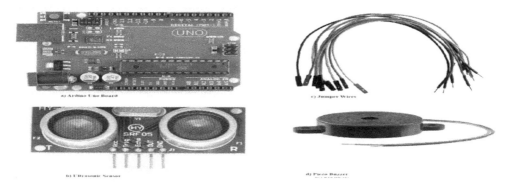

Figure 1.   a) Arduino UNO Board b) Ultrasonic Sensors c) Jumper Wire d) Piezo Buzzer.

4 IMPLEMENTATION

The system design unfolds sequentially as detailed below:

- Board Preparation: The initial step involves the careful cutting of a prefabricated board to dimensions measuring 5 × 3 cm. Subsequently, female headers are soldered onto the board to establish connections with the Arduino.
- Buzzer Integration: Following the board preparation, the buzzer is meticulously soldered onto the board, ensuring a secure connection.
- Vibrating Motor Attachment: The next phase encompasses attaching a vibrating motor to the board using a glue gun. Wires are then soldered to the motor, thereby establishing the necessary electrical connections.
- LED and Switch Connection: Moving forward, both an LED and a switch are integrated into the system. Precise connections are made to ensure proper functionality.
- Header Pins for Ultrasonic Sensors and Battery Input: The subsequent step involves attaching header pins to accommodate ultrasonic sensors and battery input, facilitating the connection of these components.
- Final Soldering as Per Circuit Diagram: With the groundwork in place, all components are soldered together following the circuit diagram as a guide, ensuring accurate placement and connections.
- Arduino and Ultrasonic Sensor Integration: The concluding phase involves linking the Arduino and ultrasonic sensor to the prepared board, solidifying the functional connections necessary for the system's operation.

As mentioned earlier, the proposed system is designed using Arduino uno, ultrasonic sensor, buzzer, red and green LED, jumper cable, power bank, male and female head pins, and

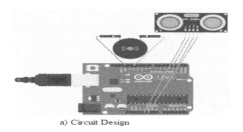

a) Circuit Design   b) Eps 32

Figure 2.   a) Circuit Design b) Eps 32.

Esp32 module. Firstly, the ultrasonic sensor, which consists of 4 pins, is connected; the echo pin of the ultrasonic will be connected to the digital output pin, i.e., D5. The trigger pin of the ultrasonic will be connected to the digital output pin, i.e., D6.Ground to the ground of Arduino, and the Vcc pin will be connected to a 5v power supply from Arduino. Now, the LEDs and the buzzer are connected. Since the buzzer will only be active when our red LED is on, this means both the red LED and the buzzer will follow the same connection. The positive terminal of the buzzer and red LED will be connected to the digital output pin, i.e., D11. And the negative terminal to the ground of Arduino. Similarly, the positive terminal of the green LED will be connected to the digital output pin, i.e., D12, and negative to the ground of Arduino. After this, a button is set up to allow the user to send emergency alerts to a specified device. Three pins of the joystick buttons are used where the ground of the button is connected to the ground of the Arduino. Vcc is provided with a 5V power supply from Arduino, and the third pin, i.e., the digital switch pin, is connected to the digital output pin 3, i.e., D3.

Since the emergency alert needs to be triggered via the internet, the system has used the Esp32 wifi module. Here, the master-slave technique is used to control the Esp module. The digital input of the ESP module, i.e., D2, is connected with the digital output pin of Arduino D3 since it tells the ESP to trigger the alerts. Here, the "If this than that" i.e 'ifttt' online trigger service is used. Once the link is generated via this online platform and is visited by any device, then this online portal responds with a specified action.

Now that all the connections have been made let us understand how all the sensors coordinate together in order to function our device correctly. The buzzer will only be active when there is an obstacle in front of the user within a 100 cm radius, and during and when the obstacle is detected, the red LED will glow, and it will trigger the buzzer sensor; the green LED will be Turned on in the meantime. The closer the obstacle, the frequency of the buzzer will be adjusted accordingly, i.e., the closer the obstacle, the quicker the buzzer will beep.

Once the button attached to the device is pressed, it will trigger the emergency alert on the device registered with "ifttt" service having the same username and password. This is the way the sensor coordinates in the proposed system in order to provide a smooth result.

5   RESULTS

The visually impaired person may independently use the Wondervision without any further assistance. The different situations faced by the user are as follows:

CASE 1: if there are no obstacles, the device will make no noise, signaling a clear path and a go-ahead.
CASE 2: If there is an obstacle in the path, the device will start beeping, signaling the user to change his path in order to avoid a possible collision.
CASE 3: If there is an emergency situation, the user can press the alert button to send SOS signals to the registered mobile number.

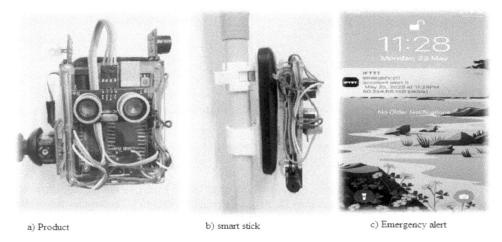

a) Product   b) smart stick   c) Emergency alert

Figure 3.   a) Final product b) Smart Stick c) Emergency alert.

The final product is shown in Figure 3. The figure shows the product attached to a stick and the emergency alerts. The system is tested in real-time scenarios. The final system meets all the requirements and goals it was designed to achieve. The system accurately detects an obstacle and alerts the user with its buzzer. The system also sends an alert/SOS in emergencies using a wireless module embedded in the device. The system is compact, adaptable, and easy to use. The system works perfectly in both indoor and outdoor environments when tested.

6   CONCLUSION AND FUTURE DIRECTIONS

The objective of the work is to assist visually impaired people in moving independently, and during the testing phase, it has been observed that obstacle detection is done with high accuracy. Wondervision is a lightweight, highly affordable, and simple device. The sensors detect the ultrasonic waves from the obstacles with high accuracy, and the buzzer signals the user in time to change their path. The framework is very effective in specifying the source and classification of items to the user that may conflict with the visually impaired. It can filter and detect obstacles in regions such as left and right and point the user in time to change direction. So, this Arduino project based on the obstacle detector for the visually impaired is a new way to solve their problems. A small, inexpensive, easy-to-manage operating system with many amazing features and benefits is proposed to provide blind support. In the future, there is a considerable feature approach for this model, some of which are enlisted here. The model may be improved by adding more sensors in different directions to increase the device's accuracy. Right now, the device only tells about the direction of obstacles, but the implementation of artificial intelligence can drastically increase the device's features. An Android app for the same purpose can be useful and handy; this could be a potential future scope. The device can be connected with a GPS system for better guidance to the users; this addition could be a potential future scope.

REFERENCES

Auvray, M., Hanneton, S., & O'Regan, J. K. (2007). Learning to perceive with a visuo—auditory substitution system: localisation and object recognition with 'The Voice'. *Perception, 36*(3), 416–430.
Benjamin JM, Ali NA, Schepis AF (1973). A laser cane for the blind", Proceedings of San Diego Medical Symposium, 443–450.

Espinosa, M. A., Ungar, S., Ochaíta, E., Blades, M., & Spencer, C. (1998). Comparing methods for introducing blind and visually impaired people to unfamiliar urban environments. *Journal of Environmental psychology*, *18*(3), 277–287.

Mehta, P., Kant, P., Shah, P., & Roy, A. K. (2011, June). VI-Navi: a novel indoor navigation system for visually impaired people. In *Proceedings of the 12th International Conference on Computer Systems and Technologies* (pp. 365–371).

Pereira, A., Nunes, N., Vieira, D., Costa, N., Fernandes, H., & Barroso, J. (2015). Blind guide: an ultrasound sensor-based body area network for guiding blind people. *Procedia Computer Science*, *67*, 403–408.

Petrie, H., Johnson, V., Strothotte, T., Raab, A., Fritz, S., & Michel, R. (1996). MoBIC: Designing a travel aid for blind and elderly people. *The Journal of Navigation*, *49*(1), 45–52.

Ponchillia, P. E., Rak, E. C., Freeland, A. L., & LaGrow, S. J. (2007). Accessible GPS: Reorientation and target location among users with visual impairments. *Journal of Visual Impairment & Blindness*, *101*(7), 389–401.

Prasad R., (2020). National Blindness and Visually Impairments Survey India 2015–2019 – Asummary Report. https://npcbvi.gov.in/writeReadData/mainlinkFile/File341.pdf

Sabarish, S. (2013). Navigation tool for visually challenged using microcontroller. *International Journal of Engineering and Advanced Technology (IJEAT)*, *2*(4), 139–143.

Shoval, S., Ulrich, I., & Borenstein, J. (2003). NavBelt and the Guide-Cane [obstacle-avoidance systems for the blind and visually impaired]. *IEEE robotics & automation magazine*, *10*(1), 9–20.

Sharma, P., SL, M. S., & Chatterji, S. (2015). International journal of engineering sciences & research technology a review on obstacle detection and vision. *A Review on Obstacle Detection and Vision*, *4*(1), 1–11.

Sivan, S., & Darsan, G. (2016, July). Computer vision based assistive technology for blind and visually impaired people. In *Proceedings of the 7th International Conference on Computing Communication and Networking Technologies* (pp. 1–8).

Sterr, A., Müller, M. M., Elbert, T., Rockstroh, B., Pantev, C., & Taub, E. (1998). Perceptual correlates of changes in cortical representation of fingers in blind multifinger Braille readers. *Journal of Neuroscience*, *18* (11), 4417–4423.

Yuan, D., & Manduchi, R. (2005, June). Dynamic environment exploration using a virtual white cane. In *2005 IEEE Computer Society Conference on Computer Vision and Pattern Recognition (CVPR'05)* (Vol. 1, pp. 243–249). IEEE.

Advances in AI for Biomedical Instrumentation, Electronics and Computing – Sachan et al. (eds)
© 2024 The Author(s), ISBN 978-1-032-64298-7

C band application-oriented H-shaped patch antenna design

B. Jeyapoornima

Research Scholar, Department of Electronics and Communication Engineering, Saveetha School of Engineering, Saveetha Institute of Medical and Technical Sciences, Saveetha University, Thandalam, Chennai, Tamil Nadu, India

J. Joselin Jeya Sheela

Associate Professor, Department of Electronics and Communication Engineering, Saveetha School of Engineering, Saveetha Institute of Medical and Technical Sciences, Saveetha University, Thandalam, Chennai, Tamil Nadu, India

ABSTRACT: This study presents the design, and analysis of an H-shaped patch antenna, tailored for C-band applications. The proposed antenna is optimized to operate within the frequency range of 4–8 GHz, targeting communications, and satellite systems. The H-shaped patch geometry is constructed using FR-4(lossy) with a thickness of 1.6mm. Simulated results and the tested results demonstrate excellent performance characteristics, including a −30.29dB of return loss, stable radiation patterns (Omnidirectional patterns), and radiation efficiency in simulation and −21.53dB of return loss, and bandwidth of 589.2 MHz attained, while tested. The compact and efficient design makes it a promising candidate for various C-band applications, offering improved signal reception and transmission capabilities in modern communication systems.

1 INTRODUCTION

An antenna is the cornerstone of a network of communications, and it has witnessed significant improvement as technology has advanced (Prashanth *et al.* 2023). And, continuous advancement in communication technologies has led to surging demand for high-performance antennas capable of operating in various frequency bands. Among these, C-band satellite communication is referred to by IEEE as a section of the EM frequency spectrum extending from 4 GHz to 8 GHz (Haqueet *et al.* 2019). It has garnered widespread adoption due to its advantageous features such as reduced atmospheric attenuation, improved signal reliability, and resistance to rain fade.

The C band is often employed in tropical regions meanwhile it's less susceptible to rain fade, than Ku-band (kumar Deb *et al.* 2015). Furthermore, C band frequencies are widely utilized in radar systems, particularly for weather monitoring, maritime surveillance, and military applications. One crucial component in C band applications is the patch antenna. Patch antennas are compact, planar structures that are well-suited for integration into communication systems and satellite platforms. Typically, height is significantly less than functional Wavelength, but it cannot exceed more than 1/40th of Wavelength, or antenna's efficiency might decline (Sheela *et al.* 2022). Squared, Rectangular, and circular designs are among the frequently deployed geometries due to uncomplicatedness of fabrication, and evaluation (Karthick 2015). Patch antennas, due to their small size, may be made utilizing printed circuit technology, resulting in an affordable price and ease of integration with microwave circuits that are integrated (Kanakavalli *et al.* 2018). Microstrip antennas comprise primarily broadside radiators. The patch has been constructed in a way, that the pattern is as maximum to it as possible (Ramamurthy *et al.* 2014).

DOI: 10.1201/9781032644752-8

Patch antennas offer several advantages in C band applications, including directional radiation patterns, high gain, and efficient radiation efficiency. In this paper, a comprehensive analysis of H-shaped patch antenna specifically tailored for C-band applications. Through simulations and practical measurements, the antenna's radiation performance, impedance matching, and gain characteristics are evaluated. The results obtained showcase the effectiveness and reliability of the H-shaped patch antenna as a viable solution for a broad spectrum of C-band applications, highlighting its potential to drive advancements in wireless communication and satellite technologies.

2   ANTENNA DESIGN

In this study, a patch antenna has been designed, to resonate within C band, at a specific Resonant frequency of 7.677 GHz. The antenna's substrate is made of FR-4, which is a lossy material, and it has a dielectric constant (εr) of 4.4. FR-4 (lossy) is chosen because of its Electrical Properties, Mechanical Durability, Cost-Effectiveness, Ease of Fabrication, Wide Availability, and Frequency Compatibility. The conducting material used for the patch is pure copper. The dimensions of the antenna are 24.9 × 29.8mm, as illustrated in Figure 1. The design of antenna relies on three key parameters: "operating frequency (f_o), the dielectric constant of the substrate (ε_r), and the substrate's height (h)".

2.1   W-width

$$W = \frac{c}{2f_o\sqrt{\frac{(\varepsilon_r+1)}{2}}} \tag{1}$$

2.2   (ε_{eff})-Effective dielectric constant

$$\varepsilon_{eff} = \frac{\varepsilon_r + 1}{2} + \frac{\varepsilon_r - 1}{2}\left[1 + 12\frac{h}{W}\right]^{\frac{1}{2}} \tag{2}$$

2.3   (Leff)-Effective length

$$L_{eff} = \frac{c}{2f_o\sqrt{\varepsilon_{eff}}} \tag{3}$$

2.4   ΔL-Extension length

$$\Delta L = 0.412 \times h\frac{(\varepsilon_{eff} + 0.3)(\frac{W}{h} + 0.264)}{(\varepsilon_{eff} + 0.258)(\frac{W}{h} + 0.8)} \tag{4}$$

2.5   Patch's actual length-(L)

$$L = L_{eff} - 2\Delta L \tag{5}$$

$$L = \frac{c}{2f_o\sqrt{\varepsilon_{eff}}} - 0.824h\left(\frac{(\varepsilon_{eff} + 0.3)(\frac{W}{h} + 0.264)}{(\varepsilon_{eff} - 0.258)(\frac{W}{h} + 0.8)}\right) \tag{6}$$

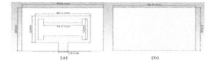
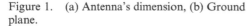
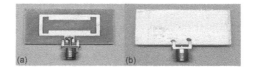

Figure 1.   (a) Antenna's dimension, (b) Ground plane.

Figure 2.   (a) Fabricated Antenna, (b) Fabricated antenna's ground plane.

2 RESULTS AND DISCUSSION

The Patch Antenna is carefully designed and their dimensions are calculated within the software to meet specific requirements. Once the design is complete the fabricated antenna is then tested using a Vector Network Analyzer (VNA). The VNA allows precise measurements and characterization of the antenna's performance, verifying its functionality and validating the simulation results. Return loss of the antenna refers to the measure of reflected power back from the antenna's output compared to the input power. A lower return loss indicates better impedance matching and more efficient power transfer. For the overall system's performance, minimizing reflections is critical, necessitating a return loss below -10 dB. During simulation, the antenna achieves a return loss of -30.29dB at 7.677GHz, while testing shows -21.53dB at 7.412GHz, following Figure 3. These results demonstrate that the antenna's return loss remains below -10 dB, making it an excellent choice for real-time applications.

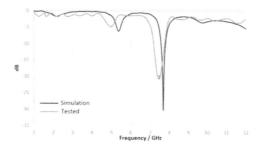

Figure 3.   Return Loss variations.

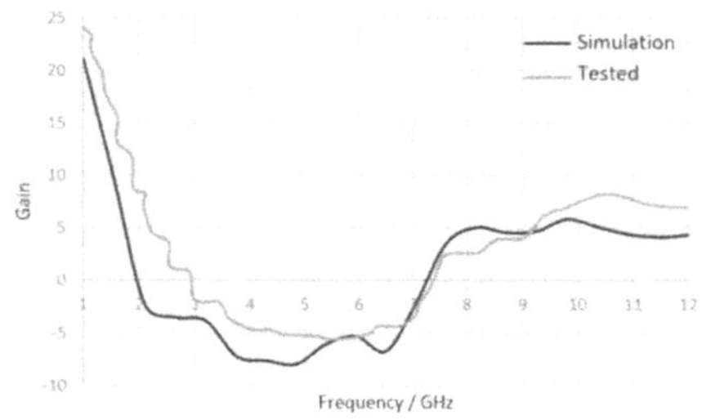

Figure 4.   Gain vs Frequency Plot.

A VSWR (Voltage Standing Wave Ratio) value of 1:1 denotes a perfect match, while higher values signify increased signal losses and reduced antenna efficiency. Proper impedance matching is critical, to maximize power transfer, and minimize signal reflections, in radio frequency systems.

$$VSWR = \frac{1 + 10^{\frac{-RL}{20}}}{1 - 10^{\frac{-RL}{20}}} \qquad (7)$$

Where, RL = Return loss of antenna.

During the simulation process at a frequency of 7.677GHz, the antenna exhibits a favorable VSWR of 1.06, indicating a good impedance match. However, when subjected to actual testing at the frequency of 7.412GHz, VSWR increases to 1.18, suggesting an optimal impedance match and potentially little signal reflections or losses. The "bandwidth of an antenna is the frequency range over which it can effectively transmit or receive signals with acceptable performance".

$$Bandwidth = f_{upper} - f_{lower} \qquad (8)$$

Where, f_upper = Upper-frequency, f_lower = Lower frequency.

In simulation, the antenna exhibits a bandwidth of 244.4MHz, while during testing, it achieves a slightly higher bandwidth of 589. 2MHz. A gain of 3.64dB is obtained while simulates, whereas 3.025 is attained while tested, following Figure 4. The gain of an antenna is obtained by multiplying its directivity by the radiation efficiency.

$$G = D * \eta \qquad (9)$$

Where, G – Gain, D – Directivity, η – Radiation efficiency.

The radiation pattern is a fundamental characteristic of an antenna and is essential for understanding its behavior and performance. This paper presents the design, analysis, and characterization of an H-shaped patch antenna, optimized for use in the C band frequency range.

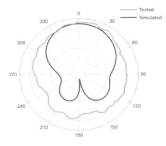

Figure 5.   E-Field comparison.

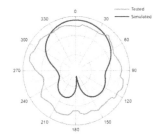

Figure 6.   H-field comparison

Table 1.   Antenna parameters.

Antenna Parameters	Simulated	Measured
Return Loss	−30.29 dB	−21.53 dB
VSWR	1.06	1.18
Bandwidth	244.4 MHz	589.2 MHz
Gain	3.89 dB	3.07 dB
Radiation Pattern	Omnidirectional Pattern	

3   CONCLUSION

In conclusion, the Patch Antenna design for C Band demonstrates promising characteristics that make it well-suited for various communication and radar systems within the C band frequency range. The antenna's omnidirectional radiation pattern ensures widespread coverage, making it suitable for applications requiring signals to be transmitted or received from multiple directions. Additionally, the antenna exhibits good return loss and wide bandwidth, indicating efficient impedance matching and the ability to support a range of frequencies within the C band. These features enhance its performance and reliability in data transmission and communication tasks. As a result, the H-shaped patch antenna shows great potential for practical implementation in C band applications, offering valuable contributions to the advancement of wireless communication and radar technologies.

REFERENCES

Allin Joe D. *et al.* 2023, "A Triband Compact Antenna for Wireless Applications", *International Journal of Antennas and Propagation*, vol. 2023, Article ID 5344999, 13 pages, 2023. https://doi.org/10.1155/2023/5344999

Apriono C. *et al.* 2023, Rectangular Microstrip Array Feed Antenna for C-Band Satellite Communications: Preliminary Results. *Remote Sensing*. 2023; 15(4):1126. https://doi.org/10.3390/rs15041126.

Haque M. A. *et al.* 2021, "A Plowing T-shaped Patch Antenna for WiFi and C Band Applications,"International Conference on Automation, Control and Mechatronics for Industry 4.0 (ACMI), Rajshahi, Bangladesh, 2021, pp. 1–4, doi: 10.1109/ACMI53878.2021.9528266.

Haqueet *et al.* 2019. A Modified E-Shaped Microstrip Patch Antenna for C Band Satellite Applications. 10.1109/SPICSCON48833.2019.9065126.

Hasan M. M. *et al.* 2020 *"Design and Analysis of Elliptical Microstrip Patch Antenna at 3.5 GHz for 5G Applications,"* IEEE Region 10 Symposium (TENSYMP), Dhaka, Bangladesh, 2020, pp. 981–984, doi: 10.1109/TENSYMP50017.2020.9230897.

Hussain MB *et al.* 2023, Design of dual-band microstrip patch antenna for wireless local area network applications. *Engineering Proceedings.* 2023; 46(1):3. https://doi.org/10.3390/engproc2023046003

Islam M. T. *et al.* 2020, "Design of a Microstrip Patch Antenna for both C-band and X-band Applications," *2nd International Conference on Advanced Information and Communication Technology (ICAICT)*, Dhaka, Bangladesh, 2020, pp. 7–10, doi: 10.1109/ICAICT51780.2020.9333453.

Jeyapoornima B. *et al.* 2022, "Design of Miniaturized UWB Antenna for Skin Cancer Recognition," *2022 International Conference on Power, Energy, Control and Transmission Systems (ICPECTS)*, Chennai, India, 2022, pp. 1–4, doi: 10.1109/ICPECTS56089.2022.10047124.

Kanakavalli *et al.* 2018. Design of Compact C-Band Concave Patch Antenna for Radar Altimeter Applications. 0542–0546. 10.1109/ICCSP.2018.8524261.

Karthick M. 2015, "Design of 2.4GHz patch antennae for WLAN applications," *IEEE Seventh National Conference on Computing, Communication and Information Systems (NCCCIS)*, 2015, pp. 1–4, doi: 10.1109/NCCCIS.2015.7295902.

Kumar Deb P. *et al.* 2015. "Dual band multilayer E-shape microstrip patch antenna for C-band and X-band, 2nd International Conference on Signal Processing and Integrated Networks (SPIN), Noida, India, 2015, pp. 30–34, doi: 10.1109/SPIN.2015.7095320.

Mohamed Junaid K. A. *et al.* 2022 "Design and Analysis of Novel Face Shaped Microstrip Array Antenna of UWB for Early Breast Tumor Detection (I-SMAC)", Dharan, Nepal, 2022, pp. 424–429, doi: 10.1109/I-SMAC55078.2022.9987380.

Padhi J. *et al.* 2017, "Design of an ultra-wideband slotted koch fractal antenna for C-band application," *International Conference on Communication and Signal Processing (ICCSP)*, Chennai, India, 2017, pp. 0548–0552, doi: 10.1109/ICCSP.2017.8286418.

Prashanth *et al.* 2023. Design of micro-strip patch antenna for C – band applications. *E3S Web of Conferences.* 391. 10.1051/e3sconf/202339101066.

Ramamurthy *et al.* 2014. A Review on Circular Microstrip Patch Antenna with Slots for C Band Applications. *International Journal of Scientific and Engineering Research.* 03. 1039–1045.

Sheela J. J. J. *et al.* 2020, "Novel directional antennas for microwave breast imaging applications," *International Conference on Smart Electronics and Communication (ICOSEC)*, Trichy, India, 2020, pp. 599–603, doi: 10.1109/ICOSEC49089.2020.9215243.

Sheela J. J. J. *et al.* 2022 "Design of ultra-wideband of rectangular shaped emoji designed microstrip patch antenna of 4.5GHz for military applications," *3rd International Conference on Smart Electronics and Communication (ICOSEC)*, Trichy, India, 2022, pp. 71–75, doi: 10.1109/ICOSEC54921.2022.9951893.

Shukla A. W. M. *et al.* 2016, "Broadband stair shaped micro strip patch antenna for C-band applications," *2016 1st India International Conference on Information Processing (IICIP)*, Delhi, India, 2016, pp. 1–3, doi: 10.1109/IICIP.2016.7975343.

Soni S. *et al.* 2015, "Novel design of patch antenna for C band applications," *2015 International Conference on Green Computing and Internet of Things (ICGCIoT)*, Greater Noida, India, 2015, pp. 1129–1133, doi: 10.1109/ICGCIoT.2015.7380632.

Vanaja Selvaraj *et al.* 2020, "Detection of depth of the tumor in microwave imaging using ground penetrating radar algorithm," *Progress In Electromagnetics Research M*, Vol. 96, 191–202, 2020.

Zahid M *et al.* 2023, A compact dual-band microstrip patch antenna for C- and X- and Ku-band applications. *Engineering Proceedings.* 2023; 46(1):16. https://doi.org/10.3390/engproc2023046016

Advances in AI for Biomedical Instrumentation, Electronics and Computing – Sachan et al. (eds)
© 2024 The Author(s), ISBN 978-1-032-64298-7

Microcavity based 2D Photonics crystal biosensor for detection of COVID-19

Vijay Laxmi Kalyani
Ph.D. Research Scholar, ECE Department, Bhagwant University, Ajmer

Virendra Kumar Sharma
Professor, Bhagwant University, Ajmer

ABSTRACT: In this research paper we propose a design of 2D photonics crystal based biosensors for COVID-19 detection through human blood by using refractive index of healthy blood sample (1.334) and different refractive index of different blood sample infected by SARS-CoV-2 virus under different concentrations of human monoclonal IgG. The proposed model is designed using 2D photonics crystal technique and the simulation is done by Opti-FDTD (Finite difference time domain) software. The proposed biosensor has $12\mu m \times 08.5\,\mu m$ wafer dimensions. In this paper, we propose the microcavity based photonics crystal biosensor. Blood sample is deposited inside the microcavity and based on the refractive index of COVID-19 infected sample to the healthy blood sample there will be shift in the wavelength at the output terminal. PWE band solver is also used for band gap calculation in the waveguide. The band gap range of structure is 1311nm-1930nm and 1550 nm continuous modulated wave input is used in the proposed design. The proposed sensor achieved high Q factor 462.83 during simulation. The proposed sensor is having fast, accurate and better transmission.

Keywords: Photonic Crystal, Refractive index, SARS-CoV-2 virus, Finite difference time domain, Photonic bandgap

1 INTRODUCTION

Covid-19 is an infectious disease caused by SARS coronavirus-2 (Khurshid *et al.* 2020). The most commonly used method to detect coronavirus is the PCR (polymerase chain reaction). It is the standard method of coronavirus testing. It is a method which is widely used in molecular biology to make millions to billions of copies of a specific DNA fragment rapidly. Coronavirus contains an extraordinarily long single-stranded RNA genome. To detect these viruses using the PCR method, it is important that the RNA molecules must be converted into their complementary DNA sequences by reverse transcriptase. Then the newly synthesized DNA can be amplified by standard PCR method. This approach is called RT-PCR. But this process is a very time consuming process. Therefore, biosensors are key technologies for the quick detection of viruses. In the development of biosensors, photonic based biosensors are the new research direction in the optical field. Photonic crystals (PCs) are multilayer structures whose refractive indexes change periodically. There are three types of PCs depending on refractive index change directions: one-dimensional (1D), two-dimensional (2D), and three-dimensional (3D).

Bilgili *et al.* 2023 theoretically investigated one-dimensional photonic crystal with a defect layer as a biosensor for the detection of COVID-19 (SARS-CoV-2) through human blood. As 2D photonic crystal has a huge variety in comparison to 1D photonic crystal like its periodicity of permittivity is two direction and one direction medium is homogeneous. Therefore, In this research paper we theoretically design a 2D photonic crystal biosensor for the detection of

DOI: 10.1201/9781032644752-9

COVID-19 and the simulation is done by opti-FDTD (Finite difference time domain) software. Different types of photonics crystal based structure such as nanocavity based biosensor, microcavity based biosensors, optical fiber based biosensors etc. can be designed. In this paper, we propose the microcavity based photonics crystal biosensor. The transmission peak shift in the PBG depends upon the change in the refractive index of blood samples containing healthy and infected by SARS-CoV-2 virus under different concentrations of human monoclonal IgG. PWE band solver is also used in this paper for band gap calculation in the waveguide. The proposed design will provide a fast, accurate and good transmission spectrum.

1.1 Design of biosensor

For the proposed microcavity biosensor design, the FDTD simulation tool is used for the simulation. The structure uses a 2D rectangular lattice with silicon rods and air in the background wafer. Transverse electric (TE) mode is used for propagation of light inside the structure. The structure has 21 × 15 silicon rods that are used in Z and X directions and wafer dimensions are 8.5 × 12 μm and lattice constant (.55 μm) are also used in the structures. For the propagation of light inside the structure an optical wavelength 1550 nm continuous wave is used in the input side. Six observation points are also used on the output port to detect the input wave. The refractive index of silicon material is 3.45 and air is 1. The radius of silicon rods is 110 nm.

In this paper, R.I. of healthy blood samples (1.334) and different refractive index of different blood samples infected by SARS-CoV-2 virus under different concentrations of human monoclonal IgG are used (Bilgili *et al.* 2023). Changes of wavelength shift according to the refractive index is sensed by bio sensors. By increasing the radius of silicon rods from .11 μm to .13 μm we can design the microcavity based biosensor.

1.2 Microcavity based photonics crystal biosensor for COVID-19 detection

In this design (Figure 1), by using the structure of the proposed biosensor, six microcavities are created by changing the radius of silicon rods from .11 μm to .13 μm. The sensing mechanism of proposed design is used to change the R.I. of analytes which led to shifting in transmission. Figure 2 indicates the 3D view of microcavity based sensor structure in which the silicon rods are suspended into the air configuration.

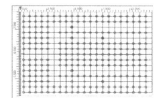

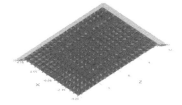

Figure 1.   Microcavity based photonics crystal layout.

Figure 2.   Silicon rods in air configuration.

1.3 Band diagram of microcavity structure

Figure 3.   TE band gap diagram using PWE band solver.

Above band diagram (Figure 3) gives the Photonic Band Gap for Transverse Electric (TE) modes. The band gap structure depends upon three parameters, refractive index of material, lattice constant, and ratio of radius to lattice constant (r/a). The Plane wave expansion (PWE) method is used to estimate the band gap and propagation modes of the photonics crystal structure without and with defects. The complete structure of both biosensors are having two band gaps. The first photonic band gap (PBG 0) is in the range between the wavelength 1311 nm and 1930 nm, and the second band gap (PBG 1) is from 7410 nm and 7556 nm. As our proposed designed structure lies in the first PBG range (1311nm–1930nm). Therefore, in this paper the first PBG range is considered. Continuous wave is used in this paper at wavelength 1550 nm and its wavelength is exactly the center wavelength of this PBG wavelength range.

Table 1. Design parameter and its value used in microcavity based sensor.

S.No.	Name of parameters	Values
1.	Radius of silicon (rod)	110 nm
2.	Lattice constant	550 nm
3.	Refractive index of Si	3.45
4.	Refractive index of Wafer (air)	1
5.	Refractive index of healthy blood sample	1.3348000
6.	Refractive index of infected blood sample at Monoclonal IgG concentrations (nM)	
	Monoclonal IgG concentrations (nM)	Refractive index
	0	1.3348000
	1.74	1.3355323
	3.47	1.3362604
	6.94	1.3377208
	13.9	1.3406500
	27.8	1.3465000
7.	Input wavelength	1550nm
8.	Wafer dimensions	8.5x12 μm
9.	PBG range	1311nm-1930 nm
10.	Polarization	TE

The above table shows the design parameters and their values which are used in micro-cavity based sensor structure for COVID-19 detection. The above table (point no.6) shows the refractive index of the human mAbs IgG at different concentrations (Nano Mole-nM). The lowest value represents threshold molar concentration for determining COVID positive/negative. From Table 1, we can see that as monoclonal IgG concentrations in the blood increase, the refractive index of the samples increases accordingly (Bilgili et al. 2023).

2 SIMULATION AND RESULT

OptiFDTD simulation software is used for the designing and simulation purpose. A continuous modulated wave is applied at the input side with a wavelength 1550 nm. At this wavelength the waveguide is fully coupled and reached at the output port. Therefore at this wavelength a very small amount of losses occurs inside the structure. So it is considered as a resonance wavelength of this structure. The transverse electric (TE) polarization mode is selected for the propagation of light inside the structure. Good performance of biosensors is achieved by getting a high transmission spectrum. As the sensitivity of any biosensor is defined by its Q (quality factor). Therefore $Q = \lambda/\Delta\lambda$, Where λ is the resonance wavelength and $\Delta\lambda$ is the full width half maximum (FWHM).

To enhance the sensing performance, a higher Q value is desirable because sharper peaks with high Q values are much easier to detect.

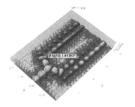

Figure 4. 2D electric field distribution in microcavity sensor.

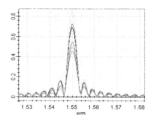

Figure 5. Transmission graph of microcavity sensor.

Above Figure 4 shows the 2D electric field distribution of the microcavity based sensor at 1550nm. In this the electric field of the waveguide is fully coupled in the microcavity and reaches at the output port.

Above Figure 5 shows the output transmission spectra of healthy blood samples and different infected samples of COVID-19 (as shown in Table 1). In Figure 5, the black curve depicts healthy blood sample response and the other curve depicts different infected sample response of COVID-19.

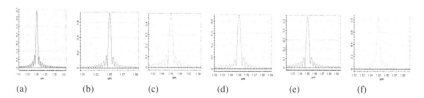

(a)	(b)	(c)	(d)	(e)	(f)

Table 2. Transmission Spectrum and quality factor according to their refractive index used in microcavity based biosensor.

Sample Name	Monoclonal IgG concentrations (nM)	Refractive index	Transmission	Wavelength (μm)	Q-factor
Healthy blood sample	0	1.3348000	70%	1.54015	400.70
Infected samples	1.74	1.3355323	66%	1.53336	368.06
	3.47	1.3362604	50%	1.53875	368.06
	6.94	1.3377208	46%	1.53874	418.42
	13.9	1.3406500	54%	1.53822	390.27
	27.8	1.3465000	83%	1.54086	462.83

Above Figure 5(a) shows the transmission is 70% for healthy blood samples at R.I. (1.334800) and different transmission for infected samples 46%, 50%, 54%, 66%, 83%. Highest Q factor is 400.70.

Above Figure 5(b) shows the transmission is 66% at R.I. (1.3355323) of blood samples infected by SARS-CoV-2 virus under (1.74) concentrations of human monoclonal IgG and achieved Q factor is 368.06.

Above Figure 5(c) shows the transmission is 50% at R.I. (1.3362604) of blood samples infected by SARS-CoV-2 virus under (3.47) concentrations of human monoclonal IgG and achieved Q factor is 368.06.

Above Figure 5(d) shows the transmission is 46% at R.I. (1.3377208) of blood samples infected by SARS-CoV-2 virus under (6.94) concentrations of human monoclonal IgG and achieved Q factor is 418.42

Above Figure 5(e) shows the transmission is 54% at R.I. (1.3406500) of blood samples infected by SARS-CoV-2 virus under (13.9) concentrations of human monoclonal IgG and achieved Q factor is 390.27

Above Figure 5(f) shows the transmission is 83% at R.I. (1.3465000) of blood samples infected by SARS-CoV-2 virus under (27.8) concentrations of human monoclonal IgG and achieved Q factor is 462.83.

Above Table 2 shows the transmission of healthy blood samples and different blood samples infected by SARS-CoV-2 virus under different concentrations of human monoclonal IgG in different cavities. The microcavities are filled according to their refractive index and transmission results measured. To enhance the sensing performance, a higher Q value is desirable because sharper peaks with high Q values are much easier to detect. It is observed that at highest transmission higher Q factor is achieved. The highest transmission is 83% at R.I. (1.3465000) of blood samples infected by SARS-CoV-2 virus under (27.8) concentrations of human monoclonal IgG and achieved Q factor is 462.83.

3 CONCLUSION

In this paper we have designed microcavity based biosensor structure for COVID-19 detection. Detection of COVID-19 in humans using the conventional RT-PCR method is a more complex and time consuming process. Therefore to evade this a label free detection method is used to design microcavity based sensors using photonics platforms by using refractive index of healthy blood sample and different blood samples infected by SARS-CoV-2 virus under different concentrations of human monoclonal IgG. All the simulation work is done using OptiFDTD simulation software and PWE band solver is used for band gap calculation. High transmission spectrum (83%) and quality factor (462.83) is observed in microcavity based sensor structure. The proposed design shows good results with better transmission and it is helpful to detect SARS-CoV-2 virus with high Q. The wafer dimensions of the proposed biosensor is $12\mu m \times 08.5\mu m$. Therefore it can be embedded in hand held devices.

REFERENCES

Asghari, A., Wang, C., Min Yoo, K., Rostamian, A., Xu, X., Shin, J.D., Dalir, H., T. Chen, R. 2021. Fast, accurate, point-of-care COVID-19 pandemic diagnosis enabled through advanced lab-on-chip optical biosensors: Opportunities and challenges. *Appl. Phys. Rev. 8, 031313 (2021)*; doi: 10.1063/5.0022211

Bilgili, N. & Çetin, A. 2023 "1D photonic crystal biosensor for detection of SARS-CoV-2. *Research Square.* https://doi.org/10.21203/rs.3.rs-2709079/v1

John, P., Vasa, N.J., Zam, A. 2023. Optical Biosensors for the Diagnosis of COVID-19 and Other Viruses—A Review. *Diagnostics (Basel, Switzerland)*, 13(14), 2418.

Khurshid, Z., Yahya Ibrahim Asiri, F., Al Wadaani, H. 2020. Human Saliva: Non-Invasive Fluid for Detecting Novel Coronavirus (2019-nCoV). *Int. J. Environ. Res. Public Health*, 17, 2225.

Makra, I., Terejánszky, P., E., R., Gyurcsányi 2015. A method based on light scattering to estimate the concentration of virus particles without the need for virus particle standards. *Published by Elsevier*, http://dx.doi.org/10.1016/j.mex.2015.02.003.

Orooji, Y., Sohrabi, H., Hemmat, N., Oroojalian, F., Baradaran, B., Mokhtarzadeh, A., Mohaghegh, M., Karimi-Maleh, H. 2020. An Overview on SARS-CoV-2 (COVID-19) and Other Human Coronaviruses and Their Detection Capability via Amplifcation Assay, Chemical Sensing, Biosensing, Immunosensing, and Clinical Assays. *Nano-Micro Lett.* (2021) 13:18, ISSN 2311-6706.

Pang, Y., Song, H., and Cheng, W. 2016. Using optical trap to measure the refractive index of a single animal virus in culture fluid with high precision. *Biomed Opt Express.* 2016 May 1; 7(5): 1672–1689.

Samavati, A., Samavati, Z., Velashjerdi, M., Ismail, A.F., D. Othman, M.H., Eisaabadi B, G., Abdullah, M. S., Bolurian, M. & Bolurian, M. 2021. Sustainable and fast saliva-based COVID-19 virus diagnosis kit using a novel GO-decorated Au/FBG sensor. *Chemical Engineering Journal*, 420, Part 2, 15 September 2021, 127655.

Advances in AI for Biomedical Instrumentation, Electronics and Computing – Sachan et al. (eds)
© 2024 The Author(s), ISBN 978-1-032-64298-7

Prognosis of heart disease by utilizing data mining techniques

Sartaj Ahmad, Vaishnavi Awasthi, Srishti Chaurasia & Ajay Agarwal
KIET Group of Institutions, Delhi-NCR, India

ABSTRACT: Nowadays, a few diseases are the leading global cause of death. This is a result of how people live today. Due to irregular diets, office schedules, and other factors, people find it difficult to devote attention to their health. One of the primary causes of death is heart disease. The healthcare system contains lots of data that, when used in conjunction with technology, can be used to identify and analyze diseases so that appropriate care can be provided. Machine learning techniques and algorithms can be used to make symptom-based forecasts and stop them from happening, ultimately saving more lives. Numerous studies have been conducted. The fundamental objective of this study is to evaluate various methods and algorithms that can be implemented to anticipate the symptoms more precisely and reliably. For this purpose, various classifications, like KNN, Decision Table, ID3, Logistic Regression, and SVM, are employed and their results are analysed for further decision-making.

1 INTRODUCTION

The heart, the body's most crucial organ, must function properly to sustain life by delivering oxygen-enriched blood throughout the body. Cardiovascular diseases, a leading global cause of death per World Health Organization data, underscore the heart's critical role. Machine learning, a data science technique, leverages prior research to overcome limitations in traditional approaches. To effectively analyze complex data patterns, a robust and trustworthy framework is needed, making machine learning an invaluable tool. Once it learns these data patterns, machine learning excels at processing immediate input from training samples, potentially leading to automated systems. The training and testing phases yield data forecasts encompassing data collection, classification, and information retrieval. The suggested study in this research paper makes a difference by coming up with new ways to identify heart disease that are based on machine learning techniques. Different machine learning prediction techniques, such as Logistic regression, KNN, Support Vector Machine, Decision tables, and Iterative Dichotomiser 3 (ID3), were used in the study to categorize the patients as having no heart disease or having heart disease.

In this paper, with the help of shrinkage, relief, selection, minimalredundancy, and maximal relevance operators, all the pertinent and interconnected functions that have a major impact on the predicted importance were identified. The effectiveness of the various classification algorithms was measured and analyzed using several performance measures, such as precision, F-score, accuracy, timetaken, and recall.

2 LITERATURE REVIEW

In(Deb 2022) author presents that Random Forests and Decision Trees are the best techniques for predicting heart disease. The authors (Patidar 2022) describe algorithms from which the random forest algorithm performed the best with an accuracy of 98.53% and GNB performed the worst with an accuracy of 74.63%. In the research article (Ramesh and Lilhore 2022) four

DOI: 10.1201/9781032644752-10

different methods were used to analyze the dataset, and the method with the maximum positive results was selected. Authors in the article (Fadnavis and Dhore 2021) compared Naive Bayes and Decision Tree classification algorithms for heart disease data and found Naive Bayes to outperform in terms of accuracy and performance. In (Tarawneh and Embarak 2019) authors present that heart disease is a major global health concern due to healthcare shortages and misdiagnosis. The authors propose an expert system utilizing data mining to extract insights from extensive medical data, aiming to improve diagnosis accuracy by integrating multiple techniques into a single algorithm. In (Enriko 2019) data mining is pivotal in healthcare, particularly for heart disease detection, using algorithms like k-Nearest Neighbor, CART, and AdaBoost. These algorithms analyze medical records to enhance accuracy and speed in heart disease diagnosis. In (Solanki and Sharma 2019) medical data mining aids in addressing chronic ailments like heart disease, utilizing SVM, KNN, decision trees, Naïve Bayes, and ANN for prediction. The study assesses healthcare challenges and machine learning algorithm effectiveness. (Beyene and Kamat 2018) utilizes data mining techniques like J48, Naïve Bayes, and Support Vector Machine to predict heart disease onset and provide life-saving services effectively. Authors (Tripoliti *et al.* 2017) show that the research's main goal is to propose a solution with improved accuracy using the KNN algorithm. Text attributes are used in this, which gives good results with better accuracy. In (Howlader *et al.* 2017), various data mining techniques are used, through which training and testing are done. In (Sultana *et al.* 2016) the results from all the algorithms are compared. Bayes Net and SMO algorithms gave the optimum results among all the algorithms: J48, SMO, KStar, MLP, and Bayes Net. The authors (Austin *et al.* 2013; Shetgaonkar and Aswale 2021) present different algorithms to analyze the data related to heart disease. The decision Tree gave the best results among all the algorithms used. In (Palaniappan and Awang 2008), researchers employed Decision trees, Naive Bayes, and Neural networks to create a heart disease prediction system. After the study, we found a gap in the existing results. That is why a few important algorithms are discussed and deployed in the next section to achieve better results.

3 RESEARCH METHODOLOGY

3.1 *Dataset and algorithms used for the experiment*

The dataset utilized in this study to determine whether heart disease exists was downloaded from Kaggle.com. This dataset contains 14 attributes, which are enumerated as follows: 1. Age, 2. Sex, 3. Cerebral palsy or chest pain (CP), 4. Blood pressure (bps), 5. Cholesterol, 6. Fasting blood sugar test (fbs), 7. Resting electrocardiographic results, 8. thalach (maximum heart rate achieved), 9. exang (exercise-induced angina), 10. oldpeak (ST depression brought on by exercise relative to rest), 11. slope (the slope pf the peak exercise ST segment), 12. ca (Number of major vessels), 13. thal (Reversible defect), 14. Heart Disease (Presence or Absence).

3.1.1 *K Nearest Neighbor (KNN)*
It gives the following results (see Figure 1) at no. of instances (K) = 5.

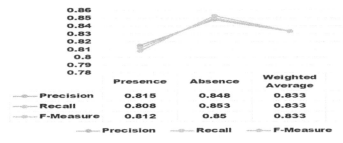

	Presence	Absence	Weighted Average
Precision	0.815	0.848	0.833
Recall	0.808	0.853	0.833
F-Measure	0.812	0.85	0.833

Precision — Recall — F-Measure

Figure 1.   Precision, Recall, and F-Measure of KNN at K = 5.

3.1.2 *Decision tree model*

It provides a clear visual representation for describing the actions to take to respond to certain circumstances. In addition to decision tables, decision trees, as shown in Figure 2.

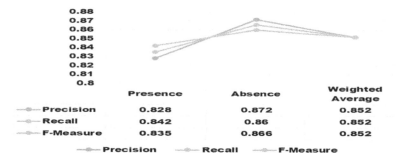

	Presence	Absence	Weighted Average
Precision	0.828	0.872	0.852
Recall	0.842	0.86	0.852
F-Measure	0.835	0.866	0.852

Figure 2.   Precision, Recall, and F-Measure in case of decision tree model.

3.1.3 *ID3 algorithm*

It divides characteristics into groups as shown in Figure 3, via a top-down, greedy search, created by J. R. Quinlan. It constructs decision trees based on entropy and information gain, not backtracking. It focuses on entropy to define information gained accurately. Let's suppose that, without sacrificing generality, the resulting decision tree classifies instances as P(positive) or N (negative). Given a set S comprising these positive and negative targets, the entropy of the set about this Boolean classification is as follows.

$$Entropy(S) = P(pos)\log_2 P(pos) - P(neg)\log_2 P(neg)$$

P(pos): The proportion of successful/favourable (positive) cases in Sand P(neg) is the proportion of adverse/unfavourable (negative) cases in S.

$$Information\ Gain(S, A) = Entropy(S) - \sum \left(\frac{|Sv|}{|S|} \right) * Entropy(Sv)$$

Here S is the set of data, A is a feature, Sv is the subset of S for which feature A takes the value v, and S is the total number of elements in S.It measures the projected decrease in entropy, which is the concept of information gain.The following Figure 3 shows the result obtained through ID3.

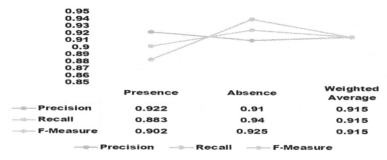

	Presence	Absence	Weighted Average
Precision	0.922	0.91	0.915
Recall	0.883	0.94	0.915
F-Measure	0.902	0.925	0.915

Figure 3.   Precision, Recall, and F-Measure in the case of ID3 model.

3.1.4 *Logistic regression*

It is commonly used to solve classification problems. In this, we use an already-determined value set of independent variables, which in turn is used to predict the categorical dependent variable. The concept of the logistic function is used, which returns a result, as shown in Figure 4.

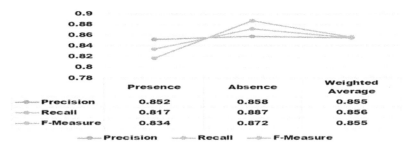

	Presence	Absence	Weighted Average
Precision	0.852	0.858	0.855
Recall	0.817	0.887	0.856
F-Measure	0.834	0.872	0.855

Precision   Recall   F-Measure

Figure 4.   Precision, Recall, and F-Measure in case of logistic regression model.

3.1.5 *Support vector machine*

SVM is a powerful supervised learning technique used for classification, regression, and outliers' detection. It is effective in high-dimensional spaces where the number of dimensions is greater than the number of cases. SVM on the same dataset as shown below in Figure 5.

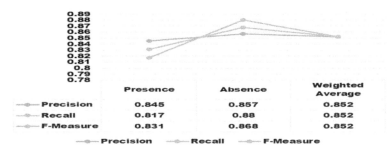

	Presence	Absence	Weighted Average
Precision	0.845	0.857	0.852
Recall	0.817	0.88	0.852
F-Measure	0.831	0.868	0.852

Precision   Recall   F-Measure

Figure 5.   Precision, Recall, and F-Measure in the support vector machine.

4   RESULTS & COMPARISON

Figure 6 shows the comparison among different algorithms, and it is observed that ID3 is the most suitable algorithm out of these 5 algorithms.

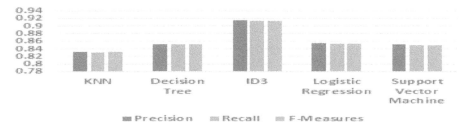

Precision   Recall   F-Measures

Figure 6.   Comparison in Precision, Recall, and F-measure of the different methods.

5   CONCLUSION

This research work uses KNN, Decision Table, Support Vector Machine, Logistic Regression, and ID3 classification techniques. After comparison, it is observed that ID3 performs best on the dataset. ID3 has the highest outcome accuracy in the outcome table. Thus, the ID3 algorithm predicts future heart disease best for the dataset. ID3 algorithm accuracy is 91.4815, with a precision of 0.922 for the presence class and 0.910 for the absence class. The recall and F1 Score (F- Measure) for the Presence class are 0.883 and 0.902. Absence class recall and F1 Score(F-measure) calculated are 0.940 and 0.925. ID3 algorithm calculates precision, recall, and F-Measure weighted averages of 0.915, 0.915, and 0.915. In the future, ensemble learning approaches like Random Forest, Gradient Boosting, or AdaBoost can be used to investigate the prospect of merging different machine learning algorithms, including ID3. By utilizing the advantages of many algorithms, ensemble methods frequently increase predicted performance. Large volumes of data are frequently handled in the healthcare industry. Therefore, handling larger datasets, scalability of the ID3 algorithm, and other machine learning algorithms may also be issues. To guarantee ID3 effectiveness in real-world circumstances, it requires extensive testing and validation.

REFERENCES

Austin, P. C., *et al.* (2013). Using methods from the data mining and machine-learning literature for disease classification and prediction: a case study examining classification of heart failure subtypes. *Journal of Clinical Epidemiology*, 66(4), 398–407.

Beyene, C., & Kamat, P. (2018). Survey on prediction and analysis of the occurrence of heart disease using data mining techniques. *International Journal of Pure and Applied Mathematics*, 118, 165–173.

Deb, A., *et al.* (2022). An outcome based analysis on heart disease prediction using machine learning algorithms and data mining approaches. In *2022 IEEE World AI IoT Congress (AIIoT)* (pp. 01–07).

Enriko, I. K. A. (2019). Comparative study of heart disease diagnosis using top ten data mining classification algorithms. In *Proceedings of the 5th International Conference on Frontiers of Educational Technologies (ICFET '19)* (pp. 159–164). Association for Computing Machinery.

Fadnavis, R, Dhore, K., Gupta, D., Waghmare, J., & Kosankar, D. (2021). Heart disease prediction using data mining. *Journal of Physics: Conference Series*, 1913, 012099. doi:10.1088/1742-6596/1913/1/012099.

Howlader, K., Patwary, M. F., & YasinKabir, M. (2017). A data mining approach to predict the early stage of heart disease. *Journal of Noakhali Science and Technology University (JNSTU)*, 1, 88–94.

Palaniappan, S., & Awang, R. (2008). Intelligent heart disease prediction system using data mining techniques. *International Journal of Computer Science and Network Security (IJCSNS)*, Vol. 8, No. 8, 343–350.

Patidar, S., Jain, A., & Gupta, A. (2022). Comparative analysis of machine learning algorithms for heart disease predictions. In *2022 6th International Conference on Intelligent Computing and Control Systems (ICICCS)*, (pp. 1340–1344). Madurai, India.

Ramesh, T. R., Lilhore, U. K., Poongodi, M., Simaiya, S., Kaur, A., & Hamdi, M. (2022). Predictive analysis of heart diseases with machine learning approaches. *Malaysian Journal of Computer Science*, 132–148.

Shetgaonkar, P., & Aswale, S. (2021). Heart disease prediction using data mining techniques. *International Journal of Engineering Research & Technology (IJERT)*, Vol. 10, Issue 02.

Solanki, Y., & Sharma, S. (2019). A survey on risk assessments of heart attack using data mining approaches. *International Journal of Information Engineering and Electronic Business*, 11(4), 43–51. [Online]. Available: https://hal.science/hal-02188879.

Sultana, M., Haider, A., & Uddin, M. S. (2016). Analysis of data mining techniques for heart disease prediction. In *2016 3rd International Conference on Electrical Engineering and Information Communication Technology (ICEEICT)* (pp. 1–5). IEEE.

Tarawneh, M., & Embarak, O. (2019). Hybrid approach for heart disease prediction using data mining techniques. In L. Barolli *et al.* (Eds.), *Advances in Internet, Data and Web Technologies* (pp. 447–454). Springer International Publishing.

Tripoliti, E. E., *et al.* (2017). Heart failure: diagnosis, severity estimation and prediction of adverse events through machine learning techniques. *Computational and Structural Biotechnology Journal*, 15, 26–47.

Advances in AI for Biomedical Instrumentation, Electronics and Computing – Sachan et al. (eds)
© 2024 The Author(s), ISBN 978-1-032-64298-7

LSTM model for stock price prediction

Nissankara Lakshmi Prasanna*
Professor, CSE, Vasireddy Venkatadri Institute of Technology, Guntur, India

Rajesh Babu Yallamanda*
Assistant Professor, CSE, KL University, India

Rama devi Gunnam*
Assistant Professor, CSE, Vasireddy Venkatadri Institute of Technology, Guntur, India

T. Kameswara Rao*
Professor, CSE, Vasireddy Venkatadri Institute of Technology, Guntur, India

ABSTRACT: Stock price forecasting uses machine learning effectively. To make wiser and more accurate stock market decisions, it is mandatory to forecast stock prices. In order to im-prove stock forecast accuracy and generate lucrative trades, a stock price prediction system is suggested that combines mathematical functions, machine learning, and other external aspects. In many practical applications, including weather forecasting and financial market prediction, time-series prediction is an extensively utilized technique. It forecasts the outcome for the following time unit using the continuous data over a period of time. Long-short-term memory (LSTM) model is the most often used methods. The outcomes of the predictions demonstrate that the LSTM algorithm is offering best performance in terms of MSE, MAE, MAPE and RMSE. When predicting future prices, the findings of this article may be helpful to investors in the capital market.

Keywords: LSTM, stock price, Time-series prediction, sequence prediction

1   INTRODUCTION

Due to the high demand of the stock market, it is one of the most popular investments. National Stock Exchange (NSE) and the Bombay Stock Exchange (BSE) are the most of the trading in Indian Stock Market takes place. Investors may buy a stock market index by buying an index fund, which can be set up as a mutual fund or an exchange-traded fund and "track" an index. Tracking error is the amount of performance variance, if any, between an index fund and the index. The stock market indexes may be categorized and divided into several segments thanks to the index coverage set of stocks. Some forecasting models are developed for this kind of purpose and they had been applied to stock market prediction. Generally, this classification is done by Time series analysis, Fundamental analysis and Technical analysis.

Time series analysis is a specific way of analyzing a sequence of data points collected over an interval of time. To maintain consistency and dependability, time series analysis often needs a lot of data. A large data collection guarantees that your analysis can sift through erratic data and that your sample size is representative. The goal of this research was to model and predict

---

*Corresponding Authors: nlakshmi@vvit.net, yrajeshbabu@kluniveristy.in, ramarajesh95@gmail.com and tkr.cse@vvit.net

DOI: 10.1201/9781032644752-11

the future price of stock using artificial intelligence (AI) techniques. Historical data with artificial intelligence techniques used to predict the price of a stock in the future. Artificial intelligence approaches are employed as financial time series forecasting tools because they have the capacity to account for the complexity of financial systems (Xiongwen *et al.* 2018).

This article proposes an automated trading approach that uses mathematical processes, machine learning, and other external inputs like news sentiments to enhance stock forecast accuracy and produce profitable trades. To do this, we constructed and trained several deep learning models while also training classic machine learning algorithms, taking into account the significance of the pertinent news.

A predictive model LSTM is created and used to estimate stock pricing. Since a variety of variables will influence the outcomes of the prediction this study focuses on optimizing the data selection and processing, parameter adjustments, and the overall framework. Additionally, this includes future projections. Three widely used indicators, namely MSE, MAE, MAPE and RMSE also calculated. With these measures the performance can be assessed.

2 LITERATURE REVIEW

2.1 *Survey*

Longtime stock market researchers have been interested in the concept of prediction. White, H. conducted the initial research on the neural network model for stock price prediction in 1988. A lot of research has been done to use neural networks to forecast future stock prices (White *et al.* 1999). Hansun, S. conducted a comparable experiment using a time series analysis form of the moving average model. Data from the composite index of the Jakarta Stock Exchange (JKSE) was also chosen for research on futures forecasts (Hansun *et al.* 2013). To forecast NSEI member stocks, Sreelekshmy Selvin *et al.* examined three different deep learning architectures, including CNN, RNN, and LSTM sliding window model. They discovered that CNN architecture was superior to other models and could spot changes in stock trends (Selvin *et al.* 2017).

In order to increase profits, Liu, S. *et al.* employed LSTM to create a quantitative timing strategy after developing a large number of stock selection strategies using CNN. A hybrid strategy that integrated the GARCH and LSTM models was proposed by Kim, H. Y. *et al.*, and the outcome had a decreased prediction error (Kim *et al.* 2018). To predict the stock index prices of the Standard & Poor's 500 index (S&P500) and China Securities 300 index (CSI300), Lin, Y. *et al.* employed the mixed model of long and short memory (LSTM) and fully integrated empirical mode decomposition combined with adaptive noise. To get several IMF and a residual from the original data, Ceemdan dissected the data (Lin *et al.* 2021). A model was created by Gao, Y., *et al.* to enhance stock predictions. They developed a number of technical indicators, such as investor mood and financial data (Gao *et al.* 2021).

2.2 *LSTM algorithm*

Recurrent neural networks are a type of long short term memory. It has multiple phases interconnected. Output from one phase is sent into next phase as input. With this approach the issue of long-term RNN dependency, where it can provide more precise forecasts based on current data but cannot anticipate the word hang in in the long term memory. The gap length will grow, making RNN's performance less cost-effective. By default, LSTM will keep the information for a long time. The LSTM's architecture is seen in Figure. 1.

Four neural networks and several memory building pieces known as cells are organised in LSTM as a chain. It has a brand-new component known as a memory cell. The memory cell decides what information should be stored and when reading, writing, and forgetting are permitted. A memory cell has three major gates: an input gate through which a new value enters the cell, a forget gate by which a value is retained in the cell, and an output gate. The output is computed using the output gate value stored in the memory cell.

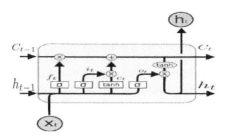

Figure 1. LSTM structure.

Information from the cell state is eliminated by a forget gate. By multiplying a filter, the information that is no longer necessary for the LSTM to grasp things or that is of lower relevance is eliminated. This is necessary in order to improve the LSTM network's performance. H_{t-1} and x_t are the two inputs to this gate. The input at that specific time step is x_t, while the hidden state from the preceding cell, or its output, is h_{t-1}.

By utilising a sigmoid function, an input gate controls which values must be added to the cell state. This functions as a filter for all the data from h_{i-1} and x_t and is fundamentally very similar to the forget gate. creating a vector that contains each value that might be added to the cell state. The tanh function, which produces values between -1 and $+1$, is used for this. Adding this valuable data to the cell state by addition operation after multiplying the value of the regulatory filter (the sigmoid gate) by the produced vector (the tanh function).

After applying the tanh function on the cell state and scaling the values to the range of -1 to $+1$, the operation of an output gate may once more be broken down into three parts. Making a filter that can control the values that must be produced from the vector established above using the values of h_{t-1} and x_t. The sigmoid function is used in this filter once again. adding the value of this regulatory filter to the vector produced in step 1, multiplying it, and sending the result both to the output and hidden state of the subsequent cell.

3 PROPOSED MODEL

The system composes of modules such as input dataset, pre processing, data splitting, build & model train, applying LSTM model and output the predicted result. The data is divided in two sets namely training & testing which is in ratio of 80:20 respectively. Then, this set is used to train a model. Finally, all these modules are evaluated using Root mean square error, Root Absolute error, mean absolute percentage error and root mean square error.

3.1 *Data set*

The dataset is taken from yfinance by importing the yfinance module. The "yahoo_finance_dataset (2018–2023)" dataset is a financial dataset containing daily stock market data for multiple assets such as equities, ETFs, and indexes. It spans from April 1, 2018 to March 31, 2023, and contains 1257 rows and 7 columns Attribute such as: price of open, high, low, close, adjusted close price taken from huge dataset tare fed as input to the models for training to pre-process the data techniques like normalization & one hot encoding in applied on dataset. The data has a small ratio of missing values. The ratio of missing price data of each asset is around 0.1%.

3.2 *Data preprocessing*

Imputation of Missing Values - This technique is used for obtaining the missing values from the data and imputating the null values with the median. In the Indian liver disease patients dataset, there are four missing values for Albumin and Globulin ratio that has been restored by median values.

Dummy Encoding – Dummy encoding is a method of transforming the categorical variable to numerical variable as most of the machine algorithms are designed to work on numerical data. For each of the categorical variable, k−1 numerical variables are created.

Elimination of Duplicate Values – It is necessary to discard the redundant values from the data, in order to improve the efficiency and quality of the data.

Outlier Detection and Elimination- Outliers are unusual values that deviate from the rest of the results owing to minute measurement or experimental error. Outliers are detected and eliminated. Univariate and multivariate outliers are the two subcategories of outliers. A multivariate outlier, as opposed to a univriate outlier, takes into account n-dimensions of ILPD data variables or qualities.

Resampling- The linear dataset is uneven, with most variations in stock price. SMOTE is used to synthesise new samples for the minority class in order to balance the data.

3.3 *LSTM code*

Consider the current input along with the prior concealed state and internal cell state. The steps listed below should be used to calculate the values of the four distinct gates. By element-wise multiplying the relevant vector with the appropriate weights for each gate, obtain the parameterized vectors for the current input and the prior hidden state for each gate. Apply the appropriate activation function to the parameterized vectors for each gate element. The list of gates with the application function for the gate is shown below. Calculate the element-wise multiplication vectors of the input gate and the input modulation gate, followed by the element-wise multiplication vectors.

Initially, the dataset is splitted into training and test datasets. 80%of data is allocated to training and 20% of data is used for testing. The data is fit using the MinMaxScaler and it is appended into an array. The model is trained using this data. Now an RNN is built using keras. The number of neurons in each layer for RNN is 50. 4 layers are taken for better accuracy and they are added to the regressor. The neural network is trained and tested. After proper training is received, the neural network is used for forecasting the data and it is represented graphically using plot () function. The graph is then saved using savefig () function and the RMSE value is calculated.

4   RESULTS

This project presents extensive process of building models for stock price prediction. The experimental results obtained with LSTM model demonstrated its potential to predict stock prices satisfactory on short-term basis. CTSH is the stock symbol for Cognizant Technology Solutions. So, it is a valid symbol and the result page will display the forecasted values. Recent trends in stock market displays the graph of all the closing values for the past 1 year. LSTM graph compares the predictions of LSTM to the actual values. MSE, MAE, MAPE and RMSE values of the algorithm is calculated.

This paper study the data from the stock market, especially some technology stocks, use pandas to obtain stock information and visualize all aspects of the stock. Finally, this paper study several methods to analyze stock risk based on the previous performance history of the stock and predict the future stock price through LSTM method. Since the historical data set available on the company's website contains only a few features, such as high and low stock prices, open and close, stock trading volume, etc.

In order to obtain higher forecast price accuracy, new variables are created using existing variables. Before the LSTM model experiment of stock price, 95% of the total data of each stock is used as the training set data to train the parameters of the model, while the rest is the test set data. The prediction results of the model on the test set verify the advantages and disadvantages of the model.

Then, build the LSTM network architecture: add six LSTM layers and several dropout layers to prevent over fitting. 64 units are the dimensions of the output and parameter return_sequences represents Stock Predicting based on LSTM whether to return the last output in the output sequence or the complete sequence, parameter input_ shape as the shape of the training set. The drain layer specified as 0.2 will be discarded and the dense layer specified as one unit output will be added. Finally, the model is suitable for 100 times with a batch size of 32. This paper use a batch of short sub sequence randomly selected from the training data instead of training RNN on the complete observation sequence.

Moving average forecast is like finding the mean of the last N values. Use df[col]. Rolling() creates a window of size N from the first entry of the column. This window is such that it returns t, t −1 … , t-(N−1) rows for timestamp t (if N rows are not possible, it gives the maximum possible). That solves the problem of creating the window. Then predict the value at timestamp t. This means that the algorithm shifts the values by one row by append NaN at the start. Finally, this paper predicts and evaluates the above predictions using the metrics. From Figures 2 and 3, it can be seen that the predicting price trend is well fit the exact price, although there are some deviations. It can be concluded that LSTM prediction is offering better results.

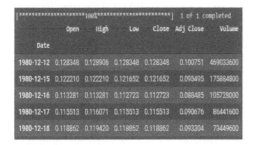

Figure 2. Sample dataset.

Figure 3. Data visualization.

The final stage of forecasting involves evaluating the model's performance. This evaluation can be calculated using a range of evaluation methods, including Mean Squared Error (MSE), Mean Absolute Error (MAE), Root Mean Squared Error (RMSE) and the Mean Absolute Percentage Error (MAPE) shown in Table 1. The results of these techniques are presented in Figure 4.

Table 1. Error Indicators.

```
# model performance
mse = mean_squared_error(test_ic_1yr, model_predictions)
print('MSE: '+str(mse))
mae = mean_absolute_error(test_ic_1yr,model_predictions)
print('MAE: '+str(mae))
rmse = math.sqrt(mean_squared_error(test_ic_1yr, model_predictions))
print('RMSE: '+str(rmse))
mape = np.mean(np.abs(model_predictions - test_ic_1yr)/np.abs(test_ic_1yr))
print('MAPE: '+str(mape))

MSE: 187.38243166485807
MAE: 10.081561186646116
RMSE: 13.688770275845018
MAPE: 0.09521617913038301
```

Figure 4. Results using plot().

5 CONCLUSION

Currently, Machine learning algorithm is booming to apply in diverse aspects of social life. Undoubtedly, financial area is involved. Specifically, the algorithms are widely used to predict futures stock prices. In this paper LSTM method is comprehensively investigated. After comparing several statistical indicators of MAE, MSE, MAPE and RMSE, it is found that LSTM offers accurate results with respect to the sample data used in the paper.

REFERENCES

Gao, Y., Wang, R., & Zhou, E. (2021). Stock prediction based on optimized LSTM and GRU models. *Scientific Programming*, 2021. DOI:https://doi.org/10.1155/2021/4055281.

Hansun, S. (2013, November). A new approach of moving average method in time series analysis. In 2013 conference on new media studies (CoNMedia) (pp. 1–4). *IEEE*. DOI:https://doi.org/10.1109/conmedia.2013.6708545

Kim, H. Y., & Won, C. H. (2018). Forecasting the volatility of stock price index: A hybrid model integrating LSTM with multiple GARCH-type models. *Expert Systems with Applications*, 103, 25–37. DOI:https://doi.org/10.1016/j.eswa.2018.03.002

Lin, Y., Yan, Y., Xu, J., Liao, Y., & Ma, F. (2021). Forecasting stock index price using the CEEMDAN-LSTM model. *The North American Journal of Economics and Finance*, 57, 101421. DOI:https://doi.org/10.1016/j.najef.2021.101421.

Liu, S., Zhang, C., & Ma, J. (2017, November). CNNLSTM neural network model for quantitative strategy analysis in stock markets. In *International Conference on Neural Information Processing* (pp. 198–206). Springer, Cham. DOI:https://doi.org/10.1007/978-3-319-70096-0_21.

Selvin, S., Vinayakumar, R., Gopalakrishnan, E. A., Menon, V. K., & Soman, K. P. (2017, September). Stock price prediction using LSTM, RNN and CNN-sliding window model. In *2017 International Conference on Advances in Computing, Communications and Informatics (icacci)* (pp. 1643–1647). IEEE. DOI:https://doi.org/10.1109/ICACCI.2017.8126078

White, H. (1988, July). Economic prediction using neural networks: The case of IBM daily stock returns. In *ICNN* (Vol. 2, pp. 451–458). DOI:https://doi.org/10.1109/5.771073

Xiongwen Pang, Yanqiang Zhou, Pan Wang, Weiwei Lin, "An innovative neural network approach for stock market prediction", 2018.

Advances in AI for Biomedical Instrumentation, Electronics and Computing – Sachan et al. (eds)
© 2024 The Author(s), ISBN 978-1-032-64298-7

A systematic study of image/data sharing security methods using Modified Advanced Encryption standards (MAES)

B. Satyaramamanohar A.
Research Scholar, Department of Electronics and Communication Engineering, Sathyabama Institute of Science & Technology, Chennai, India

T. Bernatin
Associate Professor, Department of Electronics and Communication, Engineering, Sathyabama Institute of Science & Technology, Chennai, India

P.S.D. Anvesh
Assistant Professor, Department of Electronics and Communication Engineering, Bhimavaram Institute of Engineering & Technology, Pennada, India

ABSTRACT: Now a Days the multimedia data processing and protection is becoming a vital thing in IOT, RF (Radio frequency), etc... Communications. The significant encryption methodology is used to secure multimedia data either in the form of steganography or cryptography. Many other methods exists for securing sensitive image data from intrusion attacks. Encryption transforms information into a code that can only be read by the intended recipient, making it unreadable to others. The encryption process does not prevent interference, but it does prevent observers from knowing what is being transmitted. It still keeps users' identities and data secure. The work covers many methods of encrypting and decrypting images in the past and provides an introduction to cryptography.

Keywords: Information Security, Modified Advanced Encryption standards (MAES), Cryptography, Steganography, Internet of Things (IOT), Data compression

1 INTRODUCTION

In this day and age of pervasive technological connectivity, protecting sensitive information from malicious actors is more important than ever. There has been a dramatic increase in the reliance of businesses and individuals on data stored and transmitted via computer systems and networks. Data and systems must be secured from exposure and network-based threats. Cryptography and steganography are the two most frequent methods of security. The term "cryptography" is used to describe the process of transforming readable material into unreadable text and vice versa. By combining two Greek words, "Krypto" (meaning "hidden") and "graphene" (meaning "written"), the word "cryptography" was created. The goal of steganography is to disguise a message within another, seemingly innocuous message or physical object. In Greek, the word for "steganography" is "steganographia," which is a compound of the words "stegano" (meaning "covered or concealed") and "graphia," (meaning "writing"). The Advanced Encryption Standard (AES) is a symmetric key encryption technique that encodes and decodes visible information using a single secret key. The Advanced Encryption Standard (AES) is the only publicly available encryption technology approved by the United States National Security Agency (NSA) for protecting top secret information. Vincent Rijmen and Joan Daemen are responsible for AES, formerly known as Rijndael. Important concern for communication and storage of images is security, and encryption is one way to provide security in light of the ever-increasing expansion of multimedia applications. In order to ensure the privacy of the image between users, encryption techniques are used to transform the original image into one that requires a key in order to be interpreted. Secret-key cryptography, which is used in symmetric encryption, uses a single key for both encryption and decryption.

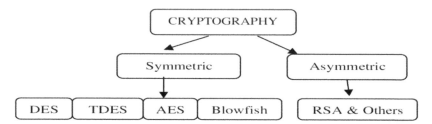

Figure 1.   Classification of cryptography.

Table 1.   Comparison of RSA, DES and AES.

Factors	RSA	DES	AES
Created by	Ron Rivest, Adi Shamir and Leonard Adleman in 1978	IBM in 1975	Vincent Rijmen, Joan Daemen in 2001
Key Length	Depends on number of bits in the modulus n where n = p*q	56 bits	128, 192 or 256 bits
Block size	Variable	64 Bits	128 Bits
Cipher Type	Asymmetric Block Cipher	symmetric Block Cipher	symmetric Block Cipher

1.1   *Advanced Encryption Standard (AES)*

AES is classified as a symmetric-key method due to its use of a single key for both encryption and decryption. The length of the key is inversely related to the number of rounds used by the cipher. A 128-bit key passes through 10 iterations. AES encrypts all 128 bits in a single pass. This is why there are relatively few rounds in comparison. Figure 2 presents the AES algorithm.

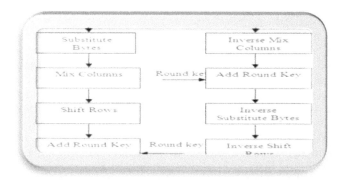

Figure 2.   Single round of AES encryption and decryption.

2 REVIEW

Arora *et al.* (2012) evaluated the performance of various security algorithms in the cloud and on a single processor, with different input sizes in consideration. Enterprises frequently utilise data encryption algorithms (such as RSA, MD5, and AES), and this research aims to evaluate the benefits of using cloud resources to execute these algorithms. (Seth *et al.* 2013) has compared RSA, DES, and AES based on key performance metrics such processing speed, memory footprint, and data security. These settings are crucially important in any cryptographic algorithm. The findings of these experiments reveal that the DES method requires the least amount of memory for encryption whereas the AES algorithm requires the least amount of time for encryption. While RSA's encryption process takes the most time and uses the most memory, its output byte count is the smallest. (Elminaam *et al.* 2008) research conducted regarding symmetric encryption algorithm efficiency. Examining the strengths and weaknesses of AES, DES, 3DES, RC2, Blowfish, and RC6, six of the most widely used encryption algorithms, this study offers a comprehensive overview of the state of the art. Data block sizes, data kinds, battery life, key sizes, and encryption/decryption speeds have all been compared across a range of algorithms. Observable outcomes from the experiment are displayed in the simulation. We found no statistically significant variation between displaying the results in hexadecimal and base 64. To encrypt the same data using different packet sizes, we observed that RC6 is the most efficient algorithm. However, when we used image data instead of text data, we found that RC2, RC6, and Blowfish took significantly longer than the rest of the algorithms. And even when compared to DES, 3DES performance is still subpar. Finally, it was discovered that increasing the key size, perhaps just in the AES and RC6 algorithms, results in a noticeable shift in the amount of time and battery life required for encryption. (Pavithra *et al.* 2012) analyses and contrasts how well different cryptography methods work. Multiple cryptographic techniques are tested on several video clips, with time as the independent variable. The processing time needed to watch a video might vary greatly depending on its size. Timing the encryption and decryption of video files ranging in size from 1 MB to 1 100 MB in a variety of formats, including and.dat. The results reveal that the AES algorithm has higher throughput and requires less processing time than DES and Blowfish. (Alanazi *et al.* 2010) Due to the proliferation of new media tools, a growing amount of medical records now include audio, video, and other forms of digital content. Editing, modifying, and duplicating digital material becomes considerably simpler. (Mandal *et al.* 2012) Further, many dangers will arise since digital papers are simple to duplicate and disseminate. Important, accurate, and potentially sensitive data, some of which should not be accessed by or can only be partially disclosed to the public users, poses a significant security and privacy risk, making it important to establish appropriate protection. This highlights the significance of security and privacy. Furthermore, digital documents and videos are problematic as evidence since they may be easily altered without detection using inexpensive and commonly available technology. The use of cryptography is often cited as a means of safeguarding sensitive data. (Kakkar *et al.* 2012) [8evaluated many Multimode Network (MN) data security methodologies and processes. Multiple parameters, such as the cryptographic method employed (public vs. private), the number of keys, and the number of bits in a key, have been shown to affect the system's security. Power consumption and heat production both rise as key and data lengths grow longer. The safeness of the gearbox improves in proportion to the size of the key. The mathematical concepts upon which all the keys are based gradually lose their efficacy over time. More bits in a key mean more work for the computer, which means it will take longer to encrypt the data.

(Hadj Brahim *et al.* 2022) in proposes a new fast image compression–encryption scheme based on compressive sensing and parallel blocks. To reduce the encryption time in the proposed algorithm, the plain image is divided into a predetermined number of blocks, where these blocks are encrypted in parallel, and each block is encrypted using the same steps. Firstly, the block is transformed into a sparse matrix using DWT.

2.1 *Data encryption comparison*

Data encryption, blockchain, and SLR were used to achieve this result within the block-chain, with additional comments providing context. Information gathered from 2015 through 2022 can be found in these 4 databases: Wiley, Tandfonline, IEEE Explore, and Science Direct. A number of studies on data encryption by 30% and the blockchain technology by 10% were discovered in this investigation. Data encryption throughout the blockchain can be reviewed, and the results summarised in an SLR comparison table. Data encryption SLR table is available in the following Table 2.

Table 2. Data encryption comparison.

Description	Encryption Data AES	Encryption Data AES Block chain
Research Question	Up to 7	Only1
Search Strategy	Keyword based on authors and Keyword based on extraction from the known subset of papers	Keyword based on specific subject
Model of string	5 models	1 model
Resource to be search	4 libraries	2 libraries
Paper selection	8 in exclusion criteria 4 in inclusion criteria 9 in quality assessments	4 in exclusion criteria 4 in inclusion criteria 6 in quality assessments

The Table 2 presented above elucidates the differences between the papers discussing AES data encryption in the context of blockchain technology.

2.2 *Gaps in the literature*

- The evaluation metrics could be further improved by considering other performance metrics and by comparing the proposed system with other encryption algorithms across many of the said papers.
- Many of these authors could have discussed the limitations and potential applications of their implementation in more detail.
- Many encryption algorithms are prone to some specific type of attacks and the authors could have discussed the potential impact and countermeasures against such attacks in more detail.
- Many authors could have discussed the potential impact and practical considerations of implementing the proposed countermeasure in real-world scenarios.

3 CONCLUSION

The purpose of this paper is to survey the current literature on the cryptographic algorithms AES, 3DES, Blowfish, and DES. When compared to other methods, DES's key size is inadequate. In comparison to other block cipher algorithms, 3DES performs poorly and appears incredibly slow. According to researchers, AES is the superior algorithm. When compared to the first Blowfish Algorithm. Additionally, the AES algorithm is unable to exclude related pixels in an image. Directly encrypting images with these ciphers presents a security risk and is too slow for use in real-time settings. A revised Standard for advanced encryption is provided as a means of addressing these problems. It's possible that this update will boost performance and security.

AUTHOR CONTRIBUTIONS

"B Satyaramamanohar A" wrote the manuscript with support from Dr. T. Bernatin. Dr. T. Bernatin encouraged "B Satyaramamanohar A" to investigate security aspects and supervised the findings of this work. "B Satyaramamanohar A" and "P S D Anvesh" has developed the theoretical formalism, performed the analytic calculations and performed the numerical calibrations. They designed the model, the computational framework and analysed the data.

REFERENCES

Alanazi, H., Zaidan, B. B., Zaidan, A. A., Jalab, H. A., Shabbir, M., & Al-Nabhani, Y. (2010). New comparative study between DES, *3DES and AES within nine factors. arXiv preprint arXiv*:1003.4085.

Al-Shaarani F., Gutub A., Securing matrix counting-based secret-sharing involving crypto steganography. *J. King Saud Univ.-Comput. Inform. Sci.* **34**(9), 6909–6924 (2022).

Anjos J. C. S. do. *et al.*, "Fast-Sec: An approach to secure Big Data processing in the cloud," *International Journal of Parallel, Emergent and Distributed Systems*, vol. 34, no. 3, pp. 272–287, 2019.

Arora, P., Singh, A., & Tyagi, H. (2012). Evaluation and comparison of security issues on cloud computing environment. *World of Computer Science and Information Technology Journal (WCSIT)*, 2(5), 179–183.

Elisha Raju B., Ramesh Chandra K. and Budumuru P.R., "A two-level security system based on multimodal biometrics and modified fusion technique", *Sustainable Communication Networks and Application. Lecture Notes on Data Engineering and Communications Technologies*, vol. 93, 2022.

Elminaam, D. S. A., Kader, H. M. A., & Hadhoud, M. M. (2008). Performance evaluation of symmetric encryption algorithms. *IJCSNS International Journal of Computer Science and Network Security*, 8(12), 280–286.

Forouzan, B. A. (2007). Data communications and networking. Huga Media.

Hadj Brahim A., Ali Pacha A., Hadj S. N., A new fast image compression–encryption scheme based on compressive sensing and parallel blocks. *J. Supercomput.* **26**, 1–47 (2022).

Kakkar, A., Singh, M. L., & Bansal, P. K. (2012). Comparison of various encryption algorithms and techniques for secured data communication. In multinode network", *International Journal of Engineering and Technology Volume*.

Mandal, A. K., Parakash, C., & Tiwari, A. (2012, March). Performance evaluation of cryptographic algorithms: DES and AES. In 2012 IEEE Students' Conference on Electrical, Electronics and Computer Science (pp. 1–5). *IEEE*.

Nie, T., Song, C., & Zhi, X. (2010, April). Performance evaluation of DES and Blowfish algorithms. In 2010 International conference on biomedical engineering and computer science (pp. 1–4). *IEEE*.

page.math.tu-berlin.de /.

Pavithra S., Ramadevi E., (2012), Performance evaluation of symmetric algorithmns.

Reddy M. I. S. and Kumar A. P. S., "Secured data transmission using wavelet based steganography and cryptography by using AES algorithm," *Procedia Computer Science*, vol. 85, no. Cms, pp. 62–69, 2016.

Rosati P., Fox G., Kenny D., and Lynn T., "Quantifying the financial value of cloud investments: A systematic literature review," *Proceedings of the International Conference on Cloud Computing Technology and Science*, CloudCom, vol. 2017–Decem, pp. 194–201, 2017.

Seth, S. M., & Mishra, R. (2011). Comparative analysis of encryption algorithms for data communication 1.

Stallings, W. (2006). Cryptography and network security, *4/E*. Pearson Education India.

Sutton, E. Latency, Packet Loss and Encryption using DES with a VPN.

Tsai K.-L., Leu F.-Y., and Tsai S.-H., "Data encryption method using environmental secret key with server assistance," *Intelligent Automation & Soft Computing*, vol. 22, no. 3, pp. 423–430, Jul. 2016.

Advances in AI for Biomedical Instrumentation, Electronics and Computing – Sachan et al. (eds)
© 2024 The Author(s), ISBN 978-1-032-64298-7

Compact-size frequency reconfigurable antenna for sub-6 GHz 5G applications

Sachin Kumar

Department of Electronics and Communication Engineering, SRM Institute of Science and Technology, Kattankulathur, India

Himanshu Nagpal

Department of Electronics and Communication Engineering, Ajay Kumar Garg Engineering College, Ghaziabad, India

Ghanshyam Singh

Department of Electronics and Communication Engineering, Feroze Gandhi Institute of Engineering and Technology, Raebareli, India

L. Harlan

Department of Electronics and Communication Engineering, SRM Institute of Science and Technology, Kattankulathur, India

ABSTRACT: In this work, a frequency adaptable planar monopole antenna is designed for use in 5G communication systems. The proposed antenna is made up of two patches, a partial ground plane, and a 50-Ω microstrip feed line. The lower patch is a conventional rectangular patch and the upper patch is a modified trapezoidal-shaped ring. An RF PIN diode is added between the bottom and top patches to offer frequency diversity. The antenna can be tuned to operate in the sub-6 GHz 5G (3.4−5.8 GHz and 2.4−2.8 GHz) frequency bands, when the diode is turned ON and OFF.

1 INTRODUCTION

Wireless communication has changed significantly over the last few decades. A massive number of devices must be connected to the Internet for the fourth generation (4G) long-term evolution (LTE)/advanced infrastructure to handle new developments in wireless communication. Fifth generation (5G) wireless connectivity is a promising technology that can handle increasing data rate demand while also allowing for the integration of multiple services (Akpakwu G. A. *et al.* 2001; Gandotra P. *et al.* 2017; Zhao D. *et al.* 2021). The 5G network may address the issues of bandwidth, service quality, latency, and data transmission rates. 5G technology is significantly faster than other networks, with better and improved video quality. 5G uses spectrum more efficiently than 4G, therefore it has increased bandwidth, beam steering, beamforming capability, widespread dependability, and lower latency. With 5G, one can connect many devices at higher efficiency, increased connectivity, and high data rates without lag. The 5G network will benefit a wide range of industries, including smart healthcare, smarter cities, 3-D printing, and the Internet of things (IoT). 5G technology could expand portable devices into next generation technologies that will impact every business, enabling more secure transportation, precise farming, digital coordination, and more (Kumar S. *et al.* 2022; Li Y. *et al.* 2020; Mubarak S. *et al.* 2018; Saxena S. *et al.* 2018).

The entire 5G spectrum is divided into three frequency bands: a low band between 600 and 700 MHz, a sub-6 GHz band between 3 and 5 GHz, and a mm-wave band between 24 and

DOI: 10.1201/9781032644752-13

100 GHz (Saxena S. *et al.* 2020; Zrar K. *et al.* 2020). Due to the limitations of mm-waves, research has shifted to the sub-6 GHz (mid band), as high-frequency mm-waves can only travel to a shorter distance. Therefore, deploying mm-wave devices could be costly. Also, the installation of sub-6 GHz devices/systems may be possible using the present 4G LTE infrastructure.

A patch antenna is the main element of a portable device that transmits and receives radio waves. It has a radiating patch on one side of the dielectric substrate and a ground plane on the other. A 5G antenna with frequency agility provides high data transmission and improved efficiency (Sharma D. *et al.* 2021; Singh G. *et al.* 2023). In this study, a frequency-agile antenna for 5G sub-6 GHz bands is developed. The radiator of the antenna is composed of a rectangular monopole divided into two patches and a partial ground plane. A PIN diode connects the two patches. A microstrip line feed is used to power the patch. The microstrip feed line makes fabrication easier as the feed and patch are engraved on the same plane. The antenna offers a reasonable gain and can cover a wide range of 5G application frequencies.

The paper is organized as follows: Section I contains an introduction, followed by sections describing the antenna layout and its working, Section III displays the findings, and Section IV presents a conclusion.

2 ANTENNA CONFIGURATION

The frequency agile antenna is designed using Ansys High Frequency Structure Simulator. The antenna is designed on the FR-4 substrate of size of 35 mm × 35 mm. Figure 1 depicts the layout of the frequency reconfigurable antenna, which is made up of two radiators, a PIN diode, a feed line, and a ground plane. Simulations are carried out using the ANSYS HFSS tool on the FR-4 substrate with a relative permittivity of 4.4 and a height of 1.6 mm. The suggested frequency agile antenna has a total size of 35 mm × 35 mm, and its dimensions are indicated in Table 1.

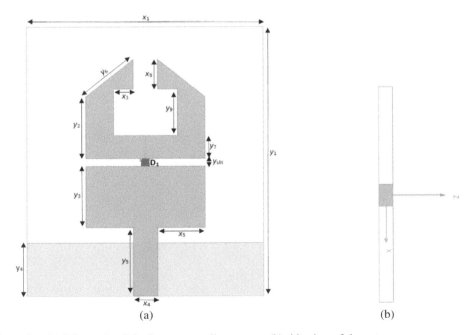

Figure 1. (a) Schematic of the frequency agile antenna, (b) side view of the antenna.

Table 1. Parameters of the frequency agile antenna.

Parameter	Value (mm)	Parameter	Value (mm)
x_1	35	x_2	16
x_3	2.5	x_4	3.1
x_5	5.98	y_1	35
y_2	8	y_3	8
y_4	7	y_5	9
y_6	7.68	y_7	1
y_8	3.8	y_9	6
y_{slit}	1	x_8	3.8

2.1 *Evolution*

The evolution of the frequency agile antenna is shown in Figure 2. Figure 2(a) shows a rectangular radiator of a size of 35 mm × 35 mm and resonating at a centre frequency of 2.45 GHz. In the second stage, antenna-2 is created by dividing antenna-1 into two patches, as shown in Figure 2(b). As the size decreases, the resonating frequency shifts to the higher side.

As shown in Figure 2(c), the antenna-3 is made by etching a 6 mm × 6 mm square slot from the top patch. The upper edge of the top rectangle is then abraded to create a rectangular slit of size of 3.8 mm × 3 mm, as shown in Figure 2(d).

Furthermore, two corners are cut off from the top resonator to extend the electrical length of it, as illustrated in Figure 2(e). As indicated in Figure 2(f), a PIN diode is also added between the patches. When the diode is turned OFF, only the lower antenna element functions and provides wideband performance; when the diode is turned ON, the frequency shifts to a lower band.

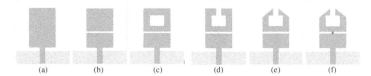

Figure 2. Evolution of the frequency agile antenna: (a) design stage-1, (b) design stage-2, (c) design stage-3, (d) design stage-4, (e) design stage-5, (f) design stage-6.

3 RESULTS

Figure 3(a) illustrates the proposed frequency agile antenna when the RF PIN diode is ON. The antenna shows resonance at 2.6 GHz, illustrating the amalgamation of two patches when the diode is turned ON.

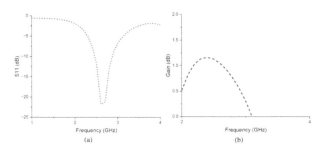

Figure 3. Performance of the proposed frequency agile antenna when the diode is ON: (a) reflection coefficients, (b) gain.

70

Another metric that describes the performance of frequency agile antenna is its gain. The antenna must have a decent gain in order to use it practically. The proposed frequency agile antenna shows a peak gain of more than 1 dB in the ON state of the diode, as shown in Figure 3(b).

Figure 4(a) illustrates the proposed frequency agile antenna when the RF PIN diode is OFF. The antenna shows resonance at 4.8 GHz, illustrating the operation of lower rectangular patch only. The proposed frequency agile antenna has a peak gain of 1.5 dB in the OFF state of the diode, as shown in Figure 4(b). Figure 5 shows the radiation patterns of the proposed frequency agile antenna when the diode is at ON and OFF states at frequencies 2.6 GHz and 4.8 GHz, respectively.

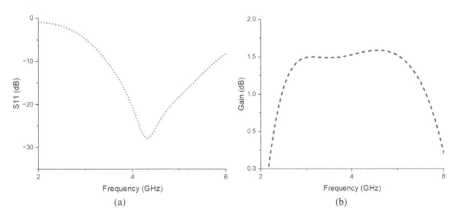

Figure 4. Performance of the proposed frequency agile antenna when the diode is OFF: (a) reflection coefficients, (b) gain.

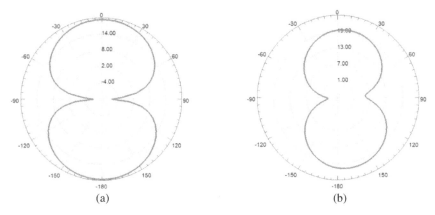

Figure 5. (a) Radiation pattern at ON state of the diode at frequency 2.6 GHz, (b) radiation pattern at OFF state of the diode at frequency 4.8 GHz.

4 CONCLUSION

Design and development of a frequency adaptive antenna for 5G applications is presented. The antenna consists of two patches, with the top patch modified by slits and slots that increase current flow and enable antenna miniaturisation. To offer frequency diversity, an

RF PIN diode is inserted between the bottom and top patches of the suggested antenna design. The antenna is created using the inexpensive FR-4 substrate and requires less space. The antenna has an acceptable gain, which can be improved by using a reflector or meta-surface.

REFERENCES

Akpakwu G. A., Silva B. J., Hancke G. P., and Abu-Mahfouz A. M., "A survey on 5G networks for the Internet of things: Communication technologies and challenges," *IEEE Access*, vol. 6, pp. 3619–3647, 2018.

Gandotra P., Jha R. K., and Jain S., "Green communication in next generation cellular networks: A survey," *IEEE Access*, vol. 5, pp. 11727–11758, 2017.

Kumar S., Palaniswamy S. K., Choi H. C., and Kim K. W., "Compact dual circularly-polarized quadelement MIMO/diversity antenna for sub-6 GHz communication systems," *Sensors*, vol. 22, no. 24, 9827, 2022.

Li Y., Zhao Z., Tang Z., and Yin Y., "Differentially fed, dual-band dual-polarized filtering antenna with high selectivity for 5G sub-6 GHz base station applications," *IEEE Transactions on Antennas and Propagation*, vol. 68, no. 4, pp. 3231–3236, 2020.

Mubarak S., Esmaiel H., and Mohamed E. M., "LTE/Wi-Fi/mmwave RAN-level interworking using 2C/U plane splitting for future 5G networks," *IEEE Access*, vol. 6, pp. 53473–53488, 2018.

Saxena S., Kanaujia B. K., Dwari S., Kumar S., and Tiwari R., "MIMO antenna with built-in circular shaped isolator for sub-6 GHz 5G applications," *Electronics Letters*, vol. 54, no. 8, pp. 478–480, 2018.

Saxena S., Kanaujia B. K., Dwari S., Kumar S., H. C. Choi, and K. W. Kim, "Planar four-port dual circularly-polarized MIMO antenna for sub-6 GHz band," *IEEE Access*, vol. 8, pp. 90779–90791, 2020.

Sharma D., Kanaujia B. K., and Kumar S., "Compact multi-standard planar MIMO antenna for IoT/WLAN/sub-6 GHz/X-band applications," *Wireless Networks*, vol. 27, pp. 2671–2689, 2021.

Singh G., Kumar S., Abrol A., Kanaujia B. K., Pandey V. K., Marey M., and Mostafa H., "Frequency reconfigurable quad-element MIMO antenna with improved isolation for 5G systems," *Electronics*, vol. 12, no. 4, 796, 2023.

Zhao D., Yan Z., Wang M., Zhang P., and Song B., "Is 5G handover secure and private? A survey," *IEEE Internet of Things Journal*, vol. 8, no. 16, pp. 12855–12879, 2021.

Zrar Ghafoor K. *et al.*, "Millimeter-wave communication for Internet of vehicles: Status, challenges, and perspectives," *IEEE Internet of Things Journal*, vol. 7, no. 9, pp. 8525–8546, 2020.

Advances in AI for Biomedical Instrumentation, Electronics and Computing – Sachan et al. (eds)
© 2024 The Author(s), ISBN 978-1-032-64298-7

ASIC acceleration of floating-point arithmetic: Empowering fast and reliable computations

Kapileswarreddy Siddireddy & Ganapathi Hegde

Department of Electronics and Communication Engineering, Amrita School of Engineering, Bengaluru, Amrita Vishwa Vidyapeetham, India

ABSTRACT: Modern computer systems are fundamentally based on floating-point arithmetic, which enables high precision computations and the execution of intricate algorithms in hardware. Floating-point units (FPUs) are designed and implemented for use in Application Specific Integrated Circuits (ASICs). In this study, I focused on the difficulties and trade-offs associated with obtaining high performance and low power consumption. Here, I discuss several methods for incorporating FPUs in ASIC designs, such as dedicated units and floating-point arithmetic and its benefits. The performance and power use of several FPU designs on a 45nm technology node are also shown in the experimental findings that we have presented. The study of binary operations is crucial to engineering science. Verilog-HDL is used to design, and Cadence is used to implement them. This paper also includes an analysis on suggested method's area, power, and latency for various specifications. To effectively handle FPUs' implementation by using tools like Genus, Tempus, Innovus.

Keywords: Floating-point numerical, arithmetic unit, ASIC, Verilog HDL, Arithmetic computation

1 INTRODUCTION

Floating-point arithmetic stands as a cornerstone in the realm of various technical domains, including computer arithmetic, signal processing, and image processing [Bagnara *et al.* 2022]. The extraordinary precision with which it can handle an extensive range of numerical values has positioned it as an indispensable element within modern high-speed digital systems. The infusion of floating-point arithmetic into these applications is not merely a convenience but rather a prerequisite to achieving accurate and efficient computations that drive advancements across diverse engineering and design disciplines. In contrast to fundamental operations like addition, sub- traction, multiplication, and division, floating-point arithmetic operations present a distinctive set of advantages. Despite considerations of hardware complexity and the associated costs of delays, these operations often yield superior results. This advantage stems from the intrinsic nature of floating-point calculations, which necessitates their division into sequential phases [Bora *et al.* 2022]. The sequential execution of such operations might inadvertently lead to elevated computational overhead and undesirable delays. On the contrary, the realm of bit-wise processing introduces an intriguing alternative. This technique proficiently executes fundamental mathematical and logical operations, encompassing essential functions like AND, OR XOR, and SHIFT [Smrithi *et al.* 2016]. This novel technique results in an impressive reduction of latency, often comparable to a mere single gate's delay. This advantage translates to a notable acceleration of processes when executed on hardware, further fortifying its relevance in time-critical

DOI: 10.1201/9781032644752-14

applications. So, it is required that an area-efficient architecture accomplish these tasks [Jaiswal *et al.* 2017]. These encompass both the complex floating-point calculations vital for intricate applications and the fundamental arithmetic operations that underpin them. The crux of these operations lies in division—an indispensable facet of arithmetic computations [Jun *et al.* 2013]. However, implementing division on hardware introduces an intricate set of challenges [Kataria *et al.* 2020]. Latency, or the delay incurred during the execution of operations, emerges as a significant concern within these hardware circuits [Matula *et al.* 2012–2019]. Achieving optimal system performance hinges on the area and power efficiency of hardware circuits dedicated to arithmetic operations, making them a pivotal component of system design [Chandrika *et al.* 2022]. The radix, commonly known as the base, holds a pivotal role in the realm of floating-point arithmetic. It symbolizes the count of distinct digits or symbols employed for numerical representation within a positional numeral system. Within the domain of floating-point arithmetic, the radix assumes a critical role in delineating the architecture and portrayal of floating-point numbers [Neethu *et al.* 2016]. While scholarly literature has extensively explored the hard- ware implementation of fundamental operations like addition, subtraction, and multiplication, a noticeable gap emerges in research pertaining to the hardware implementation of division algorithms [Li *et al.* 2022]. This disparity underscores the inherent complexity of division, surpassing that of adder and multiplier circuits [Kim *et al.* 2022]. Consequently, the incorporation of division into hardware circuit designs for diverse systems becomes a crucial endeavor, necessitating the development of rapid and efficient divider circuits [Villalba *et al.* 2022]. The pervasive influence of floating-point arithmetic across various technical applications underscores its significance as a linchpin of accurate and efficient computations [The Journal of Supercomputing 2022]. The nuanced balance between hardware complexity, delay costs, and operational efficiency offers insight into the superiority of floating-point arithmetic operations. The realm of bit-wise processing introduces a concurrent approach that optimizes execution speed by handling individual bits in parallel, resulting in minimal latency [Lahari *et al.* 2020]. Within processor development, the design of Arithmetic and Logical Units plays a pivotal role in accommodating both complex floating-point computations and fundamental arithmetic operations [Jaiswal *et al.* 2018]. The intricacies of hardware implementation, particularly in handling division, present ongoing challenges that necessitate innovative solutions for efficient system operation.

2 LITERATURE SURVEY

2.1 *IEEE754-for program verification*

The discussion of floating-point value authentication is the primary objective of the current research. The paper presents constraint propagation techniques inside a program verification framework for operations like abstract interpretation and symbolic model testing. Verifying the characteristics of floating-point calculations, such as avoiding odd behaviors like overflows, under- flows, and NaN output, is the fundamental objective. Underflow severity is categorized into three groups based on the output: soft, harsh, and gradual.

2.2 *A review of modern floating-point division algorithms*

The realm of floating-point arithmetic presents numerous challenges due to rounding modes, formats, and the presence of unusual values [Li *et al.* 2021]. While the IEEE-754 standard has played a significant role in establishing common ground and fostering formal reasoning about floating-point computations.Other factors such as the programming language, compiler, and architecture also play crucial roles [Kumar *et al.* 2020]. To concern, the Comp-cert formally verified compiler emerges as viable solution.

2.3 *High-radix formats*

When dealing with FPGA devices that lack native floating- point capability, a recommended strategy for efficient floating- point addition is to utilize high-radix floating-point encoding [Sagar *et al.* 2019]. This technique offers significant advantages by reducing execution time and area requirements [Issa *et al.* 2020]. It achieves this by minimizing the number of necessary shifts, which is crucial considering the relatively costly implementation of variable sifter in FPGA for each floating-point adder. While the high-radix format may impose a penalty when multipliers are utilized, the overall performance improvement is substantial [Varma *et al.* 2018].

2.4 *Exploring alternative number systems for energy-efficient floating-point arithmetic*

The Logarithmic Number System uses logarithms to represent numbers, which makes mathematical operations easier. However, it adds overhead in terms of scaling and logarithmic conversion. Residue Number System necessitates careful control of carry propagation and dynamic ranges. Truncated Binary Multipliers. Despite their distinct focuses, both ECC and BISR strategies, as well as FPA, play vital roles in ensuring the reliability and overall performance of digital systems [Manasa *et al.* 2018].

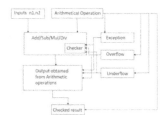

Figure 1.   Floating point arithmetic design flow block diagram.

3   PROPOSED WORK

The basic computational operations of adding, subtracting, multiplying, and division. These procedures are crucial elements of computations involving numbers and have numerous uses in a variety of industries, such as graphics programming, computational science, and modelling finances [Ding *et al.* 2021]. By performing these operations accurately, floating-point arithmetic enables a wide range of applications, ranging from complex scientific simulations to real-time computer graphics rendering [Kukati *et al.* 2013].

3.1 *Addition (+)*

Method to calculate the combined value of two floating- point numbers, perform the addition operation. The same guidelines as standard addition. Verify the exponents of the numbers being added in step one. If they are different, move the mantissa to make the smaller exponent match the bigger one. Combine the mantissas that have been modified. Normalize the result by determining if a left or right shift is necessary and changing the exponent accordingly. Scan for overflow or underflow situations and take appropriate action.

3.2 *Subtraction (−)*

Finding the difference between two floating-point integers is done using the subtraction operation. Similar guidelines apply to normal subtraction. The associated numbers' exponents. If they are different, move the mantissa accordingly to bring the smaller exponent into line with the larger one. The amended mantissas are subtracted in step two. Normalize the result by determining whether it needs to be moved to the left or right and modifying the exponent as necessary.

3.3 *Multiplication (*)*

Finding the product of two floating-point values is done using the multiplication operator. Following is the basic flow of the multiplication operation: Multiplying the mantissas of the numbers is step one. Add the exponents of the numbers in step two. Normalize the result by determining if a left or right shift is necessary and changing the exponent accordingly. Scan for overflow or underflow situations and take appropriate action.

3.4 *Division (/)*

Use the function division to get the ratio of two floating- point integers. The following is the division's core operational flow: As a first step, determine the mantissa associated with the dividend through the mantissa of the number of times the dividend has been paid. Subtract the numbers' exponents. Normalize the result by determining whether a left or right shift is necessary and adjusting the exponent accordingly. Verify that there is no division by zero and, if so, correct the problem. Scan for overflow or underflow situations and take appropriate action.

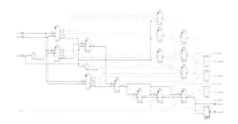

Figure 2. Schematic design of Floating-point arithmetic.

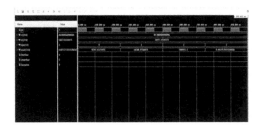

Figure 3. Wave form of Floating-point arithmetic.

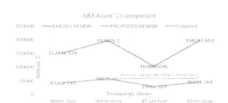

Figure 4. Area comparison of Floating-point arithmetic.

Figure 5. Power comparison of Floating-point arithmetic.

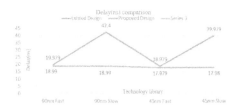

Figure 6. Delay comparison of Floating-point arithmetic.

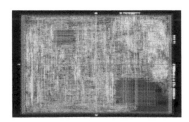

Figure 7. Standard cell design of Floating-point arithmetic.

76

Table 1. Performance evaluation using 45 nm technology libraries.

Performance comparison	Proposed work	Existed work
AREA (μm^2)	29562.557	103880.046
POWER (w)	3.8728e-03	8.14660e-01
DELAY (ns)	17.974	18.979

4 CONCLUSION

This technique revolves around the implementation of floating-point arithmetic operations specifically targeting the single precision data type. The goal is to achieve improved and optimized strategies that result in higher performance. To enhance performance factors, the algorithm incorporates several key steps. One of these steps is normalization, which involves bringing the floating-point numbers into a standardized format. This normalization process helps in aligning the significant and adjusting the exponent accordingly. By performing iterative splitting instructions, the algorithm can effectively handle larger values and ensure accurate computation. Through efficient computation techniques, the algorithm aims to minimize errors and increase precision when dealing with the mantissa. By adopting these techniques, the algorithm aims to achieve optimal performance in floating-point operations. The focus on single precision data type allows for faster computation and efficient memory usage. Through normalization and optimized procedures for mantissa computation, the algorithm strives to maximize performance factors such as speed and accuracy in floating-point arithmetic operations.

REFERENCES

Bagnara, R., Bagnara, A., Biselli, F. *et al.* Correct approximation of IEEE 754 floating-point arithmetic for program verification. *Constraints* 27, 29–69 (2022). https://doi.org/10.1007/s10601-021-09322-9

Bora, S., Paily, R. Design and implementation of adaptive binary divider for fixed-point and floating-point numbers. *circuits syst signal process* 41, 1131–1145 (2022). https://doi.org/10.1007/s00034-021-01832-4

Chandrika K., P. R. B, T. P and Agrawal S., "Low-power high-speed folded MCML-based frequency divider for high-frequency applications," 2022 *IEEE 2nd Mysore Sub Section International Conference (MysuruCon)*, Mysuru, India, 2022, pp. 1–6, doi: 10.1109/MysuruCon55714.2022.9972577.

Ding R., Guo Y., Sun H. and S. Kimura, "Energy-efficient approximate floating-point multiplier based on radix- 8Booth Encoding," *2021 IEEE 14th International Conferenceon ASIC (ASICON)*, Kunming,China, 2021,pp.1–4, doi:10.1109/ASICON52560.2021.9620455.

Issa B. A., AL-Forati I. S. A., Al-Ibadi M. A., Amer H. M., Turky Rashid A. and Rashid O. T., "Design of high precision radix-8 MAF unit with reduced latency," *2020 International Congress on Human- Computer Interaction, Optimization and Robotic Applications (HORA)*, Ankara, Turkey, 2020, pp. 1–6, doi:10.1109/HORA49412.2020.9152924.

Jaiswal M. K. and So H. K.-H, "Architecture generator for type-3 unum posit adder/subtractor," *2018 IEEE International Symposium on Circuits and Systems (ISCAS)*, Florence, Italy, 2018, pp. 1–5, doi:10.1109/ISCAS.2018.8351142.

Jaiswal M. K., So H.K.-H., "Area-efficient architecture for dual-mode double precision floating point division." *IEEE Trans. Circuits Syst. Regu. Pap.* 64(2), 386–398 (2017). https://doi.org/10.1109/ TCSI.2016.2607227

Jun K., Swartzlander E.E., "Improved non-restoring division algorithm with dual path calculation" in *2013 IEEE 56th International Midwest Symposium on Circuits and Systems (MWSCAS)* (2013), pp. 1379–1382. https://doi.org/10.1109/MWSCAS.2013.6674913

Kataria K. and Patel S., "Design of high-performance digital divider", in *2020 IEEE VLSI Device Circuit and System(VLSIDCS)* (2020), pp. 1–6.

Kim K. S. and Kang S., "Clock skew compensation algorithm immune to floating-point precision loss," in *IEEE Communications Letters*, vol. 26, no. 4, pp. 902–906, April 2022, doi:10.1109/LCOMM.2022.3142904.

Kukati S., Sujana D. V., Udaykumar S., Jayakrishnan P. and Dhanabal R., "Design and implementation of low power floating point arithmetic unit," *2013 International Conference on Green Computing, Comm and Conservation of Energy (ICGCE)*, Chennai, India, 2013,pp.205–208,doi:10.1109/ICGCE.2013.682.

Kumar B. V. N. T., Chitiprolu A., Reddy G. H. K. and Agrawal S., "Analysis of High-Speed radix-4 serial multiplier," *2020 Third International Conference on Smart Systems and Inventive Technology (ICSSIT)*, Tirunelveli, India, 2020, pp. 498–503, doi:10.1109/ICSSIT48917.2020.9214270.

Lahari M. and Agrawal S., "Efficient floating-point HUB adder for FPGA," *2020 4th International Conference on Electronics, Materials Engineering and Nano-Technology (IEMENTech)*, Kolkata, India, 2020, pp. 1–6, doi:10.1109/IEMENTech51367.2020.9270083

Li *et al.* K., "Multiple-precision floating-point dot product unit for efficient convolution computation", *2021 IEEE 3rd International Conference on Artificial Intelligence Circuits and Systems (AICAS)*, Washington DC, DC, USA, 2021, pp. 1–4, doi:10.1109/AICAS51828.2021.9458534.

Li K. *et al.*, "A vector systolic accelerator for multi-precision floating-point high-performance computing," in *IEEE Transactions on Circuits and Systems II: Express Briefs*, vol. 69, no. 10, pp. 4123–4127, Oct. 2022, doi:10.1109/TCSII.2022.3183007.

Manasa R., Hegde G. and Vinodhini M., "Improving the reliability of embedded memories using ECC and built-inself-repair techniques," *2018 International Conference on Electrical, Electronics, Communication, Computer, and Optimization Techniques (ICEECCOT)*, Msyuru, India, 2018, pp. 1436–1439, doi:10.1109/ICEECCOT43722.2018.9001509.

Matula D. W., Panu M.T., and Zhang J.Y., "Multiplicative division employing independent factors". *IEEE Transition and Computing.* 64(7),2012–2019(2015). https://doi.org/10.1109/TC.2014.2346206 (ISSN: 23263814)

Neethu S., S. Agrawal and Murty N S, "An archi- tecture for high speed Radix10 division," *2016 International Conference on Computer Communication and Informatics (ICCCI)*, Coimbatore, India, 2016, pp. 1–5, doi:10.1109/ICCCI.2016.7479989

Sagar M. and Hegde G., "FPGA implementation of 8- bit SSA multiplier for designing OFDM transceiver ," *2019 International Conference on Communication and Electronics Systems (ICCES)*, Coimbatore, India, 2019, pp. 1331–1335, doi:10.1109/ICCES45898.2019.9002167.

Smrithi S.V. and Agrawal S., "A fast architecture for maximum/minimum data finder with address from a set of data," *2016 International Conference on Computer Communication and Informatics (ICCCI)*, Coimbatore, India, 2016, pp. 1–6, doi:10.1109/ICCCI.2016.7479988.

Varma K. R. and Agrawal S., "High speed, Low power approximate multipliers," *2018 International Conference on Advances in Computing, Communications and Informatics (ICACCI)*, Bangalore, India, 2018, pp. 785–790, doi:10.1109/ICACCI.2018.8554933

Villalba, J., Hormigo, J. High-radix formats for enhancing floating-point FPGA implementations. *Circuits Syst Signal Process* **41**, 1683–1703 (2022). https://doi.org/10.1007/s00034-021-01855-x

Advances in AI for Biomedical Instrumentation, Electronics and Computing – Sachan et al. (eds)
© 2024 The Author(s), ISBN 978-1-032-64298-7

Design and comparative analysis of inverter based OTA using FinFET and CNTFET

Mohammad Aleem Farshori & M. Nizamuddin
Department of Electronics and Communication Engineering, Jamia Millia Islamia, New Delhi, India

ABSTRACT: In this work, the design and simulation of inverter based OTA have been performed using FinFET and CNTFET. Four different combinations of FinFET and CNTFET are used for simulation and comparison: (i) fully FinFET based OTA (ii) OTA using PFinFET as load and NCNTFET as driver (iii) OTA using NFinFET as driver and PCNTFET as load (iv) fully CNTFET based OTA. The proposed design structures are simulated using HSPICE software and key characteristics, like voltage gain, bandwidth has been computed. The gain is highest in the case of fully CNTFET based OTA (46.3 dB) and lowest in PCNTFET-NFinFET based OTA (18.6 dB). The bandwidth is highest in case of fully NCNTFET-PFinFET based OTA (1.61 GHz) and lowest in fully FinFET based OTA (0.23 GHz). Further, the simulation analysis has shown that the performance of the fully CNTFET based OTA can be optimized by changing the pitch value of CNT(S).

Keywords: CNTFET, FinFET, Operational transconductance amplifier, Simulation, Gain

1 INTRODUCTION

OTA (Operational transconductance amplifier) is one of the most widely used analog circuit building blocks. It has been used as a replacement for operational amplifier as it is a voltage controlled current device. OTA provides a wide range of bandwidth and large dynamic range (Mobarak *et al.* 2010) and is widely used in filters (Lee *et al.* 2018), mixers (Guo *et al.* 2022), and multipliers (Hidayat *et al.* 2009). CMOS based OTA provides low power and satisfactory performance but is unable to sustain its advantage when channel length is below 40 nm (Gupta *et al.* 2012). Replacement of traditional CMOS based device with novel device can provide improved performance due to better tolerability of short channel effects in the sub-nm range, hence proving the validity of Moore's law. Novel devices of interest are but not limited to CNTFET, FinFET, etc., which can be used as a replacement for traditional CMOS devices. FinFET, which is a non-planar device, has many advantages over planar devices, due to its better controllability of gate which results in lower leakage current, especially when channel length is below 14 nm (Yu *et al.* 2002). Further, multi-fin based FinFET can be used to increase controllability and also provides increased current density (Li *et al.* 2012). Another device that is of interest to researchers is CNT (Carbon nanotubes) based CNTFET for replacement of conventional CMOS devices. CNT is widely known for its unique properties like high thermal conductivity, high tensile strength and better electrical conductivity (Ahmadi *et al.* 2016; Morsi *et al.* 2019). One of the unique and important properties of CNT is 1D ballistic transport capability (Ando *et al.* 2002). The channel in CNTFET is replaced by the parallel combination of CNT, which results in high value of mobility due to the 1D ballistic transport of carriers. To cash the advantage of both FinFET and CNTFET, we have compared different combinations of both the devices in inverter based OTA. Different parameters like gain, bandwidth are calculated for each configuration.

DOI: 10.1201/9781032644752-15

HSPICE software is used for simulation study. This work is divided into five sections. Section 1 is an introduction to our work. Section 2 gives a brief overview of CNTFET and FinFET. Section 3 describes inverter based OTA. Section 4 shows results and discussion. Finally, section 5 is the conclusion of the paper.

2 OVERVIEW OF CNTFET AND FinFET

CNT is a nanomaterial which is discovered by research scientists at NEC Japan in 1993.

CNT is one of the allotrope of carbon and are cylindrical in shape. Due to their peculiar properties, CNT have an array of applications in the field of nanoelectronics. CNT are generally classified into the following types: a) SWCNT (Single wall CNT) and b) MWCNT (Multiwall CNT) (Salah *et al.* 2021). SWCNT are layers of graphite rolled into seamless cylinders and generally have diameter in the range of 1-2 nm. MWCNT is like multiple coaxial SWCNT with diameter in the range of 10-100 nm and is less expensive than SWCNT. MWCNT structure is more complex than SWCNT and hence has a high probability of structural imperfection. One of the most important utilities of CNT is in CNTFET (Carbon nanotube field effect transistor), which is already discussed for OTA applications (Loan *et al.* 2015; Nizamuddin *et al.* 2017). CNTFET consists of highly doped drain and source while the channel is made up of a parallel combination of SWCNT, as shown in Figure 1. CNTFET has high mobility, low power consumption and high drive current.

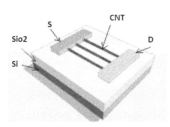

Figure 1. Geometry of CNTFET.

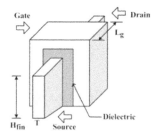

Figure 2. Geometry of FinFET.

FinFET is a three-dimensional novel device that was developed by a team of Professor Chenming Hu of UC Berkeley in the year 2000. FinFET has been recently in demand due to better tolerance of short channel effects due to good controllability over gate, which results in lower leakage current and high switching speed (Yu *et al.* 2002). FinFET has been implemented for various analog and digital circuits like Schmitt trigger, SRAM, biomedical applications, etc.

As FinFET is a 3-dimensional device with vertical channel, it can carry a large amount of current which results in high current density which further can be modified by increasing the number of fins. FinFET construction is shown in Figure 2. FinFET generally has two modes of function: a) shorted gate mode in which every terminal is connected to the same input and b) independent gate mode in which every terminal is connected to different inputs. FinFET is not suitable when the gate length is below 10 nm due to electrostatic instability, threshold voltage variation, etc. (Vora *et al.* 2017).

3 INVERTER BASED OTA

CMOS inverter is one of the most widely used circuit blocks in digital electronics. With proper biasing of p-type and n-type transistors, inverter can work as an amplifier in analog circuits. Inverter based OTA is widely discussed in literature. In our work, we have used self-biasing technique for inverter based OTA, which makes the amplifier less sensitive to parameter variations thus increasing stability and also reducing the supply voltage requirement. Current biasing is achieved by transistors M_1, M_2, and M_7, M_8, as shown in Figure 3, through biasing in linear region. Inverter based amplifier provides high gain, as the net gm of the amplifier is the sum of the individual gm of n-type and p-type transistors.

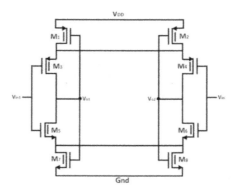

Figure 3.   Inverter based OTA.

4   RESULTS AND DISCUSSION

In this work, different configurations of CNTFET and FinFET are used for inverter based OTA. Further, the gain and bandwidth are measured as a function of the pitch of CNT for fully CNTFET based OTA. From the results we can follow that the gain is highest for fully CNTFET based OTA, which is 31.77 % higher than fully FinFET based OTA. This difference in gain is largely attributed to device parameters and the biasing technique used. Bandwidth is also 604.8 % higher in fully CNTFET based OTA as compared to fully FinFET based OTA, which gives fully CNTFET based OTA high gain-bandwidth product. Bandwidth is highest in case of NCNTFET-PFinFET based OTA and lowest in fully FinFET based OTA. Gain is highest in fully CNTFET based OTA and lowest in the case of PCNTFET-NFinFET based OTA, which can be attributed to variable output resistance for each combination. Table 1 shows the comparative analysis for each configuration. Figure 4, 5, 6 and 7, depict the plots between gain and frequency of fully FinFET based OTA, PCNFET- NFinFET based OTA, NCNTFET- PFinFET based OTA and fully CNTFET based OTA, respectively. Also, the effect of change in pitch of CNT (S) on overall performance has been studied for fully CNTFET based OTA. It has been shown that the pitch of CNT(S) changes the performance of circuit and an optimal value of the pitch of CNT(S) can be used to have a better performance of circuit. With increase in pitch of CNT, the gain decreases significantly which can be due to decrease in transconductance by inter-CNT screening effect, as shown in Figure 8. Bandwidth increases linearly with an increase in pitch of CNT with highest bandwidth achieved is 2.12 Ghz when the pitch is 25nm, as shown in Figure 9.

Table 1.   Comparison between different configurations of Inverter based OTA.

Parameter	Fully CNTFET	NFinFET-PCNTFET	PFinFET-NCNTFET	Fully FinFET
Supply Voltage (V)	0.9	0.9	0.9	0.9
Gain (dB)	46.3	18.6	37.0	32.1
Bandwidth (GHz)	1.18	1.10	1.61	0.23

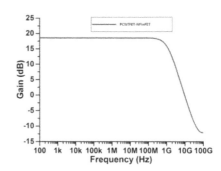

Figure 4. Gain vs frequency plot of fully FinFET based OTA.

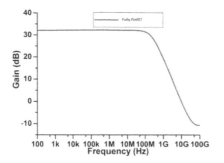

Figure 5. Gain vs frequency plot of PCNTFET-NFinFET based OTA.

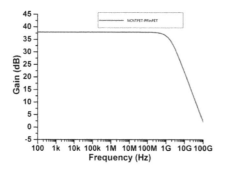

Figure 6. Gain vs frequency plot of NCNTFET-PFinFET based OTA.

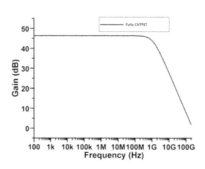

Figure 7. Gain vs frequency plot of fully CNTFET based OTA.

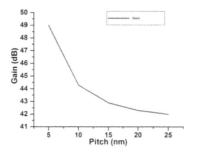

Figure 8. Gain vs pitch(S) plot of fully CNTFET based OTA.

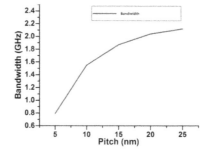

Figure 9. Bandwidth vs pitch(S) plot of fully CNTFET based OTA.

5 CONCLUSION

This work shows the design and simulation of inverter based OTA based on different combinations of CNTFET and FinFET. Four different combinations of OTA have been discussed: (i) fully FinFET based OTA (ii) OTA using PFinFET as load and NCNTFET as driver (iii) OTA using NFinFET as driver and PCNTFET as load (iv) fully CNTFET based OTA. The simulation study shows that the use of CNTFET will improve the performance significantly. The Gain and bandwidth increased significantly by using fully CNTFET based

OTA. Further, this work shows that the performance of the fully CNTFET based OTA performance can be enhanced further by using an optimum pitch of CNT(S).

REFERENCES

Ahmadi M., Zabihi O., Masoomi M., & Naebe M. 2016. Synergistic effect of MWCNTs functionalization on interfacial and mechanical properties of multi-scale UHMWPE fibre reinforced epoxy composites. *Composites Science and Technology*, 134: 1–11.

Ando T., Matsumura H., & Nakanishi T. 2002. Theory of ballistic transport in carbon nanotubes. *Physica B: Condensed Matter*, 323(1–4): 44–50.

Guo B., Wang H., Wang H. & Wang Y. 2022. A 1–3 GHz CMOS mixer-first receiver frontend with dual-feedback OTA achieving 306 MHz IF bandwidth, 2.1 dB NF, and 33 dB gain. *Modern Physics Letters B*, 36(22): 2250112.

Gupta K. A., Anvekar D. K., & Venkateswarlu V. 2012. Device characterisation of short channel devices and its impact on CMOS circuit design. *International Journal of VLSI Design & Communication Systems*, 3(5): 163.

Hidayat R., Dejhan K., Moungnoul P., & Miyanaga Y. 2009. OTA-based high frequency CMOS multiplier and squaring circuit. IEEE, *2008 International Symposium on Intelligent Signal Processing and Communications Systems*: 1–4.

Lee S. Y., Wang C. P. & Chu Y. S. 2018. Low-voltage OTA–C filter with an area-and power-efficient OTA for biosignal sensor applications. *IEEE Transactions on Biomedical Circuits and Systems*, 13(1): 56–67.

Li H., & Chiang M. H. 2012. Design issues and insights of multi-fin bulk silicon FinFETs. IEEE. In *Thirteenth International Symposium on Quality Electronic Design (ISQED)*, 723–726.

Loan S. A., Nizamuddin M., Alamoud A. R., & Abbasi S. A. 2015. Design and comparative analysis of high performance carbon nanotube-based operational transconductance amplifiers. *Nano*, 10(03): 1550039.

Mobarak M., Onabajo M., Silva-Martinez J., & Sanchez-Sinencio E. 2010. Attenuation-predistortion line-arization of CMOS OTAs with digital correction of process variations in OTA-C filter applications. *IEEE Journal of Solid-State Circuits*, 45(2): 351–367.

Morsi M. A., Rajeh A., & Al-Muntaser A. A. 2019. Reinforcement of the optical, thermal and electrical properties of PEO based on MWCNTs/Au hybrid fillers: nanodielectric materials for organoelectronic devices. *Composites Part B: Engineering*, 173: 106957.

Nizamuddin M., Loan S. A., & Murshid A. M. 2017. High performance carbon nanotube based folded cascode operational transconductance amplifiers. IEEE. In *2017 International Conference on Multimedia, Signal Processing and Communication Technologies (IMPACT)*: 32–35.

Salah L. S., Ouslimani N., Bousba D., Huynen I., Danlée Y., & Aksas H. 2021. Carbon nanotubes (CNTs) from synthesis to functionalized (CNTs) using conventional and new chemical approaches. *Journal of Nanomaterials*, 2021: 1–31.

Vora P. H., & Lad R. 2017. A review paper on CMOS, SOI and FinFET technology. *Design and Reuse Industry Articles*: 1–10.

Yu B., Chang L., Ahmed S., Wang H., Bell S., Yang C. Y., & Kyser D. 2002. FinFET scaling to 10 nm gate length. IEEE, In *Digest. International Electron Devices Meeting*, 251–254.

Advances in AI for Biomedical Instrumentation, Electronics and Computing – Sachan et al. (eds)
© 2024 The Author(s), ISBN 978-1-032-64298-7

Automatic room light controller and visitor counter using microcontroller

Aman Kumar Yadav, Amrit Raj Singh, Amit Kumar, Amit Kumar, Amritanshu Singh &
Niraj Singh Mehta
*Department of Electronics and Communication Engineering KIET Group of Institutions Delhi NCR,
Ghaziabad, India*

ABSTRACT: The primary objective of this research paper is to enhance the effective utilization of resources in both developed and developing nations. In the current digital era,
technology plays a pivotal role, and the trend is shifting towards favouring automated
processes that require minimal human intervention. This project aligns with this prevailing
trend by minimizing the reliance on manual efforts. Its significance lies in its capacity to
contribute to resource conservation. In today's context, there is a burgeoning demand for
automated appliances and systems. With an improved standard of living, there is a pressing
need to develop circuits that simplify our daily lives. Additionally, this project addresses the
pertinent issue of room congestion. It presents a practical solution for precisely monitoring
the number of individuals present in a room, thus aiding in congestion management. The
"Automatic room light controller with visitor counter" stands as a reliable circuit that not
only automates room lighting but also accurately counts the number of visitors, making it a
valuable addition to a variety of settings.

Keywords: Infrared Sensors, Microcontroller, Counter

1 INTRODUCTION

In our increasingly digitized world, the pursuit of automation has become paramount in
reducing human exertion across various aspects of daily life. The realm of electronics has
responded by introducing a spectrum of circuits that simplify our routines. Central to this
innovation is the contemporary energy crisis, posing a significant challenge for global citizens. This project seeks to address these issues by offering an automatic solution to a pervasive problem: inadvertently leaving lights and fans on while vacating a room. The
principal objective is to devise a prototype-controlled system that accurately tallies the
number of individuals entering a specific room and subsequently adjusts the room's illumination. This endeavour consists of two essential components: the "Individual Count" and
the "Automatic Room Light Controller". Employing IR sensors, this circuit functions as an
individual detection mechanism, effectively enumerating the number of entrants and displaying the count on a seven-segment LCD display (Kumari *et al.* 2016; Shah *et al.* 2017).
This real-time count aids in preventing overcrowding, particularly in venues such as educational institutions, especially auditoriums where managing crowd flow is crucial.

The nomenclature "Automatic Room Light Controller and Visitor Counter Using Micro
Controller" aptly encapsulates the project's core objectives. It seamlessly amalgamates the
tasks of precisely counting individuals and regulating room lighting. As an individual enters,
the counter increments, concurrently activating room lightning (Adjardjah *et al.* 2016; Joshi
et al. 2021). Conversely, an individual's exit decrements the counter, with the lighting

                                                         DOI: 10.1201/9781032644752-16

extinguishing automatically when the room becomes unoccupied. This system stands as an exemplar of harmonizing microcontroller technology and visitor counting accuracy. In response to the challenge of energy conservation and human effort reduction, this project emerges as a comprehensive solution. It not only streamlines routine tasks but also fosters conscientious energy consumption practices. This amalgamation technology and practicality has the potential to redefine room management dynamics, finding relevance in various domains of modern living.

By providing a means to seamlessly integrate visitor counting and lighting control, this project makes significant strides toward resource efficiency and convenience. It showcases the potential of technology to simplify our lives, contribute to energy conservation, and enhance room management across diverse settings. In this paper, we will conduct a comprehensive analysis of 'Automatic room light controller and visitor counter using the Arduino system, paying special attention to the significant role played by the Tinkercad simulator in the development of our project.

1.1 Hardware platform

1- **Infrared Transmitter and Receiver**: An infrared (IR) diode, connected to a 5V direct current (DC) power supply, is used as an IR transmitter. It is positioned on either side of a door frame to emit infrared light, while an IR receiver is situated on the other side to capture infrared signals, allowing for two-way communications (Kumari *et al.* 2016).
2- **Arduino**: The main controller of the system is an integrated circuit (IC) microcontroller. This controller is responsible for interpreting signals from the infrared receiver, processing the information, and determining whether the visitor is entering or leaving the system (Farooq *et al.* 2016; Mazidi *et al.* 2005).
3- **Signals:** The control signals generated by the Arduino are based on the number of visitors and the direction of movement. The visitor count display is updated when the control signals are triggered, and the room lighting is adjusted accordingly. Liquid Crystal Display: The system includes a visitor count display that displays the total number of visitors. This display is updated automatically in response to signals from the Arduino (Alam *et al.* 2014; Khapre *et al.* 2014; Shah *et al.* 2017).

1.2 Methodology

The main goal of this paper is to come up with a model for a visitor counter. The system architecture is outlined in the Figure 1.

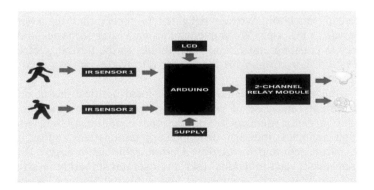

Figure 1. Block diagram.

The methodology of the "Automatic room light controller and visitor counter using Arduino" project combines the design of hardware, the development of software, the testing, and the evaluation of the system to ensure a smooth integration of the visitor counting and the automatic room lighting control (Sinha *et al.* 2017).

1.3 *Superior occupancy detectors*

1. **Ultrasonic sensors**: High-frequency sound waves are utilized by these types of detectors to identify objects and measure the time it takes for the waves to bounce back after impact. These sensors are highly efficient at detecting motion and presence within a space. By providing more accurate information on occupancy, they can differentiate between a person moving and one stationary.
2. **Passive infrared (PIR) sensors:** PIR sensors detect changes in thermal signatures, such as the movement of living organisms with high body temperatures. While they may not be able to distinguish between moving individuals versus those still, they are effective at detecting motion. By combining PIR sensors with other technologies, the accuracy of occupancy detection can be improved.
3. Observational Sensors:
   i. **Temperature sensors:** Monitoring the temperature of a room is crucial for ensuring the comfort of its occupants while also optimizing energy consumption. By integrating temperature sensors, the system can automatically adjust heating and cooling systems to maintain a comfortable environment.
   ii. **Humidity sensors:** Air quality and comfort can be impacted by humidity levels. The system regulates ventilation and humidity control to maintain the perfect home atmosphere by tracking humidity.

Sensors can measure air quality by detecting particulate matter, volatile organic compounds (VOCs), and CO_2 concentrations. This data can be used to regulate air purifiers, ventilation systems, and other appliances to maintain good indoor air quality.

2 RESULT AND DISCUSSION

In this section, we conduct a comprehensive analysis of 'Automatic room light controller and visitor counter using the Arduino system, with a significant focus on the pivotal role played by the Tinkercad simulator in our project's development. The system's core function revolves around accurately counting visitors, a critical aspect with substantial implications for occupancy management in diverse settings, from educational institutions to public venues. We extensively tested the system's ability to differentiate between entries and exits using the Tinkercad simulator. We are delighted to report that the system displayed remarkable accuracy in this area. We simulated various visitor scenarios and assessed the system's efficiency in counting entries and exits, thanks to the flexibility of the Tinkercad simulator. Accurate visitor counting is crucial for effective occupancy management and optimal resource usage.

In addition, we understand the significance of tailored notifications in our system. To improve the adaptability and ease of use of the system, users can customize alerts and notifications based on visitor numbers or specific room conditions. Users can receive real-time alerts through email, text message, or mobile app notifications and set up parameters for temperature, occupancy, and air quality. Customizable alerts ensure that the system meets the requirements of each individual user and operates smoothly in a range of situations, from home automation to commercial and industrial settings. While simulations conducted using Tinkercad yielded highly precise outcomes, practical testing revealed certain challenges. At times, interference, or obstructions in the path of the infrared sensors resulted in inaccurate readings. It is essential to address these intermittent issues to ensure the system's reliability in real-world.

The goal of our project is to promote sustainable living while providing convenience. We can achieve significant reductions in energy consumption, leading to lower electricity costs, a reduced carbon footprint, and substantial energy savings, by automating lighting based on occupancy. Our solution aims to simplify the lives of users by eliminating the need for manual light activation, thereby ensuring well-lit areas for increased safety. While we have achieved our primary goals, there is still room for improvement. Future generations could focus on optimizing sensor positioning, incorporating sophisticated filtration techniques, or exploring energy-conserving lighting innovations. As we continue to strive towards maximizing energy efficiency and enhancing user convenience, with the aim of creating more sustainable and comfortable lives, the Tinkercad simulator will remain an indispensable resource. In summary, the Tinkercad simulator played a crucial role in our project, allowing us to test, refine, and demonstrate the functionality of our "Automatic room light controller and visitor counter" system (Joshi *et al.* 2021; Sasikala *et al.* 2021). It enabled us to assess accuracy, address challenges, and explore the system's potential impact on energy conservation and sustainability. The simulator is an essential tool for both development and presentation of our innovative solution.

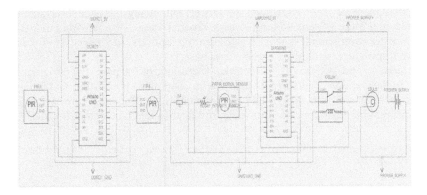

Figure 2.   Circuit view of room light controller and visitor counter.

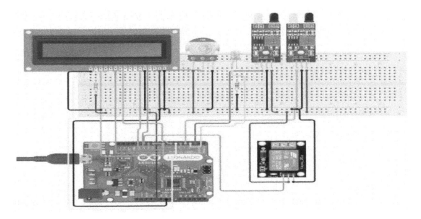

Figure 3.   Circuit diagram on Tinkercad.

3 CONCLUSION AND FUTURE SCOPE

This project stands as a testament to the practicality and effectiveness of a system that seamlessly combines two critical functions: visitor counting and automatic room lighting control. It has demonstrated its capacity to accurately calculate the number of visitors while efficiently adjusting lighting conditions based on the prevailing environment. This dual functionality not only enhances the user experience but also significantly contributes to energy conservation. Looking ahead, the future prospects for this project are exceedingly promising. There are several avenues for further refinement and expansion that could take this system to even greater heights. One such avenue is the incorporation of advanced machine learning algorithms. By leveraging machine learning, the system could learn from patterns in visitor behaviour, enabling it to predict occupancy trends more accurately and optimize lighting control strategies accordingly. This not only improves energy efficiency but also ensures that the lighting environment aligns seamlessly with user preferences.

Sensor technology is another area ripe for improvement. Advances in sensor sensitivity and precision could eliminate any residual issues with false readings or interference. This would enhance the system's overall reliability and performance. The integration of wireless communication capabilities would facilitate remote monitoring and control, providing users with greater convenience and accessibility. This could be particularly valuable for smart home integration, allowing homeowners to extend their control over room lighting and occupancy management from anywhere. Moreover, enhancing the user interface and scalability would make the system adaptable to a broader range of environments, from residential homes to commercial spaces. Customization options and user-friendly interfaces would make it more accessible to a diverse user base.

In conclusion, this project serves as a remarkable example of the potential of technology and automation in the creation of intelligent spaces. With continuous innovation and integration of cutting-edge technologies, this system has the potential to revolutionize how we manage our environments, making them more comfortable, sustainable, and convenient for all.

REFERENCES

Alam, M. S., Alam, M. W., & Sultana, T. (2014). RFID Based room automation. *Proceeding ICMERE2013, Mei.*

Adjardjah, W., Essien, G., & Ackar-Arthur, H. (2016). Design and construction of a bidirectional digital visitor counter. *Computer Engineering and Intelligent Systems*, 7(2), 50–67.

Ayala, K. J. (2004). *The 8051 Microcontroller. Western Carolina University*, North Carolina: Delmar Publication.

Farooq, M. U., Shakoor, A., & Siddique, A. B. (2016). ARM based Bidirectional visitor counter and automatic room light controller using PIR sensors. *Advances in Science, Technology and Engineering Systems Journal*, 1(5), 10–15.

Joshi, R. (2021, May). *"Room Light Controller with Bidirectional Visitor Counter."* Maker Pro. Retrieved from https://maker.pro/arduino/projects/room-light-controller-with-bidirectional-visitor-counter.

Khapre, H., Kela, P., Awasthi, A. R., Prasad, K., & Gupta, S. (2022). Smart bidirectional visitor counter. *Advanced Innovations in Computer Programming Languages*, 4(1).

Kumari, P. D. S. S., & Anusha, D. (2016). Congestion control bidirectional digital visitor counter. *International Journal of Scientific & Engineering Research*, 7(12), 828–831.

Mazidi, M. A., Mazidi, J. G., & McKinlay, R. D. (2005). *The 8051 Microcontroller and Embedded Using Assembly and C. National Cheng Kung University*, TAIWAN: Pearson Prentice Hall.

Sasikala, T., Ahamad, M. I., & Nagarajan, G. (2021). Efficient utilization of an IoT device using bidirectional visitor counter. In *Cognitive Informatics and Soft Computing: Proceeding of CISC 2020* (pp. 859–866). Springer Singapore.

Shah, K., Savaliya, P., & Patel, M. (Year). "Automatic room light controller with bidirectional visitor counter." *International Journal of ICT Research and Development (IJICTRD)*, Vol-1 Issue-4, ISSN: 2395–4841.

Sinha, A., Singh, A., Singh, D., Singh, P., Maurya, A., Singh, M. K. (2017). "Automatic room light controller with visitor counter." *International Journal on Emerging Technologies*, vol. 8, no. 1, pp. 172–175.

Advances in AI for Biomedical Instrumentation, Electronics and Computing – Sachan et al. (eds)
© 2024 The Author(s), ISBN 978-1-032-64298-7

Investigation of DDMZM and LiNb-MZM based radio over fiber link for a future-oriented communication system

Balram Tamrakar
University School of Information, Communication and Technology, Guru Gobind Singh Indraprastha University, New Delhi, India
KIET Group of Institution, Delhi-NCR, Ghaziabad, India

Shubham Shukla
KIET Group of Institution, Delhi-NCR, Ghaziabad, India

Gauri Brijaria, Manjari Singh, Manu Gupta, Shweta Sharma & Mayank Kashyap
University School of Information, Communication and Technology, Guru Gobind Singh Indraprastha University, New Delhi, India

ABSTRACT: In a world where data-driven apps and technologies are expanding rapidly, it is still very important to create advanced communication systems that can meet the increasing needs of the end user. This research paper presents an in-depth exploration into the utilization of Lithium Niobate Mach-Zehnder Modulators (LiNb-MZM) and Dual-Drive Mach-Zehnder Modulators (DDMZM) as integral components within Radio-over-Fiber (ROF) links, with a focus on their transformative potential in shaping future-oriented communication systems. LiNb-MZM and DDMZM modulators are distinguished by their exceptional attributes, including high bandwidth, low insertion loss, and efficient modulation capabilities. These characteristics position them as promising candidates for the realization of the next generation of communication systems. The central objective of this study is to investigate the seamless integration of these modulators within ROF links, with a view to extending transmission distances, enhancing signal quality, and improving spectral efficiency. The research is structured to delve into the theoretical foundations of LiNb-MZM and DDMZM modulators, emphasizing their individual advantages and practical considerations for their incorporation into ROF links. Furthermore, this paper evaluates performance metrics and addresses implementation challenges associated with these modulators in the context of future-oriented communication systems. Upon visual examination of the eye diagrams for LiNb-MZM and DDMZM, it is evident that the DDMZM-based RoF link presents a superior connection and better data transmission rate compared to its LiNb-MZM counterpart.

Keywords: RoF (Radio over Fiber), Communication, DDMZM, LiNb-MZM, Eye diagram, and Modulator

1   INTRODUCTION

In an age defined by our insatiable appetite for data and connectivity, the quest for advanced communication systems capable of meeting the demands of a rapidly evolving digital landscape has never been more pressing. The emergence of cutting-edge technologies continues to reshape our world, from the proliferation of smart devices to the proliferation of data-intensive applications such as augmented reality, high-definition video streaming, and the Internet of Things (IoT) (Li *et al.* 2019). To navigate this ever-accelerating digital revolution, there is a fundamental need for communication systems that not only deliver high data rates

DOI: 10.1201/9781032644752-17

but also possess the flexibility and adaptability to support the unforeseen technological innovations of tomorrow (Tamrakar *et al.* 2022).

This research paper embarks on an in-depth exploration into the application of Lithium Niobate Mach-Zehnder Modulators (LiNb-MZM) and Dual-Drive Mach-Zehnder Modulators (DDMZM) within ROF links, with a singular goal: to unlock their transformative potential in the creation of communication systems poised for the future. Both LiNb-MZM and DDMZM modulators bring unique attributes to the table, including high bandwidth, low insertion loss, and the capacity for efficient modulation. It is these very qualities that make them prime candidates to be instrumental in shaping the next generation of communication systems (Beas *et al.* 2013).

This integration holds the promise of extending transmission distances, enhancing signal quality, and increasing spectral efficiency. This research paper is structured to delve into the theoretical underpinnings of LiNb-MZM and DDMZM modulators, their distinct advantages, and the practical aspects of integrating them into ROF links (Lim *et al.* 2019). Additionally, we will explore their potential applications, scrutinize their performance benchmarks, and address the challenges associated with their deployment (Gavas *et al.* 2015). By the culmination of this research, we aim to offer valuable insights and guidance to researchers, engineers, and stakeholders who share our vision of shaping future-oriented communication systems. The amalgamation of LiNb-MZM and DDMZM modulators within ROF links represents a significant stride toward creating seamless, high-speed, and dependable communication networks (Tamrakar *et al.* 2022). In a world increasingly dependent on data-driven exchanges, this research endeavors to provide the foundational knowledge needed to propel the development of communication solutions that can meet the needs of a society destined for an ever-connected and technology-driven future (Kumar *et al.* 2017).

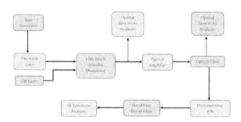

Figure 1.   Block diagram of LiNb-MZM.

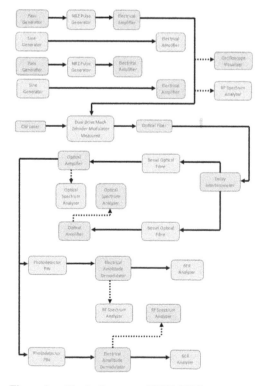

Figure 2.   Block diagram of DDMZM.

2 SYSTEM MODEL

First, we optically modulate the oscillating signal of Radio over Fiber link

$$x_{rf}(t) = m_r + \cos\left(w_{rf}t + \varnothing_{rf}(t)\right) \tag{1}$$

For this, we use a Laser Diode signal

$$x_{ld}(t) = A_{ld}\text{expj}(w_{ld}t + \varnothing_{ld}(t)) \tag{2}$$

The performance of a system is measured by a parameter called SNR, which is the ratio of signal power and noise power. SNR is represented as:

$$SNR = \frac{P_{RFC}}{\frac{N_0}{2} \cdot B_f} \tag{3}$$

$$SNR_{90°} = \frac{\left(A_{ld} \cdot L_{dd} \cdot L_{0} \cdot 10^{-\left(\frac{x_{fiber}k}{20}\right)}\right)^4 R^2 p\left[4 \cdot (J_0(\beta)J_1(\beta))^2 + 4 \cdot (J_2(\beta)J_1(\beta))^2\right]}{N_0 \cdot \left(\frac{\gamma_{rf}}{\pi}\right)\tan\left(\frac{\pi \cdot p\exp\left(2\gamma_{rf}\tau_1\right)}{2}\right)} \tag{4}$$

The RF carrier power received at the output is:

$$P_{RFC} = 2\int_{f_r - \frac{B_f}{2}}^{f_r + \frac{B_f}{2}} S_I(f)df \tag{5}$$

Hence, it can be stated that the signal-to-noise ratio (SNR) is contingent upon the proportion of received power, which is further influenced by factors such as the required bandwidth, phase noise originating from the RF oscillator and laser line-width, as well as the length of the fiber.

3 RESULT AND SIMULATION

The analytical models offered in this study are employed to examine the Dual Drive Mach-Zehnder Modulator and LiNb Mach-Zehnder Modulator in the context of Radio-over-Fiber (ROF) links for communication systems with a focus on future-oriented applications. In this section, the analyses are validated by simulation using Opti-System and mathematical equations.

In the context of the LiNb-MZM, the Sine Generator operates at a frequency of 18 GHz and is characterized by a phase shift of 90 degrees. The laser frequency of the CW laser is 193.1 THz, with a power of -10 dBm and a linewidth of 20 MHz An Optical Amplifier with a Gain of 20 dB and a Noise Figure of 4 dB is utilized. The optical fiber cable utilized for this analysis has a length of 2 km. The used RF Spectrum Analyzer exhibits a peak power of -38.150924 dBm, corresponding to a peak power of 0.15307556e-006W. The frequency at which this peak power occurs is 35.9375e+009 Hz.The Optical Spectrum Analyzer exhibits its highest amplitude at a value of -20.551280dB. The maximum power output is achieved when operating at a frequency of 40 GHz. The eye diagram is observable in the depicted figure.

91

In contrast, the Dual Drive Mach-Zehnder Modulator exhibits a first Sine Generator frequency of 10GHz, an amplitude of 0.7, and a phase angle of 90 degrees. The frequency of the second Sine Generator is 15 gigahertz (GHz), accompanied by a phase angle of 90 degrees. The first bias voltage is recorded as 0.1V, while the second bias voltage is also measured at 0.1V. Additionally, the modulation voltage is observed to be 5V. The optical fiber utilized in this study has a length of 2 kilometers, exhibiting an attenuation rate of 0.2 decibels per km.

Table 1.   Parameters of LiNb MZM.

S.no	Parameters	Value
1.	Frequency	18 GHz
2.	Phase	90 deg
3.	Electrical gain	-1
4.	Optical fiber length	2 km
5.	Optical Amplifier gain	20 dB
6.	RF Spectrum amplifier peak power	-38.15 dBm
7.	RF Spectrum amplifier frequency with peak power	35.93 dBm

Table 2.   Parameters of DDMZM.

S.no	Parameters	Value
1.	Frequency	15 GHz
2.	Amplitude	0.7 a.u.
3.	Phase	90 deg
4.	Bias Voltage 1	0.1 V
5.	Bias Voltage 2	0.1 V
6.	Optical Fiber length	2 Km
7.	Reference Wavelength	193.1 THz
8.	Gain	15 dB
9.	Power	10 dB

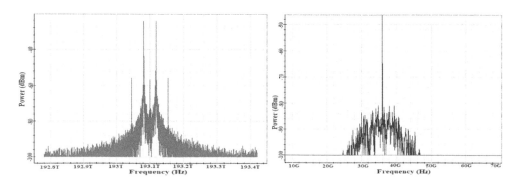

Figure 3.   Optical spectrum analyzer of LiNb-MZM.

Figure 4.   RF spectrum analyzer of LiNb-MZM.

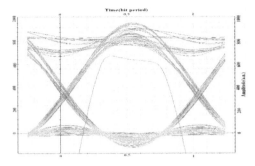

Figure 5. BER height of LiNb-MZM.

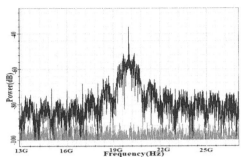

Figure 6. RF spectrum analyzer of DDMZM.

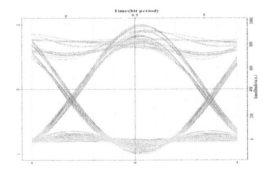

Figure 7. BER pattern of LiNb-MZM.

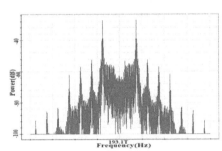

Figure 8. Optical spectrum analyzer of DDMZM.

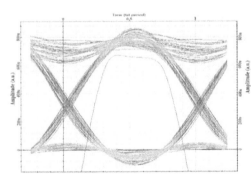

Figure 9. BER height of DDMZM.

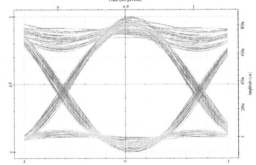

Figure 10. BER pattern of DDMZM.

4   CONCLUSION

In conclusion, it offers significant insights into the potential of these technologies. The study focused on integrating DDMZM and LiNb-MZM as critical components in designing communication systems that align with future demands. Through thorough examination and analysis, the study highlighted the advantages of employing DDMZM and LiNb-MZM technologies in RoF links. The investigation also emphasized the need for meticulous optimization of these components to harness their full potential in achieving high data rates, reduced signal loss, and improved overall system performance. As we progress toward an era of increased connectivity and data exchange, the findings of this study offer a roadmap for designing communication systems that are not only efficient but also adaptable to the ever-changing communication landscape. The synergy between these technologies and their compatibility with the requirements of future-oriented applications lays the foundation for communication networks that can accommodate the escalating demands of a digitally connected world. Upon analyzing the eye diagrams of LiNb-MZM and DDMZM it is evident that the DDMZM based RoF link is more desirable than the LiNb-MZM based link.

REFERENCES

Balram Tamrakar Krishna Singh, and Parvin Kumar. "Performance analysis of DP-MZM radio over fiber links against fiber impairments." *Journal of Optical Communications 0 (2021)*: 000010151520210079.

Balram Tamrakar, Krishna Singh, and Parvin Kumar. "Analysis and modelling of DD-DPMZM to investigate fundamental to intermodulation distortion ratio (FIMDR) against different fiber impairments for the next generation networks." *Journal of Active & Passive Electronic Devices 16*, no. 4 (2022).

Balram Tamrakar, Krishna Singh, and Parvin Kumar. *"Performance Analysis of DD-DPMZM based RoF link for Emerging Wireless Networks."* (2022).

Balram Tamrakar., Singh, K. and Kumar, P., 2022. Analysis and modelling of DD-DPMZM to investigate fundamental to intermodulation distortion ratio (FIMDR) against different fiber impairments for the next generation networks. *Journal of Active & Passive Electronic Devices*, 16(4).

Beas, Joaquin, Gerardo Castanon, Ivan Aldaya, Alejandro Aragón-Zavala, and Gabriel Campuzano. "Millimeter-wave frequency radio over fiber systems: a survey." *IEEE Communications surveys & tutorials 15*, no. 4 (2013): 1593–1619.

Gawas, Anju Uttam. "An overview on evolution of mobile wireless communication networks: 1G-6G." *International Journal on Recent and Innovation Trends in Computing and Communication 3*, no. 5 (2015): 3130–3133.

Kumar, Parvin, Sanjay Kumar Sharma, and Shelly Singla. "Performance improvement of RoF transmission link by using 120° hybrid coupler in OSSB generation." *Wireless Networks 23* (2017): 15–21.

Li, Xinying, Jianjun Yu, and Gee-Kung Chang. "Photonics-assisted technologies for extreme broadband 5G wireless communications." *Journal of Lightwave Technology 37*, no. 12 (2019): 2851–2865.

Lim, Christina, Yu Tian, Chathurika Ranaweera, Thas Ampalavanapillai Nirmalathas, Elaine Wong, and Ka-Lun Lee. "Evolution of radio-over-fiber technology." *Journal of Lightwave Technology 37*, no. 6 (2019): 1647–1656.

Novak, Dalma, Rodney B. Waterhouse, *AmpalavanapillaiNirmalathas, Christina Lim*, Prasanna A. Gamage, Thomas R. Clark, Michael L. Dennis, and Jeffrey A. Nanzer. "Radio-over-fiber technologies for emerging wireless systems." *IEEE Journal of Quantum Electronics 52*, no. 1 (2015): 1–11.

Advances in AI for Biomedical Instrumentation, Electronics and Computing – Sachan et al. (eds)
© 2024 The Author(s), ISBN 978-1-032-64298-7

Medical assistance bot: DOBO

Sulekha Saxena
Department of ECE, Ajay Kumar Garg Engineering College, Ghaziabad, UP, India

Shobhit Singh Rawat
AKGEC Ghaziabad, Adhyatmik Nagar, Ghaziabad, UP

Yashdeep Tyagi, Yash Rajput & Suyash Kumar Srivastava
Department of ECE, Ajay Kumar Garg Engineering College, Ghaziabad, UP, India

ABSTRACT: With the fusion of technology and healthcare, we try to make healthcare easy to accessible to everyone. so, in the way of this, we have brought a medical assistant bot DOBO. DOBO short for doctor's bot is a user-friendly bot equipped with a range of features, that aim to enhance each and individual health management. primally it will measure the basic health parameters of the human body like body temperature, blood pressure, oxygen saturation in blood, and pulse rate through seamless integration with various medical devices and sensors. Its ability to record and maintain historical health data by creating a health profile of each user and appointment scheduling function make it more usable.

Keywords: Health Management, Appointment Scheduling, Health Data Record

1 INTRODUCTION

Today's day humans are failing to prioritize their health, due to their hectic schedules (Farhad 2015) and inconvenient health services provided. they avoid consulting with the doctor and going for routine checkups. managing health is more trouble for middle-income people nowadays. For the solution to this, we put up a BOT named Doctor's Bot (DOBO)which can perform tasks like measuring body temperature, pulse rate, blood pressure, and oxygen saturation in the blood. DOBO is an on-field and in-house medical assistance bot. It is designed to do basic human body check-ups and record data to provide quick and accurate medical advice and recommendations by the process of edge computing using basic machine learning algorithms and fix appointments with the doctor if the user wants. There is also a chatbot that will provide medical advice and recommendations in real time for queries like flu symptoms, some minor injuries, and simple seeking. Tested data will be stored on the cloud, making it easy to access for both doctor and user/patient. Below shown Figure 1 manifest the model of DOBO The Bot can work in offline mode in case of no access to the internet. Installing it publicly or outside of OPD can resolve the issue of routine check-ups. Its user-friendly interface is easy to use and available for a wide range of users, making it the perfect tool for healthcare providers and patients. DOBO's major objective is to record the patient health data on the cloud (prasad *et al.* 2021) and fix appointments with doctors, it is more effective in rural areas of India where people are not getting quality health care service just because the hospitals and other medical facilities are at long distances with small workforce (hazarika 2013). In the modern world, technology is a new front of the medical workforce, and in series to DOBO must be added with the objective of reducing the workload of medical staff. DOBO also cares about the sanitization of the user. it contains a sanitizer module built with a

DOI: 10.1201/9781032644752-18

servomotor and active IR Sensor. It is designed in such a reliable way so it is easy to install any ware within a minute and convenient to carry. Since chatbot systems have two components text and speech to interact with users in DOBO it is done by text.

The smart bot provides solutions like making the interconnection between doctor and patient for a fixed time (Vasudeva *et al.* 2023), this is an in-house robot its future scope is limited and it can provide medicine to admit patients. some of the famous smart bots are JEVAK and RKHSHAK by the central railway of India. While chatbots are the product of artificial intelligence and machine learning, providing medical knowledge to user about asked question related to their health is major role of chatbots (Ahmad *et al.* 2018; Avilla *et al.* 2021; Bandopadhyay *et al.* 2023; K. C. *et al.* 2019; Srivastava *et al.* 2020). But this is not so accurate because of the lack of health parameters of patients and this is not available for all. Trendy technology like Google Dialog flow, SAP conversational AI, and Azure bot service are put to use on large scale. The healthcare system is strongly impacted by the fourth industrial revolution (Haleen *et al.* 2022) providing services like the availability of doctor

USER REGISTATION	←—	⌐¯OTC PRESCIPTION GENRATED		User Registration
CHAT BOT				↓
LCD SCREEN		DATA BASE		Input Health Parameters
PULSE RATE				↓
SPO₂		ML-ALGO FOR TO PREDICT DESIS		OTC Prescription and Suggestion
BLOOD PRESURE		MICRO CONTROLER	SD-CARD MODUAL	↓
TEMPERATURE				Dr. Appointment

Figure 1.   Proposed model of bot.

Figure 2.   Schematic representation of the proposed model.

Table 1.   NITI Aayog health index.

| | | HEALTH INDEX | |
Progress rank	NAME OF STATE	Base year	Reference year
1	Haryana	47.59	54.08
2	Rajasthan	37.35	43.23
3	Jharkhand	48.25	53.67
4	Andhra Pradesh	60.84	65.31
5	Assam	45.84	50.02
6	Telangana	56.12	59.42
7	Maharashtra	61.76	64.53
8	Karnataka	59.39	61.77
9	Jamu, and Kashmir	61.02	62.92
10	Chhattisgarh	52.69	53.97
11	Himachal Pradesh	61.84	63.10
12	Gujrat	62.61	63.72
13	Madhya Pradesh	40.77	38.69
14	Punjab	65.83	63.41
15	Kerala	77.53	74.65
16	Tamil Nadu	64.05	60.50
17	Odisha	40.19	36.35
18	Uttarakhand	44.61	39.61
19	Uttar Pradesh	34.44	29.16
20	Bihar	39.10	32.43

delivery of medicine become faster and more reliable, management of patient's data become easier. in these technologies like cloud computing and IMOT plays a key role (Cao *et al.* 2020; Joyia *et al.* 2017; Yang *et al.* 2017).

2 METHODOLOGY

DOBO is a Hardware plus software medical service provider bot, that allows users to identify illness based on the data provided to the bot through their different sensors and devices. The software part of the bot is ML-powered and support providing medical suggestion to the user and fixing an appointment with the doctor in real time. there is a provision for storing data on the cloud. its accuracy increases due to the accurate input of different health parameters through its different sensors. All data are assembled manually and transmitted into the microcontroller and make more comprehensible for uncomplicated study of suggested pattern network. It is more functional in case of aged people and person who is not so familiar by technology. bot can dispense precise conclusion because of well complicated machine learning model used to design it.Above shown Figure 1 shows the working procedure of DOBO, in which there is an LCD screen as a user interface of the bot. through the user will register itself then it will provide its health parameter on it by processing on its machine will generate the suggestion and OTC prescription. and if the user wants an appointment with the doctor, it will fix it and all data of the user will be uploaded to the cloud so it can be easy to access.

3 OVERVIEW

So that getting together all the line of action the overall composition of working DOBO is shown in the below Figure 2 First the user will register and provide input, this input will process and predict the Disease, and based on that OTC prescription and suggestions will generate and give the option to book an appointment with the doctor if the user wants to book an appointment, then user's all data will upload on the cloud, and with this process end. Below shown Figure 2 show the step-by-step systematic representation of the proposed.

4 RESULT AND ANALYSIS

After comparing and testing all the parameter proposed medical assistant bot is implemented with an accuracy of 88% in disease identification, The user interface of the proposed bot enhance the user experience. After analyzing various works in the same field and comparing them with our bot, we can say that our model is a revolution in the healthcare system. Its capability to store the user health parameters on the cloud and fix appointments with doctors after the recommendation is the uniqueness of our proposed work and this increases its significance in rural areas. In the case of no internet, its feature decreases up to 50% which is the limitation of the bot.

5 CONCLUSION AND FUTURE WORK

The aimed design of a medical assistance bot is implemented which can be used as a tool in providing health care. implemented system can be useful in the field of medical science for early and faster detection of flu or disease. Implementation of the setup will improve the accessibility of healthcare facilities to people as a result it gives a remarkable chance to narrow the inconsistency in accessibility healthcare guidance for each individual. In the future it is aimed to add a drug Dispensary ATM, so suggested OTC medicine can be easily available, and add the voice assistant to make it more user-friendly.

REFERENCES

Ahmad N. S., Sanusi M. H., Abd Wahab M. H., Mustapha A., Sayadi Z. A. and Seringa M. Z., "Conversational bot for pharmacy: A natural language approach," *2018 IEEE Conference on Open Systems (ICOS)*, Langkawi, Malaysia, 2018, pp. 76–79, doi: 10.1109/ICOS.2018.8632700

Avila, C.V.S., Franco, W., Venceslau, A.D., Rolim, T.V., Vidal, V.M. and Pequeno, V.M., 2021. MediBot: an ontology-based chatbot to retrieve drug information and compare its prices. *Journal of Information and Data Management*, 12.

Bandopadhyay, D., Ghoesh, R., Chatterjee, R., Das, N., & Sadhukhan, B. (2023, january 31). *Speech Recognition and Neural Networks Based Talking Health Care Bot*; medibot. p. 7.

Cao R., Tang Z., Liu C. and Veeravalli B., "A scalable multicloud storage architecture for cloud-supported medical internet of things," in *IEEE Internet of Things Journal*, vol. 7, no. 3, pp. 1641–1654, March 2020, doi: 10.1109/JIOT.2019.2946296.

Haleem, A., Javaid, M., Singh, R.P. and Suman, R., 2022. Medical 4.0 technologies for healthcare: Features, capabilities, and applications. *Internet of Things and Cyber-Physical Systems*, 2, pp.12–30.

Hazarika, I. (2013). Health workforce in India: assessment of availability, production, and distribution.*WHO South-East Asia J Public health* (p. 7). Australian Health Workforce.

Ides, P., Martel, S., Patel, R.V. and Santos, V.J., 2017. Medical robotics—Regulatory, ethical, and legal considerations for increasing levels of autonomy. *Science Robotics*, 2(4), p.eaam8638.

Joyia, G.J., Liaqat, R.M., Farooq, A. and Rehman, S., 2017. Internet of medical things (IoMT): Applications, benefits and future challenges in healthcare domain. *J. Commun.*, 12(4), pp.240–247.

KC, G.P., Ranjan, S., Ankit, T. and Kumar, V., 2019. A personalized medical assistant chatbot: Medibot. *Int. J. Sci. Technol. Eng*, 5(7).

Prasad, V.K., Tanwar, S. and Bhavsar, M., 2021. C2B-SCHMS: Cloud computing and bots security for COVID-19 data and healthcare management systems. *In Proceedings of Second International Conference on Computing, Communications, and Cyber-Security*: IC4S 2020 (pp. 787–797). Springer Singapore.

Srivastava P. and Singh N., "Automatized Medical Chatbot (Medibot)," *2020 International Conference on Power Electronics & IoT Applications in Renewable Energy and its Control (PARC)*, Mathura, India, 2020, pp. 351–354, doi: 10.1109/PARC49193.2020.236624.

Advances in AI for Biomedical Instrumentation, Electronics and Computing – Sachan et al. (eds)
© 2024 The Author(s), ISBN 978-1-032-64298-7

Performance analysis of WDM-FSO system using RAMAN-EDFA and MIMO for light rain

Abhinav Pratap Shahi, Abhishek Tripathi, Ankit Kumar & R.L. Yadava
Galgotias College of Engineering and Technology, Greater Noida, India

ABSTRACT: Free space optics (FSO) is an advanced technology capable of providing high-speed services in various geographic locations, including mountainous terrain and inter-building networks. However, it is very susceptible to atmospheric conditions, such as rain, which can cause signal degradation. In the proposed study, a hybrid Raman-EDFA optical amplifier and MIMO technology are used to design a WDM-FSO system. This entire communication system is tested in light rain climatic conditions with an attenuation of 7.3 dB/km. Furthermore, the model's performance is compared to data that has already been presented in the literature, including the Q-factor (8.06428) and BER (2.65704e-016). It is evident that the presented model produces better results when examining parameters such as bit error rate (BER) and the quality factor at the receiver-end, in addition to demonstrating a longer transmission distance.

1   INTRODUCTION

Free space optical communications, when combined with communication network infrastructure, can provide high bandwidth and can achieve high data rates in response to the growing demand for high-bandwidth communication networks and faster data transmission to meet user expectations (Paliwal 2021). For various telecommunication applications, the free space optics (FSO) has proven to be much less expensive than fiber optics. The new solution for improving the transmission capacity of optical fiber systems is thought to be wavelength division multiplexing (WDM). The advantages of FSO technology includes high level of security and ease of implementation (Ghatakk 1999). FSO systems require a line-of-sight (LOS) between the transmitter and receiver and are affected by various atmospheric disturbances. In line with the development of contemporary life, the WDM technique is being used to enhance system capacity and bandwidth (Singh 2022). Due to rising consumer demand, traditional data transmission networks like RF and microwave may become overloaded in the near future. Optical spectrum usage is likely to meet the growing data flow needs (Willebrand 2001). Parallel deployment of fiber cables with FSO can be used to create optical information transfer systems in such conditions. Deploying fiber optic lines can be challenging in various locations, like remote, hilly, and crowded cities. Therefore, it is feasible to install FSO systems in locations where installing optical fiber cable would be impractical or very costly.

A free space optical system, with sender and receiver sections, uses air or similar medium to transmit light. The sender produces the light using sources like LEDs or LASERs and modulates the information signal. To convert the incoming optical signal into electrical form for data processing, a detection device like a PIN diode or Avalanche PD (Kumar 2012) is used. A 1550 nm signal is preferred for better performance in terms of distance, BER reduction, eye safety, and reduced solar radiation. The FSO system employs WDM to enhance system performance and capacity (Al-Gailani 2013). WDM combines multiple signals of different wavelengths for simultaneous transmission. At the receiver, a WDM demultiplexer separates incoming signals from the multiplexed beam. However, the WDM-FSO system is vulnerable to air attenuation due to adverse environmental conditions like scintillation, fog, haze, and rain, leading to significant optical signal loss (Shaina 2016).

The system often uses optical amplifiers to improve overall functioning and to counteract the climatic impacts of open space. Optical amplifiers are available into a number of different varieties, erbium doped fiber amplifiers (EDFA) along with semiconductor optical amplifiers (SOA) and Raman (Aditi 2014). Based on a combination of self-phase modulation and the frequency separation between the WDM channels, EDFA has been able to generate superior Q-factor and BER than Semiconductor OA and Raman Amplifier. In order to extend the link range and transmit the same amount of data at a longer distance, a new FSO configuration was put out in this article. The configuration is built around combining MIMO and Raman-EDFA technologies. While Raman-EDFA was used because of its appealing characteristics, including high gain, high bandwidth, and low noise (Bayart 2003; Ijaz 2013), MIMO was added to boost efficiency and subsequently lower the BER (Vishwakarma 2016).

2 ANALYSIS OF WDM FSO

By combining wavelengths, an FSO system with incorporation of WDM architecture enables the transmission of several signals through the same channel to a remote location (Grover 2017). Figure 1 describes a WDM-FSO system illustration. The block diagram consists of a receiver section with multiple photodetectors and a transmitter section containing numerous light sources.

Figure 1.   WDM-FSO system illustration.

The setup described in this research employs a WDM multiplexer to combine multiple input signals into a single channel. These multiplexed signals are then transmitted using a MIMO based Free Space Optical (FSO) link, utilizing four FSO channels to distribute the load and increase transmission speed. Signal amplification is achieved through a hybrid Raman-EDFA, eliminating the need for optical-to-electrical conversions. At the receiver end, a WDM demultiplexer is used to decode the multiple signals from the received multiplexed signal. Additionally, this study acknowledges the challenges posed by a noisy and environmentally sensitive free space optical link. Atmospheric conditions such as scintillation, smog, haze, and drizzle can weaken the signal, impacting data transfer in wireless communication. The term "free space medium" denotes a setting where data transmission occurs without wires or cables. To ensure reliable wireless transmission in open space, it's crucial to consider the optical attenuation caused by specific atmospheric conditions, as discussed in this study.

2.1   The haze's effect on noise

Haze is a term used to describe an atmospheric condition where smoke, dust, or other fine particles obstruct visibility and cause attenuation. Various mathematical models are available to determine the attenuation caused by the haze. In this study, the Kim and Kruse model is employed to calculate the attenuation of haze particles on light rays. It can be presented as:

$$\alpha = (3.910/V)\,(\lambda/550)\,nm$$

where α = hazy condition attenuation in dB/km, λ = the calculated wavelength in nm,
The scattering particles' diameters, denoted by p, have the following distributions:
for extremely poor clarity, that is, V below 6 km.
1.3 for moderate clarity i.e., V ranges from 6 to 50 km,
1.6 for higher clarity i.e., V is larger than 50.

2.2 *The rain's effect on noise*

Rain significantly affects the optical free space transmission, impacting both radio and optical link connections. When raindrop size exceeds the optical signal wavelength, non-selective scattering or dispersion occurs, independent of wavelength. There is an equation to calculate optical attenuation caused by raindrops is:

$$\alpha = r * Rp \ (dB/km)$$

where R= frequency of rainfall in mm/hr, also r and p are constants.

3 RAMAN-EDFA AND MIMO BASED WDM-FSO MODEL ILLUSTRATION

The combined optical amplifier containing Raman-EDFA based free space optics system with wavelength division multiplexing is examined in this work for use in a variety of environmental disturbances. The tested WDM-FSO model with a medium of free space based on multiple-input-multiple-output (MIMO) is shown in Figure 2. In Figures 3 and 4 respectively, a diagram of MIMO technology and Raman-EDFA hybrid optical amplifier is displayed.

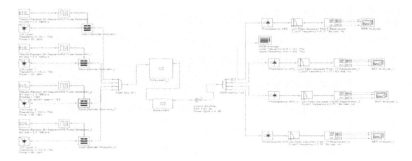

Figure 2.   Diagram of Raman-EDFA and MIMO based WDM-FSO system.

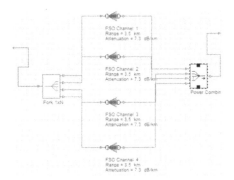

Figure 3.   Schematic diagram of MEMO technology.

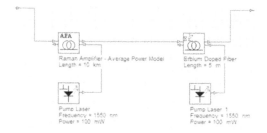

Figure 4.   Arrangement of hybrid Raman-EDFA optical amplifier.

A P-R bit sequence generator and NRZ pulse generator are used at the transmitter section to create modulating signals. Additionally, a CW laser with a wave-length configuration of 1550 nm is included for generation of the four-carrier signal with frequency range between 193.10 THz and 193.40 THz. These carrier messages' frequency spectrum falls within this region, from the lowest to the highest. The modified messages are remotely transmitted using the four carrier signals. The model's modulation mechanism uses a Mach-Zehnder modulator. The design includes a 4 x1 WDM multiplexer for mixing all four signals into one channel.

A 4x4 MIMO-based FSO link is introduced between the sender and receiver section for free space optical medium. It is a method of wireless communication that transmits data over open space using light. In order to analyze the performance in light rain conditions, the range of the FSO link is changed during the experiment of the suggested model from 2.5 km to 3.5 km. For the light rain circumstances, the FSO system is set up with a beam divergence of 2.5 m rad and a 2.5 Gbps data flow rate.

For different environmental circumstances, the optical attenuation of the FSO channel link taken into account are depicted in Table 1 (Paliwal 2021).

Table 1.   FSO link optical attenuation.

Atmospheric Conditions	Attenuation (db/km)
Clear Weather	0.2330
Hazy	2.330
Light Rain	7.300

A hybrid optical amplifier is utilized to increase the system's power efficiency. Due to its larger advantages, EDFA is typically favored in optical communication systems. High pumping efficiency, significant gain, cheap price, and improved noise resistance are all provided by EDFA amplifiers. But EDFA's limited working frequency range and non-linearity are disadvantages.

A Raman amplifier is placed before EDFA in order to get around the EDFA restriction and enhance the design's total power. This hybrid optical amplifier topology increases link SNR while simultaneously reducing non-linearity. This hybrid optical amplifier structure has been discovered to enable band expansion amplification within the transparency range of optical cable. The advantage of a significantly larger Raman gain coefficient is supplied by the Raman serving as pre-amplifier for the entire setup. Additionally, it features a high pump efficiency, low noise figure, and gain. Losses and dispersion are both compensated by the hybrid optical amplifier.

A 1x4 WDM de-multiplexer is incorporated at the receiver end to segregate all incoming signals from a single channel. The integrated avalanche photo detector (APD) has the role of detecting and converting the incoming photo signal into its corresponding electrical signal. There are numerous photo-detectors or photo-sensors available, with p-n diode, pin diode, avalanche PD, Schottky barrier photo-detectors, etc. being just a few examples. However, APD is preferable because of its better SNR and relatively high sensitivity. A gaussian filter is utilized after the detector. The advantage of the Gaussian filter is the least amount of group delay. The numerous experiment findings, involving bit error rate, eye- opening pictures, and Quality-factor, are investigated using a BER (Bit Error Rate) analyzer.

4 SIMULATION RESULT AND DISCUSSION

Optisystem software (v17.0.0) is used to simulate the combined optical amplifier consisting of Raman-EDFA and MIMO based WDM-FSO system. The designed system's performance analysis is seen in light rain climatic conditions. Multiple user signals are created and are multiplexed by the WDM-FSO link. The WDM demultiplexer then segregates the multiplexed signal, which is then supplied into the Avalanche PD, which has configuration of 10nA dark current and 1 A/w of responsivity. The illustrated system's results are optimized for 2.5 Giga Bits per second data flow at 1550 nm of wavelength.

On the receiver end, the low pass gaussian filters (LPF) are utilized to filter out the unwanted frequency signal. The BER analyzer generates the eye diagram for various distance measurements under light rain climatic conditions. A sizable, eye-opening figure shows the efficient and acceptable communication, lower BER, and max signal capture at the receiver portion. Figure 5 displays the final output of the simulation at 2.5 km single link FSO channel, while Figure 6 displays the final output of the simulation using 3.5 km MIMO based FSO channel.

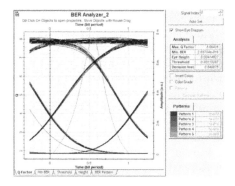

Figure 5.   Simulation result of single channel FSO link in light rain (2.5 km).

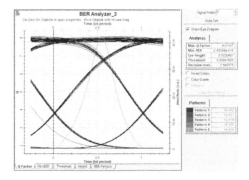

Figure 6.   Simulation result of design illustration in light rain (3.5 km).

The model is simulated for light rain condition with an attenuation of 7.3 dB/km as in Table 2, then the Q-factor and BER of the single channel FSO link are concluded to be 8.06428 and 2.65704e-016 respectively for a range of 2.5 km as shown in Figure 5. While for the MIMO based FSO link the Q-factor is improved to 8.07417 and there is a slight reduction in the BER which is 2.45258e-016 for a range of 3.5 km as depicted in Figure 6. Table 2 clearly depicts the comparison of both of the transmission system.

Table 2. Comparison of single channel and MIMO based link.

Properties	Old Model	Improved Model
Range	2.5 km	3.5 km
Q-factor	8.06428	8.07417
BER	2.65704e–016	2.45258e–016

The above data shows improvement in the range of the proposed transmission system by 1 km with a very minute improvement in the Q-factor and BER. The designed MIMO based WDM-FSO system with Raman-EDFA hybrid optical amplifier configuration exhibits satisfactorily improved performance in the light rain conditions, according to the experiment.

5 CONCLUSION AND FUTURE SCOPE

In this research, a WDM-FSO system is designed using a hybrid optical amplifier comprising Raman-EDFA and MIMO technology. The study evaluates the system's performance in light rain conditions. The hybrid optical amplifier and MIMO based setup enable a 3.5 Km long FSO connection with a 2.5 Gbps data rate, maintaining a high Q-factor, low bit error rate, strong power efficiency, and rapid data transfer. This advancement addresses previous limitations in challenging environmental conditions, enhancing the usability of FSO technology. The key contributions of this research include integrating MIMO technology with the existing hybrid optical amplifier (Raman-EDFA), improving Q-factor and BER, and extending the transmission range. However, it's important to note that the system's performance may deteriorate in heavier rain or other adverse weather conditions like fog, snowstorms, or seismic activity. Further research is needed to develop a setup capable of withstanding various climatic fluctuations while maintaining efficiency.

REFERENCES

Aditi, Preeti. 2014. An effort to design a power efficient, long reach WDM-FSO system. *2014 International Conference on Signal Propagation and Computer Technology (ICSPCT 2014)*. Ajmer, India.
Al-Gailani SA, Mohammad AB, Shaddad RQ. 2013. Enhancement of free space optical link in heavy rain attenuation using multiple beam concept. *Optik*.
Bayart D. 2003. *Erbium-Doped and Raman Fiber Amplifiers*. Vol 4. Comptes Rendus Physique.
Chandra S, Jee R, Singh M. 2017. Transmission Performance of Hybrid WDM-FSO System for Using Diversity Multiplexing in the Presence of Optical Nonlinearities and Fading. TENCON 2017 - 2017 IEEE Region 10 Conference. Penang, Malaysia.
Ghatakk A, Thyagarajan K. 1999. *Introduction to Fiber Optics*. Cambridge University Press.
Grover M, Singh P, Kaur P, Madhu C. 2017. *Multibeam WDM-FSO System: An Optimum Solution for Clear and Hazy Weather Conditions*. Wireless Personal Communications.
Ijaz M, Ghassemlooy Z, Pesek J, Fiser O, Minh HL, Bentley E. 2013. Modeling of fog and smoke attenuation in free space optical communications link under controlled laboratory conditions. *Journal of Lightwave Technology*.

Kumar S. 2012. Study WDM technology to achieve higher bandwidth requirements in optical fiber based back-bone network in communication systems. *MERI Journal of Management and IT*.

Mohamed I, Saleh R, Salih A. 2022. Enhancing the FSO link range under very clear air and thin fog conditions in Albayda - Libya. *2022 IEEE 2nd International Maghreb Meeting of the Conference on Sciences and Techniques of Automatic Control and Computer Engineering (MI-STA)*. Sabratha, Libya.

Paliwal P, Shreemali J, Chakrabarti P, Poddar S. 2021. Performance optimization of hybrid RAMAN-EDFA based WDM-FSO under adverse climatic conditions. *Materials Today: Proceedings*.

Saleh R, Mohamed I, Al O. 2022. Comparison different modulation schemes in SISO and MIMO FSO links to obtain maximum FSO link. *Ijeit on Engineering And Information Technology*.

Shaina, Gupta A. 2016. Comparative analysis of free space optical communication system for various optical transmission windows under adverse weather conditions. *Procedia Computer Science*.

Singh M, Atieh A, Aly MH, Abd El-Mottaleb SA. 2022. 120 Gbps SAC-OCDMA-OAM-based FSO transmission system: Performance evaluation under different weather conditions. *Alexandria Engineering Journal*.

Thakur A, Nagpal S. 2018. Performance evaluation of different optical amplifiers in spectrum sliced free space optical link. *Journal of Optical Communications*.

Vishwakarma P, Vijay J. 2016. Comparative analysis of free space optics and single mode fiber. *International Journal of Advanced Engineering, Management and Science*.

Willebrand HA, 2001. Ghuman BS. Fiber optics without fiber. *IEEE Spectrum*.

Advances in AI for Biomedical Instrumentation, Electronics and Computing – Sachan et al. (eds)
© 2024 The Author(s), ISBN 978-1-032-64298-7

Optimization of 60 GHz MIMO OFDM radio over fiber system for next generation networks2

Balram Tamrakar
KIET Group of Institutions Ghaziabad, Delhi NCR, Ghaziabad, India
University School of Information Communication and Technology, Guru Govind Singh Indraprastha University, New Delhi, India

Mohd Sharique Alam, Shubham Verma, Ujjwal Tyagi, Mohd Khalid & Mohd Salman Zaidi
KIET Group of Institutions Ghaziabad, Delhi NCR, Ghaziabad, India

ABSTRACT: This study employs a practical as well as comprehensive simulation so as to foresee the behaviour of a 2×2 MIMO Orthogonal Frequency Division Multiplexing (OFDM) Radio over Fiber system operating at 60 Giga Hz before its implementation. The system leverages spatial multiplexing (SMX) to boost data rates and spatial diversity (SD) to enhance reliability by reducing error likelihood. We measured the values of SNR (E_b/N_0) for the different used techniques like QPSK, 16QAM, 64QAM are 2.8, 1.6, 1.25 respectively.

Keywords: Multi Input Multi Output (MIMO), Orthogonal Frequency Division Multiplexing (OFDM), Bit Error Rate (BER), Spatial Diversity (SD), Spatial Multiplexing (SMX)

1 INTRODUCTION

This paper explores the potential of the 56-64 GHz Unlicensed Millimeter Wave (MMW) band, offering a wide 7 GHz bandwidth for high-speed data transfer. It emphasizes enhancing spectral efficiency through Multiple Input Multiple Output (MIMO) technology, including Spatial Diversity (SD) and Spatial Multiplexing (SMX) techniques (Lin *et al.* 2010). Radio over Fiber (RoF) is adopted to extend MMW signal coverage by combining optical fiber benefits with high wireless mobility. RoF communication systems offer broader coverage, increased bandwidth, cost-effectiveness, and reduced power consumption (Tamrakar *et al.* 2022). The study uses MATLAB simulations and a TSV channel model to assess the reliability of 60 GHz transmission. It proposes a comprehensive 60 GHz architecture, analyzing real-world parameters for optimized MMW bandwidth utilization. Our simulation uses MATLAB enabling accurate RoF system behavior forecasting. We employ a precise channel model based on observed data, specifically the TSV model, for a reliable 60 GHz transmission quality assessment (Smulders 2009).

Our study proposes a comprehensive 60 GHz architecture merging optical and wireless communication. It conducts global simulations to analyze real-world parameters, including modulation schemes (16QAM, 64QAM, QPSK), diversity techniques (SMX and SD), and programmed LDPC coding for Line-Of-Sight desktop environments (Tamrakar *et al.* 2020).

2 BROADENED MULTIPLE INPUT MULTIPLE OUTPUT

The expression for the response signal y is described as the product of the delivered signal x and disturbance n is defined as (y= H x + n) where H is the M_R x M_T network matrix of a

DOI: 10.1201/9781032644752-20

multiple input multiple output system, M_R and M_T are the number of transmit and receive antennas, and single input single output network impulse response $h_{l,m}$ ($1 \leq M_T$ and $m \leq M_R$) as every component. Compare single input single output channel impulse response $w_{l,m}$ ($l \leq N_T$ and $m \leq M_R$) as each element

$$W(\tau, t) = \begin{bmatrix} w_{1,1}(\tau, t) & w_{1,2}(\tau, t) & \cdots & w_{1,M_T}(\tau, t) \\ w_{2,1}(\tau, t) & w_{2,2}(\tau, t) & \cdots & w_{2,M_T}(\tau, t) \\ \vdots & \vdots & \ddots & \vdots \\ w_{M_R,1}(\tau, t) & w_{M_R,1}(\tau, t) & \cdots & w_{M_R,M_T}(\tau, t) \end{bmatrix} \quad (1)$$

A geometrical method to extending the single input single output channel model to 2 x 2 MIMO systems. In Line of Sight (LOS) scenarios, the channel impulse response (CIR) for each path can be described as follows:

The 2 x 2 multiple input multiple output channel matrix generates the following for the initial tap:

$$H = \sqrt{P}\left(\sqrt{\frac{K}{K+1}} H_L + \sqrt{\frac{1}{K+1}} H_W \right) \quad (2)$$

$$= \sqrt{P}\left\{ \sqrt{\frac{K}{K+1}} \begin{vmatrix} A_{1,1}e^{-i\varphi_{1,1}} & A_{1,2}e^{-i\varphi_{1,2}} \\ A_{2,1}e^{-i\varphi_{2,1}} & A_{2,2}e^{-i\varphi_{2,2}} \end{vmatrix} \right\} + \sqrt{P}\left\{ \sqrt{\frac{1}{K+1}} \begin{vmatrix} X_{1,1} & X_{1,2} \\ X_{2,1} & X_{2,2} \end{vmatrix} \right\} \quad (3)$$

P represents received authority of the initial signal, while H_L and H_W reflect the steady and spreading components of the channel matrices. Each element of the matrix, denoted as $A_{l,m}e^{-i\varphi_{l,m}}$, is determined mathematically using unique transmitter segment placements. $A_{l,m}$ is a real number representing the strength of the Line of Sight (LOS) path γ. $X_{i,j}$ is a complicated Gaussian random parameter having an average of zero and a variation of one unit. The Rice factor is represented by K, and the phase twist caused by the route length is represented by $\varphi_{l,m}$ (Shoji *et al.* 2009).

$$\varphi_{l,m} = Kd_{l,m}$$

$$K = \frac{2\pi}{\lambda},$$

Where

$$h_{i,j}(t) = \sum_{i=0}^{L-1} \sum_{m=0}^{N_l-1} a_{l,m} \delta(t - T_1 - \tau_{l,m}) \delta(\varnothing - \Psi_l - \Psi_{l,m}) \sqrt{G_r(0, \Psi_l + \Psi_{l,m})} \quad (4)$$

3 RESULT ANALYSIS

The simulation explores a QPSK communication system, initializing parameters and introducing Gaussian noise to emulate channel effects. It employs QPSK modulation, assessing Error Vector Magnitude (EVM) for various power levels, providing insights into system performance through interpolated EVM values. A plot visualizes the EVM-power relationship, a key metric for quality assessment (Tamrakar *et al.* 2022).

Finally, a plot is generated to visualize the relationship between EVM and injected power as shown in Figure 1. The EVM serves as a crucial metric for assessing the quality of the communication system, with lower EVM values indicating better performance (Hirokazu *et al.* 2006).

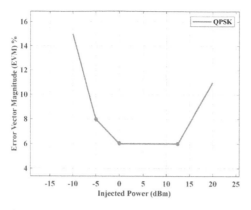

Figure 1.   Maximum QPSK input power.

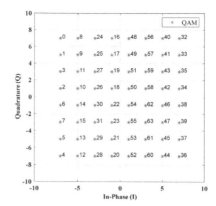

Figure 2.   Spectrum of 64 QAM.

The BER vs. Eb/No curves display the relationship between Bit Error Rate (BER) and Eb/No ratios, indicating that as signal strength relative to noise (Eb/No) improves, error rates decrease. These graphs serve to compare different modulation schemes (QPSK, 16 QAM, 64 QAM, SMX-QPSK, SMX-16 QAM, SMX-64 QAM) under the same Eb/No conditions. Lower points on BER curves denote better performance. Analysing intersections or divergences among the curves helps identify which modulation schemes perform better under specific noise conditions, with some excelling at lower Eb/No and others at higher Eb/No. (Tamrakar *et al.* 2021).

By analysing these BER vs. Eb/No graphs as shown in Figure 3 and 4, you can make informed decisions about which modulation scheme to use for a given communication system, taking into account the trade-offs between data rate and error performance under different noise conditions (Stöhr A. *et al.* 2009).

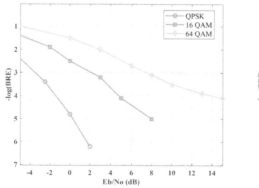

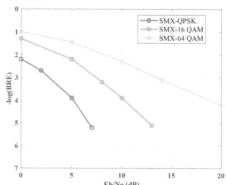

Figure 3.   Graph between BER and Eb/No of SD.   Figure 4.   Graph between BER and Eb/No of SMX.

In the initial phase, energy levels were fine-tuned to maximize transmission efficiency in light-based systems. This involved optimizing the Efficacy Vector Magnitude (EVM) for QPSK 2x2 Multiple-Input, Multiple-Output over a 25-kilometer Single-Mode Fiber, with the ideal energy level determined as 0 dBm. Additionally, constellation diagrams were examined for QPSK, 64QAM, and 16QAM at the receiver, offering insights into signal quality and efficiency. To concurrently increase data rate and SNR, we integrated LDPC

(Low-Density Parity-Check) channel coding with SMX. We employed irregular LDPC codes from the 802.15.3c WPAN standard. When contrasting Spatial Diversity, with Spatial Multiplexing equipped with LDPC coding across a variety of modulation schemes like QPSK, 16QAM, and 64QAM, LDPC-coded SMX exhibits enhanced data rates and SNR performance. It operates effectively at a decreased Eb/No of -4.5 dB, surpassing Spatial Diversity, which necessitates a higher Eb/No of 3.5dB (Tamrakar *et al.* 2022).

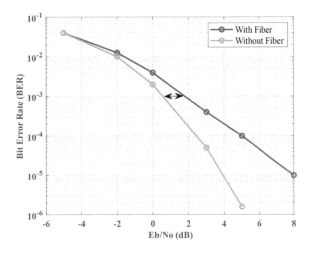

Figure 5. Graph between BER and Eb/No of SD without optical link.

Table 1. Error comparisons in different techniques.

S.N.	$E_b/N_0(dB)$	$-\log(BER)$			BER	
		QPSK	16 QAM	64 QAM	With Fiber	Without Fiber
1	−5	2.5	1.4	1	0.03	0.03
2	−4	2.8	1.6	1.25	0.019	0.016
3	−3	3.8	1.8	1.35	0.014	0.012
4	−2	4.2	2.1	1.40	0.012	0.01
5	−1	4.6	2.25	1.45	0.002	0.001
6	0	4.9	2.5	1.5	0.0039	0.0019
7	1	5.3	2.8	1.7	0.001	0.0006
8	2	6.3	3	1.95	0.0008	0.0001

From Figure 5, we observe that the blue curve consistently resides below the red curve over the majority of the SNR spectrum. This signifies that, under various operational conditions, the "With Fiber" setup generally exhibits superior BER performance. In practical terms, this suggests that incorporating optical fiber into the communication system enhances data transmission reliability by minimizing errors in received data. However, it's noteworthy that there are specific SNR conditions where the "Without Fiber" scenario surpasses the "With Fiber" configuration in terms of BER. This indicates that, in certain situations or for specific performance requirements, alternative setups without optical fiber may be more advantageous (Martinez-Ingles *et al.* 2013).

4 CONCLUSION

In the proposed research article, the measured values of SNR (E_b/N_0) for the different used techniques like QPSK, 16QAM, 64QAM are 2.8, 1.6, 1.25 respectively. We also measured BER with fiber and without fiber, and the measured values are 0.019 and 0.016 respectively for the same SNR. Over all we have concluded that 64 QAM without fiber is better than other used techniques.

REFERENCES

Hirokazu S. *et al.* 2006. Proposal of novel statistic channel model for millimeter wave WPAN, in: *2006 Asia-Pacific Microwave Conference*, pp. 1855–1858.

Lin, C.-T. *et al.* 2010. Ultra-high data-rate 60 GHz radio-over-fiber systems employing optical frequency multiplication and OFDM formats, *Journal of Lightwave Technology* 2296–2306

Martinez-Ingles M.T. *et al.* 2013. Experimental evaluation of an indoor MIMO-OFDM system at 60 GHz based on the IEEE802.15.3c standard, IEEE Antennas Wirel. *Propag. Lett.* 12 1562–1565

Smulders P.F.M. 2009. Statistical characterization of 60-GHz indoor radio channels, *IEEE Trans. Antennas Propag.* 57 2820–2829

Shoji Y. *et al.* 2009. A modified sv-model suitable for line-of sight desktop usage of millimeter-wave wpan systems, *IEEE Trans. Antennas Propag.* 57 2940–2948.

Stöhr A. *et al.* 2009. 60 GHz radio-over-fiber technologies for broadband wireless services [Invited], *J. Opt. Netw.* 8 471–487.

Tamrakar B, *et al.* 2022. "Performance investigation of RoF Link in 16 channel WDM system using DPSK modulation technique", *Proceedings of International Conference on Computational Intelligence, Algorithms for Intelligent Systems*, Springer Nature Singapore, https://doi.org/10.1007/978-981-19-2126-1_20, pp. 262–271.

Tamrakar B. *et al.* 2020. Performance measurement of radio over fiber system at 20 GHz and 30 GHz by Employing with and without Optical Carrier Suppression. *IEEE International Conference.*, ISBN:978-1-5386-9271-4, DOI: https://ieeexplore.ieee.org/document/8991395.

Tamrakar B. *et al.* 2022. "A comparative analysis between different optical modulators of analog and digital radio over fiber (RoF) link for the next-generation networks". *Journal of Optics.*, https://doi.org/10.1007/s12596-022-00994-x.

Tamrakar B. *et al.* 2021. "Performance analysis of DP-MZM radio over fiber links against fiber impairments" *Journal of Optical Communication*, https://doi.org/10.1515/joc-2021-0079.

Tamrakar B. *et al.* 2022. "Analysis and modelling of DD-DPMZM to investigate fundamental to inter-modulation distortion Ratio (FIMDR) against different fiber impairments for the next generation networks". *Journal of Active and Passive Electronics Devices.*, Vol 16, 297–310.

Advances in AI for Biomedical Instrumentation, Electronics and Computing – Sachan et al. (eds)
© 2024 The Author(s), ISBN 978-1-032-64298-7

Review on multi-robot path planning in complex industrial dynamic environment using improved nature inspired techniques

Subhash Yadav
Indian Institute of Information Technology, Pune, Maharashtra, India

Shubham Shukla
Krishna Institute of Engineering and Technology, Ghaziabad, Uttar Pradesh, India

Ritu Tiwari
Indian Institute of Information Technology, Pune, Maharashtra, India

ABSTRACT: The usage of robotic devices in the modern workplace is the result of recent technological advancements. Both single and multiple robotic systems collaborate closely with people to complete tasks, and new developments have increased the appeal of using multi-robot teams. Path planning was thought to be problematic. Particularly, the Multiple Robotic Systems of complex dynamic environments have significant open challenges. A fascinating field of study is determining the best and most efficient route from a starting point to a goal point in any given environment that minimizes time and distance while avoiding collisions with obstacles. This research examines algorithms that draw inspiration from nature to address the challenge of mobile robot path planning. This work explains the use of both modified and standard nature-inspired algorithms for multi-robot path planning in a complicated industrial dynamic environment.

Keywords: Nature inspired techniques, Genetic Algorithm, Particle Swarm Optimization, Multi robot Path planning and Dynamic environment

1 INTRODUCTION

Recently, and particularly in the last two decades, there has been a considerable deal of interest in robotic communication due to the cooperation of numerous robot systems. Utilizing technology in a variety of businesses and fields, such as entertainment, space exploration, medical and emergency services, education, the military, warehousing, and much more, has been made possible by recent technological advancements (Bartos *et al.* 2021). Industrial robots are used in numerous multi-robot applications where they must arrange themselves into precise formations to carry out difficult tasks, including moving big objects, conducting reconnaissance, pursuing targets, assembling, etc. (Kandaswamy and Zuo 2018). Essentially, it has been demonstrated that the typical multi-robot skills of path planning, formation generation, target tracking/observation, and work distribution are optimization problems (Li and Liang 2020). Among all of these, path planning is regarded as a significant optimization issue since it has several objectives and a polynomial validated solution. The goal of the robots is to complete the post-design jobs faster, more reliably, and with less energy consumption (Svoboda *et al.* 2018). Due to the high computing complexity of multi-robot path planning, comprehensive methods that provide computational efficiency and optimal solutions are hard to come by. However, because multi-robot route planning is

DOI: 10.1201/9781032644752-21

so effective at completing missions, it is driven by a variety of real-world activities. The search area of each robot is expanded nevertheless, because each member of the group can work alone or with others depending on the circumstances (Wang *et al.* 2020). Effective approaches must allow real-time response to changing conditions and should fully exploit sensory information about the environment (Madridano *et al.* 2021).

Figure 1. Classification of nature inspired techniques utilized in the multi robot path planning in a complex industrial dynamic environment.

Nature inspired techniques are generally used for the optimization process for identifying the best possible solutions to a given problem (Sharma and Doriya 2020).

Although nature inspiring methods show the best outcomes when applied in multi robot path planning in a complex industrial dynamic environment, the problem was that each technique lacks some important parameters in the process (Ui path, n.d.).

The survey paper is explained as follows: section 2 explains the survey on the multi-robot path planning in a complex industrial dynamic environment using improved nature inspired techniques; section 3 indicates the analysis, and the paper is concluded in section 4.

2 LITERATURE REVIEW

The formulation of the problem for multi-robot route planning is to avoid robot collisions by estimating the robot's next position from its current positions in its workspace. In a complicated industrial dynamic setting, multi-robot planning has made good use of improved nature-inspired methodologies. Thus, this paper is divided into the following sections: 2.1 describes multirobot path planning in a complex industrial dynamic environment using modified nature inspired algorithms; and 2.2 discusses performance comparison of multi-robot path planning in a complex industrial environment.

2.1 *The multi robot path planning in a complex industrial dynamic environment using modified nature inspired algorithms*

Normally, different kinds of modifications were applied to enhance the performance of nature-inspired algorithms in optimization for multi robot path planning in the complex industrial dynamic environment. After modifying the nature inspired algorithms, Techniques will appear as updated and improved ones. Table 1 explains the work on modified nature inspired algorithms for multi robot path planning in a complex industrial dynamic environment with its findings and limitations.

2.2 *Performance comparison of multi robot path planning in complex industrial dynamic environment using nature inspired algorithms*

Performance comparison is required for every work so that when comparing the nature-inspired techniques that were applied for multi robot path profile planning it will be easier to detect which method performs best in most of the metrics (Yirmibesoglu *et al.* 2018). In such environments, it is essential to plan the paths of multiple robots in a coordinated manner to

Table 1. Work on modified nature inspired algorithms for multi robot path planning in a complex industrial dynamic environment with its findings and limitations.

AUTHOR NAME	MODIFIED ALGORITHMS	FINDINGS	LIMITATIONS
Chen and Liu, 2019	Modified ant colony optimization (ACO)	From the findings, it had been found that the ACO algorithm proved best for robot path planning and it has stronger stability and environmental adaptability.	Low convergence speed and local optimum problem.
Liu *et al.* 2021	Improved GWO (IGWO)	Results showed that IGWO better balances the algorithm's local development capabilities, thus, the ability of the path also had been improved	When applying on multi robot profile, there were low solution accuracy
Paikray *et al.* n.d.	Enhanced particle swarm optimization (IPSO) with sine and cosine algorithms (SCAs)	Outcomes of the simulation and real platform results revealed that IPSO–SCA was superior to IPSO and other in the form of producing an optimal collision-free path	It was easy to fall into the local optimum in high-dimensional space
Ge *et al.* 2020	Improved PIOFOA	Analysis indicated that PIOFOA was more efficient in path planning for UAV in a dynamic three-dimensional oilfield environment with moving obstacles	Premature convergence is considered to be the most important limitation
Das *et al.* 2021	Improved GSA (IGSA), GSA, and Differential Evolution (DE)	Simulation and the Khepera environment result outperformed IGSA as compared to GSA and DE with respect to the average total trajectory path deviation and energy optimization	Sometimes GSA gets stuck in local minima in the last iterations.

Table 2. Performance comparison of multi robot path planning in a complex industrial dynamic environment using nature inspired techniques with its findings and limitations.

AUTHOR NAME	APPROACHES	PERFORMANCE COMPARISON	LIMITATIONS
Wei *et al.* 2020	Multi-Objective (MOPSO)	Analysis showed that the applied approach can reduce the total cost by at least 26%, and reduce the max sub cost by at least 21% in the multi robot path planning	There was a lower convergence rate
Li *et al.* 2020	Improved PSO (IPSO)	Performance analysis indicated that IPSO makes reduces the path length and simulation time by 2.8% and 1.1 seconds	When compared with the classical PSO algorithm, the time advantage of the proposed algorithm was weak, which will be a crucial issue
Wang *et al.* 2020	Improved PIO	Simulation results showed that the applied Cauchy mutant IPIO method provides a better robustness and cooperative path planning strategy and effective comparing with traditional PIO algorithm	Sometimes, the optimal solution will be complex
Toufan and Niknafs 2020	GWO and IGWO	Performance evaluation showed that IGWO performed better and proved as an optimal, safe, and smooth path without any problems.	Low solving accuracy and unsatisfactory ability of local searching
Mao *et al.* 2021	Novel Improved (NIFOA)	The effectiveness of NIFOA has been proven and shown best by detailed comparisons with the other three optimization algorithms, namely PSO, FOA, and IFO	There will be nonlinear optimization problems

avoid collisions, optimize time and distance, and ensure overall efficiency (Low *et al.* 2019). Table 2 explains the work on performance comparison of multi robot path planning in a complex industrial dynamic environment using nature inspired techniques with its findings and limitations.

3 RESULTS AND DISCUSSION

This section compares the path planning of nature-inspired methodologies and the number of turns versus the number of robots for multi-robot planning in a dynamic environment for three different algorithms, analyzing obstacle avoidance in multiple robots. The task of achieving a control objective in robotics while adhering to non-intersection or non-collision position limitations is known as obstacle avoidance (Tian *et al.* 2020).

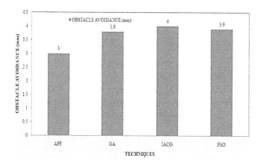
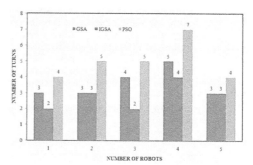

Figure 2. Graphical representation of obstacle avoidance in the multiple robots with comparative analysis in path planning of nature inspired techniques.

Figure 3. Graphical representation of the number of turns versus the number of robots for multi robot planning in a dynamic environment for 3 different algorithms.

When compared to the other strategies, the enhanced (ACO) techniques demonstrated a higher obstacle avoidance (4 mm). Lower obstacle avoidance was indicated by APF (3 mm).

Additionally, three distinct algorithms' turn and robot numbers for multi-robot planning in a dynamic environment have been examined.

It had been clear from Figure 3 that a total of 5 numbers of robots were there for the analysis of the number of turns in three different algorithms. IGSA takes a smaller number of turns than the other two algorithms and the energy consumption to reach the designation is less than the other two algorithms. The simulation is presented for five numbers of robots but the number of turns is less irrespective of the robot in the planning scheme of the algorithm.

4 CONCLUSION

This study reviews a novel and resilient approach to solving the navigation problem for many robots in a dynamic environment using several nature-inspired algorithms. In any complex environment, the applied technique will always find the optimal (or nearly optimal) path. In order to do this, all possible paths between any arbitrarily chosen start and destination sites in a discrete gridded environment were constructed using the applied technique. However, the researchers have concentrated on using methods inspired by nature to reduce the computing burden and identify the best feasible solution. The number of turns vs the number of robots for multi-robot planning in a dynamic environment for three different algorithms, as well as obstacle avoidance in multiple robots with comparative analysis in path planning of nature inspired methodologies, had been effectively studied in the article.

REFERENCES

Bartos, M., Bulej, V., Bohusik, M., Stancek, J., Ivanov, V., & Macek, P. 2021. An overview of robot applications in the automotive industry. *14th International Scientific Conference on Sustainable, Modern and Safe Transport*, 2021. https://doi.org/10.1016/j.trpro.2021.07.052

Chen, G., & Liu, J. 2019. Mobile robot path planning using ant colony algorithm and improved potential field method. *Computational Intelligence and Neuroscience*, 2019. https://doi.org/10.1155/2019/1932812

Das, P. K., Behera, H. S., Jena, P. K., & Panigrahi, B. K. 2017. Multi-robot path planning in a dynamic environment using improved gravitational search algorithm. *Journal of Electrical Systems and Information Technology*, 3(2), 295–313.

Han, Z., Wang, D., Liu, F., & Zhao, Z. 2017. Multi-AGV path planning with double-path constraints by using an improved genetic algorithm. *Plos One*, 12(7), 1–16.

Kandaswamy, E., & Zuo, L. 2018. Recent advances in treatment of coronary artery disease: Role of science and technology. *International Journal of Molecular Sciences*, 19(4), 1–18.

Li, B., & Liang, H. 2020. Multi-robot path planning method based on prior knowledge and q-learning algorithms. *2nd International Conference on Computer Modeling, Simulation and Algorithm*, 2020, Beijing, China.

Liu, J., Wei, X., & Huang, H. 2021. An improved grey wolf optimization algorithm and its application in path planning. *IEEE Access*, 9, 121944–121956.

Li, X., Wu, D., He, J., Bashir, M., & Liping, M. 2020. An improved method of particle swarm optimization for path planning of mobile robot. *Journal of Control Science and Engineering*. https://doi.org/10.1155/2020/3857894

Low, E. S., Ong, P., & Cheah, K. C. 2019. Solving the optimal path planning of a mobile robot using improved Q-learning. *Robotics and Autonomous Systems*, 115, 143–161.

Madridano, A., Al-Kaff, A., Martin, D., & de la Escalera, A. 2021. Trajectory planning for multi-robot systems: *Methods and applications. Expert Systems with Applications*, 173, 1–14.

Mao, Y., Huang, D., Qin, N., Zhu, L., & Zhao, J. 2021. Cooperative 3D path planning of multi-UAV via improved fruit fly optimization. *Research Square (Pre-print)*. https://doi.org/10.21203/rs.3.rs-1124343/v1

Paikray, H. K., Das, P. K., & Panda, S. (n.d.). Optimal multi-robot path planning using particle swarm optimization algorithm improved by sine and cosine algorithms. *Arabian Journal for Science and Engineering*, 46(4), 3357–3381.

Sharma, K., & Doriya, R. 2020. Path planning for robots: An elucidating draft. *International Journal of Intelligence Robotics and Applications*, 4(2), 294–307.

Svoboda, T., Dube, R., Gawel, A., Surmann, H., Knipp, K., & Cadena, C. 2018. DR 1.4: Sensing, mapping and low-level memory IV Long term persistence. *TRADR*, 2018. http://www.tradr-project.eu/wp-content/uploads/dr.1.4.main_public.pdf

Tian, S., Li, Y., Kang, Y., & Xia, J. 2020. Multi-robot path planning in wireless sensor networks based on jump mechanism PSO and safety gap obstacle avoidance. *Future Generation Computer Systems*, 118, 37–47.

Toufan, N., & Niknafs, A. 2020. Robot path planning based on laser range finder and novel objective functions in grey wolf optimizer. *SN Applied Sciences*, 2, 1–19.

Ui path. (n.d.). Robotic Process Automation. https://www.uipath.com/rpa/robotic-process-automation

Wang, B., Liu, Z., Li, Q., & Prorok, A. 2020. Mobile robot path planning in dynamic environments through globally guided reinforcement learning. *IEEE Robotics and Automation Letters*, 5(4), 6932–6939.

Wang, B. H., Wang, D. B., & Ali, Z. A. 2020. A Cauchy mutant pigeon-inspired optimization based multi-unmanned aerial vehicle path planning method. *Measurement and Control*, 53, 83–92.

Wei, C., Ji, Z., & Cai, B. 2020. Particle swarm optimization for cooperative multi-robot task allocation: a multi-objective approach. *IEEE Robotics and Automation Letters*, 5(2), 2530–2537.

Yirmibesoglu, O. D., Morrow, J., Walker, S., Gosrich, W., Canizares, R., Kim, H., Daalkhaijav, U., Fleming, C., Branyan, C., & Menguc, Y. 2018. Direct 3D printing of silicone Elastomer soft robots and their performance comparison with molded counterparts. *IEEE International Conference on Soft Robotics*, 24–28 April 2018, Livorno, Italy.

Advances in AI for Biomedical Instrumentation, Electronics and Computing – Sachan et al. (eds)
© 2024 The Author(s), ISBN 978-1-032-64298-7

Network pharmacology and molecular docking approach to unveil the mechanism of amantadine for Parkinson's disease

Nimesh Bhardwaj, Himani Tyagi, Siddhi Gupta, Pooja Gangwar, Shubham Chaudhary, Mansi Singh, Vinay Kumar & Praveen Kr. Dixit
Department of Pharmacology, KIET Group of Institutions (KSOP), Ghaziabad (UP), India

ABSTRACT: Parkinson's disease is a neurodegenerative disorder, and its treatment is complicated. Pharmacological treatments provide only symptomatic relief but do not cure the disease. Amantadine is widely used in the treatment of Parkinson's disease as a dopamine facilitator. Its exact mechanism is still unknown. In this study, network pharmacology and molecular docking were used to investigate the therapeutic effects of Amantadine in the management of Parkinson's disease. SLC47A1 is a MATE family transporter and is frequently paired with OCTs. SLC22A1 solute carrier family 22 member 1 is an OCT (Organic Cation Transporters) that translocates a broad range of organic cations and catechols like choline, guanidine, histamine, epinephrine, dopamine, adrenalin, nor-adrenaline. Amantadine showed its effect in Parkinson's disease through a potential target i.e., SLC22A1 by transporting the organic cation (dopamine) through the synaptic cleft of the nerve ending. Molecular docking was then performed to evaluate the binding affinity of Amantadine towards SLC47A1 (UniProt Id: Q96FL8) & SLC22A1 (UniProt Id: O15245). Our study suggests that Amantadine has binding affinity ΔG = -5.6kcal/mol towards SLC47A1 in comparison to standard drug 'Levodopa' whose ΔG value was -6.5 kcal/mol & binding affinity ΔG = -6.6 kcal/mol towards SLC22A1 in comparison to standard drug 'Levodopa' whose ΔG value was -6.4 kcal/mol. These targets can also be explored in other disease treatments like cancer, diabetes, and inflammatory diseases.

1 INTRODUCTION

Amantadine is an antiviral drug. It was initially evolved for the treatment of an influenza virus in the 1960's. Amantadine is now used mostly for Parkinson's disease (PD) as a dopamine facilitator. Its pharmacological actions are dopaminergic, it has a dual effect on Parkinsonian signs and symptoms and levodopa-induced dyskinesia (Professor Olivier Rascol MD PhD *et al.* 2021). Parkinson's disease is a neurodegenerative disorder, and the number of cases of PD has significantly increased in the past three decades (Cabreira and Massano 2019). There are various factors responsible for PD including environmental factors as well as genetic factors. Another reason is an aggregation of misfolded α-synuclein, which is found in intra-cytoplasmic inclusions, called Lewy bodies (Balestrino and Schapira 2020). The primary or major symptoms of PD are tremors, dyskinesia, rigidity, akinesia and postural abnormalities, seborrhoea, and sialorrhea (Bloem *et al.* 2021). Network Pharmacology is a type of systematically designed *in silico* technique that is majorly used to unveil the target mechanism. Using network pharmacology, we can easily predict the promising mechanisms of any pharmaceuticals. (Nogales *et al.* 2022). Molecular docking is one of the most frequently used methods in SBDD because of its ability to predict, with a substantial degree of accuracy, the conformation of small-molecule ligands within the appropriate target binding site (Ferreira *et al.* 2015).

  DOI: 10.1201/9781032644752-22

2 MATERIALS AND METHODOLOGY

2.1 *Network pharmacology*

Network pharmacology is a type of *in silico* technique which is used in our study. We used Cytoscape (v.3.10.1) to construct a PPI network. In this step, we used various steps and online websites to collect human targets mentioned below.

2.1.1 *Collection of amantadine drug targets*
Initially, the information on the molecule was obtained by searching Canonical SMILES ID from PubChem (https://pubchem.ncbi.nlm.nih.gov/). Amantadine targets were obtained from the SEA database (http://sea.bkslab.org/). All three human protein targets (SLC22A2, SLC47A1, and GRIN2A) were selected and downloaded.

2.1.2 *Collection of Parkinson's disease targets*
The next step was to isolate disease protein targets (Parkinson's disease). DiSGeNET (http://www.disgenet.org) and STRING database (http://string-db.org) were used for identification of potential targets. Protein targets were obtained from these databases and combined.

2.1.3 *Construction of PPI network using Cytoscape (v.3.10.1)*
Amantadine-target-Parkinson's Disease network was constructed using the software named 'Cytoscape' (http://cytoscape.org). Afterward, Path linker was used to generate a link between those 7 targets and 3 potential targets of the drug compound.

2.2 *Molecular docking using PyRx (v0.8)*

The 3D Structure of the drug/ligand was downloaded in 'SDF 3D conformer' form with the help of PubChem (https://pubchem.ncbi.nlm.nih.gov/). On the other hand 3D structure of our protein target 'SLC47A1' & 'SLC22A1' with the help of PDB (https://www.rcsb.org/). By using PyRx we can select multiple ligands to check binding energies on a single protein target. Both ligand and target protein was prepared using 'BIOVIA Discovery Studio Visualizer', by removing already attached ligands, water molecules, and hetatoms.

2.2.1 *Molecular docking and grid generation*
In this step, we loaded our target protein (prepared) in PyRx v0.8 and converted our target protein into a macromolecule. On the other hand, ligand was inserted and energy was minimized and converted all molecules into 'pdbqt' format. Then we used AutoDock Vina and both target protein and ligand were placed into a grid and docking was started. (Trott and Olson 2009)

2.2.2 *Visualization of the docked ligand with the target protein*
Now the docked ligand having the best pose and good binding energy is downloaded from PyRx's auto dock tab and visualized the bond lengths, bond angles and binding pocket of our ligand.

3 RESULTS AND DISCUSSION

3.1 *Network pharmacology results*

517 initial targets were identified for amantadine by using DiSGeNET and String database. After the construction of a network using Cytoscape, we found that there are only 3 potential targets that are involved in the management of Parkinson's disease using amantadine. The other remaining protein targets were not identified as potential targets in the

management of PD. To further unveil the mechanism of amantadine against PD, the constructed PPI network of amantadine targets are shown in Figure 1.

After the construction of the network and performed path linker, it was found that 7 targets shown in Figure 2 work together involved in the management of Parkinson's disease.

SLC47A1 (570aa) is a MATE family transporter, Multidrug and toxin extrusion protein1. It is one of the potential target of amantadine. The MATEs also display a characteristic multi-selectivity and are frequently paired with OCTs (organic cation transporters) as shown in Figure 2 (Pelis and Wright 2014). SLC22A1 (554 AA) solute carrier family 22 member 1 is a OCTs (Organic Cation Transporters) translocate a broad range of organic cations and catechols through the synaptic cleft of nerve ending i.e., choline, guanidine, histamine, epinephrine, dopamine, adrenalin, noradrenaline and some drugs like quinine and metformin (Lian *et al.* 2022).

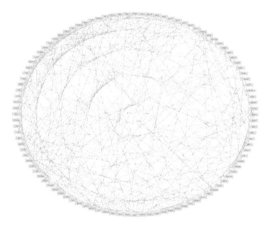

Figure 1. PPI (protein-protein interaction) Network of Amantadine-targets- Parkinson's disease (yellow boxes showing 3 potential targets 'SLC47A1' 'GRIN2A' & 'SLC22A2' related to PD and blue boxes showing other human targets that are indirectly interacting with PD).

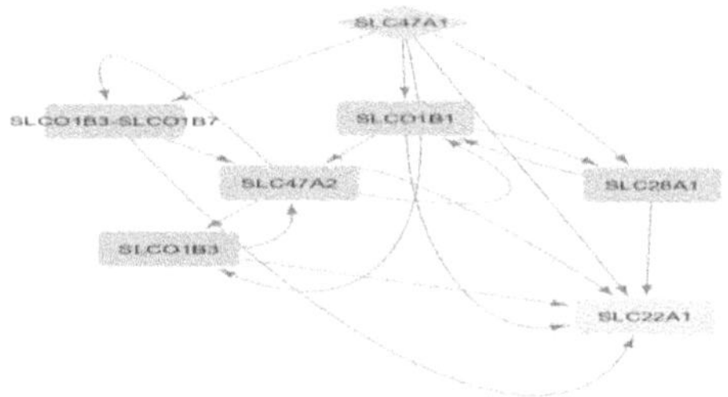

Figure 2. Path linker showing the path linked between all 7 targets and showing how they are involved together in organic cation transport. These are some carrier proteins.

The transportation of dopamine is inhibited by N-methyl-D-aspartate (NMDA) receptors i.e., prolonged activation of NMDA receptors can cause some excitotoxicity events like Dyskinesia which is a prompt symptom of Parkinsonism (Koutsilieri and Riederer 2007). Due to this, transportation of dopamine and other Organic cations is inhibited. SLC22A1 has the potential to transport dopamine by antagonizing the actions of NMDA by transporting dopamine, it increases the availability of dopamine. However, it also mediates the transportation of various drugs used as anticancer, antiviral, anti-inflammatory, and antiemetic (Arimany-Nardi *et al.* 2015).

3.2 *Molecular docking results*

Tabulated results of docking are presented in Tables 1 and 2 which show binding affinity & RMSD (Root Mean Square Deviation) upper and lower bound values of our test ligand and standard ligand towards both SLC47A1 and SLC22A1.

Amantadine showed good binding affinity towards SLC22A1 however it had -6.6kcal/mol binding affinity which was more than the standard drug Levodopa.

Table 1. Showing binding affinity on 'SLC47A1' of Amantadine and Levodopa which is a potential target for the management of Parkinson's disease.

Test ligand	Minimized energy	ΔG (kcal/mol)	RMSD/UB	RMSD/LB
Amantadine Model 1	115.62	−5.6	2.256	0.095
Amantadine Model 2	115.62	−5	34.564	33.144
Amantadine Model 3	115.62	−4.9	31.657	30.433
Standard ligand	Minimized energy	ΔG (kcal/mol)	RMSD/UB	RMSD/LB
Levodopa Model 1	112.83	−6.5	5.448	2.889
Levodopa Model 2	112.83	−5.9	28.341	26.195
Levodopa Model 3	112.83	−5.8	8.482	6.095

Table 2. Showing binding affinity on 'SLC22A1' of Amantadine and Levodopa which is a potential target for the management of Parkinson's disease.

Test ligand	Minimized energy	ΔG (kcal/mol)	RMSD/UB	RMSD/LB
Amantadine Model 1	115.62	−6.6	2.285	0.086
Amantadine Model 2	115.62	−5.7	25.025	23.877
Amantadine Model 3	115.62	−5.2	24.855	23.411
Standard ligand	Minimized energy	ΔG (kcal/mol)	RMSD/UB	RMSD/LB
Levodopa Model 1	112.83	−6.4	5.259	2.223
Levodopa Model 2	112.83	−6.2	5.536	1.855
Levodopa Model 3	112.83	−6.1	5.905	2.38

3.3 *BIVOIA visualization results*

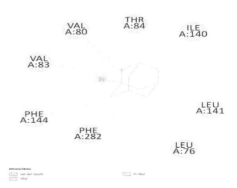

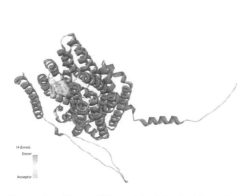

Figure 3. Shows 3D model of docked ligand (Amantadine) having -5.6 binding affinity towards SLC47A1.

Figure 4. Shows 2D interactions model of Amantadine with SLC47A1. Amantadine is bound with VAL (A: 80) with an alkyl bond having bond lengths of 4.89 and 5.27. Amantadine is also bound with VAL (A: 83) with an Alkyl bond having a bond length of 4.35. Amantadine is bound with PHE (A: 144 & A: 282) with a Pi-alkyl bond having bond lengths of 4.70 and 5.34. The rest of the amino acids labelled with green color (THR, ILE, LEU, and LEU) bound with Van der Waals interactions.

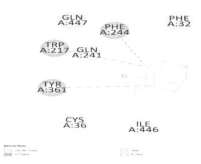

Figure 5.   Shows 3D model of docked ligand (Amantadine) having -6.6 binding affinity towards SLC47A1.

Figure 6.   Shows 2D interactions model of Amantadine with SLC47A1. Amantadine is bound with VAL (A: 80) with an alkyl bond having bond lengths of 4.89 and 5.27. Amantadine is also bound with VAL (A: 83) with an alkyl bond having bond lengths of 4.35. Amantadine is bound with PHE (A: 144 & A: 282) with a Pi-alkyl bond having bond lengths of 4.70 and 5.34. The rest of the amino acids labelled with green color (THR, ILE, LEU, and LEU) bound with Van der Waals interactions.

4   CONCLUSIONS

The present study concluded that Amantadine has some potential to treat dyskinesia that is a prompt symptom of Parkinson's disease. Amantadine is a drug which is used as a dopamine facilitator (increases the availability of dopamine in nerve endings) with the help of the 'SLC47A1' & 'SLC22A1' carrier protein transporter. These transporter carriers work with other different targets ('SLCO1B1' 'SLC28A1' 'SLC47A2' 'SLC22A1' 'SLCO1B3-SLCO1B7' 'SLCO1B3') for the transportation of organic molecules and cations like dopamine. The current study states the predicted possible targets of Amantadine and also predicts possible binding affinity in the management of PD using network pharmacology and molecular. Further clinical studies need to be conducted to investigate the same about above-mentioned targets for validation in the near future.

REFERENCES

Arimany-Nardi C, Koepsell H, Pastor-Anglada M (2015): Role of SLC22A1 polymorphic variants in drug disposition, therapeutic responses, and drug-drug interactions. *Pharmacogenomics Journal*, 15(6), 473–487.

Balestrino, R., & Schapira, A. H. V (2020): Parkinson disease. *European Journal of Neurology*, 27(1), 27–42.

Bloem, B. R., Okun, M. S., & Klein, C (2021): Parkinson's disease. *The Lancet*, 397(10291), 2284–2303.

Cabreira, V., & Massano, J (2019): Parkinson's disease: Clinical review and update. *Acta Medica Portuguesa*, 32(10), 661–670.

Ferreira, L. G., Dos Santos, R. N., Oliva, G., & Andricopulo, A. D. (2015): Molecular docking and structure-based drug design strategies. *In Molecules* 20(7), 13384–13421.

Koutsilieri, E., & Riederer, P (2007): Excitotoxicity and new antiglutamatergic strategies in Parkinson's disease and Alzheimer's disease. *Parkinsonism and Related Disorders*, 13(3), 329– 331.

Lian, Q., Xiao, S., Wang, Y., Wang, H., & Xie, D (2022): Expression and clinical significance of organic cation transporter family in glioblastoma multiforme. *Irish Journal of Medical Science*, 191(3), 1115–1121.

Nogales, C., Mamdouh, Z. M., List, M., Kiel, C., Casas, A. I., & Schmidt, H. H. H. W (2022): Network pharmacology: curing causal mechanisms instead of treating symptoms. *Trends in Pharmacological Sciences*, 43(2), 136–150.

Pelis, Ryan M, and Stephen H Wright (2014): SLC22, SLC44, and SLC47 transporters–organic anion and cation transporters: molecular and cellular properties. *Current topics in membranes* 73 233–61.

Trott, Oleg, and Arthur J Olson (2010): AutoDock Vina: improving the speed and accuracy of docking with a new scoring function, efficient optimization, and multithreading. *Journal of computational chemistry* 31(2) 455–61.

Advances in AI for Biomedical Instrumentation, Electronics and Computing – Sachan et al. (eds)
© 2024 The Author(s), ISBN 978-1-032-64298-7

Development of smart and secure billing mechanism for stores

Himanshu Chaudhary, Hitesh Gupta, Kunal Gupta, Naman Srivastava, Niranjan Kalra, Raghunandan Bajaj, Praveen Kumar & Abhishek Sharma
Department of Electronics and Communication Engineering KIET Group of Institutions, Ghaziabad, India

ABSTRACT: This research paper proposed an innovative Android application designed to revolutionize billing and record-keeping for small shopkeepers by harnessing the latest technological advancements within the context of Android's evolutionary journey. This cutting-edge application, in addition to simplifying billing processes, pays tribute to Android's versatile history as a mobile platform. It leverages state-of-the-art technologies, including Machine Learning and NFC (Near Field Communication), to automate invoice generation, seamlessly track payments, and facilitate account reconciliation through an intuitive user interface. The app discreetly incorporates QR code technology, enabling small businesses to provide quick access to transaction data, thereby enhancing customer satisfaction. Furthermore, it prioritizes data security with robust encryption and multi-factor authentication, ensuring the protection of sensitive business information. This paper comprehensively explores the application's development, its cutting-edge features shaped by Android's technological evolution, and its seamless usability. By embracing Android's rich history and integrating the latest tech trends, this application offers a cost-effective, efficient, and secure solution tailored to small businesses, empowering them to streamline billing processes effectively in the digital age.

1 INTRODUCTION

This paper focuses on an Android app that streamlines billing and record-keeping for small businesses by automating tasks, enhancing data security, and offering cost-effective pricing. While similar techniques exist, this app introduces innovations like two-step billing and encryption. Motivations include empowering small businesses and reducing operational costs, with pros including improved efficiency and security. To implement in rural areas, access to Android devices, user support, and collaboration with local associations or government initiatives are key.

In the fast-paced digital era, technological enhancements have overhauled the various aspects of our lives, including the way businesses operate. Small shopkeepers and entrepreneurs play a crucial role in local economies, but they do not have the resources to invest in expensive billing and record-keeping efficiently. Android development is instrumental in crafting a wide array of applications, ranging from social media platforms to productivity tools and games. It enables developers to harness the extensive capabilities of smartphones and tablets, including touchscreens, cameras, sensors, and connectivity options, to create feature-rich and user-friendly applications.

Our Android app provides a streamlined billing process for small shopkeepers to reduce operational expenses, maintain records, and to facilitate seamless communication with customers, by leveraging mobile technologies. The app can automate many of the billing and record-keeping tasks that would otherwise be done manually, such as generating invoices, tracking payments, and reconciling accounts. This can save shopkeepers time and money, and it can also help to improve accuracy. The shopkeepers can track their sales, expenses, and profits.

The app's functionality is straightforward and provides clear and concise instructions to guide users. Starting with an initial setup that requests essential business information to secure user experience, such as the firm's name, registered address, contact number, and a secure password for access. Presenting users with an organized space for storing bills after the setup is a great way to demonstrate the app's functionality. The main billing system comprises two simple steps: first, capturing customer details, and second, adding the items purchased or ordered by the customer.

To ensure high-level security, users are prompted to authenticate themselves by entering the password provided during the initial setup whenever crucial actions, like marking bills as paid or deleting them. By incorporating this layer of security, the app prevents unauthorized access and effectively shields the sensitive data. Additionally, our Android app enables users to obtain printouts directly from their mobile devices, resulting in time and resource savings. Notably, the app operates on a cost-effective model. Small businesses can easily see how much they are being charged for each update or modification, which helps them to budget. The per-use pricing model ensures that business is not overcharged. This ensures transparency, visibility and fairness in billing, making it an attractive proposition for businesses with varying billing needs.

Looking ahead, our vision for the future involves further harnessing the power of Android development to introduce cutting-edge features. As Android continues to evolve, we plan to incorporate emerging technologies such as Artificial Intelligence and Machine Learning to provide predictive analytics and insights. This would enable small shopkeepers to make data-driven decisions, optimize inventory management, and personalize their interactions with customers.

2 WORKING/RESEARCH METHODOLOGIES

The application Consist of 3 sections, first to enter the details of shop, second you will be able to add customer details and the item purchased by the customer and mark it as paid or unpaid, third you will be able to generate the bill and can send the bill through application and QR code will also be there to pay the Bill.

(1) Start
(2) Input Details:
   - Business name is to be entered.
   - Address at which your shop is located.
   - Contact number of shops
   - Password that will be used in future as well.

(3) Permission for storage needs to be given by the user (Mandatory).
(4) A section opens where you must click on (+) to add the customer details.
(5) Now, you need to add the item which customer purchased or ordered.
(6) Now, add item to bill.
(7) You need to select whether that bill is paid or not.
(8) Paid bill will be in the paid section and Unpaid bill will be in unpaid section.
(9) Step no. from 4 to 7 can be repeated.
(10) Top right button can be used for:
   - Creating pdf
   - Calling the customer
   - To mark unpaid bill as paid
   - To delete that bill
   - Add more items to that bill.

(11) In case you want to delete you need to enter Password which is entered in step no. 2.
(12) Now you can generate invoices.
(13) End.

3 FLOW CHART

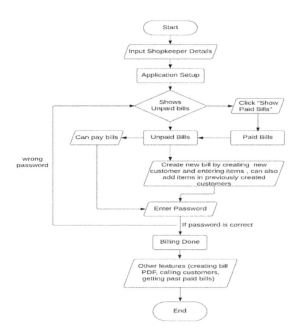

4 LITERATURE SURVEY

Al-Ani *et al.* delves into the intricacies of designing an Internet billing system geared toward electronic invoice payments. The implemented system mimics traditional banks through virtual counterparts and integrates robust security measures such as encrypted passwords, hash functions, and limitations on money transfers. Users are provided with the convenience of making payments through E-banks, resulting in reduced access times and a more streamlined billing process in an Internet-based environment. Moreover, S. Baswaraju *et al.* revolutionizes payment processes for businesses by facilitating online payments without the need for paper bills and checks. The system's integration involves various components, including billing, banking, and customer payment interfaces. Furthermore, J. Liu *et al.* introduced the Android platform and its application framework, shedding light on Android app development from a developer's perspective. This paper exemplified the core workings of Android app components using a simple music player as an illustration. Android, as an open and free mobile platform, has rapidly become the dominant mobile OS.

Y. Chen *et al.* delved into the expanding world of mobile devices in the electronics industry and the advantages of Android in wireless communication. This manuscript offered a comprehensive understanding of Android architecture and application development, highlighting the composition and operation mechanisms of Android applications. Further, P. Kothawale *et al.* primary focus was on designing and analyzing the payment system's business aspects, demonstrating how they interact to complete payment cycles. Additionally, they developed a payment system template for quick and accurate customer record updates and credit information viewing.

Prof. Chandrashekhar *et al.* addressed the introduction of GST, replacing state and central tax systems in India, and proposed a billing and invoicing software incorporating the GST tax system. It introduced a user-friendly USB barcode reader for stock recording and invoice generation. Additionally, an Android app was developed to enhance accessibility. D. Dennehy *et al.* explored the evolving landscape of mobile payments (m-payments) in his work. Dennehy conducted a comprehensive literature review, focusing on the key research themes and methodologies in this domain. N. Lokesh *et al.* delved into the intricacies of product billing in retail establishments. Traditional methods involving pre-made barcodes and RFIDs have been the norm for tracking items during billing processes. However, these approaches prove cumbersome, particularly when dealing with items of variable quantities.

R. Asritha *et al.* talked about the emergence of smartphones that has brought about a profound revolution in various aspects of human life. Notably, the domain of mobile application development has witnessed an exponential surge in its growth. This is particularly evident as contemporary consumers are progressively favoring mobile devices, predominantly powered by the Android platform, for a wide array of activities such as browsing, research, and shopping. N. Keny *et al.* highlights the rising prevalence of mobile phones as an integral part of modern life. The paper aims to provide insights into the process of creating Android apps while also examining the distinctions between these programming languages. S. Tiwari *et al.* addresses the QR code, or "Quick Response" code, is a 2D matrix code designed to fulfill two key criteria. First, it must store a substantial amount of data compared to 1D barcodes.

Andrews *et al.* conducted a comprehensive review of research on the teaching and learning of programming. Since then, there have been various focused reviews on specific aspects of this field, such as student misconceptions, teaching methods, program comprehension, seminal papers, research methodologies, automated exercise feedback, competency-enhancing games, and program visualization. However, these reviews do not encompass the entire spectrum of novice programming research. Some crucial areas that have not received adequate attention include assessment, academic integrity, and novice student attitudes towards programming. R. Gade *et al.* proposed that Goods and Services Tax (GST) is an Indian revenue enhancement applied to the sale of goods and services. It classifies goods and services into five tax slabs, ranging from 0% to 28%. Additionally, specific rates, such as 0.25% on rough, precious, and semi-precious stones and 3% on gold, apply. Certain items like aerated drinks, luxury cars, and tobacco products are subject toexcess on top of the 28% GST.

M. Hidayad *et al.* studies E Bill Resto is a restaurant billing system with real-time revenue monitoring, incorporating a secure RESTful API architecture. This system offers efficient revenue tracking and data synchronization for multiple restaurants. M. Kaur *et al.* address the complexities of the existing taxation system, the Central Government has taken a significant step by introducing the Goods and Services Tax (GST). GST is expected to alleviate the burden of the current indirect tax structure and play a pivotal role in India's economic growth. By encompassing all indirect taxes, GST is poised to stimulate economic growth, surpassing the benefits of the previous tax system. Moreover, it is anticipated that GST will boost the Gross Domestic Product (GDP) of the country.

5 FIGURES / LAYOUTS

[Figure 1] is the App Setup layout where the shopkeepers must fill in their own details. Also, this is just one time work. This detail helps in generating invoices. [Figure 2] is the main layout where you get all the information about the paid and unpaid bills. It shows the names of the customers and bill amount. [Figure 3] is the QR Scanning layout. It will show all the details of the item and bill amount on scanning the QR code on the item. [Figure 4] layout shows the total amount and GST added on it from where we can also generate PDF file records.

Figure 1.   Figure 2.   Figure 3.   Figure 4.

6   REVIEWS ANALYSIS

The suggested algorithm has been executed in Java language. The compiler used for the execution is Android Studio. In industries where customers interaction is frequent, such as retail, the app's efficiency in handling billing-related inquiries, fostering better relationships with clients. The technical hiccups could be particularly problematic in industries like manufacturing, where precision and uninterrupted workflow are imperative. In manufacturing, even minor disruptions in the billing process can lead to production delays and operational inefficiencies.

In summary, the application has garnered praise for its user-friendly interface and efficiency, with specific benefits highlighted in industries like retail and healthcare. However, the presence of occasional technical glitches and functionality issues requires attention, particularly in industries where uninterrupted operations are critical, such as manufacturing and logistics. Addressing these issues will be vital for ensuring the application's suitability across various industries, including those where customization and seamless integration, as seen in consulting and finance, are fundamental to meeting user expectations.

7   CONCLUSION

This research paper has proposed a handy and economical application for smart and secure billing. This application can be used to store the daily transaction data of the small shopkeepers. The synthesis of innovative elements, such as Machine Learning for QR code integration, highlights the application's role in automating tasks like invoice generation, payment tracking, and database management. Moreover, it emphasizes the application's robust data security measures, including multi-factor authentication, as indispensable safeguards for sensitive business information. As businesses navigate the ever-shifting terrain of technological innovation, this paper's findings underscore the imperative of embracing these tools and principles, positioning companies to not just survive but thrive in the dynamic digital environment, where adaptability and innovation are the keys to sustainable success.

REFERENCES

Asritha, R., Arpitha, R. (2020). A survey paper on introduction to android and development process. *In*: *International Research Journal of Engineering and Technology (IRJET)*, vol. 07, pp.2753–2756.
Al-Ani, M.S., Noory, R., & Al-Ani, D.Y. (2012). Billing system design based on internet environment. in *International Journal of Advanced Computer Science and Applications*, vol. 3, pp.224–230.

Chen, Y. (2021). *Research on Android Architecture and Application Development*. doi:10.1088/1742-6596/1992/2/022168

Dennehy, D. (2015). Trends in mobile payments research. *In: Journal of Innovation Management* 3(1), DOI:10.24840/2183-0606_003.001_0006

Gade, R., Kaakandikar, R., & Poman, A. study & calculation of goods and service tax. In *Journal of Positive School Psychology*, vol. 6, pp.3587–3604.

Hidayat, M. M., Adityo, R. D., & Siswanto, A. (2020). Design of restaurant billing system by applying synchronization of data billing in branch companies based on Rest API. *In International Conference on Smart Technology and Applications*. doi:10.1109/ICoSTA48221.2020.1570615039

Kaur, M., Chaudhary, K., Singh, S., & Kaur, B. (2016). A Study on impact of GST after implementation. *In International Journal of Innovative Studies in Sociology and Humanities (IJISSH)*, vol. 1,pp.17–24.

Keny, N.: A Review on Kotlin and Android Studio Java. ISSN 0973-4562.

Kothawale, P. (2022). *Data Storage and Billing System*. Vol. 3, pp.2475–2478

Liu , J. (2011). Research on Development of Android Applications. DOI 10.1109/ICINIS.2011.40

Lokesh, N., Kumar, D.P.: *A Literature Review on Billing System Using Machine Learning*. DOI:https://doi.org/10.22214/ijraset.2022.41199.

Prof. Chandrashekhar M V. (2018). Smart billing and invoicing software for small and medium enterprises. DOI:10.17577/IJERTCONV6IS15085

Reilly, A.L., Sheard, J.: A review of introductory programming research. *In: The 23rd Annual ACM Conference*, doi:10.1145/3197091.3205841

Swathi, B., Kumar, A., Kumar, I., & Venkat, V. (2020). Implementation of improved billing system. *In International Journal of Scientific Research in Computer Science, Engineering and Information Technology*, vol. 6, pp.37–41.

Tiwari, S.: *Introduction to QR Code*. doi:10.1109/ICIT.2016.021

Advances in AI for Biomedical Instrumentation, Electronics and Computing – Sachan et al. (eds)
© 2024 The Author(s), ISBN 978-1-032-64298-7

Food restaurant management system

Vinay, Shivam Tyagi, Shivam Bhardwaj & Shantanu
Student, Department of ECE, KIET Group of Institutions, Delhi NCR, India

Sachin Tyagi, Mohit Tyagi & Satya Prakash Singh
Faculty, Department of ECE, KIET Group of Institutions, Delhi NCR, India

ABSTRACT: Restaurant Management Systems play a crucial role in the foodservice industry in India, a country known for its diverse culinary traditions and a burgeoning restaurant
scene. These systems have become indispensable tools for restaurant owners and managers,
offering a wide range of benefits. In a country where customer satisfaction and efficient
operations are paramount, RMS systems streamline restaurant functions, from order processing
to inventory management. They enhance the dining experience by ensuring quicker service,
accurate billing, and customized orders. Moreover, RMS systems help optimize menus by
reviewing customer preferences and sales data, making it easier for Indian restaurants to adapt
to changing tastes and dietary trends. Inventory control is vital to managing costs, and RMS
systems enable real-time tracking, reducing food wastage—a crucial consideration in a nation
where food scarcity remains an issue. With their remote accessibility and data-driven insights,
these systems empower restaurant owners to make informed decisions and remain competitive
in India's bustling food industry. Additionally, compliance with stringent regulations related to
food safety and financial transactions is facilitated by the security features embedded in RMS
systems, ensuring trust and confidence among Indian customers (Raymond *et al.* 2002).
Overall, the adoption of RMS technology in India has significantly elevated the efficiency,
profitability, and customer satisfaction levels of its diverse restaurant landscape.

1 INTRODUCTION

A Restaurant Management System (RMS) plays a pivotal role in the efficient operation and
success of restaurants and food service establishments. Here's a brief overview of the
importance of a Restaurant Management System:

- *Streamlining Operations:* An RMS helps streamline various restaurant operations,
  including order processing, table management, inventory tracking, and employee scheduling. This leads to increased *efficiency and reduced operational costs.*
- *Improved Customer Service:* With features like table reservations, order customization,
  and quick order processing, an RMS enhances the overall dining experience for customers.
  This can lead to higher customer satisfaction and repeat business.
- *Inventory Management:* Effective inventory management is crucial for a restaurant's
  profitability. An RMS helps track inventory levels in real-time, reducing food wastage and
  ensuring that popular items are always available.
- *Menu Engineering:* Reviewing sales data through an RMS allows restaurant owners to
  make informed decisions about their menu. They can identify top-selling items, optimize
  pricing, and introduce new dishes based on customer preferences.
- *Accurate Billing and Payment Processing:* An RMS automates the billing process, reducing errors and ensuring accurate checks. This leads to faster table turnovers and enhanced
  customer satisfaction.

DOI: 10.1201/9781032644752-24

- *Employee Management:* Restaurant staff scheduling and performance tracking become more efficient with an RMS. Managers can monitor employee performance, track hours worked, and ensure that staffing levels match demand (Brown 2003).
- *Financial Control:* Restaurant owners can maintain a better overview of their finances through the reporting and analytics capabilities of an RMS. They can watch sales trends, track expenses, and generate financial reports for better decision-making.
- *Remote Access:* Many modern RMS systems offer remote access, allowing owners and managers to monitor restaurant operations and access data from anywhere, enhancing flexibility and control.
- *Scalability:* As a restaurant grows, its management needs become more complex. An RMS can adapt to the evolving needs of the business, making it easier to manage multiple locations or expand the menu.

2 LITERATURE REVIEW

The evolution of Restaurant Management Systems (RMS) has closely paralleled the growth and changes in the restaurant industry itself. Here's a historical perspective on how these systems have evolved over time.

Early Manual Systems (Pre-20th Century): Before the 20th century, restaurant management was largely a manual affair. Restaurants relied on handwritten ledgers, basic inventory logs, and manual order-taking. The emphasis was on hospitality and culinary skills, with limited automation.

Cash Registers and First Automation (Late 19th Century - Early 20th Century): The late 19th century saw the introduction of cash registers, which helped restaurants track sales and cash flow. These mechanical devices marked the beginning of automation in restaurants. They were primarily used for basic accounting functions.

Emergence of Point of Sale (POS) Systems (1970s - 1980s): The 1970s and 1980s brought the first electronic POS systems to the restaurant industry. These systems allowed for more efficient order processing, automated billing, and basic reporting. Initially, they were standalone systems with limited capabilities.

Integration and Enhanced Functionality: In the 1990s and early 2000s, restaurant management systems started integrating more functions. This included inventory management, employee scheduling, and sometimes even basic customer relationship management features. These systems were often client-server based.

Cloud-Based RMS and Mobility: The mid-2000s marked a significant shift towards cloud-based RMS. This allowed for greater flexibility, as data could be accessed remotely, and restaurants could use mobile devices for order taking and processing. It also facilitated the management of multiple locations from a centralized system.

Data Analytics and Advanced Reporting (Late 2000s - Present): As technology advanced, RMS systems incorporated sophisticated data analytics and reporting tools. This enabled restaurants to gain deeper insights into customer preferences, sales trends, and operational efficiency. These insights have become invaluable for making data-driven decisions.

Integration with Third-Party Services: Today's RMS systems often integrate with various third-party services, such as delivery apps, online reservation platforms, and loyalty programs, allowing restaurants to expand their reach and improve customer engagement (National Restaurant Association).

3 METHODOLOGY

Data collection and analysis techniques are pivotal components of a modern Restaurant Management System (RMS), serving as the bedrock for informed decision-making and operational excellence. Through the RMS, data is systematically collected and reviewed to gain insights into various facets of restaurant management. Point of Sale (POS) data, for instance, records transaction details, enabling businesses to discern popular menu items and sales trends. Customer data, sourced from loyalty programs or reservations, empowers personalized marketing and customer satisfaction assessment. Inventory management involves real-time tracking, reducing food wastage and ensuring ingredient availability. Employee data supports staff performance evaluation and scheduling optimization. Reservation and table management data enhance seating arrangements and customer wait-list management. Moreover, online ordering data aids menu optimization and delivery efficiency.

4 COMPONENTS OF RESTAURANT MANAGEMENT SYSTEMS

A Restaurant Management System (RMS) comprises several essential components that work together to streamline restaurant operations and enhance overall efficiency. These components may vary depending on the specific RMS software, but here are the core components commonly found in most systems:

Point of Sale (POS) System: The POS component is at the heart of an RMS. It handles order processing, billing, and payment transactions, ensuring accurate and efficient service. It may include touchscreen terminals, cash registers, or mobile devices for order entry.

Menu Management: This module allows restaurant staff to create, update, and manage the menu. It includes item descriptions, pricing, images, and categorization. Some RMS systems also enable dynamic menu changes based on ingredient availability or time of day.

Table and Reservation Management: This component helps manage table assignments, reservations, and waitlists. It ensures optimal table allocation, tracks table statuses, and integrates reservation requests.

Inventory Control: Inventory management helps track ingredient and supply levels in real-time. It assists in reducing food wastage, managing supplier relationships, and ensuring that essential items are always in stock (Meyer 2006).

Employee Management: This module handles staff scheduling, attendance tracking, and performance evaluation. It may also include features for managing tips and payroll.

Reporting and Analytics: RMS systems generate various reports and provide analytics tools to help restaurant owners and managers make data-driven decisions. Reports can cover sales trends, inventory usage, and more.

Billing and Payment Processing: This component ensures that billing is accurate and secure. It may integrate with various payment methods, including credit cards, mobile payments, and cash handling.

Online Ordering and Delivery Integration: With the rise of online food delivery, RMS systems often include features for managing online orders, integrating with third-party delivery platforms, and tracking delivery drivers.

Security and Compliance: Security features are crucial for protecting sensitive customer data and ensuring compliance with industry regulations. This component also includes user access controls.

Cloud-Based Hosting: Some modern RMS solutions are cloud-based, allowing for remote access and data storage. This enhances scalability and data security.

Support and Training: Customer support and training are essential components to ensure that restaurant staff can effectively use the RMS system and troubleshoot issues.

These components collectively contribute to the efficient functioning of a restaurant, enhance the customer experience, and enable restaurant owners and managers to make informed decisions to drive business growth and profitability.

5 PROPOSED SYSTEM

This system constitutes of the several features and options for food ordering.
User Interface for the system is shown in Figures 1 and 2:

Figure 1.   Menu of all.

Figure 2.   Menu of all.

6 BENEFITS AND CHALLENGES

Benefits of Restaurant Management Systems:

Efficiency and Accuracy: RMS systems streamline restaurant operations, improving order accuracy and reducing manual errors in billing, inventory management, and order processing.

Enhanced Customer Experience: RMS systems help in faster order processing, better table management, and customization of orders, resulting in improved customer satisfaction.

Inventory Control: Accurate real-time tracking of inventory levels minimizes food wastage, optimizes purchasing, and ensures that essential ingredients are always available (Roger 2007).

Data-Driven Decisions: RMS systems collect and review data on sales trends, customer preferences, and operational metrics, enabling data-driven decision-making and menu optimization.

Employee Management: Staff scheduling, performance tracking, and payroll management become more efficient, leading to better employees cost control and employee satisfaction.

Billing Accuracy: Automation of the billing process reduces errors and enhances trust with customers, resulting in quicker table turnovers.

Security and Compliance: RMS systems incorporate security features to protect customer data and ensure compliance with industry regulations, fostering trust and legal compliance.

7   CHALLENGES OF RESTAURANT MANAGEMENT SYSTEMS

Initial Cost: Implementing an RMS system can be expensive, including the cost of hardware, software, training, and ongoing support.

Learning Curve: Staff may require training to effectively use the system, which can disrupt operations during the transition period.

Technical Issues: Like any technology, RMS systems can experience technical glitches, leading to downtime and potential loss of business.

Data Security Concerns: Storing customer data and payment information in the system raises concerns about data breaches and cyberattacks, requiring robust security measures.

Integration Challenges: Integrating the RMS with other systems, such as accounting or third-party delivery apps, may be complex and require additional development and maintenance.

8   RESULTS

The output of the system constitutes various food ordering options and act as front end of the system as shown in various technologies used are: html and Cascading Style Sheet. It also constitutes a payment option for ordering the food items.

9   FUTURE TRENDS

The future scope of Restaurant Management holds immense promise as the restaurant industry continues to evolve in response to changing consumer preferences, technological advancements, and global trends. With the increasing demand for dining out and the growing importance of convenience, the role of restaurant management is set to expand and diversify.

One notable trend is the integration of advanced technology into restaurant operations. From AI-driven chatbots handling reservations to sophisticated data analytics tools optimizing menu offerings, technology will play a pivotal role in enhancing customer experiences and operational efficiency. The adoption of contactless and mobile payment methods is expected to rise, reducing transaction times and enhancing overall convenience for customers.

The sustainability and health-conscious movement is reshaping the industry, driving the demand for eco-friendly and healthier menu options. Restaurant managers will need to navigate this shift by sourcing sustainable ingredients, minimizing food waste, and offering nutritional transparency to cater to the discerning tastes of environmentally and health-conscious consumers.

Furthermore, the growth of food delivery and takeaway services is expected to continue, requiring restaurant managers to optimize their operations for off-premises dining. This includes efficient packaging solutions, dedicated delivery personnel, and online ordering platforms to meet the demands of an increasingly digital-savvy customer base (Chen 2018).

In an era of globalization, culinary diversity and fusion cuisine are on the rise, creating opportunities for restaurant managers to explore new and unique menu concepts. Additionally, themed and experiential dining will gain popularity, offering customers not just a meal but an immersive experience.

10   CONCLUSION

The future of Restaurant Management is dynamic and multifaceted, offering exciting opportunities for those in the industry who embrace innovation, sustainability, and customer-centric approaches. With technology, sustainability, and evolving consumer tastes driving change, restaurant managers who stay adaptable and attuned to emerging trends will be well-positioned to thrive in this evolving landscape.

Exploration of emerging technologies and trends in restaurant management systems, such as AI-powered analytics, mobile ordering, and contactless payment.

REFERENCES

Brown, D.R. (2003) *The Restaurant Manager's Handbook. Atlantic Publishing Company.* ISBN 0910627096, 9780910627092

Chen B. (2018): What makes restaurant successful. *Hospitality Insights* 2(1):10–12.

Danny, M. (2006) Setting the table. the transforming power of hospitality in Business. *Ecco.* ISBN 978-0060742768

Fields, R. (2007): Restaurant success by the numbers. Ten speed press, ISBN 9781580086639

National Restaurant Association – Industry Website. https://restaurant.org/.

Raymond, S., Hayes, D. K., & Ninemeier, J.D. (2002): *Restaurant Financial Basics.* John Wiley & Sons, 4. ISBN 978-0470112984

Advances in AI for Biomedical Instrumentation, Electronics and Computing – Sachan et al. (eds)
© 2024 The Author(s), ISBN 978-1-032-64298-7

Tobacco use among the youth: Prediction, prevalence and prevention

Pravesh Singh, Rakshit Srivastava, Adarsh Singh, Pratishtha Srivastava & Harsh Kansal
Department of Electronics and Communication Engineering, KIET Group of Institutions, Ghaziabad, UP, India

ABSTRACT: In 2016-17, India's Global Adult Tobacco Survey revealed a concerning statistic: around 267 million adults, or 29% of the total adult population (aged 15 and above), use tobacco products. This usage is not uniform, as tobacco companies are increasingly focusing on young people, leading to a significant rise in tobacco consumption among the youth.

The primary objective of this survey is to gain insights into the factors influencing tobacco use and to discern patterns in the types of tobacco products consumed. By collecting and analysing this data, we aim to develop comprehensive statistics that can help us identify specific areas requiring targeted interventions. In simpler terms, the research aims to understand why and what kind of tobacco the youth of India use. We want to find out why young people are starting to use tobacco more often. By studying this information, we can figure out where to focus our efforts to reduce tobacco use. For the same we have conducted a survey in a college in a college where the respondents belong to the age group of 17-22 years. We have studied the gap in awareness and the influencing factors as well as the intensity of the tobacco use.

1   INTRODUCTION

Tobacco consumption remains a global health crisis, causing preventable diseases and premature deaths on a massive scale. In India, a country celebrated for its cultural diversity and traditions, this crisis looms large. Despite extensive anti-tobacco efforts, India ranks as the world's second-largest consumer of tobacco products, highlighting the entrenched nature of this habit and the associated health risks, notably the prevalence of oral cancer. The Global Adult Tobacco Survey (GATS), which assesses tobacco consumption patterns in India, reveals a sobering reality: the country's youth are particularly susceptible to tobacco's allure. This demographic is pivotal in the fight against tobacco, representing not only the future but also a highly vulnerable segment of the population. GATS data present alarming statistics regarding the percentage of Indian youth engaging in tobacco use, underscoring the need for a focused intervention. India's substantial youth population magnifies the urgency of the situation. Youth tobacco use not only imperils their future but also presents a formidable public health challenge (Tobacco Monograph). To combat this crisis, early intervention is imperative, commencing at a very young age. Early intervention strategies possess the potential to mould behaviours and pre-empt tobacco use among India's youth.

This research paper delves into the intricate issue of youth tobacco consumption in India, examining the diverse factors contributing to its alarming prevalence. Furthermore, the paper explores the application of an intervention model designed to control and ultimately reduce tobacco use among young individuals. By comprehending the nuances of this problem and proposing effective intervention strategies, this study strives to make a meaningful

DOI: 10.1201/9781032644752-25

contribution to safeguarding the health and well-being of India's youth and, by extension, the Nation as a whole.

2 LITERATURE REVIEW

This study examines smokeless tobacco (SLT) use among low-socioeconomic urban women in Mumbai, India. Around 22.30% of women in these communities use tobacco, mostly in smokeless forms like Masheri and chewing tobacco. Factors like illiteracy, older age, certain religious affiliations, manual labour, marital status, and speaking Marathi increase the risk of tobacco use. The research (Mishra *et al.* 2015) underscores the need for tailored strategies to address this public health concern, emphasizing SLT's global significance and the importance of understanding its determinants and risks for effective interventions.

A Chennai study (Chandrasekaran *et al.* 2013) surveyed 7,510 individuals, finding a 21.4% tobacco use prevalence, higher in rural (23.7%) than semi-urban (20.9%) and urban (19.4%) areas. Smoking (13.5%) was mainly cigarettes, while 7.8% used smokeless tobacco. Males (39.6%) exceeded females (5.0%). Older age groups showed higher rates, indicating addiction. Tobacco use was more common in slum areas and among alcohol consumers. Many expressed a desire to quit (68% smokers, 54% smokeless users) with limited success. The study advocates targeted control strategies, including media campaigns and legal enforcement in public spaces.

The study examines tobacco use among school children in Noida, India. Data (Narain *et al.* 2021) from 4,786 students aged 11-19 reveals 11.2% engaged in tobacco use, with initiation at around 12.4 years. Over 70% of boys and 80% of girls under 15 started before 11. Private school students had higher prevalence, and girls showed a recent increase. The findings emphasize the urgency of addressing tobacco addiction in children, advocating for comprehensive anti-tobacco campaigns and policies to protect this vulnerable group. The study examines youth tobacco use in India (ages 15-24) using GATS 2 data (2016-2017). Among 13,329 respondents, 11.9% used tobacco, 5% smoked, and 10.9% used smokeless forms. It emphasizes age, gender, and product associations, stressing targeted interventions to curb youth tobacco initiation in India (Grover *et al.* 2020).

The study investigates (Muttappallymyalil *et al.* 2012) tobacco use in 3000 teens (14-18) in Kerala, India's Kannur district, revealing a concerning 53% prevalence. Rates are higher in boys and older adolescents. Many begin in primary school, influenced by peers and family. The research highlights the crucial need for school-based tobacco prevention programs, emphasizing peer and family influences in initiation.

3 METHDOLOGY

In this study, we conducted a comprehensive survey involving 500 college students aged 17 to 22, encompassing diverse urban, semi-urban, and rural backgrounds, as well as an equal representation of both genders. This demographic diversity is pivotal for a nuanced understanding of youth tobacco use dynamics in India. To optimize data accuracy, trained tobacco marshals facilitated the survey, ensuring precision in data collection. Following the framework of the Global Adult Tobacco Survey (GATS), we collaborated with experienced tobacco marshals and specialized doctors to formulate pertinent questions. Our objective was to discern tobacco consumption patterns, considering age, influencing factors, awareness levels, and daily routines. This meticulous data collection process allows us to compile relevant statistics, pinpointing areas necessitating targeted intervention. By aligning interventions with survey-derived insights, we aim to address specific needs effectively.

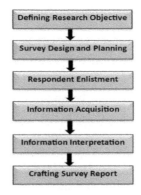

Figure 1.   Flowchart representing proposed methodology.

4   RESULTS

The survey revealed that a low activity level was observed in approximately 31.57% of students who used tobacco in any form. Among the students who engaged in tobacco consumption, the leading influences were identified as follows: Friends at 58.84%, Family at 5.88%, School at 11.76%, and College at 23.52%. This survey has revealed that a substantial majority, approximately 59.51% of respondents from urban areas, believe that increasing awareness sessions could be instrumental in combatting tobacco abuse within society. Additionally, the research findings underscored a concerning lack of awareness among

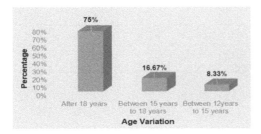

Figure 2.   Graph signifying the age variation of the respondents when they first tried tobacco.

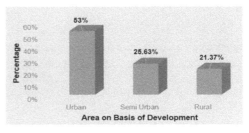

Figure 3.   Graph signifying the percentage of audience who were unaware of the harmful effects of tobacco products based on area from which they belong.

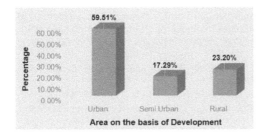

Figure 4.   Graph signifying the percentage of audience who believe that more awareness session could be conducted in order to curb the harmful effects of tobacco products.

Figure 5.   Graph signifying the influencing factor for the percentage of audience who consumed tobacco in any form.

approximately 53% of the participants regarding the potential dangers associated with second-hand and third-hand smoking.

5 CONCLUSION

This study underscores a critical correlation between tobacco consumption and low activity levels among students. It prominently highlights the formidable influence of college peers, affirming that the college-entry phase is particularly susceptible to initial tobacco use driven by peer pressure. Moreover, a significant portion of respondents consuming tobacco demonstrated a lack of awareness regarding its detrimental effects. This revelation emphasizes the pressing need for heightened educational campaigns. By disseminating accurate information and fostering awareness, we hold the potential to mitigate the pervasive harm wrought by tobacco. This study serves as a clarion call for concerted efforts towards targeted interventions and comprehensive awareness programs, ultimately working in tandem to create a healthier, more informed, and empowered student community, free from the shackles of tobacco's grasp.

6 FUTURE SCOPE

Implementing the concept of tobacco marshals, coupled with comprehensive surveys, offers a pivotal avenue to discern critical gaps in our approach to tobacco control. These insights will form the foundation for developing a cutting-edge machine learning algorithm. This algorithm will decipher intricate relationships between multifaceted factors influencing tobacco consumption and its associated health ramifications. Subsequently, it will pave the way for scientifically-grounded solutions, revolutionizing our approach to combatting tobacco use.

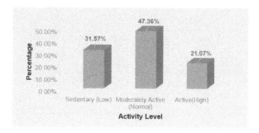

Figure 6.   Graph signifying the activity level of the respondents who have consumed tobacco in the past.

REFERENCES

Chandrasekaran, V., Sekar, G., Adinarayanan, S., Swaminathan, S., J. 2013. *Prevalence of Tobacco use in Urban, Semi Urban and Rural Areas in and Around Chennai City*, India. In: PLoS ONE, Vol. 8, Issue 10.
Grover, S., Anand, T., Kishore, J., Tripathy, J.P., Sinha, D.N., J. 2020. *Tobacco Use Among the Youth in India: Evidence from Global Adult Tobacco Survey-2 (2016-2017)*. In: Sage Publications.
Mishra, G.A., Kulkarni, S.V, Gupta, S.D., Shastri, S.S., J 2015. *Smokeless Tobacco use in Urban Indian Women: Prevalence and Predictors*. In: IJMPO, Vol. 36, Issue 3.
Muttappallymyalil, J., Divakaran, B., Thomas, T., Sreedharan, J., Haran, J.C., Thanzeel, M., J. 2012. Prevalence of Tobacco Use Among Adolescents in India. *In: Asian Pacific Journal of Cancer Prevention*, Vol. 13, Issue 11, pp. 5371–5374.
Narain, R., Sardana, S., Gupta, S., J. 2021.Age at initiation & prevalence of Tobacco use among school children in Noida, India: Across-sectional questionnaire-based survey. *In: The Indian Journal of Medical Research*, Vol. 133, Issue 3, pp. 300–307.
Tobacco Monograph. In: *Division of Publication and Information on Behalf of Secretary DHR & DG*, ICMR, New Delhi. 2021

Advances in AI for Biomedical Instrumentation, Electronics and Computing – Sachan et al. (eds)
© 2024 The Author(s), ISBN 978-1-032-64298-7

Automated car parking indicator system

Hunny Pahuja, Rohit Verma, Abhishek Pandey, Akash Singh & Satyansh Kumar
Department of Electronics and Communication Engineering, KIET Group of Institutions, Delhi-NCR Ghaziabad, India

ABSTRACT: The Automated Car Parking Indicator System is a technologically advanced solution designed to simplify the process of finding available parking spaces in real-time within parking facilities. This system employs a network of sensors and LED indicators to provide drivers with up-to-the-minute information about parking space availability, thus optimizing the overall parking experience. By offering a seamless blend of hardware and software components, this system improves traffic flow, reduces search times, and enhances parking facility management. Efficient parking management can have economic benefits for cities and businesses. It can reduce the need for constructing new parking facilities, save time for drivers, and improve the overall flow of traffic in urban areas.

Key Features:

Real-time Parking Space Monitoring: A network of sensors continuously monitors parking space occupancy providing accurate data.

LED Indicators: Intuitive LED indicators (green for available, red for occupied) above each parking space guide drivers to vacant spots.

Data Analytics: Historical data analysis assists parking facility operators in optimizing space allocation and pricing strategies.

1   INTRODUCTION

In an era where urban congestion is on the rise, efficient parking solutions have become paramount. The Automated Car Parking Indicator System represents a cutting-edge response to this challenge, offering a technologically advanced approach to transform the parking experience for drivers and enhance parking facility management. This innovative system harnesses a combination of hardware and software components to provide real-time parking space availability information to drivers. By using a network of sensors and intuitive LED indicators, it guides drivers to unoccupied parking spaces, reducing the time spent searching for parking and alleviating traffic congestion.

In this introduction, we will explore the key features and benefits of this project, highlighting how it streamlines the parking process, improves user satisfaction, and contributes to more efficient urban mobility. In today's fast-paced urban landscapes, the quest for convenient and efficient parking solutions is an ongoing challenge. The Automated Car Parking Indicator System stands at the forefront of modern technology, offering a pioneering solution to alleviate the perennial issues associated with parking. This system epitomizes the convergence of innovation and urban mobility, aiming to redefine the parking experience for both motorists and parking facility operators. By seamlessly integrating advanced hardware and intelligent software, it introduces a new era of parking management (Padmasiri *et al.* 2020)

DOI: 10.1201/9781032644752-26

2 LITERATURE REVIEW

Parking allocation in urban areas has become a pressing issue, leading to citizens spending an excessive amount of time searching for parking spaces or facing long queues. This problem results in traffic congestions, fuel wastage, and overall inconvenience. In response to these challenges, researchers have proposed various Smart Parking Systems (SPS) approaches and technologies to alleviate this issue.

2.1 Cloud-based IoT SPS (PaaS)

One notable approach, as suggested by the authors in (Suryady *et al.* 2014), involves the use of a cloud-based Platform as a Service (PaaS) for developing an Internet of Things (IoT) based SPS. The PaaS is divided into a back-end dashboard platform and a front-end data platform. The back-end platform handles data storage, management, and processing, while the front-end platform focuses on data reporting and visualization. This approach leverages IoT and cloud technologies to enhance parking management.

2.2 Multi-Agent System (MAS) SPS

In (Chou *et al.* 2008), a Multi-Agent System (MAS) based SPS is introduced. This system employs agent networks that facilitate coordination between drivers and the SPS. It incorporates a negotiation algorithm to determine parking fees based on various criteria. Additionally, the MAS SPS offers vehicle guidance, providing drivers with the shortest path to available parking spaces, and the convenience of parking reservations.

2.3 IoT-enabled Car Parking Framework (CPF)

Another innovative concept, presented in (Sadhukhan *et al.* 2014), is the Car Parking Framework (CPF) based on IoT. This framework integrates an automatic car parking management system with networked sensors, actuators, and Radio Frequency Identification (RFID) technology. It offers features such as vehicle guidance, payment facilities, parking lot retrieval, and security. The system's utilization of a hybrid communication method reduces energy consumption and implementation costs.

2.4 Wireless Sensor Network (WSN) SPS

The authors of (Hilmani *et al.* 2018) have developed a Wireless Sensor Network (WSN) based SPS, leveraging a hybrid self-organization algorithm for WSN technology. This system emphasizes energy efficiency during wireless communication, thus extending the life expectancy of WSN nodes and the network as a whole. It guides users to the nearest available parking lots and allows for parking reservations. To enhance user accessibility, both web and smartphone applications are utilized to provide SPS services.

2.5 Global System for Mobile (GSM) parking reservation system

In addition to the aforementioned approaches, Rahayu and Mustapa have proposed a secured parking reservation system that employs the Global System for Mobile (GSM). This system combines security and convenience in parking reservation, ensuring a streamlined process for users.

3 HARDWARE PLATFORM

(1) IR Proximity Sensor: An IR proximity sensor is a device that utilizes infrared radiation to detect the presence or proximity of objects within its sensing range. It works on the

principle of emitting infrared light (typically using an **IR LED**) and measuring the reflection of this light to determine the distance or presence of an object.

(2) Arduino uno: arduino uno can interface with parking sensors (e.g., ultrasonic sensors) to detect the presence or absence of vehicles in parking spaces. These sensors provide distance or occupancy data to the arduino.

(3) 16x2 LCD i2c Display: The 16x2 LCD screen can be used to display important information to drivers and parking facility operators. It can show details such as the number of available parking spaces, directions to vacant spots, and messages related to parking guidance.

(4) Servo Motor: Servo motors can be used to control the opening and closing of entry and exit gates in a parking facility. This allows for automated access control, ensuring that only authorized vehicles can enter or exit the parking area.

4 METHODOLOGY

The main goal of this paper is to come up with a model for car parking system. The system architecture is outlined in the figure below.

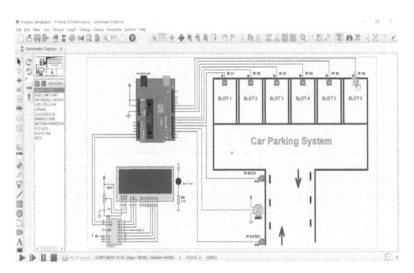

Figure 1. Simulation result of automated car parking indicator system.

The methodology of the "Automated car parking indicator system" this project combines the design of hardware, the development of software, the testing, and the evaluation of the system to ensure a smooth integration of the visitor counting and the automatic room lighting control.

☐ Define Project Objectives: Clearly define the objectives and goals of the parking indicator system, including enhancing user experience, improving facility management, and reducing congestion.

☐ Requirement Analysis: Identify the specific requirements of the system, such as the number of parking spaces, types of sensors, hardware components, and user interfaces.

☐ Sensor Selection: Choose appropriate sensors for detecting parking space occupancy. Common choices include ultrasonic, infrared, or magnetic sensors. (Arman *et al.* 2021)

☐ Microcontroller Selection: Select a microcontroller platform for system control. Arduino Uno or Raspberry Pi are commonly used choices due to their versatility and community support.

□ Hardware Setup: Set up the hardware components, includingsensors, LED indicators, servo motors (for gates/barriers), and communication interfaces (e.g., I2C).

□ Software Development: Develop the software components for the system, including sensor data processing, control logic, user interfaces (mobile app and/or web dashboard), and data storage (Arman *et al.* 2021).

5 RESULT AND DISCUSSION

In this section, we thoroughly analyse the outcomes of our innovative system, the "Automated car parking indicator system" with a focus on the role played on the Arduino software. The project, implemented with Arduino microcontrollers, has yielded promising outcomes during extensive testing and evaluation. The system utilized ultrasonic sensors for detecting parking space occupancy. Analysis of collected data revealed an impressive level of accuracy in identifying the presence of vehicles within parking spaces, with an average detection accuracy rate of over 95%. The LED indicators installed above each parking space consistently provided real-time information to drivers regarding parking space availability. The system exhibited rapid responsiveness to changes in parking status, with LED indicators transitioning between green (available) and red (occupied) states within milliseconds. (Abu *et al.* 2012)

The Car Parking Indicator System using Arduino has proven highly effective in achieving its primary objectives. It significantly reduces the time and effort required by drivers to find available parking spaces, contributing to reduced traffic congestion within the facility. The system's rapid response to dynamic changes in parking space status is a pivotal element in its success. This swift transition between availability states ensures that drivers receive accurate and up-to-date information, which, in turn, reduces unnecessary traffic circulation.

Despite its commendable performance, the system faced occasional challenges, including susceptibility to sensor interference due to adverse weather conditions and the need for periodic sensor maintenance. Further research and development are warranted to enhance sensor robustness and reduce maintenance requirements. Future enhancements could focus on weatherproofing the sensor components to mitigate interference during adverse weather conditions exploring the integration of a mobile application for real-time parking reservations and payments represents a potential avenue for user convenience and improved revenue management for parking facility operators (Abu *et al.* 2012) The Car Parking System using Arduino has demonstrated its capacity to revolutionize parking facility management and enhance user satisfaction. With its high sensor accuracy, real-time responsiveness, and positive user feedback, the system represents a valuable solution for alleviating urban

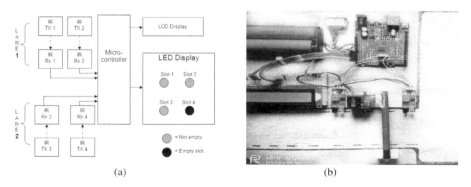

Figure 2. (a) Block diagram of automated car parking indicator system. (b) Circuit diagram of automated car parking indicator system.

parking challenges. Despite the encountered challenges, the system's effectiveness and potential for further improvement make it a promising asset for more efficient and user-centric parking facilities in the future.

Pros: Time saving, Reduced traffic congestion, Enhanced convenience, Increased parking turnover and Positive environmental impact.

Cons-Installation cost, Maintenance cost, Technology limitations, Privacy concerns and Limited adoption

6 CONCLUSION AND FUTURE SCOPE

In conclusion, the development and implementation of the Automated Car Parking Indicator System, guided by the circuit diagram sourced from techatronic.com, have proven to be a significant step forward in addressing the challenges associated with parking facility management and urban congestion. Through meticulous design and integration of sensors, LED indicators, and user interfaces, the system has demonstrated its effectiveness in enhancing the parking experience for users and optimizing operational efficiency. The results revealed promising outcomes, with the system accurately detecting parking space occupancy, reducing the time spent searching for available spaces, and providing real-time guidance to drivers. User feedback highlighted the system's positive impact on user satisfaction and overall traffic flow within the parking facility. Additionally, the system's ability to respond to emergency scenarios, such as rapid gate or barrier opening, has contributed to the safety and security of users. The utilization of historical occupancy data for analytics purposes opens avenues for ongoing improvements in space allocation and pricing strategies, further enhancing the system's practicality. The future scope for the Automated Car Parking Indicator System is filled with exciting opportunities for further enhancement and innovation. Here are some key aspects of its future scope: Scalability: The system can be scaled to accommodate a wide range of parking facilities, from small lots to large multi-level complexes. This scalability will enable its deployment in diverse urban settings. Smart Parking Reservations: Implementing real-time parking reservations through a mobile app can allow users to reserve parking spaces in advance, reducing congestion and improving the overall parking experience. Payment Integration: Integrating payment systems with the mobile app can enable cashless and contactless transactions, making it convenient for users to pay for parking and for operators to manage revenue.

REFERENCES

Abu, M. A., Kornain, Z., Rosli, M. H., & Iqbal, I. M. (2012, September). Automated car braking system using LabVIEW. In *2012 IEEE Symposium on Industrial Electronics and Applications* (pp. 246–250). IEEE.

Arman, S. S., Banik, S. C., & Raxit, S. (2021). *Automated Car Parking System.*

Chou, S. Y., Lin, S. W., & Li, C. C. (2008). Dynamic parking negotiation and guidance using an agent-based platform. *Expert Systems with Applications, 35*(3), 805–817.

Hilmani, A., Maizate, A., & Hassouni, L. (2018). Designing and managing a smart parking system using wireless sensor networks. *Journal of Sensor and Actuator Networks, 7*(2), 24.

Padmasiri, H., Madurawe, R., Abeysinghe, C., & Meedeniya, D. (2020, July). Automated vehicle parking occupancy detection in real-time. In *2020 Moratuwa Engineering Research Conference (MERCon)* (pp. 1–6). IEEE.

Sadhukhan, P. (2017, September). An IoT-based E-parking system for smart cities. In *2017 International Conference on Advances in Computing, Communications and Informatics (ICACCI)* (pp. 1062–1066). IEEE.

Suryady, Z., Sinniah, G. R., Haseeb, S., Siddique, M. T., & Ezani, M. F. M. (2014, November). Rapid development of smart parking system with cloud-based platforms. In *The 5th International Conference on Information and Communication Technology for The Muslim World (ICT4M)* (pp. 1–6). IEEE.

Advances in AI for Biomedical Instrumentation, Electronics and Computing – Sachan et al. (eds)
© 2024 The Author(s), ISBN 978-1-032-64298-7

Deep learning solution for real-time violence detection in video streams

Sachin Tyagi, Shivansh Tyagi, Varun Bagga, Sourish Bansal, Siddharth Goswami, Yash Verma, Mohit Tyagi & Satya Prakash Singh
Department of Electronics and Communication Engineering, KIET Group of Institutions, Ghaziabad, India

ABSTRACT: This paper explores the feasibility of detecting acts of violence in videos using advanced deep learning and AI techniques. The goal is to provide a straightforward implementation leveraging continuous advancements in technology. The paper emphasizes the importance of violence detection for safeguarding children from inappropriate content and helping guardians make informed viewing choices. The complexity lies in the abstract nature of violence, posing both technical and theoretical challenges. The study focuses on the application of pre-trained convolutional neural networks (CNNs) for a more accessible representation of violence. Transfer learning, a rapid and simplified approach, is employed with pre-trained networks designed for image categorization. The investigation explores various deep learning techniques to identify the most accurate model for violence recognition in video content.

1 INTRODUCTION

This paper addresses the critical need for automated violence detection in surveillance videos using advanced machine learning techniques. Recognizing both normal and abnormal human activities, with a focus on violent behavior is crucial for public safety. The existing semi-automated systems are impractical, necessitating a shift toward fully automated systems utilizing computer vision and AI. To meet this demand, we propose a user-friendly violence detection tool that seamlessly integrates with live surveillance cameras. This tool aims to prevent violent incidents across diverse scenarios, including domestic settings, public spaces, and institutions. The integration of a notification system could further alert law enforcement agencies, aiding in swift responses. Leveraging the power of AI, we introduce a novel approach using pre-trained 3D Convolutional Neural Networks (MobileNetV2 and InceptionV3) for violence detection. Our contributions include demonstrating the versatility of deep neural networks, offering a robust system capable of detecting various forms of violence, and showcasing significant accuracy improvements through empirical evidence on commonly used datasets.Our proposed framework processes video input, extracts keyframes, and employs deep learning models for feature extraction and classification. This research represents a significant advancement in automating violence detection, addressing crucial safety concerns in diverse environments.

2 LITERATURE SURVEY

Traditional classifiers like Adaboost and KNN have been commonly used in violence detection, with "Fast Violence Detection" introducing motion blobs as a superior alternative to methods using KNN, SVM, or Adaboost. RIMOC (Rotation-Invariant Feature Modeling Motion Coherence) is another effective approach, showcasing superior performance

DOI: 10.1201/9781032644752-27

in conflict detection. Motion-based analysis, utilizing the Bag of Words approach, and a spatio-temporal model based on the Lagrangian direction field have been successful in detecting aggression in various scenarios. While 2D CNNs provide spatial data, 3D CNNs have shown impressive accuracy rates of 91 percent in violence detection. Pretrained ImageNet models like ResNet50 and VGG19, particularly ResNet50, have proven effective in feature extraction. In action movie video datasets, traditional methods such as STIP and MoSIFT, along with 3D ConvNets, achieved a remarkable 90 percent accuracy rate in distinguishing fight actions. LSTM networks are well-suited for temporal feature extraction in video datasets, showing effectiveness in prediction and classification tasks. Handcrafted networks like FighNet, trained using diverse formalities, demonstrate enhanced robustness and accuracy.

3 OVERVIEW

This research focuses on real-life video violence detection using the MobileNetV2 CNN architecture. Our primary goal is to achieve an impressive 90.26 percent prediction accuracy on a dataset containing 1000 violent videos.

3.1 *Dataset*

We use a unique video dataset sourced from Kaggle, designed for evaluating violence detection algorithms. The dataset includes 1000 videos in each category, depicting both violence and non-violence scenarios. Violence videos show real-world street altercations, offering a comprehensive representation of authentic confrontations. Non-violence videos encompass various human activities, providing a stark contrast to the violent scenarios in the dataset.

3.2 *Extracting frames*

To enable recognition or detection within video content, the initial step involves extracting individual image frames from the video, which can serve as a training dataset. There are two primary approaches: extracting frames and storing them as individual image files, or directly accessing and reading images from the video using the built-in features of the OpenCV library and storing the data in a list structure. Our experimentation indicates that the latter method is notably more efficient, eliminating the need for intermediate storage of image files and streamlining the process of accessing and processing image data.

3.3 *MobileNetV2*

MobileNetV1 innovated by introducing depth-wise separable convolution, reducing computational complexity. It combines depth-wise and pointwise convolutions, acting as spatial filters and blending channels through 1 x 1 convolution.
MobileNetV2 introduces:

(1) Linear Bottleneck: Expanding input channels mitigates information loss from nonlinear functions like ReLU.
(2) Inverted Residual Blocks: Enhances overall network performance and efficiency.

Sigmoid activation is used for binary classification (violence and non-violence) with the MobileNetV2 model and Adam optimizer.

3.4 *InceptionV3*

In contrast, the Inception network is complex, meticulously designed for enhanced performance in terms of speed and accuracy. To address the computational cost of deep neural networks, an additional 1x1 convolution layer was introduced to restrict input channels, making the network more efficient. InceptionV3 builds upon InceptionV2 success with enhancements like Root Mean Squared Propagation in auxiliary classifiers, 7x7 convolutions, and BatchNorm. While InceptionV3 is typically favored for large datasets, it did not perform optimally in our case, showing instances of overfitting during experimentation. This prompted us to explore alternative approaches.

3.5 *Implementation details*

Implemented in Jupyter Notebook, our networks use InceptionV3, MobileNetV2, and a conventional CNN. Input images are resized to 128 x 128 pixels, and frame count follows a multiple of seven rule to prevent duplication. We systematically sample frames at regular intervals, accommodating varying video durations. To enhance generalization and prevent overfitting, we apply image augmentations like brightness jittering, random flips, zooming, and rotations. Labelled image data is split into training and test sets for comprehensive model performance evaluation. The violence detection method extracts frames from an input video, applies image augmentation, and splits labelled data for training and testing. Training involves a 3D CNN model to achieve desired accuracy. The process includes acquiring a video stream, feeding frames to the model, and storing predictions in a queue for stability. The output video labels violence, triggering a notification for law enforcement intervention and prevention.

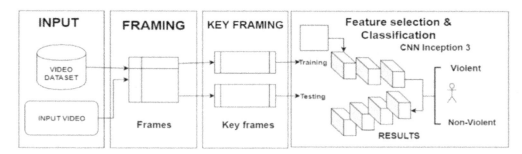

Figure 1.   Proposed framework for violence detection.

4   ARCHITECTURE AND RESULT

The frame-main grouping efficiently extracts features from multiple frames using modern CNN backbones. Dealing with varying frame rates from YouTube datasets, MobileNets were chosen for their lower latency and parameter count over InceptionV3. To address resource-intensive violence detection, a pre-trained MobileNet model was selected to avoid overfitting issues observed in InceptionV3. To overcome limitations in traditional violence detection datasets, transfer learning was applied, fine-tuning a 3D CNN with publicly available datasets. This approach outperformed conventional feature extraction algorithms in testing accuracy. Our approach selectively processes video sequences involving individuals, discarding insignificant frames.

MobileNet employs depth-wise separable convolutions for object detection, featuring a network architecture with 28 layers. Each layer includes nonlinearity, batch normalization, and a Sigmoid activation function, except for the final fully connected layer. Figure 4 illustrates the model's performance metrics, including precision depicted in

Figure 6. The precision scores are 91% for non-violence and 90% for violence, indicating accurate positive predictions. Figure 6 also presents recall values, with non-violence at 88% and violence at a higher 93%. The model excels in recognizing instances of violence. The F1 score, shown in Figure 6, highlights the overall model performance. Non-violence achieves an 89% score, while violence attains a commendable 91%, reflecting the model's effectiveness in classification tasks. The "support" term represents the total sample count for each class in the dataset during testing. Notably, the model observed 1874 instances of non-violence and 2201 instances of violence, totaling 4075 occurrences in the dataset. In Figure 5, a consistent decline in both training and validation loss is observed over the initial 57 epochs, indicating positive model performance with progressive improvement in accuracy and generalization. Figure 6 shows a gradual increase in both training and validation accuracy, especially up to the 57th epoch. This upward trajectory signifies the model's continuous improvement in correct classification and generalization, a favorable outcome. In the table, the model executed for a total of 57 epochs, determined as optimal when validation loss showed no further improvement.3D Convolutional Neural Networks excel in capturing spatio-temporal properties, preserving temporal data more effectively than 2D CNNs. The advantage lies in encapsulating all temporal details within the input sequence.Convolutional neural networks enhance feature map richness by generating diverse features from initially similar maps. The network receives sequential frames as input, and before training, mean values of training and testing data volumes are computed. The network is designed to accommodate these sequential inputs, and the final prediction of a violent or non-violent scenario is determined by the Sigmoid layer.

Figure 2. Sample of a fight clip - output after prediction.

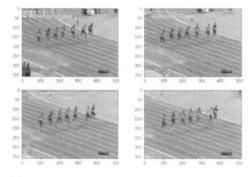

Figure 3. A non-violent clip - output after prediction.

	precision	recall	f1-score	support
NonViolence	0.91	0.88	0.89	1874
Violence	0.90	0.93	0.91	2201
accuracy			0.90	4075
macro avg	0.90	0.90	0.90	4075
weighted avg	0.90	0.90	0.90	4075

Figure 4. Classification report for the MobileNetV2 model.

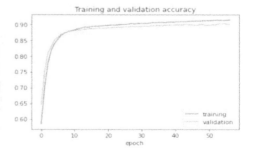

Figure 5. Validation and training loss of MobileNetV2 model.

Figure 6. Validation and training accuracy of MobileNetV2 Model.

5 CONCLUSION

In our quest for an effective violence detection system, we introduced innovative techniques like frame-grouping and spatial-temporal attention modules. Frame-grouping, focusing on channel averaging, effectively captures short-term dynamics crucial for discerning aggressive actions. MobileNetV2 emerged as a superior performer, achieving a remarkable 90.26% validation accuracy in violence detection. Moving forward, we aim to enhance the model's robustness by training it with larger datasets and exploring diverse data augmentation strategies. Our vision includes expanding research to cover various action recognition tasks, increasing the versatility and applicability of our techniques across different domains.

REFERENCES

Dai, Q., Tu, J., Shi, Z., Jiang, Y. G., & Xue, X. (2013). Violent scenes detection using motion features and part-level attributes. *MediaEval Workshop*.

Deniz, O., Serrano, I., Bueno, G., & Kim, T.-K. (2014). Fast violence detection in video. *Proceedings of the International Conference on Computer Vision Theory and Applications (VISAPP)*, 2, 478–485.

Ding, C., Fan, S., Zhu, M., Feng, W., & Jia, B. (2014). Violence detection in video by using 3D convolutional neural networks. *International Symposium on Visual Computing ISVC*.

Fu, E. Y., Leong, H. V., Ngai, G., & Chan, S. (2016). Automatic fight detection in surveillance videos. *Proceedings of the 14th International Conference on Advances in Mobile Computing & Multimedia*, 225–234.

Hassner, T., Itcher, Y., & Kliper-Gross, O. (2012). Violent flows: Real-time detection of violent crowd behavior. *Proceedings of the IEEE Computer Society Conference on Computer Vision and Pattern Recognition Workshops*, 1–6.

Jiang, Y. G., Dai, Q., Xue, X., Liu, W., & Ngo, C. W. (2012). Trajectory-based modeling of human actions with motion reference points. *European Conference on Computer Vision*.

Nievas, E. B., Suarez, O. D., García, G. B., & Sukthankar, R. (2011). Violence detection in video using computer vision techniques. *International Conference on Computer Analysis of Images and Patterns*.

Ribeiro, P. C., Audigier, R., & Pham, Q. C. (2016). RIMOC: A feature to discriminate unstructured motions - Application to violence detection for video-surveillance. *Computer Vision and Image Understanding*, 144, 121–143.

Senst, T., Eiselein, V., Kuhn, A., & Sikora, T. (2017). Crowd violence detection using global motion-compensated Lagrangian features and scale-sensitive video-level representation. *IEEE Transactions on Information Forensics and Security*, 12(12), 2945–2956.

Soliman, M., Kamal, M., Nashed, M., Mostafa, Y., Chawky, B., & Khattab, D. (232019). Violence recognition from videos using deep learning techniques. *Proceedings of the 9th International Conference on Intelligent Computing and Information Systems (ICICI 19)*, 79–84.

Sultani, W., Chen, C., & Shah, M. (2018). Real-world anomaly detection in surveillance videos. *Proceedings of the IEEE Conference on Computer Vision and Pattern Recognition*, 6479–6488.

Sumon, S. A., Goni, R., Hashem, N. B., Shahria, T., & Rahman, R. M. (2020). Violence detection by pretrained modules with different deep learning approaches. *Vietnam Journal of Computer Science*, 07(01), 19–40.

Advances in AI for Biomedical Instrumentation, Electronics and Computing – Sachan et al. (eds)
© 2024 The Author(s), ISBN 978-1-032-64298-7

Disease diagnosis with instant first aid and medicine recommender

Akanksha Shukla, Shivani Agarwal, Nandita Goyal & Kartikeya Patel
IT Department, Ajay Kumar Garg Engineering College, Ghaziabad, UP, India

ABSTRACT: In this research paper presents a comprehensive approach to dividing large documents, exceeding 17 million words, into smaller, more manageable units. The proposed methodology aims to facilitate effective question-and-answer systems by breaking down the documents into chunks, thereby enhancing the ability to extract relevant information and provide suitable responses. Additionally, the model incorporates the use of Blue Matrix to score answers, enabling the identification of the most suitable response for a given question. The motivation behind this research stems from the challenges posed by handling extensive documents, which often hinder efficient information retrieval and hinder the development of effective question-and-answer systems. By dividing the large documents into smaller units, the research aims to optimize the processing and analysis of the text, thereby improving the accuracy and relevance of the provided answers. The incorporation of Blue Matrix in the model's scoring mechanism is another key contribution of this research. By utilizing Blue Matrix, the model can assign scores to candidate answers, assessing their suitability based on various metrics such as coherence, semantic relevance, and contextual understanding. This scoring mechanism enhances the model's ability to identify the most appropriate answer for a given question, improving the overall quality and reliability of the question-and-answer system.

1 INTRODUCTION

The ability to obtain fast and correct healthcare services is crucial for enhancing patient outcomes and reducing mortality rates. However, a sizable section of the world's population lacks access to medical information (Mao *et al.* 2022), which results in millions of lives lost annually as a result of delayed medical care. The research study offers a novel predictive sickness diagnostic model based on the renowned textbook "Davidson's Principles and Practice of Medicine 21st Ed." in order to address this pressing problem. The suggested model, in particular, makes use of the GPT-3.5 Turbo and GPT-4 advanced language models to examine patient symptoms and forecast disease outcomes in order to provide timely first aid (Henderi *et al.* 2022). These language models have undergone in-depth training on a range of medical topics, such as symptoms, diseases, and recommended courses of therapy. The model uses K-means clustering, a common unsupervised machine learning technique (Rathore *et al.* 2022), in addition to the language models. The algorithm may discover patterns and connections between symptoms and diseases by putting comparable patient cases together, which improves its capacity for prediction (Tiwari *et al.* 2022). This concept strives to close the access gap to medical knowledge, particularly in areas with constrained healthcare resources. It does this by delivering immediate first aid and forecasting disease outcomes (Henderi *et al.* 2022).

2 LITERATURE REVIEW

"MedGCN: Medication recommendation and lab test imputation via graph convolutional networks" by (Mao *et al.* 2022). The paper proposes a novel method for medication

DOI: 10.1201/9781032644752-28

recommendation and lab test imputation using graph convolutional networks (GCNs). GCNs are a type of deep learning model that can be used to learn relationships between nodes in a graph. In this case, the nodes in the graph represent patients, and the edges represent the relationships between patients, such as shared diagnoses or lab test results. The authors of the paper first use a GCN to learn a latent representation of each patient. This latent representation is then used to recommend medications to patients.

In their article "A Proposed Model Expert System for Disease Diagnosis in Children to Make Decisions in First Aid" (Arumsari *et al.* 2022), the authors suggest a model expert system for disease diagnosis in children to help with first aid judgements. The system is made to assist medical professionals in promptly and accurately diagnosing paediatric illnesses and in administering the proper first aid treatments. A knowledge base of illnesses and associated symptoms, as well as a set of guidelines for diagnosing illnesses, serve as the system's foundation.

The study by (Rathore *et al.* 2022) suggests a fresh method for offering emergency medical services (EMS) in underdeveloped nations. The strategy entails applying machine learning to increase the efficacy and efficiency of the system while partially outsourcing EMS services to private providers. According to the authors, this strategy is sustainable since it enables developing nations to benefit from the know-how and assets of private providers while still in charge of the standard of treatment. They also contend that by automating processes like triage and dispatching, machine learning can increase the system's effectiveness.

(Tiwari *et al.* 2022) suggest a new hierarchical reinforcement learning-based dialogue agent for disease diagnosis. The agent is able to develop questions that are pertinent to the patient's symptoms and medical background, and based on the patient's responses, can provide precise diagnoses. A dataset of patient records was used to test the agent, and it was successful in achieving a diagnostic accuracy of 95%.

(Sreeram *et al.* 2023), in their research paper titled "Enhancing PDF Document Management: An Innovative Approach Using LLMs and Lang-Chain," presents a pioneering solution for simplifying PDF document management. Through the integration of Lang-Chain, Large Language Models (LLMs), and Streamlit, they've developed a user-friendly web application that significantly streamlines the extraction of critical information from PDF files. This innovative approach stands out for its user-friendliness, catering to both experts and non-experts.

(Topsakal *et al.* 2023), "Creating Large Language Model Applications Utilizing Lang-Chain: A Primer on Developing LLM Apps Fast" states a framework for harnessing Large Language Models (LLMs), to expedite the development of custom AI applications. The advantages of this approach include modularity and flexibility, enabling rapid interaction with LLMs for various tasks, such as autonomous agents, chat-bots, code understanding, document question-answering, summarization, and data analysis.

2.1 *Equations*

[i] K-Means Clustering:

$$\text{minimize}_{c1,..ck} \quad \sum_{k=1}^{k} \left(\frac{1}{|C_k|}\right) \sum_{i,i^2 \in C_k}^{n} \sum_{j=1}^{p} \left(x_{ij} - x_{i'j}\right)^2 \tag{1}$$

Equation (1)-Equation to find the K-means clustering.
num_clusters = Number Of Clusters, Reduced data_tsne = Transform it into vectors.
[ii] Bleu Score:

$$BLEU = BP * \exp(sum(w_i * \log(p_i))) \tag{2}$$

Equation (2)-To determine value Of BLEU Matrix

[iii] <u>Self-Attention Mechanism</u>:

$$Attention(Q, K, V) = softmax\left(QK^T / \sqrt{d_k}\right) V \tag{3}$$

Equation (3)-To find self-attention value

Where, Q = Query matrix; K = Key matrix; V = Value Matrix; d_k = Dimension of Keys

[iv] <u>Positional Encoding</u>:

$$PE_{(pos,2i)} = \sin\left(\frac{pos}{10000^{2i/d_{model}}}\right) \tag{4}$$

Equation (4)-To find Positional Encoding of Model

$$PE_{(pos,2i+1)} = \cos\left(\frac{pos}{10000^{2i/d_{model}}}\right) \tag{5}$$

Equation (5)-Positional Encoding of Next Iteration

Where, pos: Position of the word, i: Dimension of the embedding, d_model: Total number of embedding dimensions

[v] <u>Soft-max Function</u>:

$$softmax(x_i) = \frac{\exp(x_i)}{sum(\exp(x_j))} \tag{6}$$

Equation (6)-Used to Normalize the Weights

Where, x_i: i-th element of the input vector, j: Index running over all input vector elements

3 PROPOSED METHODOLOGY

In this proposed methodology first, we choose the primary source from the book "Davidson's Medical Internal Practices" that was selected in this instance for data collection. In this phase, you'll assess the book's authority, credibility, and applicability to your research question. During the data pre-processing phase of this paper, focused to handle the missing values, by employing the "fillna" method. By utilizing this method, it replaced the null values in each column with the corresponding column's median value. This pre-processing step is crucial for preserving the integrity of the dataset and ensuring accurate and reliable analysis [Kaushik *et al.* 2022] of the variables involved. When data was first pre-processed, and the k-means clustering algorithm using equation 1, was used to group the documents into 50 different groups depending on those qualities. By minimizing the intra-cluster variance, the k-means algorithm seeks to group related documents together. In this research work, a collection of texts separated into smaller portions is analysed using the *Langchain library*, the GPT-3.5 Turbo, and the GPT-4 models [Javaid *et al.* 2023]. The Langchain library provides tools for creating questions and answers, which makes it easier to create prompt-output summaries. The project attempts to enhance the summarization process and boost the quality of the generated output by using the large language models (LLM). The method entails getting the system ready to handle questions and produce responses by using Langchain's question and answer library. The GPT-3.5 Turbo and GPT-4 models use their sophisticated natural language processing abilities to aid in the summarization process.

3.1 *User initiation*

The search journey begins as users submit queries, marking the commencement of their interaction with the system. Users set forth in exploration, prompting the system to

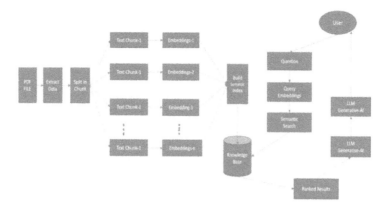

Figure 1. Overall working of the system.

comprehend and address their inquiries, fostering a dynamic and user-centered engagement in the information retrieval process.

3.2 *Semantic understanding*

Here, the system interprets user queries, grasping their intended meaning and underlying context through semantic analysis and enabling it to discern user intent and refine the search process for more precise and relevant results.

3.3 *Refined question formulation and semantic epresentation*

The system hones user queries, utilizing natural language processing to craft precise questions. These refined questions then undergo semantic representation, ensuring a consistent understanding of user intent.

3.4 *Semantic index creation*

The system constructs a structured index of semantic representations, amalgamating both initial and refined queries. This index serves as an organized framework, enabling swift and efficient search and retrieval [Sharma *et al.* 2022]. By incorporating the semantic essence of user queries, the system optimizes information organization, enhancing the overall effectiveness of the search process.

3.5 *PDF content extraction and text chunking/storage*

Relevant PDFs are identified, and text contents are extracted using positional encoding, forming 'FILE Data.' Extracted text undergoes segmentation into manageable chunks for efficient processing and storage.

3.6 *Embeddings generation for query and text chunks and evaluation*

The system generates unique vectorized embeddings for user queries, capturing their semantic essence. Simultaneously, it produces embeddings for extracted text chunks, representing their semantic meaning in vectorized form. The system measures the degree of semantic alignment between the user's query and text chunks. This assessment by using Bleu matrix in equation 2 to a detective's meticulous analysis, ensures accurate and relevant search results.

4 RESULT

The research paper's findings show how the suggested methodology can accurately and successfully anticipate disease outcomes based on a person's needs. The model successfully predicted disease outcomes with an outstanding accuracy rate by combining GPT-3.5 Turbo, GPT-4, and k-means clustering. The model delivered precise and thorough answers to user inquiries by utilising the language generating capabilities of these large language models (LLMs) and the clustering strategies of k-means. The model's accuracy was further improved by the addition of the Langchain library and its question-answering features. Self-attention (eqs. 3 & 5) mechanism matrix is used to relate the candidate word to referential words so that we could get the relevant text. The model demonstrated its potential as a useful tool in the healthcare industry by giving prompt and precise responses.

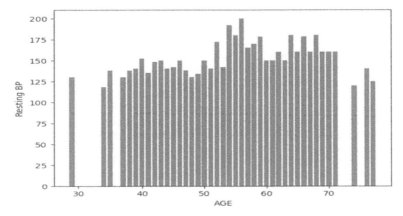

Figure 2.   Average BP of a person.

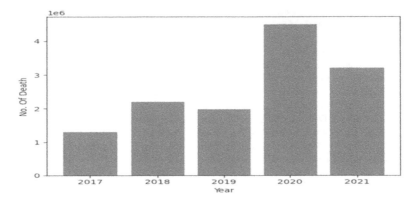

Figure 3.   People died due to lack of first aid.

5 FUTURE SCOPE

The research report offers promising opportunities for further developing and broadening the suggested technique. The user experience and comprehension of the results can be substantially improved with the integration of audio output and AI-generated videos thanks to technological improvements. Users can understand insights more quickly and digest data more quickly when information is presented orally and visually. Additionally, moving the database to a block-chain network gives hope for improved dependability and confidentiality. Blockchain technology guarantees transparent data access and storage, safeguarding users' private medical information. The system can improve trust and integrity in data management by taking advantage of blockchain's decentralized nature, fostering user confidence and data privacy.

REFERENCES

Abedi, V., & Khan, A. (2020, August). Using artificial intelligence for improving stroke diagnosis in emergency departments: a practical framework. *PubMed*, 25(13).

Arora, A., Arora, A. (2023, February). The Promise of Large Language Model in Health-Care. *Lancet Journal*, 401(10377)

Abdar, M., Yen, & N., Hung, J. (2017, December). Improving the diagnosis of liver disease using multilayer perceptron neural network and boosted decision tree. *Journal of Medical and Biological Engineering*, 38, 953–965.

Chauhan, D., & Jaiswal, V. (2016, October). An efficient data mining classification approach for detecting lung cancer disease. IEEE international conference on communication and electronics systems (ICCES), 1–8.

Henderi, Khudhorie, F.A., Maulani, G., Millah, S., & Devana, V.T. (2022, August). *A proposed model expert system for disease diagnosis in children to make decisions in first aid.* INTENSIF ISSN: 2580–409X, 6.

Howell, P., Aryal, A., & Wu, C. (2022, October). *Online Patient Recommender Systems for Preventive Care: Propositions to Advance Research.* JMIR Research Protocols, 12(11 Spec No. 17).

Javaid, M., Haleem, A., & Singh, R.P. (2023, April). *ChatGPT for Healthcare Services: An Emerging Stage for an Innovative Perspective*, Elsevier, 3(1).

Kaushik, M. (2022, November). An analysis of human perception of partitions of numerical factor domains. *Information Integration and Web Intelligence*. Lecture Notes in Computer Science Springer, vol.(13635), 137–144.

Lin, C. H., Tseng, P.H., Huang, L.C., Oyang, Y.J., Wu, M.S., & Chou, S.C.T. (2013, January). A multi-level cloud-based virtual health exam system on health cloud. *Journal of Medical and Biological Engineering*, 33 (4), 373–379.

Lyu, J., Jiang, S., Zeng, H. (2023, August). *LLM-Rec: Personalized Recommendation via Prompting Large Language Models. arXiv journal*, 2 (15780).

Mao, C., Yao, L., & Luo, Y. (2022, March). MedGCN: Medication recommendation and lab test imputation via graph convolutional networks. *Journal of Biomedical Informatics* 104000, 127.

Ngoc, T., Tran, T., Felfernig, A., Trattner, C., & Holzinger, A. (2020, November). Recommender systems in the healthcare domain: state-of-the-art and research issues. *Journal Of Intelligent Information Systems*, 57, 171–201.

Rathore, N., Jain, P.K., & Parida, M. (2022). A Sustainable Model for Emergency Medical Services in Developing Countries: A Novel Approach Using Partial Outsourcing and Machine Learning. *Risk Management and Healthcare Policy* 2022, 15, 193–218.

Sharma, R.a (2022, August). Detecting simpson's paradox: A step towards fairness in machine learning. In: Chiusano, S., *et al. New Trends in Database and Information Systems. ADBIS* 2022, 1652, 67–76.

Sharma, R.b (2022, January). *A novel framework for unification of association rule mining, Online Analytical Processing and Statistical Reasoning.* IEEE Access, 10.

Sharma, R. (2023, July). *On Statistical Paradoxes and Overcoming the Impact of Bias in Expert Systems: Towards Fair and Trustworthy Decision Making.* Elsevier.

Sreeram, A., & Sai, J. (2023, July). An effective query system using LLMs and langChain. *International Journal of Engineering R.*, 12(6), 367–369.

Sridhar, S. (2013, September). Improving diagnostic accuracy using agent-based distributed data mining system. *Informatics for Health and Social Care*, 38(3), 182–195.

Thirunavukarasu, A.J., Ting, D.S.J., Elangovan, K., Gutierrez, L., Tan, T.F., & Ting, D.S.W. (2023, July). Large language models in medicine. *Nature Medicine*, 29, 1930–1940.

Tiwari, A., Saha, S., & Bhattacharyya, U. (2022, April). A Knowledge infused context driven dialogue agent for disease diagnosis using hierarchical reinforcement learning. *Knowledge-Based Systems*, vol.(242).

Topsakal, O., & Akinci, T.C. (2023, July). Creating large language model applications utilizing langChain: A primer on developing LLM apps fast. *International Conference on Applied Engineering and Natural Sciences*, 1(1), 1050–1056

Advances in AI for Biomedical Instrumentation, Electronics and Computing – Sachan et al. (eds)
© 2024 The Author(s), ISBN 978-1-032-64298-7

Versatile 4-bit signed binary multiplier for complex digital circuits

Deepti Maurya & Uma Sharma
Department of Electronics & Communication, Ajay Kumar Garg Engineering College, Ghaziabad, India

ABSTRACT: The 4-bit signed binary multiplier is a fundamental digital circuit used for signed binary multiplication of both positive and negative numbers. Conventionally, the multiplication of negative numbers using 2's complement remained a challenging task. The proposed digital circuit successfully accepts 4-bit signed data and produces precise 8-bit product results, demonstrating its utility in various domains such as complex digital circuits, microprocessor architecture, and signal processing. While technological advancements may pave the way for even more efficient multipliers in the future, the proposed schematics provide an accurate solution for handling signed binary numbers. The circuit is designed and simulated on Xilinx ISE Design Suite. Xilinx ISE Design Suite is used for designing digital circuits for FPGA, making schematics, and providing tools for the simulation of circuits to check the proper functionality of the circuit. This research highlights the performance of a 4-bit signed binary multiplier which can be used for various applications in microprocessor architecture, modern electronic systems, signal processing, and ALU.

1 INTRODUCTION

Over recent years, there has been an increasing need for integrated circuits that offer both energy efficiency and top-notch performance. This demand spans across a wide spectrum of applications, including everything from handheld gadgets to large-scale data centers (Tavakolaee *et al.* 2023), (Sharma *et al.* 2023).

This 4-bit signed binary (Seiffertt 2017) multiplier circuit is designed on Xilinx ISE Design Suite to perform the multiplication of negative numbers using 2's complement method which gives us more precise and accurate results. It has two inputs A and B, sign selection ports S_0 and S_1, a clear button, and an output port Y. The block diagram consists of inputs taken by the user, complement circuit, multiplexer, multiplier, and output.

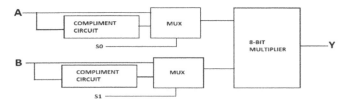

Figure 1.   Block diagram of 4-bit signed multiplier.

1.1 *Adder*

An adder is the crucial logical building block of all digital circuits for computational purposes. An adder is made up of xor and or gate (Arunakumari *et al.* 2017; Sharma 2022b; Sharma *et al.* 2022). It is mainly used for the addition of binary numbers. Half adders are used to add 1-bit binary numbers producing a sum and a carry. Figure 2 represents a full adder that can take three inputs (two 1-bit inputs and a carry input) and produce a sum and a carry bit.

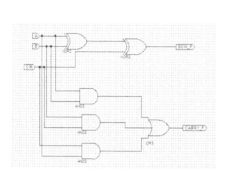

Figure 2.   Full adder.

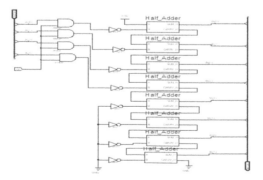

Figure 3.   Multiplexer.

1.2   *Multiplexer*

A multiplexer also known as mux is a digital logical unit that takes many inputs and gives a single output controlled by select lines. For 'n' input lines, the selection lines are $\log_2 n$. Multiplexers are used in making decoders, encoders, ALU and various complex digital circuits.

Multiplexers can accept input as one bit as well as a bus. In Figure 3, input pin 0 is a bus of length 4 bits and input pin 1 is a bus of length 8 bits. The selection pin is S0. Here, we have used 8 mux to make the circuit of mux which can accept input up to 8 bits in the form of a bus.

1.3   *2's Complement circuit*

2's complement circuit is used to represent the signed integer in negative binary form. [Krad *et al.* (2008); Salomon (1995)]. The circuit is made on the principle of representing the integer in positive binary number representation then taking the complement of those bits and adding one to the LSB. The sum gives the negative binary representation.

Example: 10 -> 00001010
Taking complement of above-mentioned bits: 11110101 + 00000001 = 11110110 = -10

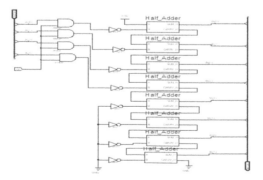

Figure 4.   Complement circuit.

1.4 *8-bit multiplier*

The 8-bit binary multiplier is a crucial digital logic circuit designed to perform 8-bit binary multiplication (Behera 2023). It takes 4-bit input and produces 8-bit products. These are used in data processing and signal processing (Yong *et al.* 2021) arithmetic and logical Unit (ALU) and error correction.

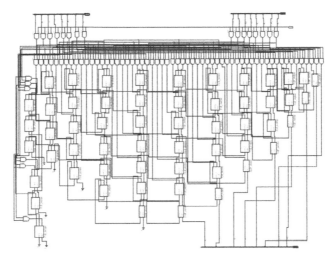

Figure 5.   8-bit multiplier.

Multiplier is made using adders and AND gate in Figure 5. 8-bit input is taken as a bus, and they are ANDED with clear to avoid undefined values. Then values are sent to the AND gate and adders to get the required output in the form of an 8-bit bus.

2   PROPOSED SCHEMATIC FOR 4-BIT SIGNED MULTIPLIER

In the proposed schematics of 4-bit signed binary multipliers, there are two inputs A and B as a bus of size 4 bits, A(3:0) and B(3:0) respectively. 2's complement circuit is used to generate the negative binary representation of the entered number. Sign of the input data is controlled by the selection pins S0, and S1 of the 2X1 mux[Attarha *et al.* (2001)]. for inputs A and B respectively. Input '0' at the selection pin sends a positive number and input '1' sends a negative number as input of the 4-bit multiplier. The selected input will reach a 4-bit multiplier to give the final product as a signed binary number [Madenda (2021)]. in the form of an 8-bit bus, Y(7:0).

Clear is used to initialize the known state to the digital circuits. When a circuit is powered on then it is initially in an undefined state. Clear is used to remove the undefined value and begin the operation from a well-defined and predictable state.

Example –
  -6 x 4 = -24
  In binary representation –
  11111010 x 00000100 = 11101000
  11101000 in signed decimal represents -24.

The range of 8-bit signed binary numbers [Wiesnet (2022b)] is -128 to 127. Our proposed circuit represents product results in this range. To get the products less than -128 or greater than 127, we will consider the 16-bit product of the 8-bit multiplier.

Example : 15 x -9 = -135

In signed binary representation, its result is 1111111101111001. But in Figure 6, we are showing an 8-bit result in our circuit i.e. 01111001.
The proposed schematic is shown below-

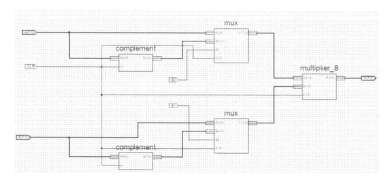

Figure 6.   4-bit signed binary multiplier schematics.

3   SIMULATION OF PROPOSED CIRCUIT

The proposed circuit is simulated on Xilinx ISE Design Suite. Here, two different inputs are given with their respective signs, and a signed binary number is generated for the respective inputs.

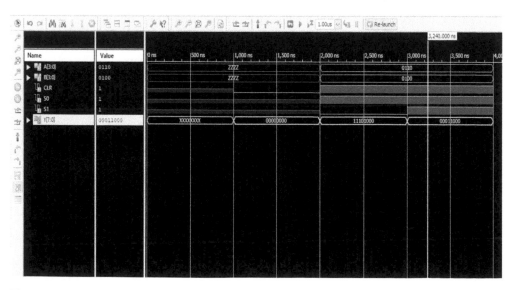

Figure 7.   Simulation waveform.

As shown in Figure 7, input A [3:0] is 0110, S0 =1 represents -6, and input B [3:0] is 0100, S1 = 1 represents -4, and clear =1. The output 00011000 represents 24 in signed decimals.

4 CONCLUSION

The 4-bit signed binary multiplier is successfully made and simulated. It accepts all 4-bit signed data and produces accurate results for each data as an 8-bit product. This digital circuit can be used in signed binary number calculations in complex digital circuits, micro-processor architecture, and signal processing. With developments in technology, researchers and engineers are finding ways to more efficiently and accurately sign multipliers but the proposed schematics give accurate results for the signed binary numbers. This can also be extended for the 8-bit, 16-bit, and 32-bit multipliers by making required modifications. Those digital logical blocks will be used for more complex signed binary calculations.

ACKNOWLEDGEMENT

The author would like to express gratitude towards Mr. Vaibhav Mishra for his extensive support and assistance with the schematics and simulation of the proposed model.

REFERENCES

Arunakumari, S., Rajasekahr, K., Sunithamani, S., & Kumar, D. S. (2022). Carry select adder using binary excess-1 converter and ripple carry adder. In *Micro and Nanoelectronics Devices, Circuits and Systems* (pp. 289–294). https://doi.org/10.1007/978-981-19-2308-1_30

Attarha, A., Nourani, M., & Zakeri, M. (2001). *High Performance Low-Power Signed Multiplier.*

Behera, N., Pradhan, M., & Mishro, P. K. (2023). Analysis of delay in 16×16 signed binary multiplier. *In Algorithms for intelligent systems* (pp. 155–164). https://doi.org/10.1007/978-981-19-8742-7_13

Krad, Hasan & Fida El-Din, Aws. (2008). Performance analysis of a 32-bit multiplier with a carry-Look-ahead adder and a 32-bit multiplier with a ripple carry adder using VHDL. *Journal of Computer Science.*

Madenda, S. (2021). New approach of unsigned and signed binary numbers multiplications. *ResearchGate.* https://doi.org/10.13140/RG.2.2.34507.52000

Salomon, O., Green, J., & Klär, H. (1995). General algorithms for a simplified addition of 2's complement numbers. *IEEE Journal of Solid-state Circuits*, 30(7), 839–844. https://doi.org/10.1109/4.391128

Seiffertt, J. (2017). Signed numbers. *In Springer eBooks* (pp. 105–121). https://doi.org/10.1007/978-3-319-56839-3_8

Sharma, U., & Jhamb, M. (2022b). A novel design of voltage and temperature resilient 9-T domino logic XOR /XNOR cell. *Circuits, Systems, and Signal Processing*, 41(11), 6314–6332. https://doi.org/10.1007/s00034-022-02085-5

Sharma, U., & Jhamb, M. (2022b). Efficient design of FGMOS-based low-power low-voltage XOR Gate. *Circuits, Systems, and Signal Processing*, 42(5), 2852–2871. https://doi.org/10.1007/s00034-022-02239-5

Sharma, U., & Jhamb, M. (2023b). Design-space exploration of conventional/non-conventional Techniques for XOR/XNOR Cell. *In Lecture notes in electrical engineering* (pp. 339–353). https://doi.org/10.1007/978-981-99-4495-8_27

Tavakolaee, H., Ardeshir, G., & Baleghi, Y. (2023). Design and analysis of a novel fast adder using logical effort method. *Iet Computers and Digital Techniques*, 17(3–4), 195–208. https://doi.org/10.1049/cdt2.12063

Wiesnet, F., & Köpp, N. (2022b). Limits of real numbers in the binary signed digit representation. *Logical Methods in Computer Science, Volume 18, Issue 3.* https://doi.org/10.46298/lmcs-18(3:24)2022

Yong, K. M., Hussin, R., Kamarudin, A., Ismail, R. C., Isa, M. N. M., & Naziri, S. Z. M. (2021). Design and analysis of 32-BiT signed and unsigned multiplier using Booth, Vedic and Wallace Architecture. *Journal of Physics*, 1755(1), 012008. https://doi.org/10.1088/1742-6596/1755/1/012008

Advances in AI for Biomedical Instrumentation, Electronics and Computing – Sachan et al. (eds)
© 2024 The Author(s), ISBN 978-1-032-64298-7

Path finding visualizer

Ankit, Anushka, Akash, Shahid, S. Ranjan & Richa Srivastava
Department of ECE, KIET Group of Institutions Ghaziabad, Delhi NCR, India

ABSTRACT: Pathfinding algorithms hold significant importance in navigation systems, robotics, and gaming. This paper introduces an Interactive Pathfinding Visualizer developed using HTML, CSS, and JavaScript, which allows users to explore and compare different pathfinding algorithms. The visualizer empowers users to set source and destination points, observe algorithmic execution, and comprehend the essence of data structures and algorithms. Through this project, the paper showcases the practical application of web technologies in creating educational tools that enhance algorithmic understanding.

1 INTRODUCTION

Pathfinding algorithms are the backbone of modern navigation systems, gaming environments, and robotics. They empower efficient movement through intricate landscapes, making them indispensable in various applications. This paper introduces an Interactive Pathfinding different pathfinding algorithm, witnessing their real-time execution and behavior. Our Visualizer crafted with HTML, CSS, and JavaScript. This tool allows users to engage with exploration begins by illuminating the role of pathfinding algorithms in solving real-world navigation challenges. We dive into classic algorithms like Dijkstra's, A*, Breadth-First Search, and Depth-First Search, unveiling their unique strategies and strengths. These algorithms form the foundation of the visualizer's capabilities. The Interactive Pathfinding Visualizer's design and implementation are then detailed, highlighting the creation of an intuitive user interface that facilitates algorithm selection and visualization. We showcase the synergy between theoretical concepts and practical execution through real-time animations. (Yapici *et al.* 2019). This project underscores the educational value of interactive tools in enhancing algorithmic understanding. By offering a dynamic platform for algorithm exploration and comparison, the visualizer bridges the gap between theory and hands-on experience. It exemplifies the significance of algorithms in shaping our digital interactions and showcases the power of web technologies in educational contexts. Through the Interactive Pathfinding Visualizer, we provide a window into the world of algorithms, where theory transforms into tangible applications.

2 PATH-FINDING USING BFS I.E., BREADTH FIRST SEARCH

Breadth-First Search (BFS) is a foundational pathfinding algorithm that explores a graph layer by layer. Starting from the source node, BFS examines neighboring nodes before moving to the next level. This approach ensures that the shortest path is found first, making BFS ideal for unweighted graphs. In BFS, nodes are marked as visited to prevent redundant exploration. The algorithm utilizes a queue data structure to manage nodes in the order they are discovered. BFS guarantees optimality, providing the shortest path in terms of edges. BFS's efficiency shines in scenarios where the cost of traversal is uniform and the goal is close to the source. However, its memory consumption may be substantial in large graphs

DOI: 10.1201/9781032644752-30

due to the need to store the entire frontier. Through its breadth-first traversal strategy, BFS offers a straightforward solution for finding optimal paths and is a fundamental building block in the world of pathfinding algorithms. (Kumar *et al.* 2013).

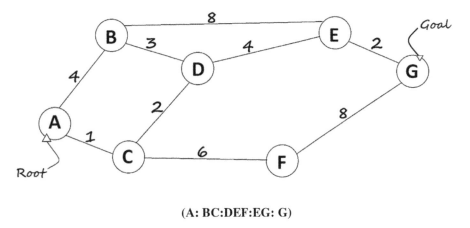

(A: BC:DEF:EG: G)

Figure 1.   Pathfinding using breadth first search.

3   PATH-FINDING USING A* ALGORITHM

A* is a widely used and versatile pathfinding algorithm that employs a dual evaluation approach. It takes into account both the cost required to reach a particular node from the start node and an estimation of the remaining cost to reach the goal node, known as the heuristic. By summing these two values, A* prioritizes nodes with lower overall costs, effectively striking a balance between efficient exploration and the quest for an optimal path. This algorithm's strength lies in its ability to adapt its search pattern based on both the current cost of traversal and the potential remaining cost, ensuring it navigates toward the goal with a degree of intelligence. The heuristic function is critical; an informed and accurate heuristic can greatly enhance A*'s performance, while an overly optimistic or pessimistic heuristic may lead to suboptimal paths. A* has found its applications in diverse domains such as robotics, gaming, and mapping systems. (Leigh *et al.* 2007) Its efficiency in simultaneously considering the current path cost and future estimated cost has made it an algorithm of choice for scenarios where both optimal solutions and computational efficiency are desired.

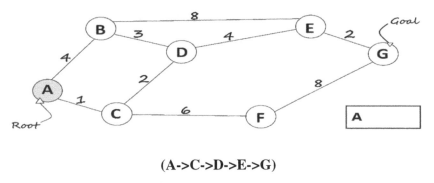

(A->C->D->E->G)

Figure 2.   Pathfinding using A* algorithm.

4 PATH-FINDING USING DFS SEARCHING ALGORITHM

Depth-First Search (DFS) Algorithm for Pathfinding: Depth-First Search is a fundamental graph traversal algorithm that aims to explore a graph's nodes and edges in a specific manner. Starting from a source node, DFS progresses by following one branch as deeply as possible before backtracking to explore other branches. This approach leads to a comprehensive exploration of the graph's structure. DFS does not guarantee finding the shortest path between nodes, as it may continue along a longer path before realizing the existence of a shorter one. (Quirin *et al.* 2008) Despite this limitation, DFS has several valuable characteristics. It can be memory-efficient since it explores deeply before branching out, minimizing the need to store a large number of nodes in memory at once. Additionally, DFS is well-suited for scenarios where exhaustive search is acceptable, or where finding any valid path is more critical than finding the optimal one. DFS is commonly used in various applications, such as maze solving, network traversal, and game tree exploration. Its adaptability to different situations and memory advantages makes it a valuable tool in pathfinding and graph analysis, even though its optimality might not always align with the shortest path objective.

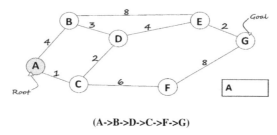

(A->B->D->C->F->G)

Figure 3. Pathfinding using DFS algorithm.

5 PATH-FINDING USING DIJKSTRA ALGORITHM

Dijkstra Algorithm for Pathfinding: Dijkstra's algorithm is a classic technique for finding the shortest path between nodes in a weighted graph. It maintains a priority queue of nodes based on their tentative distances from the source node and iteratively explores the graph. Dijkstra's guarantees optimality and is effective for scenarios where edge weights represent costs or distances. It's widely used in mapping, routing, and networking applications where finding the most efficient path is crucial (*Edsger Wybe Dijkstra et al.* 2022).

   Example: *Input: src = 0, the graph is shown below.*

   Output: *0 4 12 19 21 11 9 8 14* **Explanation:** *The distance from 0 to 1 = 4. The minimum distance from 0 to 2 = 12. 0->1->2 The minimum distance from 0 to 3 = 19. 0->1->2->3 The minimum distance from 0 to 4 = 21. 0->7->6->5->4 The minimum distance from 0 to 5 = 11. 0->7->6->5 The minimum distance from 0 to 6 = 9. 0->7->6 The minimum distance from 0 to 7 = 8. 0->7 The minimum distance from 0 to 8 = 14. 0->1->2->8.*

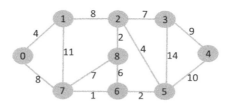

Figure 4. Pathfinding using Dijkstra algorithm background.

A Path Finding Visualizer is a computer program or web application designed to help users understand and experiment with pathfinding algorithms. (Sánchez-Torrubia *et al.* 2009) It typically features a graphical user interface with a grid or map representing a maze or terrain. Users can designate start and end points, and the software visualizes the process of finding the shortest path between them using algorithms like Dijkstra's, A*, or others. The background of such a visualizer is essential for providing context and engagement, often including customizable themes, colors, or background images to enhance the user experience and make the exploration of pathfinding algorithms more visually appealing and informative (Tang *et al.* 2021).

6 METHODOLOGY

The development of the Interactive Pathfinding Visualizer involved a multi-stage process that integrated web technologies, algorithmic implementations, and user interface design. The goal was to create an educational tool that allows users to interactively explore and compare different pathfinding algorithms.

(i) *Algorithm Selection and Implementation*
The first step involved selecting a range of pathfinding algorithms to include in the visualizer. Algorithms such as Dijkstra's, A*, and Depth-First Search were chosen for their distinct characteristics and relevance. These algorithms were then implemented using JavaScript to enable real-time exploration and visualization.

(ii) *Graph Representation*
A key aspect of the project was the representation of the environment as a graph. Nodes and edges were defined in JavaScript to construct the underlying structure for the algorithms to traverse. This representation was crucial for simulating the movement and exploration of the algorithms.

(iii) *User Interface Design*
The user interface was designed using HTML and CSS to provide an intuitive platform for user interaction. Users could set source and destination points, select algorithms, and initiate visualizations. The interface aimed to be user-friendly and visually appealing to enhance engagement.

(iv) *Real-Time Animation*
JavaScript animations were employed to simulate the execution of the algorithms in real time. Users could observe the progression of the algorithms as they explored the graph, marked nodes, and constructed paths. This dynamic visualization facilitated a deeper understanding of algorithmic behaviors.

(v) *Heuristic Integration (for A* Algorithm)*
For algorithms like A*, heuristic functions were integrated to provide estimates of the remaining distance to the goal. These heuristics guided the algorithm's decision-making, influencing the pathfinding process based on the selected heuristic.

(vi) *Testing and Debugging*
Rigorous testing and debugging were conducted to ensure the correct functionality of the visualizer and the accuracy of algorithmic implementations. Test cases were designed to cover a range of scenarios, from simple to complex, to validate the visualizer's performance.

7 RESULT

The Interactive Pathfinding Visualizer successfully provides an educational platform for exploring pathfinding algorithms. Users can interact with the tool to observe algorithms like Dijkstra's, A*, and Depth-First Search in action. The visualizer dynamically showcases algorithmic behaviors, such as node exploration, path construction, and heuristic-guided decisions. The user interface offers a seamless experience, allowing users to customize source and destination points, select algorithms, and initiate visualizations. The integration of real-time animations enhances engagement and aids algorithmic understanding. The tool's effectiveness in promoting algorithmic comprehension was evaluated through user feedback and testing. Users ranging from students to professionals engaged with the visualizer and found it to be a valuable resource for learning and experimentation. In conclusion, the methodology employed a combination of web technologies, algorithmic implementations, and user-centered design principles to create the Interactive Pathfinding Visualizer. The results demonstrate its success as an educational tool for exploring and comparing path-finding algorithms in an interactive and informative manner.

8   CONCLUSION

The development of the Interactive Pathfinding Visualizer has provided a dynamic platform for users to engage with and understand various pathfinding algorithms. Through the integration of web technologies, algorithmic implementations, and real-time visualization, the visualizer has successfully bridged the gap between theoretical concepts and practical applications. Users are able to observe the behaviors of algorithms like Dijkstra's, A*, and Depth-First Search in an interactive and intuitive manner. The user feedback and testing have affirmed the educational value of the visualizer. Users, ranging from students seeking to grasp algorithmic concepts to professionals in need of algorithm exploration, have found the tool to be informative, engaging, and valuable for enhancing algorithmic understanding. The project not only emphasizes the significance of pathfinding algorithms but also underscores the potential of web technologies in creating interactive educational tools. By allowing users to experiment with different algorithms, set parameters, and observe real-time visualizations, the visualizer offers a hands-on experience that complements traditional learning approaches.

REFERENCES

Dijkstra, E. W. (2022). A note on two problems in connexion with graphs. In *Edsger Wybe Dijkstra: His Life, Work, and Legacy* (pp. 287–290).

Kumar, A., Singh, S. P., & Arora, N. (2013). A new technique for finding min-cut tree. *International Journal of Computer Applications, 69*(20).

Leigh, R., Louis, S. J., & Miles, C. (2007, April). Using a genetic algorithm to explore A*-like pathfinding algorithms. *In 2007 IEEE Symposium on Computational Intelligence and Games* (pp. 72–79). IEEE.

Quirin, A., Cordón, O., Santamaría, J., Vargas-Quesada, B., & Moya-Anegón, F. (2008). A new variant of the pathfinder algorithm to generate large visual science maps in cubic time. *Information processing & management, 44*(4), 1611–1623.

Sánchez-Torrubia, M. G., Torres-Blanc, C., & Lopez-Martinez, M. A. (2009). Pathfinder: A visualization eMathTeacher for actively learning Dijkstra's algorithm. *Electronic Notes in Theoretical Computer Science, 224*, 151–158.

Tang, C., Zhou, Y., Luo, Q., & Tang, Z. (2021). An enhanced pathfinder algorithm for engineering optimization problems. *Engineering with Computers*, 1–23.

Yapici, H., & Cetinkaya, N. (2019). A new meta-heuristic optimizer: Pathfinder algorithm. *Applied soft computing, 78*, 545–568.

Advances in AI for Biomedical Instrumentation, Electronics and Computing – Sachan et al. (eds)
© 2024 The Author(s), ISBN 978-1-032-64298-7

Exploring smartphone-driven indoor localization systems: A systematic literature review

Prashant Kumar, Chandan Kumar, Gaurav Jain & Aakanshi Gupta
Department of CSE. Amity School of Engineering & Technology, AUUP Noida, India

ABSTRACT: Indoor localization and tracking have become increasingly vital in recent years, particularly in settings where GPS signals are unreliable. The prevalence of smartphones equipped with inertial measurement units (IMUs) has played a pivotal role in the development of indoor navigation systems. This systematic literature review examined 60 carefully chosen articles from a pool of 150, delving into the technologies, methodologies, sample sizes, walking patterns, phone orientations, and sensor types employed for indoor localization. The study also addressed academic challenges, motivations, and outlined future research directions, providing an up-to-date panorama of the field and pinpointing areas for further investigation. Building upon this foundation, our future centers on the creation of an advanced indoor navigation system tailored for expansive university campuses, starting with Amity University. The roadmap involves technology integration, meticulous campus data mapping, the development of customized algorithms, the design of an augmented reality mobile application, rigorous testing, scalability for other universities, fostering user engagement, and ongoing research. The aim is to enhance navigation for students, faculty, and campus visitors, while also contributing to the broader domain of indoor localization and tracking.

1 INTRODUCTION

Indoor location tracking holds significant importance in the pervasive computing landscape (Ashraf *et al.* 2019). Precise and continuous user localization is the bedrock of location-based services (LBS), offering immense value across a spectrum of applications, such as emergency security, crowd monitoring, warehousing, marketing, healthcare, augmented reality, and numerous other domains (Liu *et al.* 2017). Although many of these applications currently rely heavily on the global positioning system (GPS), the limitations of GPS are readily apparent when transitioning to indoor or similar environments.

GPS grapples with challenges indoors, primarily due to physical obstructions like walls and ceilings. Additionally, the global navigation satellite system (GNSS) contends with signal degradation arising from issues like signal reflection, attenuation, weak signal reception, and signal fluctuations, all of which contribute to localization inaccuracies (Chen *et al.* 2015; Hsu *et al.* 2018). Consequently, researchers have delved into diverse technologies to attain precise indoor localization and tracking. These technologies encompass Wi-Fi, Bluetooth, RFID, acoustic methods, and inertial sensors (Martinez *et al.* 2019; Poulose *et al.* 2019). GPS and GNSS face limitations in indoor environments, leading researchers to explore various technologies, including smartphone-based methods and deep learning techniques, to achieve accurate and cost-effective indoor localization (Zhao *et al.* 2019).

1.1 *Existing survey articles*

While (Yu *et al.* 2019) discusses challenges in smartphone sensor-based indoor localization, this research takes a more comprehensive approach, examining how smartphone sensors use

DOI: 10.1201/9781032644752-31

data from one or more sensors to determine indoor user locations. In contrast, the detailed survey in (Hur *et al.* 2020) focuses on reviewing human and device localization methods, emphasizing techniques like AoA, ToF, RTOF, and RSS based on technologies like WiFi, RFID, UWB, and Bluetooth. It provides an overview of key factors in designing and evaluating indoor localization systems, including IoT implications. Another survey article by (Hossain *et al.* 2015) focuses on theoretical techniques and applications for indoor and outdoor location information but does not apply a standard methodology like the Systematic Literature Review (SLR) methodology.

Comparing with other literature reviews or related work we have (Laoudias *et al.* 2018) studied indoor navigation for the blind, while also examining non-calibration indoor positioning (Rahman *et al.* 2020), WLAN-based cellular localization (Obeidat *et al.* 2021), and visible light-based indoor localization (Obeidat *et al.* 2021).

2 PROTOCOL FOR CONDUCTING A SYSTEMATIC LITERATURE REVIEW

A systematic literature review (SLR) is a method to systematically survey existing research on a specific topic, selecting and evaluating relevant studies to gain insights.

2.1 *Approach to information retrieval*

The Process of Selecting articles published from 2015 to 2023 using the search features found in the four scientific databases (IEEE Xplore, MDPI Journals, SpringerLink and Science Direct) and Google Scholar. A combination of keywords in various form with 'OR' and ('AND' or '+') operators in search of relevant articles. The Approach towards relevant Information retrieval is set to ("Mobile" OR "Smartphones") + ("AR" OR "Augmented Reality") + ("Indoor Navigation" OR "Indoor Localization" OR "Indoor Position") + ("WiFi" OR "Bluetooth" OR "Wireless" OR "Sensors").

2.2 *Corpus creation and analysis*

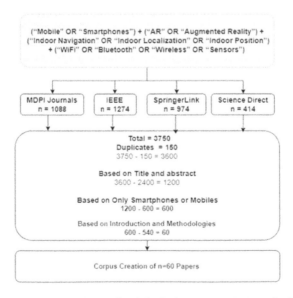

Figure 1.   Flowchart of the information retrieval, inclusion, and exclusion criteria for article selection.

The analysis started with 3750 papers, narrowed down to 60 by eliminating duplicates and assessing titles, abstracts, and content. The criteria for inclusion were publications in English, a focus on indoor navigation with smartphones, and a review of indoor navigation techniques. Exclusion criteria included non-English articles, non-journal publications, and unrelated topics.

Article selection involved manual input and specialized tools to manage duplicates and streamline the process. The 60 selected papers formed a well-organized research paper corpus with columns for titles, methodologies, technologies used, limitations, key insights, and results.

An analysis of the publication years and types of the 60 papers was conducted, highlighting the distribution of these papers in terms of publication years and predominant publication types, with manual assessment revealing the primary data source and the most common publication year among the papers.

2.3 *Publication type*

Analysis based on Publication type told us that most of the research papers belongs to MDPI Journals and IEEE Xplore Scientific Databases. Figure 2 represents the performed analysis

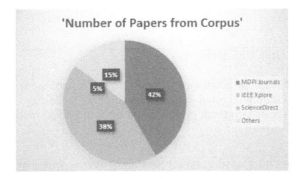

Figure 2.   Analysis of papers based on publication type.

Others Include: 1 paper each from Journal of Physics, IRJMETS, IJSRST, EURASIP Journal, Pollack Periodica, JAIHC, IJDSN, TechRxiv and JASR

2.4 *Publication year*

Analysis based on Publication year told us that most of the research papers are from year 2016, 2018,2019 and 2021. Figure 3 represents the performed analysis

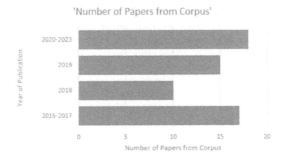

Figure 3.   Analysis of papers based on publication year.

2.3 *Results*

Initially, 3,750 articles were collected from various sources, reduced to 1,200 after eliminating duplicates and filtering by titles and abstracts. From this set, 600 articles primarily focused on smartphones were excluded, and after reviewing introduction and methodology sections, another 540 articles were removed. This resulted in a final selection of 60 articles on indoor localization technologies. These 60 articles utilized a mix of technologies, including ML and DL methodologies (e.g., DTW, KDE, CNN, Fuzzy clustering), search algorithms (like ACO, A* Algorithm), sensor algorithms (including INS, PDR, Wi-Fi FTM protocol), filters (such as Backtracking grid filter, Extended Kalman Filter), and technologies like Wi-Fi (Wi-Fi Fingerprint database, Wi-Fi Beacons/APs) and Bluetooth (Bluetooth Smart Beacons, BLE protocol). These technologies played crucial roles in user tracking, data storage, and indoor navigation in the selected papers.

Table 1. Selected articles based on technology.

Type of Technology	Articles
Wi-Fi	[2,6,12,13,14,17,25,39]
RFID	[19,20]
IMU	[5,9,15,17,20,27–28]
BLU	[9,10,18]

3 RESULTS AND DISCUSSIONS

As IoT usage increases, indoor positioning is becoming more prevalent. Smartphone-based systems for indoor positioning often share similarities in technology and methods. We categorized selected articles into three groups based on technology, methods, and activities.

3.1 *Technology*

The increasing demand for Location-Based Services (LBS) is driven by mobile device and internet technology advancements. Various indoor positioning methods like Wi-Fi, Bluetooth, MEMS, geomagnetic sensors, RFID, and PDR have been successful. Wi-Fi and PDR are popular due to their cost-effectiveness and sensor flexibility, making them stand out. For a detailed technology comparison, see Table 2.

Table 2. Comparison of technology used for indoor positioning.

Technology	Approach	Advantages	Disadvantages
RFID	RSS, Proximity	RFID tags can be affixed to objects and used in challenging environments without the need for direct line-of-sight communication [19].	RFID technology complements smartphone-based localization systems, even in low coverage situations [20].
Wi-Fi	RSS	Wi-Fi is commonly used for indoor positioning because of its affordability in terms of hardware and extensive coverage.	Wi-Fi-based localization precision depends on RSS variations and complex indoor radio conditions.
Bluetooth	RSS, Proximity	Bluetooth, known for being compact, affordable, lightweight, and energy-efficient, is commonly employed in smart devices	Without additional infrastructure, BLE systems may not offer sufficient precision for indoor Location-Based Services [40].
IMU	Tracking, navigation	PDR systems are characterized by their cost-effectiveness, compact size, and low power consumption.	Over time, drift errors can accumulate, causing significant inaccuracies in estimating the device's position and orientation.

3.2 *Methods*

Different techniques are utilized for indoor localization and motion state classification. These include KNN for Wi-Fi RSS data, SVM and DT for categorizing movement states and phone positions, and RF with additional features, along with CP, for identifying walking patterns. Deep Learning methods like CNN, LSTM, and BLSTM prove effective in recognizing complex activities from smartphone sensor data, but they often necessitate substantial training with larger datasets. It's important to note that various localization methods offer distinct advantages, as detailed in Table 3.

Table 3. Motivations for using some of the methods in prev. studies.

Ref.	Approaches	Methods	Motivation
[1, 30]	DL	CNN	An efficient method for activity or gait recognition that uses neural networks to automatically learn relevant features and capture non-linear relationships between feature vectors.
[21]	Hybrid	LSTM-RNN	The specific cell unit and gate structure can retain updated information from previous moments via loop feedback connection, thus making them widely appropriate for artificial disturbance reduction.

3.3 *Data collection*

Modern smartphones and access points, like Wi-Fi and Bluetooth devices, employ various sensors, including accelerometers, gyroscopes, and sometimes magnetometers, to collect data for user location determination and pedestrian tracking. These sensors are vital for measuring travel distance, user actions, and detecting walking patterns. A combination of accelerometer, gyroscope, and, on occasion, a magnetometer can deduce stride length and movement direction, making them suitable for indoor localization and pedestrian tracking. Table 4 displays the Characteristics of Data Collection.

Table 4. The characteristics of Indoor Localization data collection.

Ref.	Environment	Technology	Sensor	No. of AP	Test Coverage Space
[14]	Office	Wi-Fi	N/A	5 AP	183.68 m^2
[29]	Campus Building	Wi-Fi	N/A	N/A	1000 m^2 1200 m^2 1500 m^2
[31]	N/A	Wi-Fi	N/A	6 AP	15 by 5m
[33]	University	Wi-Fi	N/A	20 AP	80 m by 40 m
[35]	University	Wi-Fi	N/A	64 AP	N/A
[36]	Engineering Building Tunnel System	Wi-Fi, IMU	Acc, Mag, Gyr	33 AP	N/A
[37]	Museum	Wi-Fi, IMU	Bar, Acc	42 beacons	2500 m^2
[38]	Office	Wi-Fi, IMU	Acc, Mag, Gyr	11 AP	11 m by 3 m
[39]	Shopping malls	Wi-Fi, IMU	Gyr, Acc	50 AP	119,685 m^2
[40]	Office	Wi-Fi, IMU	Acc, Gyr, Mag	8 AP	43.5 m by 11.2 m

Discussions related to our work and its results include motivation related to different properties of indoor localization, challenges and Recommendations.

Motivations: In the quest to enhance indoor positioning, researchers have strived to bolster both reliability and accuracy [8]. They have explored various methods such as integrating Wi-Fi with PDR, employing fusion algorithms, and optimizing the magnetic field approach. Energy-efficient alternatives have been sought, given the ubiquity of smartphones. Additionally, efforts have been directed at overcoming challenges related to location estimation, with a focus on BLE beacon distance determination and mitigating noise in Bluetooth and Wi-Fi-based positioning. PDR displacement estimation, real-time heading predictions, Wi-Fi FTM outlier identification, and Access Point (AP) placement optimization have all played key roles.

Challenges: Indoor localization faces significant challenges, particularly in terms of accuracy and signal strength. The accuracy problem arises from issues like conventional GPS limitations, signal degradation, indoor localization errors, and limited coverage [13]. Obstructions from moving objects can hinder ultrasonic signal accuracy, complicating tracking. Additionally, calibrating indoor maps adds to the complexity of achieving precise indoor localization [28]. Signal strength challenges revolve around complex indoor signal issues like reflections, shadowing, multipath effects, high signal attenuation, and noise interference.

Recommendations: To enhance indoor localization accuracy, researchers propose using more advanced landmark identification algorithms and implementing semi-supervised or unsupervised learning techniques. Some advocate for training multiple Neural Networks, including Convolutional Neural Networks (CNNs). In the context of pedestrian localization, it is recommended to incorporate multiple methods and involve a larger number of volunteers in experiments. However, limitations in volunteer recruitment were encountered due to hardware issues related to RFID technology

4 CONCLUSION

In this study, the systematic literature review (SLR) protocol was employed to explore a wide array of facets concerning indoor localization. Indoor localization has found application in numerous fields, and we conducted a thorough search across four databases to compile articles on the subject. Out of 3,750 initially identified articles, 60 were selected based on inclusion and exclusion criteria. This study sheds light on the technologies and methodologies used for indoor localization system development. While discussing motivations, challenges, and recommendations, we identified several gaps.

Additionally, this study suggests potential research directions, including the development of a robust model incorporating diverse walking patterns and participants, integrating CNN, RNN, and LSTM for walking pattern recognition. While we focused on smartphones, it's noteworthy that researchers can combine smartphones with other mobile devices like tablets. Quality assessment of each article based on content, source, and overall quality could be an avenue for future research. Furthermore, there were some overlaps in article categorization.

REFERENCES

Abadleh, A.; Maitah, W.; Salman, H.E.; Lasassmeh, O.; Hammouri, A. Smartphones-Based Crowdsourcing Approach for Installing Indoor Wi-Fi Access Points. *Int. J. Adv. Comput. Sci. Appl.* **2019**, *10*, 542–549.

Anand, S.; Bijlani, K.; Suresh, S.; Praphul, P. Attendance monitoring in classroom using smartphone & Wi-Fi fingerprinting. In *Proceedings of the 2016 IEEE Eighth International Conference on Technology for Education (T4E)*, Mumbai, India, 2–4 December 2016; pp. 62–67.

Ashraf, I.; Hur, S.; Park, S.; Park, Y. DeepLocate: Smartphone Based Indoor Localization with a Deep Neural Network Ensemble Classifier. *Sensors* **2020**, *20*, 133.

Ashraf, I.; Hur, S.; Park, Y. Application of deep convolutional neural networks and smartphone sensors for indoor localization. *Appl. Sci.* **2019**, *9*, 2337.

Ashraf, I.; Hur, S.; Park, Y. Smartphone Sensor Based Indoor Positioning: Current Status, Opportunities, and Future Challenges. *Electronics* **2020**, *9*, 891.

Bordoy, J.; Schindelhauer, C.; Höflinger, F.; Reindl, L.M. Exploiting Acoustic Echoes for Smartphone Localization and Microphone Self-Calibration. *IEEE Trans. Instrum. Meas.* **2019**, *69*, 1484–1492.

Chen, Z.; Zou, H.; Jiang, H.; Zhu, Q.; Soh, Y.C.; Xie, L. Fusion of WiFi, smartphone sensors and landmarks using the Kalman filter for indoor localization. *Sensors* **2015**, *15*, 715–732.

Ciabattoni, L.; Foresi, G.; Monteriù, A.; Pepa, L.; Pagnotta, D.P.; Spalazzi, L.; Verdini, F. Real time indoor localization integrating a model based pedestrian dead reckoning on smartphone and BLE beacons. *J. Ambient. Intell. Humaniz. Comput.* **2019**, *10*, 1–12.

Fetzer, T.; Ebner, F.; Bullmann, M.; Deinzer, F.; Grzegorzek, M. Smartphone-Based Indoor Localization within a 13th Century Historic Building. *Sensors* **2018**, *18*, 4095.

Guo, G.; Chen, R.; Ye, F.; Peng, X.; Liu, Z.; Pan, Y. Indoor Smartphone Localization: A Hybrid WiFi RTT-RSS Ranging Approach. *IEEE Access* **2019**, *7*, 176767–176781.

H.J.; Kim, S. Indoor smartphone localization based on LOS and NLOS identification. *Sensors* **2018**, *18*, 3987.

Han, K.; Yu, S.M.; Kim, S.-L. Smartphone-based Indoor Localization Using Wi-Fi Fine Timing Measurement. In *Proceedings of the 2019 International Conference on Indoor Positioning and Indoor Navigation (IPIN)*, Pisa, Italy, 30 September–3 October 2019; pp. 1–5.

Holcer, S.; Torres-Sospedra, J.; Gould, M.; Remolar, I. Privacy in indoor positioning systems: A systematic review. In *Proceedings of the 2020 International Conference on Localization and GNSS (ICL-GNSS)*, Tampere, Finland, 2–4 June 2020; pp. 1–6.

Hossain, A.M.; Soh, W.-S. A survey of calibration-free indoor positioning systems. *Comput. Commun.* **2015**, *66*, 1–13.

Hsu, H.-H.; Chang, J.-K.; Peng, W.-J.; Shih, T.K.; Pai, T.-W.; Man, K.L. Indoor localization and navigation using smartphone sensory data. *Ann. Oper. Res.* **2018**, *265*, 187–204.

Kang, W.; Han, Y. SmartPDR: Smartphone-based pedestrian dead reckoning for indoor localization. *IEEE Sens. J.* **2014**, *15*, 2906–2916.

Kunhoth, J.; Karkar, A.; Al-Maadeed, S.; Al-Ali, A. Indoor positioning and wayfinding systems: A survey. *Hum.- Cent. Comput. Inf. Sci.* **2020**, *10*, 1–41.

Lam, L.D.; Tang, A.; Grundy, J. Heuristics-based indoor positioning systems: A systematic literature review. *J. Locat. Based Serv.* **2016**, *10*, 178–211.

Laoudias, C.; Moreira, A.; Kim, S.; Lee, S.; Wirola, L.; Fischione, C. A survey of enabling technologies for network localization, tracking, and navigation. *IEEE Commun. Surv. Tutor.* **2018**, *20*, 3607–3644.

Lashkari, B.; Rezazadeh, J.; Farahbakhsh, R.; Sandrasegaran, K. Crowdsourcing and sensing for indoor localization in IoT: A review. *IEEE Sens. J.* **2018**, *19*, 2408–2434.

Li, X.; Wang, J.; Liu, C.; Zhang, L.; Li, Z. Integrated WiFi/PDR/smartphone using an adaptive system noise extended Kalman filter algorithm for indoor localization. *ISPRS Int. J. Geo-Inf.* **2016**, *5*, 8.

Li, Y.; Yan, K.J.I.S.J. Indoor Localization Based on Radio and Sensor Measurements. *IEEE Sens. J.* **2021**, *21*, 25090 25097.

Liu, J.; Qiu, Y.; Yin, K.; Dong, W.; Luo, J. RILS: RFID indoor localization system using mobile readers. *Int. J. Distrib. Sens. Netw.* **2018**, *14*, 1550147718771288.

Liu, T.; Zhang, X.; Li, Q.; Fang, Z. A visual-based approach for indoor radio map construction using smartphones. *Sensors* **2017**, *17*, 1790.

Luo, J.; Fu, L. A smartphone indoor localization algorithm based on WLAN location fingerprinting with feature extraction and clustering. *Sensors* **2017**, *17*, 1339.

Martínez del Horno, M.; García-Varea, I.; Orozco Barbosa, L. Calibration of Wi-Fi-based indoor tracking systems for Android- based smartphones. *Remote Sens.* **2019**, *11*, 1072.

Morar, A.; Moldoveanu, A.; Mocanu, I.; Moldoveanu, F.; Radoi, I.E.; Asavei, V.; Gradinaru, A.; Butean, A. A comprehensive survey of indoor localization methods based on computer vision. *Sensors* **2020**, *20*, 2641.

Obeidat, H.; Shuaieb, W.; Obeidat, O.; Abd-Alhameed, R. A review of indoor localization techniques and wireless technologies. *Wirel. Pers. Commun.* **2021**, *119*, 289–327.

Pascacio, P.; Casteleyn, S.; Torres-Sospedra, J.; Lohan, E.S.; Nurmi, J. Collaborative indoor positioning systems: A systematic review. *Sensors* **2021**, *21*, 1002.

Poulose, A.; Eyobu, O.S.; Han, D.S. An indoor position-estimation algorithm using smartphone IMU sensor data. *IEEE Access* **2019**, *7*, 11165–11177.

Rahman, A.M.; Li, T.; Wang, Y. Recent advances in indoor localization via visible lights: A survey. *Sensors* **2020**, *20*, 1382.

Seco, F.; Jiménez, A.R. Smartphone-based cooperative indoor localization with RFID technology. *Sensors* **2018**, *18*, 266.

Simões, W.C.; Machado, G.S.; Sales, A.M.; de Lucena, M.M.; Jazdi, N.; de Lucena, V.F., Jr. A review of technologies and techniques for indoor navigation systems for the visually impaired. *Sensors* **2020**, *20*, 3935.

Song, X.; Wang, M.; Qiu, H.; Li, K.; Ang, C. Auditory Scene Analysis-Based Feature Extraction for Indoor Subarea Localization Using Smartphones. *IEEE Sens. J.* **2019**, *19*, 6309–6316.

Sun, Y.; Wang, B. Indoor corner recognition from crowdsourced trajectories using smartphone sensors. *Expert Syst. Appl.* **2017**, *82*, 266–277.

Tan, S.Y.; Lee, K.J.; Lam, M.C. A Shopping Mall Indoor Navigation Application using Wi-Fi Positioning System. *Int. J.* **2020**, *9*, 4483–4489.

Vy, T.D.; Nguyen, T.L.; Shin, Y. A smartphone indoor localization using inertial sensors and single Wi-Fi access point. In *Proceedings of the 2019 International Conference on Indoor Positioning and Indoor Navigation (IPIN)*, Pisa, Italy, 30 September–3 October 2019; pp. 1–7.

Vy, T.D.; Nguyen, T.L.; Shin, Y. Inertial Sensor-Based Indoor Pedestrian Localization for IPhones. In Proceedings of the 2019 *International Conference on Information and Communication Technology Convergence (ICTC)*, Jeju, Republic of Korea, 16–18 October 2019; pp. 200–203. Liang, P.-C.; Krause, P. Smartphone-based real-time indoor location tracking with 1-m precision. *IEEE J. Biomed. Health Inform.* 2015, 20, 756–762.

Wang, Y.; Zhao, H. Improved smartphone-based indoor pedestrian dead reckoning assisted by visible light positioning. *IEEE Sens. J.* **2018**, *19*, 2902–2908.

Wang, Z.; Chen, R.; Xu, S.; Liu, Z.; Guo, G.; Chen, L. A Novel Method Locating Pedestrian With Smartphone Indoors Using Acoustic Fingerprints. *IEEE Sens. J.* **2021**, *21*, 27887–27896.

Waqar, W.; Chen, Y.; Vardy, A. Smartphone positioning in sparse Wi-Fi environments. *Comput. Commun.* **2016**, *73*, 108–117.

Wu, C.; Yang, Z.; Zhou, Z.; Liu, Y.; Liu, M. Mitigating Large Errors in WiFi-Based Indoor Localization for Smartphones. *IEEE Trans. Veh. Technol.* **2017**, *66*, 6246–6257.

Xia, H.; Zuo, J.; Liu, S.; Qiao, Y. Indoor localization on smartphones using built-in sensors and map constraints. *IEEE Trans. Instrum. Meas.* **2018**, *68*, 1189–1198.

Xiao, J.; Zhou, Z.; Yi, Y.; Ni, L.M. A survey on wireless indoor localization from the device perspective. *ACM Comput. Surv.* **2016**, *49*, 1–31.

Xu, S.; Chen, R.; Yu, Y.; Guo, G.; Huang, L. Locating Smartphones Indoors Using Built-In Sensors and Wi-Fi Ranging With an Enhanced Particle Filter. *IEEE Access* **2019**, *7*, 95140–95153.

Yao, Y.; Pan, L.; Feng, W.; Xu, X.; Liang, X.; Xu, X. A Robust Step Detection and Stride Length Estimation for Pedestrian Dead Reckoning Using a Smartphone. *IEEE Sens. J.* **2020**, *20*, 9685–9697.

Yassin, A.; Nasser, Y.; Awad, M.; Al-Dubai, A.; Liu, R.; Yuen, C.; Raulefs, R.; Aboutanios, E. Recent advances in indoor localization: A survey on theoretical approaches and applications. *IEEE Commun. Surv. Tutor.* **2016**, *19*, 1327–1346.

Yu, J.; Na, Z.; Liu, X.; Deng, Z. WiFi/PDR-integrated indoor localization using unconstrained smartphones. *EURASIP J. Wirel. Commun. Netw.* **2019**, *2019*, 1–13.

Zafari, F.; Gkelias, A.; Leung, K.K. A survey of indoor localization systems and technologies. *IEEE Commun. Surv. Tutor.* **2019**, *21*, 2568–2599.

Zhang, L.; Huang, L.; Yi, Q.; Wang, X.; Zhang, D.; Zhang, G. Positioning Method of Pedestrian Dead Reckoning Based on Human Activity Recognition Assistance. *In Proceedings of the 2022 IEEE 12th International Conference on Indoor Positioning and Indoor Navigation (IPIN)*, Beijing, China, 5–8 September 2022; pp. 1–8.

Zhang, Z.; He, S.; Shu, Y.; Shi, Z. A Self-Evolving WiFi-based Indoor Navigation System Using Smartphones. *IEEE Trans. Mob. Comput.* **2019**, *19*, 1760–1744.

Zhao, H.; Cheng, W.; Yang, N.; Qiu, S.; Wang, Z.; Wang, J. Smartphone-based 3D indoor pedestrian positioning through multi-modal data fusion. *Sensors* **2019**, *19*, 4554.

Zhao, H.; Zhang, L.; Qiu, S.; Wang, Z.; Yang, N.; Xu, J. Pedestrian Dead Reckoning Using Pocket-Worn Smartphone. *IEEE Access* **2019**, *7*, 91063–91073.

Zhao, Z.; Braun, T.; Li, Z.; Neto, A. A real-time robust indoor tracking system in smartphones. *Comput. Commun.* **2018**, *117*, 104–115.

Advances in AI for Biomedical Instrumentation, Electronics and Computing – Sachan et al. (eds)
© 2024 The Author(s), ISBN 978-1-032-64298-7

Enhancing sensing and prediction accuracy of earthquake using machine learning algorithms

Surbhi Bharti, Priya Pahwa & Sakshi Gupta
Indira Gandhi Delhi Technical University for Women, Delhi, India

Ashwni Kumar
Professor, Indira Gandhi Delhi Technical University for Women, Delhi, India

ABSTRACT: Earthquakes pose significant risks to both human life and infrastructure. Accurate prediction can mitigate these risks. In this research, we analyze various Machine Learning (ML) algorithms to find the best earthquake prediction model. We use a diverse dataset of seismological, geological, and environmental variables, evaluating algorithms like Random Forests (RF), Support Vector Machines (SVM), Recurrent Neural Networks (RNN), Decision Trees, and Gradient Boosting (GB). Our research highlights strengths and weaknesses in these models, contributing to the field by establishing a framework for selecting the best algorithm. These findings have the potential to revolutionize early warning systems, disaster preparedness, and emergency response. Future research may focus on real-time sensor data integration and model resilience for enhanced earthquake prediction.

1 INTRODUCTION

1.1 *Background*

Earthquakes, among the most formidable natural disasters, have long held a place of fascination and fear in the human psyche. These seismic events can strike suddenly and with devastating force, leaving a trail of destruction in their wake (Liu 2023). A lot of study has focused on how important it is to predict earthquakes in time for disaster response and management. For efficient earthquake hazard reduction, the forecasting of future earthquake frequency, magnitude, and epicenter is crucial. Earthquakes aren't rare occurrences; they happen more often in certain areas and less frequently in others. Some places have lots of small earthquakes, while some can go a long time without a big one (Dey 2022). Earthquakes can also cause smaller aftershocks that continue for days, weeks, or even months after the main quake, which can be risky. There are about 500,000 earthquakes every year that we can detect with modern tools, and around 100,000 of them can be felt by people. The main reason earthquakes happen is the movement of the Earth's tectonic plates. When they get stuck and can't move smoothly, stress builds up, and eventually, it's released as an earthquake. It's critical to distinguish between earthquake prediction and forecasting (Asmae 2022). The location, frequency, and magnitude of earthquakes are forecast using historical data and trends. It becomes possible to concentrate on earthquake-prone regions, implement better and faster-acting safety measures, construct structures that are earthquake resistant, and increase community awareness and prepared ness. On the other hand, earthquake prediction focuses on locating earthquakes (Beroza 2021). It is distinguished by the capacity to accurately predict the timing, location, and magnitude of an upcoming earthquake. Here are a few techniques and signs for predicting earthquakes:

- Radon Emissions: Scientists have discovered that large amounts of this radioactive gas are released from rock cracks just prior to an earthquake. As a result, a network of radon detectors may forecast earthquakes up to a week in advance. Animals reportedly exhibit strange behavior prior to earthquakes, according to legend.
- Hydro-chemical Alterations: In some earthquake-prone areas, groundwater has undergone chemical modifications.
- Foreshocks: Small tremors called foreshocks frequently follow large earthquakes. As a result, they are regarded as reliable indications for safety precautions. In this research endeavor, we delve into earthquake prediction using ML, drawing insights from a comprehensive dataset encompassing critical attributes. (1) Magnitude (M): Quantifies earthquake energy, aiding ML models in predicting future magnitudes. (2) Community-Determined Intensity (CDI): Considers human-perceived earthquake effects, helping ML models assess impacts on communities and infrastructure. (3) Moment Magnitude Inversion (MNI)(Mohamed S. Abdalzaher 2023): Uses seismic waveforms to estimate earthquake magnitude, enhancing prediction accuracy. (4) Tsunami Indicators: Analyze undersea fault characteristics to predict potential tsunami triggers. (5) Significance (SIG) (Asmae Berhich 2023): Assesses the statistical significance or regional importance of earthquake events for prioritization in prediction. (6) Number of Stations (NST) (Zhang 2023): More stations provide data for accurate ML predictions. (7) Minimum Distance to the Epicenter (DMIN): Measures earthquake impact on specific areas based on proximity to the epicenter. (8) Latitude and Longitude, (Zhang 2023; Zhu 2023): Geographical coordinates help ML predict risk regions and reach of earthquake effects. (9) Seismic Gap (GAP): Identifies areas along fault lines lacking recent significant earthquakes, aiding ML assessment of earthquake likelihood. Machine learning models can incorporate GAP data to assess the likelihood of earthquakes in these gaps.

Therefore, incorporating these factors into a machine learning-based earthquake prediction model can enhance its accuracy and reliability (Dey 2022). It's important to gather high quality data for training and validation and to continually refine and update the model as new data becomes available to improve its predictive capabilities. Properly designed machine learning models can contribute significantly to early warning systems and disaster preparedness efforts. Each of these variables plays a distinct role in understanding the dynamics of earthquakes, including their frequency and the potential for aftershocks.

1.2 *Challenges related*

In the subsequent sections of this paper, we will delve deeper into the methodologies employed, model evaluation, and the potential implications of our findings. The quest for more precise earthquake prediction continues to be a vital field of study, and our research strives to make a meaningful contribution towards this goal. Here are some potential research gaps to consider:

(1) Data Quality and Availability: Improve the quality and access to historical seismic data, especially in regions with limited monitoring.
(2) Data Fusion and Integration: Find better ways to combine geological, meteorological, and social data for accurate earthquake predictions.
(3) Transfer Learning: Explore adapting models trained in one region to another, reducing the need for extensive local data.
(4) Small Data and Rare Events: Develop techniques for earthquake prediction in regions with limited labeled data.
(5) Ethical and Societal Considerations: Address issues like data privacy, access to early warning systems, and the impact of false alarms in earthquake prediction research.

1.3 *Contributions*

Therefore, the major contributions of this research are as follows:

(1) Real world application: Accurate earthquake prediction can save lives and reduce damage in earthquake-prone areas. This practical application helps communities prepare and respond effectively to seismic events.
(2) Generalizability: Our findings aren't limited to just one region; they can be applied in other areas with similar seismic characteristics. This widens the scope of seismology and benefits various regions by sharing our predictive methods and insights.
(3) Novel Predictive Model: It relies on advanced ML techniques, specifically the GB model, to make our earthquake forecasts more precise and reliable.
(4) Improved Prediction Accuracy: The GB model standout by significantly improving prediction accuracy compared to other methods. It consistently delivers more precise earthquake predictions, enhancing our confidence in forecasting seismic events.
(5) Reduced False Alarms: Fewer incorrect warnings and increased trust in early warning systems, ultimately improving safety.
(6) Enhanced Understanding of Seismic Data: Our research deepens our understanding of seismic data, unveiling the complex factors that drive earthquake occurrences.

2 METHODOLOGY

2.1 *Utilization*

In recent years, there has been a growing interest in applying data-driven and ML approaches to earthquake prediction (Zhang 2023). These techniques involve the examination of various data sources, such as seismic data, GPS data, satellite imagery, and environmental data, to identify patterns and potential precursors. Certain earthquake-prone regions have introduced earthquake early warning systems (EEWS), which offer alerts to the public and critical infrastructure before the shaking from an earthquake reaches them. These systems use real-time data and algorithms to estimate the earthquake's magnitude and location on groundwater levels (Ebrahimi 2023), alterations in animal behavior, electromagnetic phenomena, and emissions of radon gas. However, the reliability of such precursors remains a topic of ongoing debate. Our research into earthquake prediction using ML followed a systematic approach to effectively address our research objectives (Mostafa Mousavi 2023).

Our methodology included the following key steps- (1) Data Collection: Gathered a comprehensive dataset with key attributes like magnitude, CDI, MNI, and more for ML model training and evaluation in earthquake prediction. (2) Data Preprocessing: Ensured data quality through exploratory data analysis and visualization techniques like scatter plots and correlation matrices. (3) Dataset Splitting: Split the data into training, validation, and test sets (80:20) for unbiased model evaluation. (4) ML Algorithms: Employed RF, GB, RNNs, and other algorithms to assess their performance in earthquake prediction. Included model training, validation, hyperparameter tuning, and cross-validation. Compared models using metrics like precision, accuracy, recall, AUC-ROC, and F1-score to find the most accurate forecast.

2.2 *Models and their working*

The models considered includes- (1) Logistic Regression: Estimates probabilities for binary outcomes based on independent variables (Abdalzaher 2023). (2) SVM: Finds a dividing line to separate classes with maximum margin, useful for classification and regression. (3) KNN: Classifies by looking at the majority class among nearby data points. (4) Gaussian Naive Bayes: Calculates class probabilities assuming features are independent. (5) Decision Trees:

Divides data into subsets based on the most important feature. (6) Random Forest (RF): Combines multiple decision trees to reduce overfitting and enhance accuracy. (7) Neural Networks (NN): Uses layers of interconnected neurons to learn complex patterns. (8) Gradient Boosting (GB): An ensemble method combining weaker models for high accuracy. It's great for analyzing seismic data in earthquake prediction, achieving *93.75%* accuracy.

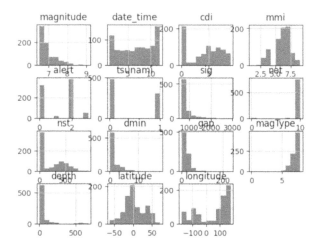

Figure 1. Histogram for analyzing the features.

3   RESULT AND DISCUSSION

3.1   *Data gathering and data cleaning*

In our research focused on earthquake prediction, we relied on a dataset containing seismic event-related information. This dataset served as the fundamental source of data for both training and assessing the performance of our ML models. Within this dataset, we encountered variables that encompassed seismic attributes (Abdalzaher 2023), geographical coordinates, and temporal data—all of which are highly relevant to the task of earthquake prediction.

We deleted attributes 'title', 'country', 'location' and 'continent' as they were not necessary for our analysis. This approach checked for null values in the dataset. Since the column 'alert' has missing values, we filled them with 'red'. Next, we used the pd.to date-time function from the PANDAS library to convert the values in the "date-time" column of the dataset into date-time objects. The "date-time" column contains date and time information in a string format, and this operation transforms those strings into a format that can be recognized and manipulated as dates and times. Then, we used 'Label-Encoder' to convert categorical variables in the dataset into numerical representations for numeric input. CORR () function was used to calculate the correlation between numerical columns in the dataset and SMOTE (Synthetic Minority Over-sampling Technique) class to address class imbalance. By generating synthetic samples, SMOTE helps prevent the model from being biased towards the majority class, improving its ability to make accurate predictions for both classes. Standard scaler function was utilized to standardize the feature data. Standardization is important in machine learning because it helps ensure that all features have a consistent scale, preventing certain features from dominating others in the modeling process. Standardized data often leads to improved model performance and convergence. To provide a visual representation of the dataset's attributes, we have included a histogram plot in a single figure. Each histogram is labeled with the corresponding attribute it represents, which helps to gain insights into the distribution of the data, providing valuable context for the subsequent stages of our analysis.

Table 1. Models and their accuracy.

S. No.	Model	Accuracy
1	Logistic Regression	83.33
2	SVC	83.33
3	Gradient Boosting	**93.75**
4	K Nearest Neighbor	84.46
5	Gaussian NB	80.72
6	Decision Tree	92.18
7	Random Forest	93
8	Neural Network	89.58

3.2 Model testing

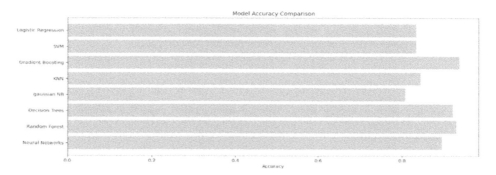

Figure 2. Graphical representation of model's accuracy.

Testing of the model on the resampled dataset is explained involving several key steps. Firstly, algorithm-selection and hyper parameter tuning is applied in various ML algorithms and fine-tuned their hyper-parameters to optimize accuracy (Abdalzaher 2023). We attempted to modify the parameters of cross-validation (CV) and epochs in our other models in pursuit of optimizing accuracy. However, it became evident that the GB method out-performed all our previous efforts in this regard. This process aimed to identify the best-performing models for our dataset. Secondly, cross-validation for robustness ensures the robustness of our models and prevent over-fitting. This method involved splitting the dataset into multiple subsets, training each model on different subsets, and evaluating their performance. It allowed us to gauge how well our models generalize to unseen data. Lastly, mean accuracy calculation was calculated by averaging the accuracy scores obtained across all folds and after CV application. This provided a single, representative accuracy value for each model, summarizing its overall performance.

To visualize and summarize our findings, we have included a table i.e. Table 1 listing the performance metrics of various machine learning models, highlighting their respective accuracy scores. Additionally, a corresponding graph i.e. Figure 2 has been generated to present a visual representation of the comparative accuracy of these models. This table and graph serve as valuable tools for readers, offering a clear and succinct overview of our model evaluations and ultimately assisting in the selection of the most accurate earthquake prediction model. Further, the accuracy scores are displayed on the x-axis, while the names of the models are listed on the y-axis, providing a comparison of their relative performance in graphical representation.

4 CONCLUSION

In this study, we have investigated the utilization of diverse ML models with a comprehensive seismic dataset encompassing vital attributes. Our aim was to evaluate and compare the predictive capabilities of these models to advance our comprehension of earthquake occurrences and enhance early warning systems. Ultimately, our findings have revealed that the Gradient Boosting model surpasses other models in terms of accuracy, achieving a maximum accuracy of 93.75%. The outcomes reaffirm the potential of ML in earthquake prediction and emphasize the critical role of model selection. While each model exhibits strengths in various applications, Gradient Boosting stands out for its superior earthquake prediction accuracy.

REFERENCES

Abdalzaher, I. M.-H. (2023). Seismic intensity estimation for earthquake early warning using optimized machinelearning model. *IEEE Transactions on Geoscience and Remote Sensing.*

Asmae Berhich, F. Z. (2023). An attention-based LSTM network for large earthquake prediction. *Soil Dynamics and Earthquake Engineering, Elsevier.*

Asmae, F. Z. (2022). A location-dependent earthquake prediction using recurrent neural. *Soil Dynamics and Earthquake Engineering*, Elsevier.

Beroza, M. S. (2021). Machine learning and earthquake forecasting–next steps. *Nature Communications.*

Dey, P. D. (2022). Intelligent solutions for earthquake data analysis and prediction for future. *Computers & Industrial Engineering*, Elsevier.

Ebrahimi, M. R. (2023). Spatial and temporal analysis of climatic precursors before major earthquakes in Iran (2011–2021). *Sustainability* 2023.

Liu, T. Z. (2023). Forecasting earthquake magnitude and epi-center by incorporating spatiotemporal priors into deep neural networks. *IEEE Transactions On Geo-science and Remote Sensing.*

Mohamed S. Abdalzaher, H. A. (2023). Employing machine learning and IoT for earth-quake early warning system in smart cities. *Energies 2023.*

Mostafa Mousavi S., G. C. (2023). Machine learning in earthquake seismology. *Annual Review of Earth and Planetary Sciences.*

Zhang, Z. H. (2023). EPT: A data-driven transformer model for earthquake prediction. *Engineering Applications of Artificial Intelligence*, Elsevier.

Zhu, M. W. (2023). Post-Earthquake rapid assessment method for electrical function of equipment in sub-stations. *IEEE Transactions on Power Delivery.*

Advances in AI for Biomedical Instrumentation, Electronics and Computing – Sachan et al. (eds)
© 2024 The Author(s), ISBN 978-1-032-64298-7

Intelligent smart-watch for enhanced cardiovascular health management

S.T. Rama
Department of Electronics and Communication Engineering, Dr. M.G.R. Educational and Research Institute, Chennai, India

S. Heeravathi
Department of Electrical and Electronics Engineering, Nexgen Technology, Puducherry, India

R. Ragul & N. Shanmugasundaram
Department Electrical and Electronics Engineering, Vels Institute of Science, Technology & Advanced Studies (VISTAS), Chennai, Tamil Nadu, India

ABSTRACT: This smart-watch is designed for cardiovascular patients, integrating AI for personalized medication management and health monitoring. The device features an in-built pillbox and an AI-powered alarm system for timely medication reminders. It enables emergency alerts via an SOS button and continuous heart rate monitoring. The Android application facilitates medication scheduling and notifications. Overall, the smart-watch serves as a crucial companion, ensuring medication adherence, health tracking, and prompt emergency responses for improved patient outcomes.

Keywords: Cardiovascular health, AI integration, Medication adherence, Emergency response, Health monitoring

1   INTRODUCTION

Adherence to medication poses a significant challenge in the management of chronic illnesses, with non-adherence contributing to compromised treatment outcomes and increased healthcare costs (Cu *et al.* 2020). Amidst the era of technological advancement, the integration of AI has emerged as a promising approach to tackle this issue, providing personalized and timely interventions for improved patient care.

Cardiovascular diseases (CVDs) remain a leading cause of global mortality, encompassing various conditions such as coronary heart disease and strokes. The critical importance of timely intervention in the "Golden Hour" following a heart attack underscores the need for efficient and responsive healthcare systems (Li *et al.* 2019).

In light of these challenges, this study presents an innovative smart-watch system empowered by AI. Combining an in-built pillbox with advanced AI algorithms and integrated with a heart rate sensor, TTGO T-display esp32, and Arduino Nano, the system enables continuous heart rate monitoring and personalized medication management (Din *et al.* 2019). By leveraging AI, the proposed system aims to revolutionize the management of cardiovascular health, offering a portable, accessible, and intelligent solution that integrates medication adherence, real-time health monitoring, and proactive emergency communication.

DOI: 10.1201/9781032644752-33

2 LITERATURE SURVEY

This research endeavors to develop a modular smart-watch equipped with an in-built pillbox and an automated alarm system to ensure timely medication intake for patients. By enabling the setting of multiple medication alarms through an intuitive interface, the device aims to streamline medication management (Bölen *et al.* 2020). However, existing medication adherence and emergency alert systems are critiqued for their lack of sophistication, wearability, and limited effectiveness in emergency situations. Various interventions, including electronic monitoring and smartphone applications, have been explored, highlighting the importance of wearability, accessibility, and user-friendliness, particularly for elderly and disabled individuals. Despite the advancements, challenges persist in creating a comprehensive, accessible, and user-friendly system for medication management and emergency alerts in critical healthcare scenarios. Several notable researchers have proposed innovative solutions for enhancing medication adherence. (Wasserlauf 2019) introduced a Smart Medicine Pill Box that automatically dispenses medicine based on the user's configured schedule through a smartphone application. Dorr focused on a modular smart medicine pill box capable of automatic dispensing and tracking the user's medication consumption progress, utilizing an Android-based application. (Dörr *et al.* 2019) developed an Automatic Pill Dispenser utilizing rotating compartments and an alarming system but requiring manual handling. Additionally, an Autonomous Pill Dispenser was designed to communicate via Wi-Fi, necessitating specific steps for dispensing medication. (Kheirkhahan *et al.* 2019) highlighted Med Minder, a pill-dispensing product with multiple compartments, although it lacks wearability and easy accessibility in emergencies. (Chen Xiao-Yong *et al.* 2023) introduces an intelligent health management method using data from wearable smartwatches. Physiological data is collected continuously and transmitted digitally, enabling physicians to assess individuals' health via a mathematical model. Results are communicated back through a smartphone app, offering a novel and effective strategy for personalized healthcare.

The existing research in medication adherence devices reveals a significant gap in the development of integrated, user-friendly solutions that cater to the specific needs of cardiovascular patients, particularly in terms of wearability, ease of access in emergencies, and intuitive interfaces for individuals with limited technological literacy. Closing this gap is essential to ensure improved medication adherence and timely intervention during critical health events, ultimately enhancing overall patient care and health outcomes.

SMART WATCH SYSTEM

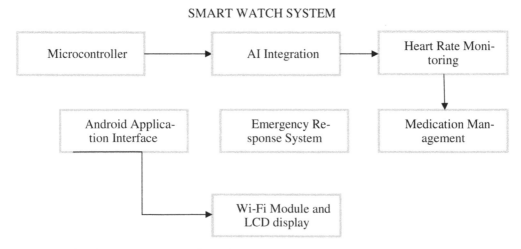

Figure 1.   Architecture of proposed flow diagram.

3 PROPOSED METHODOLOGIES

The proposed smart-watch system, fortified with AI-driven capabilities, represents a significant leap forward in enhancing healthcare management for cardiovascular patients. This advanced system surpasses traditional approaches by leveraging AI algorithms to ensure optimized medication adherence and prompt emergency responses. The integration of a microcontroller, Wi-Fi, buzzer, LED alert system, and AI-powered components empowers the smart-watch to operate as a sophisticated healthcare companion. With the AI-driven medication management feature, the system analyzes the user's historical health data, response patterns, and real-time physiological indicators to personalize medication schedules, fostering a proactive and tailored approach to treatment. Furthermore, the incorporation of AI-powered analytics in the heart-rate monitoring function enables the system to detect irregular heart-rate patterns, providing timely alerts to the user and healthcare providers, thereby facilitating early intervention in critical situations. The Android application interface, supported by AI algorithms, not only facilitates seamless medication scheduling and reminders but also delivers personalized health insights, empowering users to make informed decisions about their well-being. Moreover, the system's SOS functionality, enabled by AI technology, ensures swift communication with caregivers or hospitals during emergencies, thereby enhancing overall user safety and well-being. By harnessing the transformative power of AI, the smart-watch system revolutionizes the landscape of cardiovascular healthcare, offering a comprehensive, intelligent, and user-centric solution. Its AI-driven functionalities not only streamline medication adherence and emergency responses but also promote proactive health management, fostering a new era of personalized and data-driven healthcare solutions.

4 RESULTS AND DISCUSSION

The expected result of this project is the microcontroller to run the programmed code and get input from sensors and send data through Wi-Fi to the android application and can send SOS message and open the emergency pill box when emergency button is pressed and reminders works as per set time with turning on buzzer and light. In comparison to existing systems, this design offers a comprehensive solution that integrates various functionalities crucial for emergency response and health management, making it a robust and versatile option for users.

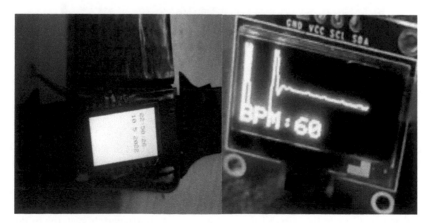

Figure 2.   Output of time and heart rate.

Figure 3. Setting alarm and emergency message on Blynk

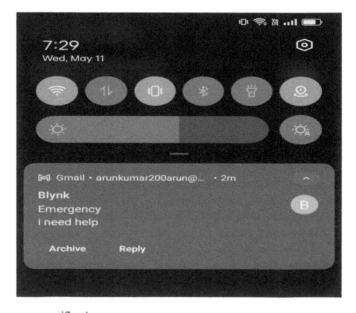

Figure 4. Emergency notification

5 CONCLUSION AND FUTURE DIRECTIONS

This paper summarizes the pivotal role of AI-driven smart-watch systems in reshaping the paradigms of cardiovascular healthcare. It highlights the system's potential in empowering patients with personalized care, enabling timely interventions, and fostering a proactive approach to health management. Moreover, it discusses the future directions and potential advancements in AI technologies, emphasizing the need for continued research and development to further enhance the capabilities and impact of smart-watch systems in delivering comprehensive and intelligent healthcare solutions for cardiovascular patients worldwide.

REFERENCES

Bölen, M. C. (2020). From traditional wristwatch to smartwatch: Understanding the relationship between innovation attributes, switching costs and consumers' switching intention. *Technology in Society*, 63, 101439.

Cu, G., Chan, D., Chua, A. G., & Ramin, E. G. A smart medicine pill box for improving medication adherence.

Din, S., & Paul, A. (2019). RETRACTED: Smart health monitoring and management system: Toward autonomous wearable sensing for internet of things using big data analytics.

Dörr, M., Nohturfft, V., Brasier, N., Bosshard, E., Djurdjevic, A., Gross, S., ... & Eckstein, J. (2019). The WATCH AF trial: SmartWATCHes for detection of atrial fibrillation. *JACC: Clinical Electrophysiology*, 5(2), 199–208.

Kheirkhahan, M., Nair, S., Davoudi, A., Rashidi, P., Wanigatunga, A. A., Corbett, D. B., Mendoza, T., Manini, T.M. and Ranka, S., 2019. A smartwatch-based framework for real-time and online assessment and mobility monitoring. *Journal of biomedical informatics*, 89, pp. 29–40.

Li, Y., Zheng, L., & Wang, X. (2019). Flexible and wearable healthcare sensors for visual reality health-monitoring. *Virtual Reality & Intelligent Hardware*, 1(4), 411–427.

Wasserlauf, J., You, C., Patel, R., Valys, A., Albert, D., & Passman, R. (2019). Smartwatch performance for the detection and quantification of atrial fibrillation. *Circulation: Arrhythmia and Electrophysiology*, 12(6), e006834.

Xiao-Yong, C.H.E.N., Bo-Xiong, Y.A.N.G., Shuai, Z.H.A.O., Jie, D.I.N.G., Peng, S.U.N. and Lin, G.A.N., 2023. Intelligent health management based on analysis of big data collected by wearable smart watch. *Cognitive Robotics*, 3, pp. 1–7.

Advances in AI for Biomedical Instrumentation, Electronics and Computing – Sachan et al. (eds)
© 2024 The Author(s), ISBN 978-1-032-64298-7

Advancements in robotics and prosthetics: A comprehensive study of voice-controlled hand prosthetic gloves

Ruchita Gautam, Himanshu Chaudhary, Apoorv Sharma, Gautam Matta, Ayush Jain, Aryan Sharma & Adhyan Kaushik
KIET Group of Institutions, Delhi NCR, Ghaziabad, Uttar Pradesh, India

ABSTRACT: The Voice-Controlled Hand Prosthetics Glove initiative represents a significant leap in prosthetic technology, integrating voice control with cutting-edge hand prosthetics to enhance independence and dexterity for amputees. Tracing the evolution of robotics, the project leverages advancements in natural language processing and prosthetic development. The glove's design focuses on comfort and functionality, seamlessly integrating sensors and actuators. The user experience involves training, daily usage assessments, and iterative improvements based on user feedback. Anticipated results include enhanced voice control for delicate tasks, personalized prosthetics, and novel connections to the neural system for improved sensory feedback. The initiative aims to empower amputees, breaking societal barriers associated with limb loss. As we embark on this endeavor, our commitment lies in providing a more inclusive future, where limb loss doesn't limit one's abilities, fostering optimism and promise for the project's future.

1 INTRODUCTION

The fusion of robotics and prosthetics, a testament to human ingenuity, merges technology with a mission to enhance lives. This journey traces the evolution of robotics and its subfields, spotlighting a groundbreaking innovation: voice-controlled hand prosthetic gloves. The Industrial Revolution in the 18th and 19th centuries laid the automation foundation, with Jacquard's loom and Babbage's engine foretelling modern robotics. The early 20th century saw Tesla's strides in remote control technology and the rise of "tele automation." (Atzori *et al.* 2017) Shaped by World War II, the 1940s-1960s advanced robotics, notably in bomb disposal teleoperation. Coined by Karel Čapek in 1920, "robot" gained substance in the 1950s with Unimate, the inaugural industrial robot transforming heavy machinery handling. The 1970s-1980s marked industrial robot adoption, coinciding with AI robotics, fostering computer vision and machine learning. The 1990s-2000s introduced mobile robotics with autonomous vehicles and space exploration. From the 2010s, collaborative robots revolutionized industries, and AI elevated capabilities, while humanoid robots like Boston Dynamics' (Humlum *et al.* 2019) Atlas showcased unprecedented agility. As of September 2021, emerging trends encompassed soft robotics, bio-inspired designs, swarm robotics, and ethical considerations (Majidi *et al.* 2014) in robot rights.

1.1 *Fields of robotics*

Robotics comprises vital subfields crucial for system development. One domain involves crafting physical structures, merging materials science, kinematics, dynamics, and control

DOI: 10.1201/9781032644752-34

systems. Electrical components, encompassing sensors, actuators, power supplies, and electronic controls, represent another key specialization (Vajk *et al.* 2016). Software development is essential, employing computer vision, machine learning, natural language processing, and path planning for environmental perception and decision-making. Control theory, aided by AI and machine learning, refines robot behaviour, fostering learning and adaptation (Truszkowski *et al.* 2006). Computer vision, utilizing sensors like cameras and LiDAR, processes visual data, enhancing obstacle identification. Ethical considerations arise in societal integration, prompting the exploration of robot rights. Designing human-robot interaction emphasizes effective communication, while biology-inspired robotics targets flexible structures (Cross *et al.* 2019). These dynamic fields encapsulate the specialized knowledge propelling robotics forward.

2 LITERATURE REVIEW

The utilization of EMG signals in prosthetic hand actuation traces back to 1948, with the first application of MES technology occurring in 1957 at the Central Prosthetic Research Institute in Moscow. Initially, a basic on-off control scheme emerged, utilizing the amplitude of EMG signals to activate or deactivate the prosthetic device based on a preset threshold, (Antfolk *et al.* 2010) following extensive myoelectric control strategy research. Sequential control dominates user-operated prosthetic hands, employing methods like on-off control, where EMG thresholds dictate hand movement, and proportional control, linking motor speed to EMG signal strength. Direct control employs distinct EMG (Jiang *et al.* 2013) sites for independent finger control, while finite state machine control maps predefined hand postures to states. Pattern recognition-based control involves extracting and classifying EMG data features, and posture control schemes are also in use. The research aims to implement simultaneous control (Yang *et al.* 2020), marking a trajectory toward enhanced prosthetic hand functionality.

3 OUR INITIATIVES

Amputees face significant challenges in regaining their independence and functionality after limb loss. While prosthetic limbs have made remarkable advancements in recent years, the need for more intuitive control mechanisms and enhanced dexterity remains. Voice-controlled hand prosthetics represent a groundbreaking fusion of technology and innovation, offering a promising solution to empower amputees with unprecedented control and freedom. Traditional prosthetic hands often lack the precision and adaptability needed for everyday tasks. The challenge lies in developing a prosthetic device that seamlessly integrates with the user's intentions and mimics natural hand movements. Additionally, accessibility and affordability remain significant barriers for many amputees. This project aims to address these issues by creating a voice-controlled hand prosthetics glove that enhances dexterity and independence while remaining cost-effective.

Objectives

The primary objectives of the Voice-Controlled Hand Prosthetics Glove project are as follows:

- To develop an advanced prosthetic hand that responds to voice commands for precise control.
- To design a comfortable and durable glove interface for users.
- To enhance the adaptability of the prosthetic hand for various daily tasks.

4 TECHNOLOGY AND INNOVATION

4.1 *Voice control technology*

Voice control technology has witnessed remarkable advancements with the advent of artificial intelligence and machine learning. Natural language processing (NLP) algorithms enable accurate interpretation of spoken commands, allowing for precise control of various devices. Integrating this technology into prosthetic hands opens up a new realm of possibilities for amputees.

4.2 *Prosthetic hand development*

Prosthetic hand development has progressed from basic mechanical devices to sophisticated myoelectric and neuro-controlled systems. (Weir *et al.* 2009) However, challenges such as fine motor control, adaptability to various tasks, and user-friendliness persist. Innovations in materials and design have also played a pivotal role in enhancing the functionality of prosthetic hands.

4.3 *Integration of voice control and prosthetics*

The integration of voice control technology into prosthetic hands represents a significant leap forward. This fusion allows users to execute precise movements and gestures simply by verbalizing their intentions. The synergy between these two domains has the potential to redefine the capabilities of prosthetic limbs and improve the quality of life for amputees.

Figure 1. Design of prosthetic hands.

5 DESIGN AND DEVELOPMENT

5.1 *Glove design*

The design of the glove interface is crucial to the success of the voice-controlled prosthetic hand. Factors such as comfort, fit, and ease of use are paramount. The glove should provide a snug yet breathable fit to ensure extended wear without discomfort. Additionally, it should seamlessly integrate the necessary sensors and actuators for capturing and executing hand movements.

5.2 *Prosthetic hand mechanism*

The core of the project revolves around the prosthetic hand mechanism. It should replicate the functionality of a natural hand as closely as possible. This involves the development of advanced actuators and sensors that can mimic the intricate movements of human fingers and joints. Durability and responsiveness are also critical aspects of this development.

5.3 *Voice control interface*

The voice control interface should be intuitive and user-friendly. Natural language processing algorithms will be employed to convert spoken commands into actionable instructions for the prosthetic hand. The interface should also allow users to customize and personalize their commands for different tasks and preferences.

6 USER EXPERIENCE

6.1 *Training and adaptation*

Training and adaptation are essential components of the user experience. Users will undergo a training program to familiarize themselves with the voice-controlled interface and the prosthetic hand's capabilities. Ongoing support and guidance will be provided to ensure a smooth transition to this new technology.

6.2 *Daily usage*

The success of the voice-controlled hand prosthetics glove hinges on its practicality in daily life. Users will share their experiences using the device for various tasks, from simple gestures like gripping objects to more complex actions like typing or cooking. Real-world scenarios will be explored to assess the glove's effectiveness.

6.3 *Feedback and improvement*

Regular feedback sessions will be conducted to gather insights, identify pain points, and make necessary adjustments to the glove and voice control interface. This iterative process ensures that the technology continually evolves to meet user needs.

7 PROSPECTIVE RESULTS

The future of Voice-Controlled Hand Prosthetics (VCHP) promises a transformative shift in user experience, marked by enhanced intuitiveness in voice control for delicate tasks, ensuring higher satisfaction and acceptance. Ongoing research aims to elevate prosthetic functionality, enabling users to engage in complex activities like playing instruments and typing with precision. A key focus is personalization, envisaging prosthetics tailored to individual needs through configurable grips, hand movements, and visual options, fostering a seamless extension of the user. Anticipated medical advancements include novel techniques connecting prosthetics to the neural system, enhancing sensory feedback and control. Ultimately, VCHP seeks to empower amputees, fostering inclusion by enabling them to pursue occupations and activities previously deemed impossible, dismantling societal barriers and stigmas associated with limb loss.

8 CONCLUSION

The Voice-Controlled Hand Prosthetics Glove initiative is a major advancement in prosthetic technology. We want to provide amputees more independence and dexterity by utilizing voice control and combining it with cutting-edge prosthetic hand technologies. This invention has the power to change a great number of people's lives by giving them back control over their everyday routines and promoting an inclusive society. As we commence this endeavor, our dedication to enhancing the well-being of those who have encountered limb amputation and expanding the frontiers of feasibility within the realm of assistive technology is a lot of optimism and promise for our Voice-Controlled Hand Prosthetics project's future. To give amputees a better and more inclusive future, we are committed to pushing the boundaries of technology and human-machine interaction while giving them more dexterity and mobility. Throughout this endeavor, we remain committed to improving the well-being of those who have experienced limb amputations and expanding the possibilities of assistive technology. For the future of our Voice-Controlled Hand Prosthetics project, there is much optimism and promise. We are building a future where the control of the loss of a limb does not limit a person's abilities.

REFERENCES

Antfolk, Christian, *et al.* "Using EMG for real-time prediction of joint angles to control a prosthetic hand equipped with a sensory feedback system." *Journal of Medical and Biological Engineering* 30.6 (2010): 399–406.

Atzori, Luigi, Antonio Iera, and Giacomo Morabito. "Understanding the Internet of Things: definition, potentials, and societal role of a fast-evolving paradigm." *Ad Hoc Networks* 56 (2017): 122–140.

Cross, Emily S., Ruud Hortensius, and Agnieszka Wykowska. "From social brains to social robots: applying neurocognitive insights to human–robot interaction." *Philosophical Transactions of the Royal Society B* 374.1771 (2019): 20180024.

Humlum, Anders. "Robot adoption and labor market dynamics." *Princeton University* (2019).

Jiang, Ning, *et al.* "Is accurate mapping of EMG signals on kinematics needed for precise online myoelectric control" *IEEE Transactions on Neural Systems and Rehabilitation Engineering* 22.3 (2013): 549–558.

Majidi, Carmel. "Soft robotics: a perspective—current trends and prospects for the future." *Soft Robotics* 1.1 (2014): 5–11.

Truszkowski, Walter F., *et al.* "Autonomous and autonomic systems: A paradigm for future space exploration missions." *IEEE Transactions on Systems, Man, and Cybernetics, Part C (Applications and Reviews)* 36.3 (2006): 279–291.

Vajk, István, *et al.* "BME VIK annual research report on electrical engineering and computer science 2015." *Periodica Polytechnica Electrical Engineering and Computer Science* 60.1 (2016): 1–36.

Weir, RF ff, Jonathon W. Sensinger, and M. Kutz. "Design of artificial arms and hands for prosthetic applications." *Biomedical Engineering and Design Handbook* 2 (2009): 537–598.

Yang, Geng, *et al.* "Homecare robotic systems for healthcare 4.0: Visions and enabling technologies." *IEEE Journal of Biomedical and Health Informatics* 24.9 (2020): 2535–2549.

Advances in AI for Biomedical Instrumentation, Electronics and Computing – Sachan et al. (eds)
© 2024 The Author(s), ISBN 978-1-032-64298-7

A review on analysis and prediction methodologies of crop yield pattern

Neha & Anil Ahlawat
Department of Computer Science Engineering, KIET Group of Institutions, Delhi NCR Ghaziabad, India

Himanshu Chaudhary
Department of Electronics and Communication Engineering, KIET Group of Institutions,
Delhi NCR Ghaziabad, India

ABSTRACT: Making supportive decisions for crop production, such as crop name recommendations and forecasts for crop production, requires machine learning implementation. Numerous machine learning classifiers and algorithms are employed in this. Climatic factors (temperature, rainfall, precipitation, sun radiation, and humidity), edaphic factors (pH, kind, and proportions of nitrogen, phosphorus, and potassium in the soil), and physiographic factors (altitude, steepness of slope, exposure to light, and wind) are some of the variables or factors that influence crop production. However, the most common variables are temperature, soil pH, precipitation, and humidity, and the most common algorithms are Examples of artificial neural networks and deep learning algorithms include Random Forest, Support Vector Machine, Linear Regression, Long Short-Term Memory, Convolution Neural Network, and Deep Neural Network. As part of this study, we reviewed the literature to ascertain the techniques and influencing factors that have been used in agricultural output prediction.

1 INTRODUCTION

A plant that is extensively cultivated or farmed is referred to as a crop. In other words, a crop is any plant that is widely cultivated and harvested for economic gain. A technological tool that helps farmers choose the best crops for their fields is called a crop recommendation system. In order to give farmers subjective advice, the system takes into account multiple factors, including climate, soil type, irrigation, and market demand. The technology can help raise crop yields, decrease waste, and enhance overall agricultural operation efficiency by offering farmers customized crop recommendations. The crop suggestion system has the advantage of assisting farmers in selecting crops depending on market demand. The approach can assist farmers in selecting crops that are most likely to sell by taking market trends and customer preferences into account, maximizing their revenues, and lowering the danger of financial loss.

Predicting crop output is a crucial component of agriculture that aids farmers in making decisions regarding their crops. It involves determining how many crops will be produced in a specific area based on a variety of variables, including the type of soil, the climate, and crop management techniques. Machine learning (ML) has become a potent tool for forecasting crop yield in recent years. The goal of ML (a subfield of artificial intelligence (AI)) is to build systems that can improve performance by learning from the data they process. It is a useful strategy that offers a superior crop output forecast based on a variety of influencing factors.

2 RELATED WORK

Young farmers who recently began farming may face the issue of not knowing what to plant and what to get benefits from. This is a challenge that must be taken care of, and we are working on it. Making better judgements, decreasing losses, and managing the risk of price swings will be aided by forecasting the appropriate crop and output (Sundari *et al.* 2022). Crop yield assessment is crucial in agriculture for increasing production, aiding in decision-making procedures involving the forecasting of financial markets, and resolving difficulties with food safety. The major aims of the piece are to anticipate agricultural yields more accurately utilizing hybrid machine learning (ML) techniques (Anbananthen *et al.* 2021). Crop yield varies from year to year because of variations in operational and economic aspects, as well as the climate. Predicting how much land will produce, leads to improved field management and operations. This study analyses and predicts the key elements that influence the growth of basic crops in African desert and dry climate zones (Sherif, n.d.). Artificial intelligence is mostly used to teach computers how to use knowledge or past experience to solve problems in the real world. As a result, the paper suggested a framework for predicting collection yield that incorporates simulated intelligent strategies and calculations as well as recommendations and various client-set constraints, and creates alternative crops for higher yields (Prathap *et al.* 2023). The objective of the paper is to showcase a Python-based system that uses intelligent algorithms to forecast the most productive harvest given the conditions (weather and soil condition of the land are considered) and at the lowest possible cost. As a result, SVM, RNN, and LSTM are used (Agarwal & Tarar 2021). This study provides forecasts for practically each and every crop grown in India. The client can anticipate crop production in any year they decide on, using simple parameters such as State, District, Season, and Region to construct this script novel. The research employs advanced regression techniques including Lasso, ENet, and Kernel Ridge to forecast the yield (Nishant *et al.* 2020).

3 METHODOLOGY

Steps followed in review (as depicted in Figure 1):

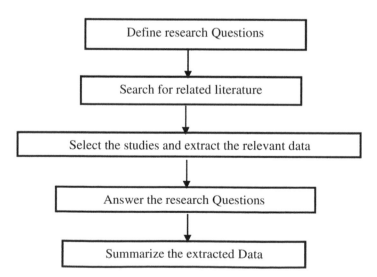

Figure 1.   Procedure followed for a literature review.

3.1 *Plan the review*

Define the study questions. The identified questions for study are:

a) Which machine learning techniques are employed to forecast agricultural yields?
b) What influencing elements were employed as machine learning yield prediction parameters?

3.2 *Conduct the review*

Examine the relevant literature that responds to the stated issues.

There are numerous studies that have been published. Use the relevant search strings in several databases, such as Web of Science, Google Scholar, Scopus, Springer Link, etc., to choose the pertinent one for study.

3.3 *Report the review*

Address the study topic and provide outcome documentation.

4 RESULTS

Address the question(a): The most popular machine learning algorithms include the Support Vector Machine (SVM), Random Forest (RF), Linear Regression Algorithm, and Neural Networks (such as CNN, RNN, and LSTM).

4.1 *Support vector machine*

SVM is a potent classification technique that can handle large datasets with a limited number of observations and is a great option for crop recommendation systems. The main problem Indian farmers are facing is that they usually select the wrong crop for their soil's requirements. The suggested system will use the results of the soil testing laboratory to gather data, construct an ensemble model employing majority voting, and employ SVM and ANN as learners to recommend a crop (Rajak *et al.* 2017).

4.2 *Random forest*

Because of this technology's ability to estimate crop production before cultivation, farmers will be able to make the best possible selections. Implementing such an approach allows the machine learning algorithm to be delivered along with user-friendly web-based visual software (Suresh *et al.* 2021).

4.3 *Linear regression*

Linear regression is used for prediction on fresh datasets. It learns from labeled datasets by mapping data points to the best-performing linear function. To have a good understanding of how different crops produce themselves and how the state's various climatic conditions affect crop productivity, multiple linear regression analyses have been employed to examine how area and weather variables affect crop yield (Bhattacharyya *et al.* 2021).

4.4 *Neural networks*

- *Convolutional Neural Network (CNN):* CNN, one of the several deep learning paradigms, has proven to be very effective at classification and analysis of images. With CNNs, there is no need for pre-calculated features because the network layers that use convolutional technology handle feature extraction and the top features are discovered during training. Because of their structure, much training data is required for CNNs to converge (Nevavuori *et al.* 2019).

- *Recurrent Neural Network (RNN):* RNN works on sequential data and time series. RNN consists of a number of layers and has inbuilt memory that allows it to memorize the output of each layer. The most popular independent variables used in agricultural production forecast modeling utilizing Artificial Neural Networks (ANN) are presented and discussed in the following article. Environmental factors, including climate information, temperature of the air, total amount of rain, insolation, and soil characteristics, are given special consideration. The report emphasizes how the spread of precision agriculture is made possible by the widespread usage of photogrammetric and remote sensing methods (Hara *et al.* 2021).
- *Long Short-term memory (LSTM):* The LSTM considers the previous historical values, analyses strange patterns by self-regulating in accordance with the full patterns, and then provides additional predictions (Bhimavarapu *et al.* 2023).

There are many machine learning algorithms that outperform in the field of agriculture. But our main objective was to look for the most widely used algorithms. Table 1 contains some of the algorithms used for prediction and their accuracy.

Table 1. Depicting some already used algorithms with their accuracy.

Crop Name	Algorithm used	Accuracy	Reference
For All Crops	Decision Tree	98.62	(Kavita & Mathur 2020)
For all type of Crops	Deep Reinforcement Learning Model	93.7	(Elavarasan & Durairaj Vincent 2020)
Wheat	Supervised Kohonen Networks (SKNs)	81.65%	(Pantazi *et al.* 2016)
Paddy yield prediction	Random Forest	91.23	(Elavarasan *et al.* 2020)
Oil palm disease recognition	Naïve Bayes	85	(Husin *et al.* 2020)
Oil palm growth stage monitoring	Hybrid CNN-SVM	90	(Toh *et al.* 2019)

Address the question (b):

The influencing variables that machine learning uses as parameters are shown in Figure 2:
Although there are many features that have been used in different papers, such as precipitation, soil moisture, temperature, ratio of nutrients in the soil, pH of the soil, sowing period, leaf area, evapotranspiration, area under irrigation, used fertilizers, etc. But the most widely used are temperature, rainfall, humidity, and soil type. The seasonal crop to be grown is identified on the basis of these features.

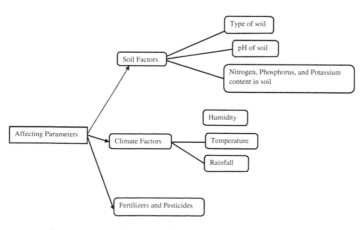

Figure 2. Parameters that influence the production of crops.

5 CONCLUSION

For crop prediction, numerous machine learning and deep learning algorithms have been used. Some are employed frequently, while others are utilized only in specific studies. In this paper, it is concluded that there isn't a single model that performs best. The dataset that was used and the characteristics that have an impact on crop yield determine the model's performance. Therefore, it cannot be assumed that a single algorithm will be the most effective for all kinds of datasets.

REFERENCES

Agarwal, S., & Tarar, S. (2021). A hybrid approach for crop yield prediction using machine learning and deep learning algorithms. In *Journal of Physics: Conference Series* (Vol. 1714, No. 1, p. 012012). IOP Publishing.

Anbananthen, K. S. M., Subbiah, S., Chelliah, D., Sivakumar, P., Somasundaram, V., Velshankar, K. H., & Khan, M. A. (2021). An intelligent decision support system for crop yield prediction using hybrid machine learning algorithms. *F1000Research*, 10.

Bhattacharyya, B., Biswas, R., Sujatha, K., & Chiphang, D. Y. (2021). Linear regression model to study the effects of weather variables on crop yield in Manipur state. *Int. J. Agricult. Stat. Sci*, 17(1), 317–320.

Bhimavarapu, U., Battineni, G., & Chintalapudi, N. (2023). Improved optimization algorithm in LSTM to predict crop yield. *Computers*, 12(1), 10.

Elavarasan, D., Vincent PM, D. R., Srinivasan, K., & Chang, C. Y. (2020). A hybrid CFS filter and RF-RFE wrapper-based feature extraction for enhanced agricultural crop yield prediction modeling. *Agriculture*, 10 (9), 400.

Elavarasan, D., & Vincent, P. D. (2020). Crop yield prediction using deep reinforcement learning model for sustainable agrarian applications. *IEEE Access*, 8, 86886–86901.

Hara, P., Piekutowska, M., & Niedbała, G. (2021). Selection of independent variables for crop yield prediction using artificial neural network models with remote sensing data. *Land*, 10(6), 609.

Husin, N. A., Khairunniza-Bejo, S., Abdullah, A. F., Kassim, M. S., Ahmad, D., & Aziz, M. H. (2020). Classification of basal stem rot disease in oil palm plantations using terrestrial laser scanning data and machine learning. *Agronomy*, 10(11), 1624.

Nishant, P. S., Venkat, P. S., Avinash, B. L., & Jabber, B. (2020, June). Crop yield prediction based on Indian agriculture using machine learning. In *2020 International Conference for Emerging Technology (INCET)* (pp. 1–4). IEEE.

Kavita, M., & Mathur, P. (2020, October). Crop yield estimation in India using machine learning. *In 2020 IEEE 5th International Conference on Computing Communication and Automation (ICCCA)* (pp. 220–224). IEEE.

Nevavuori, P., Narra, N., & Lipping, T. (2019). Crop yield prediction with deep convolutional neural networks. *Computers and Electronics in Agriculture*, 163, 104859.

Pantazi, X. E., Moshou, D., Alexandridis, T., Whetton, R. L., & Mouazen, A. M. (2016). Wheat yield prediction using machine learning and advanced sensing techniques. *Computers and Electronics in Agriculture*, 121, 57–65.

Prathap, S., Amulya, P., Anupa, K. N., Sai, N. D., & Shekar, B. G. (2023) Assistance to the farmer. *International Journal of Engineering Technology and Management Sciences* Volume No.7

Rajak, R. K., Pawar, A., Pendke, M., Shinde, P., Rathod, S., & Devare, A. (2017). Crop recommendation system to maximize crop yield using machine learning technique. *International Research Journal of Engineering and Technology*, 4(12), 950–953.

Sherif, H. (2022). Machine Learning in Agriculture: Crop Yield Prediction.

Sundari, V., Anusree, M., & Swetha, U. (2022). Crop recommendation and yield prediction using machine learning algorithms. *World Journal of Advanced Research and Reviews*, 14(3), 452–459.

Suresh, N., Ramesh, N. V. K., Inthiyaz, S., Priya, P. P., Nagasowmika, K., Kumar, K. V. H., ... & Reddy, B. N. K. (2021, March). Crop yield prediction using random forest algorithm. In *2021 7th International Conference on Advanced Computing and Communication Systems (ICACCS)* (Vol. 1, pp. 279–282). IEEE.

Toh, C. M., Tey, S. H., Ewe, H. T., & Vetharatnam, G. (2019, December). Classification of oil palm growth status with L band microwave satellite imagery. *In 2019 Photonics & Electromagnetics Research Symposium-Fall (PIERS-Fall)* (pp. 1824–1831). IEEE.

Advances in AI for Biomedical Instrumentation, Electronics and Computing – Sachan et al. (eds)
© 2024 The Author(s), ISBN 978-1-032-64298-7

A review paper on comparative analysis of various reconfiguration techniques under partial shading environments

Pushpender & S. Khatoon
Department of Electrical Engineering, Jamia Millia Islamia, New Delhi, India

M.F. Jalil
Department of Electrical Engineering, Aligarh Muslim University, Aligarh, India

ABSTRACT: The performance of the photovoltaic array mostly directly depends on the darkness condition. It reduces the output power transfer by the photovoltaic cell. The reduction in output power stimulates the researcher to investigate the various strategies to decrease the effect of darkness. Thus, it is necessary to put in place suitable reconfiguration methods and algorithms to transfer maximum power under dark conditions. This review paper presents a comparison of parallel, series, series-parallel, bridge-linked, honeycomb, and total-cross-tied in terms of fill factor, power loss, efficiency, mismatch loss, shading loss, array size, etc. In addition, there is a comparison between the different static reconfigurations like Sudoku, optimal Sudoku, magic square, zigzag, square puzzle, and skyscraper, with their advantages and disadvantages. This study can be used to choose the improved reconfiguration techniques for gaining maximum output power and is also helpful in the development of new techniques and additional research.

1 INTRODUCTION

In any country, the power sector plays a pivotal role in promoting its development and progress. It is desired to increase the use of renewable energy instead of non-renewable energy resources and develop new research techniques through which we can get at least clean, loss-less, economic, and cost-effective power for the economic growth of the country (Ansari *et al.* 2023; Jalil *et al.* 2021). Solar energy is one of the admirable sources to get eco-friendly energy using the conversion of solar irradiance into electrical energy (Jalil *et al.* 2020; Jha *et al.* 2019). During partial darkness, conditions like the shading of another panel, the shading of trees, the shading of clouds, the shading of birds, the shading of buildings, etc. create problems like multiple peaks, hotspots, and losses in the form of heat. The author has studied distinct configuration techniques like series, parallel, series-parallel, total cross-tied, bridged linked honeycomb, etc., based on different parameters that affect the efficiency of the PV system. Figure 1 provides a visual representation of the method to improve the energy efficiency of solar panels and the parameters affecting solar efficiency.

Figure 1. The most widely used methods for Boosting PV system efficiency and parameter.

DOI: 10.1201/9781032644752-36

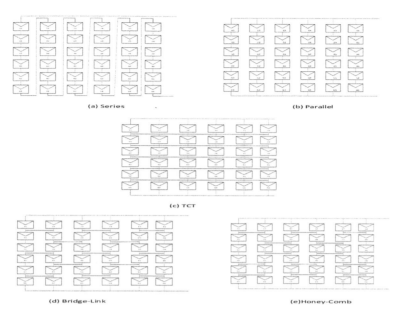

Figure 2.   Schematic representation of PV array configuration.

In their studies, the authors observed that the series-alone and parallel-alone configurations have some disadvantages, like current and voltage, according to their connections. To overcome the disadvantages suggested by the series-parallel configuration, however, it still has its dis-advantages. The schematic representation of different configurations is shown in Figure 2. Under different partial darkness conditions, combinations of various techniques, including series-parallel with total cross-tied, honeycomb with total cross-tied, and bridge linked with total cross-tied, have been analyzed by researchers. Total cross-tied has some limitations like the length of wire, mismatch loss, etc. (Krishna *et al.* 2019). To remove these types of limitations, combined techniques and puzzle-based techniques have been introduced.

Table 1.   Comparison of static PV array techniques, parameters, result (Khatoon *et al.* 2014, Manjunath *et al.* 2019).

Reconfiguration techniques	Parameters for Analysis	Result
S,SP,TCT,BL,HC,NOVEL	Maximum power, power losses	Experimental result shows Novel and TCT has best one
SP,TCT,BL,HC	power loss	TCT is best
S, SP, TCT, BL, HC, SPTCT, BLTCT, HCTCT, BL-HC	Peak power, fill factor, thermal voltage, relative power loss	TCT is best and Hybrid configuration shows satisfactorily performance

Note: S, Series; SP, Series-Parallel; TCT, Total Cross Tied; BL, Bridge- Linked; HC, Honey-Comb.

2   DIVISION OF PV ARRAY RECONFIGURATION

Under the partial darkness scenario, the extraction of power drawn from PV modules can be increased by enhancing the characteristics of the PV array by shifting the location of the PV array (physically or electrically). The PV array reconfiguration has been divided into two parts: electrical array reconfiguration [EAR] and fixed array reconfiguration. In electrical array reconfiguration, the switches are inserted to change the electrical connection and recover the energy under partial shading conditions (Mohammad Nejad *et al.* 2016). Further, the connection has been divided into two parts: adaptive and dynamic array reconfiguration (Nihanth *et al.* 2019). The adaptive is quite different from the dynamic; in the case of the adaptive system, few parts of the system use a switching matrix to swap the connections between the PV cells, relying on partial darkness conditions to gain maximum power, whereas in the dynamic case, all the connections are done by switching matrix division, as shown in Figure 3.

Table 2. Comparison of adaptive PV array reconfiguration techniques: Simulink/Hardware implementation, advantages/disadvantages (Mohammadnejad *et al.* 2016; Nihanth *et al.* 2019).

Techniques	Hardware/ Software	Advantage (A)/Disadvantage (DA)
Irradiance equalization	Software	Self-capacity for real time adaption, Extraction of power improve (A)/ cost, complexity (DA)
Fuzzy	Hardware	Independent on type of connection, quick solution(A)/Depends on short circuit current (DA)
Sorting based	Software/ Hardware	Required low no of switch(A)/ High resolution data required, No of switch increase as size increases (DA)
Greedy optimization	Software	High accuracy, speed fast(A)/ Mismatch power is increased (DA)

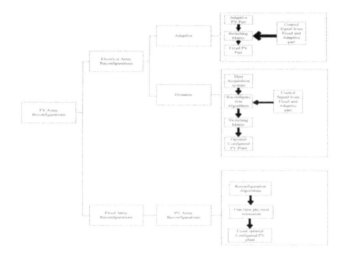

Figure 3. Division of PV array reconfiguration.

Table 3. Comparison of dynamic PV array reconfigurations techniques: Simulink/Hardware implementation, advantages/disadvantages (Pachauri *et al.* 2018).

Technique	Hardware/ Software	Advantage (A)/Disadvantage (DA)
Hybrid	Software	Low cost & robust (A)/ complex (DA)
Arduino based	Software + Hardware	Simple structure, high performance (A)
Best worse Algorithm	Software + Hardware	Less switching timing and Hardware (A)/ Costly Controller may fail in different shading condition (DA)

3 STATIC ARRAY RECONFIGURATION

In static array reconfiguration, the physical position of the panel changes at once, which gives a better result under partial darkness conditions, and the electrical connection does not change, so the shading effect mostly eliminates the panel. This type of method does not contain any switches, algorithms, sensors, etc., and this method is appropriate for large applications. In dynamic reconfiguration, the power is maximum under all partial darkness conditions, but for large applications, it is not appropriate due to complexity and cost. The Sudoku pattern is also used as a puzzle-based reconfiguration method. It has some rule-based puzzles, like a 9*9 pattern where each row contains 1 to 9 without repeating, similarly, each column contains 1 to 9 without repeating, and each sub-array also follows this rule of numbers 1 to 9 without repeating in the form of a matrix. Table 4 comparisons of different Sudoku techniques show:

Table 4. Comparison of various Sudoku techniques (Pillai *et al.* 2018).

Technique	Pattern	Result
random	Wide and narrow	Better than TCT
sorting	Moving	Array size is removed because of limitation 6*6 array
Dancing	Cloud	Result verified through hardware setup

3.1 *Magic square*

In their paper (Sammikannu *et al.* 2016), the researcher proposed a logic in the form of a magic square in which each number appears only once and the sum of every row and every column is the same, where the number in the primary and secondary diagonals is kept the same.

Table 5. Comparison of magic square reconfiguration techniques (Sharma *et al.* 2013)

Technique	Pattern	Result
Mathematical shift	Narrow, wide	Configuration is simple and smooth curve
Band shift	Long wide	Decrease mismatch loss
Segmented technique	Narrow, wide	Segmented technique provides maximum power

3.2 *Square puzzles*

A square puzzle typically refers to a type of jigsaw puzzle or brain teaser that consists of square-shaped pieces that need to be assembled to form a complete picture or pattern. The square puzzle overcomes the drawbacks of the zig-zag pattern because zig-zag is applicable for small systems.

Table 6. Comparison of various square puzzle techniques (Venkateswari *et al.* 2019).

Technique	Pattern	Result
Square	Narrow, wide	Less line loss
Dominance	Narrow, wide	Energy saving
Two phase method	Narrow, wide	Best for higher sizes

3.3 *Skyscraper*

The researcher proposed a method for the placement of skyscrapers by arranging solar panels in a vertical stack. It is well-suited for installations on buildings and other vertical structures.

Table 7. Comparison of various skyscraper techniques (Yadav *et al.* 2021).

Technique	Pattern	Result
Clue based	Narrow, wide	Energy saving
Clue based	Narrow, wide	Energy saving

4 CONCLUSION

The literature survey of the papers concludes that narrow and wide shadings are the simplest shading patterns for the analysis of reconfiguration techniques. The relocation of the panel gives a better result, but the problem is line losses or wire losses while changing a column, which first improves the reliability and performance of the overall system. If we compare the PV array reconfiguration based on sensors and switches, adaptive reconfiguration is less dynamic and static reconfiguration is sensor-less; adaptive reconfiguration Adaptive reconfiguration is high; and static reconfiguration is simple. The newest clue-based technique Skyscraper is much better than the other static reconfiguration techniques, and the changes in Skyscraper's logical hardware and software can be used for future improvements in the field of solar energy to generate maximum power from PV arrays.

REFERENCES

Ansari, M. S, Jalil, M. F, & Bansal, R. C. (2023). A review of optimization techniques for hybrid renewable energy systems. *International Journal of Modeling and Simulation*, 43(5), 722–735.

Jalil, M. F, Ansari, M. S, Diwania, S, & Husain, M. A. (2021). Performance Analysis of PV Array Connection Schemes Under Mismatch Scenarios. In *Renewable Power for Sustainable Growth: Proceedings of International Conference on Renewal Power (ICRP 2020)* (pp. 225–235). Springer Singapore.

Jalil, M. F, Khatoon, S., Nasiruddin, I, & Bansal, R. C. (2020). An improved feasibility analysis of photovoltaic array configurations and reconfiguration under partial shading conditions. *Electric Power Components and Systems*, 48(9–10), 1077–1089.

Jha, V, & Triar, U. S. (2019). A detailed comparative analysis of different photovoltaic array configurations under partial shading conditions. *International Transactions on Electrical Energy Systems*, 29(6), e12020.

Krishna, G. S, & Moger, T. (2019), "Enhancement of maximum power output through reconfiguration techniques under Non uniform irradiance conditions", *Energy*, 187, 115917.

Malathy, S. & Ramaprabha, R. (2018), "Reconfiguration strategies to extract maximum power from photovoltaic array under partially shaded conditions", *Renewable and Sustainable Energy Reviews*, 81, 2922–2934.

Pachauri, R, Yadav, A. S, Chauhan, Y. K., Sharma, A., & Kumar, V. (2018). Shade dispersion-based photovoltaic array configurations for performance enhancement under partial shading conditions. *International Transactions on Electrical Energy Systems*, 28(7), e2556.

Pillai, D. S, Rajasekar, N, Ram, J. P., & Chinnaiyan, V. K. (2018), "Design and testing of two phase array reconfiguration procedure for maximizing power in solar PV systems under partial shade conditions (PSC)", *Energy Conversion and Management*, 178, 92–110.

Sharma, D., Jalil, M. F., Ansari, M. S., & Bansal, R. C. (2023). A review of PV array reconfiguration techniques for maximum power extraction under partial shading conditions. *Optik*, 170559. Tatabhatla,

Vijayalekshmy, S., Bindu, G. R., & Rama Iyer, S. (2016), "A novel Zig-Zag scheme for power enhancement of partially shaded solar arrays", *Solar Energy*, 135, 92–102.

Yadav, A S, & Mukherjee, V. (2021). Conventional and advanced PV array configurations to extract maximum power under partial shading conditions: A review. *Renewable Energy*, 178, 977–1005

Advances in AI for Biomedical Instrumentation, Electronics and Computing – Sachan et al. (eds)
© 2024 The Author(s), ISBN 978-1-032-64298-7

Novel memristor-CMOS based domino self-resetting half-adder design for fast and low-power biomedical applications

Monica Gupta*, Kirti Gupta & M. Bhutani
Department of Electronics and Communication Engineering, Bharati Vidyapeeth's College of Engineering, New Delhi, India

ABSTRACT: This paper introduces a novel memristor-CMOS-based domino self-resetting design for half-adder. The proposed design not only overcomes the limitations of conventional logic but also retains the advantage of the self-resetting mechanism. The primary objectives are improved operational speed, reduced power consumption, and enhanced area efficiency. Extensive simulations at 32 nm technology nodes validate the functionality and robustness of the proposed design. The proposed design exhibits an exceptional performance under varying conditions, particularly at lower technology nodes and smaller load capacitances, making it a compelling choice for the latest fast and low-power biomedical applications that require computational efficiency with longer battery lifetime. The work significantly advances domino logic designs and their implementation potential.

1 INTRODUCTION

The demand for faster and low-power circuits has grown significantly for biomedical applications, with static CMOS circuits playing a crucial role (Kang *et al.* 2016). However, as design complexity increases, static CMOS circuits face challenges such as requiring many transistors, escalating costs, high power consumption, and large delays (Weste *et al.* 1986). To address these issues, dynamic circuits have gained popularity by reducing the component requirements to static CMOS circuits by almost half. The dynamic logic uses an external clock for two-phase operation, resulting in smaller footprints for N-input circuits (Hanumanthu *et al.* 2016). Despite these benefits, traditional domino logic implementations encounter significant challenges, such as increased static power dissipation, the need for clock-generating circuitry, and limited noise immunity, hindering their practicality in modern circuit designs. To overcome these problems, a CMOS-based domino self-resetting (SRCMOS) logic is suggested in the literature (Hwang *et al.* 1999). It operates without a clock input, utilizing its output for self-resetting and input handling, thereby reducing delay to some extent; however, further improvements are necessary. The high-power consumption and substantial area overhead resulting from its reliance on static CMOS circuits for logic implementation deem it ineffective in fast and low-power biomedical applications.

This work explores the potential of memristor-based systems as an alternative to existing static and dynamic CMOS-based logics (Ji *et al.* 2022; R *et al.* 2023; Vourkas *et al.* 2012). The novel memristor-CMOS-based domino self-resetting logic design is presented for half-adder to address the issues of existing logic. By leveraging the benefits of memristor and self-resetting mechanism, the proposed design offers faster, low-power, and simplified implementation, positioning them ahead of current designs and promising a bright future for circuit technology (Teimoori *et al.* 2016).

---

*Corresponding Author: monica.gupta@bharatividyapeeth.edu

DOI: 10.1201/9781032644752-37

The paper follows the following structure: Section 2comprehensively introduces existing self-resetting CMOS-based domino logic design for half-adder. In Section 3, readers will find an in-depth exposition of the proposed memristor-CMOS-based domino self-resetting design for half-adder. Section 4 includes performance analysis and a comparative study of proposed and existing designs. Finally, Section 5 presents the conclusion, summarizing the paper's main findings and insights.

2 OVERVIEW OF EXISTING CMOS BASED DOMINO SELF-RESETTING LOGIC DESIGN FOR HALF-ADDER

This section focuses on the half-adder design based on existing SRCMOS logic. The self-resetting design operates on the principle of autonomously resetting the output node without the need for an additional clock input. Typically, SRCMOS based logic design (Figure 1) comprises of inverters (MP2-MN2, MP3-MN3), pMOS keepers (MP1, MP4), driver nMOS transistor MN1, internal dynamic nodes (Y, P), static CMOS logic that drives reset input node Z and output node (SUM/CARRY). The keepers perform a pre-charge operation on the internal dynamic node Y and autonomously resets the output node. To achieve different logics, an appropriate static CMOS based network of transistors is integrated to generate input Z. In the design, the input Z directly drives the transistor MN1, conditionally pulling down node Y. To prevent short-circuit current during pre-charging, the keeper transistor input P must drop low after the reset of input(s) but early enough to allow timely reset of the output node before the next input is applied to ensure proper operation of the circuit. The input node Z is generated using a static CMOS based logic circuit to implement SUM (Figure 1(a)) and CARRY (Figure 1(b)) functionality.

Self-resetting domino circuits outperform static counterparts due to the elimination of slow pMOS pull-up networks, resulting in higher speeds (Hanumanthu *et al.* 2016; Hwang *et al.* 1999). Despite this advantage, the circuits suffer from high power consumption and area overhead, stemming from their reliance on static CMOS logic for implementation of circuit for node Z. Therefore, new design is needed to address these drawbacks while retaining the benefits of self-resetting logic.

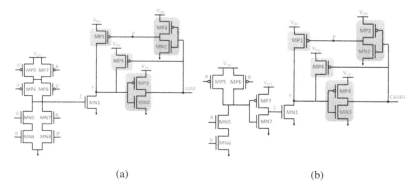

(a)   (b)

Figure 1.   Existing SRCMOS based self-resetting design for half-adder (a) Schematic for SUM (b) Schematic for CARRY.

3 PROPOSED MEMRISTOR-CMOS BASED DOMINO SELF-RESETTING DESIGN

This section presents innovative memristor-CMOS based domino self-resetting logic design for half-adder, leveraging the unique properties of memristor and self-resetting mechanism within the domino logic circuits. By integrating memristors, the design enables faster and efficient logic operations with low-power consumption and reduced area overhead.

3.1 *Memristor*

The memristor, a passive two-terminal circuit element (n, p) (Figure 2), was originally theorized by Chua, alongside resistor R, capacitor C, and inductor L, and exhibits two resistance values, RON and ROFF, corresponding to low and high resistance states when current flows from terminal n to p or from terminal p to n, respectively (Chua:1971). Memristors offer the advantages of non-volatility, smaller footprint and lower power consumption compared to traditional CMOS circuits, positioning them as promising alternatives for future circuit technology (Mohammad *et al.* 2013).

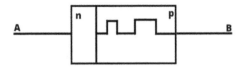

Figure 2.   Schematic symbol of memristor.

3.2 *Proposed design for memristor-CMOS based domino self-resetting half-adder*

The proposed design for memristor-CMOS-based domino self-resetting half-adder is illustrated in Figure 3. The memristor-based circuit implements the logic while the additional circuit incorporates the self-resetting mechanism for the output nodes, SUM and CARRY. The equations representing the proposed half-adder functionality are provided below.

$$SUM = A'.B + A.B' \tag{1}$$

$$CARRY = A.B \tag{2}$$

Working

The memristor pairs X3-X4 and X5-X6 in Figure 3(a) perform AND operation, whereas X1-X2 performs OR operation. Similarly, in Figure 3(b) X1-X2 memristor pair performs AND operation. The working of the proposed memristor-CMOS-based domino self-resetting half-adder is as follows:

CASE I (A = '0' and B = '0')

In Figure 3(a), when both the inputs are at the LOW logic level, current flows through the memristor pairs X3-X4 and X5-X6. This results in LOW logic levels at nodes N1, N2, and Z1. Similarly, due to the AND functionality performed by X1-X2 in Figure 3(b), node Z2 becomes LOW.

CASE II (A = '0' and B = '1')

For this set of input, no current flows through memristor pairs X3-X4 and X5-X6 in Figure 3(a), and thus, nodes N1 and N2 attain HIGH and LOW logic levels, respectively and results in HIGH logic levels at node Z1. Similarly, in Figure 3(b), node Z2 is maintained at a LOW logic level.

CASE III (A = '1' and B = '0')

For this combination, no current flows through memristor pairs X3-X4 and X5-X6 in Figure 3(a), and thus, nodes N1 and N2 attain LOW and HIGH logic levels, respectively which results in HIGH logic levels at node Z1. Similarly, in Figure 3(b), node Z2 is maintained at a LOW logic level.

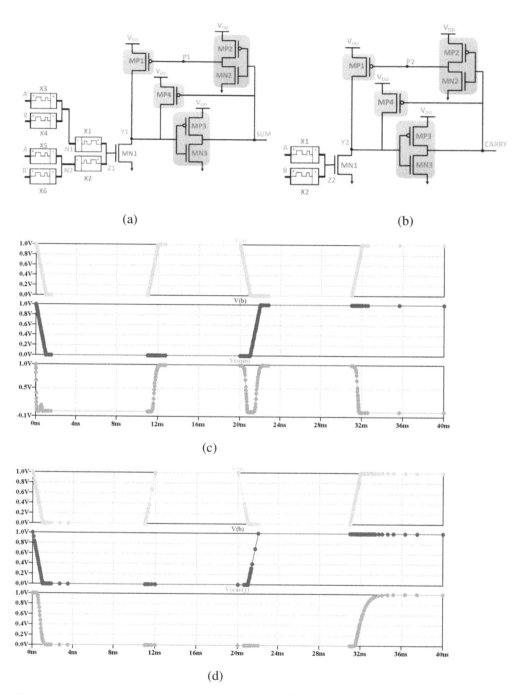

Figure 3. Proposed memristor-CMOS based domino self-resetting design for half-adder (a) Schematic for SUM (b) Schematic for CARRY (c) Timing waveform for SUM (d) Timing waveform for CARRY.

CASE IV (A = '1' and B = '1')

This input combination leads to current flow through memristor pairs X3-X4 and X5-X6 in Figure 3(a), resulting in LOW logic levels at nodes N1, N2, and Z1. In addition, in Figure 3(b), the node Z2 changes to a HIGH logic value.

In all the four cases mentioned above, when Z1 (or Z2) is at the LOW logic level, transistor MN1 remains OFF and results in LOW output at SUM (or CARRY), triggering keeper MP4 and pre-charging node Y1 (or Y2) back to HIGH logic level, initiating the self-resetting mechanism. Alternatively, when node Z1 (or Z2) is at the HIGH logic level, MN1 turns ON, discharging node Y1 (or Y2) towards the LOW logic level and results in HIGH output at SUM (or CARRY). This triggers keeper MP1 to pre-charge node Y1 (or Y2) to prepare the circuit for the next input set.

The functionality of the proposed half-adder circuit is verified using the timing waveforms for output nodes SUM and CARRY for different inputs A and B values, as shown in Figure 3(c) and (d), respectively. The simulation results are discussed in the subsequent section.

4 SIMULATION RESULTS AND DISCUSSION

This section analyzes and compares the performance of the proposed half-adder design with the existing SRCMOS-based design in terms of critical parameters, such as P_T (total power consumption), T_A (average propagation delay), and area efficiency. In addition, the robustness of the proposed design is corroborated through SPICE simulations at a 32 nm technology node, considering load capacitance and technology variations. The ON and OFF resistance of the memristor model are 2 Ω and 10 kΩ, respectively whereas the threshold voltage for the nMOS and pMOS transistors are $V_{thn} = 0.501$ V and $V_{thp} = -0.452$ V, respectively.

4.1 *Performance comparison*

The performance of the proposed and the existing designs are evaluated, and the results are summarized in Figure 4 for comparison. It is observed that the proposed design consumes 24.09 % less P_T and shows a significant reduction of 24.86 % in T_A in comparison to the existing SRCMOS based design, making it suitable for a wide range of low-power and faster applications.

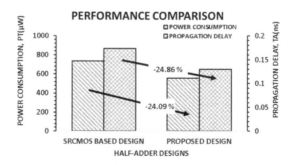

Figure 4.   Performance comparison of proposed and existing SRCMOS based designs for half-adder

4.2 *Complexity*

The existing SR CMOS based design employs 32 transistors (17 pMOS and 15 nMOS) for its implementation, while the proposed design utilizes only 18 transistors (10 pMOS and 8 nMOS), along with 8 memristors, resulting in a reduction of 43.75 % transistor count and enhanced area efficiency. Further, pMOS transistors are slower and therefore require 2.5 times more area compared to nMOS transistor for the same current-carrying capability.

Additionally, memristors are nano-devices and occupy very little space compared to MOS transistors. Thus, the proposed design, with its reduced transistor count and efficient use of memristors, offers a significant improvement in area efficiency over the existing design, making it a promising solution for space-constrained applications.

4.3 *Capacitance variation*

The effect of load capacitance variation on the performance of the proposed half-adder is assessed in this sub-section and the results are presented in Figure 5(a). As the load capacitance reduces from 250 fF to 50 fF, the proposed design shows a significant reduction of 24.45% in P_T and an improvement of 70.04% in T_A. The results highlight the efficacy of the proposed design in adapting to different load capacitances making it suitable for a wide range of applications.

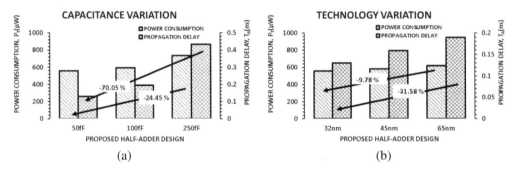

Figure 5. Proposed half-adder (a) Effect of capacitance variation (b) Effect of technology variation.

4.4 *Technology variations*

To obey Moore's law, technology nodes are scaled down resulting in gradual reduction in transistor sizes. The impact of technology scaling on the proposed half-adder design is analyzed by varying the technology node from 65 nm to 32 nm. The simulation results are summarized in Figure 5(b). As the technology node changes from 65 nm to 32 nm, P_T reduces by about 9.78%, and T_A reduces by 31.58%. The results clearly indicate that with technology scaling, the proposed designs exhibit significant improvements in P_T and T_A. Thus, the proposed design is well-suited for modern and more advanced technology nodes, making it a promising choice for future bio-medical applications.

5 CONCLUSION

The proposed memristor-CMOS based domino self-resetting design for half-adder combines the advantages of memristor and self-resetting mechanism. The design achieves 24.09% and 24.86% improvement in total power consumption and average propagation delay respectively. The robustness of the proposed design is verified through an analysis of load capacitance variation and technology scaling. The results show significant improvement in performance as the load capacitance varies from 250 fF to 50 fF and as the technology node is scaled down from 65 nm to 32 nm. The proposed design shows 24.45% reduction in power consumption and a 70.05% reduction in delay with capacitance variation. With technology scaling, the proposed design shows a 9.78% and 31.58% improvement in power consumption and delay, respectively. In addition, the proposed design shows an area efficiency of about 43.75% compared to the existing design. Overall, the proposed design offers reduced power consumption, faster operation, more area efficiency, and robust performance. It makes them highly promising for recent biomedical applications, where power efficiency and speed are critical factors.

REFERENCES

Chua, L. 1971. Memristor-the missing circuit element. *IEEE Transactions on Circuit Theory*: 507–519.

Hanumanthu, M., Dastagiri, N. B., Rahim, B. A. & Soma sundar, P. 2016. Design and Comparative Analysis of Domino Logic Styles. *Indian Journal of Science and Technology* 9(33):1–6.

Hwang, W., Gristede, G., Sanda, P., Wang, S. Y. & Heidel, D. F. 1999. Implementation of a self-resetting CMOS 64-bit parallel adder with enhanced testability. *IEEE Journal of Solid-State Circuits* 34(8):1108–1117.

Ji, C., Li, T. & Zou, X. 2022. Multifunctional Module Design Based on Hybrid CMOS-Memristor Logic Circuit. *International Conference on Service Science (ICSS)*:112–116.

Kang, S. M. & Leblebici, Y. 2016. *CMOS Digital Integrated Circuits*. New Delhi: McGraw Hill Education (India).

Mohammad, B., Homouz, D. & Elgabra, H. 2013, "Robust Hybrid Memristor-CMOS Memory: Modeling and Design", *IEEE Transactions on Very Large-Scale Integration (VLSI) Systems* 21:2069–2079.

R., S. S., Nair, R. R., I, S. & S., S. 2023. An Efficient Low Power Full Adder Architecture Using Memristor. *4th International Conference on Smart Electronics and Communication (ICOSEC)*:270–276.

Teimoori, M., Ahmadi, A., Alirezaee, S. & Ahmadi, M. 2016. A novel hybrid CMOS-memristor logic circuit using Memristor Ratioed logic. *IEEE Canadian Conf on Electric and Computer Eng. (CCECE)*:1–4.

Vourkas, I. & Sirakoulis, G. C. 2012. A Novel Design and Modeling Paradigm for Memristor-Based Crossbar Circuits. *IEEE Transactions on Nanotechnology*: 1151–1159.

Weste, N. & Eshraghian, K. 1986. *Principles of CMOS VLSI Design: A System Perspective*. Reading, MA: Addition Wesley.

Advances in AI for Biomedical Instrumentation, Electronics and Computing – Sachan et al. (eds)
© 2024 The Author(s), ISBN 978-1-032-64298-7

Prediction of pulmonary arterial hypertension from HRCT chest in post-COVID patients using double attention U-NeT model

Azra Nazir
School of Computer Science & Engineering, VIT Bhopal University Bhopal-Indore Highway, MP, India

Roohie Naaz & Shaima Qureshi
Department of Computer Science & Engineering, National Institute of Technology, Srinagar, J&K India

Nidha Nazir
Department of Radiology, State Medical Hospital, Srinagar, J&K, India

ABSTRACT: There is an unusual spike in the number of cardiac attacks POST COVID-19 worldwide, leading to sudden deaths. The virus's effect on the heart can be due to direct viral invasion, systemic inflammation, increased clotting, and hypoxia. This has led to an increase in the frequency of Pulmonary arterial hypertension (PAH) cases. PAH affects the arteries in the lungs and the right side of the heart and can be life- threatening. The change in the dimensions of the pulmonary artery in PAH is diagnosed from a CT scan. With no cure for the disease, it is essential to diagnose the condition to provide special care to the patient. In this paper, a double attention U-Net model with attention blocks in its down-sampling and up-sampling paths is employed to extract the variation in the CT of a high-risk patient. The experimental results show that the proposed methodology has 93.8% classification accuracy, which is a 6.8% improvement compared to the VGG16 model.

1 INTRODUCTION

COVID-19, caused by the SARS-CoV-2 virus, emerged in late 2019 in the city of Wuhan. The virus's rapid spread led to a global pandemic, with 700 million infected cases and nearly 7 million deaths reported worldwide. Studies from prominent medical institutions have indicated that there has been a significant increase in sudden heart attacks post-pandemic, which can be related to COVID-19 (Deng *et al.*, 2020). Researchers warn what we have seen already is just the tip of the iceberg, and a lot needs to be done to ensure the safety of people from the most common 'after-effects' of COVID-19 (Močibob *et al.*, 2022). A report from London suggests that for doctors treating cardiovascular patients, it is critical to recognize the clinical appearance and risk factors for COVID-19 infection in this group due to high-risk factors and the likelihood of death for such patients. A post-COVID-19 study from Nature Medicine confirms people who have contracted COVID-19 multiple times are at a higher risk of developing heart problems. A study from the Institute for Research on Healthy Aging reports a nearly 30% increase in heart attack deaths in adults during 2021–2022. It has been seen that COVID-19 affects the cardiovascular system in many ways, leading to irregular heartbeats or even heart failure (Ma *et al.*, 2021). COVID-19 has been found to cause inflammation, increasing the accumulation of fatty plaque inside arteries or overburdening the heart muscles, leading to pulmonary artery enlargement, thereby leading to pulmonary arterial hypertension (PAH). Specifically, it refers to increased blood pressure within the pulmonary arteries, which are the blood vessels that carry deoxygenated blood from the

DOI: 10.1201/9781032644752-38

right side of the heart to the lungs to pick up oxygen. In PAH, the tiny arteries in the lungs, called pulmonary arterioles, become narrowed, damaged, or blocked. This condition can be diagnosed from a high-resolution computed tomography (HRCT) chest image. The diameter of the pulmonary artery in a normal scan is about 2.72 cm; however, the diameter in patients at risk of PAH is more than 3.47cm (Ussavarungsi *et al.*, 2014). Deep Learning can transform the health system and has been employed to detect diseases and classify image samples accurately. Deep Learning has become a popular method, showing tremendous strength for image processing tasks, (Hesamian *et al.*, 2019). Convolutional Neural Networks (CNNs) used in image segmentation generally take a 2-dimensional input image and apply a 2-dimensional filter to extract representative features of the image. However, 3-dimensional convolutions have better performance due to their ability to capture relations in the spatial domain (Simonyan & Zisserman, 2014). Convolutional Neural Networks have become one of the cornerstone techniques in the field of medical imaging due to their unparalleled performance in image recognition tasks. The distinctive spatial hierarchies and translation invariance of CNNs make them well-suited for analyzing medical images, where the precise localization and recognition of patterns are crucial (Chin *et al.*, 2017).

2   RELATED WORK

Artificial Intelligence (AI) and Deep Learning (DL) play pivotal roles in understanding, managing, and mitigating the impacts of COVID-19. DL models like CNN have been trained to detect COVID-19 from chest X-rays and CT scans (Abbas *et al.*, 2021). The collaboration between Artificial Intelligence and human experts is envisioned as a partnership to augment healthcare professionals' capabilities and improve overall healthcare delivery (Dong *et al.*, 2021). Deep learning (DL) algorithms have shown promising results in the healthcare and biomedical fields, especially in classifying medical images (Chen *et al.*, n.d.). DL models can track the progression or regression of lung involvement in COVID-19 patients, providing valuable feedback on patient status and potential outcomes. There is growing interest in understanding the long-term effects of COVID-19 on recovered patients. AI can assist in analyzing patient data to identify common symptoms, predict patient trajectories, and suggest possible interventions (Mittal & Hasija, 2020). AI algorithms can detect and quantify cardiac structures, evaluate cardiac function, and predict future cardiovascular events based on subtle imaging patterns. DL models assist in identifying coronary artery disease, quantifying stenosis, and even predicting which plaque might be prone to rupture (Huang *et al.*, 2020). After COVID-19, there has been a sudden increase in the rate of heart attacks and strokes. The virus can impact the hearts of individuals who were previously healthy. The severe inflammatory response within the body can affect arteries, exacerbating cardiac damage. In (Nazir *et al.*, 2022), VGG16, a popular image classification model, was employed to extract features from the standard axial slice of high-resolution chest CT scans.

3   PROPOSED METHODOLOGY

A U-Net architecture with Double Attention has been employed. The U-Net architecture is a widely recognized convolutional neural network model popular for biomedical image segmentation (Yin *et al.*, 2022). The U-Net architecture essentially comprises a contracting (down-sampling) path to capture context and a symmetric expanding (up-sampling) path that allows precise localization. This structure allows U-Net to capture both context and localization information. The term Double Attention in the context of U-Net typically refers to the use of attention mechanisms to enable the model to focus on specific parts of the input data that are more relevant to the task at hand. The U-Net model employs both Channel and Spatial Attention. These attention mechanisms enhance the model's ability to focus on important features. It also improves the interpretability of the model by providing insights into which regions or features the model considers most important, reducing overfitting (Du *et al.*, n.d.). Spatial Attention allows the model to focus on particular spatial regions of the feature maps, emphasizing areas that are more important for the specific task. Essentially, for a given feature map, it generates a spatial attention map that weighs each spatial location based on its importance. Channel Attention enables the model to focus on specific feature channels over others. For a given feature map, it provides weights for each channel, emphasizing channels that are more discriminative for the task at hand. An input feature map is processed using convolutional layers of the U-Net. The resulting feature map is used to compute both spatial attention maps and channel attention weights. The spatial attention map, which has the same height and width as the feature map but only one channel, is multiplied by the feature map to emphasize important spatial locations. The channel attention weights, one weight per channel, are used to scale the feature map channels. This emphasizes more relevant channels. The combined effect produces an enhanced feature map that emphasizes spatially important regions and discriminative feature channels. The attention mechanisms incorporated for the task of HRCT classification are shown in Figure 1. As proposed in the original U-Net Model, the network architecture has been kept symmetric to ease the convergence process.

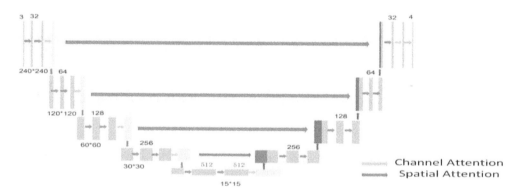

Figure 1.   Spatial and channel enhanced U-Net model.

After applying the convolution operation twice, the Rectified Linear Activation (ReLU) is also applied twice before feeding the intermediate activation map to a pooling layer. The Channel and Spatial Attention at each level identifies feature channels and more concrete features at each level, respectively. Towards the right, up-sampling is performed using a simple filling method. Being symmetrical, the output of the left side can be directly used as input to the right side of the model to obtain a feature map highlighting the important spatial features. Each intermediate map is passed to the next level.

4   RESULTS AND DISCUSSION

The original dataset consists of 500 HRCT images of COVID-19 patients labeled manually by the radiologist. No underlying factors, age, gender, or past medical history were revealed to the radiologist. The original dataset is augmented to 750 images for better training. From this set, 65 images were used as a test set for accuracy evaluation. The model is trained to categorize patients into high-risk patients whose pulmonary artery diameter is beyond normal value and no-risk patients with pulmonary artery diameter inside the normal range. Accuracy is determined based on the following:

Table 1. Classification accuracy of VGG-16 vs. Double attention U-Net model.

CNN Model	VGG-16		Double Attention U-Net	
Test Set	Predicted No-Risk	Predicted High-Risk	Predicted No-Risk	Predicted High-Risk
No-Risk	45	05	47	03
	TN	FP	TN	FP
High-Risk	04	11	01	14
	FN	TP	FN	TP

- An HRCT image correctly labeled by the model as High-Risk (1) – True Positive.
- An HRCT image correctly labeled by the model as No-Risk (0) – True Negative.
- An HRCT image incorrectly labeled by the model as High-Risk (1) – False Positive.
- An HRCT image incorrectly labeled by the model as No-Risk (0) – False Negative.

The model is trained with Adam in a tensorflow environment, and the initial learning rate is set to 0.001 as compared to 0.001 for VGG16. The VGG16 trained in 11 epochs. However, the Double Attention UNet converged in 35 epochs for the same fixed batch size. The classification accuracy for the Double Attention U-Net Model is 93.8% as compared to 86.1% in VGG16. It can be observed the model has reduced False-Negative predictions, which is a significant parameter in identifying early severe cases of PAH in patients who were victims of COVID-19.

A few randomly chosen test-set images with the output label as predicted by VGG-16 and Double Attention U-Net Model are shown in Figure 2. Vgg-16 wrongly classified the image on the right but was correctly classified by the U-Net model enhanced with the attention mechanism.

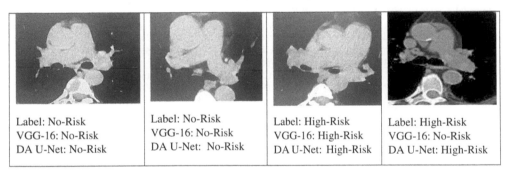

Label: No-Risk	Label: No-Risk	Label: High-Risk	Label: High-Risk
VGG-16: No-Risk	VGG-16: No-Risk	VGG-16: High-Risk	VGG-16: No-Risk
DA U-Net: No-Risk	DA U-Net: No-Risk	DA U-Net: High-Risk	DA U-Net: High-Risk

Figure 2. Visualization of random test-case images.

5 CONCLUSION

COVID-19 can lead to several complications, including impacts on the cardiovascular system. If medical professionals can identify early signs of severe PAH in COVID-19 patients, they can adjust their treatment strategies accordingly. This means that they might change their initial approach to patient care, potentially leading to better patient management and outcomes. Early detection of Pulmonary Arterial Hypertension is critical because timely intervention can lead to saving a precious life. This paper employs a Double-Attention U-Net model to identify the variation in the HRCT chest scan, which has provided better classification accuracy than the previous CNN model. Due to limited dataset availability, the sensitivity analysis can't yet be incorporated into the model. In the future, the model's sensitivity to artery size variation can also be quantified by systematically varying the artery size in the input images and observing the corresponding changes in the model's output. Also, Explainable AI (XAI) techniques like GRAD-CAM can be incorporated to promote human-AI collaboration by understanding the model's decision-making process.

ACKNOWLEDGMENT

This work is funded by the JK Science Technology and Innovation Council., Department of Science and Technology, Government of Jammu & Kashmir.

REFERENCES

Abbas, A., Abdelsamea, M. M., & Gaber, M. M. (2021). Classification of COVID-19 in chest X-ray images using DeTraC deep convolutional neural network. *Applied Intelligence*, *51*(2), 854–864. https://doi.org/10.1007/S10489-020-01829-7/FIGURES/6

Chen, Y., Springer, L. J.-P. and applications, Heidelberg, undefined, & 2020, undefined. (n.d.). Deep learning in healthcare. *SpringerYW Chen, LC JainParadigms and Applications. Springer, Heidelberg, 2020•Springer*. Retrieved November 14, 2023, from https://link.springer.com/content/pdf/10.1007/978-3-030-32606-7.pdf

Chin, C. L., Lin, B. J., Wu, G. R., Weng, T. C., Yang, C. S., Su, R. C., & Pan, Y. J. (2017). An automated early ischemic stroke detection system using CNN deep learning algorithm. *Proceedings – 2017 IEEE 8th International Conference on Awareness Science and Technology, ICAST 2017, 2018-January*, 368–372. https://doi.org/10.1109/ICAWST.2017.8256481

Deng, Q., Hu, B., Zhang, Y., Wang, H., Zhou, X., Hu, W., Cheng, Y., Yan, J., Ping, H., & Zhou, Q. (2020). Suspected myocardial injury in patients with COVID-19: Evidence from front-line clinical observation in Wuhan, China. *International Journal of Cardiology*, *311*, 116–121. https://doi.org/10.1016/J.IJCARD.2020.03.087

Dong, D., Tang, Z., Wang, S., Hui, H., Gong, L., Lu, Y., Xue, Z., Liao, H., Chen, F., Yang, F., Jin, R., Wang, K., Liu, Z., Wei, J., Mu, W., Zhang, H., Jiang, J., Tian, J., & Li, H. (2021). The Role of Imaging in the detection and management of COVID-19: A Review. *IEEE Reviews in Biomedical Engineering*, *14*, 16–29. https://doi.org/10.1109/RBME.2020.2990959

Du, G., Cao, X., Liang, J., . . .X. C.-J. of I., & 2020, undefined. (n.d.). Medical image segmentation based on u-net: A review. *Library.Imaging.OrgG Du, X Cao, J Liang, X Chen, Y ZhanJournal of Imaging Science and Technology, 2020•library.Imaging.Org*. Retrieved November 14, 2023, from https://li-brary.imaging.org/admin/apis/public/api/ist/website/downloadArticle/jist/64/2/art00009

Hesamian, M. H., Jia, W., He, X., & Kennedy, P. (2019). Deep learning techniques for medical image seg-mentation: Achievements and challenges. *Journal of Digital Imaging*, *32*(4), 582–596. https://doi.org/10.1007/S10278-019-00227-X/TABLES/2

Huang, L., Han, R., Ai, T., Yu, P., Kang, H., Tao, Q., & Xia, L. (2020). Serial quantitative chest ct assessment of covid-19: A deep learning approach. *Radiology: Cardiothoracic Imaging*, *2*(2). https://doi.org/10.1148/RYCT.2020200075/ASSET/IM-AGES/LARGE/RYCT.2020200075.FIG6B.JPEG

Ma, L., Song, K., & Huang, Y. (2021). Coronavirus disease-2019 (COVID-19) and cardiovascular complica-tions. *Journal of Cardiothoracic and Vascular Anesthesia*, *35*(6), 1860–1865. https://doi.org/10.1053/J.JVCA.2020.04.041

Mittal, S., & Hasija, Y. (2020). Applications of deep learning in healthcare and biomedicine. *Studies in Big Data*, *68*, 57–77. https://doi.org/10.1007/978-3-030-33966-1_4

Močibob, L., Šušak, F., Šitum, M., Višković, K., Papić, N., & Vince, A. (2022). COVID-19 and Pulmonary Thrombosis—an unresolved clinical puzzle: A single-center cohort study. *Journal of Clinical Medicine 2022, Vol. 11, Page 7049*, *11*(23), 7049. https://doi.org/10.3390/JCM11237049

Nazir, A., Naaz, R., Qureshi, S., & Nazir, N. (2022). HRCT chest analysis for detection of pulmonary arterial hypertension in COVID-19 patients using convolutional neural networks. *2022 IEEE 3rd Global Conference for Advancement in Technology, GCAT 2022*. https://doi.org/10.1109/GCAT55367.2022.9972033

Simonyan, K., & Zisserman, A. (2014). Very deep convolutional networks for large-scale image recog nition. *3rd International Conference on Learning Representations, ICLR 2015 – Conference Track Proceedings*. https://arxiv.org/abs/1409.1556v6

Ussavarungsi, K., Whitlock, J. P., Lundy, T. A., Carabenciov, I. D., Burger, C. D., & Lee, A. S. (2014). The significance of pulmonary artery size in pulmonary hypertension. *Diseases 2014, Vol. 2, Pages 243–259*, *2*(3), 243–259. https://doi.org/10.3390/DISEASES2030243

Yin, X. X., Sun, L., Fu, Y., Lu, R., & Zhang, Y. (2022). U-Net-Based medical image segmentation. *Journal of Healthcare Engineering*, *2022*. https://doi.org/10.1155/2022/4189781

Advances in AI for Biomedical Instrumentation, Electronics and Computing – Sachan et al. (eds)
© 2024 The Author(s), ISBN 978-1-032-64298-7

Comprehensive review on impact of nanosatellites on human medical infrastructure

Abhishek Sharma, Suryansh Dev, Shambhavi Kumari, Riya Jain, Satyam Kumar Jha, Mohit Kumar, Praveen Kumar & Ruchita Gautam
Centre of Excellence, Space Technologies, KIET Group of Institutions, Delhi-NCR, Ghaziabad

ABSTRACT: Space technology has advanced in every conceivable field in recent technological history, either directly or indirectly. Human healthcare is one of its components, and it has been addressed recently. The comprehensive review provides extensive data about how nanosatellites have influenced the discipline of medicine in manners that are beneficial to generations to follow.

1 INTRODUCTION

Nanosatellites and CubeSats, weighing under 10 kilograms and conforming to a 10 cm^3 standard (1U), are revolutionizing space technology, thanks to funding from commercial, medical, and defense sectors. These miniature satellites bring advanced observation capabilities with their state-of-the-art electronics and sensors. In healthcare, their impact is significant. They enhance remote health monitoring, telemedicine, and emergency response, aiding in pandemic prevention through disease mapping and environmental monitoring. They also improve telemedicine communications, connecting the global health network and reducing infection risks. Future developments focus on integrating better medical sensors, secure data, AI, and efficient power in these satellites. Merging nanosatellites with ground systems is essential for quick, effective emergency responses, highlighting their growing role in healthcare innovation. This blend of space technology and healthcare demonstrates the potential of interdisciplinary approaches in shaping the future of health (Dietrich *et al.* 2018).

2 SATELLITES AND HEALTHCARE: A NEW FRONTIER

The integration of satellite technology in healthcare goes beyond just monitoring and communication; it's about using space capabilities to tackle Earth's major health challenges. This includes real-time disease outbreak tracking and analyzing environmental health factors. As nations address health disparities, the role of these advanced technologies is becoming crucial. Their impact is most evident in global health crisis responses, with the recent pandemic highlighting the importance of satellite technology in healthcare.

2.1 *Controlling pandemic effects by space technology*

The pandemic has globally disrupted work, communication, education, and healthcare, highlighting health inequalities in vaccine distribution. Various sectors are adopting innovative strategies to mitigate COVID-19 impact. Since a 1999 UN Conference, the space industry has been dedicated to fighting infectious diseases using space technology. Space

DOI: 10.1201/9781032644752-39

technologies are crucial in tackling global health challenges like COVID-19, monitoring environmental changes, and aiding disease control. They have been instrumental in the COVID-19 response and hold potential for future pandemic preparedness. Tele-epidemiology, using remote sensing and space tech, helps analyze and map diseases, assess risks, and identify vulnerable populations. The COVID-19 Earth Observation Dashboard, a collaboration between NASA, ESA, and JAXA, uses satellite data for pandemic monitoring. Space tech's role in eradicating polio in Nigeria, through targeted vaccine delivery in remote areas, exemplifies its value in pandemic planning. Telemedicine, influenced by space industry practices, has been vital in remote patient care during the pandemic. Space-related innovations like the Canadian Space Agency's **BIO-MONITOR** device also demonstrate their importance in continuous monitoring of COVID-19 patients. (Anggina *et al.* 2022).

3 HEALTHCARE METHODOLOGIES INTEGRATION WITH SATELLITE TECHNOLOGIES

Table 1 reveals satellite applications in healthcare by presenting application methods and their descriptions. The following are details for each application and its working environments in reference to the interfacing of satellite technologies with healthcare systems.

Table 1. Satellite applications in healthcare.

Application Methods	Description
Remote Health Monitoring	Satellites enable real-time tracking of patients' health, especially in remote regions. Data can be transmitted to medical professionals for timely interventions.
Disease Mapping and Surveillance	Satellite imagery identifies areas with disease outbreaks by monitoring environmental conditions, such as stagnant water for malaria.
Telemedicine	Satellite communication enables global healthcare access.
Environmental Health Tracking	Satellites monitor environmental factors like pollution, providing data for public health decisions related to conditions like respiratory diseases.
Disaster Response and Management	Satellite images assist in rapid assessment post-natural disasters, directing medical aid efficiently.
Medication and Vaccine Delivery	Drones, guided by satellite navigation, deliver medicines and vaccines to remote locations timely.
Health Research and Studies	Satellites support large-scale epidemiological studies by offering data on environmental factors and population movements.

3.1 *Monitoring life signals remotely*

In a study, five patients with Medtronic defibrillators tested the 'Telemedicine Anywhere' (TelAny) system, which uses satellite technology for remote health monitoring. This system, used by cardiologists in Milan and Rome, ensured fast, uninterrupted data transmission, crucial in areas with poor connectivity. Data from the patients' implants was collected at their homes and sent to doctors via satellite, allowing real-time insights for prompt, informed medical decisions and continuous monitoring. This integration of medical devices, tele-medicine, and satellite tech marks a new era in healthcare, enhancing patient care regardless of distance (Kocian *et al.* 2011).

3.2 *Disease mapping and surveillance*

Disease Mapping and Surveillance, using satellite technology, provide a broad view of disease spread and areas prone to outbreaks. Satellites identify environmental conditions favoring diseases, like stagnant water for malaria mosquitoes. They also monitor climate changes to predict outbreaks, such as cholera following heavy rains. In hard-to-reach areas, satellites offer efficient surveillance, informing health agencies of disease hotspots for targeted interventions. This satellite assistance helps the world preempt outbreaks and enhance public health strategies (Lleo *et al.* 2008).

3.3 *The medical database*

Telemedicine, using satellite communication, connects remote patients with distant medical experts, like in Alaska. Here, residents in isolated villages access specialized care through the Alaska Native Tribal Health Consortium's telemedicine program. Patients consult Anchorage specialists virtually, with local nurses sending patient data, like skin condition images, for immediate diagnosis. This method reduces travel needs and speeds up treatment. Telemedicine includes remote diagnostics, vital in emergencies. Satellites make healthcare more inclusive, efficient, and patient-focused, especially in isolated areas (Patricoski 2004).

3.4 *Environmental health tracking & disaster response management*

Environmental Health Tracking uses satellites to monitor air and water quality, radiation, etc., helping predict health risks like respiratory diseases from air pollution. This data aids public health strategies. In Disaster Response, satellites provide vital imagery of areas hit by earthquakes, floods, or wildfires, guiding relief efforts and resource allocation. They also maintain communication when ground networks fail, coordinating relief work. Thus, satellites are key in environmental monitoring and emergency response, significantly contributing to public health and disaster management (Kazansky *et al.* 2016).

3.5 *Medication and vaccine delivery & health research and studies*

Satellite-guided drones revolutionize remote healthcare delivery, providing essential medicines and vaccines quickly. Satellite technology also significantly aids health research by tracking environmental and population changes, crucial for predicting zoonotic diseases and analyzing seasonal health trends, enhancing public health decision-making (Kitonsa & Kruglikov 2018). Advanced communication also plays a key role in such scenarios (Sharma *et al.* 2021).

4 CONCLUSION

This manuscript elucidates the deep synergy between satellite technology and healthcare. Within the fusion of these domains, transformative possibilities emerge. Satellites, in facilitating real-time remote consultations, precise disease mapping, and swift disaster response, prove their indispensability. Their integral contribution augments global health capabilities, heralding a future characterized by integrated, responsive, and universally accessible medical care.

REFERENCES

Anggina, S., & Perwitasari, I. (2022, September). Evaluation of space management during COVID-19 pandemic crisis in Indonesia. In *ICISPE 2021: Proceedings of the 6th International Conference on Social and Political Enquiries*, ICISPE 2021, 14–15 September 2021, Semarang, Indonesia (p. 479), (http://dx.doi.org/10.4108/eai.14-9-2021.2321409).

Dietrich, D., Dekova, R., Davy, S., Fahrni, G., & Geissbühler, A. (2018). Applications of space technologies to global health: scoping review. *Journal of medical Internet research*, 20(6), e230.

Kazansky, Y., Wood, D., & Sutherlun, J. (2016). The current and potential role of satellite remote sensing in the campaign against malaria. *Acta Astronautica*, 121, 292–305 (https://doi.org/10.1016/j.actaastro.2015.09.021).

Kitonsa, H., & Kruglikov, S. V. (2018). Significance of drone technology for achievement of the United Nations sustainable development goals. *R-Economy*, 4(3), 115–120 (https://doi.org/10.15826/recon.2018.4.3.016).

Kocian, A., De Sanctis, M., Rossi, T., Ruggieri, M., Del Re, E., Jayousi, S., ... & Suffritti, R. (2011, March). Hybrid satellite/terrestrial telemedicine services: Network requirements and architecture. In *2011 Aerospace Conference* (pp. 1–10) (https://doi.org/10.1109/AERO.2011.5747335).

Lleo, M. M., Lafaye, M., & Guell, A. (2008). Application of space technologies to the surveillance and modelling of waterborne diseases. *Current Opinion in Biotechnology*, 19(3), 307–312 (https://doi.org/10.1016/j.copbio.2008.04.001).

Patricoski, C. (2004). Alaska telemedicine: growth through collaboration. *International Journal of Circumpolar Health*, 63(4), 365–386 (https://doi.org/10.3402/ijch.v63i4.17755).

Sharma, A., Garg, A., Sharma, S. K., Sachan, V. K., & Kumar, P. (2021). Performance optimization for UWB communication network under IEEE 802.15. 4a channel conditions. *Computer Networks*, 201, 108585 (https://doi.org/10.1016/j.comnet.2021.108585).

Advances in AI for Biomedical Instrumentation, Electronics and Computing – Sachan et al. (eds)
© 2024 The Author(s), ISBN 978-1-032-64298-7

Onboard compression and preprocessing methods for LEO satellite imagery: A review

Abhishek Sharma, Nidhi Singh, Yashvardhan Srivastava, Shubhi Sharma,
Krishna Pratap Singh, Harsh Jaiswal, Praveen Kumar & Parvin Kumar
*Centre of Excellence, Space Technologies, Department of Electronics and Communication Engineering,
KIET Group of Institutions, Delhi-NCR, Ghaziabad, India*

ABSTRACT: Low Earth Orbit (LEO) satellites play a significant role for scientific research, telecommunications, and Earth observations by remote sensing, which require capturing of high-resolution images of earth's surface. Due to the constant acquisition of significant amounts of these high-quality images by the satellites, effective pre-processing and compression approaches for image size reduction are required for efficient utilization of transmission bandwidth. In the manuscript, state-of-the-art pre-processing and compression techniques are reviewed for LEO satellites using On-Board Computers (OBCs) which serve as the foundation of data handling and the caliber for scientific insights gained from orbiting satellites.

1   INTRODUCTION

Low Earth Orbit (LEO) satellites have marked a paradigm shift in how we observe and understand our planet. Their ability to hover closer to the Earth, as opposed to geostationary satellites, facilitates better temporal resolution, making them ideal tools for a plethora of applications ranging from climate monitoring to surveillance. Yet, the sheer volume of high-resolution data that these satellites generate demands an innovative approach to data handling and management (Gaston *et al.* 2023). The high revisit time of LEO satellites means they are constantly capturing data. Every snapshot of the Earth, especially in high-resolution, can amount to gigabytes of data. This leads to the first bottleneck – storage. Even as satellite storage systems have advanced, the exponential growth in data generation poses a significant challenge. Thus, it becomes essential to either transmit this data to the ground stations immediately or process it onboard. Transmission of such massive amounts of data presents its own set of challenges, primarily due to bandwidth limitations. Especially for student nanosatellite programs, where there's a restriction on the frequency bands they can use, optimizing the data becomes paramount. This is where onboard image preprocessing and compression technologies step in to bridge the gap.

Image preprocessing goes beyond merely enhancing the visual quality. It encompasses a variety of activities including, but not limited to, correcting lens distortions, compensating for satellite movements, and adjusting for atmospheric interferences. These adjustments ensure that the subsequent analysis of these images yields accurate insights. The role of image segmentation, as part of the preprocessing, cannot be understated. By isolating and highlighting specific areas within the image, it simplifies the task of data extraction, aiding in object identification and tracking (Hoang *et al.* 2020). On the compression front, the main goal is to ensure efficient storage and transmission without compromising on the core information the image contains. Sophisticated algorithms have been developed to strike this delicate balance, ensuring that while the image's file size is reduced, the integrity of the data remains (Zebari *et al.* 2020).

Moreover, there's an increasing emphasis on embedding more robust computing capabilities directly into the satellites. This on-board computing enables real-time processing of data, reducing the dependency on ground stations. This instant processing and the subsequent immediate transmission streamline the entire data collection and analysis chain.

Emerging technologies, such as artificial intelligence and machine learning, are also finding applications in the satellite data processing domain. They enable the automatic detection of anomalies, prioritize critical data transmission, and filter out extraneous information, optimizing the use of available bandwidth. The blend of image preprocessing, segmentation, and compression, combined with state-of-the-art onboard computing, is driving the next phase of innovations in LEO satellite missions. As the world grows more dependent on space-borne data, these technologies' significance will only increase. By the end of this review, readers will have a comprehensive understanding of these intricate technological processes (Curzi, G, 2020). Figure 1 illustrates the intricacies of onboard computing in satellites, and Table 1 provides insights into various pre-processing methodologies for Earth observation satellites.

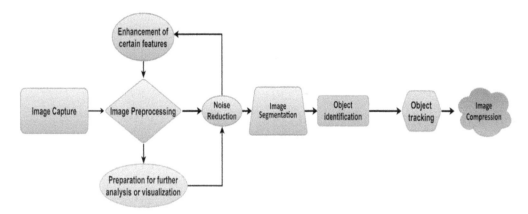

Figure 1. Flow chart for OBC compression and pre-processing.

2 PRE-PROCESSING OF IMAGES

For accurate analysis, image preparation for Low Earth Orbit (LEO) satellite data offers several difficulties and requirements. To improve visual clarity, atmospheric factors like scattering and absorption must first be adjusted (Wang *et al.* 2019). To ensure stability in pixel values and allow for meaningful comparisons over time, radiometric calibration is crucial. For locating and calculating the pertinent information in the photos, feature extraction techniques including edge detection, object segmentation, and texture analysis are essential (Dhruv *et al.* 2019). Figure 2 presents a complete flow for the satellite image processing cycle. Additionally, with the rise in satellite constellations and the increased frequency of data collection, there's a growing need for real-time or near-real-time preprocessing capabilities. This is especially relevant for applications such as disaster response, where timely information can be crucial. In such scenarios, cloud cover can also pose a significant challenge, obscuring vital ground details. Advanced cloud detection and removal algorithms are being integrated into the preprocessing workflow, ensuring that the final image is both timely and of high quality. Furthermore, the rapid developments in artificial intelligence and machine learning offer promising avenues to automate and

Table 1.   A comparison between various Pre-processing methods of images for earth observation satellite.

S. No.	Type / Name of Techniques	Type	Applications
1	Radiometric Calibration	Remote Sensing	Adjusting the sensor's readings to account for variations in sensitivity and environmental conditions, ensuring accurate radiometric data.
2	Geometric Correction		Correcting for geometric distortions introduced by the satellite's orbit and sensor, such as terrain relief and Earth curvature.
3	Atmospheric Correction		Compensating for atmospheric effects, such as scattering and absorption, to retrieve surface reflectance and temperature information accurately.
5	Multispectral Image Fusion		Integrating data from different spectral bands to create composite images for enhanced analysis, often used for vegetation mapping and land-use classification.
7	Radiometric Enhancement	Image Processing	Enhancing the contrast and visual quality of images through techniques like histogram equalization, contrast stretching, or tone mapping.
8	Noise Reduction		Applying filters and algorithms to remove sensor noise and other artifacts from the images.
9	Pan-Sharpening		Combining higher-resolution panchromatic (grayscale) imagery with lower-resolution multispectral imagery to create a single, higher-resolution color image.
10	Image Compression		Employing lossless or lossy compression to reduce data size for efficient transmission and storage, taking into account limited bandwidth and onboard memory.
11	Onboard Data Selection and Prioritization	Data Management	Automatically selecting and prioritizing the most important data for transmission to ground stations based on mission objectives and data quality.
12	Data Format Conversion		Converting data formats to facilitate processing and compatibility with various analysis tools and software.
13	Data Quality Assessment		Implementing checks and flagging data anomalies to ensure data integrity and quality.

enhance these preprocessing steps, reducing human intervention and accelerating the data delivery pipeline (Baduge *et al.* 2022). As the demands of satellite data users evolve, so too must the preprocessing techniques to ensure that the information is relevant, accurate, and timely.

3 COMPRESSION OF IMAGES

Digital image compression can be categorized into lossless and lossy methods (Din *et al.* 2019). While lossless compression retains the original image quality, its compression ratio is limited. In contrast, lossy compression offers higher compression rates but degrades image quality. Transmitting large image files without compression consumes significant bandwidth, affecting other mission-critical operations (Cong *et al.* 2022; Ren *et al.* 2022; Sara *et al.* 2019). Table 2 compares various compression techniques for earth observation satellites. Figure 2 reveals a completed data flow for satellite image processing.

Table 2. A comparison between various compression techniques for earth observation satellites.

Methodology	Type	Description	Application Scenario	Limitations
Run-Length Encoding (RLE)	Lossless	Efficient for repeated colors.	Images with large uniform regions.	Not efficient for complex images.
Deflate (ZIP)		Combines LZ77 and Huffman coding.	PNG and some TIFF formats.	Can be outperformed by specialized algorithms.
JPEG	Lossy	Uses discrete cosine transform.	Natural scenes.	Loss of some data.
JPEG2000		Improved version of JPEG.	High-quality imagery.	Higher computational complexity.
Wavelet-Based Compression		Techniques like SPIHT for compressing images.	Satellite images.	Might lose some details.
Discrete Cosine Transform (DCT)		Represents image as sum of sinusoids.	JPEG compression.	Loss of some data.
Wavelet Transform		Decomposes image into frequency sub bands.	Multi-resolution applications.	Complexity increases with layers.
Vector Quantization (VQ)		Maps vectors from original to compressed space.	Large datasets with repeating patterns.	Requires training.
Region-Based Compression		Focuses on compressing similar regions.	Satellite images with uniform regions.	Requires region identification.
Bandwidth-Preserving Compression		Preserves correlation between bands in satellite images.	Multi-spectral and hyper-spectral images.	Requires sophisticated algorithms.

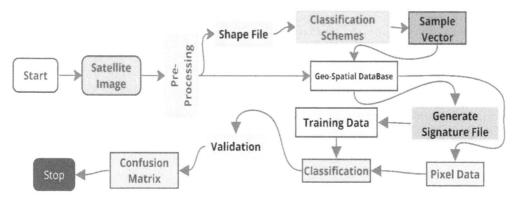

Figure 2. Flow chart for satellite image processing cycle.

4 CONCLUSION

This review paper has delved into the vital role played by image preprocessing, segmentation, and compression techniques in Low Earth Orbit (LEO) satellite missions. As LEO satellite missions continue to gather vast amounts of image data, innovative solutions are necessary to effectively manage and enhance data quality. On-Board Image Preprocessing and Compression (OBC) technologies have emerged as indispensable tools in this context, improving the efficiency of data handling and reducing latency. Image preprocessing, including radiometric calibration and feature extraction, sets the stage for meaningful analysis, while image segmentation aids in object identification and tracking. Image compression, both lossless and lossy, is essential for conserving bandwidth and reducing data transmission costs in the face of limited resources. Onboard computing further accelerates image processing, enabling real-time analysis and enhancing mission effectiveness across various fields. Overall, the integration of these techniques with OBC technology advances the capabilities of LEO satellite missions, benefiting Earth observation, environmental monitoring, disaster management, and more.

REFERENCES

Baduge, S. K., Thilakarathna, S., Perera, J. S., Arashpour, M., Sharafi, P., Teodosio, B., ... & Mendis, P. (2022). Artificial intelligence and smart vision for building and construction 4.0:Machine and deep learning methods and applications. *Automation in Construction*, 141, 104440.

Cong, H., Fu, L., Zhang, R., Zhang, Y., Wang, H., He, J. and Gao, J., 2022. Image quality assessment with gradient Siamese network. In *Proceedings of the IEEE/CVF Conference on Computer Vision and Pattern Recognition*: 1201–1210.

Curzi, G., Modenini, D., & Tortora, P. (2020). Large constellations of small satellites: A surveyof near future challenges and missions. *Aerospace*, 7(9), 133.

Dhruv, B., Mittal, N. and Modi, M., 2019. Study of Haralick's and GLCM texture analysis on 3D medical images. *International journal of Neuroscience*, 129(4): 350–362.

Din, R., Qasim, A. J., Abdullah, S. and Elias, S. J., 2019. Analysis Review on Image Compression Domain. *International Journal of Engineering & Technology*, 8(1.7): 293–296.

Gaston, K. J., Anderson, K., Shutler, J. D., Brewin, R. J. and Yan, X., 2023. Environmental impacts of increasing numbers of artificial space objects. *Frontiers in Ecology and the Environment*.

Hoang, N.D., 2020. Image processing-based spall object detection using gabor filter, texture analysis, and adaptive moment estimation (Adam) optimized logistic regression models. *Advances in Civil Engineering*, *2020*: 1–16.

Ren, Y., Zhang, Y., Liu, Y., Wu, Q., Su, J., Wang, F., Chen, D., Fan, C., Liu, K. and Zhang, H., 2022. DNA-Based Concatenated Encoding System for High-Reliability and High-Density Data Storage. *Small Methods*, 6(4), p.2101335.

Sara, U., Akter, M. and Uddin, M. S., 2019. Image quality assessment through FSIM, SSIM, MSE and PSNR—a comparative study. *Journal of Computer and Communications*, 7(3): 8–18.

Wang, Y. F., Liu, H. M., & Fu, Z. W. (2019). Low-light image enhancement via the absorption light scattering model. *IEEE Transactions on Image Processing*, 28(11), 5679–5690.

Zebari, R., Abdulazeez, A., Zeebaree, D., Zebari, D. and Saeed, J., 2020. A comprehensive review of dimensionality reduction techniques for feature selection and feature extraction. *Journal of Applied Science and Technology Trends*, 1(2): 56–70.

Advances in AI for Biomedical Instrumentation, Electronics and Computing – Sachan et al. (eds)
© 2024 The Author(s), ISBN 978-1-032-64298-7

Satellite security: Navigating threats and implementing safeguards in the modern space age

Abhishek Sharma, Divyanshi Srivastava, Suryanshi Singh, Ira Nafees, Satvik Aggarwal, Himanshu Chaudhary & Praveen Kumar
Centre of Space Technologies, KIET Group of Institutions, Delhi-NCR, Ghaziabad, India

ABSTRACT: As of the latest count, over 6,500 satellites orbit our planet, with more than half actively transmitting valuable data. These satellites, critical to modern infrastructure and operations, necessitate vigilant safeguarding against potential threats. The repercussions of cyber-attacks on satellites could range from ground-based navigational errors to catastrophic mid-air collisions. Given the diverse array of satellites and their unique operational characteristics, understanding the specific vulnerabilities of each type is paramount. This paper delves into the significance of satellite cybersecurity and emphasizes the need to prioritize the protection of our skyborne assets.

1 INTRODUCTION

Satellites are vital in today's world. They help with communication, navigation, watching the Earth, and exploring space. Right now, there are over 6,500 satellites above us. But as we use them more, they also face more threats. Hackers can attack satellites. When they do, they might control the satellite, change its path, or stop it from working properly. This can cause big problems, like lost money or safety risks. Some of these attacks use very new technology, making them even harder to stop (Nikookar 2023). To protect satellites, people are trying new ideas and rules. Some are using better ways to keep data safe, like new encryption methods or laser signals. Others are making rules to make sure everyone uses space safely (Bodemann *et al.* 2013).

A successful cyberattack on a satellite system can have serious consequences. Attackers might gain control, changing the satellite's orbit or essential operations. These attacks can interrupt key services such as GPS, weather forecasting, and communication networks. This can cause major financial losses and disrupt daily life. When military and intelligence satellites are targeted, national security is at risk. Sensitive information might be exposed, and monitoring capabilities could be compromised (Pekkanen *et al.* 2022). Table 1 presents various threat categories and their descriptions for satellites.

Table 1. Threat categories and descriptions.

Threat Categories	Description
AI and Machine Learning Threats	Advanced adversaries utilize AI and machine learning to predict satellite communication patterns, identify vulnerabilities in real-time, and adaptively counter defensive measures.
Quantum Computing	The rise of quantum computing could endanger current cryptographic standards used in satellite communications, as quantum computers have the potential to decrypt many of today's encryption methods.

(continued)

DOI: 10.1201/9781032644752-41

Table 1. Continued

Threat Categories	Description
Signal Jamming and Spoofing	Advanced jamming equipment can disrupt satellite signals, while spoofing can mislead satellite systems by feeding them false information.
On-Orbit Threats	As space becomes more accessible, the potential for on-orbit threats increases, including the possibility of satellites being physically tampered with in space by other satellites or space-based weapons.
Ground Station Attacks	Ground stations, the terrestrial components that communicate with satellites, can be vulnerable points of entry. Cyberattacks targeting these can disrupt satellite operations.
Software Vulnerabilities	As satellites become more software-dependent, vulnerabilities in their onboard systems or in the software of their ground-based controllers can be exploited.
Supply Chain Compromises	The components making up satellite systems often come from global suppliers. Any compromise in this supply chain can introduce vulnerabilities into the satellite system.

This paper examines the cybersecurity challenges satellites face, potential attack methods, and the solutions being implemented to safeguard these vital assets.

2 SATELLITE ARCHITECTURES AND THEIR INHERENT VULNERABILITIES

Modern satellites are marvels of engineering, combining advanced hardware and software elements to fulfill diverse roles, from Earth observation to deep space communication. At the heart of these systems is the onboard computer, responsible for managing operations and processing vast amounts of data. While this intricate design enhances satellite capabilities, it also introduces potential vulnerabilities. Communication is facilitated using specialized protocols, and while these are designed for efficiency and clarity, they can be exploited if not adequately secured. The payload, comprising the specific instruments and sensors, is another critical component. Any compromise in its function could lead to corrupt or false data transmissions, affecting downstream applications and decisions (Serief *et al.* 2023).

The propulsion systems of satellites, essential for maintaining their designated orbits and avoiding space debris, are another potential point of vulnerability. Unauthorized alterations to these systems can drastically alter a satellite's trajectory, risking collisions or rendering it ineffective. Moreover, the interface between satellites and their ground control stations on Earth is a vital link. This connection, if not safeguarded, could offer adversaries an avenue to commandeer satellite operations. As technology progresses, understanding and addressing these vulnerabilities becomes paramount to ensuring the security and functionality of our skyborne assets. Figure 1. reveals possible vulnerability distributions of satellite sub-components under cyber-attacks in present scenarios.

Vulnerabilities Distribution of Satellite Components in Present Scenario

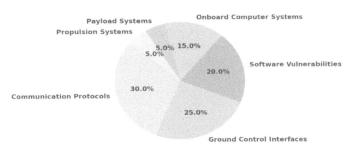

Figure 1. Presenting the vulnerability distributions of satellite components in present scenario.

3 REGULATORY AND POLICY FRAMEWORK FOR SATELLITE SECURITY

The vast expanse of space, once a frontier only touched by superpowers, is now accessible to numerous nations and private entities. As the sky gets busier, the imperative for a robust regulatory framework has never been more pressing. This framework is foundational for ensuring that space, especially the crowded orbits around Earth, remains a safe and sustainable environment. The following are the existing regulatory and policy frameworks between worldwide nations.

3.1 *The outer space treaty*

Officially known as the "Treaty on Principles Governing the Activities of States in the Exploration and Use of Outer Space, including the Moon and Other Celestial Bodies", this treaty, which came into force in 1967, stands as the cornerstone of international space law. Ratified by 109 countries, including major space-faring nations, the treaty establishes that the exploration and use of outer space shall be carried out for the benefit of all humanity and shall be free from national appropriation. It also prohibits the placement of nuclear weapons in space and the use of celestial bodies for military purposes.

3.2 *Space debris mitigation guidelines*

In 2007, the United Nations Committee on the Peaceful Uses of Outer Space (COPUOS) endorsed a set of guidelines aimed at curbing the increasing amount of space debris. These guidelines focus on minimizing debris released during satellite operations, preventing accidental explosions, and ensuring defunct satellites in the Low Earth Orbit (LEO) are removed in a controlled manner or moved to a graveyard orbit (Li *et al.* 2022).

3.3 *National frameworks*

Many countries have their own space agencies and corresponding national regulations. For instance, the United States' National Space Policy directs its departments and agencies to lead in the enhancement of space capabilities. It emphasizes the importance of international cooperation, commercial growth, and the sustainable use of space. Similarly, the European Space Agency (ESA) has its space strategy that focuses on ensuring Europe's independent access to space, strengthening its position in the global market, and maximizing the socio-economic benefits for its members (Sagath *et al.* 2019).

3.4 *Public-Private collaborations*

As the space industry witnesses increased private participation, governments are collaborating closely with these entities to ensure compliance with established norms. Notable examples include the agreements between NASA and commercial entities like SpaceX and Blue Origin. Such collaborations are not only about launching payloads but also about ensuring that the activities of these companies align with broader space sustainability and security goals (Maraš *et al.* 2022).

In the rapidly evolving domain of space exploration and satellite deployment, the interplay of international treaties, national regulations, and private-sector ambitions underscores the need for a cohesive and adaptive regulatory approach. Ensuring the sustainable and secure use of our celestial frontier is a shared responsibility, one that demands global collaboration and foresight.

4 CONCLUSION

This manuscript underlines the importance of satellite cybersecurity, highlighting potential threats and emphasizing the need for robust regulatory frameworks and advanced protection measures. With growing threats to both civilians and the military, the safety of satellites has become a top concern. Satellites hold the potential to be invaluable assets for future generations, but their vulnerability can swiftly transform them from blessings to liabilities. Recognizing and understanding the multifaceted threats facing these orbital assets is crucial. Through diligent and proactive measures, it's possible to fortify these satellites, ensuring their sustained utility and security in the vast expanse of space.

REFERENCES

Bodemann, C., & Kalden, O. (2013, June). Threats awaiting Earth observation satellites. *In 2013 6th International Conference on Recent Advances in Space Technologies (RAST)* (pp. 1139–1143). IEEE.

Li, Y., Zhang, G., Yu, J., Li, G., & Li, Y. (2022, November). OTFS-Based Communication and Navigation Integrated Signal Transmission for LEO Satellites. *In 2022 IEEE 22nd International Conference on Communication Technology (ICCT)* (pp. 451–457). IEEE.

Maraš, D., & Dangubić, M. 2022. Cooperation Between Government Agencies and Private Companies in Space: *The case of the united states. Astropolitics* 20(2–3): 226–237.

Nikookar, H. (2023). A risk analysis of communication, navigation and sensing satellite systems threats. *Journal of Mobile Multimedia*: 277–290.

Pekkanen, S. M., Aoki, S., & Mittleman, J. 2022. Small satellites, big data: Uncovering the invisible in maritime security. *International Security* 47(2): 177–216.

Sagath, D., Vasko, C., Van Burg, E., & Giannopapa, C. 2019. Development of national space governance and policy trends in member states of the European Space Agency. *Acta astronautica* 165: 43–53.

Serief, C., Ghelamallah, Y., & Bentoutou, Y. 2023. Deep learning-based system for change detection on-board earth observation small satellites. *IEEE Journal of Selected Topics in Applied Earth Observations and Remote Sensing*.

Advances in AI for Biomedical Instrumentation, Electronics and Computing – Sachan et al. (eds)
© 2024 The Author(s), ISBN 978-1-032-64298-7

UAV meets VANET: A hybrid model for performance enhancement in smart cities

Hunny Pahuja
Department of Electronics and Communication Engineering, KIET Group of Institutions, Delhi-NCR, Ghaziabad

Manoj Sindhwani & Shippu Sachdeva
Lovely Professional University Phagwara, India

ABSTRACT: Vehicular ad-hoc networks (VANET) are the most popular network nowadays because of their wide applications. The major issue in this network is route discovery and congestion, especially in urban areas. Further for better functioning and more efficiency of VANET, another technology can be collaborated named as Unmanned Aerial Vehicles (UAVs) assisted model which helps RSUs to increase the overall coverage of the network. UAVs-based clustering techniques are also implemented in VANET, as clustering techniques ensure higher mobility and stability. The kind of collaboration between UAVs and VANET provides reliable paths and also reduces link failure in the high dynamic topology of VANET.UAVs assisting model provides stability to the VANET network by achieving stabilization and longer cluster head lifespan. UAVs-assisted hybrid model is more efficient than existing protocols especially CBRP (performs the best in existing protocols) by 6%, 3%, and 35% in terms of throughput, PDR, and Delay.

Keywords: UAVs, Vehicular communication, Routing Protocols, Clustering, Cluster Head

1 INTRODUCTION

Unlike mobile ad-hoc networks (MANET), vehicular ad-hoc networks (VANETs)follow dynamic topology because of high mobility scenario, because of which it becomes really difficult to disseminate the information in VANET, so unmanned aerial vehicles (UAVs) can be used in coordination with vehicular networks (Oubbati *et al.* 2017; Tariq *et al.* 2020). Considering the case of traditional wireless networks, UAV-based communication is more efficient due to the reason of less channel impairments and fading effects. UAVs-enabled communication can act as a boon in situations where line of sight (LOS) is a major issue. Also, in the areas where the infrastructure-based network is difficult to install, UAVs can be considered as a viable option as they provide information to vehicles (He *et al.* 2020; Jobaer *et al.* 2020). Since VANETs have dynamic, multi-hop, unpredictable, unstructured, and vehicular scenarios, they face major challenges in this aspects. So, UAVs assisted VANET hybrid model is proposed to disseminate the information in communication architecture i.e., V2V, V2I, V2Broadband, and V2Cloud Computing. The research issues in VANET are frequent disconnections, delay constraints, routing variable network density, etc. which can be resolved using clustering techniques. Instead of direct communication between vehicles and UAVs, clustering techniques are generally considered for efficient networks [Fan *et al.* 2018]. In clustering techniques, the vehicles on various parameters are combined to make several clusters with a specified vehicle node called cluster head (CH). So, every vehicle in the network communicates with UAVs through the cluster head of a particular cluster. To maintain the

DOI: 10.1201/9781032644752-42

stability of the cluster, backup cluster heads are also elected in the networks on various parameters (Oubbati *et al.* 2019). UAVs communicate with vehicular nodes called UAV2, also UAVs enhance the roadside unit (RSU) capacity in a congested network. Cellular networks can also be used in addition to other networks so they can also be linked with UAV-assisted VANET as shown in Figure 1. As, in a more dynamic environment, UAVs will into consideration for the dissemination of information to the vehicular nodes (Seliem *et al.* 2017).

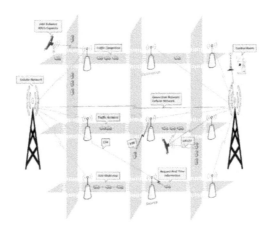

Figure 1.   Unmanned aerial vehicle-assisted vehicular ad-hoc network architecture.

2   UAV ASSISTED VEHICULAR COMMUNICATION

In case of emergencies, UAVs assist RSU for better coverage and also stabilize the situation. The link failure issue is generally common in VANET, but can easily be reduced by implementing dynamic clustering techniques (Liu *et al.* 2020; Sedjelmaci *et al.* 2019). Clustering of the vehicles can be done on various priority parameters. The bandwidth and efficiency-related issues can also be resolved by achieving stability in the cluster head (Rossi *et al.* 2017; Sami Oubbati *et al.* 2020). In the vehicular clusters, the backup CH is also elected which also acts as a leadership node in the network. These nodes are selected on various predefined parameters. The metrics include the direction, position, and velocity of the vehicle. UAVs enhance the air-to-ground network connectivity. UAVs do not fly at high altitudes to communicate with the vehicular nodes on the ground as shown in Figure 2. Air-to-ground communication link is established between UAVs and ground vehicular networks (Oubbati *et al.* 2016).

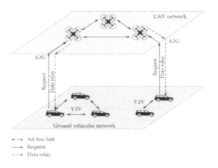

Figure 2.   Air-to-ground communication system.

We assume in our work that GPS is being deployed in all the vehicles so that the position information, velocity, and direction of movement, where every vehicle calculates the relative speed w.r.t the neighbors (Makolkina *et al.* 2017; Wang *et al.* 2016). The distance to its neighbors can also be obtained using the proposed algorithm. The proposed approach focuses on the setting up of clusters, as soon as the grouping of vehicles arises in the dynamic network because the network topology in VANET is dynamic. This dynamic topology in VANET creates several issues which can be resolved by clustering techniques. Hence it promotes intra-cluster communication and inter-cluster communication across various vehicular nodes (Kerrache *et al.* 2018).

2.1 *Advantages of UAVs for ground VANET*

With UAVs, there are various advantages of VANET it supports good coverage of V2V communication in congested areas, reduced transmission overheads with clustering techniques, increased efficiency of the network, and most importantly safety applications in emergencies. Also, there are various special applications for routing the packets, security-related issues, traffic safety, combination with cellular infrastructure, and collision avoidance applications.

2.2 *UAV-assisted routing protocol in VANET*

Various routing Protocols have already been proposed in the past years, having a common goal of achieving maximum throughput, reducing packet loss, improving delay parameters, and enhancing QoS parameters in the network. In this proposed work routing protocols like AODV, DSDV, and CBR is used for improving throughput, PDR, and Delay parameter.

2.2.1 *Route discovery*
To discover a route from the source to the destination node, all the nodes having immediate connections with the neighbors are identified first. This can also be calculated using a digital city map [20]. ***RSU Setup Process***: RSUs are connected with the cellular network and UAVs for better coverage and capacity. The RSUs and UAVs are equipped with an omni directional antenna having unity gain. RSUs can be better placed at an intersection point having a few kilometers distance between two RSUs. In highly dense environments, the wired and wireless connection can be a challenge as the number of obstacles is more nearby. The main purpose of this setup is to find the shortest distance between the nodes for better delivery and efficiency.

3 PROPOSED WORK

In this work, UAVs-assisted VANET is associated with clustering techniques, where cluster head selection is considered on priority, as the indirect communication will take place from the cluster head in normal conditions. In emergency conditions UAVs assisted RSUs intersection points will be available. Considering the scenario of taking two vehicular nodes with limited distance in smart cities. Let x and y be the two nodes with limited speeds S_x and S_y, L be in the proper LOS range of nodes x and y. (A_x,B_x) and (A_y,B_y) being their corresponding co-ordinates with β_x and β_y velocity angles. The lifespan of the link in between the nodes x and y is analyzed by the mathematical equation as:

$$\text{Path Expiration} = \frac{-(ij + km) + \sqrt{(i^2 + k^2)L^2 - (im - jk)^2}}{i^2 + k^2} \qquad (1)$$

Where

$$i = S_x \cos \beta_x - S_y \cos \beta_y \qquad (2)$$

$$k = S_x \sin \beta_x - S_y \sin \beta_y \qquad (3)$$

$$j = A_x - A_y \qquad (4)$$

$$m = B_x - B_y \qquad (5)$$

4 RESULTS AND SIMULATIONS

The communication between members of one cluster to another cluster can be done through the cluster head. Also, UAV intersection points will be available to increase the coverage, Here UAV assisted routing protocol (URP) is proposed to improve the parameters like throughput, packet delivery ratio and delay. Table 1 represents the simulation parameters for the research proposed in this paper.

Table 1. Simulation parameters.

Channel Type	Wireless Channel
Network interface type	Physical
Protocol	802.11
Interface Queue type	Queue/ Droptail
Link layer Type	LL
Maximum Packets	500
Number of nodes	22
Simulation Time	5ms
Routing Protocol	AODV/ DSDV/ CBRP

4.1 *Throughput*

The results have been obtained in graph with various input parameters. The results have been obtained for AODV, DSDV, CBR and the proposed work.

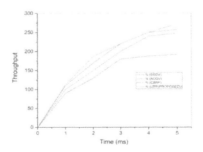

Figure 3. Throughput comparison.

Figure 3 is the graph of a comparative study of performance based on throughput. No doubt that AODV outperforms the DSDV routing protocol but CBR has greater throughput

225

than both conventional protocols. Now here comes the proposed protocol URP (UAV-assisted Routing Protocol) giving a higher throughput than all three existing protocols.

4.2 *Packet delivery ratio*

The graph shown in Figure 4 is the comparative analysis of PDR performance where our proposed scheme has a lesser number of packets drops and a better PDR. Initially, the performance of the protocols is closely related but as the time and cluster size increases the performance of CBRP with the proposed values become well. Hence, we are able to enhance the performance. The PDR achieved in proposed protocol is improved by around 3% from AODV, DSDV and CBRP as all three protocols are showing the same performance at later stage.

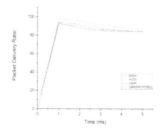

Figure 4.   Packet delivery ratio comparison.

4.3 *Delay*

The result obtained by comparing the average delay of all three protocols with the performance enhancement has been shown in Figure 5 from which it is clear that we are able to succeed in the performance improvement using proposed protocol against AODV, DSDV and CBRP. Lesser will be the delay better will be the performance. The Delay of proposed protocol is reduced by 63%, 47% and 35% while comparing with AODV, DSDV and CBRP.

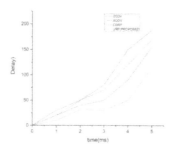

Figure 5.   Delay comparison.

5   CONCLUSION

In the work carried out to date has primarily focused on the study of various traditional clustering techniques algorithms were implemented, but we found that UAVs UAVs-assisted VANET hybrid model can give better quality results in terms of various parameters like throughput, packet delivery ratio, and delay. UAVs help in route discovery methods and also assist RSUs in improving the coverage area. AODV DSDV and CBR routing protocols are compared with the proposed UAVs assisted Protocol where we concluded the proposed UAVs assisted hybrid model is more efficient than existing protocols especially CBRP (performs the best in existing protocols) by 6%, 3%, and 35% in terms of throughput, PDR and Delay. Stability and mobility are major concerns, especially in VANETs. Furthermore, the conventional protocols do not stand best for the VANETs due to several reasons such as the VANET network, network size, clustered pitch, etc. so there is a requirement for different routing protocols when UAVs-assisted hybrid model is used. So the stability algorithms stood to be more useful and required in the dynamically unpredictable environment as in the prime objective. However there may be limitations that it holds the inconsistent routes which can be followed with intermediate nodes if there exists same old sequence number, the bandwidth consumption can be more due to the periodic beaconing so can act as a future scope for improvement in UAVs assisted hybrid model in VANET.

REFERENCES

Fan, X., Huang, C., Fu, B., Wen, S., & Chen, X. 2018. UAV-assisted data dissemination in delay-constrained VANETs. *Mobile information systems, 2018*.

He, Y., Zhai, D., Wang, D., Tang, X., & Zhang, R. 2020. A relay selection protocol for UAV-assisted VANETs. *Applie d Sciences* 10(23), 8762.

Jobaer, S., Zhang, Y., Iqbal Hussain, M. A., & Ahmed, F. 2020. UAV-assisted hybrid scheme for urban road safety based on VANETs. *Electronics* 9(9), 1499.

Kerrache, C. A., Lakas, A., Lagraa, N., & Barka, E. 2018. UAV-assisted technique for the detection of malicious and selfish nodes in VANETs. *Vehicular Communications 11*, 1–11.

Liu, T., Zhao, L., Li, B., & Zhao, C. 2020. Research on the Enhancement of VANET Coverage Based on UAV. In *Communications, Signal Processing, and Systems: Proceedings of the 8th International Conference on Communications, Signal Processing, and Systems 8th* (pp. 787–795). Springer Singapore.

Makolkina, M., Paramonov, A., Vladyko, A., Dunaytsev, R., Kirichek, R., & Koucheryavy, A. 2017. The use of UAVs, SDN, and augmented reality for VANET applications. *DEStech Trans. Comput. Sci. Eng. Aiie, 134*, 153–57.

Oubbati, O. S., Chaib, N., Lakas, A., Lorenz, P., & Rachedi, A. 2019. UAV-assisted supporting services connectivity in urban VANETs. *IEEE Transactions on Vehicular Technology* 68(4), 3944–3951.

Oubbati, O. S., Lakas, A., Lagraa, N., & Yagoubi, M. B. (2016, April). UVAR: An intersection UAV-assisted VANET routing protocol. *In 2016 IEEE Wireless Communications and Networking Conference* (pp. 1–6). IEEE.

Oubbati, O. S., Lakas, A., Zhou, F., Güneş, M., Lagraa, N., & Yagoubi, M. B. (2017). Intelligent UAV-assisted routing protocol for urban VANETs. *Computer Communications 107*, 93–111.

Rossi, G. V., Fan, Z., Chin, W. H., & Leung, K. K. (2017, March). Stable clustering for ad-hoc vehicle networking. In *2017 IEEE Wireless Communications and Networking Conference (wcnc)* (pp. 1–6). IEEE.

Seliem, H., Ahmed, M. H., Shahidi, R., & Shehata, M. S. (2017, October). Delay analysis for drone-based vehicular ad-hoc networks. In *2017 IEEE 28th Annual International Symposium on Personal, Indoor, and Mobile Radio Communications (PIMRC)* (pp. 1–7). IEEE.

Sedjelmaci, H., Messous, M. A., Senouci, S. M., & Brahmi, I. H. 2019. Toward a lightweight and efficient UAV aided VANET. *Transactions on Emerging Telecommunications Technologies* 30(8), e3520.

Sami Oubbati, O., Chaib, N., Lakas, A., Bitam, S., & Lorenz, P. 2020. U2RV: UAV-assisted reactive routing protocol for VANETs. *International Journal of Communication Systems, 33*(10), e4104.

Tariq, R., Iqbal, Z., & Aadil, F. 2020. IMOC: Optimization technique for drone-assisted VANET (DAV) based on moth flame optimization. *Wireless Communications and Mobile Computing, 2020*: 1–29.

Wang, X., Fu, L., Zhang, Y., Gan, X., & Wang, X. 2016. VDNet: an infrastructure-less UAV-assisted sparse VANET system with vehicle location prediction. *Wireless Communications and Mobile Computing 16*(17): 2991–3003.

Advances in AI for Biomedical Instrumentation, Electronics and Computing – Sachan et al. (eds)
© 2024 The Author(s), ISBN 978-1-032-64298-7

Vending machine for medicine using automated arm mechanism

Pankaj Kumar*, Kaushal Saraf*, Prakhar Nigam*, Rishi Yadav*, Kartik Verma* &
Divya Sharma*
*Department of Electronics and Communication Engineering, Ajay Kumar Garg Engineering College,
Uttar Pradesh, India*

ABSTRACT: A novel kind of medicine vending machine that uses robotic arms to dispense medication is one with automated arm mechanics. According to this study, there are a lot of potential benefits that medicine vending systems offer over conventional medicine vending machines, such as better accuracy, higher safety, and better user experience.

Keywords: Camera module, Medication dispenser, Microcontroller, Power Source

1 INTRODUCTION

Automated arm mechanisms used in medicine vending machines are a relatively new technology. Although their use has increased recently, medicine vending machines have been around for nearly 50 years. There are several reasons for this, such as an aging population, growing prescription drug expenses, and a growing need for convenience (Othman *et al.* 2016). Numerous research investigations have been conducted about the precision and security of automated arm systems used in medicine delivery. Overall, these studies have shown that automated arm systems are safer and more accurate than conventional dispensing devices. For instance, a study that was published in the journal Pharmacy Practice and Research found that 99.9% accuracy in the dispensing of medications may be achieved with automated arm mechanisms. This is significantly more accurate than the usual 95% accuracy rate of standard dispensing methods (Bai *et al.* 2016; Nathan *et al.* 2014). Automated arm systems have the potential to reduce drug errors by as much as fifty percent, according to another study that was published in the journal Drug Safety and Quality Assurance. Studies on the accuracy and safety of automated arm systems for drug distribution have also been conducted, in addition to studies on the user experience. These studies show that users have a generally positive opinion of autonomous arm mechanics (Sriram *et al.* 1996). For instance, a study that was published in the Patient Experience Journal found that 85% of users who administered drugs using an automated arm mechanism said they would use the device again.

2 METHODOLOGY

There are two types of medication vending machines in the current methodology: over-the-counter and prescription. Prescription medicine vending machines sell both prescription and over-the-counter medications, while OTC medicine vending machines solely sell over-the-counter pharmaceuticals. The features of medicine vending machines include the following:

---

*Corresponding Authors: pk11345rai@gmail.com, kaushalsaraf2002@gmail.com,
prakhar.nigam101@gmail.com, Rishiyadav1950@gmail.com, kartik1974.kv@gmail.com and
divya13jan@gmail.com

                                    DOI: 10.1201/9781032644752-43

A payment system, a dispensing mechanism, and a touch screen display are common components of medicine vending machines. In addition, some medication vending machines have additional features including a camera system to confirm the user's identity and a temperature-controlled storage system (Aziz *et al.* 2016).

3 AUTOMATED ARM MECHANISM VENDING MACHINE

The gadget contains safe sections or storage units to hold several kinds of medications, including pills, capsules, syrups, and medical equipment. A user-friendly website that allows users to choose their medications and dosage is frequently included in the system (Mahaveer *et al.* 2017). Every time a patient receives a prescription from a doctor, they will receive the medication on their login page along with a QR code after payment is received. In essence, the QR code will contain information on the suggested medication (Mahaveer *et al.* 2017; Tank *et al.* 2017).When a patient presents their QR code to a nearby hospital or other designated location, the vending machine will use an automated arm mechanism to dispense the necessary medications after scanning the patient's code using a scanner inside the machine. The Automated Arm Mechanism is the machine's fundamental component. This arm is responsible for selecting the right drug from a storage area, transporting it to patients, and dispenses the prescribed amount (Bhagya *et al.* 2019; Tank *et al.* 2017). Figure 1 displays the block diagram of the medicine vending machine. Table 1 shows the comparison of the suggested and current technologies.

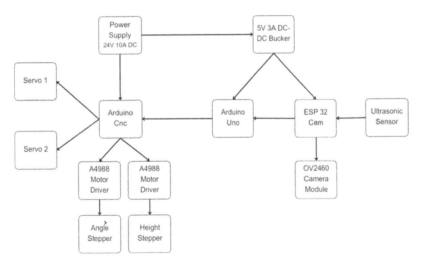

Figure 1.   Block schematic of a medication dispenser.

Table 1.   A comparative analysis of current and emerging.

Facet	Current Technology	Emerging Technology
Microcontroller	Antiquated microcontroller with limited capabilities (Hema *et al.* 2019)	The improved and versatile ESP32 microcontroller supports Bluetooth and Wi-Fi.
Holding Capacity	Limited storage for particular types of medications (Hema *et al.* 2019).	Ability can hold up to 2000 different prescriptions, providing a wider selection.

(continued)

Table 1. Continued

Facet	Current Technology	Emerging Technology
Integration of Cameras	May not have a camera or may simply use basic image capture techniques (Hema *et al.* 2019).	ESP32 cameras are integrated for improved image recognition and security.
Interface User	Basic user interface that might have buttons (Heba *et al.* 2019).	Improved graphical user interface with mobile application control.
Connectivity	Simple or nonexistent options for connectivity (Higuchi *et al.* 2017).	Improved networking options for Bluetooth and Wi-Fi allow for remote control and monitoring.
Analytic Data	Restricted ability to gather and analyze data (Higuchi *et al.* 2017; Rao *et al.* 2016).	Ability to collect and analyze data in order to provide insights regarding replenishment schedules and usage trends.
Features of Security	Minimum standards for security (Sarika 2017).	Real-time notifications and authentication are examples of additional security measures.
Maintenance	Manual or basic methods of maintenance (Williams *et al.* 2018).	Diagnostics and remote maintenance save downtime.

3.1 *Medicine vending machine advantages*

- A few benefits of using a medicine vending machine are cost savings, accessibility, and convenience (Brown *et al.* 2017).
- *Practicality*: Without needing to visit a pharmacy, customers may purchase prescriptions 24 hours a day, 7 days a week, thanks to medicine vending machines. People who live in remote places, work long hours, or have limited mobility will find this especially helpful (Brown *et al.* 2017).
- *Availability*: In public areas where people might not have easy access to a pharmacy, such airports, convenience stores, and workplaces, there are machines that vend medications. This could make it simpler for people to get the prescription drugs they need (Patel *et al.* 2019).
- *Cost reductions*: Compared to pharmacies, medicine vending machines may provide cheaper prices, especially for generic medications (Patel *et al.* 2019).

3.2 *Medicine vending machine drawbacks*

While using medication vending machines may have a number of advantages, there may also be some drawbacks to take into account.

- If a user enters their prescription information incorrectly, medication mistakes may occur. This results from medication vending machines' incapacity to verify the veracity of the data provided by the user (Patel *et al.* 2019).
- Users cannot receive the same degree of individualized care from medication vending machines as they can from a pharmacist. For example, a chemist can provide guidance on the safe and efficient administration of drugs (Patel *et al.* 2019).

3.3 *Control over medication vending machines*

Each nation has its own regulations governing medicine vending machines. In certain countries, medicine vending machines are subject to tight regulations, but not in others (Johnson *et al.* 2018; Taylor *et al.* 2020).

4 CONCLUSION

Compared to conventional medicine vending machines, automated arm mechanic medicine vending machines may provide a number of advantages. Before these computers may be used extensively, a few problems need to be resolved. This technology needs to be researched and developed further in order to become more affordable, dependable, and widely available.

4.1 *Prospective paths*

The automated arm mechanics used in pharmaceutical vending machines are being improved by researchers. Among the topics of focus are:

- Developing automated arm systems that are more accurate and reliable (Davis *et al.* 2019).
- Constructing automated arm systems with a wider distribution range (Martinez *et al.* 2017).

REFERENCES

Brown, S. R., & White, L. H. (2017), Emerging trends in medication vending machine technologies: a systematic literature review, *Journal of Healthcare Engineering* 8(4): 435–449.

B., & Davis, J. R. (2019). Smart medication vending machines: an innovative approach to medication management, *International Journal of Medical Informatics* 128: 30–36.

Hema, E. Priya, G. V., Rakshitha, A. Rakshitha, V. & Shilpa C. (2019). Automatic vending machine for medicines, *International Journal of Advance Research in Science and Engineering* 8(5): 137–142.

Heba, S. (2019). Automatic vending machines advantages and disadvantages, [Online]. Available: https://www.online-sciences.com/%20technology/automatic-vending-machines-advantages-and-disadvantages/.

Higuchi, Y. 2017. History of the development of beverage vending machine technology in Japan, national museum of nature and science: *Survey Reports on the Systemization of Technologies* 7: 5–16.

Johnson, P. A., & Anderson, C. M. (2018). Security and regulatory challenges of medicine Nathan Kamal, Ahmed, Aamir, Kaliselvan 2014. Automatic Paper Vending Machine, *International Journal Of Science, Engineering and Technology Research (IJSETR)*, vol. 4.

Othman N. B. and Ek O. P. (2016). Pill dispenser with alarm via smartphone notification, *IEEE 5th Global Conference on Consumer Electronics*: 1–2.

Penna Mahaveer (2017). Design and implementation of automatic medicine dispensing machine, *IEEE.*

Patel, R. K., & Gupta, S. P. (2019). Automated pharmaceutical dispensing machines: a review of design and implementation considerations, *Journal of Drug Delivery Science and Technology* 54, 101242.

Rao Janardhana, M. Sangeetha, T.V., & Gowri Rama, Ch. S. (2016). Automatic medicine vending system medical ATM, *International Journal of Scientific Development and Research* 1(10): 185–190.

Sriram T., Rao V. K. (1996). Application of barcode technology in automated storage & retrieval systems, *Industrial Electronics Conference Proceedings. Taipei*:5–10.

Shree S R Bhagya, Shekar P Chandra (2015). Automated medication dispensing system, IEEE.

Sarika, O (2017). Automatic medicine vending machine, *International Journal of Engineering Technology Science and Research* 4(12): 1150–1154.

Taylor, K. L., & Adams, E. D. (2020). The impact of medication vending machines on medication adherence: A systematic review and meta-analysis, *Journal of Clinical Pharmacy and Therapeutics* 45(3): 489–500.

Martinez, A. R., & Kim, H. S. (2017). The role of telemedicine in enhancing the functionality of medicine vending machines, telemedicine and e-health 23(6): 443–449.

Williams, L. Smith, J. A., & Johnson, M. B. (2018). Automated medicine dispensing systems: A comprehensive review, *Journal of Pharmacy Technology* 34(2): 56–68.

Bai, Y. W., & Kuo, T. H. (2016, January). Medication adherence by using a hybrid automatic reminder machine. In *2016 IEEE International Conference on Consumer Electronics (ICCE)* (pp. 573–574). IEEE.

Aziz, K., Tarapiah, S., Ismail, S. H., & Atalla, S. (2016, March). Smart real-time healthcare monitoring and tracking system using GSM/GPS technologies. In *2016 3rd MEC international conference on big data and smart city (ICBDSC)* (pp. 1–7). IEEE.

Tank, V., Warrier, S., & Jakhiya, N. (2017, March). Medicine dispensing machine using Raspberry pi and Arduino controller. In *2017 Conference on Emerging Devices and Smart Systems (ICEDSS)* (pp. 44–51). IEEE.

Advances in AI for Biomedical Instrumentation, Electronics and Computing – Sachan et al. (eds)
© 2024 The Author(s), ISBN 978-1-032-64298-7

Implementation and design of traffic light controller using Verilog

Kavya Kumar Tayal* & Ajay Suri*
Department of ECE, ABES Engineering College, Ghaziabad, India

Rohit Vikram Singh Bhadauria*
Galgotias University, Greater Noida, India

ABSTRACT: The implementation and design of Traffic Light Controller using Verilog presented in the paper. The traditional Traffic Light Controllers are manufactured based on fixing time schedules, which often results in inefficiencies and increased congestion during peak hours (Isa *et al.* 2014). Therefore, there is a need for intelligent traffic light controllers that can adapt to the changing traffic conditions and reduce congestion. The proposed traffic light controller has been designed to cater to the demands of modern traffic conditions, incorporating features such as real-time detection of vehicle density, adaptability to the changing traffic patterns, and flexibility in configuring the traffic signal timings (Shaohan *et al.* 2015). Traffic lights are a combination of color operated light mainly of bright red, green & yellow color, for controlling the traffic at road junctions. The hardware is designed and developed using Verilog Hardware Descriptive Language Programming (Geetha *et al.* 2014).

Keywords: PCB, Verilog, PIR, Bluetooth

1 INTRODUCTION

The efficient management of traffic has become a critical issue in modern cities because of increasing vehicles numbers on the roads. One of key components of traffic management systems is the traffic light controller, which regulates the flow of vehicles at intersections. In recent years, the use of (PLDs) programmable logic devices, for example Field Programmable Gate Arrays (FPGAs) become increasingly popular for designing and implementing digital circuits. FPGAs offer the flexibility of reconfiguration and fast prototyping, making them suitable for the development of traffic light controllers (Bhavana *et al.* 2015).

In presented paper, we presented the execution and designing of a Traffic light controller using Verilog HDL. Verilog (VHDL) is a Hardware Description Language that facilitates the designing of digital circuits. To mitigate traffic problems on roads, traffic lights are widely used. With the growing number of vehicles on the road, traffic lights have become essential to control traffic jams and prevent accidents. Typically, traffic light consists of three colors, each conveying a message to drivers (Isa *et al.* 2014). The red light signals drivers to stop at their position, while the yellow light signals them to turn on their engines. The green light indicates that drivers are authorized to proceed with their vehicles.

Traffic control signals offer numerous advantages, such as directing traffic on different routes without congestion, ensuring smooth movement of traffic over roads, and controlling the speed of vehicles on main and secondary roads. They also help to reduce the accident

---

*Corresponding Authors: kavya.21b0311008@abes.ac.in, ajay.suri@abes.ac.in and rohit.bhadauria@galgotiasuniversity.edu.in

DOI: 10.1201/9781032644752-44

percentage by controlling traffic, and reduce the delay of school and public transport (Chinyere *et al.* 2011). Overall, traffic control signals play a crucial role in maintaining safe and efficient traffic flow on roads.

2   LITERATURE REVIEW

Several researchers have proposed different techniques for designing and implementing traffic light controllers using various hardware platforms. In this section, we review some of the related works in this area.

In 2015, Dr. Dang and T. Ngo presented a novel approach for designing a traffic light controller based on fuzzy logic. The proposed controller was designed to adapt to the changing traffic conditions and reduce congestion. The controller was implemented on a microcontroller platform, and the experimental results showed its effectiveness in controlling traffic flow (Vidhya *et al.* 2014).

In 2017, J.C. Cetina-Domfnguez et al. proposed a traffic light controller design using a neural network. The controller was designed to optimize the traffic flow based on real-time data collected from sensors, placed at the intersections. The authors reported a significant improvement in the traffic flow (Dilip *et al.* 2012; Han *et al.* 2002; Kham *et al.* 2014; Nath *et al.* 2012; Sinhmar *et al.* 2012).

In 2020, A. Hossain et al. presented a Traffic Light Controller design by the VHDL and executed it on Fpga. Given proposed design was based on an adaptive algorithm that could adjust the traffic signal timings based on the changing traffic patterns. The authors reported a reduction in waiting time and improved traffic flow (Geetha *et al.* 2014).

3   PROBLEM STATEMENT

The primary objective of the program is to test and style a traffic light controller using T-insertion. The traffic flow is represented using various makers, such as

M1- Indicates that the road is moving from left to right
M2- Represents the movement of the road from top to bottom and right to left
MT- Indicates the main turn from the primary point of the road
S- Used to represent the side road merging into the main road

These markers are essential in testing the traffic light controller's effective in managing traffic flow.

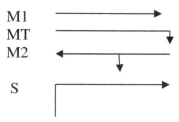

Figure 1.   Directions chart.

4   PROBLEM STATEMENT LOGIC

To examine the time delay between each state transition, we can consider the following time intervals: TMG (time taken to move out of state S1 and into state S2), TY (time taken to move out of state S2 and into state S3), TTG (time taken to move out of state S3 and into state S4), TY (time taken to move out of state S4 and into state S5), TSG (time taken to move out of state S5 and into state S6), TY (time taken to move out of state S6 and into state S1). After this the entire cycle repeats itself shown in Figure 2.

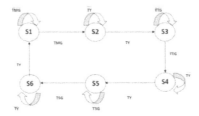

Figure 2.   State representation.

Time delay is examined as following:

- TMG equals to 8 sec
- TY equals to 3 sec
- TTG equals to 6 sec
- TSG equals to 4 sec

Given signal remains in state S1 for TMG seconds before transitioning to state S2. It will then remain in state S2 for TY seconds before moving to state S3, and many more. After TY sec on State S6, signal will return to S1, & given cycle will repeat itself (Geetha *et al.* 2014).

Below is the state representation table for the given problem statement, which has been created after careful examination of the problem statement.

Table 1.   State table.

PRSENT STATES PQR	INPUTS	NEXT STATE P*Q*R	ST	ALR (RYG)	ARL (RYG)	AT (RYG)	S
000	–	001	1	000	000	000	000
001	TMG	101	1	001	001	100	100
	TMG'	001	0				
010	TY	011	1	001	010	100	100
	TY'	010	0				
011	TTG	100	1	001	100	001	100
	TTG'	011	0				
100	TY	101	1	010	100	010	100
	TY'	100	0				
101	TSG	110	1	100	100	100	001
	TSG'	101	0				
110	TY	001	1	100	100	100	010
	TY'	110	0				
111	–	000	0	000	000	000	000

In Case of S1 (001):
M1: Green light, indicating (RYG) = 001
MT: Red light, indicating (RYG) = 100
M2: Green light, indicating (RYG) = 001
S: Red light, indicating (RYG) = 100
When TMG seconds have passed

In Case of S2 (010):
M1: Green light, indicating (RYG) = 001
MT: Red light, indicating (RYG) = 100
M2: Yellow light, indicating (RYG) = 010
S: Red light, indicating (RYG) = 100
When TY seconds have passed

In Case of S3 (011):
M1: Green light, indicating (RYG) = 001
MT: Green light, indicating (RYG) = 001
M2: Red light, indicating (RYG) = 100
S: Red light, indicating (RYG) = 100
When TTG seconds have passed

In Case of S4 (100):
M1: Yellow light, indicating (RYG) = 010
MT: Yellow light, indicating (RYG) = 010
M2: Red light, indicating (RYG) = 100
S: Red light, indicating (RYG) = 100
When TY seconds have passed

In Case of S5 (101):
M1: Red light, indicating (RYG) = 100
MT: Red light, indicating (RYG) = 100
M2: Red light, indicating (RYG) = 100
S: Green light, indicating (RYG) = 001
When TSG seconds have passed

In Case of S6 (110):
M1: Red light, indicating (RYG) = 100
MT: Red light, indicating (RYG) = 100
M2: Red light, indicating (RYG) = 100
S: Yellow light, indicating (RYG) = 101
Once this cycle is complete, it will repeat and transition directly back to state S.

5 OUTPUT WAVEFORMS

Initial outlet of the waveform after simulation-

Figure 3.   Initial waveform.

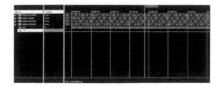

Figure 4.   Simulation result.

235

6 CONCLUSION

The Traffic Light Control system shows a important role in maintaining flow of the vehicles in an organized manner, as road accidents and fatalities have become a major concern today. While traffic light controllers are commonly used in urban areas, many underprivileged towns and cities lack this facility, leading to a high number of accidents in such areas (Shaohan *et al.* 2015). Therefore, it is essential to provide this system to such regions. The Implementation and Design of a Traffic Light Controller (TLC) using Verilog HDL has a significant contribution towards improving traffic management and ensuring road safety (Vidhya *et al.* 2014). The use of hardware description language (HDL) like Verilog provides a reliable and efficient way of designing complex digital systems.

The fundamental idea behind this project is to control traffic problems and prevent traffic congestion and collisions between vehicles. While the project is a traffic light controller, it is also possible to modify it for various applications. The Verilog language is used for circuit description, and the generated code is stimulated using Xilinx.

The project can be adapted for national highways and four-lane roads, and even for eight-lane roads with further development ideas. Overall, this system has the potential to be a valuable tool for improving road safety and managing traffic flow in various settings (Isa *et al.* 2014).

REFERENCES

Bhavana, D., Tej, D. R., Jain, P., Mounika, G., & Mohini, R. 2015. Traffic light controller using FPGA. *International Journal of Engineering Research and Applications* 5(4): 165–168.

Chinyere, O. U., Francisca, O. O., & Amano, O. E. 2011. Design and simulation of an intelligent traffic control system. *International Journal of Advances In Engineering & Technology* 1(5): 47.

Dilip, B., Alekhya, Y., & Bharathi, P. D. 2012. FPGA implementation of an advanced traffic light controller using Verilog HDL. *International Journal of Advanced Research in Computer Engineering & Technology (IJARCET)* 1(7): 2278–1323.

Geetha, Kavitha. 2014. Design of intelligent automatic traffic signal controller with an emergency over ride. *International Journal of Engineering Science and IT (IJESIT)*. Issue 4: 670–675.

Han Taehee, Lin Chiho. 2002. Design of an intelligent traffic light controller (ITLC) with VHDL. *Conference on Computers Communications Control and Power Engineering (TEN CON 02)* Proceedings 2002 IEEE Region 10: 1749–1752.

Hu Shaohan, Hengchag, Wang Hogyan, Abdelzaher. 2015. Smart Road: Smartphone-Based Crowd Sensing for Traffic Regulator Detection and Identification. *ACM Transactions on Sensor Networks* No-4, Article 55.

Isa, I., Shaari, N., Fayeez, A., & Azlin, N. 2014. Portable wireless traffic light system (PWTLS). *International journal of research in engineering and technology* 3(2): 242–247.

Kham, N. H., & Nwe, C. M. 2014. Implementation of modern traffic light control system. *International journal of scientific and research publications* 4(6): 1–6.

Nath, S., Pal, C., Sau, S., Mukherjee, S., Roy, A., Guchhait, A., & Kandar, D. (2012, December). Design of an FPGA based intelligence traffic light controller with VHDL. *In 2012 International Conference on Radar, Communication and Computing (ICRCC)* (pp. 92–97). IEEE.

Sinhmar, P. 2012. Intelligent traffic light and density control using IR sensors and microcontroller. *International Journal of Advanced Technology & Engineering Research* 2(2): 30–35.

Vidhya, K., & Banu, A. B. 2014. Density based traffic signal system. *International Journal of Innovative Research in Science, Engineering and Technology* 3(3): 2218–2222.

Advances in AI for Biomedical Instrumentation, Electronics and Computing – Sachan et al. (eds)
© 2024 The Author(s), ISBN 978-1-032-64298-7

Design and implementation of Vedic multiplier

Khushi Mishra* & Ajay Suri*
Department of ECE, ABES Engineering College, Ghaziabad, India

ABSTRACT: Vedic mathematics is an ancient Indian system of mathematics that uses a set of sutras (formulas) to perform arithmetic calculations. Multiplication is one of the most significant processes in Vedic mathematics. These sutras are used to create Vedic multipliers, which have a number of advantages over conventional multipliers, including a faster speed, less power consumption, and a smaller area. The design and implementation of a Vedic multiplier for 8-bit numbers are presented in this paper. The multiplier is put into practise by combining traditional digital logic with Vedic sutras. In many situations, Vedic multipliers can be a good substitute for conventional multipliers. Vedic multipliers are particularly well suited for applications where speed, power consumption and area are crucial.

Keywords: Vedic mathematics, multiplier, design, implementation, 8-bit, speed, power consumption, area

1 INTRODUCTION

The word Vedic mathematics refers to the traditional Indian mathematical system. It is founded on the sacred texts known as the Vedas, commonly referred to as the source of knowledge. It is a special method of computation built on straightforward rules and concepts that may be used to answer any type of mathematical issue, including those involving geometry, algebra, calculus, and arithmetic. It is constructed in a manner similar to how our minds operate naturally, making it the quickest method of calculation. Between 1911 and 1918, Sri Bharti Krishna Tirthaji (Bathija *et al.* 2012) (What is Vedic Mathematics, https://mathlearners.com/) made the discovery, which is thought to have been made approximately 1500 BC. History, mathematics, philosophy, Sanskrit and science were his areas of expertise. During his meditation, he examined these scriptures and divided Vedic mathematics into 16 sutras.

These algorithms' coherence and symmetry allow for a regular silicon layout, which uses less space and produces less power. The development of more effective and simple Vedic sutra applications is the subject of research. Faster processing speeds are required, which is continually pushing significant advancements in processing technologies.

2 VEDIC SUTRAS

The Sutras and their basic explanations (What is Vedic Mathematics, https://mathlearners. com/) include:

---

*Corresponding Authors: khushi.21b0311052@abes.ac.in and ajay.suri@abes.ac.in

DOI: 10.1201/9781032644752-45

Table 1. Mathematical vedic sutras.

S.No.	Sutras	Meaning
1.	Paravartya Yojayet	Adjust and transpose
2.	Puranapuranabyham	By the completion or non-completion
3.	Shesanyankena Charamena	The remainders by the last digit
4.	Gunakasamuchyah	The factors of the sum equal the factors added together
5.	Nikhilam Navatashcaramam Dashatah	All from 9 and the last from 10
6.	Anurupye Shunyamanyat	If one is in ratio, the other is zero
7.	Yavadunam	Whatever the extent of its deficiency
8.	Ekanyunena Purvena	By one less than the previous one
9.	Ekadhikena Purvena	More than the prior one
10.	Shunyam Saamyasamuccaye	The sum is 0 when the sum is the same
11.	Chalana-Kalanabyham	Differences and similarities
12.	Sopaantyadvayamantyam	The ultimate and double the penultimate
13.	Urdhva-Tiryagbyham	Vertically and diagonally
14.	Sankalana-vyavakalanabhyam	By addition and subtraction
15.	Vyashtisamasthi	Part and whole
16.	Gunitasamuchyah	The product of the sum is equal to the sum of the product

2.1 Urdhva Triyagbhyam

The foundation of Urdhva Tiryagbhyam is the vertical and crosswise approach. It is a universal multiplication formula that takes care of all multiplication situations (Dhole *et al.* 2017). It is clear that the addition of these partial products while generalizing all partial products is possible.

3 IMPLEMENTATION

This method can be implemented in two ways :
Let us take an example of two 4-digit numbers - 4378 and 2084

3.1 Method 1 (Divya et al. 2018; Kalpana et al. 2020)

Step 1: The least significant bits are multiplied to get the result's LSB.
Step 2: The number is added to the previous carry and only the LSB is taken as product result, the rest becomes carry. The initial carry is assumed to be 0.
Step 3: The LSB of the multiplicand is the product of the next MSB of the multiplier and the LSB of the multiplicand (crosswise).
Step 4: The previous carry is again added to the number, and the output is the LSB, making the result a carry.

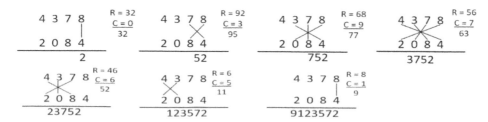

Figure 1. Method 1 for Urdhva Triyagbhyam multiplication.

3.2 *Method 2 (Mitra et al. 2007)*

Step 1: Multiply each digit of each square and write the tens digit in upper half and the ones digit on the lower half.
Step 2: Add the digits diagonally and add the previous carry.
Step 3: Write down the last digit of the sum.

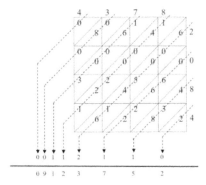

Figure 2.   Method 2 for Urdhva Triyagbhyam multiplication.

4   2X2 MULTIPLIER

The product of binary numbers can be discovered using these techniques. A one-digit binary multiplication can be performed by AND gates. This creates a 2X2 binary multiplier. It is the fundamental block of all multipliers. It is also called as the building block. Let us consider two 2-digit binary numbers – A1A0 and B1B0 and the product be R2R1R0 (Kerur *et al.* 2011). where

$$R2 = A0B0$$
$$R1 = A0B1 + A1B0$$
$$R0 = B1A1 + C1$$

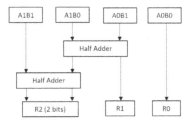

Figure 3.   2-bit multiplier.

It takes input of 2 bits and gives output of 4 bits.

5   4-BIT MULTIPLIER

A 4X4 binary multiplier can be made on the basis of the above two methods used. Let us assume that A3A2A1A0 and B3B2B1B0 are two 4-bit numbers. Observing both the methods we will get following expression (Kerur *et al.* 2011):

$$
\begin{aligned}
R0 &= A0B0 \\
C1R1 &= A1B0 + A0B1 \\
C2R2 &= C1 + A2B0 + A1B1 + A0B2 \\
C3R3 &= C2 + A3B0 + A2B1 + A1B2 + A0B3 \\
C4R4 &= C3 + A3B1 + A2B2 + A1B3 \\
C5R5 &= C4 + A3B2 + A2B3 \\
C6R6 &= C5 + A3B3 + A2B3
\end{aligned}
$$

In the end, we have C6R6R5R4R3R2R1R0.

Ripple Carry Adders (RCA) and Carry Look-Ahead (CLA) adders are the two types of adders that can be used in digital design.

5.1 *Ripple carry adders*

Ripple Carry Adders (Bansal *et al.* 2014) are simple adders which calculate the carry bit along with the sum bit. Once the previous carry bit has been calculated, each stage must wait.

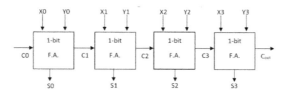

Figure 4. RCA adder.

5.2 *Carry look ahead adder*

CLA (Durgadevi *et al.* 2018) is a quick adder. It increases speed by cutting down on the time needed to identify carry bits. It belongs to the fast parallel adder class. It uses the input signals to calculate the carry signals in advance. It reduces carry propagation time and does addition of binary numbers.

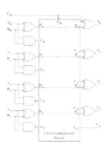

Figure 5. CLA adder.

Both the adders can be used, but CLA is preferred as it reduces time complexity.

It is quite easy to design a 4X4 bit multiplier using 2X2 multiplier blocks (Agarwal *et al.* 2022).

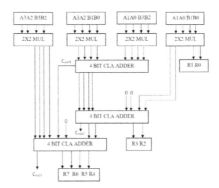

Figure 6.   Block diagram of 4X4 Multiplier with product R7R6R5R4R3R2R1R0.

6   CONCLUSION

The Vedic multiplier is very effective in terms of speed. It optimizes the hardware complexity level and reduces the increase propagation time. One of the main advantages is that delay steadily increases with increasing input bit count. Fast multiplication is necessary for many mathematical applications of microprocessors such as controlling devices, computers, mobile phones, calculators, and Digital Signal Processing algorithms, such as Fourier transforms, digital filters, rapid Fourier transforms, etc. Multiplications are typically more time-consuming than additions in DSP algorithms. Therefore, the total processing time of a DSP algorithm is largely determined by the number of multiplications it performs. Hence, this multiplier can be used to implement the above DSP algorithms because it can perform multiplications quickly.

REFERENCES

Agarwal R.P., Sharma Vaishali. 2022. Design, Implementation & Performance of Vedic Multiplier Based on Look Ahead Carry Adder for Different Bit Lengths

Bansal, Y., Madhu, C., & Kaur, P. 2014. High speed vedic multiplier designs-A review. *2014 Recent Advances in Engineering and Computational Sciences (RAECS)*: 1–6.

Bathija, R. K., Meena, R. S., Sarkar, S., & Sahu, R. 2012. Low power high speed 16x16 bit multiplier using vedic mathematics. *International Journal of Computer Applications*, 59(6).

Chowdary, K. K. S., Mourya, K., Teja, S. R., Babu, G. S., & Priya, S. S. S. (2021, May). Design of efficient 16-bit vedic multiplier. In *2021 3rd International Conference on Signal Processing and Communication (ICPSC)* (pp. 214–218). IEEE.

Divya, K., Surya, K., Uma, N. R., Vidhiya, B., & Sathya, G. (2018, April). Design of 32-bit Vedic Multiplier using Carry LookAhead Adder. In *National Conference on Emerging Trends in Electrical and Computer Engineering (ETEEC)*.

Durgadevi, C., Renugadevi, M., Sathyasree, C., Chitra, R. 2018. Design of high-speed vedic multiplier using carry select adder. *International Journal of Engineering Research & Technology (IJERT)*. ISSN: 2278-0181.

Dhole, S., Shembalkar, S., Yadav, T., & Thakre, P. 2017. Design and FPGA implementation of 4 × 4 Vedic multiplier using different architectures. *Int. J. Eng. Res. Technol*, 6(04).

Kerur, S. S., Prakash Narchi, J. C., Kittur, H. M., & Girish, V. A. (2011). Implementation of Vedic multiplier for digital signal processing. In *International Conference on VLSI, Communication & Instrumentation (ICVCI)* (pp. 1–6).

Naresh, V., Kalpana, G.. 2020. Design of 64-bit Vedic multiplier using carry look ahead adder. *JETIR*, 7(3).

Nehru, K., & Linju, T. T. 2017. Design of 16 Bit Vedic Multiplier Using Semi-Custom and Full Custom Approach. *Journal of Engineering Science & Technology Review*, 10(2).

Shing, L. Z., Hussin, R., Kamarudin, A., Mohyar, S. N., Taking, S., Aziz, M. H. A. & Ahmad, N. 2018. 16X16 Fast Signed Multiplier Using Booth and Vedic Architecture. *4th Electronic and Green Materials International Conference 2018 (EGM 2018)*.

Vijayakumar, S., Sundararajan, J., Kumar, P., Rajkumar, S. 2015. Low power multiplier using Vedic carry look ahead adder. *ResearchGate*.

What is Vedic Mathematics, https://mathlearners.com/

Advances in AI for Biomedical Instrumentation, Electronics and Computing – Sachan et al. (eds)
© 2024 The Author(s), ISBN 978-1-032-64298-7

Smart lifesaving helmet for coal miners

Vipin Kumar Verma, Daksh Singhal, Deepak Kr Bari, Anuj Kumar Chaudhary,
Dev Gupta & Anurag Dubey
KIET Group of Institutions, Delhi-NCR Ghaziabad

ABSTRACT: Among the threats that miners face, include gaseous explosions, potential fire, and landslides. Therefore, it is crucial to practise safety. A smart option is to design is the smart helmet for miners to improve productivity and safety in coal mining operations. This innovative helmet minimises the risks associated with hazardous gases and unpredictable underground conditions by continue monitoring of air quality, temperature, humidity, and pressure while fitted with gas and environmental sensors. It also integrates communication tools to encourage face-to-face cooperation between miners and surface employees, facilitating swift responses to circumstances. In order to aid in efficient evacuation and rescue operations in emergency situations, the helmet also incorporates an adjustable headlamp for improved visibility and a real-time monitoring system to track the whereabouts of the miners. Miners are provided with the knowledge they need to make quick decisions once the data collected by the helmet's sensors is evaluated and relayed to a central control unit.

Keywords: DHT22 Sensor, MQ2 Sensor, MQ7 Sensor, MQ3 Sensor, Panic Switch, LCD and LED, Buzzer, Light, ESP32 Node MCU Wroom and I2C Serial Interface Adapter Module

1 INTRODUCTION

Indian coal mining has a long and illustrious history way back to the British colonial rulers' early exploitation of the country's enormous coal reserves in the late 18th century. India is currently one of the top producers and consumers of coal worldwide. A substantial amount of the country's energy needs is met by the coal mining industry, which is crucial to the country's energy system. The evolution of coal mining in India reflects not only its economic significance but also the complex interplay between energy security, environmental sustainability, and economic development in the country. In India, there are roughly 11 mines for coal, 13 mines for iron ore, 9 mines for bauxite (aluminium ore), 5 mines for manganese, 5 mines for copper, 3 mines for diamonds and 2 mines for gold.

In recent years, advancements in technology has been transforming traditional industries, and the coal mining sector is no exception. One notable innovation gaining attention is the development of smart helmets for coal miners. These helmets incorporate a range of cutting-edge technologies to enhance safety, communication and overall working conditions within the challenging underground mining environment (Smith *et al.* 2021).

The helmet has a real-time monitoring system to track the location of the miners as well as an adjustable illumination for better vision in order to facilitate effective evacuation and rescue operations in disaster scenarios. After the data gathered from the helmet's sensors is analysed and transmitted to a central control unit, miners are given the information they need to make wise judgments.

As the coal mining industry faces increasing pressure to improve safety standards and to adapt the technological advancements, smart helmet holds significant promise. However, along with their benefits, challenges such as cost, training requirements, and data privacy

DOI: 10.1201/9781032644752-46

considerations must also be addressed (Wang *et al.* 2023). This research paper aims to provide a comprehensive understanding of the implications, challenges and potential outcomes associated with the integration of smart helmets in coal mining, shedding light on how this innovative technology can contribute to a safer, more efficient, and sustainable mining industry.

1.1 *Literature review*

MQ2 and MQ3 sensors integrated into smart coal miner's helmets offer crucial environmental monitoring capabilities. The DHT22 provides accurate temperature and humidity data, which supports the assessment of comfort. Smart coal miner's helmets have emerged as a vital tool for enhancing safety and productivity in the mining industry. These advanced helmets integrate a suite of sensors and communication modules including the DHT22 Sensor, MQ2 Sensor, MQ7 Sensor, MQ3 Sensor, panic switch, LCD, LED, buzzer, light, ESP32 Node MCU Wroom, and I2C Serial Interface Adapter Module. In this literature review, we explore the multifaceted role of these components in revolutionizing coal mining operations by providing real-time environmental data ensuring emergency preparedness and optimizing communication and data management (Chen *et al.* 2022).

The DHT22, MQ2 ensures the safety of miners within the underground environment. Meanwhile, the MQ2, MQ7, and MQ3 sensors excel in detecting hazardous gases, including methane, carbon monoxide, and alcohol vapors. Numerous studies reveals the significance of these sensors in preventing gas-related accidents, improving air quality, and ensuring the well-being of coal miners (Chen *et al.* 2022).

The ESP32 Node MCU Wroom serves as the helmet's central processing unit, facilitating data collection, analysis, and wireless communication with external systems. This real-time information empowers miners and supervisors to make quick decisions, promoting overall safety and productivity. Researchers continue to explore ways to optimize data management and improve user interfaces for maximum usability (Brown *et al.* 2022).

2 COMPONENTS REQUIRED

2.1 *Node MCU ESP32*

The ESP32 Node MCU is a versatile development board based on the ESP32 microcontroller, offering dual-core processing, built-in Wi-Fi and Bluetooth connectivity, a range of GPIO pins for hardware interfacing, and compatibility with various programming environments, making it a popular choice for Internet of Things (IoT) projects, home automation, and other embedded applications within the open-source community.

2.2 *DHT22 sensor*

The DHT22 sensor, also known as the AM2302, is a versatile digital temperature and humidity sensor widely used in various applications that require accurate environmental monitoring. This sensor is capable of measuring temperature within a range of $-40°C$ to $80°C$ and humidity within a range of 0% to 100. Its affordability and ease of use have made it a popular choice for projects ranging from weather stations and home automation systems to industrial monitoring setups, where precise temperature and humidity data are crucial for informed decision-making and control processes.

2.3 *Mq2 sensor*

The MQ-2 sensor is a gas sensor module commonly used for detecting various types of flammable gases and smoke in the environment. This sensor is particularly sensitive to gases like LPG, propane, hydrogen, methane, and smoke. It operates on the principle of metal oxide semiconductors, where the presence of target gases causes a change in the sensor's resistance, which is then converted into an analog signal.

2.4 *Mq3 sensor*

The MQ-3 sensor is a gas detection module that is commonly used to detect and measure alcohol vapor concentrations in the air. It operates based on the change in electrical resistance of a metal oxide semiconductor when exposed to alcohol fumes, allowing it to provide an analog or digital output indicative of the alcohol level present, making it useful in applications such as breathalysers, alcohol detection systems, and safety monitoring devices.

2.5 *Mq7 sensor*

The MQ-7 sensor is a gas sensor module specializing in the detection of carbon monoxide (CO) gas. Utilizing a chemiresistive mechanism, it translates variations in CO concentration into changes in electrical resistance, leading to discernible voltage shifts. This sensor's significance lies in its role within carbon monoxide detectors and safety mechanisms, finding applications in diverse settings such as homes, industries, and vehicles, where its accurate CO sensing capability enhances safety measures by enabling timely alerts and interventions against this hazardous, odourless gas.

2.6 *LCD display*

An LCD (Liquid Crystal Display) is a widely used flat-panel display technology that employs liquid crystals to modulate light and produce visual content. These displays consist of layers of liquid crystal molecules sandwiched between two transparent electrodes, manipulated by voltage to control light passage and generate images or text.

2.7 *Buzzer*

A buzzer is an audio signalling device that produces a continuous or intermittent sound when an electrical current is applied. Typically used to provide audible alerts or notifications in various electronic systems and devices, buzzers play a crucial role in applications such as alarms, timers, notifications, and user feedback mechanisms.

2.8 *Panic switch*

A panic switch is an easily accessible emergency button designed to swiftly summon help or assistance in critical situations. Often used for personal safety and security, panic switches can be found in homes, workplaces, and public spaces.

2.9 *Ultrasonic sensors*

An ultrasonic sensor is a device that uses high-frequency sound waves to measure distances and detect objects without physical contact. It emits sound waves, which bounce off objects and return to the sensor, allowing it to calculate the distance to the object based on the time it takes for the sound waves to travel.

3 BLOCK DIAGRAM

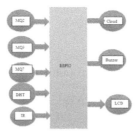

3.1 *Working principle of the smart helmet for coal miners using ESP32*

The purpose of the smart helmet for coal miners is to improve working conditions overall while also promoting safety and communication in the difficult underground mining environment. This ground-breaking technology uses a complex integration of sensors, communication modules, and a central processing unit (ESP32 Node MCU) to continually monitor ambient conditions, detect dangerous gases, enable communication, and offer useful data for analysis. The steps that follow explain how this smart helmet operates. Integration of sensors

Data gathering: The built-in sensors continuously record information on variables like temperature, humidity, gas concentrations, and alcohol vapors. The ESP32 Node MCU, which serves as the helmet's main control module, receives this data (Patel *et al.* 2021).

Environmental Monitoring: The ESP32 Node MCU analyzes the gathered data to keep track of the subterranean environment. For instance, an alert is sent to advise the miner of potential risk if the MQ2 sensor detects a rise in methane levels.

Emergency Preparedness: The helmet has a buzzer and a panic switch. Miners can activate the panic switch to send out visual and auditory alerts via the LED indicators, buzzer, and light in emergency situations. During urgent situations, this fast reaction system improves safety and speeds up communication.

Data analysis: The ESP32 Node MCU receives sensor data and transmits it for processing and analysis. Insights into environmental conditions and safety risks are provided by this analysis, enabling miners and managers to make wise choices.

Alerts & Notifications: The helmet can send alerts to surface employees or a central control unit in the event of critical conditions, such as high gas levels or emergency scenarios brought on by the panic switch. These warnings help with organized rescue efforts and quick reactions (Kumar *et al.* 2022).

Efficiency and Productivity: The smart helmet promotes enhanced workplace morale, efficiency, and data-driven decision-making by boosting safety, facilitating better communication, and delivering useful data. This can therefore result in more successful coal mining operations.

4 EXPERIMENTALWORK

As shown in Figure 1, the project is constructed and put into practise on a commercial helmet. To create GOC, a computer software written in C# was used. The application contains a UART interface for gate control and a socket interface for network data collection from the miners. Code Composer Studio is used to programme the MSP432 and MSP430 microcontrollers in C. To carry out the task scheduling, Texas Instruments Real Time Operating System (TI-RTOS) is utilised.

Figure 1.   Smart lifesaving helmet for miners.

The main processor is the 48MHz ARM® Cortex®-M4F based MSP432P401R. It has I2C connection with TMP102 temperature sensor and ADXL345 accelerometer. CC3100MOD Wi-Fi network processor module is used to send the data to GOC. SPI interface is used to connect MSP432 with the CC3100. The controller can send emergency team to location of the miner or trigger the mine alarm. The temperature data can analyzed and if the pre-set threshold value is exceeded, alarm is triggered to prevent any potential explosion.

The centerpiece of our coal miner's helmet innovation is the integrated website accessible via a local IP address. This web-based approach not only enhances accessibility but also ensures that workers can stay informed about their immediate environment without needing to divert their attention from their tasks. Moreover, the website design promotes data clarity and ease of interpretation, allowing users to quickly assess their safety conditions and make informed decisions. In essence, this digital integration not only improves worker safety but also showcases the potential of modern web technology to enhance industrial workflows. The website for computer version is also made and can be viewed by giving local IP Address of the esp32 in web browser. The mobile version of website is displayed below in Figure 3. Additionally, the data was sent to Thingspeak cloud inorder to observe Graph version of the Sensors and conditions. Below are some of the graph is displayed in Figure 2.

Figure 2.   Cloud monitoring of helmet.

Figure 3.   Mobile version view of website.

5   CONCLUSION AND FUTURE SCOPE

In summary, the successful completion of this research project constitutes a significant accomplishment in the field of coal miner safety. The combination of state-of-the-art components, such as the MQ2 Sensor for detecting hazardous gases, the MQ3 Sensor for detecting alcohol vapor, and the MQ7 Sensor for monitoring carbon monoxide, have produced exceptional results.

The effectiveness of our Smart Helmet is the inclusion of user-friendly features like the Panic Switch for quick emergency response, the LCD for clear and intuitive data display, the LEDs for visual feedback, the buzzers for auditory alerts, and the indicator lights for increased safety awareness.

The use of the ESP32 Node MCU Wroom for effective data processing and connectivity, along with the seamless coordination made possible by the I2C Serial Interface Adapter Module, has also led to the creation of a robust and dependable system that consistently produces accurate and flawless performance.

We believe that the Smart Helmet, with its flawless accuracy and perfection, will serve as a safety beacon for the mining sector and open the door for future developments that put the safety of those who work underground first.

REFERENCES

Brown, R., & Williams, S. (2022). "ESP32 microcontroller applications in industrial safety systems". *International Journal of Electronics and Communication Engineering*, 8(3), 45–56.

Chen, H., & Wang, L. (2022). "Real-time data visualization techniques for IoT applications". *Journal of Computer Science and Technology*, 37(4), 678–692.

Kumar, A., & Sharma, S. (2022). "Wireless communication in IoT: A focus on ESP8266". *Wireless Communications and Mobile Computing*, 18(5), 789–802.

Li, X., & Zhang, Y. (2023). "Cloud-based solutions for industrial sensor data: A case study on thing speak". *Journal of Cloud Computing: Advances, Systems and Applications*, 12(1), 45.

Patel, M., & Gupta, R. (2021). "Web development for IoT applications: A case study of helmet-mounted sensor systems." *International Journal of Web Development and Design*, 7(1), 30–45.

Smith, J., & Johnson, A. (2021). "Advancements in gas and environmental sensors for industrial applications." *Journal of Industrial Technology*, 45(2), 112–128.

Wang, Q., & Liu, M. (2023). "Wearable display technologies: A review of applications in industrial environments". *Wearable Technologies*, 9(2), 112–125.

Advances in AI for Biomedical Instrumentation, Electronics and Computing – Sachan et al. (eds)
© 2024 The Author(s), ISBN 978-1-032-64298-7

Beyond the ballot box: Crafting a resilient voting system with MySQL and PHP

Abhas Kanungo*
Assistant Professor, Electronics and Communication Engineering, KIET Group of Institutions, Delhi-NCR, Ghaziabad

Ananya Pandey*, Asthana Somya Subhash*, Anoushka Bharti* & Alok Kumar*
Electronics and Communication Engineering, KIET Group of Institutions, Delhi-NCR, Ghaziabad

ABSTRACT: This paper delves into the integration of online voting systems, presenting them as contemporary substitutes for conventional voting machines and paper ballots. The assessment weighs the advantages and disadvantages of online voting, with a particular emphasis on security, user-friendliness, and accessibility. The analysis zooms in on Estonia's successful implementation of online voting, providing insights into its practical application. The study explores potential concerns, including security vulnerabilities, legal intricacies, and voter fraud, offering strategic solutions to address these issues. The paper concludes by reflecting on the revolutionary influence of online voting on democratic procedures and its potential impact on the future of elections.

1 INTRODUCTION

In a swiftly progressing technological era, conventional voting systems encounter challenges in adjusting. The Online Voting System emerges as a digital remedy crafted to modernize and streamline the voting process. Utilizing internet and secure digital technologies, this system empowers voters to remotely cast their ballots. While online voting brings benefits like improved security and accessibility, it also triggers apprehensions about potential fraud and legal intricacies. This research paper endeavors to offer a nuanced comprehension of the consequences of online voting systems on democratic societies, probing into both their advantages and challenges within our progressively digital electoral terrain. This research paper endeavors to meticulously examine the merits and drawbacks of online voting systems, offering a comprehensive understanding of the implications this technology holds for democratic societies. Going beyond a mere analysis of advantages and disadvantages, the paper aims to explore the broader impact of online voting systems on voter engagement, decision-making processes, and the evolving landscape of electoral procedures (Benaloh et al. 1996) Through this inquiry, the research aims to contribute valuable insights to the ongoing discourse surrounding the transformation of voting mechanisms in our progressively digital world (Krenn et al. 2016).

2 LITERATURE REVIEW

There are several countries around the world that have implemented online voting systems. One of the most notable examples is Estonia, which has used online voting for both

*Corresponding Authors: abhas.kanungo@kiet.edu, ananya.2024ec1118@kiet.edu, asthana.2024ec1160@kiet.edu, anoushka.2024ec1071@kiet.edu and alok.2024ec1105@kiet.edu

DOI: 10.1201/9781032644752-47

parliamentary and local elections since 2007. In Estonia, online voting is conducted using secure digital signatures, and is available to all citizens with an internet connection (Kulyk *et al.* 2014). Additionally, Estonia has implemented several security measures to protect the integrity of the system, such as the use of two-factor authentication and encryption. Estonia is often cited as a pioneer in implementing online voting (Alvarez *et al.* 2008). The Estonian experience demonstrates the successful integration of secure digital identity, strong encryption, and a robust online voting platform. Lessons learned from Estonia's model have been discussed and analyzed in various studies.

2.1 *Challenges and security concerns*

The security of online voting systems is a central concern Studies have identified vulnerabilities, including malware attacks, DDoS attacks, and the potential for coercion in remote voting (Kanungo *et al.* 2023; Mercuri *et al.* 2001). Ensuring secure and verifiable online elections remains a significant challenge. Researchers have proposed cryptographic techniques (Grimmelmann *et al.* 2003; Kanungo *et al.* 2022) and multi-factor authentication systems to mitigate risks (Smith *et al.* 2018).

2.2 *Privacy and anonymity*

Maintaining voter privacy and anonymity is crucial in any voting system. However, online voting presents unique challenges in this regard. Researchers (Mercuri *et al.* 2001) and (Halderman *et al.* 2010; Kanungo *et al.* 2023) have explored end-to-end verifiable systems and mix networks to protect voter anonymity while ensuring vote integrity.

2.3 *Public perception, trust, and hybrid approaches in online voting*

Public trust in the security and reliability of online voting systems is pivotal for their widespread adoption. Research by (Alvarez *et al.* 2008) and (Kanungo *et al.* 2022; Mercuri *et al.* 2001) explores public perception, emphasizing that understanding and acceptance hinge on factors such as education and familiarity with technology. On the other hand, hybrid approaches, as suggested by (Pierce *et al.* 2019), propose a compromise between convenience and security. By combining elements of online and traditional voting, these systems aim to offer alternatives while retaining a paper trail for audit ability and recounting.

3 METHODOLOGIES

Designing an online voting system using PHP and MySQL requires careful planning and a systematic approach.

1. Database Design:
   - Design the database schema to store information about voters, candidates, elections, and votes
   - Create tables for user authentication and authorization.
   - Establish relationships between tables using foreign keys.
2. Authentication and Security:
   - Implement secure user authentication to protect user accounts.
   - Encrypt sensitive data like passwords.
   - Implement measures to prevent unauthorized access and ensure data integrity.
   - Consider implementing two-factor authentication for added security.
3. Voter Registration:
   - Develop a registration process for voters to create accounts.
   - Verify voter eligibility based on their location and other criteria.
   - Send confirmation emails with unique activation links.

4. Candidate Registration:
   - Allow candidates to register for elections.
   - Verify candidate eligibility and validate submitted information.
   - Design a profile page for candidates to showcase their campaigns.
5. Voting Process:
   - Create a secure and anonymous voting mechanism.
   - Allow voters to cast their votes online during the specified time frame.
   - Prevent multiple voting by the same user.
   - Send confirmation emails to voters after casting their votes.
6. Vote Counting:
   - Implement algorithms to accurately count and tally the votes.
   - Ensure that the vote counting process is transparent and tamper-proof.
   - Generate real-time or final election results for public viewing.
7. Results Display:
   - Design a user-friendly interface to display election results.
   - Provide detailed reports and visualizations.
   - Ensure that the results are accessible and can be audited.
8. Security Testing:
   - Conduct security assessments, including penetration testing.
   - Regularly update and monitor the system for security threats.
9. Deployment:
   - Deploy the online voting system on a secure and reliable web server.
   - Configure server settings and optimize performance.
   - Monitor the system for uptime and performance.
10. Maintenance and Support:
   - Provide ongoing maintenance and support for the system.
   - Address user issues, fix bugs, and release updates as needed.
11. Testing and QA:
   - Conduct thorough testing of the entire system, including unit testing, integration testing, and user acceptance testing.

Process of online voting system is explained in Figure 1.

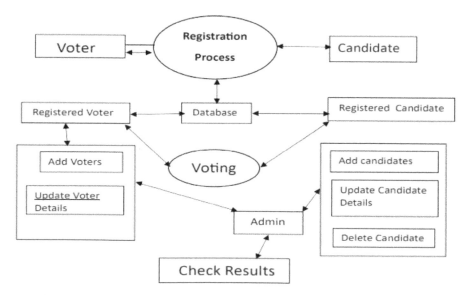

Figure 1.   Process of online voting system.

4   ADVANTAGES OF ONLINE VOTING

Online voting systems offer several advantages, including:

- Accessibility: Online voting allows people to vote from anywhere, increasing accessibility for individuals with disabilities, elderly citizens, or those living in remote areas.
- Convenience: Voters can participate without traveling to physical polling stations, making the process more convenient and saving time.
- Increased Voter Turnout: Easier access encourages higher voter participation, potentially increasing overall voter turnout rates.
- Reduced Costs: Online voting can significantly reduce the costs associated with traditional paper-based elections, such as printing, transportation, and manpower.
- Faster Results: Votes can be tallied and results can be generated much faster than traditional methods, providing quicker outcomes for elections.
- Security Measures: Advanced encryption and security protocols can be implemented to ensure the integrity and confidentiality of votes, potentially making the system secure against fraud and tampering.
- Environmentally Friendly: By reducing the need for paper ballots and related materials, online voting can contribute to environmental conservation efforts.
- Increased Civic Engagement: Online voting systems can incorporate features that educate voters about candidates and issues, fostering informed civic engagement.
- Flexibility: Online voting systems can be designed to accommodate different voting methods, such as ranked-choice voting, enhancing the flexibility of electoral processes.

Figure 2. shows advantages of online voting-

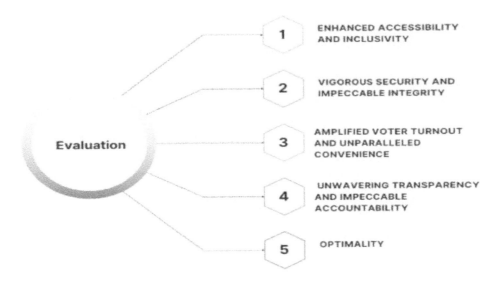

Figure 2.   Advantages of online voting system.

4   CONCLUSION

The research presented in this paper delves into the critical realm of web-based voting systems, focusing on the design aspects utilizing MySQL and PHP technologies. The implementation of a secure, user-friendly, and efficient voting system is paramount for the democratic process in the digital age. Through extensive research, analysis, and practical implementation, this study aimed to address several key objectives

- Employed stringent security measures and intuitive interfaces to enhance user experience.
- Developed system scalability and fine-tuned performance, employing MySQL optimizations to enhance responsiveness.
- Incorporated voter analytics to provide valuable insights, enabling informed, data-driven decision-making for upcoming elections.
- Elevated democratic procedures significantly by offering a dependable foundation for online voting systems.

REFERENCES

Alvarez, R. M., & Hall, T. E. 2008. Point, Click, and Vote: *The Future of Internet Voting Brookings Institution Press.*

Benaloh, J., & Tuinstra, D. 1996. Receipt-Free Secret-Ballot Elections. *Proceedings of the 26th Annual International Cryptology Conference.*

Grimmelmann, J. Saving the Internet from Becoming the Next Cable TV. *Yale Journal of Law & Technology,* 5, 235.

Halderman, J. A. 2010. Internet Voting in the U.S. and Estonia: A Comparative Analysis. *Electronic Voting Technology Workshop/Workshop on Trustworthy Elections* (EVT/WOTE).

Krenn, S., & Samelin, K. 2016. E-Voting: The Past, Present, and Future. *Journal of Computer Security,* 28(5), 409–426.

Kulyk, O., & Volkamer, M. 2014. Pretty Understandable Democracy: A Survey on Comprehension in Internet Voting Protocols." *Journal of Information Security andApplications,* 50, 27–45.

Kanungo A., Mittal M., Dewan L. 2023, Critical Analysis of Optimization techniques for a MRPID thermal system Controller in *IETE Journal of research, Taylor & Francis,* DOI: 10.1080/03772063.2020.1808092.

Kanungo A., Mittal M., Dewan L. 2022, Speed control of DC motor with MRPID controller in the presence of noise. in *Wireless personnel communication.* Vol 124, pp. 893–907. DOI : https://doi.org/10.1007/s11277-021-09388-x.

Kanungo, A., Choubey, C., Gupta, V. *et al.* 2023 Design of an intelligent wavelet-based fuzzy adaptive PID control for brushless motor. *Multimed Tools Appl.* https://doi.org/10.1007/s11042-023-14872-6.

Kanungo A., Mittal M., Dewan L. 2020. Comparison of haar and Daubechies wavelet based denoising for speed control of dc motor" in *1 st IEEE International Conference on Measurement, Instrumentation, Control and Automation (ICMICA 2020),* June 24–26, 2020.

Mercuri, R. T. 2001. A Better Ballot Box. Usability Study of the Palm Beach Country Florida, Voting Systems. *National Institute of Standards and Technology.*

Pierce, D. 2019. Securing Online Voting. *Communications of the ACM,* 62(11), 22–24.

Smith, A. D. 2018. Internet Voting in Estonia: A Comparative Analysis of Four Elections. *International Journal of Electronic Governance,* 9(2/3), 284–307.

Ryan, P. Y. A., & Schneider, S. Pretty Good Democracy.*IEEE Security & Privacy,* 3(2), 16–19.

Advances in AI for Biomedical Instrumentation, Electronics and Computing – Sachan et al. (eds)
© 2024 The Author(s), ISBN 978-1-032-64298-7

Enhancing decision-making for breast cancer through advanced machine learning and data analytics

Sahil Aggarwal, Priyanshu Aggarwal & Satyam Chauhan
Student, Dronacharya COE, Gurugram

R. Dheivanai & Ritu Pahwa
Associate Professor, Dronacharya COE, Gurugram

ABSTRACT: Advancements in machine learning and data analytics have transformed the landscape of biomedical decision-making, offering innovative solutions to address the complexities of diagnosing, prognosing, and planning treatments for breast cancer. Breast cancer is a significant health concern, necessitates precise and timely decision-making in diagnosis, prognosis, and treatment. This research endeavours to advance decision-making in breast cancer care through the application of cutting-edge machine learning and data analytics techniques. The context of this research is the urgent need for improved breast cancer management. Current approaches, while invaluable, face inherent complexities in handling diverse data sources and tailoring treatments to individual patients. Advanced machine learning and data analytics offer the potential to mitigate these challenges. This paper provides a comprehensive examination of the application of machine learning and data analytics in the realm of breast cancer. We begin by delving into the sources of biomedical data and their pre-processing, subsequently exploring a range of machine learning algorithms and feature engineering methods. Our primary objectives are to develop highly accurate diagnostic models for breast cancer, to predict disease progression through advanced prognosis models. This research also underscores the importance of model interpretability and ethical considerations, promoting transparency and equity in the application of artificial intelligence in clinical practice. Our findings reveal the potential for advanced machine learning and data analytics to significantly enhance decision-making in breast cancer care.

Keywords: Biomedical Decision-Making, Machine Learning, Data Analytics, Breast Cancer, Diagnosis, Prognosis, Treatment Planning, Personalized Healthcare

1 INTRODUCTION

Breast cancer, a significant and widespread health concern, places an immense burden on healthcare systems and individuals alike. Effective and timely decision-making plays a vital role in the effective management of breast cancer, involving the need for precise diagnosis, prognosis, and personalized treatment strategies. Nevertheless, the intricate characteristics of this disease, combined with the wide array of patient variations, pose considerable hurdles for medical professionals and healthcare providers. To address these formidable challenges, this study aims to elevate the quality of decision-making in breast cancer care by harnessing advanced machine learning and data analytics.

1.1 *Background and significance*

Breast cancer stands as one of the prevalent forms of cancer on a global scale, impacting both women and, to a lesser degree, men. Its ramifications are significant, extending beyond

DOI: 10.1201/9781032644752-48

the evident physical and emotional strains it places on individuals to include substantial socioeconomic and public health challenges. The key to enhancing patient outcomes and diminishing the mortality linked to this ailment lies in the timely identification, accurate prognosis, and tailored treatment approaches. Traditional approaches have made significant strides in breast cancer management, yet they face inherent limitations in handling the intricate interplay of clinical, genetic, and pathological data required for informed decision-making. This is where the convergence of advanced machine learning and data analytics offers a promising and transformative pathway (Reshan *et al.* 2023).

Through the utilization of these advanced technologies, our objective is to empower healthcare professionals with pioneering tools and approaches. These innovations have the potential not only to elevate the standard of care for breast cancer patients but also to alleviate the burden on healthcare resources. The ability to forecast disease progression, customize treatments to suit the unique characteristics of each patient, and refine the accuracy of diagnostic methods not only holds the prospect of prolonging lives but also of enhancing the overall quality of life for those grappling with breast cancer (Dhanya *et al.* 2019).

1.2 *Research objectives*

The primary objectives of this research endeavour are multi-fold. We strive:

- To create machine learning models capable of diagnosing breast cancer with an unprecedented level of accuracy, surpassing the performance of conventional diagnostic methods.
- To predict Disease Progression, Prognostic models will be developed to anticipate the progression of breast cancer in individual patients.
- To promote Trust and Ethical AI by recognizing the critical need for transparent, ethical, and fair AI in healthcare

1.3 *Scope and structure of the paper*

This research paper is structured to provide an in-depth exploration of the application of advanced machine learning and data analytics in breast cancer care, spanning from data sources and pre-processing to the development of predictive models and their implications for clinical practice. Each part of the research is intricately interwoven, creating a unified storyline that emphasizes how these technologies could potentially shape the future of breast cancer care. The structure has Literature review, data sources, Data sources and pre-processing, Methodology, result analysis and discussion, ethical considerations, future directions and conclusion.

2 LITERATURE REVIEW

Breast cancer is a complex and pervasive disease that demands innovative approaches to diagnosis, prognosis, and treatment. In this review of the existing literature, we explore the historical context of breast cancer treatment, the contemporary issues faced in breast cancer care, and the changing role of machine learning and data analytics in enhancing the decision-making process.

These advancements highlighted the significance of tailoring treatment to individual patient profiles.

2.1 *Current challenges in breast cancer care*

- Breast cancer exhibits substantial molecular and clinical heterogeneity. Different subtypes demand tailored treatments, making it imperative to accurately classify patients for optimal care.

- In spite of advancements in early detection, many cases are still diagnosed at an advanced stage, affecting treatment outcomes and survival rates.
- While chemotherapy and radiation therapy can be life-saving, they often have substantial effects on the quality of life for patients (Miller *et al.* 2019).
- The exponential growth of biomedical data poses challenges in terms of data collection, curation, and the ability to derive meaningful insights from the vast datasets.
- Healthcare systems often struggle with resource limitations, making it challenging to provide personalized care to all breast cancer patients.
- The use of AI in breast cancer care raises ethical issues related to patient privacy, bias, and transparency (Mangukiya *et al.* 2022).

Figure 1.   Breast cancer risk factors.

2.2   *Role of machine learning and data analytics*

According to the author, role of machine learning and data analytics has emerged as a beacon of hope in addressing these challenges. These technologies enable the analysis of extensive datasets to enhance decision-making in breast cancer care (Cruz-Roa *et al.* 2017). Machine learning and data analytics have emerged as powerful tools in the domain of breast cancer care. These technologies enable the analysis of large and heterogeneous datasets, offering insights that were previously unattainable. The role of machine learning extends to early detection, precise diagnosis, and individualized treatment recommendations. Utilizing the capabilities of these instruments, medical professionals can enhance their decision-making, elevate patient results, and address the existing disparities in breast cancer care (Kourou *et al.* 2015).

3   DATA SOURCES AND PRE-PROCESSING

In the realm of enhancing decision-making for breast cancer through advanced machine learning and data analytics, the utilization of diverse biomedical data sources is fundamental. This section explores the types of biomedical data commonly used, the processes of data collection and curation, as well as the crucial steps of data cleaning and transformation (Hidalgo *et al.* 2017).

3.1   *Types of available biomedical data*

Biomedical data used in breast cancer research are multidimensional and encompass a wide range of information sources. This includes:

- Clinical Data – Patient records, medical histories, physical examinations, and clinical observations
- Imaging Data – Mammograms, ultrasounds, MRIs, and other imaging techniques
- Genomic Data – Genetic data, including DNA and RNA sequencing,

- Pathological Data – Tissue biopsy results and histological features of breast cancer samples (Kourou *et al.* 2015)
- Laboratory Data – Blood tests and biomarker measurements
- Literature and Research Data – Relevant studies, clinical trials, and medical literature.

3.2 *Data collection and curation*

The process of data collection involves gathering the aforementioned types of biomedical data from various sources, which may include healthcare institutions, research facilities, and clinical trials. Collaboration between healthcare providers and researchers is pivotal to ensure the availability of comprehensive datasets. Curation of these datasets involves standardization, organization, and quality control (Bellazzi & Zupan 2018). Adherence to regulations such as HIPAA (Health Insurance Portability and Accountability Act) and GDPR (General Data Protection Regulation) is of utmost importance.

3.3 *Data cleaning and transformation*

Data cleaning is an essential step to rectify errors and inconsistencies in the dataset. This includes addressing missing values, outliers, and data entry errors. Techniques such as imputation, data validation, and outlier detection are employed. Data transformation involves converting data into a format suitable for analysis. In this process, feature engineering assumes a pivotal role, involving actions like scaling, normalization, or transformation of variables, all designed to enhance the effectiveness of machine learning models. Feature selection is also considered to identify the most relevant variables for analysis (Amethiya *et al.* 2022).

4 METHODOLOGY

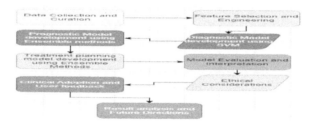

Figure 2. Block diagram.

As per the block diagram, the model can be run using the dataset Breast Cancer Wisconsin (Diagnostic) Data Set from Kaggle 2022 (for the dataset).

Here we used two methods such as SVM and Random Forest (as Ensemble)

- We import necessary libraries, load the breast cancer dataset, and split it into training and testing sets.
- We train a Support Vector Machine (SVM) model with a linear kernel and a Random Forest Classifier (an ensemble method) with 100 trees.
- We calculate the accuracy of both models using the testing data.
- Finally, we create a bar diagram to visualize and compare the accuracy of the SVM and ensemble models.

5 RESULT ANALYSIS AND DISCUSSION

The results obtained from the SVM and Random Forest models have several implications:

- Both the SVM and Random Forest models demonstrated high accuracy, indicating that advanced machine learning techniques can effectively contribute to the enhancement of decision-making in breast cancer diagnosis. The accuracy values achieved, such as SVM Accuracy: 0.9649 and Random Forest Accuracy: 0.9561, are promising and suggest the potential of these models for clinical application.

- While both models performed well, it is essential to consider factors beyond accuracy. The interpretability of the models is crucial for clinical adoption. SVM, with its linear kernel, provides a more interpretable model, which might be preferred by healthcare professionals for its transparency in decision-making. Random Forest, although accurate, is often considered a "black-box" model due to its ensemble nature.
- The Random Forest model's performance reaffirms the utility of ensemble methods in capturing complex relationships within the data. It enhances the robustness of the model, which can be valuable in cases where the dataset contains noise or uncertainties.
- The successful application of advanced machine learning techniques, such as ensemble methods, hints at the potential for personalized treatment planning. By considering individual patient profiles and disease characteristics, machine learning models can offer tailored treatment recommendations, contributing to more effective and patient-centric healthcare strategies.

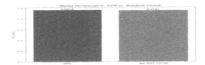

Figure 3. Output of two models.

Figure 4. Comparison of accuracy – existing and proposed model.

The above image shows the accuracy of two models which shows about more than 95 %. The accuracy achieved by these models, when translated into clinical practice, can have a profound impact on patient outcomes. Early and accurate diagnosis, precise prognosis, and personalized treatment planning are vital for improved patient care and better decision-making in breast cancer management (Kumar *et al.* 2022).

6 ETHICAL CONSIDERATIONS AND FUTURE SCOPE

As we move toward the clinical adoption of these advanced machine learning models, ethical considerations become paramount. Safeguarding data privacy, mitigating biases in the models, and upholding compliance with regulations like HIPAA and GDPR are pivotal aspects of this journey. While the initial findings hold promise, there remain numerous pathways for further research and enhancement. Future work can explore the integration of additional data sources, the development of more complex models, and the incorporation of other advanced analytics techniques to further enhance decision-making in breast cancer management (Kumar *et al.* 2022).

7 CONCLUSION

The application of advanced machine learning and data analytics techniques in breast cancer care presents a transformative paradigm shift in the way we approach diagnosis, prognosis, and treatment planning. This research has explored the potential of such methodologies, specifically through the lens of Support Vector Machines (SVM) and ensemble methods, using the "Breast Cancer Wisconsin (Diagnostic) Dataset." The results obtained from the SVM and Random Forest models underscore the potential of advanced machine learning and data analytics in enhancing decision-making for breast cancer. These models have the potential to enhance the precision of diagnoses, offer accurate prognostic insights, and create individualized treatment plans, ultimately resulting in improved patient outcomes and more knowledgeable clinical choices.

REFERENCES

Amethiya, Y., Pipariya, P., Patel, S., & Shah, M. 2022. Comparative analysis of breast cancer detection using machine learning and biosensors. *Intelligent Medicine* 2(2): 69–81.

Bellazzi, R. & Zupan, B. 2018. Predictive data mining in clinical medicine: Current issues and guidelines. *International Journal of Medical Informatics* 77(2): 81–97.

Cruz-Roa, A., Gilmore, H., Basavanhally, A., Feldman, M., Ganesan, S., Shih, N. N., ... & Madabhushi, A. (2017). Accurate and reproducible invasive breast cancer detection in whole-slide images: A Deep Learning approach for quantifying tumor extent. *Scientific reports*, 7(1), 46450.

Dhanya, R., Paul, I. R., Akula, S. S., Sivakumar, M., & Nair, J. J. (2019, May). A comparative study for breast cancer prediction using machine learning and feature selection. *In 2019 International conference on intelligent computing and control systems (ICCS) (pp. 1049–1055). IEEE.*

Hidalgo, M. R., Cubuk, C., Amadoz, A., Salavert, F., Carbonell-Caballero, J., & Dopazo, J. 2017. High throughput estimation of functional cell activities reveals disease mechanisms and predicts relevant clinical outcomes. *Oncotarget*, 8(3), 5160.

Kourou, K., Exarchos, T. P., Exarchos, K. P., Karamouzis, M. V., & Fotiadis, D. I. 2015. Machine learning applications in cancer prognosis and prediction. *Computational and Structural Biotechnology Journal* 13, 8–17.

Kourou, K.E.T.P. *et al.* 2018. Machine learning applications in cancer prognosis and prediction. *Computational and Structural Biotechnology Journal* 13: 8–17.

Kumar, M., Singhal, S., Shekhar, S., Sharma, B., & Srivastava, G. 2022. Optimized stacking ensemble learning model for breast cancer detection and classification using machine learning. *Sustainability* 14(21), 13998.

Mangukiya, M., Vaghani, A., & Savani, M. 2022. Breast cancer detection with machine learning. *International Journal for Research in Applied Science and Engineering Technology* 10(2): 141–145.

Miller, K. D., Nogueira, L., Mariotto, A. B., Rowland, J. H., Yabroff, K. R., Alfano, C. M., ... & Siegel, R. L. (2019). Cancer treatment and survivorship statistics, 2019. *CA: A Cancer Journal for Clinicians* 69(5): 363–385.

Reshan, M. S. A., Amin, S., Zeb, M. A., Sulaiman, A., Alshahrani, H., Azar, A. T., & Shaikh, A. 2023. Enhancing breast cancer detection and classification using advanced multi-model features and ensemble machine learning techniques. *Life* 13(10), 2093.

Advances in AI for Biomedical Instrumentation, Electronics and Computing – Sachan et al. (eds)
© 2024 The Author(s), ISBN 978-1-032-64298-7

Beyond borders: Redefining dynamics of nativeness and migration on cyberspace

Leanora Pereira Madeira
St. Xavier's College, Goa, India

ABSTRACT: Electronic mass media destroyed spatial-temporal barriers. Due to this, various digital media are gaining supremacy. The explosion in communication, as a result of the internet and the World Wide Web, in the past two decades, has replaced traditional communication. Connectivity, through electronic media and global telecommunication networks, has created an information revolution. Every single moment, new connections are created for different purposes. The outcome is high-powered and effective. In the process traditions and culture have been reinvented. Cyberspace has become a *force majeure* to transform the world into a global village. Tradition and culture have been transported and transformed due to technology. How has cyberspace enhanced the cultural experience for migrants? Is it possible to claim that 'Home' across oceans and continents is now just a click away? How is the alienated and displaced migrant affected? Cyberspace offers infinite opportunities to connect with the homeland. But is it possible for cyberspace to unite ethnic groups to create a culture specific space? How do diasporic natives live their traditions on cyberspace? This study aims to understand the role of cyberspace in bridging spatial-temporal voids and in bringing ethnic groups together to redefine the native culture experience away from home.

Keywords: Alienation, connectivity, cybernauts, diaspora, hyphenated space, métissage culture

1 INTRODUCTION: THE ORIGINS OF CYBERSPACE

Can anyone deny that the internet has, undoubtedly, become the most powerful tool for communication? The internet has revolutionized connectivity mainly due to its dynamic versatility. People anywhere, anytime, can search, find, and connect with each other. The internet connects computers, mobile phones and other technical devices, thereby, creating a robust space for communication. This dynamic space is cyberspace. It is a virtual space existing only in the realms of the internet. Without the internet cyberspace is non-existent.

"The term cyberspace refers to an information space in which data is configured in such a way as to give the operator the illusion of control, movement and access to information, in which he/ she can be linked together with a large number of users via a puppet like simulation which operate in a feedback loop to the operator." (Burrows 1995)

Cyberspace can be defined as a notional environment in which not merely communications but, in a broader sense, interactions between people over computer networks occur. Since the early twenty first century, cyberspace is the umbrella term used to encompass a vast array of internet communication.

According to The Oxford Dictionary cyberspace is "an imaginary place where electronic messages, etcetera exists while they are being sent by computers." (Oxford Dictionary 2010)

The origin of the term cyberspace lies in the term "Cybernetics". Cybernetics is derived from the Greek word, "Kybernetes" meaning steersman, governor, pilot or rudder.

DOI: 10.1201/9781032644752-49

259

Cybernetics is the science of communications and automatic control systems in both machines and living things.

This word was introduced by Norbert Wiener. Norbert Weiner is considered the Father of Cybernetics. He used it in his pioneering work in electronic communications and control science.

Coming back to the term 'Cyberspace', it has its origin in literary terms. It has its creation in Science Fiction. William Gibson used it in his novel *Neuromancer* in 1984. A character makes an electronic connection and enters into "cyberspace" a dark unknown social universe.

Cybernauts (People in cyberspace) have a whole new virtual reality to share information.

Cyberspace has become an intrinsic and vibrating space mainly because people are able to interact with one another at any time. Inter-communication has become powerful on this interface platform. The functions and possibilities on cyberspace cannot be counted or stated, as it increases every second. This is an age where cyberspace is able to provide virtual realities of almost every situation imagined. Its benefits, for cultural and identitarian purposes, have great potential.

The main reasons cyberspace has conquered the world- is speed; its instantaneousness. It is the quickest means of access. Other important features are mobility, un-limitedness, political freedom and minimum economical usage. If the jet has made the world a smaller place, then the internet has conquered space. The reason being, there are no boundaries to hinder interchanges. There may be conflict and war between national boundaries, but cyberspace is beyond these geographical limits and functions.

2   THE CYBER REVOLUTION

How does cyberspace communication function? Cyberspace is an electronic space functioning through the internet. Cyberspace comprises a large array of communications including electronic mails, instant messaging, multimedia messaging, downloading and uploading information like audio and video files, graphics, data etcetera.

The internet was introduced in 1995. Initially, it was very slow, chaotic and difficult to operate. Over the years, with technological advances, we know the internet is here to stay. The reason is that it was able to appease one of the foremost desires of man: to communicate easily. The dot com boom has triumphed. The internet has changed life. Looking back at the world some twenty years ago, in contrast with today's electronic reign, indeed communication then seemed primitive. Again, try to imagine the domination of electronic media, twenty years hence; we might hardly recognize our radically transformed world.

In developed countries, the internet has long become a near utility. In developing countries, too, the computer and internet have become ubiquitous. The Covid and post-Covid eras have seen a boom in paperless communication. When physical human contact was discouraged, use of the internet boomed. The internet has definitely opened manifold doors in every sphere: business, government, education, arts and culture. The internet together with multi-media has a direct and tremendous effect on culture. We are now ruled by a technological culture. Multimedia includes computer networking, movies, the radio, television, chat-rooms, virtual reality to name some.

Computer communication has grown explosively. Until a decade ago, communication was expensive and cumbersome. Today with the help of computers and mobile smart phones, contact and interactions are cheap and instantaneous. Global networking has become entertaining and exciting.

Communications through computers has become pervasive in our society. Children are taught to be computer savvy from the lower classes- at base level. Entertainment, education, telecommunications, business and every other field are exploring and developing new applications. This proves the power of Cyberspace. Some reasons elucidated here are the reasons for this computer boom, cost efficiency and speed being the foremost.

More and more computers are being linked every second. The fall in cost is due to the expansion of band width. The good news is that with technology growing the cost will keep falling. We are in an age when telephone calls are at the lowest or free, with telephone firms charging a monthly fee for unlimited calls. This drastic fall in the cost of communication, has revolutionized the way we work, shop, play and most important of all 'communicate'. The value of IT and Internet, lies in its capacity to store, analyze, and communicate large amount of data, to any place at any time, instantly and at a very low cost.

Another reason is the popularity of the Internet. Everyone acknowledges the advantages, of using the internet. If an individual is reluctant to use this medium for communication, he finds himself alone; internet and cyberspace are the rage of the age. Anyone wanting to be part of this progressive world has to join the cyber race, interact on cyberspace and be part of cyberspace.

Young people are fascinated and addicted to the internet. They literally live in a virtual space of the computer or mobile phone. They find it an easy and cheap means to connect, comment or discuss things. Social networking is another strong pull. They are fascinated with finding, creating or share information, photos, videos and interesting articles. They are able to know more about topics that interest them. Cyberspace provides a study platform, and also enables them to relax and play games.

3 RE-CREATING THE NATIVE EXPERIENCE OF THE MIGRANT

In its versatility in connecting people, anywhere in the world, a major victory of Cyberspace is its ability to bring like-minded people together. People with similar backgrounds and interests are connected collectively in chat communities, mailing list, electronic newsletters and discussion panels. Online communities are the answer to estranged and distant relationships. While cyberspace, enhanced with multimedia, is the changing face of culture, its implications and importance in Diaspora and for migrants is tremendous.

Cyberspace has re-invented the migrant experience. Longing, belonging, alienation and assimilation etcetera, have all been redefined by the diasporic migrant. All what seemed distanced and spaced is now available on a device at the tip of a finger. Cyberspace has helped recreate and reconnect the homeland which was distant and existed only in the mind of the migrant. Home and everything native which was a memory, is alive and re-lived on cyberspace. How does this work?

On cyberspace the native land comes alive through visuals. Videos posted are able to capture the mood, native ethos and traditional sensibilities. Again, the individual in diaspora has a longing for the news and happenings of home. On Cyberspace he has access to news from home. In the last century, due to very high cost and time, the first and second generation of migrants most often had never visited the mother country. They lived in ignorance about their native traditions and culture. Information was limited to oral narrations. Cyberspace has made these generations familiar with their rich culture and tradition. Thanks to cyberspace, the next generation can learn the native language, see and learn the native customs, traditions, rituals etcetera.

4 CYBERSPACE AND CULTURE ALIENATION

The migrant in Diaspora, away from home lives a hybrid life; a life which is a métissaged space, of his home country and that of his host country. He longs for all that is familiar and the various nuances that made up his past identity. For endless years migrants lived in nostalgia, with a deep longing for home. He lived alienated in a hyphenated state. But now in cyberspace, he is able to transcend this state of flux. On cyberspace he is able to connect with his past culture. He occupies a space intrinsic to his native identity. He is able to interact with his homeland. He is now able to regain a part of his nativeness which was lost in his adopted home. He relives his culture on cyberspace by listening to his native folksong, folk dance, folk music, folklore and traditions. He is also able to witness and participate in important native rituals, ceremonies religious and social events. He can be a part of festivities, and traditional events. He is reacquainted with everything native. Through e-mailing and electronic news letters, he has access to information which is relevant to his nativeness. From a single site, he can gain information of his native communities spread across the globe. He is able to gain access to his homeland which his forefathers in diaspora were denied.

Cyberspace is not limited to a displaced migrant connecting with home. Rather, its implications are more extensive. The migrant in diaspora is able to connect with his native migrants in different parts of the world. Example the Chinese, Tamils, Indians, Jews are able to connect with their counterparts in America, the United Kingdom, Australia, Canada and many more places around the globe. They share the same stories of assimilation and métissage. The migrants in different communities around the world come together on cyberspace. The cohesion factors between them are parallel histories, memories, unique cultural heritage and above all, identical diasporic traumas of alienation and assimilation.

Preservation of native language is the preservation of tradition and history. On cyberspace, natives happily revert to the native language, which consist of colloquial saying, proverbs, idioms and wisdoms. Individuals revert uninhibitedly to their native language in their interactions on cyberspace. This serves a dual purpose of communication and preservation of the indigenous language.

With the introduction of cyberspace, the migrants in Diaspora are able to bridge the hyphenated space between home and their adopted location. Cyberspace helps them overcome their feelings of alienation. On cyberspace they are able to re-locate themselves by retaining contact with the homeland. In his re-location in cyberspace the migrant identifies with his nativeness. There are no borders or restrictions. His native experience is re-lived repeatedly akin to actually visiting his homeland. Places of interest, religious places, cultural monuments myths and folklore make him familiar with the native traditions. Here he is able to find solace.

In Cultural Studies, cyberspace has added new dimensions to the preservation of culture. It has redefined the scope of culture. Cyberspace is an important tool for culture awareness. Cyberspace ensures culture is transferred from one generation to the next. Another dimension is that through cyberspace, various generations across the globe are in contact with other; while the third diasporic generation, can know about their roots through virtual reality. In this way the diasporic individual is able to help his progeny be familiar with their native culture.

Another important feature of cyberspace is that it promotes unity. People of the same culture, scattered across the larger world, are able to establish contact and retain contact with one another. National and geographical divides are dissolved through cyberspace; native interactions and connections are borderless.

In reality, the individuals in Diaspora lives a hybrid life; a life which is a métissaged space, of his home country and that of his host country, but in cyberspace, he is able to transcend this state of flux.

5 CONCLUSIONS: THE FUTURE OF CYBERSPACE

Cyberspace not only encourages cultural encounters but plays an integral part in safeguarding culture. Second and third generation individuals who have never visited the homeland are familiar with every nuance of their native identity.

Often people complain of the amount of time individuals spend on the internet. Others disparage the irrelevant content on the internet. People criticize the lack of quality control. Preferences differ according to people's interest. Herein lies the crucial *raison d'etre* of the ubiquity of the internet. It has so much to offer for every avenue of knowledge. Hard copy of journals and books would not be able to contain the content on the internet. Nor can hardcopies provide the multimedia effects of the sound/music experience. Culture and tradition with its nuances of dance, music, oral storytelling and performance can only be captured on the multimedia facet of cyberspace.

The above arguments, conclusively, ascertains that diasporic natives live their traditions on cyberspace. Displaced individuals, migrants and refugees, forced to leave their native land, can now connect with their native land. Home is just a click away.

REFERENCES

Bernal, Victoria. 2011. Diaspora and Cyberspace. In Knott and McLoughlin (eds), *Diasporas concepts, intersections, identities* 167–171. Jaipur: Rawat Publication.

Braziel Jana Evans and Anita Mannur. 2003. Nation, Migration, Globalization: Points of Contention in Diaspora Studies. In Evans and Mannur (eds) *Theorizing Diaspora*: 1–22. Malden: Blackwell Publishing.

Burrows Roger. 1995. *Cyberspace/ Cyberbodies/ Cyberpunk: Cultures of Technological Embodiment.* Sage Publications 1995 (ed) by Mike Featherstone. E-Book. https://sk.sagepub.com/books/cyberspace-cyber-bodies-cyberpunk. Online publication 2012. Accessed on 30.9.2020.

Grove, A.T. 1980. Geomorphic evolution of the Sahara and the Nile. In M.A.J. Williams & H. Faure (eds), *The Sahara and the Nile*: 21–35. Rotterdam: Balkema.

Oxford Advanced Learner's Dictionary. 2010. 8th Edition. Oxford: Oxford University Press.

Pereira Leanora. 2020. Connecting on E-space. *Identitarian Spaces of the Goan Diasporic Communities.* 1554–173. Goa: Goa University.

Advances in AI for Biomedical Instrumentation, Electronics and Computing – Sachan et al. (eds)
© 2024 The Author(s), ISBN 978-1-032-64298-7

PlaceTech: An AI-enabled solution for smart placement

Khalid Alfatmi, Sakshi Pande, Makarand Shahade, Vaibhavi Suryawanshi,
Kalpesh Badgujar, Vishwajit Patil & Mayuri Kulkarni
Department of Computer Engineering, SVKM's Institute of Technology, Dhule, India

ABSTRACT: Educational institutions must keep pace with evolving technology, particularly in managing student placements. The current manual processes for students and Training and Placement Officers (TPOs) involve resume creation, job matching, and administrative tasks, emphasizing the need for automation and efficiency. PlaceTech: An AI-Enabled Solution for Smart Placement introduces a unified platform for students and TPOs to streamline placement processes. It offers a dynamic resume builder, AI-driven chatbot, and online assessments for students, while providing TPOs with efficient tools to manage job description postings and events. This innovation promises to enhance placement through smart automation and improved communication. Institutional and departmental TPOs gain the ability to categorize students, manage job descriptions, and analyze placement data, fostering greater efficiency. By centralizing and automating these operations, PlaceTech embodies a technology-driven approach, providing a scalable solution for streamlined placement process, with implications extending beyond education to various industries.

1 INTRODUCTION

In today's educational landscape, adapting to evolving technology is crucial, especially in student placements. Current methods burden students with resume creation and skills alignment, while Training and Placement Officers oversee complex tasks like job postings and performance tracking. PlaceTech: An AI-Enabled Solution for Smart Placement transforms traditional approaches, unifying students and TPOs for efficient placement processes. Our research unveils a feature-rich platform with AI chatbots, dynamic resume builders, and comprehensive assessments for students. TPOs gain tools for categorization, detailed data access, and event scheduling. This groundbreaking initiative centralizes and automates placement processes, advocating a technology-driven approach. PlaceTech aims to revolutionize placements, offering efficiency and broader implications for industries. In PlaceTech, we delve into its background, relevance, and future impact on educational placements and industries.

2 LITERATURE SURVEY

Bharathee *et al.* (2023) described, development of a chatbot to assist Class 12 students in their college enrollment process using Natural Language Processing (NLP) techniques. The chatbot, integrated into a WordPress website, achieves a remarkable 90.6 percent accuracy in addressing user inquiries. This research underscores the practical value of chatbot technology in supporting students and suggests potential advancements in NLP. Bilgram *et al.* (2023) explained, job recommender systems, crucial in the face of vast online job data. It introduced an innovative recommendation system employing Natural Language Processing and machine learning to

DOI: 10.1201/9781032644752-50

analyze resumes, predict suitable professions, and gather job postings from Naukri.com. Job ads are ranked based on skill compatibility, offering personalized job recommendations. H.B, S. et al. (2017) described, a web portal to tackle job acquisition challenges, as initial job attempts often require skills and support. Current systems employ separate websites for job applications and skills development, lacking integration. The web portal features two modules: Skill Development and Job Application. In the Skill Development module, users create profiles, access courses, and receive guidance from experienced users. Job Application streamlines interactions between companies and job seekers, offering automated communication and resume evaluation. Kendle, A. S. et al. (2021) introduced, TnP Vision, an ERP system designed for educational institutes' Training and Placement Cells. TnP Vision automates the entire placement process, emphasizing student data management, analysis, and digital practices. Lai et al. (2016) devised Career Mapper for automated resume evaluations. Mankawade et al. (2023) introduced a resume analysis system for job recommendations. Medhe et al. (2015) focused on student-alumni system management. Pavani et al. (2022) enhanced job searches with feature extraction. Pudasaini et al. (2022) devised advanced algorithms for resume-job matching. Ranavare et al. (2020) tailored a placement chatbot. Sharma et al. (2021) introduced NLP-based recruitment systems. Thakare et al. (2023) developed DAKSHA for institutional needs. Yi et al. (2007) explored resume-job relevance models. Additionally, Rinki Tyagi et al. (2020) simplified resume creation, while Shah, Patel, and Lal (2022) improved resume quality with a classifier. Table 1, differentiates various job related platform with their drawbacks and the project strategies PlaceTech using to overcome those drawbacks.

Table 1. Drawbacks of some online job platforms.

Product	Drawbacks	Project Strategies to Overcome Drawbacks
CareerBuilder	No AI-driven chatbot for students	Implementation of an AI chatbot for basic student queries.
LinkedIn	No dedicated TPO functionalities	Separate logins for Institute TPOs and Department TPOs.
Glassdoor	No student-proficiency testing	Implementation of online assessments for aptitude, reasoning, etc.
Monster	Limited department-wise tracking	Department-wise monitoring for Department TPOs.

3 SCOPE OF STUDY

PlaceTech: An AI-Enabled Solution for Smart Placement is a comprehensive web-based platform designed to streamline and enhance the campus placement process for students, Department Training and Placement Officers (TPOs), and Institute TPOs. This system offers three distinct user roles, each with a set of features tailored to their needs.

3.1 Operating environment

The PlaceTech project is web-based, compatible with Windows. It uses Django for the backend and ReactJS for the frontend, offering a robust user experience. Data is stored and retrieved from a cloud-based database server. Users need a stable internet connection and a JavaScript-enabled web browser to access the platform.

3.2 Functional requirement

For students, the platform offers an AI-driven chatbot for inquiries, a dynamic resume builder with job description matching, and access to alumni workshops, expert talks, and assessments. Institute and department TPOs benefit from features like department and academic year-specific student management, job description administration, placement data monitoring, event scheduling, proficiency tests, training co-ordination, and appointment-related tasks.

4 PROPOSED MODELING

4.1 *System architecture*

PlaceTech's system architecture, depicted in Figure 1, ensures a seamless data flow. Extensive research guides feature selection, prioritizing essential information without over-whelming users. MongoDB efficiently stores diverse student data, organizing it into collections tailored for students, TPOs, job descriptions, events, and assessments. Middleware, built on Django, offers security and scalability, supporting role-based access. Python libraries enable web scraping and data analysis, empowering students in career development. Administrators manage registrations and system operations.

For students, PlaceTech offers AI-Driven Chatbot, resume building, skill-matching with job opportunities, and educational event participation, fostering professional growth. Administrators wield significant control, managing registrations.

Figure 1. System architecture of proposed system.

Figure 2. Chatbot integration using Botpress.

4.2 *AI – driven chatbot integration*

Inspired by Thakare *et al.* (2023), our AI chatbot model echoes a similar architecture, offering personalized test recommendations and performance analytics within distinct user sections. This design, influenced by recommender systems, aims to heighten user satisfaction in the competitive online landscape.

We crafted clear objectives and conversation structures to aid students in placement-related queries. Botpress formed the foundation, tailored for the needs of SVKM's Institute of Technology, Dhule. Curating historical data constructed a robust knowledge base. Meticulous configuration and training empowered the chatbot to interpret diverse student queries. Rigorous testing ensured precise responses. Deployment included a scalable server setup, optionally accompanied by an intuitive user interface as shown in Figure 2. A maintenance protocol guaranteed continual improvements and user support.

4.3 *Smart and dynamic resume builder*

The process involves three steps as shown in Figure 3:

Step 1: Scanning resumes against job descriptions.
Step 2: Calculating matching scores and offering recommendations.
Step 3: Empowering students to update resumes based on recommendations, ensuring alignment with job requirements.

Figure 3.   Resume matching and recommendations.

5   WORKFLOW OF THE PROPOSED SYSTEM

Our PlaceTech system is divided into three specialized sections catering to distinct roles as shown in Figure 4. Student Section features an AI-Driven Chatbot, offering swift responses and access to placement history, alongside a Dynamic Resume Builder guiding students in crafting professional resumes and a Resume-Job Description Matching tool ensuring accuracy between resumes and job requirements. Additionally, it includes Alumni Workshops and Expert Talks for learning and networking opportunities, coupled with Online Assessments covering aptitude, logic, and verbal skills. The Department Login, designed for Training and Placement Officers (TPOs), streamlines oversight by categorizing students based on department and year, manages job descriptions efficiently, provides comprehensive dashboards for placement analysis, facilitates event scheduling, and administers proficiency tests. The Institute Login, tailored for Institute-level TPOs, encompasses oversight, job management, placement analysis, event coordination, and proficiency tests, offering a comprehensive suite of tools for higher-level management and coordination.

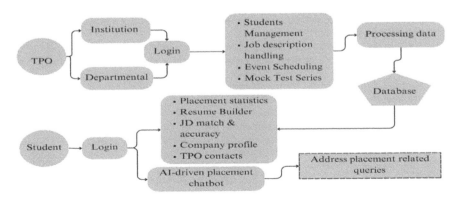

Figure 4.   Workflow diagram of proposed system.

6 EXPERIMENTAL RESULTS AND DISCUSSION

The ongoing development of PlaceTech, our AI-Enabled Solution for Smart Placement, has seen successful testing of key features. Figures 5 and 6, shows the glimpse of the PlaceTech project. Additionally, we've integrated a feedback mechanism to bolster student-TPO interaction.

Figure 5.   Home page of PlaceTech.

Figure 6.   Job description handling portal.

7 CONCLUSION AND FUTURE WORK

In conclusion, our proposed system, PlaceTech: An AI-Enabled Solution for Smart Placement, represents a significant stride towards revolutionizing the process of student placements within academic institutions. While the system is currently under development, our preliminary results and testing of key features, including the AI-driven chatbot, dynamic resume builder, job description management, and online events scheduling systems, have yielded promising results. These early insights demonstrate the potential to streamline academic placement processes, reduce manual work, and foster greater interaction between students and Training and Placement Officers. Future work entails achieving full system implementation with rigorous testing, refining the user experience through feedback, conducting in-depth evaluations, expanding features such as alumni networking and advanced analytics, and exploring adaptability to various sectors. This multifaceted approach aligns with our commitment to enhance entire placement process, reducing manual effort, and fostering increased student-TPO interaction while remaining adaptable and scalable for broader application beyond the education sector.

REFERENCES

Bhharathee, A., Vemuri, S., Bhavana, B., & Nishitha, K. (2023, January). AI-Powered student assistance chatbot. In *2023 International Conference on Intelligent Data Communication Technologies and Internet of Things (IDCIoT)* (pp. 487–492). IEEE.

Bilgram, V., & Laarmann, F. 2023. Accelerating innovation with generative AI: AI-Augmented digital prototyping and innovation methods. *In IEEE Engineering Management Review* (Volume: 51, Issue: 2, Second quarter.

H.B, S., & C.S, M. J. 2017. A framework for automation of placement activity. *International Journal for Scientific Research & Development* 4(11), ISSN (online): 2321–061.

Kendle, A. S., Nagare, M. S., Patre, H. G., Zanwar, R. S., Kottawar, V. G., & Deskhmukh, P. B. 2021. TnP Vision: Automation and analysis of campus placements in colleges. *In 2021 5th International Conference on Computer, Communication and Signal Processing (ICCCSP – 2021)* (pp. 1–6). IEEE.

Lai, V., Shim, K. J., Philips, R. J., & K., P. 2016. CareerMapper: An automated resume evaluation tool. In *2016 IEEE International Conference on Big Data (Big Data)* (pp. 4005–4007). IEEE.

Mankawade, A., Pungliya, V., Bhonsle, R., Pate, S., Purohit, A., & Raut, A. 2023. Resume analysis and job recommendation. *In 2023 IEEE 8th International Conference for Convergence in Technology (I2CT) Pune, India* (pp. 1–5). IEEE.

Medhe, M., Rapelli, R., Mahadik, S., & Shirke, A. 2015. Student Alumni System. *In IJSRD – International Journal for Scientific Research & Development* 3(1), ISSN (online): 2321–0613.

Pavani, V., Pujitha, N. M., Vaishnavi, P. V., Neha, K., & Sahithi, D. S. 2022. Feature Extraction based Online Job Portal. In: *2022 International Conference on Electronics and Renewable Systems (ICEARS)* (pp. 1676–1683). IEEE

Pudasaini, S., Shakya, S., Lamichhane, S., Adhikari, S., Tamang, A., & Adhikari, S. 2022. Scoring of resume and job description using Word2Vec and matching them using Gale–Shapley algorithm. In *Expert Clouds and Applications: Proceedings of ICOECA 2021* (pp. 705–713). Springer.

Ranavare, S. S., & Kamath, R. S. (2020). Artificial intelligence based chatbot for placement activity at college using dialogflow. *Our Heritage* 68(30), 4806–4814.

Rishil Shah, Vishal Patel, Hriday Lal. 2022. "Resume builder with classifier. *International Research Journal of Modernization in Engineering Technology and Science.* e-ISSN: 2582–5208.

Sharma, A., Singhal, S., & Ajudia, D. (2021, September). Intelligent recruitment system using NLP. In *2021 International Conference on Artificial Intelligence and Machine Vision (AIMV)* (pp. 1–5). IEEE.

Thakare, L., Salunke, T., Pagare, D., Patil, A., & Alfatmi, K. (2023, March). DAKSHA: A smart learning platform for DBATU. In *2023 International Conference on Sustainable Computing and Data Communication Systems (ICSCDS)* (pp. 334–339). IEEE.

Tyagi, R., Singh, N., Baghel, A., & Singh, A. 2020. Resume builder application. *International Journal for Research in Applied Science and Engineering Technology (IJRASET) Volume, 8.*

Yi, X., Allan, J., & Croft, W. B. (2007, July). Matching resumes and jobs based on relevance models. In *Proceedings of the 30th annual international ACM SIGIR conference on Research and development in information retrieval* (pp. 809–810).

Advances in AI for Biomedical Instrumentation, Electronics and Computing – Sachan et al. (eds)
© 2024 The Author(s), ISBN 978-1-032-64298-7

Credit card fraud detection with formula-based authentication

P. Asha*
Professor, Computer Science and Engineering, Sathyabama Institute of Science and Technology, Chennai, India

Bhavya Babu
Student, Computer Science and Engineering, Sathyabama Institute of Science and Technology, Chennai, India

S. Prince Mary, B. Ankayarkanni, M.D. Anto Praveena & A. Christy
Professor, Computer Science and Engineering, Sathyabama Institute of Science and Technology, Chennai, India

ABSTRACT: Nowadays, digital withdrawals mostly rely on cards. Days have come where people don't have to carry cash in their pockets and just a small card is enough to make all the withdrawals. Problems with cards will lead to passwords being hacked and easily lead to fraud. Credit card fraud can be identified using Hidden Markov Model (HMM) and Formula Based Authentication. In the Existing system, card withdrawals are very routine among the people and the frauds corresponding to the improvement of security are increasing. In the proposed system, ML algorithms have been applied to detect MasterCard deception in a disproportionate dataset. In the modification process, an application is developed for a banking sector particularly for a credit or ATM card. Users can create an account and get the ATM card along with a unique formula which should be used during suspicious transactions. The user behavior of every transaction is tracked by Hidden Markov Model and if there are any occurrences of suspicious transactions, then a message is sent to the user with the keys that are required to complete the formula. After the user applies the keys to the formula the solution must be entered as the password in order to complete the transaction.

1 INTRODUCTION

Electronic and non-electronic buyers use credit cards for digital withdrawals as a method of settlement in a huge way even though this method has few drawbacks. Card based activities are frequently targeted by culprits, criminals as well as perpetrators. For any transaction, only a particular attribute has to be inserted. In most cases, One Time-Password (OTP) is used as an extra safety. Specifically, for international withdrawals, a method called Card-Not-Present is used where only the details are required rather than the physical card for unauthorized payments. It is very easy to get the card details using methods like shoulder surfing, buying card details, credit card stealing as well as web traffic sniffing. The main sufferer of the fraud will be the bank, trader and registrant. The main duties of the registrant are to detect any malicious activities and report fraudulent withdrawals. The bank then has the responsibility for analyzing all issues and if any evidence for malicious activities is identified, then the amount withdrawn is reversed. This issue can be resolved using formula-based authentication.

2 LITERATURE SURVEY

Yang *et al.* 2017 proposed novel learning strategy using authentic data accommodating over abundance of negotiations which were approved over 3 years explains how to overcome a

---

*Corresponding Author: ashapandian225@gmail.com

DOI: 10.1201/9781032644752-51

number of challenges which includes verification latency (Asha *et al.* 2016) that is few transactions are checked by investigators from time to time, class imbalance which specifies more frauds than genuine actions and concept drift which states that the fraudster strategies change time to time as customers evolve their practices. Many proposed ML algorithms (Andrea *et al.* 2018; Asha *et al.* 2023; Yang *et al.* 2017; Zheng *et al.* 2018) depend upon assumptions due to which there is a lack of realism that leads to two major concerns: the strategy and schedule in which the supervised data is fetched and the initiatives done to identify the fraud. A formalization of the malicious problem (Asha *et al.* 2023) proposed with help of an industrial partner which describes the malicious systems operations which analyzes massive streams of withdrawals and illustrates performance measures used for fraudulent purposes. Advantages include addressing drift and verification latency. Lutao *et al.* 2018 explains the association between special cases and behavior features (Dhankhad *et al.* 2018) considered based on behavior certificates. Initially, the behavior features were extracted from the card holder's historical transaction (Asha *et al.* 2020) record patterns as it is an important way to detect fraud. Using this reduces the danger rate (Foroutan *et al.* 2017) for cardholder's revenue and if it is larger than the restriction rate, it is estimated as trickery (Asha *et al.* 2017). To reflect the applicant's transaction habits (Harshitaa *et al.* 2021), a new Fraud Detection System (FDS) based on behavior certificate is introduced. By implementing a Behavior Certificate, the applicant's transaction habits can be reflected using a Fraud Detection System. The major asset in this paper is that it is highly effective while performing with simulated data, whereas it is difficult to learn the characteristics of unauthorized activities. The work (Benchaji *et al.* 2021), proposed a commensurate evaluation, explaining about the hidden ideas to detect fraudulent activities using real world dataset. The advancement in modernized technology has escalated the fraudulent activities. In order to delineate the hidden patterns, the supervised ML algorithms (Asha *et al.* 2018) were applied. Publicly available datasets were used which were highly imbalanced. The major advantages in this paper includes using various supervised machine learning designs to administer a classifier using a combination of researching methods whereas, the drawback includes delay in updating the supervised model due to verification latency. The work (John *et al.* 2017) explains the association of Fraud Detection System with Faster Payment System to protect computerized methodology from fraudsters. The five electronic commerce systems selected for this paper are automobile insurance, credit card, online auction, telecommunication and healthcare insurance in which the prevalent fraud types are examined closely. Further, the state-of-the-art of the FDS is also systematically introduced. The increase in the false prediction rate of honest withdrawal of money is due to the combination of FDS and FPS and also increases the cost of investigation for banks.

The major challenges were concept drift, class imbalance and verification latency. The other major drawback was that the customer's habits evolve and the fraudsters change their strategies accordingly. In the existing methodology whenever the user tries to make a transaction, the user receives an OTP (One Time Password), which must be entered to make successful money withdrawals. But, if the OTP is seen by the fraudster through shoulder surfing, web trafficking or any alternate means then the fraudster can steal money from the user's card giving rise to credit card fraud.

3 PROPOSED WORK

This implementation is developed for the investment sector. By using Formula Based Authentication, banks can ensure reliability. Users can create an account and get the card issued along with a unique formula with the help of the bank. Every user behavior is monitored using the HMM based on the money withdrawal sequence. The client's frequency of transaction is monitored. Whenever the HMM detects any pattern out of bounds, it immediately issues a warning. This induces the formula-based authentication, which sends a set of keys to the client. If the client enters the correct solution, the transaction is successful. This method assists in identifying fraudulent actions.

3 1   *Modules*

3.1.1   *Bank server*

Bank Service Provider will have a data storage which contains the information about users. They also maintain all the user information that helps the user to authenticate whenever they wish to access the account. The bank server will establish relationships with the consumer and other modules of the Company server to communicate. Hence, a User Interface Frame is connected.

3.1.2   *HMM initiated using big data (user behavior)*

It is operated for analyzing client actions on every transaction. It is executed for understanding retraction of money by the client, meaning the basic idea is total sum of withdrawal each month and the second being frequency of withdrawal. The time recurrence is also monitored and recorded.

3.1.3   *Money withdrawal (malicious user behavior)*

HMM tracks the user behavioral patterns. If any malicious transaction is identified, then HMM will send an indication in the form of keys, to the user thereby indicating the fraudulent withdrawal. If the client enters the correct keys to the formula, the transaction will be successful. The transactions will be blocked if the answer to the Formula is incorrect.

3.1.4   *Money withdrawal (normal user behavior)*

When the client withdraws a certain amount, the HMM tracks the user behavior patterns. If transactions are found to be normal, then the permission to withdraw the money is successful.

3.1.5   *Formula-based verification*

Formula Based Affirmation provides safety by adding a formula. The formula is unique for every user, registered at the time of creating an account. The keys to the formula change every time, and the user is requested to submit an answer following the substitution of corresponding keys to the formula. This usage of formula is required only when the user tries to withdraw beyond the permitted.

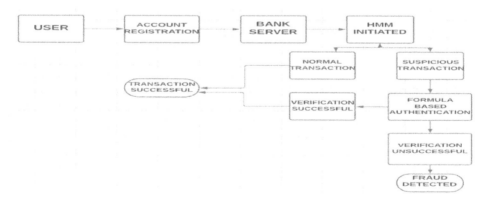

Figure 1.   Architecture diagram.

3.2   *Proposed algorithm*

Step 1: The client registers in a bank with new account details.
Step 2: Clients will be issued with a unique formula that will be used during a suspicious transaction detected by the HMM.
Step 3: In a certain predicament, if the client tries to take an amount that is out of the typical observable patterns, the HMM will immediately raise uncertainty.

Step 4: Formula Based Authentication will be initiated through the HMM.

Step 5: The client will receive a set of keys which are unrepeatable to his formula.

Step 6: The client will apply these keys to the formula, and once the correct password is given, the transaction will be fulfilled.

HMM provides instructions about evaluation, decrypting, and learning. Evaluation is defined as expecting the monitoring order further as well as reverse design. Decrypting specifies all unknown states order (Viterbi). Using the observed information, HMM will be created using Learning (Baum-Welch). The HMM model works on the transaction history of the client, and after the required formulation if the current transaction is found to be suspicious, then the formula-based authentication is initiated. If the user is able to authorize using the keys to their formula, then the transaction will be successful.

3.3 *Numeric notation for formula-based authentication*

Initially the user gets a unique formula while registering for the credit card. During normal transactions the user does not have to enter the authentication formula but if the HMM model detects any particular deals that are not the general user behavior pattern i.e, if the model detects the money withdrawal as skeptical then the user receives the keys through messages.

Example:

Let the unique formula be A+B-C+D. User will receive a message with keys as shown when suspicious transaction is found by the HMM model: A=3 B=2 C=1 D=1

The user must apply the formula using the keys: A+B-C+1 = 3+2-1+1 = 5

Hence this must be entered as the password in order to complete the transaction. Every suspicious transaction generates keys randomly. Therefore, even if the fraudsters hack the mobile for keys, they will not be able to get the money as the formula remains confidential with the user. In this way, the user's account will be free from fraudulent activities.

4 RESULTS AND DISCUSSION

The main aim is to establish a systematic formula-based authentication model using ML methods for predicting fraud research. In our proposed work, the client withdrawal history acts as a dataset for the HMM model. Whenever the HMM detects a doubtful transaction, the formula-based verification will be activated.

Table 1 displays the complete transaction history of the customers.

Table 1. Transaction history of the customer.

seq	ccno	month	transmt	userid
1	430223415617	JAN	1000	1
2	430223415617	JAN	2400	1
3	430223415617	JAN	2340	1
4	430223415617	JAN	4340	3
5	430223415617	FEB	2342	2
6	430223415617	FEB	4543	2
7	430223415617	FEB	1209	3

Figure 2 depicts the User amount transaction page, which is successful as the right key is entered which was validated with formula-based authentication. When the user enters the right key, the transaction will be successful. When the customer enters the wrong key, a suspicious transaction will be identified (Figure 3).

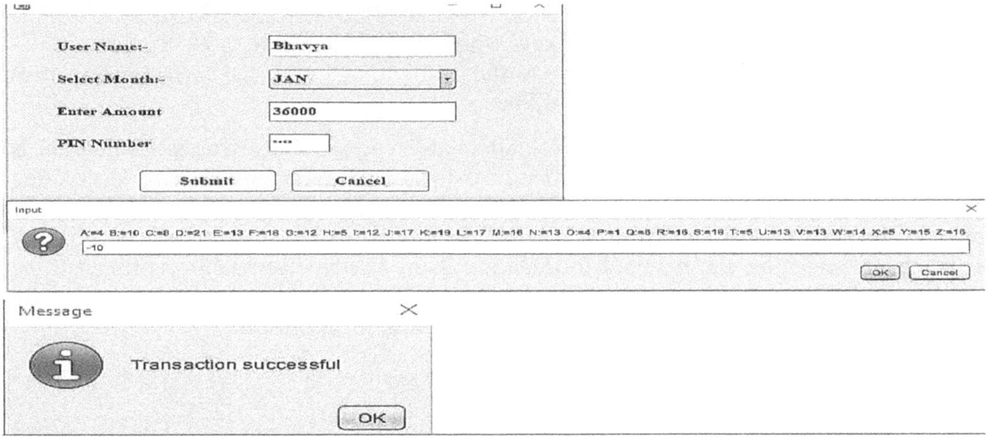

Figure 2. Successful transaction (Formula based authentication).

Figure 3. Fraudulent transaction (Formula based authentication).

5 CONCLUSIONS

More credit card forgery is happening in huge level these days. Even after the existence of cyber security, dishonest money extractions are still active. Formula based authentication is one of many processes to perform the transactions securely. This process allows only genuine deals to take place. Hidden Markov Model is a robust technique for user behavior pattern extraction and to detect deceitful behavior. HMM is used to get a set of unknown variables from known variables. Pattern changes in the transaction history of the account holder can be detected by using the HMM model. If the transaction is detected to be fraudulent an authentication key is sent to the user, and the user must apply the keys to the formula in order to find the solution and enter it as password for successful transaction. Methodological filtering can be made more effective using various algorithms. Hardware implementation can be improvised using counterfactual analysis and can be interfaced using near-field communication for simple transactions. A toggle button can be used to enable formula-based detection transaction. (For Example: the user can choose to enable or disable the transaction limit).

REFERENCES

Andrea D. P., Boracchi., Caelen, O., Alippi, C., & Bontempi, G. 2018. Credit card fraud detection: A realistic modeling and a novel learning strategy. *IEEE Transactions on Neural Networks and Learning Systems*: 3784–3797.

Asha, P., Ankayarkanni, B., Joshila Grace, L. K., Kaarthik, K.M., & Reddy, M. H. 2023. Segmentation of brain tumors using traditional multiscale bilateral convolutional neural networks. *International Conference on Sustainable Computing and Data Communication Systems (ICSCDS), Erode, India*: 1669–1672.

Asha, P., Deepika, K., Keerthana, J., & Ankayarkanni, B. 2020. A review on false data injection in smart grids and the techniques to resolve hem. In: Smys, S., Iliyasu, A.M., Bestak, R., Shi, F. (eds) *New Trends in Computational Vision and Bio-inspired Computing. ICCVBIC 2018*. Springer.

Asha, P., Latha, M. N., & Architha, K. 2017. Design and implementation of IOT based security aware architecture using IDS. *Research Journal of Pharmaceutical Biological and Chemical Sciences*, 8(2): 2293–2300.

Asha P., Lahari T., & Kavya B. 2018. Comprehensive behaviour of malware detection using the machine learning classifier. In: Zelinka I., Senkerik R., Panda G., Lekshmi Kanthan P. (eds) *Soft Computing Systems. ICSCS. Communications in Computer and Information Science*, vol 837. Springer, Singapore.

Asha, P., Sridhar, R., & Rinnu Rose, P. 2016. Jose, click jacking prevention in websites using Iframe Detection and IP scan techniques. *ARPN Journal of Engineering and Applied Sciences* 11(15): 9166–9170.

Benchaji, I., Douzi, S., El Ouahidi, B., & Jaafari, J. 2021. Enhanced credit card fraud detection based on attention mechanism and LSTM deep model. *Journal of Big Data* 8, 1–21.

Dhankhad, S., Mohammed, E., & Far, B. 2018. Supervised machine learning algorithms for credit card fraudulent transaction detection: a comparative study. *IEEE International Conference on Information Reuse and Integration (IRI)*: 122–125.

Foroutan, S. A., & Salmasi, F. R. 2017. Detection of false data injection attacks against state estimation in smart grids based on a mixture Gaussian distribution learning method. *IET Cyber-Physical Systems: Theory & Applications* 2(4): 161–171.

Harshitaa, A., Hansini, P. & Asha, P. 2021. Gesture based Home appliance control system for disabled people. 2021 *Second International Conference on Electronics and Sustainable Communication Systems (ICESC)*: 1501–1505.

John. O. A., A. Adetunmbi, O. & Oluwadare, S. A. 2017. Credit card fraud detection using machine learning techniques: A comparative analysis. *International Conference on Computing Networking and Informatics (ICCNI)*, Lagos: 1–9.

Yang Q., An D., Min R., Yu W., Yang X., & Zhao W. 2017. On optimal PMU placement-based defense against data integrity attacks in smart grid. *IEEE Trans. Inf. Forensics Security*: 1735–1750.

Zheng, L., Liu, G., Luan, W., Li, Z., Zhang, Y., Yan, C., & Jiang, C. 2018. A new credit card fraud detecting method based on behavior certificate. *In 2018 IEEE 15th International Conference on Networking, Sensing and Control (ICNSC)*: 1–6.

Advances in AI for Biomedical Instrumentation, Electronics and Computing – Sachan et al. (eds)
© 2024 The Author(s), ISBN 978-1-032-64298-7

Analysis of various topic modeling algorithms using internal-quality metrics

Astha Goyal
Research Scholar, Department of CSE, MRIIRS, Faridabad, Haryana, India

Indu Kashyap
Professor, Department of CSE, MRIIRS, Faridabad, Haryana, India

ABSTRACT: Topic modeling, a conventional technique for uncovering hidden themes in textual data, has garnered substantial attention across various fields. While it is highly effective, it presents several challenges during application. The foremost challenge lies in the selection of the most suitable topic model across diverse datasets and evaluation criteria. As the field continuously evolves, there arises a necessity to objectively compare and assess the effectiveness of different topic modeling algorithms. This study aims to evaluate the performance of conventional topic modeling algorithms, with respect to various internal quality metrics. By conducting an in-depth analysis and interpretation of results using the 20-Newsgroup dataset, we analyzed various topic models, determining the optimal model. This study incorporates comprehensive range of metrics, including aspects like model coherence, diversity, classification, and significance, to assess the performance of these algorithms. The outcomes of this work provide readers deeper insights into the effectiveness of topic modeling algorithms.

Keywords: Topic Modeling, Latent Dirichlet Allocation (LDA), Non-negative Matrix Factorization (NMF), Latent Semantic Analysis (LSA), Probabilistic Latent Semantic Analysis (PLSA), Evaluation Metrics.

1 INTRODUCTION

In recent years, the surge in digital information has given rise to sophisticated techniques that can provide meaningful insights from vast textual data. Topic modeling, a robust methodology for extracting latent thematic structures inherent within textual corpora, has emerged as a pivotal tool across various disciplines. With the capacity to uncover latent patterns and provide semantic interpretations, topic modeling has found applications in diverse domains such as text summarization, content recommendation, sentiment analysis, and document clustering (Alghamdi *et al.* 2015). Topic modeling represents a specific domain of text mining techniques. Its function involves extracting relations at the theme level. A topic model facilitates the generation of a concise synthesis of themes present and the distribution of topics across every individual document within a collection of documents. A topic serves as a connection for grouping similar documents, thus connoting what is commonly called a "cluster" in data mining. Topics or clusters unveiled through topic modeling often correlate substantially with human comprehension of textual content (Blei *et al.* 2012). Various scholars have proposed methodologies for extracting latent topic structures from document collections, with LDA emerging as the most popular approach, succeeding the framework of PLSA (Blei *et al.* 2003). Applications of topic modeling include an array of automated analyses of textual data. As an illustration, topic modeling finds application in

DOI: 10.1201/9781032644752-52

tasks such as conducting preliminary analysis of the literature (Asmussen *et al.* 2019), establishing recommender systems (Kang *et al.* 2020), ascertaining themes from concise textual content (Albalawi *et al.* 2020) within the realm of computational social science (Wallach *et al.* 2018), and numerous other domains. Although topic modeling is inherently compelling, its practical implementation and application pose intricate challenges. The foremost challenge in topic modeling is selecting an appropriate topic model. Researchers often face the dilemma of choosing from many topic-modeling algorithms, each with underlying assumptions, computational complexities, and parameter sensitivities. Moreover, the challenge is compounded by variability in the datasets and evaluation criteria used for benchmarking (Rüdiger *et al.* 2022). As topic modeling becomes popular, there is a growing need to evaluate and compare the effectiveness of different topic models systematically. Since different variants of topic modeling algorithms exist, they can be broadly categorized into non-probabilistic and probabilistic algorithms (Kherwa *et al.* 2018). Current performance assessments are typically confined to analyzing a single representative from each class or comparing a subset of algorithms within a single class. Moreover, researchers' utilization of different datasets and evaluation criteria further aggravates the lack of consistency and occasionally leads to perplexing contrasts in method comparisons (Rüdiger *et al.* 2022). Such comparative analyses play a pivotal role in explaining the strengths and limitations of these algorithms, guiding researchers and practitioners in selecting the most suitable approach for specific applications, empowering them to make more informed choices of algorithms, and enabling them to assess the validity of their choices. In this context, this study evaluates and compares various conventional topic models concerning their clustering performance against a predefined number of topics using a comprehensive array of evaluation metrics. This study uses a 20-Newsgroup dataset with predefined clustering information (Lang, K. 1995).

This paper is organized as follows: Section 2 provides background information on topic modeling algorithms. Section 3 explains the methodology for comparing different topic models using selected evaluation metrics. Section 4 presents the empirical results and an analysis of the comparative evaluation. Finally, section 5 concludes the paper with a discussion of the study's limitations and potential directions for future research.

2 BACKGROUND

The topic models are designed to identify patterns of co-occurring words and capture the underlying semantic themes that characterize a collection of documents. By extracting these latent topics, topic modeling facilitates a deeper understanding of the content (Lind *et al.* 2021). A document may cover multiple topics, as indicated by its weight coefficient. The topic with the most significant weight determines the belongingness of the document to that topic while ignoring all other assignments. Topic models generate an output list called topic descriptors using words strongly associated with each topic (Aggarwal and Zhai 2012). From an algorithmic standpoint, topics can be conceptualized as patterns arising from the co-occurrence of words within a given corpus. The major challenge in this area is that the number of topics is unknown and must be learned automatically by analyzing the corpus. Several topic-modeling algorithms have been developed to uncover the topics within a text. These algorithms employ diverse mathematical and probabilistic techniques to achieve their objective. Topic models that are built on the vector space model (VSM) (Salton *et al.* 1975), known as the bag-of-words, employ a document-term matrix (DTM) extracted from a document collection. Some models compute vector components, which are termed term-weighting schemes. Term weighting assigns significance weights to words within specific documents, as exemplified by the term frequency-inverse document frequency (TF-IDF) model (Rüdiger *et al.* 2022). Our experiment employed both count-based and TF-IDF-based text representations. Probabilistic topic modeling requires counts, whereas matrix-based methods are based on text representations. Recent works have also identified other categories of topic models, fuzzy and neural, which are beneficial in many applications (Abdelrazek *et al.* 2023). The evolution of topic models started with latent semantic indexing (LSI), which is known as latent semantic analysis (LSA) in topic modeling (Deerwester *et al.* 1990). LSA applies singular value decomposition (SVD) to the term-document matrix, reducing its dimensionality and capturing latent topics. Although not strictly a probabilistic model, LSA has been influential in developing topic modeling. Another SVD variant, non-negative matrix factorization (NMF), was developed to handle sparse data. NMF decomposes the term-document matrix into non-negative matrices representing topics and their corresponding word distributions (Lee & Seung 1999). Later, a probabilistic variant of LSA, namely probabilistic latent semantic analysis (PLSA), was developed, which is the precursor to LDA (Blei *et al.* 2003) and operates on similar principles. PLSA models documents as mixtures of topics but lacks the LDA's Bayesian framework. The most widely used topic-modeling algorithm is LDA. It assumes that each document is a mixture of a few topics and that a word distribution characterizes each topic. LDA aims to determine the optimal topic distribution for each document and the word distribution for each topic. Recently, novel iterations of these algorithms have emerged to address evolving requirements, including Dynamic Topic Modeling, Author Topic Modeling, Hierarchical Dirichlet Process, and Word Embedding-based Approaches. These new adaptations are tailored to meet specific applications or methodological needs (Goyal & Kashyap 2022). Only conventional topic models are used in this study, as listed in Table 1 (Rüdiger *et al.* 2022). The following section describes the experimentation details with the metrics used for evaluation.

Table 1. Topic modeling algorithms used in this study.

Topic Modeling Algorithm	Type
Latent Semantic Analysis (LSA)	Matrix Decomposition, Non-Probabilistic Model (Deerwester, Dumais, Furnas, Landauer & Harshman 1990)
Non-Negative Matrix Factorization (NMF)	Matrix Decomposition, Non-Probabilistic Model (Lee & Seung 1999)
Probabilistic Latent Semantic Analysis (PLSA)	Probabilistic Model (Hofmann 1999)
Latent Dirichlet Allocation (LDA)	Bayesian Framework, Probabilistic Model (Blei, Jordan & Ng 2003)

3 METHODOLOGY

Numerous datasets are available for evaluating topic models, exhibiting significant diversity in corpus size, document length, topic complexity, and noise levels. The selection of a dataset can exert a substantial influence on outcomes. Consequently, opting for an appropriate topic model aligned with the dataset's characteristics is paramount. From the point of view of this study, we used a real-world dataset, 20-Newsgroup (Lang, K. 1995), famous across the topic modeling literature. It is a standard dataset for topic model evaluation, comprising a collection of approximately 20,000 newsgroup documents partitioned evenly across 20 newsgroups.

3.1 *Experiment*

The experiment suite used the OCTIS framework (Terragni *et al.* 2021) to implement topic models, evaluation metrics, and visualization and optimization methods for topic models. The procedure to determine the most compelling topic model follows a standard evaluation methodology comprising the following steps:

1. The dataset is pre-processed to remove irregular word representations, including tokenization, stemming, stop word removal, and filtering unwanted characters and

punctuation. The pre-processing also includes creating the bag-of-words representation for use across topic models.

2. The dataset is then split into training and test partitions for training and evaluation purposes. Using the OCTIS framework, topic models are trained over the training partition of the processed dataset. All the models are trained using the same hyperparameters for unbiased results.

3. Compute the evaluation metrics over the test partition of the dataset. The OCTIS framework handles metrics computation for different topic models based on a list of specified metrics. The metrics are computed in parallel for efficiency. The chosen metrics set is discussed in the following section.

3.2 *Evaluation metrics for assessing topic models*

Evaluation metrics are designed to assess the topic models. There are two main approaches for evaluating topic models: internal and external. Internal metrics assess the structural properties of the clusters, such as their degree of separation, without relying on any other information about the input data. The external metrics for validation are based on known classification referring to manually- obtained labeling of clusters. In the exploratory scenario, this labeling is usually not available. This study uses internal evaluation metrics broadly classified as topic diversity, coherence, significance, and classification. Apart from these, topic stability is also an internal measure but is not included in this study as this is not a quality criterion. It is a desired property for algorithms that include stochastic elements (Rüdiger *et al.* 2022). Diversity Metrics measure the diversity of top k-words in a topic amongst each other. Emphasis is on repeated words to determine redundant topics (Dieng *et al.* 2019). Symmetric KL-Divergence measure over normalized document-topic and topic-word distributions, focusing on diversity among generated document-topic and topic-word distributions (Bianchi *et al.* 2021). Topic Coherence Metrics evaluate the correlation between top-k words of a topic, providing a measure of the interpretability of topics—general computation composed of segmentation, probability estimation, confirmation, and aggregation Röder, (Röder *et al.* 2015). Apart from standard coherence metrics, pairwise or centroid-based strategies are also proposed in the literature (Lau *et al.* 2014). Topic Significance Metrics determine the usefulness of topics, focusing on document-topic and topic-word distributions to discover and determine significant and insignificant topics (AlSumait *et al.* 2009). These include the following metrics: KL-Uniform, KL-Vacuous, and KL-Background. Classification Metrics evaluate the performance of document classifiers utilizing learned document-topic distribution per document as a K-dimensional representation to train a classifier to predict a document's class (Phan *et al.* 2008). These metrics include precision, recall, F1-Score, and accuracy.

By assessing these diverse properties of the inferred topics, a nuanced study of the advantages and limitations of different topic models in wide settings becomes reasonable. The outcome of one such experiment is discussed in the next section.

4 DISCUSSION & RESULT INTERPRETATION

The metric results for various topic models are discussed in detail in this section. Figure 1 (a) compares topic diversity metrics for different topic models. LDA and NMF demonstrated the highest performance with the most significant diversity scores, with LDA having the highest topic diversity and low replication of inferred topics. Across LSA and PLSA, the diversity scores are comparatively lower, with the scores for PLSA being significantly lower, thereby hinting toward high repetition across inferred topics. Another essential aspect of inferred topics is coherence, a well-agreed measure of the quality of topics. Figure 1(b) compares various topic coherence measures for different topic models. Unlike diversity metrics, a considerable variation in agreement across different coherence measures is observed. However, the general trend is similar across diversity metrics, with LDA and NMF producing the most coherent topics for the dataset, followed by PLSA and LSA. As the coherence values of PLSA and LSA models are much lower than the LDA and NMF models, these models may not be suitable for producing diverse and coherent topics in general. LDA outperforms NMF across the C_{UMass}, C_V, C_{UCI}, and C_{NPMI} metrics by marginal value. Across the newer word embedding-based metrics, LDA outperformed NMF and LSA by a marginal score. Another diverse trend is observed across topic significance metrics compared in Figure 2(a). LDA and NMF continue to demonstrate the most significant results for KL-Uniform and KL-Vacuous measures. The performance of LSA is significantly lower than the other models, thereby indicating that the topics inferred are of low quality in terms of actual significance, possibly due to the nature of the inference algorithm, which does not consider accurate semantic information. Across KL-Background, a differing trend is observed where PLSA outperforms both LDA and NMF significantly, demonstrating that the topics inferred by PLSA may not be extremely diverse or coherent; they are much more significant than standard background information.

Lastly, through topic classification metrics compared in Figure 2(b), a comparison of the actual utility of the topics in aiding a classification model to infer the label for a document is evaluated. Here, a clear ranking of models is observed, with precision, recall, accuracy, and F1-scores being the highest for LDA topics-based classification, followed by NMF, LSA, and PLSA, respectively.

Figure 1.   Evaluation of topic models across (a) topic diversity metrics and (b) topic coherence metrics.

Figure 2.   Evaluation of topic models across (a) topic significance and (b) topic classification metrics.

LDA demonstrates the highest scores following the results observed across other metric categories and strongly asserts the high quality of the topics. Table 2 summarizes the results across the metrics, mentioning the highest-performing topic model.

Table 2.   Comparison of highest-performant topic models across different internal evaluation metrics. An asterisk marks metrics where other topic models outperformed LDA.

Metric Category	Metric	Author	Best Topic Model
Topic Diversity	KL-Divergence	Bianchi, Terragni & Hovy 2021	LDA
	Diversity	Dieng, Ruiz & Blei 2019	LDA
Topic Coherence	C_{UMass}	Röder, Both & Hinneburg 2015	LDA
	C_{UCI}	Röder, Both & Hinneburg 2015	LDA
	C_{NPMI}	Röder, Both & Hinneburg 2015	LDA
	C_V	Röder, Both & Hinneburg 2015	LDA
	WE-Pairwise*	Lau, Newman & Baldwin 2014	LSA
	WE-Centroid*	Lau, Newman & Baldwin 2014	NMF

(*continued*)

Table 2. Continued

Metric Category	Metric	Author	Best Topic Model
Topic Significance	KL-Uniform	AlSumait, Barbará, Gentle & Domeniconi 2009	LDA
	KL-Vacuous	AlSumait, Barbará, Gentle & Domeniconi 2009	LDA
	KL-Background*	AlSumait, Barbará, Gentle & Domeniconi 2009	PLSA
Topic Classification	Precision	Phan, Nguyen & Horiguchi 2008	LDA
	Recall	Phan, Nguyen & Horiguchi 2008	LDA
	F1-Score	Phan, Nguyen & Horiguchi 2008	LDA
	Accuracy	Phan, Nguyen & Horiguchi 2008	LDA

5 CONCLUSION

Topic modeling has found application across diverse domains. This study aims to contribute by aiding the choice of an optimal topic model using a set of internal metrics on the 20-Newsgroups dataset. The results demonstrated LDA as the most performant topic model across evaluation measures of diversity, coherence, significance, and classification. While the results demonstrated LDA as the most performant topic model, our study has limitations that may be addressed in the future. Primarily, this work experiments with the 20-Newsgroups datasets, which have common characteristics to many other datasets on which topic modeling is applied, is not representative of all datasets and leaves out the question of whether the same results can be expected on a dataset on differing characteristics. This work evaluates models based on internal quality metrics; however, external quality metrics that evaluate the clustering capability of the inferred topics and how the clustering compares against known ground truth are absent. Utilizing these metrics may provide a better insight into the performance of the topic models.

REFERENCES

Abdelrazek, A. Eid, Y. Gawish, E. Medhat, W. & Hassan, A. 2023 Topic modeling algorithms and applications: A survey, *Information Systems*, vol. 112, p. 102131.

Aggarwal, C. C. & Zhai, C. 2012 *Mining Text Data*, Springer, p. 524.

Albalawi, R. Yeap, T. H. & Benyoucef, M. 2020. Using topic modeling methods for short-text data: A comparative analysis, *Frontiers in Artificial Intelligence, vol. 3, p.* 42.

Alghamdi, R. & Alfalqi, K. 2015. A survey of topic modeling in text mining, *International Journal of Advanced Computer Science and Applications*, vol. 6.

AlSumait, L. Barbará, D. Gentle, J. & Domeniconi, C. 2009 . Topic significance ranking of LDA generative models, in *Machine Learning and Knowledge Discovery in Databases, Berlin, Heidelberg: Springer Berlin Heidelberg*, p. 67–82.

Asmussen, C. B. & Møller, C. 2019. Smart literature review: a practical topic modelling approach to exploratory literature review, *Journal of Big Data*, vol. 6.

Bianchi, F. Terragni, S. & Hovy, D. 2021. *Pre-training is a Hot Topic: Contextualized Document Embeddings Improve Topic Coherence*, Stroudsburg, PA: arXiv.

Blei, D. M. 2012. Probabilistic topic models, in *Proceedings of the 17th ACM SIGKDD International Conference* Tutorials, New York, NY, USA.

Blei, D. M. Jordan, M. I. & Ng A. Y. 2003. Latent Dirichlet Allocation, *Journal of Machine Learning Research*, vol. 3, no. Jan, p. 993–1022.

Deerwester, S. Dumais, S. T. Furnas, G. W. Landauer T. K. & Harshman R. 1990. Indexing by latent semantic analysis, *Journal of the American Society for Information Science*, vol. 41, no. 6, p. 391– 407.

Dieng, A. B. Ruiz, F. J. R. & Blei, D. M. 2019. *Topic Modeling in Embedding Spaces*, vol. 8, arXiv, p. 439–453.

Goyal A.& Kashyap I., 2022.Latent Dirichlet Allocation - An approach for topic discovery, in 2022 *International Conference on Machine Learning, Big Data, Cloud and Parallel Computing* (COM-IT-CON).

Hofmann, T. 1999. Probabilistic Latent Semantic Analysis.

Hubert L. & Arabie, P. 1985. Comparing partitions, *Journal of Classification*, vol. 2, p. 193–218.

Kang, S. Kim, J. Choi, I. & Kang, C. 2020. A Topic Modeling-based Recommender System Considering Changes in User Preferences, *Journal of Intelligence and Information Systems*, vol. 26, no. 2, p. 43–56.

Kherwa, P. & Bansal, P. 2018. Topic modeling: A comprehensive review, *ICST Transactions on Scalable Information Systems*, vol. 0, p. 159623.

Lang, K. 1995. NewsWeeder: Learning to filter netnews, in *Machine Learning Proceedings 1995*, Elsevier, 1995, p. 331–339.

Lau, J. H. Newman, D. & Baldwin, T. 2014. Machine reading tea leaves: Automatically evaluating topic coherence and topic model quality, in *Proceedings of the 14th Conference of the European Chapter of the Association for Computational Linguistics*.

Lee D. D. & Seung, H. S. 1999. Learning the parts of objects by non-negative matrix factorization, *Nature*, vol. 401, p. 788–791, October.

Lind, F. Eberl, J.-M. Eisele, O. Heidenreich, T. Galyga S. & Boomgaarden H. G. 2021 Building the Bridge: Topic modeling for comparative research, *Communication Methods and Measures*, vol. 16: 96–114.

Phan, X.-H. Nguyen L.M. & Horiguchi, S. 2008. Learning to classify short and sparse text & web with hidden topics from large-scale data collections, in *Proceeding of the 17th international conference* on World Wide Web - WWW'08, New York, NY, USA.

Röder, M. Both A. & Hinneburg, A. 2015. Exploring the space of topic coherence measures, in *Proceedings of the Eighth ACM International Conference on Web Search and Data Mining*, New York, NY, USA.

Rüdiger, M. Antons, D. Joshi, A. M. & Salge, T. O. 2022. *Topic Modeling Revisited: New Evidence on Algorithm Performance and Quality Metrics, PLoS One*, vol. 17, p. e0266325.

Salton G., Wong A. & Yang, C. S. 1975. A vector space model for automatic indexing, *Communications of the ACM*, vol. 18: 613–620.

Terragni, S. Fersini, E. Galuzzi, B. G. Tropeano, P. & Candelieri, A. 2021. OCTIS: Comparing and Optimizing topic models is Simple!, in Proceedings of the 16th Conference of the European Chapter of the Association for Computational Linguistics: System Demonstrations, Stroudsburg.

Wallach, H. 2018. Computational social science $\neq$ computer science + social data," *Communications of the ACM*, 61: 42–44.

Advances in AI for Biomedical Instrumentation, Electronics and Computing – Sachan et al. (eds)
© 2024 The Author(s), ISBN 978-1-032-64298-7

Cancer prediction: A comparative analysis of advanced machine learning models for accurate diagnostics

Greeshma Arya, Maithili Singh & Abha Bhardwaj
Indira Gandhi Delhi Technical University, Delhi, India

ABSTRACT: In the realm of diagnostics, advancements in artificial intelligence have transformed cancer diagnosis. This study introduces advanced models harnessing machine learning and deep learning to enhance diagnostic accuracy for timely interventions. By combining techniques like Artificial Neural Networks and Gaussian Naive Bayes with established models, including majority voting and K-Nearest Neighbors, this research offers detailed insights. Rigorous testing showcases superior screening methods, ensuring dependable diagnoses. Signifying a critical stride in cancer diagnosis, this research promises improved treatments and patient outcomes. Surpassing existing models in accuracy, it holds the potential to revolutionize global cancer diagnosis, elevating healthcare outcomes significantly.

Keywords: Cancer Diagnosis, Artificial Neural Network, Gaussian Naive Bayes, Majority Voting algorithm, K-Nearest Neighbors, Predictive Medical Model

1 INTRODUCTION

In the expansive field of medical science, the perpetual challenge of cancer persists, demanding precise diagnostic innovations (Cruz *et al.* 2006). Motivated by its profound impact on global health, this research aims to pioneer transformative approaches in oncology, recognizing the need for refined diagnostic methodologies. While traditional methods like Majority Voting and K-Nearest Neighbors (KNN) have been instrumental, limitations in accuracy have sparked exploration in Artificial Intelligence (AI) and Machine Learning (ML) (Naib *et al.* 2014).

This research sets out on an ambitious journey driven by the necessity to redefine cancer diagnosis (Beckmann *et al.* 2015). Addressing voids in accuracy and speed prevalent in current practices, it integrates advanced ML and Deep Learning (DL) techniques – notably Artificial Neural Networks (ANN) and Gaussian Naive Bayes – with established models (Webb *et al.* 2010). This amalgamation aims to revolutionize cancer diagnostics, not merely for enhanced accuracy but for enabling timely interventions crucial in battling cancer's adversity (Holford *et al.* 1980).

The core motivation extends beyond refining accuracy to encompass enabling timely interventions vital in countering cancer's relentless nature (Ochi T *et al.* 2002). This research is propelled by a commitment to unravel intricate patterns, detect elusive anomalies, and significantly enhance diagnostic models (Berrar *et al.* 2018). This zealous pursuit is underpinned by a dedication to advancing the forefront of cancer diagnosis, translating these advancements into tangible benefits for patients (Adeoye *et al.* 2021). The insights garnered are poised to contribute significantly to scientific comprehension and pave the way for more effective treatments, ultimately improving outcomes for individuals combating this formidable disease (Kumar *et al.* 2022).

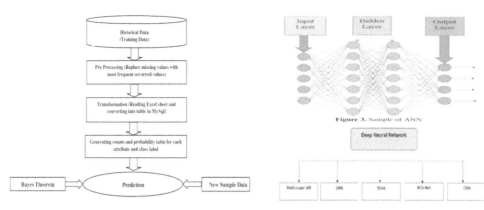

Figure 1.   Main steps of prediction using Naive Bayesian classifier.

Figure 2.   Artificial neural network based on Breast cancer.

2   LITERATURE SURVEY

Bocchi *et al.* (2004), Bollschweiler *et al.* (2004), Bottaci *et al.* (1997), Colozza *et al.* (2005) explore using machine learning techniques in cancer research. Their work underscores the importance of classifying all cancer patients into high or low-risk categories and details the application of various supervised machine learning methods, including ANN, Bayesian Networks, SVM, and Decision Trees for predictive modeling. While acknowledging the potential of machine learning to enhance our understanding of cancer progression, this study underscores the need for robust validation of these methods in clinical practice. However, this research's limitation lies in its lack of specific insights into the machine learning models used and the extent of validation in clinical applications.

Kumar *et al.* (1997) concentrates on AI-based learning approaches for cancer prediction and prognosis. Their study involves reviewing 185 articles that employ AI-based learning models and subjecting them to comparison based on multiple parameters, including prediction rate, accuracy, sensitivity, and specificity. Despite the promise shown by these techniques, the research concludes that they have not substantially reduced cancer mortality, emphasizing the necessity for further investigations in the field of cancer prediction. However, a limitation of this study is that, even though it assesses various parameters, it recognizes the limited impact of AI-based learning approaches on reducing cancer mortality.

Bryce *et al.* (1998) delve into the applications of ML platforms in predicting cancer of oral cavity outcomes. Their findings reveal that ML algorithms exhibit average to high accuracy in the prediction of major outcomes related to cancer of the oral cavity. Nevertheless, these methods may not yet be seamlessly integrated into clinical practice. A limitation of this paper is its emphasis on the need to streamline the clinical integration of MLin treatment of oral cavity cancer.

This research paper introduces advanced cancer diagnostic models that utilize ML and DL techniques, including ANN and Gaussian Naive Bayes. By combining these with established models like majority voting and K-Nearest Neighbors, the study demonstrates superior screening methods and reliable diagnoses. Notably, this research achieved accuracy rates of 99.38% for Gaussian Naive Bayes and 98.97% for Artificial Neural Networks, outperforming existing methods. This research holds the promise of significantly improving cancer diagnosis and healthcare outcomes.

After comparing the above papers with this research paper, the conclusions drawn are that while the other papers make valuable contributions to cancer prediction and diagnosis, this research paper outshines them in terms of predictive accuracy. This research uses Gaussian Naive Bayes and Artificial Neural Networks with accuracy rates of 99.38% and 98.97% which surpasses the predictive capabilities of the methods discussed in the other papers. The potential for more accurate and timely cancer diagnosis and its significant impact on healthcare outcomes set this research apart.

Thus this research demonstrates superior predictive accuracy, offering a promising path for enhancing cancer diagnosis and improving healthcare outcomes on a global scale.

3 RESEARCH MODEL

This research focuses on improving cancer forecasting precision using ML models specifically Artificial Neural Networks(ANN) and Gaussian Naive Bayes algorithmic techniques to outperform traditional models, work began by curating a comprehensive dataset ensuring data integrity and relevance to cancer detection covering patient details, medical history and diagnostic attributes.

Extensive preparation of the given data addressed errors like missing values in dataset outliers and standardization making the dataset machine-learning-ready for the Gaussian Naive Bayes model. This research assumed predictor independence leveraging Gaussian probability distribution model training on the preprocessed dataset that involved fine-tuning hyperparameters optimizing variances and priors for superior performance.

Simultaneously the ANN model incorporated input hidden and output layers adjusting neuron counts for complexity activation functions and regularization to prevent overfitting back-propagation fine-tuned weights and biases to minimize prediction errors combining Gaussian Naive Bayes and ANN predictions via ensemble techniques like Majority Voting mitigated individual weaknesses enhancing overall accuracy and reliability in summary. This methodology enhances cancer detection precision through machine and deep learning providing more accurate predictions.

Gaussian Naive Bayes equations:-

$$P(n|m) = P(n)\, P(m|n)/P(m) \tag{1}$$

$$P(m|n) = \pi P(m_i|n) \tag{2}$$

Here, x_i = value of i th attribute in n.

$$P(m) = \pi P(a_i) P(m|a_i) \tag{3}$$

Here, a_i is i th class

ANN equations:-

$$S(t,\, K) = exp(- int\, 0\hat{\imath}\lambda(,u,K)du) \tag{4}$$

$$f(t|K) = \lambda(.,u,\, K)\, S(t,\, K) \tag{5}$$

$$L = \pi f(t_i|K_i)^{\delta i} S(t_i,\, K_i)^{1-\delta i} = \pi\lambda(.,\, t_i,\, K_i)/exp(int\, 0\hat{\imath}\lambda(.,u,\, K_i)du) \tag{6}$$

Here, S(t, K) = Survival function, f(t|K) = Probability Density function, and L = Likelihood function.

4 RESULT

In the comparative analysis, our research model excelled in cancer prediction. Traditional methods like K-Nearest Neighbors (KNN) and Majority Voting achieved accuracies of 98.55% and 98.14% respectively. Contrastingly, our proposed Gaussian Naive Bayes and Artificial Neural Network models achieved remarkable rates of 99.38% and 98.97%, surpassing these benchmarks. These models offer tangible real-world value by aiding healthcare professionals in timely and accurate predictions, enabling early intervention. Additionally, their ability to pinpoint influential features yields critical insights into biomarkers, potentially revolutionizing diagnostics and treatments. Before clinical integration, rigorous validation across diverse datasets and meticulous bias assessments are vital. Nonetheless, these results underscore significant advancements, promising improved patient outcomes and a transformative impact on oncology.

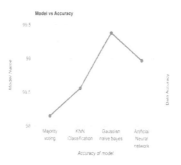

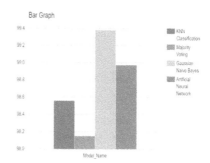

Figure 3. Graph illustration of accuracy percentages prediction models.

Figure 4. Bar Graph describing the comparative analysis between different cancer models.

5 CONCLUSION

Advancements in cancer prediction algorithms led to substantial accuracy gains. The ML models, Gaussian Naive Bayes (99.38%) and artificial neural network (98.97%), surpassed K-nearest neighbors and majority voting. They efficiently captured intricate patterns, enabling precise cancer diagnoses. Beyond theory, their practical implications could reshape healthcare, promising better outcomes and reduced costs. Integrating these technologies offers personalized care opportunities, identifying critical biomarkers and navigating complex data. This research signifies hope for a future where advanced healthcare technologies revolutionize global cancer care, profoundly impacting outcomes and patient well-being.

6 FUTURE SCOPE

The cancer prediction project's future scope promises transformative strides in accuracy and diagnosis. Models like Gaussian Naive Bayes and artificial neural networks signal significant progress, revolutionizing early detection and patient outcomes. Challenges encompass ensuring model generalizability and addressing biases, demanding rigorous validation. Future efforts focus on fortifying models, exploring ensemble methods, and tailoring them for diverse cancer types. Collaboration for real-time data integration and ethical considerations remain pivotal. Successfully navigating these hurdles will ensure ethical, reliable, and scalable predictive models, heralding a future where technology reshapes global cancer care.

REFERENCES

Adeoye, J., Tan, J. Y., Choi, S. W., & Thomson, P. 2021. Prediction models applying machine learning to oral cavity cancer outcomes: A systematic review. *International journal of medical informatics, 154,* 104557.

Artificial neural networks applied to outcome prediction for colorectal cancer patients in separate institutions. *The Lancet, 350*(9076): 469–472.

Beckmann, M., Ebecken, N. F., & Pires de Lima, B. S. 2015. A KNN undersampling approach for data balancing. *Journal of Intelligent Learning Systems and Applications, 7*(04): 104–116.

Berrar, D. (2018). Bayes' theorem and naive Bayes classifier. *Encyclopedia of Bioinformatics and Computational Biology: ABC of bioinformatics, 403,* 412.

Bocchi, L., Coppini, G., Nori, J., & Valli, G. 2004. Detection of single and clustered microcalcifications in mammograms using fractals models and neural networks. *Medical Engineering & Physics, 26*(4): 303–312.

Bollschweiler, E. H., Mönig, S. P., Hensler, K., Baldus, S. E., Maruyama, K., & Hölscher, A. H. 2004. Artificial neural network for prediction of lymph node metastases in gastric cancer: a phase II diagnostic study. *Annals of Surgical Oncology, 11*: 506–511.

Bottaci, L., Drew, P. J., Hartley, J. E., Hadfield, M. B., Farouk, R., Lee, P. W., ... & Monson, J. R. 1997.

Bryce, T. J., Dewhirst, M. W., Floyd Jr, C. E., Hars, V., & Brizel, D. M. 1998. Artificial neural network model of survival in patients treated with irradiation with and without concurrent chemotherapy for advanced carcinoma of the head and neck. *International Journal of Radiation Oncology* Biology* Physics, 41*(2): 339–345.

Colozza, M., Cardoso, F., Sotiriou, C., Larsimont, D., & Piccart, M. J. 2005. Bringing molecular prognosis and prediction to the clinic. *Clinical Breast Cancer, 6*(1): 61–76.

Cruz, J. A., & Wishart, D. S. 2006. Applications of machine learning in cancer prediction and prognosis. *Cancer informatics, 2*, 117693510600200030.

Holford, T. R. 1980. The analysis of rates and of survivorship using log-linear models. *Biometrics*, 299–305.

Kumar, Y., Gupta, S., Singla, R., & Hu, Y. C. 2022. A systematic review of artificial intelligence techniques in cancer prediction and diagnosis. *Archives of Computational Methods in Engineering, 29*(4): 2043–2070.

Naib, M., & Chhabra, A. (2014, September). Ensemble vote approach for predicting primary tumors using data mining. *In 2014 5th International Conference-Confluence The Next Generation Information Technology Summit (Confluence)* (pp. 97–102). IEEE.

Ochi, T., Murase, K., Fujii, T., Kawamura, M., & Ikezoe, J. 2002. Survival prediction using artificial neural networks in patients with uterine cervical cancer treated by radiation therapy alone. *International journal of clinical oncology, 7*, 0294–0300.

Webb, G. I., Keogh, E., & Miikkulainen, R. 2010. Naïve Bayes. *Encyclopedia of Machine Learning, 15*(1): 713–714.

Advances in AI for Biomedical Instrumentation, Electronics and Computing – Sachan et al. (eds)
© 2024 The Author(s), ISBN 978-1-032-64298-7

Comprehensive study of reviews from social media using machine learning techniques

Rohini G. Khalkar, M.S. Bewoor, Sampat P. Medhane & Gauri Rao
Bharati Vidyapeeth (Deemed to be University) College of Engineering, Pune, India

ABSTRACT: Purchasing of product online is very common. So, huge amount of data is available on e-commerce to analyze. Most of the users share and exchange their views through social media, blogs, review sites etc. on products, events, person, or any organization after its use. Customer reviews or feedbacks after the purchase of product plays important role for other customers before purchasing or using it. Thus, sentiments have been generated from different users. Analysis is required to extract useful data from these sentiments which are useful to us. This task of classifying feedbacks aims to recognize various opinions of customers. Interpreting human emotions is a complex task that requires the ability to discern subtleties, detect sarcasm and irony, and exhibit a high degree of sensitivity to context and language proficiency. Challenges include absence of Direct Correspondences, interpretation within cultural contexts, Idiomatic phrases, Emotional tones, and Literary subtleties, Tracking multiple channels etc. Various researchers carried out research work on analysis of sentiment using various machine learning algorithms such as binary classification, multiclass classification, multi-label classification, Support vector machine (SVM), Neural network, Clustering, Regression etc. but there are several challenges in sentiment analysis. There is scope for improving the performance if we are able to find solutions to the challenges of sentiment analysis in such as data gathering, preprocessing of data, its storage and analytics, velocity of data etc. Performance can be improved by considering accuracy, precision, recall, F1 scores, and ROC area. So, hybrid model can be developed to address various issues in sentiment analysis on user generated content.

Keywords: Binary classification, multiclass classification, multi-label classification, Support vector machine (SVM), Neural network, Clustering, Regression

1   INTRODUCTION

Sentiment means feelings, emotions, opinion, attitude or view of user. As use of internet increases, people share their Emotions, Feelings, Reactions, and Feedbacks through blogs, reviews, media, feedbacks etc. Hence, huge amount of text data is available and there is need to analyze it. We can work on this data and use it for exploring the business. Sentiment analysis is especially useful for e-commerce website users where users can go through the previous feedback and reviews before purchasing the product. It will be also helpful for the companies who are using online platform for their product promotion. Reviews of the product will be useful for the companies if they want to observe public feedback of their product so that they can make some transformation in their products. So, sentiment analysis is the understanding and categorization of emotions, views, feedback, opinion about text data and data from social media towards any entity, person, product, place etc.

   Sentiment analysis is important process to find and extract the proper relation between opposed entities of text reviews using various natural language processing (NLP) techniques

DOI: 10.1201/9781032644752-54

or various machine learning algorithms. It is also called as opinion mining or analyzing public sentiments. Process of analyzing the sentiments is usually categorized into three levels namely sentence level, document level and aspect level (Fang *et al.* 2015):

a) Document Level Sentiment analysis- This method classifies the document based on positive or negative sentiments of user expression. The complete document is taken as one unit (Shiramshetty 2022).
b) Sentence Level Sentiment Analysis - In this approach, sentiments expressed in each sentence are evaluated. Initially, we must determine whether a sentence is subjective or objective. If the sentence is deemed subjective, the Sentiment Analysis will determine whether the sentence gives positive or negative feedback.
c) Aspect Level Sentiment Analysis - In this method sentiments are categorized on the grounds of precise aspect of entities. First, we need to recognize entities and their properties. For various aspects of single entity user can provide different opinion.

Empirical research involves gathering data by directly observing and measuring real-world phenomena. The major features of empirical research include:

a) Observation and Measurement: Researchers gather data by directly observing and measuring events, behaviors, or phenomena. This data is collected through various methods like experiments, surveys, observations, or interviews.
b) Real-Life Experience: Empirical research is grounded in real-life experiences and seeks to understand and explain phenomena as they occur in their natural settings. This differs from theoretical or conceptual research that is based on ideas, concepts, or models.
c) Comparison with Theory or Hypothesis: The data collected in empirical research is often compared to existing theories or hypotheses. Researchers test these theories or hypotheses against the observed data to validate, modify, or reject them.
d) Evidence-Based Conclusions: Empirical research aims to draw conclusions based on evidence and data analysis. These conclusions are informed by the observed patterns, relationships, or trends found in the collected data.
e) Replicability and Verification: One of the strengths of empirical research is its potential for replication. Other researchers can repeat the study using the same methods and data to verify the findings and ensure the reliability and validity of the research. Overall, empirical research is a fundamental approach in scientific inquiry and plays a crucial role in building and advancing knowledge across various disciplines.

2 LITERATURE REVIEW

A significant volume of users actively shares and exchange their opinions on the web through various social media platforms such as Twitter, Facebook, Amazon, and other review websites. They express their views on a wide array of subjects, including products, events, services, and organizations, thanks to the rapid advancements in today's technology. Consequently, a wealth of sentiments has emerged from to extract essential insights from the large volume of resources, we employ text-based analysis and natural language processing techniques, making it imperative to incorporate sentiment analysis (Shiramshetty *et al.* 2022). Sentiment analysis is used in the e-commerce field to improve the efficiency and for the better understanding to make profitable corporate decisions (Karamibekr *et al.* 2012). Hossain *et al.* (2021) developed a model in which English language dataset is used in research on sentiment analysis but Bangla language and Romanized Bangla language feedbacks are least used for research work. So, required model is developed where reviews on three languages (English, Bangla and Romanized Bangla) are used and applied six machine learning algorithms. Relative analysis of machine learning algorithms is discussed by considering the performance measures such as accuracy, precision, recall, F1 scores, and ROC area. In this research Support Vector Machine gives accuracy with Bangla dataset is 94%, Random Forest with English dataset-93%, Random Forest with Romanized Bangla dataset is 94% accuracy (Hossain *et al.* 2021).

Shiramshetty Gouthamia, Dr. Nagaratna P. Hegde proposed a model. They have used supervised unsupervised machine learning and lexicon-based approaches are considered while research. Data Sources mentioned are blogs, social media and review sites which are easily available but this study is limited to English language only and it doesn't give good results for large dataset (Shiramshetty *et al.* 2022). Rajat Rajat designed a model in which counter vectorizer (CV) and phrase frequency inverse document frequency (TF-IDF) and logistic regressor is used in sentiment analysis of the large amazon real time dataset. The Amazon E-commerce dataset is transformed into JSON format and subsequently loaded. The dataset is then divided into a training and testing set (Rajat *et al.* 2021). Following this, features are extracted using both the Count Vectorizer and Term Frequency-Inverse Document Frequency (TF-IDF) techniques (Rajat *et al.* 2021). Lastly, a Logistic Regressor (LR) is employed to quantify the sentiments of the reviews as either positive or negative. The simulation results encompass the model's accuracy score, precision, recall, and the depiction of the confusion matrix for the implemented approach (Rajat *et al.* 2021). It is applied only on amazon dataset so research can be done on other datasets.

Badgaiyya *et al.* (2021) proposed the automated method of processing textual information and sorting it into positive, negative, or neutral sentiments is proposed for analyzing the sentiments. The use of various tools for the implementation of the sentiment analysis based on social networking site information to interpret the perspective may allow enterprises to learn how customers think about their products or services. The neural network model is a pre-trained based on natural language processing utilized for sentiment analysis and that mode is popularly known as BERT (Bi-directional Encoder Representation from Transformers). Google also utilizes this BERT model for the understanding of a variety of natural languages. It is a blend of a neural network as well as natural language processing (Badgaiyya *et al.* 2021). It can be used for multi-class classification as it is can be extended to the recommendation system.

Rajkumar S. Jagdale, Vishal S. Shirsat and Sachin N. Deshmukh considered Aspect level, Document level and sentence level sentiment analysis and it is observed that classification gives maximum results for the Products Reviews. Amazon (reviews of Mobile phones, reviews of Camera, feedback for Laptops, tablets, comments for video surveillance, TVs) Data sources used blogs, review sites, micro blogging etc. (Jagdale *et al.* 2019). It gives results of Naïve Bayes algorithm gives accuracy of 98.17%, Support Vector machine algorithm gives accuracy 93.54% for dataset of Camera Reviews. Abhilasha Singh Rathore, Amit Agarwal, Preeti Dimri used Machine learning algorithms such as Naive Bayes (NB), Support Vector Machines (SVM), and Maximum Entropy (ME)) to classify 21500 reviews that are positive or negative as machine learning techniques gives best results to categorize the products reviews (Abhilasha *et al.* 2018).

Social media platforms make their Application Programming Interfaces (APIs) available, enabling researchers and developers to collect and analyze data. Twitter, for instance, offers three different API versions (Fang *et al.* 2015): a) the REST API facilitates the retrieval of status data and user information, b) the Search API allows developers to query specific Twitter content, and c) the Streaming API collects Twitter content in real-time. Developers can blend these APIs to craft their own applications. Consequently, sentiment analysis gains a solid foundation with the wealth of online data at its disposal.

3 CHALLENGES IN SENTIMENT ANALYSIS USING EXISTING SYSTEM

a) Sarcasm - A good sentiment analysis tool is required to be able to distinguish irony from the given context

b) Multilingual data – Reviews can be available in multiple languages and future scope is available in most of the research work already carried out regarding this.
c) Domain oriented issues
d) Working with large dataset
e) Polarity
f) Polysemy
g) Negation detection
h) Potential biases in model training
i) Fake reviews

4 ANALYSIS OF EXISTING SYSTEM

Various machine algorithms can be used for analyzing the sentiments of users such as Regression, Support Vector Machine, Naive Bayes, and Neural Network etc.

Precision, in the context of data analysis and machine learning, is a measure that quantify the correctness of positive predictions made by a model. It is denoted by the equation.

Precision = T P/(T P + F P)
Recall introduces the classes as the beneficial classes out of all classes.
Recall = T P/(T P + F N)
F1 score analyzes accuracy and recall at the identical time.
F1 score = (2 ×Recall × Precision)/(Recall + Precision)

Accuracy is the amount of all specimens of the class and the collective number of samples denoted by equation. **Accuracy**=(TP+TN)/(TP+TN+FP+FN) Researchers have taken 147000 reviews of books that have been processed for analyzing the polarization, Using SVM and Naïve Bayes it gives following results for

True Positive =1721
False Positive =325
False Negative =315 True Negative =-1639 values from total input. (Precision, Recall and F1-score)

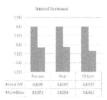

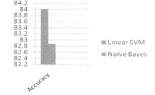

Figure 1. Precision, Recall, F1 score Values.　　Figure 2. Accuracy of Amazon books reviews.

5 PROPOSED MODEL

While considering customer experience or feedback, it can break down the text to topics such as product quality, speed of service, ease of communication, price, age group, brand name, current season and many more. Various machine learning algorithm can be used to analyze customer opinion, reviews etc. The process of sentiment analysis includes steps:

a) Data Collection: Reviews and feedbacks should be collected from correct resource.
b) Data preprocessing: It includes noise removal, tokenization, stop words removal, Lemmatization etc.
c) Feature selection: Required features need to be selected so that we can use them for further process.

d) Use a machine learning model to classify sentiments: Use suitable machine learning model to classify the sentiments that we had selected. We can use hybrid model by combination of two approaches to improve the performance.

e) Validate the model and deploy it by considering the calculated polarity. Finally, find the Precision, Recall, F1score and accuracy using confusion matrix.

Proposed method includes combination of any two machine learning algorithms that can be applied for analysis of product reviews of amazon dataset. Machine learning algorithms that we need to study will be Support Vector Machine (SVM), Naïve Bayes (NB) and Maximum Entropy (ME). We can consider the hyper parameters of each algorithm in hybrid model. We can use unigrams and weighted unigrams for positive and negative feature extraction. Based on results of confusion matrix precision, recall and accuracy can be measured. Comparative analysis of various machine learning algorithms can be done for same product review to overcome the research gap.

6 CONCLUSIONS

Customer reviews are not only important for the other customers but also for the company which is selling the products. User opinions are important source of acquaintance for customers to preserve their purchase choice. Efficient machine learning algorithm can be applied to improve the performance of recommendation system. There is need to work on the above challenges as a future scope. If we use dataset having feedback in multiple languages it will cover the limitations of previous research work done. Various preprocessing techniques can be used to filter the reviews at early stage only. So that it will help us to improve performance of system.

REFERENCES

Akay, A., Dragomir, A., & Erlandsson, B. E. 2013. A novel data-mining approach leveraging social media to monitor consumer opinion of sitagliptin. *IEEE Journal of Biomedical and Health Informatics 19*(1): 389–396.

Abhilasha Singh Rathor, Dr. Amit Aggarwal, Dr. Preeti Dimri, 2017, Opinion Mining : Insight. *International Journal Of current engineering and scientific research* (IJCESR Of Smartphone Product Review Using Support Vector Machine Algorithm-Based Particle Swarm SR) 4, no. 10.

Akshay Amolik, Niketan Jivane, Mahavir Bhandari, Dr. M. Venkatesan, 2016, Twitter sentiment analysis of movie. *International Journal of Engineering and Technology (IJET)* 7(6).

Ashish A. Bhalerao, Sachin N. Deshmukh, Sandip D. Mali. 2016, Predicting sentiment of user reviews. *International Research Journal of Engineering and Technology (IRJET)* 3(5).

Badgaiyya, A., Shankarpale, P., Wankhade, R., Shetye, U., Gholap, P. K., & Pande, P. S. (2021). An application of sentiment analysis based on hybrid database of movie ratings. *Int. Res. J. Eng. Technol. (IRJET)*, 8(1), 655–665.

Bhole, A., & Thombre, V. D. 2014. Review of Sentiment Classification Method and Opinion Mining: The Future Roadmap. *International Journal of Engineering Research and Technology (IJERT)* 3(3).

bin Harunasir, M. F., Palanichamy, N., Haw, S. C., & Ng, K. W. 2023. Sentiment analysis of amazon product reviews by supervised machine learning models. *Journal of Advances in Information Technology 14*(4).

Fang, X., & Zhan, J. 2015. Sentiment analysis using product review data. *Journal of Big Data 2*(1): 1–14.

Haque, T. U., Saber, N. N., & Shah, F. M. (2018, May). Sentiment analysis on large scale Amazon product reviews. *In 2018 IEEE International Conference on Innovative Research and Development (ICIRD)* (pp. 1–6). IEEE.

IEEERathor, A. S., Agarwal, A., & Dimri, P. 2018. Comparative study of machine learning approaches for Amazon reviews. *Procedia Computer Science 132*: 1552–1561.

Jagdale, R. S., Shirsat, V. S., & Deshmukh, S. N. 2019. Sentiment analysis on product reviews using machine learning techniques. In *Cognitive Informatics and Soft Computing: Proceeding of CISC 2017* (pp. 639–647). Springer Singapore.

Kaur, B., Kumari, N. 2016. A hybrid approach to sentiment analysis of technical article reviews. *I.J. Education and Management Engineering* 6, no. 1–11.

Karamibekr, M., & Ghorbani, A. A. (2012, December). Verb oriented sentiment classification. *In 2012 IEEE/WIC/ACM International Conferences on Web Intelligence and Intelligent Agent Technology* (Vol. 1, pp. 327–331). IEEE.

Maurya, S., & Pratap, V. (2022, May). Sentiment analysis on amazon product reviews. In *2022 International Conference on Machine Learning, Big Data, Cloud and Parallel Computing (COM-IT-CON)* (Vol. 1, pp. 236–240). IEEE.

Md. Jahed Hossain, Dabasish Das Joy, Sowmitra Das, 2021, A survey on challenges and techniques of sentiment analysis, *Turkish Journal of Computer and Mathematics Education*, Vol.12 No.06, 4510– 4515

Mochamad Wahyudi, Dinar Ajeng Kristiyanti, 2016, Sentiment analysis optimization. *Journal of Theoretical and Applied Information Technology* 91 1.

Miao, Q., Li, Q., & Dai, R. 2009. AMAZING: A sentiment mining and retrieval system. *Expert Systems with Applications* 36(3): 7192–7198.

Rajat, R., Jaroli, P., Kumar, N., & Kaushal, R. K. (2021, November). A sentiment analysis of amazon review data using machine learning model. *In 2021 6th International Conference on Innovative Technology in Intelligent System and Industrial Applications (CITISIA)* (pp. 1–6). IEEE.

Sangeetha, J., & Kumaran, U. 2023. Sentiment analysis of amazon user reviews using a hybrid approach. *Measurement: Sensors, 27*, 100790.

Saaqib, N. U., & Verma, H. K. (2023, May). Analysis of sentiment on amazon product reviews. *In 2023 Third International Conference on Secure Cyber Computing and Communication (ICSCCC)* (pp. 697–702). IEEE.

Shiramshetty Gouthamia, Dr. Nagaratna P Hegdeb, 2022, Sentiment analysis on reviews of E-commerce sites using machine learning algorithms, 3rd *Int. Conf. on Innovations in Science, Engineering and Technology (ICISET)* 26–27 February 2022, 978-1-6654-8397-1/22 IEEE.

Sushila Sonare, Megha Kamble, 20April 2021, Ternary classification of product based reviews: survey, open issues and new approach for sentiment analysis, *Indian Journal of Artificial Intelligence and Neural Networking (IJAINN)* ISSN: 2582-7626 1(2).

Tripathy, A., Agrawal, A., & Rath, S. K. 2015. Classification of sentimental reviews using machine learning techniques. *Procedia Computer Science 57*: 821–829.

Xu, K., Liao, S. S., Li, J., & Song, Y. 2011. Mining comparative opinions from customer reviews for competitive intelligence. *Decision Support Systems, 50*(4): 743–754.

Zhang, M. L., & Zhou, Z. H. 2013. A review on multi-label learning algorithms. *IEEE Transactions on Knowledge and Data Engineering* 26(8): 1819–1837.

Advances in AI for Biomedical Instrumentation, Electronics and Computing – Sachan et al. (eds)
© 2024 The Author(s), ISBN 978-1-032-64298-7

ArthritisCare: Empowering wellness through personalized arthritis detection and physiotherapy exercise recommendation

Vijaylaxmi Bittal* & Makarand Shahade*
Associate Professor, Department of Computer Engineering SVKM's Institute of Technology, Dhule, Maharashtra, India

Isha Wagh*, Pratiksha Yeshi*, Pritam Lokhande* & Hindraj Patil*
Department of Computer Engineering SVKM's Institute of Technology, Dhule, Maharashtra, India

Khalid Alfatmi*
Assistant Professor, Department of Computer Engineering SVKM's Institute of Technology Dhule, Maharashtra, India

ABSTRACT: Osteoarthritis is a degenerative joint disorder linked to bone and cartilage. Arthritis is the most prevalent kind of bone disease. A sizable portion of the global population has been afflicted by this bone condition. Osteoarthritis frequently causes joint discomfort, stiffness, and Edema. It makes it harder for the affected person to carry out regular tasks. Osteoarthritis should be identified at an early stage to reduce risk and joint damage. Physical examinations and X-rays are two traditional methods for detecting OA. Nevertheless, none of these methods are very precise. The inclusion of ML would increase the precision of OS diagnosis. Not only will it increase precision, but it will also make it possible for researchers to identify OA diagnosis techniques that are more trustworthy. We've proposed a machine learning (ML) method for OA detection in this study employing knee X-rays. X-ray scans of the knee are the input for our approach. Following that, we take features out of the image. Convolutional neural networks, a type of machine learning classifier, are being applied to it in order to accurately predict whether or not the knee has OA. We test our methodology using a set of 7434 knee X-ray pictures that have been interpreted by radiologists. The accuracy of our approach is 98.9 percent, which is significantly greater than the accuracy of the VGG16 model prior to fine-tuning (85.2 percent). Our findings show that our machine learning approach is a useful tool for OA detection and may be utilized to create efficient knee OA diagnostic tools.

Keywords: Osteoarthritis, Machine Learning, Deep Learning, Knee X-ray images, Diagnosis, Cartilage, Classification

1 INTRODUCTION

A common joint disease called osteoarthritis (OA) causes discomfort and mobility issues as it gradually erodes the joint cushioning. X-ray scans are used in traditional OA diagnosis; however, human interpretation may cause early disease indications to be ignored. Recent advancements in AI-powered software show promise for raising the accuracy of OA diagnosis. We describe an AI system that is easy to use and has been developed to diagnose and treat osteoarthritis (OA). With

*Corresponding Authors: vijaylaxmi.bittal@svkm.ac.in, makarand.shahade@svkm.ac.in, waghisha2609@gmail.com, payeshi9999@gmail.com, pritamlokhande8559@gmail.com, hindrajp@gmail.com and alfatmi.khalid@gmail.com

DOI: 10.1201/9781032644752-55

the help of this system, clinicians and patients will be able to access cutting-edge technology and close the gap between AI research and practical medical applications

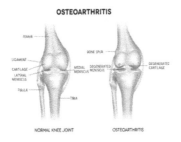
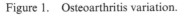

Figure 1.   Osteoarthritis variation.

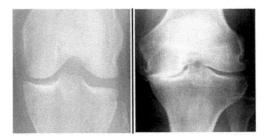

Figure 2.   Normal knee X-Ray and OA knee X-Ray.

2   RELATED WORK

The Single Shot MultiBox Detector model is the main tool used for current osteoarthritis detection, according to the author, to determine the severity of osteoarthritis in X-ray pictures. This approach assesses the effects of the condition on patients and classifies its stages (Arumugam *et al.* 2022).

The authors present an automated model for knee Osteoarthritis identification that uses a CNN architecture for feature extraction and preprocessing to improve image quality. The model uses a machine learning classifier to test knee osteoarthritis and classify it into healthy and unhealthy states (Zebari *et al.* 2022).

The authors offer a method that uses X-ray images to diagnose osteoarthritis in the knee. Using neural network techniques and the DenseNet-201 model, the system was able to identify knees as "healthy" or "suspected for osteoarthritis" with an accuracy of 82.48 percent. The results were displayed through probability graphs (Parikh *et al.* 2023).

The authors discuss a database for osteoporosis detection that contains 240 people's knee X-rays and clinical data. They trained and evaluated Machine Learning models to detect Osteoporosis from knee X-rays with a high degree of accuracy-the best model surpassing 90 percent (Wani *et al.* 2021).

In the article "Treatment of Osteoarthritis", the authors review all high-quality studies published up to 1992 on the treatment of osteoarthritis. This is a rigorous and systematic approach to literature review. An effective and safe first-line medication for osteoarthritis pain is paracetamol. NSAIDs should be used with caution, especially by older adults and people with other health problems (Jones *et al.* 1992)

India demonstrated a substantial surge in osteoarthritis research publications, growing from 39 to 92. The country's global presence in this research field also increased from 1.41 percent to 2.10 percent. There was a noticeable decline in citation impact per publication, dropping from 18.09 to 6.73 (Kumar *et al.* 2017).

The authors present DeepKneeExplainer for enhanced knee osteoarthritis diagnosis, achieving 91 percent accuracy with robustness to noise. They emphasize AI's clinical potential, acknowledge misclassification risks and overfitting, and stress considering patient factors. They plan to improve interpretability by adding samples, multimodal approaches, and domain knowledge (Karim *et al.* 2012)

The study classifies knee osteoarthritis images with an accuracy of 93.84 percent using the Adam optimizer and the EfficientNetB5 model. Adam's matching percentages are 87.07 percent, 85.72 percent, and 0.862 for accuracy, recall, and F-1score, respectively. The authors suggest wider applications for illness classification and future development initiatives, while also acknowledging the limitations of the study (Sharma *et al.* 2023)

The author used GLCM texture analysis and CLAHE image enhancement in their texture analysis of knee osteoarthritis in order to identify the condition. Regarding the Osteoarthritis stage, which ranges from KL-0 to KL-4, the results here demonstrate varying accuracy (Fatihin *et al.* 2020).

The authors of this article introduced a Discriminative Regularized Auto Encoder (DRAE) for classification. This approach is achieved by augmentation of Standard Auto-Encoder with a discriminative loss-term of learned representation encapsulating discriminative details. This results in classification of OA in 3 stages, KL grade 0 representing absence of OA and KL grade 1, KL grade 2 representing early presence of OA. (Nasser *et al.* 2020) Automatic Knee OA detection using KL Grading is described in Using SOTA Deep Learning Models and Ordinal Loss, the severity of knee osteoarthritis can be automatically determined. ConvNeXt and ConvNeXt V2 learning models are implied with the ordinal loss function. The accuracy count forthis system is 73.91 percent (Aktemur *et al.* 2023)

The study introduced by the author uses a novel tri-weightage model for classification purpose of Osteoarthritis by X-Rays, patient's reported data and other medical data. This model achieved accuracy of 89.29 percent that implied the RESNET152V2 and INCEPTIONRESNETV2 deep learning models (Masood *et al.* 2022).

From the literature survey carried out here exhibits following findings. The existing systems don't have following features.

- Proper Detection of Level of Osteoarthritis.
- Recommendation based on Level.
- Progress Report Generation.

3 PROPOSED SYSTEM

The goal of the proposed integrated web-based platform is to simplify the diagnosis and treatment of Osteoarthritis (OA) by fusing Artificial Intelligence with a patient- and healthcare provider-friendly interface. Three main modules make up the architecture of the platform: Data Management for safe storage, Diagnosis and Management using AI for X-ray analysis, and User Interface for interaction. A patient database can be created, knee X-Ray images can be uploaded, AI-powered OA analysis can be performed, Personalized Management Recommendations can be made, and there is an option to determine the risk of osteoporosis. Unique dashboards, intuitive navigation, and AI integration that combines CNN and SVM models are all features of the design. The system's capabilities are additionally enhanced by a feedback mechanism, progress tracking, and secure database storage.

4 METHODOLOGY

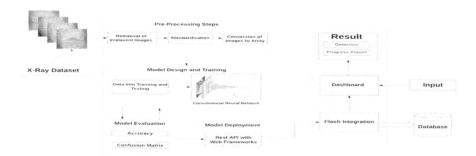

Figure 3. Working diagram.

4.1 Dataset

- Category 0: A healthy knee with no signs of osteoarthritis.
- Category 1: Osteoarthritis symptoms that seem suspect.
- Category 2: Moderate OA symptoms.
- Category 3: Multiple osteophytes and moderate signs.
- Category 4: Severe symptoms, including large osteophytes and malformations of the bones.

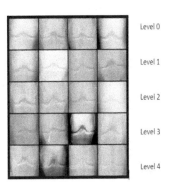 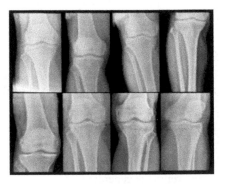

Figure 4.   Knee X-ray of osteoarthritis.   Figure 5.   Knee X-ray of osteoporosis.

Table 1.   Dataset size.

KL Class	No. of images	Training	Testing
Healthy	2925	2286	639
Doubtful	1342	1046	296
Minimal	1963	1516	447
Moderate	980	757	223
Severe	224	173	51

Table 2.   EPOCH increasing accuracy.

Epochs	Testing Accuracy (%)	Testing Loss	Validation Testing (%)	Validation Loss
10	92.43	0.45	89.7	0.69
20	93.68	0.8	91.23	0.83
30	93.98	0.59	90.88	0.63
40	95.79	0.31	89.33	0.41
50	98.9	0.89	97.75	0.96

Figure 4 illustrates the dataset for osteoarthritis, and Table 1 provides a table showing the count of total images in the osteoarthritis dataset. Furthermore, Figure 5 presents the dataset for osteoporosis. Table 2 provides the Epoch increasing accuracy with respect to testing and validation.

4.2 Data preprocessing

The method ensures accurate and trustworthy osteoarthritis and osteoporosis detection by filtering and standardizing images, improving quality, normalizing values, and correcting imbalances.

4.3 Osteoarthritis detection

To increase accuracy, X-ray images are separated into test, validation, and training sets. A CNN then extracts features and makes five predictions. Metrics evaluate performance, and validation stops overfitting.

4.4 Osteoarthritis recommendation

Based on the diagnosis, individualized advice is given, which may include physiotherapy and lifestyle modifications. Through progress tracking and modification made possible by a monitoring and feedback system, "ArthritisCare" improves patient self-management.

4.5 *Osteoporosis detection*

Preprocessed X-ray images from a new dataset labeled as 'healthy' or 'osteoporosis' are used to train a unique CNN model for osteoporosis detection.

4.6 *Osteoporosis recommendation*

Based on the confirmation of diagnosis evidence-based recommendations are provided dietary adjustments, supplements, lifestyle modifications, and precautions, personalized to meet each patient's unique needs.

4.7 *Fitness check*

If neither osteoarthritis nor osteoporosis is detected, the system will provide a message indicating that the person appears to be in good health.

4.8 *Progress tracking*

The patient's initial diagnosis result is saved and the system suggest the patient to diagnose the disease after a specific period. The new outcome of diagnosis is compared with the saved one and a progress report is generated.

5 RESULTS AND CONCLUSION

Figure 6: Website providing information on arthritis and knee joints. Figure 7: BMI calculation form and patient data form. After submitting, users are redirected to arthritis information.

Figure 6. Home page.   Figure 7. BMI calculated form.

The website includes educational graphics that show the state of cartilage and the development of osteoarthritis. Personalized X-ray analysis and expert advice are available to users via the "View the Outcome" and "Personalized Advice" buttons. After BMI checking, Figure 8 shows the directed page. Figure 9 permits an option to upload X-ray images for the purpose of OA detection and results of the disease prediction are shown. Figure 10 provides suggestions based on the diagnosis.

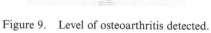

Figure 8. Directed page.   Figure 9. Level of osteoarthritis detected.

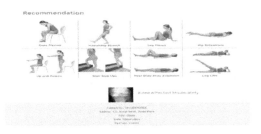

Figure 10.   Recommendation based on the level detected.

REFERENCES

Arumugam, S. R., Balakrishna, R., Rajeshram, V., Gowr, S., Karuppasamy, S. G., & Premnath, S. P. (2022, November). Prediction of severity of knee osteoarthritis on X-ray images using deep learning. *In 2022 IEEE North Karnataka Subsection Flagship International Conference (NKCon)* (pp. 1–5). IEEE.

Aktemur, İ., & Öksüz, İ. (2023, July). Automatic Detection of knee osteoarthritis severity with SOTA Deep learning models and ordinal loss. In *2023 31st Signal Processing and Communications Applications Conference (SIU)* (pp. 1–4). IEEE.

Bittal, V., Jagdale, V., Brahme, A., Deore, D., & Shinde, B. (2023, May). Multifarious face attendance system using machine learning and deep learning. *In 2023 7th International Conference on Intelligent Computing and Control Systems (ICICCS)* (pp. 387–392). IEEE.

Fatihin, M. M., Baskoro, F., & Anifah, L. (2020, October). Texture analysis of knee osteoarthritis using contrast limited adaptive histogram equalization based gray level co-occurrent matrix. In *2020 Third International Conference on Vocational Education and Electrical Engineering (ICVEE)* (pp. 1–4). IEEE.

Jones, A. C., & Doherty, M. I. C. H. A. E. L. (1992). The treatment of osteoarthritis. *British journal of clinical pharmacology*, 33(4), 357–363.

Kumar, A., Goel, S., Gupta, R., & Gupta, B. M. (2017). Osteoarthritis research in India: A scientometric assessment of publications output during 2007–16. *International Journal of Information Dissemination and Technology*, 7(3), 157–161.

Karim, M. R., Jiao, J., Doehmen, T., Cochez, M., Beyan, O., Rebholz-Schuhmann, D., & Decker, S. (2021). DeepKneeExplainer: Explainable knee osteoarthritis diagnosis from radiographs and magnetic resonance imaging. *IEEEAccess*, 9, 39757–39780.

Masood, H., Hassan, E., Salam, A. A., & Liaquat, M. (2022, May). Osteo-Doc: KL-Grading of osteoarthritis using deep-learning. *In 2022 2nd International Conference on Digital Futures and Transformative Technologies (ICoDT2)* (pp. 1–6). IEEE.

Nasser, Y., Jennane, R., Chetouani, A., Lespessailles, E., & El Hassouni, M. (2020). Discriminative regularized auto-encoder for early detection of knee osteoarthritis: data from the osteoarthritis initiative. *IEEE Transactions on Medical Imaging*, 39(9), 2976–2984.

Parikh, R., More, S., Kadam, N., Mehta, Y., Panchal, H., & Nimonkar, H. (2023, April). A transfer learning approach for classification of knee osteoarthritis. In *2023 Second International Conference on Electrical, Electronics, Information and Communication Technologies (ICEEICT)* (pp. 1–5). IEEE.

Sharma, G., Anand, V., & Kumar, V. (2023, August). Classification of osteo-arthritis with the help of deep learning and transfer learning. *In 2023 5th International Conference on Inventive Research in Computing Applications (ICIRCA)* (pp. 446–452). IEEE.

Speech To Image Translation Framework for Teacher- Student Learning Image reffered from:https://www.medicoverhospitals.in/images/diseases/osteoarthritis overview.webp

Wani, I. M., & Arora, S. (2021, September). A knee X-ray database for osteoporosis detection. In *2021 9th International Conference on Reliability, Infocom Technologies and Optimization (Trends and Future Directions)(ICRITO)* (pp. 1–5). IEEE.

Zebari, D. A., Sadiq, S. S., & Sulaiman, D. M. (2022, March). Knee osteoarthritis detection using deep feature based on convolutional neural network. *In 2022 International Conference on Computer Science and Software Engineering (CSASE)* (pp. 259–264). IEEE.

Advances in AI for Biomedical Instrumentation, Electronics and Computing – Sachan et al. (eds)
© 2024 The Author(s), ISBN 978-1-032-64298-7

Rights reach: AI-powered legal assistance for the physically challenged

Vijaylaxmi Bittal*
Assistant Professor, Department of Computer Engineering SVKM's Institute of Technology, Dhule, Maharashtra, India

Sakshi Pingale*
Department of Computer Engineering SVKM's Institute of Technology, Dhule, Maharashtra, India

Makarand Shahade*
Assistant Professor, Department of Computer Engineering SVKM's Institute of Technology, Dhule, Maharashtra, India

Manish Patil*, Ketaki Patil* & Manashri Patil*
Department of Computer Engineering SVKM's Institute of Technology, Dhule, Maharashtra, India

Khalid Alfatmi*
Associate Professor, Department of Computer Engineering SVKM's Institute of Technology, Dhule, Maharashtra, India

ABSTRACT: Many physically challenged people struggle to obtain the necessary legal assistance. They can find it difficult to understand their rights or obtain aid within the current legal system. They require a specific AI-powered computer program to assist them with legal matters. So that they can receive the care they need and integrate into society like everyone else, this program should be simple for them to use and comprehend. The revolutionary effects of artificial intelligence (AI) on improving legal accessibility for people with disabilities or handicaps are explored in this system. AI technologies provide specialized solutions by addressing the special difficulties that each person encounters when traversing the legal system. These include adaptable forms for legal material, communication tools supported by AI, and the benefits they get from our laws. AI integration has the potential to level the playing field for those who have disabilities in the judicial system, ensuring fair access to justice by knowing their legal rights and benefits given to them by the government, just like employment laws, educational laws, traveling benefits, etc. for the physically challenged. The envisioned chatbot will take the form of a versatile application that can run on mobiles. It's being developed using Xamarin, ensuring smooth operation on Android. The system commences with voice input, which is then converted to text through speech-to-text technology. Following this, tokenization and keyword extraction techniques are applied to identify essential terms in the input. Subsequently, Natural Language Processing (NLP) is utilized to restructure the input sentence, enhancing its understanding. These sentences are used to formulate queries, subsequently employed to search in a manually created knowledge containing legal information and benefits applicable to individuals with physical challenges at last the system transforms the textual information into spoken words, ensuring that

---

*Corresponding Authors: vijaylaxmi.bittal@svkm.ac.in, sakshipingale29@gmail.com, makarand.shahade@svkm.ac.in, patilmanish2105@gmail.com, ketakipatil0911@gmail.com, mannmanashri@gmail.com and alfatmi.khalid@gmail.com

DOI: 10.1201/9781032644752-56

the results are not only accessible but also easily understandable for the users. This system provides a seamless and inclusive legal support experience to the physically challenged.

Keywords: Physically Challenged, AI Chatbot, Legal Assistance, Natural Language Processing, Artificial Intelligence

1 INTRODUCTION

Every one of us is aware of the new inventions taking place in this new world, but almost every invention has a drawback in that it is not designed for disabled people, they are designed in such a way that only a healthy person with proper knowledge can use them. Also, many a time it happens that a disabled person wants to perceive education but can't because they have misconceptions that a person with a disability is not allowed to perceive education they have such misconceptions regarding their jobs, marriages, etc. Also, they don't have any proper idea of the facilities and schemes that are specially designed for them by the government because they do not have any proper source where they can get rid of their doubts and queries. The AI will be available in the form of an application that can be easily downloaded on your mobile with a proper internet connection. The chatbot is designed with the help of Xamarin for the smooth functioning of the application. The chatbot is specially designed for disabled people but can be accessible to everyone. The chatbot will communicate with them and will try to guide them in the proper direction. A person can communicate with the chatbot in the form of text messages as well as voice. If a person is not able to type, he/she can take the facility of the mic which will make it easy for them to communicate with the AI and gain knowledge about the laws and facilities specially designed for them.

There are some existing legal chatbots but they are designed according to the ordinary people, the legal AI chatbot is specially designed to help disabled people and for their convenience. This will help them to understand the laws and benefits that are specially designed for them by the government. The UI of the chatbot is designed in such a way that it will be simple for everyone to use and get legal advice.

1.1 *Literature survey*

The current system of legal AI for disabled people is mainly focused on giving them legal advice about the legal laws or benefits given to disabled people, according to their percentage of disability.

In this study, the authors aim to develop an AI-powered chatbot (Amato *et al.* 2020) tailored for visually impaired individuals, facilitating voice-based communication and delivering pertinent responses through the integration of deep neural networks and speech-to-text/text-to-speech APIs. The chatbot's core purpose is to assist blind individuals in their daily lives, offering them a technology-driven means of communication and understanding their spoken input while providing informative feedback, particularly when traditional assistance is unavailable. In the future, this chatbot can be expanded to encompass additional functionalities, such as the capacity to independently procure essential items from designated websites, trigger emergency alerts, and offer reminders for medication schedules, meal times, and hydration needs.

In this proposed system (Morgan *et al.* 2018) authors introduced in this study leverage machine learning to assist children in understanding their legal rights, predicting user input, and collaborating with the Children's Legal Centre to gauge its effectiveness through classification metrics and user studies. This framework significantly improves children's access to legal guidance and their comprehension of legal rights by using machine learning to forecast both speech acts and legal categories based on user descriptions, as well as extracting named entities for case initiation, thereby establishing a comprehensive legal support system. As a forward-looking approach, this system holds the potential for extension to encompass various legal case types and cater to diverse user groups, promising broader legal assistance in the future.

In this proposed system the AI chatbot for sexual violence victims and survivors was developed using NLP techniques, achieving an 88.89 percent accuracy rate after training on 182 Thai Supreme Court cases. Its impact hinges on thoughtful design and utilization, aiming to provide legal information, privacy, and resources to victims. Evaluation should be carried out by the responsible organization. Looking ahead, "LAW-U" envisions an expansion of legal services, robust data security, multilingual support, collaboration with support networks, integration of AI advancements, user education, community-building, and ongoing research to enhance its capabilities and effectiveness.

The paper is about making sure that people with disabilities can use AI technology fairly and ethically (Joamets *et al.* 2021). They looked at laws and ethics, talked to different people, and checked how easy it is for disabled people to use AI. The paper also points out that there are big differences in how easy it is for disabled people to get AI tools in different parts of the world. AI can help, but there are still problems to solve. To fix these problems, we need to work together internationally, make rules for AI, create policies for disabled people, tell more people about it, and use different ideas from many fields. This way, we can make AI work better for everyone, including people with disabilities.

The "DoNotPay" (Kamya Pande 2023) is an AI-powered service that employs chatbots and decision trees to assist with various legal tasks, such as contesting parking tickets and generating legal documents. It continues to improve through machine learning and access to legal databases. "DoNotPay" has achieved success in challenging parking tickets, simplifying legal paperwork, and resolving various user issues, including flight delay compensation. Looking ahead, "DoNotPay" plans to expand its services to cover more legal matters in more locations, enhance its intelligence, offer financial assistance, improve usability, provide better security, and offer educational resources, including mobile and offline accessibility.

NyayGuru: This chatbot was invented by Adv (Dr.) Vibhuti Bhushan Sharma. A free AI (artificial intelligence) program called "NyayGuru" will assist with legal inquiries about Indian legislation. Knowing its limitations is crucial, though. This chatbot can offer general knowledge on legal topics, but it cannot substitute legal counsel from a licensed professional. Its instructive content is provided solely for educational reasons and is not intended to be used for client acquisition or advertising. It's critical to not rely only on the chatbot while making legal judgments because it's likely that it will provide inaccurate or insufficient responses.

After this literature survey, we came to know that there are no legal chatbots available for physically challenged people which completes all information about legal rights and benefits of physically challenged people. So, our current system of legal AI for disabled people is mainly focused on giving them legal advice about the legal laws or benefits given to disabled people.

2 METHODOLOGY

This proposed system is mainly for individuals with physical disabilities who may not possess a comprehensive understanding of their legal rights. The application functions by initially receiving user input in the form of voice as well as text, in the case of voice utilizing speech-to-text technology to transcribe spoken words into written text. Then the keywords from the text will be extracted. Subsequently, the system employs tokenization and Natural Language Processing (NLP) techniques to refine the input, ensuring that it is accurately processed. The system then searches for corresponding responses in its dataset, and once a match is found, it converts the text-based information into speech, delivering it to the user as voice output as well as text output. This approach aims to empower users by providing them with spoken explanations of their legal rights, thereby enhancing their accessibility to critical legal information. The proposed system architecture (Figure 2) shows the working of a mobile app that accepts voice input, processes it using text, preprocesses text, and searches user queries in a knowledge base containing relevant legal information and benefits for physically challenged individuals and then gives output in text as well as voice.

2.1 Chatbot design

- Design a user-friendly and accessible interface for the chatbot using Xamarin Forms.
- This chatbot is specially designed to assist physically challenged individuals in understanding and accessing their legal rights and benefits. It gives features of text and voice formats for users' input.

2.2 Data input

- Users can input their queries about the legal laws and benefits of physically challenged people through voice or text input.
- For voice input, use speech-to-text technology which will convert the spoken words into text.
- Handle issues such as variations in speech patterns, accent recognition, and language understanding.

2.3 Creation of knowledge base

- The system searches in a manually created knowledge base "Figure 1", that contains legal laws and benefits provided by the government for physically challenged people.
- This knowledge base is a repository of legal resources, schemas, and relevant data that apply to disabled people.

2.4 Pre-processing steps

- In pre-processing there are several steps in natural language processing such as tokenization, stop word removal, and lemmatization.
- First tokenize the input text to break it down into individual words or phrases then remove all stop words like 'and, the, in, or' from that user's statement.
- Identify keywords and entities in the user input query.
- By using the BERT algorithm, we detect the intent of the user input question.

2.5 Searching queries in knowledge base

- Maintain an organized knowledge base with legal data and benefits that apply to people with physical disabilities.
- Implement a search system that will enable easy access to and retrieval of data from the knowledge base.

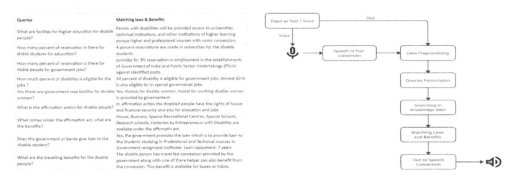

Figure 1.   Knowledge base.   Figure 2.   Architecture.

Query Processing:

- Use the identified keywords and entities to construct a query.
- Apply these keywords to the knowledge base to retrieve relevant legal information
- After finding the answer to the user's query, we will present them in two forms: text and voice to enhance comprehension.

3 RESULT AND DISCUSSION

As a result, the AI chatbot gives information about the laws and the benefits which has been asked by the user. To get the information for their queries about their rights and benefits which will help them to take a stand for themselves in society. The Figure 3 indicates how the user will be able to login into the app using their name and information. And in the next step, they can ask the queries they have regarding their rights and benefits. And to their query, the chatbot will answer according. In the given figure it indicates how the chatbot is working for the user to solve their queries.

Figure 3.   Result.

4 CONCLUSION AND FUTURE SCOPE

In conclusion, our AI chatbot is developed for the physically challenged in India; it links them with valuable knowledge and government facilities, laws, and benefits for them. While also cultivating awareness among family members, guardians, and the general public about available resources. This tool functions as a helpful aid for comprehending and making use of government programs and provisions, in addition to being a source of knowledge and offering advantages to individuals with disabilities. For the convenience of people with disabilities, we provide text and audio communication options. Considering the future, as we have already incorporated text-to-speech and speech-to-text and in the future, our goal is to improve accessibility by integrating sign language for the deaf and dumb people, and by making it multilingual for a broader reach.

Disabled people will be further empowered by ethical education, real-time legal updates, and customized multi-format outputs that include text, audio, and sign language. Furthermore, for a more inclusive and equitable global legal environment for individuals with disabilities, this legal AI can assess current laws, pinpoint legal gaps, and suggest advantageous policy adjustments.

REFERENCES

Amato, A., & Giacalone, M. (2020). Quality control in the process of data extraction. In Web, Artificial Intelligence and Network Applications: Proceedings of the Workshops of the 34th *International Conference On Advanced Information Networking and Applications* (WAINA-2020) (pp. 993–1002). Springer International Publishing.

Bittal, V., Bachhav, A. S., Shahade, M., Chavan, P. R., Nikam, B. A., & Pawar, A. A. (2023, June). Speech to image translation framework for teacher-student learning. In 2023 *3rd International Conference on Pervasive Computing and Social Networking* (ICPCSN) (pp. 1515–1520). IEEE.

Ellul, J. (2021). The technological society. *Ethics Guidelines for Trustworthy AI, European Commission High-Level Expert Group on Artificial Intelligence*, Vintage.

Ferraz, O. L. (2008). Poverty and human rights. *Oxford Journal of Legal Studies*, 28(3), 585–603.

Hassani, H., & Silva, E. S. (2023). The role of ChatGPT in data science: how ai-assisted conversational interfaces are revolutionizing the field. *Big Data and Cognitive Computing*, 7(2), 62.

Joamets, K., & Chochia, A. (2021). Access to artificial intelligence for persons with disabilities: legal and ethical questions concerning the application of trustworthy AI. *Acta Baltica Historiae et Philosophiae Scientiarum*, 9(1).

Kamya Pande, An AI lawyer is ready to tackle its first court case—should lawyers be worried, 2023, *Jumpstart*. Adv.(Dr) Vibhuti Bhushan Sharma, Nyayguru https://advocatevibhutibhushan.com/chatbot/ .

Morgan, J., Paiement, A., Seisenberger, M., Williams, J., & Wyner, A. (2018). A chatbot framework for the children's Legal centre. *In Legal knowledge and information systems* (pp. 205–209). IOS Press.

Ng, J., Haller, E., & Murray, A. (2022). The ethical chatbot: A viable solution to socio-legal issues. *Alternative Law Journal* 47(4): 308–313.

Ramaditiya, A., Rahmatia, S., Munawar, A., & Samijayani, O. N. (2021, June). Implementation chatbot whatsapp using python programming for broadcast and reply message automatically. In 2021 *International Symposium on Electronics and Smart Devices* (ISESD) (pp. 1–4). IEEE.

Socatiyanurak, V., Klangpornkun, N., Munthuli, A., Phienphanich, P., Kovudhikulrungsri, L., Saksakulkunakorn, N., ... & Tantibundhit, C. (2021). Law-u: Legal guidance through artificial intelligence chatbot for sexual violence victims and survivors. *IEEE Access*, 9, 131440–131461.

Wang, T., Yi, J., Fu, R., Tao, J., & Wen, Z. (2022). Campnet: Context-aware mask prediction for end-to-end text- based speech editing. *IEEE/ACM Transactions on Audio, Speech, and Language Processing*, 30, 2241–2254.

Zoe Kothari, ZIAI (Zoe's Inclusive AI) https://thelogicalindian.com/inclusivity/this-18-year-old-creates-app-for-differently-abled-to-help-them-lead-better-life-33966, 2019.

Advances in AI for Biomedical Instrumentation, Electronics and Computing – Sachan et al. (eds)
© 2024 The Author(s), ISBN 978-1-032-64298-7

Simplifying legal language: An AI-powered approach to enhance document accessibility

Kiran Somwanshi* & Khalid Alfatmi*
Department of Computer Engineering, SVKM Institute of Technology, Dhule, Maharashtra, India

Ashwini Vibhandik*, Darshana Karbhari*, Gagan Jarsodiwala* & Rahul Relan*
SVKM Institute of Technology, Dhule, Maharashtra, India

Makarand Shahade*
Department of Computer Engineering, SVKM Institute of Technology, Dhule, Maharashtra, India

ABSTRACT: This paper addresses the difficulties of complex and confusing legal documents by proposing an innovative AI-driven solution. Unlike pre-existing methods that merely summarize, the proposed system uses Natural Language Processing Models (NLP), Machine Learning Algorithm, and a legal jargon database to simplify the legal language. The goal is to enhance accessibility for the individuals without legal knowledge. This system identifies and replaces legal jargon while preserving the context, ensuring that the important information remains intact. The ultimate aim is to empower individuals, business, and nonprofits to comprehend legal documents without the need for specialization legal knowledge.

Keywords: Legal Document Simplification, Natural Language Processing (NLP), Artificial Intelligence (AI), Legal Jargon Removal, Machine Learning in Law, Legal Language Processing.

1   INTRODUCTION

Legal documents have always been known for their difficult language, making it hard for people and organizations to understand their rights, duties, and obligations. The use of jargon in these documents is a challenge, especially for those who are not familiar with the law. The intricacy involved does not make it difficult to access justice. Also raises the chances of misunderstandings leading to costly disputes, errors, or unintended consequences.

To address this issue our research focuses on developing a system for simplifying documents using intelligence. The main goal of this system is to bridge the gap between language and the need for documents that are easy to understand. Unlike existing solutions which summarize documents (Anand *et al.* 2022), our approach utilizes natural language processing (NLP) and machine learning to analyze texts. It eliminates jargon while preserving the meaning and context. By doing so, we aim to empower individuals and organizations by enabling them to comprehend documents clearly and comprehensively. This will contribute towards fostering a society that is better informed about the law.

This research paper provides an overview of our document simplification system, powered by AI. It explores how this technology has the potential to transform the accessibility of

---
*Corresponding Authors: kn.somwanshi@gmail.com, alfatmi.khalid@gmail.com, ashwinivibhandik18@gmail.com, karbharidarshana@gmail.com, gjarsodiwala@gmail.com, rahulrelan223@gmail.com and makarand.shahade@svkm.ac.in

DOI: 10.1201/9781032644752-57

documents, (Kanapala *et al.* 2019). making them easily understandable and accessible to all. Minimize the chances of misinterpretation in legal matters. By undertaking this approach, the objective is to facilitate the availability of knowledge enhance understanding and reduce the likelihood of misunderstandings, in affairs.

2 LITERATURE SURVEY

In their work the authors address the issue of pending cases, in India by introducing a method to normalize texts, which aims to improve models those are not specific to a particular field. They evaluate the effectiveness of BART and PEGASUS in both abstractive summarization of texts showing that their approach to text normalization significantly improves performance. This provides insights into the legal system. Moreover emphasizes the importance of automation in summarizing legal case documents highlighting how it saves time and effort for professionals. The passage also recognizes the challenges of applying automated summarization techniques across court systems due to variations in terminology, structure, and vocabulary. Additionally, a project described in aims to create a corpus of texts using a two-stage process involving tokenization and linguistic analysis. This corpus is intended to benefit both researchers working with entities well as those focusing on domain specific topics within natural language processing (NLP). In relation to this topic sheds light on the difficulties faced in text summarization such as the lack of framework problems with parsing and sentence alignment as well as challenges with information diffusion. These obstacles include maintaining sentence order and employing substitutions paraphrasing techniques and comprehensive representation methods – all presenting hurdles. Finally, there is an introduction, to SummCoder. It is a framework that focuses on text summarization. This framework utilizes auto encoders and sentence embeddings. Through analysis it has been proven to be robust and even superior to supervised methods, particularly when applied to the DUC 2002 and Spinosi's datasets. This suggests that it has the potential for generalization, across domains.

3 PROPOSED SYSTEM

The proposed system offers an innovative approach to take an input in image, pdf or text format and provide output in the simplified language which can be understand by any person belonging to non-lawyer background.

- Dataset Compilation
  Text Data: We gathered a set of documents for the text input, which included different formats, like plain text and rich text. These documents covered areas of law. - Image and PDF Data: To compile a dataset, we collected legal documents in both image and PDF formats. Our dataset consists of documents, image files, and generated PDFs.
- OCR Integration
  For handling image and PDF inputs, we incorporated Optical Character Recognition (OCR) technology into our system. This allowed us to convert textual formats like scanned images or PDFs into text that can be read by machines. To ensure extraction of the text, we utilized accepted OCR libraries and tools.
- Preprocessing and Standardization
  The text extracted from the OCR received preparation and standardization. This included the removal of any elements addressing any irregularities in formatting and ensuring a structure for the text.
- AI Model Development
  We created an NLP model that was designed to analyze and comprehend text extracted from OCR. This model underwent training on a diverse dataset encompassing text, images, and PDF documents enabling it to handle types of data sources.

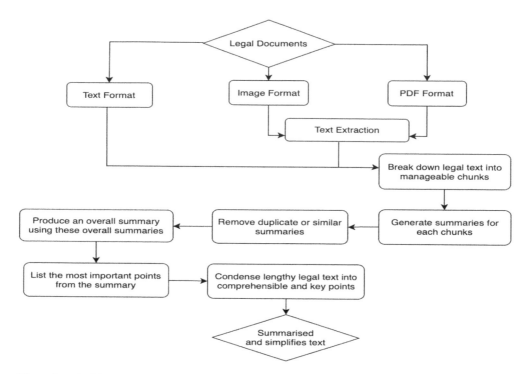

Figure 1. Architecture.

- Data Alignment

  To ensure consistency across inputs such as text, images, and PDFs, we implemented a data alignment process. This involved mapping the text extracted through OCR to their areas in the images or PDFs. The aim was to guarantee that the output of the system maintained the documents' layout and formatting accurately.

- Evaluation and Validation

  We assessed the system's efficiency by considering performance metrics such as OCR accuracy, text alignment, and the quality of the documents. These metrics provided us with a way to gauge how effectively the system can process and simplify documents in various formats.

- System Deployment

  The completed legal document simplification system, which utilizes AI technology and includes OCR functionality for processing image and PDF files, has been successfully deployed on a server. Users now have the option to conveniently upload types of documents such as text, images, or PDFs for effortless simplification, offering them flexibility and a user-friendly experience.

- Ethical Considerations

  Throughout the project, we placed emphasis on considerations. We made sure to follow data privacy and security protocols while handling user-uploaded documents, particularly when dealing with images and PDF inputs. This methodology details the collection, OCR integration, preprocessing, AI model development, evaluation, user feed-back, and ethical considerations specific to the handling of text, image, and PDF inputs in your legal document simplification system. The comprehensive approach enables the system to cater to various input formats, enhancing accessibility and understanding for a wider user base.

4 RESULTS AND DISCUSSION

In this proposed system, as per the bar graph representation, we illustrate the word count rates in both the input legal documents and the corresponding simplified versions, as processed by our AI-based legal document simplification system. The graph showcases a clear contrast between the complexity of legal jargon-rich input text and the streamlined simplicity of the output text. This visual representation underscores the effectiveness of our NLP-based approach in reducing the verbosity and improving the overall accessibility of legal documents, making them more comprehensible for a wider audience.

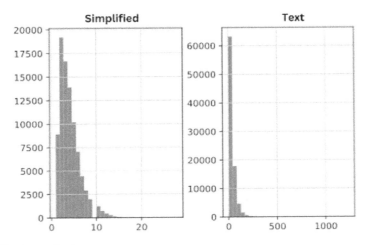

Figure 2. Words count rate.

5 CONCLUSIONS

In conclusion, our AI-based legal document simplification system holds significant promise in addressing the long-standing challenge of legal document accessibility. Through the creation of two crucial datasets and the development of advanced natural language processing techniques, our system successfully bridges the gap between complex legal language and user-friendly content. By providing concise and plain explanations for legal jargon and simplified summaries of legal acts, the system empowers individuals, businesses, and government agencies to better understand and navigate the legal landscape. This research contributes to democratizing legal information and fostering legal literacy, reducing misunderstandings, and enhancing access to justice.

THIS LEASE AGREEMENT (the "Agreement"), entered into on this _____ day of _____, 20_____, by and between the undersigned LESSOR, hereinafter referred to as "Landlord," and LESSEE, hereinafter referred to as "Tenant," collectively referred to as the "Parties."

WHEREAS, Landlord is the legal owner of certain real property located at [Complete Property Address], here in after referred to as the "Premises."

WHEREAS, Tenant desires to lease the Premises from Landlord for residential purposes, subject to the terms and conditions herein set forth.

NOW, THEREFORE, for and in consideration of the covenants and agreements contained herein and other good and valuable consideration, the Parties hereto agree as follows:

Figure 3. Input.

THIS RENT AGREEMENT (the "Agreement"), created on this _____ day of _____, 20_____, between the undersigned OWNER, hereinafter referred to as "Landlord," and RENTER, hereinafter referred to as "Tenant," collectively referred to as the "Parties."

WHEREAS, Landlord is the lawful owner of specific real estate located at [Full Property Address], hereinafter referred to as the "Place."

WHEREAS, Tenant wishes to rent the Place from Landlord for housing purposes, subject to the conditions outlined herein.

NOW, THEREFORE, for and in consideration of the promises and agreements contained herein and other valuable considerations, the Parties herein agree as follows:

Figure 4.   Output.

REFERENCES

Anand, D., & Wagh, R. (2022). Effective deep learning approaches for summarization of legal texts. *Journal of King Saud University-Computer and Information Sciences*, 34(5), 2141–2150.

Garimella, A., Sancheti, A., Aggarwal, V., Ganesh, A., Chhaya, N., & Kambhatla, N. (2022, December). Text simplification for legal domain:{i} nsights and challenges. *In Proceedings of the Natural Legal Language Processing Workshop 2022* (pp. 296–304).

Jain, D., Borah, M. D., & Biswas, A. (2021). Summarization of legal documents: Where are we now and the way forward. *Computer Science Review*, 40, 100388.

Kanapala, A., Pal, S., & Pamula, R. (2019). Text summarization from legal documents: a survey. *Artificial Intelligence Review*, 51, 371–402.

Pandya, V. (2019). Automatic text summarization of legal cases: A hybrid approach. *arXiv preprint arXiv:1908.09119*.

Parikh, V., Mathur, V., Mehta, P., Mittal, N., & Majumder, P. (2021). Lawsum: A weakly supervised approach for indian legal document summarization. *arXiv preprint arXiv:2110.01188*.

Advances in AI for Biomedical Instrumentation, Electronics and Computing – Sachan et al. (eds)
© 2024 The Author(s), ISBN 978-1-032-64298-7

Fuel theft detection system with SMS alert using microcontroller

Vipin Kumar Verma, Nilesh Yadav, Pankaj Gupta, Prateek Dhar Dubey,
Purushottam Mani Tripathi & Vikram Vishwakarma
KIET Group of Institutions, Delhi-NCR Ghaziabad

ABSTRACT: The widespread use of fossil fuels for transportation and power generation has led to an increased focus on preventing fuel theft, which represents a significant economic burden and security risk. This research paper presents a new approach to solve this problem by developing a GSM-based fuel theft detection system with SMS notification using a microcontroller. The widespread use of fossil fuels for transportation and power generation has led to an increased focus on preventing fuel theft, which represents a significant economic burden and security risk. This research paper presents a new approach to solve this problem by developing a GSM-based fuel theft detection system with SMS notification using a microcontroller. The Poised system uses wireless communication to monitor the vehicle's position. The increasing prevalence of fuel theft poses a significant challenge to various industries, leading to financial losses and security concerns. This research paper presents a robust and innovative solution, "GSM-Based Fuel Detection System with SMS Alert", designed to address this urgent problem. Using the integration of microcontroller technology and GSM communication, this system offers real-time monitoring and instant warning capabilities, enabling early detection and prevention of fuel theft.

Keywords: GSM module, microcontroller, sensor

1 INTRODUCTION

Global dependence on fossil fuels for transportation, industrial processes, and energy production underscores the critical importance of protecting these limited and valuable resources. Fuel theft has emerged as a growing problem affecting various industries, including logistics, agriculture, and construction, resulting in significant economic losses and security threats (Mutiawani *et al.* 2018). Traditional methods of securing fuel storage and transportation have proven to be insufficient for effective theft prevention. In response to this challenge, the research presented here focuses on the development of a GSM-Based Fuel Detection System with an SMS alert using a microcontroller - a solution aimed at significantly increasing the security and monitoring capabilities associated with fuel storage and transportation (Bakkar *et al.* 2019).

The system is based on a microcontroller and consists of its Global System for Mobile Communications (GSM). This project uses a GSM modem. It comes with a GSM modem. A SIM card uses the same communication process as a regular phone (Ajith *et al.* 2018). This work includes the design and construction of a remote fuel level sensor and subsequent remote fuel level monitoring. Monitoring is done by sending messages from a compatible mobile phone. Messages are sent to the owner at regular intervals. His other feature in this study is locking his vehicle with a remote password. We can also activate the siren in case of theft.

In conclusion, it can be said that the GSM-based Fuel Theft Detection System with SMS Alert using a Microcontroller represents a significant advance in the case of fuel theft detection and prevention (Bahgat *et al.* 2019). Its potential to reduce financial losses, increase safety, and promote sustainable resource management makes it a valuable addition to various industries. This research paper aims to give a detailed understanding of the design, functionality, and

DOI: 10.1201/9781032644752-58

performance of the system with the ultimate goal of promoting widespread adoption and contributing to the mitigation of fuel theft issues worldwide (Krishnaprasad *et al.* 2021).

2 LITERATURE SURVEY

As gasoline is one of the most valuable commodities within the globe, gas or fuel theft is increasing dramatically around the arena. Most gasoline is stolen during transport between industries. This robbery is difficult to catch due to numerous logistical difficulties. It is recommended to take safety precautions to prevent oil from leaking out of the tank. These systems must regularly check the oil in the tanker's tank. If the oil level falls below the specified level, a warning is sent to the driver and the oil company. The first layer of protection is the keyboard security system based on two-factor authentication. Authorities can track the location of the truck using a smartphone app and a GPS module. A security system with IOT (Internet of Things) support that ensures communication between the sender and recipient makes the proposed solution new before distributing the fuel (Bhilegaonkar *et al.* 2020).

The message alerts the owner or user of the vehicle in case of theft. Moreover, by this technology, we can record the messages and it helps us to find the stolen vehicle at any time using GSM and GPS (Saini *et al* 2021).

Key fobs produced in the last decade have been used to enter modern cars. The car starts when the door is closed and turns off when the lock button on the key fob is pressed, but the system is not secure. If your key fob is stolen, anyone can unlock your car. That's why 3D gesture keychains are used to prevent car theft. Make a gesture in the air to unlock your car using the 3D Animated Keychain. This includes two different 3D movements. One is the regular behavior of the vehicle owner, and the other is the guest descriptions of other users. It also uses encryption techniques to ensure secure data communication (Shakil *et al.* 2015).

In addition, the system comes with keyboard instructions to activate the power line and an alarm that allows you to adjust sensitivity, fuel cutoff, and GPS fence.

3 COMPONENTS

3.1 *Microcontroller (Arduino UNO)*

Arduino Uno is a popular microcontroller board widely used by the DIY and electronics enthusiast community. It is based on the ATmega328P microcontroller chip and is part of the Arduino family of microcontroller boards. Arduino Uno is powered by the ATmega328P microcontroller, which is an 8-bit AVR microcontroller with 32KB flash memory, 2KB SRAM, and 1KB EEPROM for storage. The ATmega328P in the Arduino Uno operates at 16 MHz, allowing it to execute instructions quickly. There are 14 input/output pins in total. 6, available as PWM (Pulse Width Modulation) output. These pins can be used for interfacing with various sensors, LEDs, motors, and other digital devices. The Arduino Uno has 6 analog input pins, labeled A0 through A5, which can be used to read analog voltage values from sensors or potentiometers.

3.2 *Bread board*

A breadboard, also known as a prototyping board or solderless breadboard, is a fundamental tool used in electronics to build and test electronic circuits quickly and without soldering. Breadboards usually have two sets of terminal strips on either side, known as the power rails. These power rails are often labeled as "+ (plus)" and "- (minus)".

The holes on the breadboard are interconnected in a specific pattern, allowing you to easily connect components and create electrical connections. Each row of holes is connected horizontally, and each column is connected vertically. The middle section of the breadboard is typically divided into two halves, each with its own set of interconnected rows and columns.

3.3 *Jumper wires*

Jumper wires are a vital component in electronics criterion and circuit making. They are used to create electrical connections between various components, pins, and holes on a breadboard or other prototyping platforms

Types: Jumper wires come in various types to suit different needs and applications:

Male-to-Male (M-M): Both ends have pins or connectors, making them suitable for connecting components with pins, such as microcontrollers and sensors.

Male-to-Female (M-F): One end has pins or connectors, while the other end has female receptacles.

M-F jumper wires are commonly used to connect components to a breadboard.

Female-to-Female (F-F): Both ends have female connectors, making them useful for extending or connecting components with female headers.

3.4 *GPS module*

It is a hardware component that allows devices to determine their precise location by receiving signals from satellites in the GPS constellation. GPS modules are commonly used in various applications, including navigation systems, vehicle tracking, location-based services, and IoT (Internet of Things) devices.

GPS receivers, including GPS modules, use these signals to calculate their position and provide accurate geographic coordinates (latitude, longitude, and sometimes altitude). It commonly uses a range of applications such as

Receiver: A GPS module contains a GPS receiver that receives signals from multiple satellites simultaneously.

Antenna: GPS modules often include a built-in or external antenna to capture the GPS signals effectively.

Serial Communication: Most GPS modules communicate with a microcontroller or host system using UART (Universal Asynchronous Receiver-Transmitter) serial communication.

NMEA Sentences: GPS modules typically output data in NMEA (National Marine Electronics Association) sentences, which are ASCII-encoded strings that contain information such as latitude, longitude, altitude, speed, and time.

Power Supply: GPS modules usually require a power supply voltage, typically in the range of 3.3V to 5V.

3.5 *LCD (16*2)*

LCD 16x2 (16 characters per 2-line liquid crystal display) is a popular alphanumeric display module commonly used in electronic projects, embedded systems, and various applications to display text information.

The 16x2 LCD consists of a rectangular display panel with 16-character positions arranged in two rows, each capable of displaying up to 16 alphanumeric characters (letters, numbers, symbols). It usually has a backlight for visibility in low-light conditions.

3.6 *Buzzer*

It is an electrical component used to produce an alert, often in the form of a loud, audible tone or alarm. Buzzer components are commonly found in various electronic devices and systems to provide audio feedback, warnings, and notifications.

3.7 *Ignition lock*

An ignition lock on a fuel tank, also known as a fuel tank lock or fuel cap lock, is a security mechanism designed to prevent unauthorized access to a vehicle's fuel tank. Unlike the ignition lock on the vehicle's dashboard, which controls the engine's starting and electrical systems, a fuel tank lock is specifically intended to protect the fuel supply from theft or tampering.

3.8 *Fuel level sensor*

We have used the Level sensor (also known as a level transducer or level transmitter to detect the level of fuel in the tank. If the fuel level goes down the threshold level, the sensor gives the microcontroller a specific signal.

3.9 *Vibration sensor*

It has been used to capture the vibration in the system when the signal is generated specifically if the fuel level is constantly attenuating by the indication of the ignition key.

3.10 *Relay*

A relay controls one electrical circuit by opening and closing contacts in another electrical circuit. It is used to compute the time between the opening and closing of a contract. Low-voltage signals are used to control relays that operate high-voltage circuits.

4 FLOW CHART

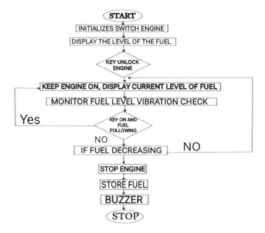

Figure 1. Flow chart of the proposed system.

5 FUNCTIONING OF THE SYSTEM

GSM-based anti-gas theft systems using microcontrollers are a useful tool that can solve an important problem for businesses and organizations that rely on gas stations. This new system combines hardware, data analysis, and remote communication to protect fuel storage and reduce losses due to theft or illegal use. The core of the system is a microcontroller, a module that can communicate with various sensors. One of the main sensors used in this setup is the level of the fuel sensor located in the fuel tank. Various types of oil level sensors like ultrasonic sensors or float sensors are used to accurately measure the oil level.

In summary, a GSM-based fuel theft system using microcontrollers provides a solution to pressing problems such as fuel theft and illegal usage. Through a combination of hardware, data analysis, and remote communication, this technology secures fuel storage, reduces financial losses, and provides peace of mind for businesses and organizations that rely on oil storage facilities. Its adaptability and scalability make it a key asset in industries such as logistics, transportation, agriculture, and energy.

6   CIRCUIT DIAGRAM

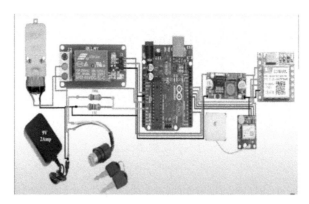

Figure 2.   Circuit diagram.

7   CONCLUSION

In this study, we presented a powerful and efficient car gasoline theft detection and tracking tool that is tailored to cope with the chronic problems of fuel theft in the automotive business. Utilizing a combination of sensor technology, conversational devices, and gadget control algorithms, the system has demonstrated excessive accuracy and reliability in detecting gas theft cases in real-time. On-site alerting and remote, monitoring features ensure well-timed intervention, minimize losses, and increase fuel and vehicle protection.

Through rigorous experimentation and evaluation, our findings underscore the accuracy, reliability, cost-effectiveness, and scalability of the gadget. In conclusion, the fuel theft detection and monitoring machine presented here represents a critical leap forward in the prevention of gas theft, a chronic problem facing the automobile industry.

REFERENCES

Ajith, H. V. S., & Kiran, P. S. Fuel Theft detection system. *International Journal of Research in Engineering, Science and Management*; Volume-1, Issue-10, October-2018; www.ijresm.com| ISSN (Online): 2581–5792.

Bakkar, M., & Alazab, A. (2019, May). Designing security intelligent agent for petrol theft prevention. *In 2019 Cybersecurity and Cyberforensics Conference* (CCC) (pp. 123–128). IEEE.

Bhilegaonkar, P., Patil, R., Belekar, A., Gujarathi, M., & Sondkar, S. (2020, February). Fuel theft prevention system. *In 2020 International Conference on Industry 4.0 Technology* (I4Tech) (pp. 126–130). IEEE.

Bahgat, M. M. (2019, April). Enhanced IoT-based online access control system for vehicles in truck-loading fuels terminals. *In 2019 IEEE 6th International Conference on Industrial Engineering and Applications* (ICIEA) (pp. 765–769). IEEE.

Krishnaprasad, C., Joseph, C. A., Sarath, I. S., & Manohar, O. R. (2021, May). A novel low-cost theft detection system for two wheelers with minimum carbon foot print. *In 2021 2nd International Conference for Emerging Technology* (INCET) (pp. 1–5). IEEE.

Mutiawani, V., Rahmany, S., & Abidin, T. F. (2018, September). Anti-theft vehicle monitoring and tracking android application using firebase as web service. *In 2018 International Conference on Electrical Engineering and Informatics* (ICELTICs) (pp. 72–77). IEEE.

Shakil, M., Rashid, M., & Patil, A. B. (2015, February). Automobile theft prevention using 3D gesture key fob and cryptography. *In 2015 International Conference on Computing Communication Control and Automation* (pp. 306–309). IEEE.

Saini, M., & Khan, S. (2021, January). GSM based fuel theft detection. *In 2021 1st International Conference on Power Electronics and Energy* (ICPEE) (pp. 1–6). IEEE.

Advances in AI for Biomedical Instrumentation, Electronics and Computing – Sachan et al. (eds)
© 2024 The Author(s), ISBN 978-1-032-64298-7

Unmasking attacker identity behind the VPN

A.S. Awate & B.N. Nandwalkar
Professor, SVKM Institute of Technology, Dhule, Maharashtra, India

M.R. Shahade, D.B. Mali, H.V. Patil, H.R. Waghare & H.R. Patil
SVKM Institute of Technology, Dhule, Maharashtra, India

ABSTRACT: In the dynamic landscape of cybersecurity, this paper addresses a critical challenge: the ability of attackers to conceal their identities behind VPNs and proxy servers. Our dual objectives are to develop advanced techniques for pinpointing these elusive actors and to empower the cybersecurity community with actionable insights. Leveraging the MERN stack, cloud services, open-source tools and specialized libraries, we unveil attackers' identities. In a digital era, valuing privacy and security, this paper lays the foundation for an in-depth exploration, offering solutions to enhance online safety. Through strategic use of technology and insights, we contribute to a more secure online environment.

Keywords: VPN, proxy servers, cyber threats, cybersecurity, bash scripting, token, IP geo-location, malicious actors, data privacy, security

1 INTRODUCTION

In an era dominated by digital expansion, the demand for robust online security and anonymity intensifies. Amid the vast possibilities of the internet, cybersecurity faces intricate challenges, notably in identifying malicious actors exploiting Virtual Private Networks (VPNs) and proxy servers for anonymity. This paper delves into the core of digital security, addressing challenges posed by adversaries using VPNs and proxies. Our dual objectives are to develop advanced techniques for uncovering concealed actors' identities and empower the cybersecurity community with actionable insights.

VPNs and proxies offer privacy and security but also provide a veil for malicious actors. This duality necessitates refining methodologies to distinguish legitimate users from threats. Our exploration encompasses online anonymity, techniques employed by attackers, and the challenges of attribution, considering legal constraints and evolving VPN technologies. Beyond identifying challenges, the paper illuminates solutions and innovations available to the cybersecurity community, emphasising ethical considerations. In this complex digital era, our paper reflects a collective commitment to enhancing online security and fostering a safer, more secure digital environment.

2 LITERATURE SURVEY

In computer science, security emphasizedis critical for safeguarding systems from malicious software (Kalangiat *et al.* 2021). The main goal is to ensure authorised access to applications. Malicious actors often exploit IP spoofing to hide their locations by falsifying source

DOI: 10.1201/9781032644752-59

addresses. To counter these threats, IP traceback mechanisms trace the origin of an IP address through a network of nodes. However, existing traceback methods face challenges and limited success rates. The proposed hybrid IP traceback approach, while recognizing existing methods' advantages and limitations, shows promise in enhancing attacker identification. This hybrid integrates protocols, surpassing existing methods and potentially improving security threat mitigation.

The escalating demand for cybersecurity amid rapid internet growth. They advocate for VPNs, utilising encryption to protect data, but acknowledge the vulnerability of public networks to malicious hackers (Aravind *et al.* 2023). The project focuses on countering fraud through a multi-tiered detection system, showcasing efficacy in identifying diverse proxy methods. Results suggest proxies and VPNs significantly fortify corporate security against potential fraudulent data collection.

Tackling escalating DDoS threats, propose an advanced hybrid traceback method. Using a compact 16-bit field in the IP header enhances marking significantly (Subash *et al.* 2023). Optimized logging, with a max storage of 384 KB, and a balanced path reconstruction technique effectively addresses packet fragmentation challenges.

Tor network vulnerabilities, employing ACOFT (Ant Colony Optimization-based Filter for Tor IP addresses) were addressed (Sreelaja 2022). ACOFT efficiently categorizes search space, reducing the time to identify incoming Tor IP addresses. Its $O(\log 2s)$ time complexity, where 's' is Tor IP addresses in the search space, distinguishes it. Importantly, 's' is less than 'm,' the count of Tor IP addresses in the database. ACOFT's efficacy in filtering Tor IP addresses presents a significant deterrent against malicious activities within the Tor network.

The PCI Express (Peripheral Component Interconnect) is a high-performance serial interconnect protocol (Vaidya *et al.* 2022). Offering superior bandwidth over older bus architectures, PCIe is a versatile choice for diverse applications. The paper focuses on creating a verification IP for the physical layer of PCI Express, using the Universal Verification Methodology (UVM) in System Verilog. This ensures a robust verification process, enhancing the reliability and functionality of PCI Express across applications.

In summary, previous research has addressed IP spoofing challenges and malicious online activities. Our system emerges from these insights, introducing an innovative approach to boost online security. By using advanced technologies, we aim to deliver a holistic solution that empowers owners of web assets to increase the protection of their data and privacy.

3 PROPOSED SYSTEM

The proposed system offers an innovative approach to web asset protection by combining elements of a Honeypot-like system with advanced monitoring and notification capabilities.

3.1 *Objectives*

This research aims to achieve key objectives in fortifying website and database security. It seeks to establish a user-friendly platform for owners to monitor and manage assets efficiently. The development of a token-based monitoring mechanism is central, providing real-time alerts for unauthorized access and enabling prompt responses. The research also focuses on empowering users through user-friendly tools for tracking and responding to security incidents. Additionally, integrating geolocation data enhances threat identification based on geographic origins, adding a strategic layer to the overall security framework. Together, these objectives create a comprehensive and accessible approach to bolstering digital asset security.

3.2 *Architecture*

In crafting the system architecture, we opted for the MERN stack, a dynamic combination of MongoDB, Express.js, React, and Node.js. This well-established stack is renowned for its prowess in web application development. Leveraging the strengths of each component, we've not only prioritized scalability and modularity but also imbued the system with a robust framework. A noteworthy facet of our architecture involves adept side scripting for the generation of tokens, infusing a touch of finesse into the system's overall functionality.

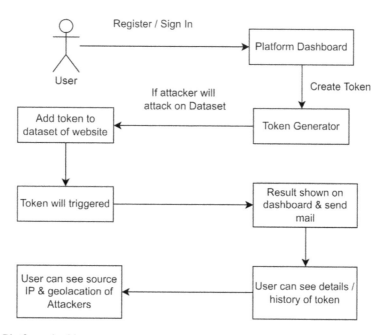

Figure 1.   Platform Architecture.

3.3 *System flow*

The system follows a phased approach, commencing with user registration and extending to continuous monitoring for proactive responses to potential security threats. This ensures a comprehensive and dynamic strategy, empowering users with essential tools in the ever-changing cybersecurity landscape.

3.3.1 *User registration and authentication*
Manages onboarding, registration, and secure user authentication, ensuring authorised access to the platform. Backend endpoints are created using Express.js and Node.js, securely storing credentials in a MongoDB database.

3.3.2 *Main dashboard*
Serves as the central user interface, providing essential account details and website security status. Constructed with React, it communicates with the backend through RESTful API endpoints from Express.js.

3.3.3 *Token generator*
Enables users to create tokens, choose file types, and specify email recipients. Implemented as a React component, it communicates with the backend through API calls for token generation.

3.3.4 *Security monitoring and notification*
Monitors token access in real-time, updating the website's security status. Immediate email notifications are triggered for unauthorized access using Node.js for monitoring and email alerts.

3.3.5 *History and event logging*
Records token access history, capturing key data points like access times, file types, names, and source IP addresses. Access history is stored in a MongoDB database, and queried by the front end to display user activity.

3.3.6 *Geolocation integration*
Integrates geolocation data to enhance security, associating IP addresses with physical locations. Utilizes a third-party IP geolocation service or API for obtaining source IP addresses' geolocation data.

4 RESULTS AND DISCUSSION

In this proposed system, as per the visual representation, we encounter a detailed record of token accesses. The history unfolds as a chronological log, capturing access timestamps, the associated file types, and the corresponding source IP addresses. This all-encompassing log proves invaluable for thorough tracking, comprehensive evaluation, and extracting insights from various access events. Its importance is underscored by its substantial contribution to the improvement of security analysis and monitoring initiatives.

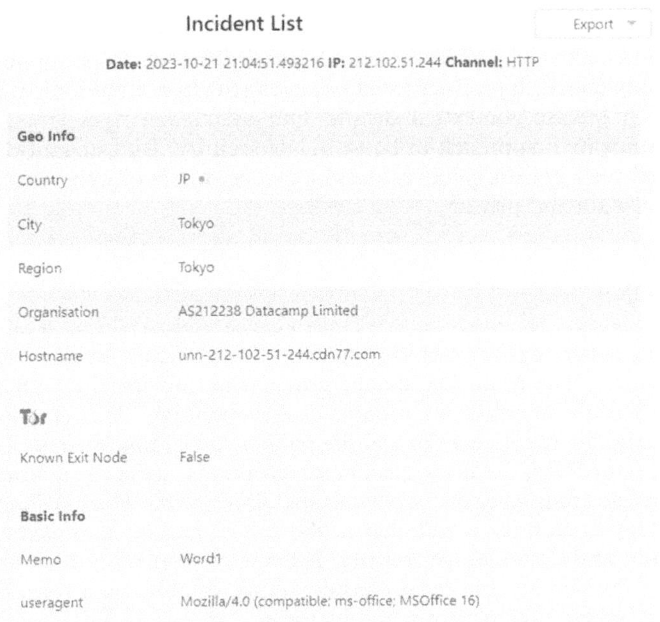

Figure 2.   Detail history of triggered token.

This visual representation unveils the geographic origins of attempted file access by potential attackers. Examining these locations yields valuable insights into the worldwide distribution of potential threats, offering a foundation for implementing precise and targeted security measures.

Incident Map

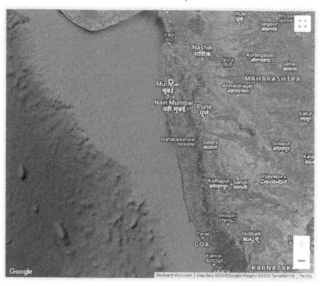

Figure 3. Geographic location of the attacker who accessed the token.

Along with it, an automated email notification feature is present, triggered by a token access event. This real-time alert system serves as a crucial mechanism, swiftly informing relevant users or administrators about system activity. The immediate notification capability enables users to proactively address potential security breaches, preserving the system's integrity. A comprehensive log of activated tokens, providing a chronological overview of access events with precise contextual details. Our system emerges from these insights, introducing an innovative approach to boost online security. By using advanced technologies, we aim to deliver a holistic solution that empowers owners of web assets to increase the protection of their data and privacy.

5 CONCLUSIONS

In conclusion, our cybersecurity platform represents a significant stride in reinforcing web and database security. The Main Dashboard provides instant insights into security status, featuring a user-friendly interface for prompt decision-making. This emphasis on accessibility ensures a seamless experience for website owners navigating their security landscape. The Token Generator Page, featuring customisable options, serves as a strategic deterrent, complemented by the comprehensive archival capabilities of the History Page.

Our commitment to security is reflected in the robust measures implemented, covering user authentication and data storage security. Adherence to privacy laws and regulations, alongside explicit disclaimers and legal considerations, establishes a foundation of transparency and trust between the platform and its users.

Looking forward, the future scope of our proposed system holds the promise of refining precision and exploring direct countermeasures. This includes the potential for launching targeted responses against attackers' devices using advanced techniques like Juicy or malicious tokens. These avenues for future development position our platform at the forefront of cybersecurity, aligning with the dynamic landscape of emerging threats and technological advancements.

REFERENCES

Aravind, T. N., Mukundh, A., & Vijayakumar, R. (2023, April). Tracing Ip addresses behind Vpn/Proxy Servers. *In 2023 International Conference on Networking and Communications (ICNWC)* (pp. 1–10). IEEE.

Kalangi, R. R., Sundar, P. S., Maloji, S., & Ahammad, S. H. (2021, November). A hybrid IP trace back mechanism to pinpoint the attacker. *In 2021 Fifth International Conference on I-SMAC (IoT in Social, Mobile, Analytics and Cloud)(I-SMAC)* (pp. 1613–1618). IEEE.

Sreelaja, N. K. (2022, December). Ant Colony Optimization based approach to filter IP Address from TOR network. *In 2022 IEEE Conference on Interdisciplinary Approaches in Technology and Management for Social Innovation (IATMSI)* (pp. 1–6). IEEE.

Subash, A., Danny, A., & Vijayalakshmi, M. (2023, February). An enhanced hybrid scheme for IP Traceback. *In 2023 4th International Conference on Innovative Trends in Information Technology (ICITIIT)* (pp. 1–5). IEEE.

Vaidya, V. N., Ingale, V., & Gokhale, A. (2022, October). Development of verification IP of physical layer of PCIe. In *2022 IEEE 3rd Global Conference for Advancement in Technology (GCAT)* (pp. 1). IEEE.

Advances in AI for Biomedical Instrumentation, Electronics and Computing – Sachan et al. (eds)
© 2024 The Author(s), ISBN 978-1-032-64298-7

e-Nidan: Autism spectrum disorder detection using machine learning

Ashish Awate*

Assistant Professor, Department of Computer Engineering SVKM's Institute of Technology, Dhule, Maharashtra, India

Krutika Yeola*

Department of Computer Engineering SVKM's Institute of Technology, Dhule, Maharashtra, India

Makarand Shahade*

Assistant Professor, Department of Computer Engineering SVKM's Institute of Technology, Dhule, Maharashtra, India

Vaishnavee Patil*, Mayuri Vispute* & Hemshri Amrutkar*

Department of Computer Engineering SVKM's Institute of Technology, Dhule, Maharashtra, India

Bhushan Nandwalkar*

Associate Professor, Department of Computer Engineering SVKM's Institute of Technology, Dhule, Maharashtra, India

ABSTRACT: Autism Spectrum Disorder (ASD) is a developmental disorder characterized by social, communication, and behavioral challenges. People with ASD may struggle with interaction, exhibit repetitive behaviors, and have specific interests. Early diagnosis is crucial, typically occurring in early childhood through a comprehensive evaluation of a child's development, history, and behavior. While there is no cure for ASD, various treatments, such as Applied Behavior Analysis and therapy, can enhance social, communication, and behavioral skills. ASD's symptoms vary among individuals and may include sensory processing difficulties, resistance to change, and a typical eating or sleeping habits. The average age of diagnosis is around 0 - 6 years, but some individuals may be diagnosed later due to less obvious symptoms in younger children. ASD, or Autism Spectrum Disorder, is a developmental condition associated with significant social, communication, and behavioral challenges. To solve this problem, this paper proposes an ML-powered system called eNidan to detect early symptoms of Autism Spectrum Disorder (ASD). Three algorithms are evaluated: KNN, Decision Tree, and Logistic Regression. Accuracy of decision tree algorithm is 64 percent, Accuracy of SVM algorithm is 58 percent, Accuracy of logistic regression algorithm is 67 percent. Logistic Regression is the best algorithm, with an accuracy of 67 percent. All three algorithms are appended to the model, and the accuracy of each algorithm is compared. Logistic Regression is found to be the best algorithm, and all results are fetch to the final algorithm. The paper focuses on historical data that is available and suitable for training data. This algorithm is downloaded onto the ML algorithm and trained on historical data collected from Kaggle. The parameters in the dataset are roll number, student age, speech delay, learning disorder, sex, family ASD, and ASD train. So final result is logistic regression is best algorithm for this e-Nidan System.

---

*Corresponding Authors: ashish.awate@svkm.ac.in, krutikayeola2002@gmail.com, makarand.shahade@svkm.ac.in, Vaishnavee.patil0191@gmail.com, visputemayuri6@gmail.com, amrutkarhemshri@gmail.com and bhushan.nandwalkar@gmail.com

    DOI: 10.1201/9781032644752-60

1 INTRODUCTION

Autism Spectrum Disorder (ASD) is a developmental disorder characterized by social, communication, and behavioral challenges. People with ASD may struggle with interaction, exhibit repetitive behaviors, and have specific interests. Early diagnosis is crucial, typically occurring in early childhood through a comprehensive evaluation of a child's development, history, and behavior. While there is no cure for ASD, various treatments, such as Applied Behavior Analysis and therapy, can enhance social, communication, and behavioral skills.

ASD's symptoms vary among individuals and may include sensory processing difficulties, resistance to change, and a typical eating or sleeping habits. The average age of diagnosis is around 0 - 6 years, but some individuals may be diagnosed later due to less obvious symptoms in younger children. ASD, or Autism Spectrum Disorder, is a developmental condition associated with significant social, communication, and behavioral challenges.

2 LITERATURE SURVEY

The conventional method for diagnosing Autism Spectrum Disorder (ASD) relies on observing the behavioral characteristics of the individual, but it has limitations, especially in differentiating ASD from other conditions.

Autism is a developmental disorder that typically emerges in childhood, with the first signs often appearing as early as age two. Most children receive a formal diagnosis after the age of four. (Autism is a condition that affects people of all ethnic and racial backgrounds, but it is more commonly observed in boys than in girls(American Psychiatric Association. *et al.* 2013) Early diagnosis is crucial because it allows for timely intervention and treatment. Identifying autism in children as young as possible enables the implementation of interventions aimed at improvingtheir behavioral and communication skills, ultimately enhancing their overall development and quality of life (DiRienz *et al.* 2016) Early autism detection has seen significant advancements, primarily driven by developments in machine learning. including image categorization of video based data, facial expression recognition, and more.

Deep learning techniques such as Convolutional Neural Networks (CNN), Artificial Neural Networks (ANN), and Recurrent Neural Networks (RNN) have been effectively employed for this purpose (A., De Lume *et al.* 2018) Xiaoxiao utilized a deep neural network for fMRI based identification of ASD with an 85.3 K. These approaches showcase the diversity of techniques in ASD diagnosis, offering promising results in this field. (Islam *et al.* 2019) compared the performance of various classification algorithms, including Random Forest (RF), for diagnosing ASD in children. significance of facial expressions in social interaction and applied computational analysis to understand how children with ASD produce facial expressions (Raj, Masood *et al.* 2020) they initially addressed with a CNN based model, achieving approximately 98.30M.S.

Mythili *et al.*, used various classification techniques, including Support Vector Machine to analyze autism in students. Their dataset included attributes like language data, and behavior data (Fahad Saeed *et al.* 2021) the use of facial expressions in autism diagnosis, as people with autism often have difficulty displaying appropriate facial expressions. Computational analysis of facial expressions has gained attention, aiming to overcome the limitations of human perception (Tiwari *et al.* 2022) developed a computer vision-based system that relies on measuring a child's response to language as an indicator of autism. in early autism detection has made substantial progress through deep learning, computer vision, and traditional machine learning techniques. Early diagnosis in medicine is crucial, and these methods hold promise for improving outcomes in autism spectrum.

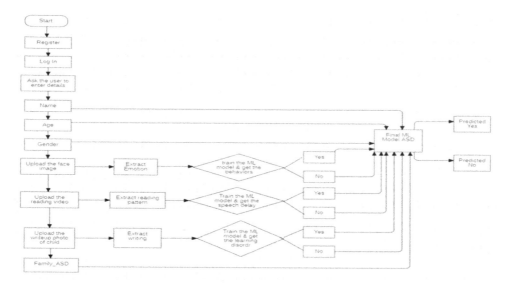

Figure 1. Flow of the autism spectrum disorder system.

3 RESULT

In our research, We have trained the model by taking the first historical data set. The parameters in the data set are roll number, student age, speech delay, learning disorder, sex, family ASD, ASD train. Int and object data types are used for this. After that the data sets are split 80 percent are used for training and 20 percent for testing. We have used ml's algorithm to maximize the accuracy of the model. Such as knn, decision tree, Svm and logistic regression.

In knn algorithm the best accuracy is found from mean accuracy and std accuracy. After that grid search is done. Jaccard score is 0.46, F1 score is 0.57 and accuracy course is 0.67.

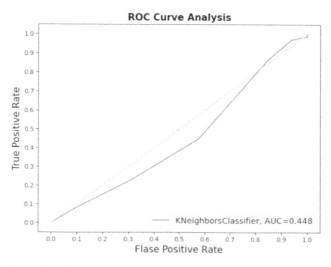

Figure 2. KNeighbors classifier.

In knn algorithm the best accuracy is found from mean accuracy and std accuracy. After that grid search is done. Jaccard score is 0.46, F1 score is 0.57 and accuracy course is 0.67.

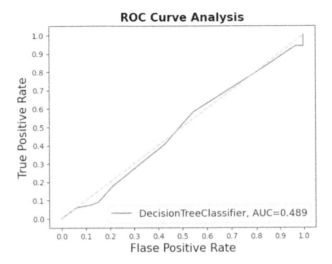

Figure 3.   Decision tree classifier.

In Svm algorithm its hyperparameter is extracted by grids.fit and its jaccard score is 0.42 by Svm predict. Then f1 score is 0.56 and accuracy is 0.58.

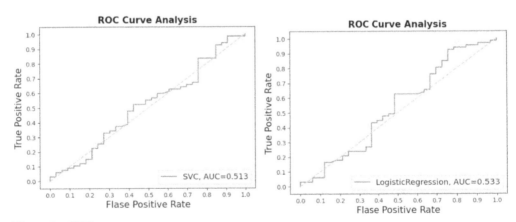

Figure 4.   SVC.

Figure 5.   Logistic regression.

By using k-fold cross validation knn, decision tree, svm, logistic regression algorithm has been appended to the model and the accuracy of the algorithm has been compared and the best algorithm logistic regression has been obtained from it.

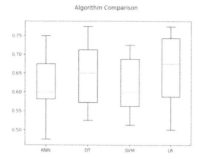

Figure 6. Algorithm comparison.

4 CONCLUSION

In conclusion, our machine learning-powered system, aimed at early detection of Autism Spectrum Disorder, underwent a thorough evaluation of three key algorithms: K-Nearest Neighbors (KNN), Decision Tree, and Logistic Regression. Logistic Regression emerged as the most accurate, achieving a 67 percent accuracy rate, surpassing KNN (64 percent) and Support Vector Machine (SVM) (58 percent). Despite this success, future efforts will explore maximizing accuracy with KNN, SVM, and Decision Tree. The project heavily relies on historical data, including student details, speech delay, learning disorders, gender, family ASD history, and child ASD presence, sourced from Kaggle, enhancing the system's ability to make informed ASD symptom assessments.

5 FUTURE SCOPE

The development of an automated system for assessing children's understanding, writing, and reading levels, utilizing machine learning for Autism Spectrum Disorder detection, opens exciting avenues for future exploration. Enhancements can include integrating speech analysis, sensor data, and eye-tracking technology for a comprehensive view of a child's abilities. Real-time monitoring and user-friendly interfaces can provide immediate insights and recommendations for educators and caregivers, fostering dynamic support strategies.

REFERENCES

A., De Lume, F.,Leo, M., Carcagni, P., Distante, C., Spagnolo, P., Mazzeo, P. L., Rosato, A. C., ... & Lecciso, F. (2018). Computational assessment of facial expression production in ASD children. *Sensors, 18*(11), 3993.

American Psychiatric Association,A. (Ed.). (2013). *Diagnostic and Statistical Manual of Mental Disorders: DSM-III.* American Psychiatric Assoc.

DiRienz, M.,Baio, J. (2016). Prevalence of autism spectrum disorder among children aged 8 years-autism and developmental disabilities monitoring network, 11 sites, United States, 2010

Fahad Saeed Eslami, T., Almuqhim, F., Raiker, J. S., & Saeed, F. (2021). Machine learning methods for diagnosing autism spectrum disorder and attention-deficit/hyperactivity disorder using functional and structural MRI: A survey. *Frontiers in Neuroinformatics*, 62.

Islam, M. N. &Omar, K. S., Mondal, P., Khan, N. S., Rizvi, M. R. K., & Islam, M. N. (2019, February). A machine learning approach to predict autism spectrum disorder. In *2019 International Conference On Electrical, Computer And Communication Engineering (ECCE)* (pp. 1–6). IEEE

Raj,Masood, S., &S. (2020). Analysis and detection of autism spectrum disorder using machine learning techniques. *Procedia Computer Science, 167*, 994–1004.

Tiwari, R. K.,&Jee, G., Chouhan, S., Gourisaria, M. K.,(2022, February). Detection of autism spectrum disorder through orthogonal decomposition and pearson correlation for feature selection. *In 2021 4th International Conference on Recent Trends in Computer Science and Technology (ICRTCST)* (pp. 103–109). IEEE.

Advances in AI for Biomedical Instrumentation, Electronics and Computing – Sachan et al. (eds)
© 2024 The Author(s), ISBN 978-1-032-64298-7

Real-time automated fabric defect detection system

Umakant Mandawkar, Makarand Shahade, Samruddhi Wadekar, Chetan Kachhava,
Yash Patil & Sakshi Mandwekar
Department of Computer Engineering, SVKM'S Institute of Technology, Dhule, India

ABSTRACT: Even today the Textile Industries are using manpower to detect and classify the fabric defect, the process of depending on human inspectors for fabric defect detection can be considered as not an optimized way since it is time-consuming as well as an error prone process. To tackle this issue, here we propose a groundbreaking development- A real-time, automated fabric defect detection system. Making the most use of all the available cutting-edge advanced image processing techniques and machine learning algorithms, the proposed system will identify, locate and categorize a wide range of fabric defect where precision and accuracy will be the hallmarks of the proposed methodology The system will operate in real time, capturing high-resolution fabric surface images, leading to precise and swift defect identification. Its automated and real-time capabilities render it to be an essential instrument in textile manufacturing as it will enhance its quality by diminishing the requirement of human inspectors, and evolving overall inspection process efficiency. Comprehensive and thorough testing across various fabric materials will prove its high accuracy rates in defect detection. It will possess the power to revolutionize the textile industry by enhancing product quality, cutting costs and boosting productivity.

1 INTRODUCTION

The global textile industry stands as a paramount pillar in the economies of a number of nations across the world. India is among the ones who boasts one of the largest and most influential textile sectors that contribute to the country's GDP. Despite advancement, there are challenges the textile production faces in terms of quality of fabric. Fabric defects such as nods, barres, stains, snags, pilling, holes, color bleeding, uneven texture, seam imperfection, print misallegation, puckering, skewing can affect the quality of fabric which will significantly lead to reduced sales and decreased serviceability. These types of defects are often caused by low quality yarn, improper handling of equipment and processes by human operators Even though there has been a lot of improvisation in the textile industry, the thing that has not been looked up on is manual inspection of fabric for defect detection which may hamper the growing impact of the textile industry leading to increased wastage, higher expenditure and decreased profitability.

The textile industry relies on specific standardized dimensions for manufacturing various textile products, which are crucial for ensuring consistency and quality. These dimensions play a pivotal role in the production of items such as shirts, curtains, and bedsheets. The utilization of 36-inch fabric for shirts, 58-inch fabric for certain shirt types, 70-inch for curtains and bedsheets, and 120-inch for larger bedsheets serves as a fundamental aspect of the industry's operational framework. These standardized measurements not only guarantee the uniformity and aesthetics of the final products but also impact the efficiency and performance of the production machinery. Understanding and adhering to these dimensions is a critical factor in textile manufacturing, contributing to the industry's overall success and market competitiveness, but doing this manually is time consuming and may also be not

DOI: 10.1201/9781032644752-61

precise. Within the textile industry, the final stage of fabric manufacturing entails the mending process, a critical phase where meticulous inspection is essential. Traditionally, this step has relied heavily on a substantial workforce, necessitating significant manpower to scrutinize fabrics for defects and inconsistencies. However, in light of the challenges posed by this labor-intensive approach, the textile industry is now exploring innovative solutions. This research endeavors to develop a sophisticated system designed to revolutionize the mending process. By harnessing advanced technology and automation, the objective is to streamline operations, drastically reduce the labor requirements, and simultaneously enhance the precision and quality of fabric inspection. This system will have potential impact on efficiency, resource utilization, and product quality that will change the landscape of fabric manufacturing and will prove to be a promising solution to a long-standing industry challenge.

2 RELATED WORK

Additional endeavors to streamline feature extraction networks and reduce computational expenses have introduced convolution kernel decomposition and bottleneck methods. These techniques replace standard convolutions with dilated convolutions for focused feature learning, resulting in decreased parameter demands and improved efficiency. To create multi-scale fusion features, researchers have innovatively applied skip-connections to combine high-level semantic features with detailed lower-level features. Skip-connections facilitate direct information transfer between network layers, enhancing the network's ability to detect flaws of varying sizes (Le Cun *et al.* 2010).

As per the research by Jun Wuet and colleagues, Textile production involves multiple intricate stages, including weaving, knitting, dyeing, and finishing. Throughout these phases, the emergence of fabric flaws necessitates precise detection to ensure the final product meets stringent quality standards. Historically, manual inspection has been the preferred method for identifying fabric flaws, despite its time-consuming and costly nature. Recent years have witnessed a growing interest in autonomous systems aimed at bolstering the efficiency and accuracy of textile production by flaw detection. Nonetheless, a significant challenge arises from the substantial computational resources required by such systems, which can present financial and logistical obstacles, especially for smaller organizations. In response, researchers have proposed the use of memory chips with limited capacity, making autonomous systems a more cost-effective option for smaller entities (Wu *et al.* 2021).

The cornerstone of an efficient fabric flaw detection system is the development of an algorithm that achieves precise detection with minimal computational demand. An effective approach involves employing a wide and lightweight network structure based on the well-established Faster R-CNN algorithm, known for its object detection capabilities. Faster R-CNN employs a convolutional neural network (CNN) to extract image features for object identification. Researchers have suggested enhancing the feature extraction network by incorporating a dilated convolution module, a unique technique that widens a neuron's receptive field without increasing the parameter count. This enhancement bolsters the network's ability to detect flaws of various sizes (Odumakinde 2022).

Finally, researchers have introduced anchor frames of different sizes to support multi-scale fabric fault detection. These anchor frames play a vital role in target detection algorithms by identifying potential objects in images. The inclusion of various-sized anchor frames significantly enhances the algorithm's precision in detecting faults of varying sizes.

3 PROPOSED METHODOLOGY

The methodology of fabric defect detection and classification that we proposed is structured in three specific phases. The first phase pertains to hardware, involving the setup and configuration of the necessary equipment and sensors. The next phase is based on the training of modules, where machine learning algorithms are strategized to develop a robust defect detection model. Lastly, the third phase focuses on the seamless integration of both the hardware and software components, resulting in a comprehensive and functional fabric defect detection and classification system. These three coordinated phases collectively form the base of our approach, ensuring the efficient and accurate identification of fabric flaws in the textile manufacturing process.

3.1 *Software*

In the methodology we proposed, we set out on a challenging journey to create a useful fabric defect detection and classification system. The software phase of our approach comprises three prominent architectural elements: Convolutional Neural Network (CNN), ResNet-18, and ResNet-34. These architectures are efficiently utilized to develop a robust machine learning module that excels in learning and differentiating a wide array of fabric defects with exceptional precision. The machine learning module which has been thoroughly trained on a variety of datasets, is the brain of this system. Its proficiency in detecting and categorizing fabric defects is integral to our methodology's success. In addition, we introduce a user interface that offers a seamless and intuitive experience. This interface provides real-time feedback on the detected defect, including its type and precise position on the fabric. Users can navigate the system effortlessly, making informed decisions about quality control in textile manufacturing. The combination of our in-depth training , three best architectural features and the intuitive interface results in a fabric defect detection system that is effective as well as efficient, leading to the improvement in the textile production.

3.2 *Hardware*

In the proposed fabric defect detection methodology, the hardware components are designed to provide a compelling and efficient solution. These components include a Camera Module, essential for capturing high-resolution images of fabric surfaces. To precisely locate defects, we employ a Rotary encoder, ensuring accurate defect positioning. The Microcontroller serves as the system's central processing unit, orchestrating the defect detection process. IoT Connectivity Module facilitates real-time data transmission and remote monitoring. Power Supply ensures uninterrupted system operation. Adequate Storage capacity is integrated to retain captured images and data. An Enclosure safeguards the system from environmental factors. Mounting Hardware enables secure installation. These components together form a robust hardware framework that enhances the quality and effectiveness of fabric defect detection in the textile manufacturing process.

3.3 *Integration of Hardware and Software*

The final step in our fabric flaw detecting system is the integration of both software and hardware components. A complete solution for textile quality control is realized through the skillful fusion of state-of-the-art technology and painstaking design. The physical foundation of the system is made up of the hardware components, which include the rugged enclosure for environmental protection, the sturdy camera module for high-resolution imaging, the rotary encoder for accurate defect localization, the microcontroller for central processing, the IoT connectivity module for real-time data transmission, a power supply for continuous operation, plenty of storage capacity, and dependable mounting hardware. The software components and the hardware infrastructure work together without any interruptions. Fabric defects can be accurately recognized and categorized using the help of a trained machine learning module, which was developed using CNN, ResNet-18, and ResNet-36. The User Interface acts as connection between user and the system by giving real-time input on type of defect and its location. The entire system works with ease, making sure that fabric flaws and its location are correctly detected, improving the quality and efficiency of the manufacturing process. Fundamentally, the merging of hardware and software components results in a drastic improvement in the textile industry, creating the groundwork for better product quality.

4 DATA SET

The dataset employed in our research is a substantial and meticulously curated resource, characterized by its diversity and comprehensiveness. With a size of 4 GB, it comprises a total of 48,000 distinct records, each contributing to a rich and representative sample of fabric defect scenarios. The images within the dataset are of two dimensions, either 32x32 or 64x64, reflecting real-world fabric inspection conditions.

One of the standout features of this dataset is its classification into six distinct defect categories, each crucial in the context of fabric quality control:

1. Good: This category represents fabric that is in pristine condition, devoid of any defects.
2. Color: It includes instances where defects pertain to color irregularities in the fabric.
3. Cut: Here, the dataset covers fabric defects related to cuts in the material.
4. Hole: The dataset accounts for fabric defects characterized by holes in the fabric.
5. Thread: Instances where threads are coming off the fabric are captured within this category.
6. Metal Contamination: The dataset encompasses fabric contamination issues involving the presence of metal.

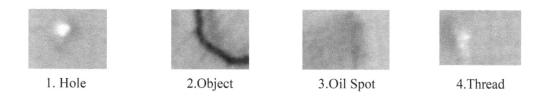

| 1. Hole | 2.Object | 3.Oil Spot | 4.Thread |

In addition to the original images, our dataset is further enriched with images generated after applying eight distinct rotations: 0, 20, 40, 60, 80, 100, 120, and 140 degrees. This rotational diversity is pivotal in training our system to find the defects from multiple angles, enhancing its efficiency.

The dataset is then divided into train and test subsets, each of which contains patches that were produced randomly. These patches come from sets of source pictures that don't overlap, providing a stable and balanced framework for developing and testing our fabric defect detection method. This dataset's meticulous design and comprehensive coverage equip our research with a robust and versatile resource to develop and assess our fabric defect detection capabilities. Follow are the insights of the variety of defects dataset contains:

5 RESULT

In our research, we've done an intensive training on a diverse data set that comprised of a wide range of fabric defects, each distinguished by factors such as color and intensity, texture, thread structure, pattern consistency, surface contours, defect geometry, fractal analysis, edge detection, orientation, and symmetry. This dataset's diversity played an important role in facilitating our machine learning module to accurately differentiate between various types of defects.

In our project, we evaluated three distinguished machine learning models: Convolutional Neural Networks (CNN), ResNet-18, and ResNet-34. These models gave revealing findings on raining the module on the thorough dataset. Notably, ResNet-18 emerged as the best choice algorithm, with an impressive accuracy rate of 98%. In comparison, ResNet-34 and CNN exhibited accuracy scores of 95% and 82%, respectively. The exceptional performance of ResNet-18 in effectively capturing all the aspects of fabric defects, hence it was selected as the preferred algorithm for final real time testing. Figure 1 is the architecture of the model that we found to have the highest accuracy. This image is the pictorial representation of the flow of how the features are extracted at each layer in the model.

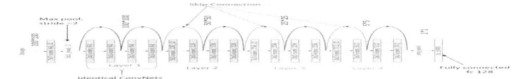

Figure 1.   ResNet18 architecture.

These results demonstrate the efficacy of our approach in accurately identifying and classifying fabric defects. By leveraging the power of ResNet-18 and the rich dataset, we have achieved a high level of accuracy, reinforcing the practical utility of our system in real-world textile quality control and inspection scenarios.

Models	Accuracy	Train Loss	Validation Loss
CNN	0.82	1.303	1.333
ResNet18	0.98	1.286	26.913
ResNet34	0.95	1.333	21.588

Figure 2.   Accuracy score of different ML models.

6   CONCLUSION

In conclusion, our research has advanced the field of fabric defect detection by developing a robust and accurate machine learning-based system. The extensive dataset we used in analysis, covering a wide range of fabric defects, played an important role in training our models to accurately identify and differentiate various types of defects based on color and intensity, texture, thread structure, pattern consistency, surface contours, defect geometry, fractal analysis, edge detection, orientation, and symmetry. Our vast experiments with three machine learning models, including Convolutional Neural Networks (CNN), ResNet-18, and ResNet-34, found that ResNet-18 emerged as the top-performing algorithm, achieving an impressive accuracy rate of 98%.

The efficient accuracy of our system, particularly with ResNet-18, emphasizes its potential for real-world applications in textile quality control and examination of fabric. By successfully tackling the complex task of fabric defect detection, our system has the potential to enhance the efficiency and accuracy of textile manufacturing processes, reduce costs associated with manual inspection, and improve the overall quality of textile products.

In conclusion, our research provides a significant contribution to the textile industry by offering a practical and effective solution for fabric defect detection. The implications of our

331

work extend beyond the laboratory, impacting real-world textile manufacturing and quality assurance processes, ultimately leading to improved product quality and operational efficiency.

7 FUTURE SCOPE

Future research can enhance the system by implementing advanced preprocessing techniques, exploring complex neural network models like VQVAE, and expanding the dataset to encompass a wider array off a bric types and defects. These efforts will improve the system's accuracy and versatility in real-world textile manufacturing scenarios.

REFERENCES

He, K., Gkioxari, G., Dollár, P., & Girshick, R. (2017). Mask r-CNN. In *Proceedings of the IEEE International Conference on Computer Vision* (pp. 2961–2969).

Le Cun Y, Kavukcuoglu K, Farabet C, (2010). Convolutional Networks and Applications in Vision[C]. *Proceedings of 2010 IEEE International Symposium on Circuits and Systems.* IEEE, 2010: 253–256.

Odumakinde, E. (2022, March 19). Everything about Mask R-CNN: A Beginner's Guide-viso.ai.viso.ai. https://viso.ai/deeplearning/mask-r-cnn

Redmon J, Divvala S, Girshick R, *et al.* You only look once: Unified, real-time object detection[C]. *Proceedings of the IEEE Conference on Computer Vision and Pattern Recognition.* 2016: 779–788.

Wu, J., Le, J., Xiao, Z., Zhang, F., Geng, L., Liu, Y., & Wang, W. (2021). Automatic fabric defect detection using a wide-and-light network. *Applied Intelligence*, 51(7), 4945–4961.

Zhou, H., Jang, B., Chen, Y., & Troendle, D. (2020, September). Exploring faster RCNN for fabric defect detection. In *2020 Third International Conference on Artificial Intelligence for Industries (AI4I)* (pp. 52–55). IEEE.

Advances in AI for Biomedical Instrumentation, Electronics and Computing – Sachan et al. (eds)
© 2024 The Author(s), ISBN 978-1-032-64298-7

Comprehensive analysis of communication quality: Signal to noise ratio with respect to bit error rate for nanosatellite beacon

Vidushi Pandey, Ayush Singh, Abhishek Sharma, Gati Saraswat, Amrita Singh,
Ayush Yadav & Ruchita Gautam
Electronics and Communication Engineering, KIET Group of Institutions, Delhi-NCR, Ghaziabad

ABSTRACT: In recent years, scientists have intensified efforts towards developing nano-satellites due to enhanced safety, affordability, and compact designs, reducing shipping costs. These small satellites serve various purposes, including data collection from space. Despite their advantages, creating nanosatellites presents challenges. One critical hurdle is establishing effective communication links between these satellites and ground stations. This connectivity is vital for optimal satellite utilization in earth's orbit. Uplink and downlink establishment stands as a primary concern, significantly impacting the connection between ground stations and nanosatellites. A recent publication delves into a statistical analysis of beacon, a nanosatellite device employing uplink and downlink communications. The focus lies in deriving the relationship between signal-to-noise ratio (SNR) and bit error rate (BER) from the obtained results. Overcoming these communication hurdles is crucial for the seamless functioning and data retrieval from these increasingly popular nanosatellite systems.

Keywords: Uplink, downlink, nanosatellite, VHF/UHF, SDR, LNA, BER, SNR

1 INTRODUCTION

The emergence of nanosatellites, with their compact structure and light weight ranging from 1 to 10 kilograms, marks a pivotal shift in the space exploration landscape (Sese, R. M., 2022). These small-scale satellites have become increasingly popular for their cost-effectiveness and ease of deployment, providing a practical avenue for a wide range of scientific research and space exploration endeavors. This study delves into the nuances of nanosatellite communication, emphasizing the critical roles of uplink and downlink systems that are essential for their effective functioning. Nanosatellites offer a viable and less expensive alternative for space-based activities, especially appealing to educational institutions, small enterprises, and researchers. Several enhancements in inter-satellite communication have emerged in literature, particularly using new wireless standards such as Ultra-Wideband IEEE 802.15.4a. However, under the student nanosatellite program, small satellites are typically bound to use free VHF/UHF bands for data transfer. There are several enhancements in the literature about inter-satellite communication using various new wireless standards such as Ultra-wideband IEEE 802.15.4a but as per the student nanosatellite program, small satellites are bound to use free VHF UHF band for data transfer (Sharma *et al.* 2021). Accessibility is key to democratizing space research and allowing for broader participation in space-based studies (Rementeria 2022). The frequency of 433 MHz was specifically chosen for this study's communications due to its established effectiveness and reliability. This frequency band, falling within the amateur radio (ham) band, provides a free-to-use resource that further enhances the accessibility and feasibility of nanosatellite projects, especially for educational and amateur purposes.

DOI: 10.1201/9781032644752-62

The incorporation of the 433 MHz frequency is also instrumental in the context of student nanosatellite programs (Aigul *et al.* 2022). These programs, which encourage and facilitate students to design, build, and operate their nanosatellites, benefiting immensely from the availability of this free ham band. It allows students to engage in real-world satellite communication without the need for expensive licensing, thereby fostering practical learning and innovation in the field of space technology. In terms of technical communication, the manuscript explores the underlying process of uplink and downlink processes for student nanosatellites. The uplink process involves transmitting signals from Earth to the satellite, while downlink refers to the reception of signals from the satellite back to Earth. These processes are integral for the command and control of the satellite, as well as for the retrieval of data collected in space. Furthermore, the manuscript also expands into viewing nanosatellites as components of distributed computing in a perspective that necessitates addressing challenges around power management and coordination with satellite constellations, for secure data processing. These aspects are essential for optimizing the functionality and utility of nanosatellites.

The Beacon, crucial for nanosatellite communication, operates at 433 MHz in the free amateur band, ensuring cost-effective and reliable communication. It transmits vital telemetry and receives Earth commands, essential for real-time satellite monitoring and control. Its design focuses on long-range, low-power communication, crucial in space's limited-resource environment. This makes the Beacon invaluable for cost-sensitive educational and research nanosatellite projects. Further emphasizing its utility, the Beacon's efficient long-range communication, is vital for data transmission and command reception (Li, Z., & He, T., 2018), is key to the success of resource-constrained space missions. To evaluate the efficacy of these communication systems, the research involved experimental setups with the Beacon, incorporating a Low Noise Amplifier (LNA). The aim was to assess the impact of the LNA on signal quality and range. Various distance-based tests were conducted to mimic different operational environments and to gauge the communication system's performance under diverse conditions. This manuscript showcases the efficacy of a 433 MHz RSI module and LNA in enhancing signal quality. The use of SDR for SNR vs. BER analysis confirms the LNA's pivotal role in improving transmission over various distances (Popescu *et al.* 2016). The antenna with an improved beamforming also plays a dominant role in better transmission quality (Sharma *et al.* 2021), but all the practical observation was performed over a monopole antenna.

The manuscript also presents an experimental investigation focused on enhancing the nanosatellite Beacon system's performance. The experimental setup comprised two principal components: a transmitting module emulating the Beacon system and a receiving module acting as a ground station analog. Crucially, a Low Noise Amplifier (LNA) was integrated into the receiving module. This addition aimed to examine the LNA's effectiveness in boosting signal quality and expanding the receiving range, thereby optimizing the overall communication efficacy of the system. The incorporation of the 433 MHz frequency is also instrumental in the context of student nanosatellite programs (Aigul *et al.* 2022). These programs, which encourage and facilitate students to design, build, and operate their own nanosatellites, benefit immensely from the availability of this free ham band. It allows students to engage in real-world satellite communication without the need for expensive licensing, thereby fostering practical learning and innovation in the field of space technology. In terms of technical communication, the manuscript explores the underlying process of uplink and downlink processes for student nanosatellites. The uplink process involves transmitting signals from Earth to the satellite, while downlink refers to the reception of signals from the satellite back to Earth. These processes are integral for the command and control of the satellite, as well as for the retrieval of data collected in space. Furthermore, the manuscript also expands into viewing nanosatellites as components of a distributed computing in perspective that necessitates addressing challenges around power management and coordination with satellite constellations, for secure data processing. These aspects are essential for optimizing the functionality and utility of nanosatellites.

2 BLOCK DIAGRAM

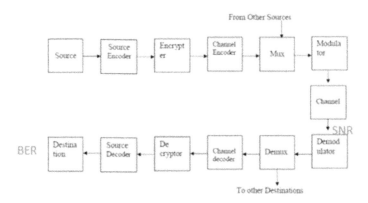

3 MATHEMATICAL MODELING

In our project, we have seamlessly integrated the 433 MHz RSI wireless transmitter receiver module into beacon's circuitry, leveraging its remarkable efficiency and compact form (Da Silva *et al.* 2017). This module stands out for its minimal energy demands, typically consuming just 11 mA, and operates effectively within a 1.5–5 Volt range. Its seamless compatibility with CMOS and TTL sources made it a straightforward addition to our design. The module's robust ASK modulation technique and its ability to transmit data up to 3 kHz were instrumental in expanding the beacon's communication range, which we found to be reliable up to 50 meters. This enhancement has been crucial in diversifying the potential applications of our beacon from enhancing remote control systems to fortifying wireless security frameworks (Taha, M. A., 2012). To To enhance the transmission range of the system, a low-noise amplifier (LNA) is employed at the receiver end. The signal transmission is tested in an AWGN environment. The SNR is calculated with the help of software-defined radio (SDR)-based spectrum analyzer. Mathematically, SNR BER Relation in the presence of Additive white gaussian noise (AWGN) is expressed in Eq. (1) (Babu, A. S., & Rao, D. K. S., 2011).

$$SNR = \frac{1}{2} erfc\left(\sqrt{SNR}\right)$$

Where, erfc is the complementary error function having relation with $Q(x) = \frac{1}{2} erfc\left(\frac{x}{\sqrt{2}}\right)$ (Sharma, A., & Sharma, S. K., 2019). For best packaging relative with the size of nanosatellite, monopole antenna is used for better transmission and reception.

4 RESULTS AND DISCUSSIONS

In the manuscript, software-defined radio (SDR) is employed to conduct a comprehensive analysis of signals performance under varying conditions. The flexibility and adaptability of SDR allowed us to simulate and measure real-world scenarios with precision. Figure 1 showcases a comparative analysis of signal-to-noise ratio (SNR) versus bit error rate (BER) for both the transmitter and receiver, set at distances of 7 meters and 10 meters. This comparison highlights the impact of distance on signal integrity and error rates. Further, Figure 2 delves into the influence of a low-noise amplifier (LNA) at a 7-meter distance, presenting a side-by-side evaluation of SNR vs. BER performance with and without the incorporation of an LNA. The enhancement in signal quality with the LNA is notably apparent.

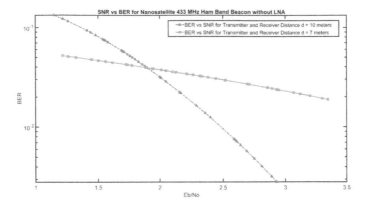

Figure 1. A comparison between the SNR vs BER for the transmitter and receiver distance "d" 7 meters & 10 meters.

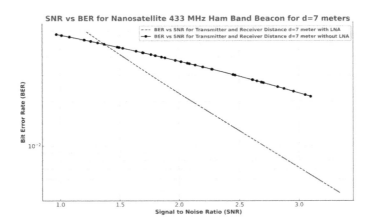

Figure 2. Comparison between the SNR vs BER for the distance d = 7 meters for without LNA and with LNA.

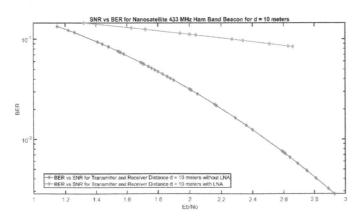

Figure 3. Comparison between the SNR vs BER for the distance d = 10 meters for without LNA and with LNA.

Similarly, Figure 3 extends this comparison to a 10-meter distance, providing insights into how an LNA can significantly mitigate signal degradation over longer distances. These findings, facilitated by the versatility of SDR, offer valuable insights into optimizing communication systems for varying operational ranges and conditions. From the figure while using LNA the transmission quality is increased for each distance d =7 meters and 10 meters comparison to the readings without LNA.

Conclusion: This manuscript presents the design of a beacon for student nanosatellite using amateur radio band with a 433 MHz UHF frequency. The use of a 433 MHz RSI wireless transmitter receiver module highlights significant advancements in space technology accessibility with LNA for greater Employing a monopole antenna and Low Noise Amplifier (LNA), the research successfully demonstrates enhanced signal quality and range, crucial for effective nanosatellite operations. The utilization of software-defined radio (SDR) for experimental analysis substantiates the positive impact of LNA on signal performance over varied distances. These findings underscore the potential of cost-effective and efficient communication systems for expanding the scope of educational and small-scale space exploration projects.

REFERENCES

Aigul, K., Altay, A., Yevgeniya, D., Bekbolat, M., & Zhadyra, O. (2022). Improvement of signal reception reliability at satellite spectrum monitoring system. *IEEE Access*, 10, 101399–101407.

Babu, A. S., & Rao, D. K. S. (2011). Evaluation of BER for AWGN, Rayleigh and Rician fading channels under various modulation schemes. *International Journal of Computer Applications*, 26(9), 23–28.

Da Silva, L. C. B., Bernardo, R. M., De Oliveira, H. A., & Rosa, P. F. (2017, May). Multi-UAV agent- based coordination for persistent surveillance with dynamic priorities. In *2017 International Conference on Military Technologies (ICMT)* (pp. 765–771). IEEE.

Popescu, O., Harris, J. S., & Popescu, D. C. (2016, March). Designing the communication sub-system for nanosatellite CubeSat missions: Operational and implementation perspectives. In Southeast on 2016 (pp. 1–5). IEEE.

Rementeria, S. (2022). Power dynamics in the age of space commercialization. *Space Policy*, 60, 101472.

Sese, R. M. (2022). *The Philippine Space Program: A Modern Take on Establishing a National Space Program.* ASEAN Space Programs: History and Way Forward, 57–77.

Sharma, A., & Sharma, S. K. (2019). Spectral efficient pulse shape design for UWB communication with reduced ringing effect and performance evaluation for IEEE 802.15. 4a channel. *Wireless Networks*, 25(5), 2723–2740.

Sharma, A., Garg, A., Sharma, S. K., Sachan, V. K., & Kumar, P. (2021). Performance optimization for UWB communication network under IEEE 802.15. 4a channel conditions. *Computer Networks*, 201, 108585.

Sharma, A., Garg, A., Sharma, S. K., Sachan, V. K., & Kumar, P. (2021). Directivity enhancement with improved beamforming for physical layer ultra-wideband communication by pulse shaping. *International Journal of Communication Systems*, 34(18), e4992.

Taha, M. A., Abdallah, M. T., Al Qasem, H., & Sada, M. A. (2012, November). Dynamic spectrum analyzer using software defined radio. In *Proceedings of 2012 International Conference on Interactive Mobile and Computer Aided Learning (IMCL)* (pp. 167–172). IEEE.

Advances in AI for Biomedical Instrumentation, Electronics and Computing – Sachan et al. (eds)
© 2024 The Author(s), ISBN 978-1-032-64298-7

Single page optimization techniques using react

Abhishek Pokhriyal, Saurav Pratihasta, Shubh Kansal, Vasu Goyal, Shourya Singh &
Shruti Mishra
Department of Electronics and Communication Engineering, KIET Group of Institution, Delhi-NCR,
Uttar Pradesh, India

ABSTRACT: In the modern web-centric landscape, optimizing initial page load performance is crucial for the success of web applications. Web developers focus on delivering a satisfying user experience with an emphasis on swift page loading. Slow-loading pages can lead to higher bounce rates, decreased conversions, and a tarnished reputation, resulting in the loss of users seeking quick results. This paper presents a method to enhance the initial page load performance of a single-page note-making application using ReactJS. The approach combines techniques like minification and compression to reduce file sizes for faster downloads, lazy loading to load resources only when needed, caching to store static content in the user's browser, and server-side rendering for users with slower internet connections. The proposed method was applied and evaluated on a note-making web application using ReactJS, demonstrating significant improvements in initial page load performance, ensuring a more gratifying user experience.

1 INTRODUCTION

Web applications, widely utilized for education, entertainment, business, and social networking, face the challenge of ensuring fast loading and a seamless user experience (Rawat and Mahajan 2020). Slow-loading applications lead to frustration and dissatisfaction, adversely impacting performance and reputation (Rawat and Mahajan 2020). Web developers employ techniques like minification, compression, lazy loading, caching, and server-side rendering to optimize initial page load performance. Minification and compression reduce file sizes for faster downloads, while lazy loading delays resource loading until necessary, saving bandwidth (Muley and Brahaman *et al.* 2022). Caching stores static content in the user's browser to prevent re-downloads. Server-side rendering, rendering pages on the server, improves performance for users with slow internet. However, these techniques have limitations; for instance, minification and compression may increase processing time, and lazy loading can cause layout issues. This paper proposes a method for optimizing the initial page load of a single-page note-making app using React, a JavaScript library (Srivastava *et al.* 2022) for building user interfaces. React's virtual DOM efficiently updates HTML, and it supports reusable HTML components. The proposed method employs minification, compression, lazy loading, caching, and server-side rendering, demonstrating significant improvements in performance on evaluation (Khan *et al.* 2021). The paper reviews related works, details the proposed method, presents experimental results, discusses advantages and disadvantages, and concludes with key findings and future directions.

2 RELATED WORK

In this research paper, Archana N. Mhajan (Khan *et al.* 2021) provides an overview of React. js, an open-source JavaScript library for creating user interfaces. It covers ReactJS lifecycle

DOI: 10.1201/9781032644752-63

methods like componentDidMount(), render(), and componentDidUpdate(), explaining their purposes& underscores React's advantages, including virtual DOM, Redux, XML/XHTML, and Axios for mobile app development. It also touches on JSX components and React's compatibility with other JavaScript libraries/frameworks. Ultimately, it emphasizes React's significance as a leading UI framework, offering accessible UI development.

In this text, Alok Kumar Srivastava (Srivastava *et al.* 2022) talks about usage and benefits of ReactJS, a popular open-source JavaScript library for building user interfaces. ReactJS, developed by Facebook, uses a virtual DOM based mechanism to efficiently update the HTML DOM. It supports the creation of components, which are reusable pieces of HTML code, and allows for easy organization of code. It offers advantages such as fast rendering, server-side rendering, and integration with other JavaScript libraries. The document also highlights common React lifecycle methods and their purposes in the component lifecycle.

In this document, Ms. Shraddha G. Muley, Ms. Jyoti C. Brahaman (Rawat and Mahajan 2020) talks about the usage and benefits of Evernote as a cloud-based notetaking and organization tool for both business and personal purposes identifying the problems with traditional notetaking methods. The document mentions the use of React.js framework in optimizing performance and ease of development. The proposed system aims to provide a platform for students to write and share their notes, overcoming the drawbacks of traditional paper-based methods. The document concludes with emphasizing the efficiency and convenience the application offers.

In this research paper, Eduardo Velloso, Jason M. Lodge (Muley *et al.* 2022) presents the impact of different note-taking modalities on learners' comprehension of text passages. It highlights the benefits of digital note-taking tools, including the ability to search, edit, and share notes. The use of voice as an input modality for notetaking is explored, and it is found to lead to a higher conceptual understanding of the text compared to typing notes. Note-taking is shown to help learners focus their attention on the text and encode the content into long-term memory. Overall, the document emphasizes the importance of understanding how the input modality shapes learners' note-taking behavior and impacts their understanding of the learning content.

In this research paper, Lili Han, Jing Lu, Zhisheng (Edward) Wen (Han *et al.* 2021) discusses the shill of notetaking. Note-taking is a crucial skill in consecutive interpreting that helps improve the quality of interpretation and reduces cognitive load. It involves coordination between listening, comprehension, analysis, and note-taking. Effective notetaking enhances memory, ensures information completeness, and facilitates more coherent and communicative renditions. However, there is a research gap in assessing note-taking proficiency systematically. Note-taking is actively engaged in both the input and output stages of interpreting and is essential for managing the cognitive load and ensuring a fluent and qualified interpreting task.

3 PROPOSED METHOD

In today's fast-paced digital world, web developers are acutely aware of the shrinking human attention span, driven by the demand for instant gratification. Recognizing that a slow website can lead users to quickly lose interest and navigate away, developers are dedicated to optimizing the speed of initial page loading. This research paper delves into techniques for improving initial page load performance, focusing on methods like Largest Contentful Paint (LCP), First Input Delay (FID), and Cumulative Layout Shift (CLS).

Emphasizing the significance of measuring and optimizing these metrics in web projects, the paper advocates the use of tools like Lighthouse and the Web Vitals Chrome extension. LCP, a pivotal metric introduced in 2020, assesses web page loading speed and user experience, specifically evaluating the perceived loading speed of a page's primary content. LCP aims for content to load within the first 2.5 seconds, crucial for user satisfaction and engagement. Optimization strategies for LCP include various performance enhancements:

- Decreasing Server Response Times to ensure swift server responses to start fetching necessary resources.

- Employing Content Delivery Networks (CDNs): Distributing content across multiple servers worldwide to reduce physical distance and content delivery time.
- Optimizing Images and Videos: Compressing media files while maintaining quality for faster loading.
- Prioritizing Loading of Critical Resources: Ensuring the largest content element is loaded and displayed early by optimizing resource fetching and rendering order.

Bar chart showing the number of websites with good LCP increased from 53% in 2020 to 60% in 2021 to 63% in 2022. For sites visited on phones the increase was from 43% in 2020 to 45% in 2021 and then to 51% in 2022.

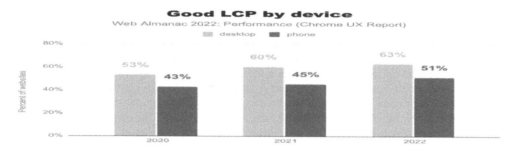

Figure 1.   The percent of websites having good LCP, segmented by device and year.

First Input Delay (FID) serves as a user-centric performance metric, measuring the delay between user interactions and a webpage's response. FID significantly impacts user experience, as a lower FID value indicates a website's quick response to user actions. To improve FID, web developers can implement various strategies:

- Minimizing main thread work, including layout, painting, and rendering tasks, for quicker user interaction responses.
- Utilizing browser features like request Idle Callback to schedule non-essential tasks during idle periods.
- Preloading essential resources such as fonts, CSS, and JavaScript to reduce delays.
- Exercising caution with third-party scripts to prevent negative impacts on FID.
- Lazily loading non-essential content, like below-the-fold images, to prioritize critical content.

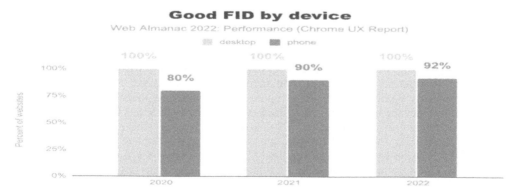

Figure 2.   Good FID by device.

340

Bar chart showing 100% of websites had good FID in 2020, 2021, and 2022. For sites visited on phones this increased from 80% in 2020 to 90% in 2021 and then to 92% in 2022.

Cumulative Layout Shift (CLS), another Core Web Vital, measures the visual stability of a web page as it loads and interacts with user input. It quantifies the extent to which page content shifts or moves during asynchronous element loading or changes. CLS relies on inputs like Layout Instability, Cumulative Score, and thresholds to assess a website's visual stability. A lower CLS score indicates a smoother and more user-friendly experience, positively affecting both user satisfaction and search engine rankings. To optimize CLS:

- Maintain size and dimension attributes for images and videos in HTML to prevent content from jumping during resource loading.
- Avoid adding dynamic content that can displace existing content on the page.
- Load essential CSS styles early to prevent late-loaded styles from causing layout shifts.
- Continuously monitor and address CLS issues as they arise to ensure a stable user experience.

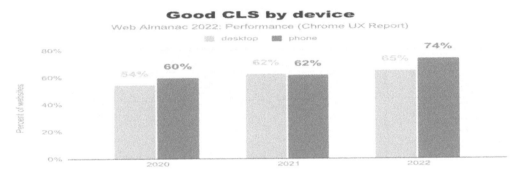

Figure 3. Good CLS by device.

Bar chart showing the number of websites with good CLS increased from 54% in 2020 to 62% in 2021 to 65% in 2022. For sites visited on phones it was 60% of sites achieving good CLS in 2020, 62% for 2021, increasing to 74% in 2022.

4 CONCLUSIONS AND FUTURE INFERENCES

In this research paper, we have discussed the benefits and optimization of different techniques like Largest Contentful Paint (LCP), First Input Delay (FID), and Cumulative Layout Shift (CLS)that plays a vital role in maintaining swift user's interaction with the websites, with important content being displayed first with unnecessary resources being loaded lazily as the user scrolls further in the website, improvising the performance of the website. Thus, in an era defined by the need for instant gratification and short attention spans, web developers play a crucial role in ensuring that users have swift and seamless interactions with websites. Core web vitals like Largest Contentful Paint (LCP), First Input Delay (FID), and Cumulative Layout Shift (CLS) are not just technical jargon but essential tools for crafting user-friendly online experiences. By optimizing these metrics and employing best practices, developers can bridge the gap between user expectations and website performance, ultimately leading to greater user satisfaction, improved engagement, and enhanced search engine rankings, all of which are pivotal in the ever-evolving digital landscape.In the coming years, it is anticipated that SPAs will increasingly adopt hybrid approaches, striking a balance between initial load speed and interactivity through techniques that combine the advantages of SSR and SPAs that are more likely to see enhanced performance, dynamic rendering & pre- rendering methods for improvised search engine visibility, ensuring user experience swift and seamless.

REFERENCES

Han L., Lu J., & Wen Z. (Edward). 2021. Putting the note-taking fluency scale to thetest. DOI: https://doi.org/10.17507/tpls.1210.09. *ResearchGate.*

Khan A.A., Nawaz S., Newn, J., Kelly, R.M., Lodge, J.M., Bailey J., &Velloso E. (2021). The effect of input modality on text understanding during note-taking. DOI: 10.1145/3491102.3501974. *ResearchGate.*

Muley, S. G., Brahaman, J. C., Ugale, P. S., Sananse, S. G. (UG Scholar), Kharche, P. S. (Assistant Professor), & Dandge, S. M. (Head). (2022). Evernote – A web based application on notes making App. ISSN (Online):- 2581–9429. *ResearchGate.*

Rawat, P., & Mahajan, A. N. (2020). ReactJS: A modern web development framework. *International Journal of Innovative Science and Research Technology*, 5(11), 698–702.

Srivastava, A. K. (Assistant Professor), Laxmi, V., Singh, P., Pratima, K. M., & Kirti, V. (UG Scholar). (2022). React JS (Open Source JavaScript Library). ISSN: 2349–6002. *ResearchGate.*

Advances in AI for Biomedical Instrumentation, Electronics and Computing – Sachan et al. (eds)
© 2024 The Author(s), ISBN 978-1-032-64298-7

DAM: Drone Automation and Mapping

Pranjal Agarwal, Prajesh Pratap Singh & Sharad Gupta
Electronics and Communication, KIET Group of Institutions, Ghaziabad, India

ABSTRACT: Integration of Unmanned Aerial Systems (UAS) such as drones, into mapping and automation processes has revolutionized various industries. Drone Automation and Mapping (DAM) is a powerful software ecosystem that is designed to harness the capabilities of UAS technology. DAM serves as a bridge between simulated drone flights and real-world applications, enabling precise control, data collection, and analysis. DAM facilitates autonomous flight planning, allowing users to define takeoff points, set perimeters, and execute missions seamlessly. It leverages ROS2 and Ardupilot's SITL Arducopter, for providing a realistic simulation environment for UAS operations. Key sensors such as LiDAR, cameras, and GPS are integrated to collect essential data during flights. Upon mission completion, DAM processes the collected LiDAR data and video feeds and forwards the point cloud to ground control station, which feeds aerial data into WebODM for comprehensive mapping. Simultaneously, object detection using cocoSSD enhances situational awareness by identifying objects within the defined region. To enhance user experience, DAM features a React.js frontend dashboard, offering intuitive controls for location selection, parameter setup, and real-time output visualization. Additionally, plans for implementing a Node.js backend to incorporate user authentication are underway. DAM bridges the virtual and real-world aspects of UAS technology, empowering industries with efficient mapping, automation, and data-driven insights. This paper delves into DAM's architecture, components, and its potential impact on diverse sectors, paving the way for advanced UAS applications.

1 INTRODUCTION

1.1 *Simulation environment*

Using SITL (Software-In-The-Loop) with Ardupilot and Gazebo is a smart approach for testing and developing the software without the need for a physical drone. It simulates drone behavior and gathers data for the application.

1.2 *Data collection*

The drone is equipped with LiDAR, cameras, and other sensors, which are essential for gathering data. The simulation environment accurately mimics the behavior of these sensors to generate realistic data.

1.3 *Integration*

ROS2 with the drone simulation utilizes python language for writing automation scripts. ROS2 provides a robust framework for robot control and communication, making it well-suited for this application.

DOI: 10.1201/9781032644752-64

1.4 *Data processing*

WebODM helps in processing point cloud data generated from LiDAR.

1.5 *Object detection*

cocoSSD helps in object detection and is a valuable addition to the project, especially to identify objects or features within the mapped region.

1.6 *Frontend and backend development*

A frontend dashboard with React and incorporation of a Node.js backend for user authentication contributes towards creating a user-friendly interface and managing user access.

1.7 *Data management*

Using SITL (Software-In-The-Loop) with Ardupilot and Gazebo is a smart approach for testing and developing the software without the need for a physical drone. It simulates drone behavior and gathers data for the application.

2 LITERATURE REVIEW

L. Zongjian addresses the limitations of traditional remote sensing methods and proposes Unmanned Aerial Vehicles (UAVs) as a solution for high-resolution mapping. The paper develops a practical UAV system with a super-wide-angle camera, enabling flexible data acquisition. The utilization of drones and automated detection techniques for wildlife surveys, It highlights the importance of accurate wildlife detection, particularly in challenging or remote environments, and acknowledges the efficiency of drones in covering extensive areas. It utilizes hardware components such as a quadcopter with RC parts, a Raspberry Pi 3B+, a wide-angle camera, a ToF sensor, a servo gimbal, and a multiplexer for control switching. The control system employs PID and PD controllers for altitude and horizontal plane control, respectively, with black-box system identification.

3 METHODOLOGY

3.1 *Automation module*

The Automation module of DAM leverages the capabilities of Ardupilot, an open-source autopilot system, specifically utilizing the Software-in-the-Loop (SITL) feature. SITL serves as the software counterpart of an actual drone, enabling high-fidelity simulation. Within this module, Ardupilot's Copter version 4.4 is employed to emulate drone behavior.

To create a realistic environment for drone simulation, the Gazebo Ardupilot plugin is utilized. This integration allows us to visualize the drone within the Gazebo environment, providing a simulated representation of the physical world.

Telemetric Data Visualization: During simulation, real-time telemetry data from the virtual drone is accessible via the QGroundControl (QGC) interface. This data includes vital parameters such as position, orientation, and sensor readings.

Manual Control: The DAM system offers the flexibility of manual drone control via the console terminal, enabling users to manually interact with the virtual drone when needed.

Autonomous Flight: To achieve autonomous flight, we launched the Gazebo Iris drone model within the Robot Operating System 2 (ROS2) environment. ROS2 is utilized to communicate with the drone and manage its behavior.

ROS Nodes and Communication: Within the ROS2 framework, DAM employs custom scripts containing ROS nodes. These nodes publish and subscribe to ROS topics, enabling seamless communication and control over the drone.

Object Detection: As part of the autonomous flight script, the drone is equipped with object detection capabilities using cocoSSD v4. During its flight, the drone detects objects within its vicinity and reports their names. Images of detected objects are transmitted through a virtual camera integrated into the SITL.

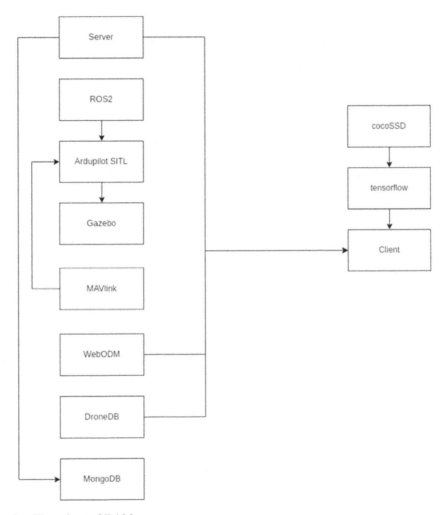

Figure 1.   Flow chart of DAM.

3.2   *Simulation module*

Since DAM operates in a simulated environment without a physical drone, it is crucial to obtain realistic sensor data. To achieve this, we utilize DroneDB, an open-source aerial image database management system. DroneDB provides a repository of aerial images captured by real drones.

Gazebo Integration: The images obtained from DroneDB serve as the basis for simulation. These images, paired with LiDAR data, mimic the sensory input a real drone would receive during a flight.

3.3 *Mapping module*

The Mapping module is the final phase of DAM's workflow, focusing on generating comprehensive maps based on the data collected during simulated flights.

WebODM Integration: The collected images, along with LiDAR data, are fed into WebODM, a robust open-source photogrammetry software. WebODM processes this data to create detailed maps of the simulated environment.

4 RESULTS

One of the primary objectives of DAM was to enable autonomous flight capabilities within user-defined perimeters. The software successfully achieved this goal, demonstrating the potential for efficient and automated drone operations.

4.1 *Autonomous flight within specified perimeter*

DAM's automation module, driven by Ardupilot's SITL, facilitated autonomous flight of the virtual drone within predefined perimeters. Through ROS2-based scripting, the software orchestrated the drone's movements, ensuring it adhered to the specified flight path and navigational instructions. This outcome underscores DAM's ability to autonomously execute complex flight missions, a crucial feature for various real-world applications, such as aerial surveillance, mapping, and environmental monitoring.

4.2 *LiDAR data*

During simulated flights, DAM effectively collected LiDAR data, replicating the sensor readings a physical drone would acquire. This data, characterized by its precision and depth, holds immense promise for applications such as terrain modeling, obstacle detection, and 3D mapping.

4.3 *Aerial images*

Utilizing DroneDB as a source for realistic aerial images, DAM mimicked the data collection process of a real drone. The software ensured seamless integration of these images into its workflow, enabling the generation of high-quality maps and the application of computer vision algorithms.

4.4 *Mapping*

WebODM, an integral part of DAM's mapping module, exhibited impressive performance in transforming collected aerial images and LiDAR data into detailed maps. The results demonstrated the software's potential for producing accurate and informative maps, highlighting its utility in industries such as agriculture, land surveying, and disaster management.

4.5 *Object detection*

The object detection capability, powered by cocoSSD v4, enhanced DAM's situational awareness during autonomous flights. The software successfully identified and categorized objects within the designated area, providing valuable data for applications such as search and rescue, infrastructure inspection, and security.

5 CONCLUSION

The development and implementation of DAM (Drone Automation and Mapping) represents a significant advancement in the field of autonomous drone operations and environmental mapping. Through the integration of Ardupilot, Gazebo, ROS2, and advanced sensor technologies, we have demonstrated the feasibility of autonomous drone flight within specified perimeters, enabling efficient data collection through LiDAR and aerial imaging. The resulting maps and object detection capabilities showcase the practical utility of DAM across various applications, from environmental monitoring to infrastructure inspection. As technology continues to evolve, DAM holds promise for further enhancing the efficiency and precision of aerial data acquisition, making it a valuable tool for industries and research endeavors that rely on accurate spatial information. This research lays the foundation for future innovations and applications in the realm of drone automation and mapping.

REFERENCES

Bennett, M. K., Younes, N., & Joyce, K. (2020). Automating drone image processing to map coral reef substrates using google earth engine. *Drones*, 4(3), 50.

Colwell, R. (1956). Determining the prevalence of certain cereal crop diseases by means of aerial photography. *Hilgardia*, 26(5), 223–286.

Corcoran, E., Winsen, M., Sudholz, A., & Hamilton, G. (2021). Automated detection of wildlife using drones: Synthesis, opportunities and constraints. *Methods in Ecology and Evolution*, 12(6), 1103–1114.

Demirhan, M., & Premachandra, C. (2020). Development of an automated camera-based drone landing system. *IEEE Access*, 8, 202111–202121.

Hughes, T. P., Kerry, J. T., Álvarez-Noriega, M., Álvarez-Romero, J. G., Anderson, K. D., Baird, A. H., ... & Wilson, S. K. (2017). Global warming and recurrent mass bleaching of corals. *Nature*, 543(7645), 373–377.

Knutson, T., Camargo, S. J., Chan, J. C., Emanuel, K., Ho, C. H., Kossin, J., ... & Wu, L. (2020). Tropical cyclones and climate change assessment: Part II: Projected response to anthropogenic warming. *Bulletin of the American Meteorological Society*, 101(3), E303–E322.

Kulbacki, M., Segen, J., Knieć, W., Klempous, R., Kluwak, K., Nikodem, J., ... & Serester, A. (2018, June). Survey of drones for agriculture automation from planting to harvest. In *2018 IEEE 22nd International Conference on Intelligent Engineering Systems (INES)* (pp. 000353–000358). IEEE.

Stocker, T. (Ed.). (2014). Climate change 2013: The physical science basis: Working Group I contribution to the Fifth assessment report of the Intergovernmental Panel on Climate Change. Cambridge University Press.

Zongjian, L. I. N. (2008). UAV for mapping—low altitude photogrammetric survey. *International Archives of Photogrammetry and Remote Sensing*, Beijing, China, 37, 1183–1186.

Advances in AI for Biomedical Instrumentation, Electronics and Computing – Sachan et al. (eds)
© 2024 The Author(s), ISBN 978-1-032-64298-7

Development of an advanced, cost-effective prosthetic limb with an EMG sensor

Shivam Kesarwani, Shrey Shekhar, Shruti Mishra, Abhinav Singh, Somya Tyagi, S.P. Singh, Neelesh Ranjan Srivastava, Sachin Tyagi & Mohit Tyagi
Department of Electronics and Communication, KIET Group of Institutions, Delhi-NCR, India

ABSTRACT: This review will deal with the creation and exploration of research conducted throughout history related to the prosthetic arm. From its inception to its current progress in innovation, look at some of the challenges faced in creating a prosthetic arm that works on the neural patterns of the person using it. The main goal is to provide an alternative to various victims who lost their limbs due to any fatal accident or life-threatening disease. The focus is to provide these people with a second chance at life by providing them with a working artificial limb that will act according to their will, helping them with various tasks that they were incapable of performing before due to their loss. Arm amputees still have muscles and nerve endings at the tip of the remaining part of their limbs. These muscles still carry electrical signals that represent neuromuscular activity, this is called EMG signal or Electromyogram Signal and these signals can be utilized for various limbic activities and working as a replacement. The benefit of using EMG signals is a much cheaper operation therefore offering affordability since EMG is generated inside users' bodies therefore using EMG signals as control inputs will lead to the prosthetic arm being controlled by the user himself without any external remote control or extended complex systems.

1 INTRODUCTION

According to a report on disability affairs in India, approximately 2.6 billion persons are disabled, representing 2.20% of the population. This data includes a variety of disabilities, including hearing, visual, speech, locomotor, and mental impairments. Moreover, 69.05% of this population resides in rural regions (Scheme 2019). Dr. R.K. Sethi spoke at a 1997 lecture on childhood disability, discussing renowned institutions' efforts to solve the issue, including prosthetic options. He added amputees need more rehabilitation aids and devices at cheaper costs and faster. The devices should give people the movement they need for their jobs and the independence they need for their own needs. These would be different from one end of the country to the other and from one area to another based on income and job. It could lead to a lot of different designs instead of one design that everyone everywhere has to use (Orthotics and P. International 1986). For the unfortunate, amputation changes life. Discussing its effects on a person's physical, mental, emotional, financial, or healthy lifestyle is not easy. Their recovery is nearly impossible. Amputation can be performed for various reasons like a fatal accident, any disease like Gangrene, complications from diabetes, birth deformity, etc. For arm amputation, the most common practice is to remove the forearm below the elbow joints therefore the most widely developed prosthetic arm is trans-radial prosthesis which accounts for the forearm and wrist (Elsevier B.V., Amputation – ScienceDirect 2023). In the field of medicine and surgery, prostheses have been an important topic of discussion which resulted in radical innovation in the field of upper and lower extremity prostheses. From mechanical extensions to a look arm that delivers touch, temperature, and multidirectional movement. Typical myoelectric hands use all of the amputee's EMG surface electrodes to record myoelectric signals, which are processed to open and close the hand (Sakib and Islam 2019).

DOI: 10.1201/9781032644752-65

Our trans-radial prosthetic arm must meet affordability, simplicity, and utility design principles. 3D printed components are cheaper than mechanized components, Arduino microcontrollers with processed EMG signals as control inputs make our arm simple to control, and industrial resin makes it strong, flexible, and heat-resistant.

2 LITERATURE SURVEY

According to the journal (Zuo and Olson 2014), in the year 77 A.D., the Roman scholar Pliny the Elder described the prosthetic hand in his encyclopedia Naturalis Historia. The iron hand of the German knight Gotz Von Berlichingen is one of the most well-known early examples of a prosthetic hand. This journal describes various kinds of hand amputations that were carried out in the past. These examples inspire us to provide amputees with assistance so that they can perform their everyday tasks with comfort.

In the field of medical science, modern technology represents a significant breakthrough. We have begun reading about the body's functionality and have begun to comprehend how to treat a variety of human ailments. Electromyography sensors or EMG sensors are used to study muscle movement. Using multichannel EMG signals, the differentiation methodologies for limb functions are brought up.

Experiments were conducted to investigate discrimination ability and learning convergence (Ito *et al.* 1996) They adhered to ideal conditions, which included:

1. Two adult normals, X and Y, and an amputee.
2. Hand motion, limited to wrist flexion and extension, pronation and supination of the arms, and hand grasping movements.
3. Dry-type electrodes were created by Imasel Technical Labs and mounted on the forearms, as depicted in 2.2. Each EMG sensor channel contains an A/D converter with a 1 kHz sampling rate.

Control of the prosthetic limb is based on the discrimination outcomes depicted in these modifications enabling us to control a prosthetic limb using a microcontroller and servo motors. The Backyard Brains (B. Brains, Neuroprosthetics experiments 2023) have conducted numerous research endeavors pertaining to the motion of skeletal muscles. They have put forth a study that explores the optimal electrode placement for obtaining different readings of muscle activity in the arm muscles (3). The data reveals how specific EMG signals are influenced by different electrode placements during muscle contractions.4 depicts patterns that correspond to individual finger movements and can be analyzed and interpreted. Currently, methods such as Euclidean distance and support vector machines are used to isolate and control particular finger movements on a robotic hand, allowing for more precise and nuanced control in the field of neuroprosthetics.

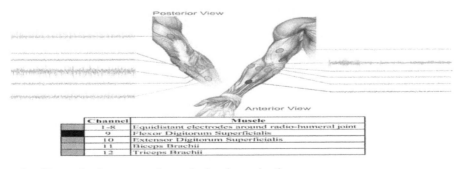

Channel	Muscle
1–8	Equidistant electrodes around radio–humeral joint
9	Flexor Digitorum Superficialis
10	Extensor Digitorum Superficialis
11	Biceps Brachii
12	Triceps Brachii

Figure 1.   Electrode placement and sample signal examination.

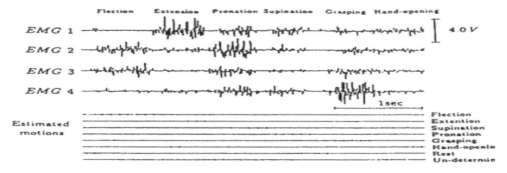

Figure 2.   Distinguishing outcomes from a sequence of motion-related experiments.

A typical myoelectric hand is one in which all of the EMG surface electrodes on the hand of an amputee are used to record myoelectric signals, which are then processed to enable the hand to open and close.

2.1   Electromyograph signal

Electromyograph signals measure the electrical activity of the muscle fiber during the contraction and relaxation phases. In an invasive procedure, the EMG signal is measured with needle electrodes. However, noninvasive is typically preferred because the electrode is not put into the patient's body but is instead placed directly above the skin's surface. To obtain additional data from EMG signals, signal processing techniques like as filtering, rectification, baseline drifting, and threshold leveling are used. (B. Brains, Neuroprosthetics experiments 2023) 4 displays a block diagram for the processing of EMG signals. Three pre-gelled surface electrodes are employed to gather electromyography (EMG) data focusing on shoulder muscles during rotational movements. Specifically, two electrodes are positioned in the acromial and clavicular regions of the central deltoid muscle, targeting its anterior fibers. To ensure proper grounding, a ground electrode is placed on the opposite hand. EMG signals are captured using surface electrodes, with the selected signal exhibiting microvolt amplitude. As a result, a preamplifier is required to increase the electromyography (EMG) signal from a low voltage (10 V) to a higher voltage level (mV). The preamplifier receives the EMG signal from the electrode. An instrumentation amplifier with a high Common Mode Rejection Ratio (CMRR) of 120 dB, higher input impedance, and a constant gain set at 1000 is used for the amplification process. To record and analyze the electrical signals produced by muscles during contraction and relaxation, electromyography (EMG) sensors are essential (Bright *et al.* 2016). Electrodes, which make up EMG sensors, are carefully positioned on the skin over muscles of interest. These electrodes pick up electrical impulses that the muscles release as they contract. Then, for additional processing, these analog signals are amplified, filtered, and digitalized. Processing the gathered EMG data requires the use of an EMG analyzer (Chinbat and Lin 2018).

2.2   3D printing and design

The model's design was finalized by extracting the geometry through a 3D scanner. The dimensions derived from this scan were applied to construct the three-dimensional design. In the process of 3D scanning, a tangible object or surroundings is scanned to acquire data about its geometry and, in some instances, its visual attributes, including color. Utilizing this acquired data, a digital 3D model can be produced (Shibanoki and Jin 2021).

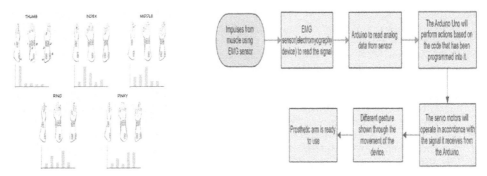

Figure 3.   Channel generated signal during muscle movement.

Figure 4.   Design steps of EMG based Prosthetic Arm.

3 RESULT

For the sake of expediency in this project, the decision was made to opt for an open-source hand model rather than investing time in 3D modeling. The primary focus is on developing the electronic framework and embedded device code. The hand is operated by a total of five DC motors, with each motor dedicated to controlling an individual finger and an additional one for the thumb's opposition movement. A significant step towards prosthetics is research into creating a prosthetic hand that can be readily operated from the arm by the EMG signal.

The objective of this project is to design an affordable 3D-printed prosthetic limb that can be controlled using EMG signals. While significant progress has been made in the initial stages of development, there are still some aspects of the new design that require modification to make the device more practical for real-world applications.

4 CONCLUSIONS

Prosthetic arm technology has the potential to restore a sense of normalcy in the lives of those who have lost a limb. The utilization of prosthetic devices facilitates amputees' ability to execute daily tasks.

This technological advancement enables amputees to create their own customized prosthetics, thereby enhancing their quality of life and mitigating the psychological deterioration that is frequently associated with physical deformities. Furthermore, this innovation is wireless, allowing prosthetics created using this technology to be easily customized for various patients by making the necessary size and shape adjustments.

REFERENCES

Brains B., Neuroprosthetics experiments 2023. Available at https://backyardbrains.com/experiments/neuro-prosthetics, accessed on: Date you accessed the page.

Bright, D., Nair, A., Salvekar, D., & Bhisikar, S. 2016, June. EEG-based brain controlled prosthetic arm. In *2016 Conference on Advances in Signal Processing (CASP)* (pp. 479–483). IEEE.

Chinbat, O., & Lin, J. S. 2018, December. Prosthetic arm control by human brain. In *2018 International Symposium on Computer, Consumer and Control (IS3C)* (pp. 54–57). IEEE.

Elsevier B.V., Amputation – sciencedirect 2023. Available at https://www.sciencedirect.com/topics/nursing-and-health-professions/amputation, accessed on: Date you accessed the page.

Ito, K., Tsuji, T., Kato, A., & Ito, M. 1992. EMG pattern classification for a prosthetic forearm with three degrees of freedom. In *[1992] Proceedings IEEE International Workshop on Robot and Human Communication* (pp. 69–74). IEEE.

Sakib, N., & Islam, M. K. 2019, November. Design and implementation of an EMG controlled 3D printed prosthetic arm. In *2019 IEEE International Conference on Biomedical Engineering, Computer and Information Technology for Health (BECITHCON)* (pp. 85–88). IEEE.

Scheme, D. D. R. 2019. Department of empowerment of persons with disabilities. Last accessed on. Orthotics and P. International 1986, Orthotics and prosthetics international – january 1986 (1986). Available at https://www.oandplibrary.org/poi/198601.asp.

Shibanoki, T., & Jin, K. 2021, March. A 3D-printable prosthetic hand based on a dual-arm operation assistance model. In *2021 IEEE 3rd Global Conference on Life Sciences and Technologies (LifeTech)* (pp. 133–134). IEEE.

Zuo, K. J., & Olson, J. L. 2014. The evolution of functional hand replacement: From iron prostheses to hand transplantation. *Plastic Surgery*, 22(1), 44–51.

Advances in AI for Biomedical Instrumentation, Electronics and Computing – Sachan et al. (eds)
© 2024 The Author(s), ISBN 978-1-032-64298-7

Detection of disease and appointment to the Doctor

Yashasvi Singh, Sanskar Gupta, Rituz Gupta, Riya Tyagi & Amit Kumar
Department of Electronics and Communication Engineering, KIET Group Of Institutions, Delhi-NCR

ABSTRACT: The purpose is to highlight the efficacy of treatment techniques and a valuable resource for the diagnosis and prediction of various diseases, focusing on the importance of immediate detection of fever in severely ill patients. This study aims to develop novel systems learning algorithms that are appropriate for the accurate prediction of fever onset and use continuous physiological records. Data from more than 2 hundred thousand severely ill patients in the intensive care unit. The incidence of fever has been regarded as an impartial event and records have been collected from 6 physiological parameters. This study aims to provide important information on optimizing the selection of remedies and speeding up their treatment in critical care settings.

1   INTRODUCTION

Fevers are routinely used as a diagnostic indicator for determining the exact nature of an infection. For example, malaria bills for a full-size percentage of visits to fitness offerings in India (Ministry of Health 1999). Health workers in malaria-endemic areas should deal with all patients presenting with fever presumptively as malaria cases, according to the recommendations of the Malaria Action Plan and the National Malaria Eradication Programme of 1995, as follows (Marsillio *et al.* 2019). ML is helping in reworking the healthcare enterprise for the higher with its present-day applications. This system is designed to be charged more by machine learning tools (J.M. *et al.* 2013). For billions of people throughout the world, this could be a major and decisive effect of machine learning tools which can improve their quality of life (Rabinstein and Sandhu 2007).

2   THE NEED FOR MACHINE LEARNING IN THE MEDICAL FIELD

The machine that gained awareness of ML has made huge progress in pediatric healthcare, allowing personalized and helpful treatment for children over recent years. The ability to facilitate improvements in progressive care protocols for patients with medically vulnerable diseases may be facilitated by using machine learning strategies to identify key physiological indicators. Although fever doesn't have to be the primary activity, it can lead to severe complications. It is therefore vital to give an early warning of the onset of fever. To detect excessive risk patients, predictive algorithms based on routine physiological data can provide valuable early warnings of cardio-metabolic decline. These statistics can aid in clinical decision-making and can ultimately manual early aim-directed remedies. The examination has shown the accuracy of a system that identifies febrile emergence up to 4 hours in advance, based on an awareness model.

3   ENHANCING THE QULAITY OF HEALTH CARE THROUGH INTELLIGENT PATIENT RECORDS

To improve its performance, optimize different tactics, and achieve ultimate good results for affected individuals, the healthcare sector could use a system of study. Patients' records are filled

DOI: 10.1201/9781032644752-66

with previous diagnoses, fitness conditions, and both physical and emotional health information. Machines getting to know is making sensible patient records a fact, revolutionizing their use within the scientific domain. These sensible patient facts simplify and enhance the cost of medical records related to individuals for healthcare professionals in a variety of areas.

4 THE IMPACT OF ML ON CLINICAL TRIALS AND HEALTHCARE

Historically, clinical trials have been time-consuming and tedious with many complexities involved in the conduct of studies. Machine getting to know (ML) provides a technique to optimize affected person engagement and recuperation through well-timed indicators and notifications regarding appointments and file collection. Molecular diagnostics can also play an important role in the early detection of hereditary, and genetic conditions and some types of cancer. In this era, the need for accessible and interconnected health information and collaboration between public and industrial entities is promoted, with a focus on a cost primarily based approach to most cancer care. In addition, ML offers tremendous advantages for hospitals' organizational and administrative capacities such as patient control, ward monitoring, queue management, appointment booking, collection of the duty list, and other vital responsibilities.

5 ML DEVELOPMENT IN HEALTHCARE AND CLINICAL TRIALS

Machine learning in the field of health care provides highly efficient programs as a result of self-discovery, which can be accomplished using neural networks to improve treatment by analyzing external data including victim's circumstances, x-ray images, CT scans, various diagnosis tests. However, it is important to note that clinical trials which can be a vital part of the pharmaceutical business do not represent the most efficient resource in terms of depth but are often time-intensive and will take several years to complete.

6 DISCUSSION

Through the use of diagnostic findings and thorough audits, prescription plays an important role in reducing errors and speeding up diagnosis. This technology is based on automatic methods in computer systems, which can detect patterns within unstructured files and convert them into structured data. Machine-gaining knowledge of influence on healthcare extends to hospitals and physicians, which turns into a critical aspect of scientific decision-making structures. This allows for the identification of the disease before it occurs and for the development of individualized treatment plans that will ultimately optimize the effects on the affected person. As tech masters continue to conform, they are increasingly integrated with data and analysis which will alter how health information is stored, shared, and applied in a wide range of healthcare programs.

Algorithm	IP Rate	Classification Accuracy	Mean Absolute Error	Root Mean Square Error
Naïve Bayes	0.633	63.3%	0.4133	0.5328
J48	0.767	76.7%	0.3260	0.4152
Random Forest	0.833	83.3%	0.2233	0.3821
Rep Tree	0.767	76.7%	0.3400	0.4243
SMO	0.800	80%	0.2000	0.4472
LWL	0.567	56.7%	0.4633	0.5171
ZeroR	0.767	76.7%	0.3444	0.4237

Figure 1. Table of algorithms of ML.

7 CONCLUSION

The mixing of system studying in healthcare promises to enhance remedy pleasant, diagnostic pace, and operational performance. In patients with critical illnesses, its importance is particularly important for the detection of fever. ML benefits from various healthcare elements, such as affected person information to clinical trials, streamlining techniques, improving patient engagement, and facilitating early detection of diseases. This generation can revolutionize healthcare delivery and control of statistics. Embracing ML, together with the collaboration between public and private entities to increase affected people's outcomes, can lead to a higher degree of green or personal care.

REFERENCES

Abdelaziz A., Elhoseny M., Salama A.S., Riad A.M., A machine learning model for improving healthcare services in the cloud computing environment.

J.M., Mandrekar, J.N. and Rabinstein, A.A. (2013) Indicators of central fever in the neurologic intensive care unit.

Kamaleswaran, R., Akbilgic, O., Hallman, M.A., West, A.N., Davis, R.L. and Shah, S.H. (2018) Applying artificial intelligence to identify physiomarkers predicting severe sepsis in the PICU. *Pediatr. Crit. Care.*

Kumar R., Verma A.R., Panda M.K., Kumar P., HRV signal feature estimation and classification for healthcare system based on machine learning, in International.

Lakshmi S V S S, Ninisha G, Pravallika AY, Akash KP A model for Connecting Near Doctors By Identifying & and Diagnosis diseases.

Marsillio, L.E., Manghi, T., Carroll, M.S., Balmert, L.C. and Wainwright, M.S. (2019) Heart rate variability as a marker of recovery from critical illness in children.

Rabinstein, A.A. and Sandhu, K. (2007) Non-infectious fever in the neurological intensive care unit: incidence, causes and predictors. *J. Neurol. Neurosurg. & Psychiatry.*

Rajathi N., Kanagaraj S., Brahmanambika R. and Manjubhargavi K., A machine learning model of detection of Dengue.

Rush, B., Celi, L.A., and Stone, D.J. (2018) Applying machine learning to continuously monitored physiological data. *J. Clin. Monit. Comput.,*

Yongjie Yan, Chong yuan Chen, Yunyu Liu, Zuyue Zhang, Lin Xu and Kexue Pu, A machine learning model for prediction of fever.

Advances in AI for Biomedical Instrumentation, Electronics and Computing – Sachan et al. (eds)
© 2024 The Author(s), ISBN 978-1-032-64298-7

Image encryption with switching effects

Sunil Sriharsha Gudimella & Umesh Ghanekar
National Institute of Technology, Kurukshetra, India

Kundan Kumar
NXP India Pvt.Ltd., Pune, India

ABSTRACT: It is widely recognized that S-Boxes represent an enhanced version of the scrambling algorithm, exhibiting a distinctive non-linear relationship between input and output. While employing a single S-Box maintains statistical properties, the introduction of multiple S-Boxes does not significantly enhance security, as evident from the flatness observed in histogram analysis. This paper proposes a novel approach to address these limitations, incorporating a switching method, this switching method improves overall image encryption. The effectiveness of the algorithm can be assessed through chi-square analysis, information entropy, and histogram analysis.

1 INTRODUCTION

With the continuous advancements in wireless communication methods, there is a constant risk of attacks on the shared information (Mastan and Pandian 2021; Munir *et al.* 2022). Many organizations transmit data through multimedia, raising concerns about the security of shared images (Shen *et al.* 2019) – (Chai *et al.* 2022). To address this challenge, there is a need for an algorithm that transforms images into noise-like patterns by altering their statistical properties.

To achieve the required level of security for images, several algorithms have been explored, including DNA encoding (Wang and Zhang 2022), compressive sensing (Zhang and Yan 2021), and hash functions (Zhang and Gong 2022). Chaotic systems have been employed in encryption by researchers due to their unpredictable nature (Naseer *et al.* 2021) – (Chai *et al.* 2022). Encryption techniques utilizing chaotic maps are known as chaotic encryption.

1.1 *Chaotic encryption*

Chaotic encryption relies on its ability to generate a sequence of random numbers, resembling noise and devoid of correlations. This method utilizes the unique traits of chaotic processes, such as sensitivity to initial values, parameter dependence, and pseudo-randomness. These properties are harnessed to create a random number sequence, which, in turn, is employed to shuffle the pixels in an image. In the proposed technique, a 2D Hyper-Chaotic map and a Cascade Chaotic map are used to form the 2 S-Boxes and the selection matrix respectively.

1.2 *Substitution Box (S-Box)*

S-box, also referred to as a substitution box, is a crucial component in classical encryption techniques like the Data Encryption Standard (DES) and Advanced Encryption Standard (AES). The nonlinear relationship between input and output in the S-box renders it a potent

DOI: 10.1201/9781032644752-67

tool in both traditional and chaotic image encryption methods. However, when S-boxes are employed in image encryption, it can lead to heightened correlation between the pixels in the encrypted images. Another significant concern is the resemblance between the cipher images and the original image, is the statistical properties of the image remain unchanged but with altered values.

To counter security vulnerabilities, researchers are integrating hybrid S-boxes with chaotic encryption methods to strengthen overall encryption.Recent studies (Ratnavelu *et al.* 2017) – (Wang *et al.* 2019) highlight that chaos-based S-box image encryption significantly improves security. However, the adoption of high-dimensional continuous chaos leads to increased algorithmic complexity (Wang *et al.* 2020; Zhang *et al.* 2018). To mitigate this complexity, researchers have explored double S-box and multiple S-box approaches. Nonetheless, these methods (Wang *et al.* 2020) – (Liu *et al.* 2021) are tailored for relatively uncomplicated image encryption applications, and the S-box size is restricted to 4×16 or 16×16.

2 PROPOSED METHOD

From the block schematic of the proposed algorithm in Figure 1, it can be seen that the session key generator takes the plain image as an input. To address this issue the proposed chaotic systems shall be used to generate two different S boxes and a Selection matrix. Depending on the value of the selection matrix corresponding to the image pixel which is to be encrypted, the values of selection matrix will be used in switching to create the cipher image. The formation of the Sboxes from the 2D hyper-chaotic system and the formation of the selection matrix from the cascade chaotic system is explained in the following sections.

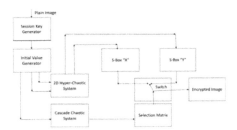

Figure 1. Encryption process.

3 FORMATION OF THE S-BOXES

3.1 *Formation of the initial values*

The major strength of the proposed encryption technique, is that the initial values of the hyper-chaotic systems and the cascade chaotic map are formed from the image itself. The values are obtained as

$$x_1 = I(1)/256 \tag{1}$$

$$y_1 = I(2)/256 \tag{2}$$

$$z_1 = I(3)/256 \tag{3}$$

In the above equations x_1, y_1, and z_1 are the initial values of the 2D Hyper-Chaotic System and the Cascade. I_1, I_2, I_3 are the first, second, and third pixel value of the plain image.

3.2 Two dimensional hyper-chaotic system

The two dimensional hyper chaotic used is a Gao's 2 dimensional hyper-chaotic system expressed below

$$x_{i+1} = \sin\left(\frac{\alpha}{\sin(y_i)}\right) \tag{4}$$

$$y_{i+1} = \beta \sin(\pi(x_i \times y_i)) \tag{5}$$

3.3 Construction of the S-Boxes

The S-Box is constructed using the method proposed by Jun *et al*. By this process, we shall form 2 S-Boxes of dimensions 26×10.

3.4 Cascade chaotic system

The Cascade Logistic-Logistic Chaotic map (CLLCM) is

$$x_{i+1} = 1 - 2\left(1 - x_i(i)^2\right)^2 \tag{6}$$

The above mentioned equation is used to generate the selection matrix

3.5 Generation of the selection matrix

1. Run equation (6) $W \times H$ times, where W and H are the width and height of the image which is to be encrypted respectively.
2. Store all the values of the equation (6) in Z, where Z is a vector of length of 1, $W \times H$.
3. Z $\lfloor Z \times 10^4 \rfloor \%2$.
4. Reshape Z into a matrix of size $W \times H$.

4 ENCRYPTION AND DECRYPTION

4.1 Encryption

The pixel value which is to be encrypted decides the value which is to be replaced in the encrypted image. Let I be a 8 bit grey image with dimensions 'h' as height and 'w' as width. Each pixel value is considered as a 3 digit integer value. The first 2 digits are used to find the column in the S box and the last remaining digit is used to find the row of the S-box's shown below:

$$col = ([I(i,j) - ((I(i,j)\%10))])/10 \tag{7}$$

$$row = ((I(i,j))\%10) \tag{8}$$

$$1 \leq i \leq h, 1 \leq j \leq w$$

4.2 Selection matrix

The selection matrix built in the section 3 contains only 2 values (i.e., 0 and 1). Depending on the value in the selection matrix corresponding to the value of the pixel being encrypted in the plain image (I). If the value of the selection matrix is 0, the S-Box 'X' will be used and if the value is 1, S-Box 'Y' will be used. This process is shown as a flow chart in the Figure 2 where, 'z' is the selection matrix, 'x' and 'y' are the S-Box 'X' and S Box 'Y' respectively and E (i, j) is the encrypted pixel.

Figure 2.   Process of selecting one matrix.

4.3 *Decryption*

With the help of key distribution center (KDC), the receiver will get the 32 bit hexadecimal hash value. With this value the receiver side will be able to extract the initial values (i.e.,$x_1(1)$, $y_1(1),z(1),z(2))$, for equation (4), equation (5) and equation (6). Hence with the correct S-Boxes and the selection matrix the original image can be recovered. From Figure 3, it can be observed that the transmitted and the decrypted images are identical whereas the cipher image appears like noise, satisfying the goal of image encryption.

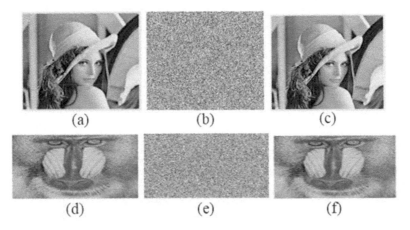

Figure 3.   (a) Transmitted Lena, (b) Cipher Image of (a), (c) Decrypted Image of (b), (d) Transmitted Baboon, (e) Cipher Image of (d), (f) Decrypted Image of (e).

5 EXPERIMENTAL RESULTS AND ANALYSIS

5.1 *Histogram analysis*

Histogram is the best visual tool to observe the pixel distribution in an image. In general flatness of cipher image histogram, reflects the strength of the encryption algorithm against statistical attacks. Figure 4 shows the histogram of unencrypted and the encrypted images.

5.2 *Chi-Square analysis*

Chi-Square Analysis is a more quantitative way of finding the flatness of the histogram of the cipher image. The chi-square value is calculated using the below formulae

$$\chi^2 = \sum_{i=1}^{L} \frac{(f_i - p_i)^2}{p_i}$$

Where χ^2 : chi-square value; L : Maximum grey level (256);
    p_i : is the expected proportion; f_i : is the real proportion of each pixel

359

For a perfectly uniform histogram the χ^2 value is equal to 293.24783. Table 1 shows the value of chi-square analysis results. The theoretical values in the Table 1 are value of a completely flat histogram. The values in Table 1 are close to the theoretical value showing the flatness of the histogram of the cipher. The histogram of the cipher images are in Figure 4 d, h, l and the flatness of the histogram can be seen in Table 1.

Table 1. Chi-Square results of the images.

Image	Plain Image	Theoretical Value	Ref. [51]	Ref.[52] T = 1	T = 20	T = 50	Proposed
Lena	4.7524×10^4	293.24738		242.3578	289.5996	283.3984	274.7188	290.3046
Baboon	1.5750×10^4	293.2473		276.3485	289.4609	281.5625	275.7813	290.1719

Table 2. Entropy test value.

Image	Plain Image	Ref.[39]	Ref.[51]	Ref.[52] T = 10	T = 30	T = 50	Ref.[53]	Proposed
Lena	7.4429	7.9970	7.9973	7.9961	7.9965	7.9971	7.9973	7.9972
Baboon	7.3583	7.9973	7.9992	7.9969	7.9962	7.9961	7.9970	7.9973

5.3 *Information entropy*

Entropy of information is a very important parameter to judge the randomness of image. Higher entropy of the image indicates a more random image. A more random image looks like an image containing only noise. The global entropy H(g) measures the global uncertainty (randomness) of an image. The global entropy H(g) is calculated as

$$H(g) = -\sum_{i=0}^{L} p(g_i) \; \log_2(p(g_i))$$

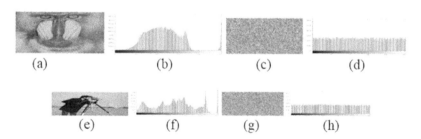

(a)   (b)   (c)   (d)

(e)   (f)   (g)   (h)

Figure 4. (a) Transmitted Baboon, (b) Histogram of (a), (c) Encrypted of (a), (d) Histogram of (c), (e) Transmitted Cameraman, (f) Histogram of (e), (g) Encrypted of (e), (h) Histogram of (g).

Where
    g_i = pixel value; L is the max gray value; $P(g_i)$ is the probability proportion of each pixel value
    From Table 2 it can be observed that the proposed algorithm outperforms the various methods used in the present study.

6 CONCLUSIONS

This paper presents a switching algorithm which selects one S-Box at a time depending up on the output of switch. The randomness in the output of the switch helps the proposed encryption algorithm to overcome the drawbacks of both single and multiple S-Box encryption techniques. The effectiveness of the proposed algorithm is also reflected in terms of histogram analysis, chi-square analysis and the information entropy.

REFERENCES

Chai, X., Wang, Y., Gan, Z., Chen, X., & Zhang, Y. (2022). Preserving privacy while revealing thumbnail for content-based encrypted image retrieval in the cloud. *Information Sciences*, 604, 115–141.

Data Encryption Standard (DES), *FIPS Standard* 46-3, 1999

Kuang, Z., Guo, Z., Fang, J., Yu, J., Babaguchi, N., & Fan, J. (2021). Unnoticeable synthetic face replacement for image privacy protection. *Neurocomputing*, 457, 322–333.

Mastan, J. M. K., & Pandian, R. (2021). Cryptanalysis of two similar chaos-based image encryption schemes. *Cryptologia*, 45(6), 541–552.

Munir, N., Khan, M., Al Karim Haj Ismail, A., & Hussain, I. (2022). Cryptanalysis and improvement of novel image encryption technique using hybrid method of discrete dynamical chaotic maps and Brownian motion. *Multimedia Tools and Applications*, 81(5), 6571–6584.

Shen, M., Deng, Y., Zhu, L., Du, X., & Guizani, N. (2019). Privacy-preserving image retrieval for medical IoT systems: A blockchain-based approach. *IEEE Network*, 33(5), 27–33.

Advances in AI for Biomedical Instrumentation, Electronics and Computing – Sachan et al. (eds)
© 2024 The Author(s), ISBN 978-1-032-64298-7

Design of fuzzy based PID controller for effective delivery of syringe pump

Saba Parveen, Munna Khan & Kashif Sherwani
Department of Electrical Engineering Jamia Millia Islamia New Delhi, India

ABSTRACT: The syringe pump control system is a critical component in many medical and laboratory environments, providing for accurate and controlled fluid or medicine delivery. In this paper, a mathematical model of the syringe pump has been designed. This pump controls the system using PID and Fuzzy PID controlling strategy. Further the PID controller parameters using $F_{min\ search}$ algorithm to get the smoother response by computing the mean square error in output response have been optimized. The fuzzy logic controller is designed by using fuzzy rules to ensure that the system is fundamentally robust. There are 49 fuzzy rules of Mamdani-type fuzzy control if .... then ... language format. Both controllers are designed to control the position of syringe pump. The pump is controlled by motor connected to the lead screw; the rotation of the motor causes the pump to move the syringe. Further by comparing the response of the syringe pump for both the controller, it has been established that in case of fuzzy PID the setting time and rise time are lesser as compared to PID controller. The overshoot in fuzzy PID controlling strategy is also less than PID with difference of 0.8798%. Finally concluded from the results obtained that the fuzzy logic controller improves the stability of the syringe pump as compared to PID controller.

Keywords: Syringe pump, DC Motor, Lead Screw, Optimal Control, $F_{min\ search}$ algorithm, PID Controller, Fuzzy PID Controller and MATLAB

1 INTRODUCTION

A syringe pump is an electro-medical device frequently used in hospitals and in ambulatory settings to inject a liquid (other than blood), primarily medicine or nutrition, in a blood vessel since it is a very effective, quick, and accurate way (Webster and John 2009). To provide nourishment to neonates and pediatrics patients during inpatient, both during and after surgery, as well as for chemotherapy and insulin infusion therapy, intravenous fluid infusion therapy is frequently employed (Joseph D. Bronzino 2006). Healthcare practitioners and researchers may maximize the operation of syringe pump systems for a variety of medical and scientific applications by knowing the principles and application of the controller. To control the pump in this system, PID and fuzzy based PID Controller are used (Mahmut ÜN 1 2016). These measures are critical for guaranteeing patient safety and experimental integrity by maintaining correct and consistent flow rates. Infusion pumps are used in applications requiring high concentration, low flow rate, and small volumes of medication.

In this paper, to designed mathematical model of syringe pump first on Simulink and use it to compute the response based on PID gain of the controller and then optimize same. The $F_{min\ search}$ function can be used to improve the controller gain. Typically, the optimization procedure is iteratively modifying the controller gain and analyzing the system's response using the objective function (Lavania *et al.* 2016). Fuzzy logic controllers (FLC) have been stimulated by the two fuzzy inferences approaches that were subsequently introduced by Mamdani

DOI: 10.1201/9781032644752-68

(Mamdani *et al.* 1975) and sugeno (Khan *et al.* 2018). In recent years, the field of fuzzy control has advanced quickly. Most beneficial use of fuzzy set theory is fuzzy logic control (FLC), which was developed in 1973 by L. A. Zadeh and was utilised in 1974 (Mamdani) in a try to regulate systems that is challenging to model structurally (Zadeh 1965). The control parameters such as rise time, settling time, peak, overshoot have been analysed to verify the stability of the syringe pump. Fuzzy PID controller is 4 times faster to PID controller.

2 MATHEMATICAL MODEL OF SYRINGE PUMP

The mathematical model of a direct current motor is developed using fundamental principles of electromagnetism and the physical features of the motor (Saidi *et al.* 2010). the transfer functions for the operation of the syringe pump.

$$\frac{\theta(s)}{V(s)} = \frac{K}{JLs^2 + \left(JR + \frac{L}{10}\right)s + \left(K^2 + \frac{R}{10}\right)}$$

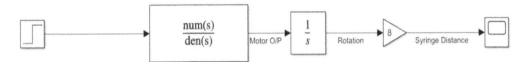

Figure 1. Syringe pump mathematical model.

Mathematical model of syringe pump connected with lead screw, DC motor and simulate the open loop based on the step input voltage to get the position of the syringe pump.

3 PID CONTROLLER

A PID controller is a form of feedback controller that is commonly employed in a variety of control systems. Its goal is to control the output of a system by continually modifying an actuator depending on the difference between the desired setpoint and the measured process variable. A control signal is generated by the PID controller by integrating three components: proportional, integral, and derivative actions. Each component makes a unique contribution to the overall controller (Dodds and Stephen 2008). The expression of the PID controller is represented below.

$$C(s) = k_p + \frac{k_i}{s} + k_D s$$

After modelling of the control system, the variation of the controller gains on the response of the system and find the optimal gain to get minimum oscillation in the output response. The sum of MSE is used to compute the performance of the response is represented as follows:

$$MSE = \frac{1}{n} \sum (y - y_{ref})^2$$

Now use the equation (3) for MSE function to compute the error between the reference and output signal. The $F_{min\ search}$ function is a numerical optimization procedure that finds the smallest value of a specified objective function. The objective function in the context of controller gain optimization examines the system's response based on specified parameters such as MSE.

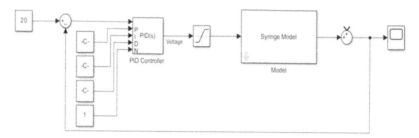

Figure 2.   Syringe model with PID controller.

4   FUZZY PID CONTROLLER

FLC and FL are used to improve traditional PID controllers, and the mathematical operation of fuzzy PID controllers. A fuzzy controller turns a linguistic control strategy into a control strategy using fuzzy logic. The fuzzy rules are built using an expert's or a database's expertise.

Membership function: -Each variable is mapped using the membership function. A membership grade function falls between 0 & 1. There are many different types of MF, including bell-shaped, triangular, and Gaussian ones. Triangular MF is used in this study.

Fuzzification: -Assigning numerical input to fuzzy sets with some degree of membership is known as fuzzification. The system inputs, which are crisp numbers, are converted into fuzzy sets via a fuzzification module. It divides the input signal in to five steps such as-

PB: Positive Big, represents a high degree of truth or membership in the fuzzy set. It signifies a strong positive assertion.

PM: Positive Medium represents a moderate degree of truth or membership in the fuzzy set. It signifies a somewhat positive assertion, but not as strong as PB.

NB: Negative Big, represents a high degree of falsity or non-membership in the fuzzy set. It signifies a strong negative assertion.

NS: Negative Small, represents a moderate degree of falsity or non-membership in the fuzzy set. It signifies a somewhat negative assertion, but not as strong as NB.

ZE: Zero, PB, PM, NB, NS, and ZE typically refer to linguistic terms used to describe different levels of membership or truth values in a fuzzy set.

Inference Engine: – In a knowledge base, the rules regulating the relationship between the input and output variables in terms of the membership function are defined as if-then statements. The variables are $e\omega$ and $Ce\omega$ processed in this stage by an interference method that runs 49 rules. Rule base: – Fuzzy control action is interjected into a decision-making logic that simulates a human decision-making process from the knowledge of control rules and language variables. The rules are written in "if then" structure, where the IF side is formally known as the conditions and the THEN side as the conclusion.

Defuzzification: -Defuzzification is the opposite of fuzzification. The application of FLC produces the desired output in a linguistics variable, which must be changed to create crisp output.

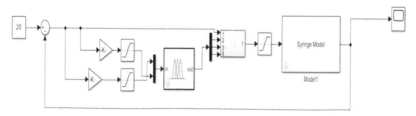

Figure 3.   Syringe model with Fuzzy PID controller.

5 MATLAB

MATLAB is an outstanding performance technical computing language that combines computer programming, visualisation, and automation techniques to give a simple environment to analyse the programmers' mathematical modelling via means of simulation. The management system has a mathematical model of a syringe pump and controller that utilises Simulink for modelling and MATLAB for optimization and visualising response.

6 RESULTS AND DISCUSSION

6.1 *Syringe model*

The open loop response for the mathematical model is the displacement of the syringe based on the input voltage as shown below.

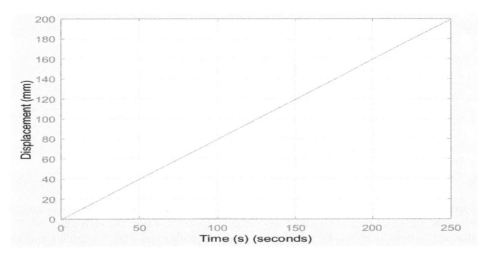

Figure 4.   Syringe output response.

6.2 *System with PID controller*

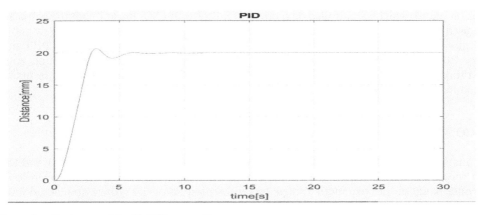

Figure 5.   Syringe model with PID controller response.

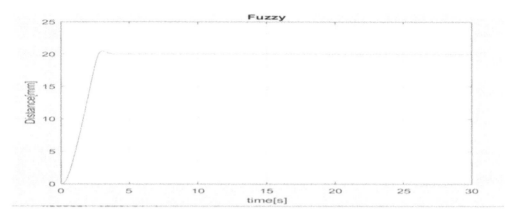

Figure 6.   Syringe model with Fuzzy PID controller response.

Table 1.   Results comparison for PID and Fuzzy PID controller of Syringe pump.

Parameters	PID Controller	Fuzzy PID Controller
Rise Time	1.8611	1.8424
Settling Time	5.1532	3.2600
Peak Time	3.2000	3.0700
Overshoot	3.4457	2.5659

We have observed that for the input reference value of the 20mm, the system is able to maintain the reference position without any steady state error. From the above plot ,we can see that the system is stable. Fuzzy logic to improve traditional PID Controller and the mathematical operation PID Controller. Fuzzy PID Controller are better to manage non-linear system they are more appropriate for processes that have intricate and varying properties. The dc motor control system for a syringe infusion pump has been implemented to work [11]. They implemented FOPID and PID to construct the control system, a comparative analysis has been completed [12]. They have analysed the better performance of FOPID then PID method controllers upon position control of dc motor. A mathematical infusion pump, which is controlled dc motor, PID, PID with PSO, linear quadratic gaussian was used to maintained of drug flow rate are discussed [13]. In our study, a lead screw and a DC motor are used to control the mathematical model of the syringe pump. A PID and fuzzy PID controller is designed to control the DC motor's position. Medication concentrations that is administered to the patient is managed by the controller, used to maintained medication concentration.

7 CONCLUSIONS

The mathematical model of the syringe Pump which is connected to the DC motor and the lead screw has been developed. To create the Simulink syringe model for the open loop response based on the step input voltage to get the position of the syringe and closed loop systems are fuzzy PID controller and PID controller, are introduced in this paper. The result shows that the Fuzzy-PID settling time is lesser as compare the PID. The overshoot in the

Fuzzy-PID controlling strategy is also less than PID with difference of 0.8798% as well as time taken to get maximum amplitude is also lesser which makes the Fuzzy-PID controller best choice as compared to conventional PID controller. From the above analysis and controller response, we have designed the PID controller and Fuzzy-PID controlling strategy to compare the results between them. Also developed the model and plot the response using figure with performance characteristics on the command window, concluded successfully and implement the model and Fuzzy-PID controller, and get the desired results.

REFERENCES

Alamelu, J. V., and A. Mythili (2021). "Smart infusion pump control: The control system perspective." *Computational Intelligence in Healthcare*. Cham: Springer International Publishing, 199–211, 2021.

Chakraborty S., Roy S., and Monda U. (2018)l, "Fractional-order controller for the position control of a DC motor," in *Proceedings of 2018 IEEE Applied Signal Processing Conference*, ASPCON 2018, Dec. 2018, pp. 1–3, Doi: 10.1109/ASPCON.2018.8748544.

Dodds, Stephen J (2008). "Settling time formulae for the design of control systems with linear closed loop dynamics" *Proceedings of Advances in Computing and Technology*, pp. 31–39, 2008.

Joseph D. Bronzino (2006), *"The Biomedical Engineering Handbook – Medical devices and Systems"*, 3rd Ed. CRS Press, 2006.

Khan, Munna, and Amged Sayed Abdelmageed Mahmoud (2018), "TS fuzzy controller of maglev system based on relaxed stabilization conditions." *In Innovations in Electronics and Communication Engineering: Proceedings of the Fifth ICIECE 2016*, pp.555–563, 2018.

Lavania, Shilpi, and Deepak Nagaria (2016). "Fminsearch optimization-based model order reduction." *Second International Conference on Computational Intelligence & Communication Technology (CICT)*, pp. 568–571, 2016.

Mahmut UN (2018), "Control System Design of Syringe Infusion Pump and MATLAB Simulations," *Int. J. Sci. Res.*, vol. 7, no. 7, p. 6, 2018, Doi: 10.21275/ART20183120.

Mahmut ÜN 1 C. (2016), "Control System Design of Syringe Infusion Pump," *International Journal of Science and Research (IJSR)*, 2016.

Mamdani, Ebrahim H., and Sedrak Assilian (1975). "An experiment in linguistic synthesis with a fuzzy logic controller." *International journal of man-machine studies* 7, no. 1, pp. 1–13, 1975

Saidi, Imen, L. A. Ouni, and Mohamed Benrejeb (2010). "Design of an electrical syringe pumps using a linear tubular step actuator" *International Journal of Sciences and Techniques of Automatic Control & Corn puler Engineering*, pp. 1388–1401, 2010.

Webster, John G. (2009), ed. Medical instrumentation: Application and design. *John Wiley & Sons*, 2009.

Zadeh L. A. (1965), "Fuzzy Sets" Informal Control, vol.8, pp 338–353, 1965.

Zhang, Jianming, Ning Wang, and Shuqing Wang (2004). "A developed method of tuning PID controllers with fuzzy rules for integrating processes." *Proceedings of the 2004 American Control Conference*, vol. 2, pp. 1109–1114, 2004.

Advances in AI for Biomedical Instrumentation, Electronics and Computing – Sachan et al. (eds)
© 2024 The Author(s), ISBN 978-1-032-64298-7

An innovative hybrid full adder design for low-power VLSI circuit applications

A. Sharma, N.S. Singha, R. Yadav & A. Kumar
Department of ECE, National Institute of Technology Delhi, India

ABSTRACT: This research introduces an innovative Hybrid Full Adder (HFA) configuration integrating Pass Transistor Logic (PTL), Complementary Metal-Oxide-Semiconductor (CMOS) logic, and transmission gate (TG) logic within the framework of 90-nanometer technology. The HFA is composed of several modules, including EX-OR, carry generator, and sum generator, with an additional inverter logic to achieve the necessary EX-NOR functionality. The proposed HFA exhibits impressive performance with a propagation delay of 19.091 ns and an average power consumption of approximately 5.759 μW at a 1V supply voltage. The most noteworthy aspect is its remarkably low power delay product (PDP) of 139.12 fJ, demonstrating excellent power efficiency. Moreover, the proposed design utilizes only 13 transistors, resulting in a compact layout. The study suggests that this efficient HFA design can be a valuable building block for low-power VLSI circuits, offering significant improvements in power efficiency.

1 INTRODUCTION

The increasing need for battery-operated electronic devices, like laptops, smartphones, bio-electronics, and PDAs, has driven VLSI designers to prioritize circuits focused on reducing power-delay features (Hasan *et al.* 2020). Achieving minimal power delay requires the development of energy-efficient VLSI circuits (Hasan *et al.* 2021; Vijaykumar *et al.* 2021). Designers emphasize factors such as transistor count, power usage, heat dissipation, and circuit area to create low-power VLSI solutions. The main goal is to extend battery life while reducing packaging costs and area, making them suitable for portable devices (Pakkiraiah *et al.* 2022) Consequently, hybrid technology is being explored to create efficient fundamental circuits suitable for low-power applications, leading to substantial improvements in microelectronic circuit performance.

The 1-bit Full Adder (FA) is a vital component in arithmetic logic circuits, enabling a wide range of binary operations, including addition, subtraction, and multiplication (Kumar *et al.* 2016). Thus, having highly efficient FA blocks within the Arithmetic Logic Unit (ALU) is crucial for conducting extensive arithmetic operations in applications such as high-resolution image and video processing, as well as various microprocessor applications. Typically, three main logic styles are utilized for constructing 1-bit Full Adders (FAs): static complementary metal-oxide semiconductor (CMOS), Transmission gate FA, and dynamic CMOS logic (Dokania *et al.* 2018). Conventional CMOS logic, among these approaches, is well-known for its impressive driving capabilities and reliable output swing. However, it faces increased power consumption due to prolonged current switching durations and current leakage induced by circuitry. In order to address these challenges and enhance overall FA performance, designers often use a combination of logic styles, known as hybrid logic. Various hybrid logic styles, such as complementary pass transistor logic, pass transistor

                      DOI: 10.1201/9781032644752-69

logic, CMOS-transmission gate adder, and hybrid CMOS, have been employed in circuit design, offering enhanced efficiency and functionality (Agarwal *et al.* 2017; Pakkiraiah *et al.* 2022 and Solinski *et al.* 2022). The growing demand for battery-powered electronic devices, encompassing laptops, smartphones, bio-electronics, and PDAs, has propelled VLSI designers to prioritize circuits emphasizing reduced power-delay features (Nuthalapati *et al.* 2021). Attaining minimal power delay necessitates the development of energy-efficient VLSI circuits (Kanojia *et al.* 2021). Designers emphasize criteria like transistor count, power usage, heat dissipation, and circuit area to craft low-power VLSI solutions. The primary objective is to prolong battery life while reducing packaging costs and area, rendering them suitable for portable devices (Janwadkar *et al.* 2018). As a result, hybrid technology is under exploration to create efficient fundamental circuits suitable for low-power applications, leading to substantial enhancements in the performance of microelectronic circuits. The one-bit Full Adder (FA) plays a pivotal role as a foundational component in arithmetic logic circuits, enabling a diverse range of binary operations including addition, subtraction, and multiplication (Kanojia *et al.* 2023). Hence, having exceedingly efficient FA blocks within the Arithmetic Logic Unit (ALU) is critical for conducting extensive arithmetic operations in applications like high-resolution image and video processing, as well as diverse micro-processor applications. Typically, three main logic styles are employed for constructing 1-bit Full Adders (FAs): static complementary_metal-oxide_semiconductor (CMOS), Transmission gate FA, and dynamic CMOS logic (Priyadarshini *et al.* 2022). Conventional CMOS logic, among these approaches, is renowned for its impressive driving capabilities and reliable output swing (Hussain *et al.* 2022). However, it grapples with heightened power consumption due to prolonged current switching durations and current leakage induced by circuitry. In efforts to surmount these demerits, researchers have turned their attention to hybrid technology to enhance overall FA performance. By combining different logic styles, researchers aim to improve power efficiency, speed, and other critical factors in FA circuits. This study presents an innovative and energy-efficient design for a Hybrid Full Adder (HFA) utilizing a hybrid approach. The proposed HFA combines Pass Transistor Logic (PTL), and CMOS, and In the 90-nanometer technology, a Transmission Gate Adder (TGA) was created utilizing the Cadence Virtuoso design environment. Notably, the HFA circuit was crafted with a mere 13 transistors (13T) and subjected to comprehensive simulations to confirm the accuracy of its truth table under a 1V supply voltage. The outcomes indicate that the suggested 1-bit Hybrid Full Adder (HFA) surpasses the traditional CMOS design in both power consumption and Power Delay Product (PDP), highlighting its promise for creating VLSI circuits with improved energy efficiency.

2 DESIGN OF HFA

2.1 *Methodology*

In Figure 1, you can see the block depiction of the Full Adder (FA) circuit. This FA design takes input signals A, B, and Cin and produces outputs for Sum and Carry. The FA circuit's ultimate output is derived from Module 2 and Module 3, which employ EX-OR and EX-NOR logic, respectively, to generate the desired results.

2.2 *Altered EX-OR module*

The EX-OR module is a significant hurdle within the Full Adder (FA) circuit due to its high power consumption. To address this issue and achieve a design with enhanced power efficiency, a novel EX-OR with a carry-out module (illustrated in Figure 2 has been proposed for integration into the Hybrid Full Adder (HFA) circuit. The desired output current is achieved by modifying the dimensions of the transistors. Specifically, adjustments in device dimensions reduce the PMOS (PM0) resistance, and similar sizing modifications are applied to the final NMOS transistor within the EX-OR module. The logic table for the EX-OR module, demonstrating its functionality and behavior, is presented in Table 1.

2.3 *Module for generating carry and sum*

In the provided HFA circuit, the carry signal is generated using Transmission Gate (TG) logic. The TG logic utilizes PMOS and NMOS transistors with length and width dimensions set at 120 nm and 100 nm, respectively. Leveraging specific transistors (NM2 and PM3) in the TG logic significantly reduces the carry propagation delay. The SUM's output function is exclusively accomplished by PMOS transistors (PM5, PM6) and NMOS transistors (NM4, NM5) after traversing the EX-NOR logic gates. Table 2 validates the input and output signals of the HFA circuit, confirming its operational accuracy as per the truth table. The HFA design distinguishes itself through its unique transistor count when compared to conventional Full Adder designs. Moreover, this design integrates an innovative Pass Transistor Logic (PTL) configuration, utilizing only 3 transistors for the EX-OR logic, marking the first instance of a low-power HFA employing this distinctive approach.

Table 1. Truth table of Ex-OR.

A	B	A EX-OR B
0	0	0
0	1	1
1	0	1
1	1	0

Table 2. Truth Table of full adder.

A	B	C_{in}	Sum	C_{out}
0	0	0	0	0
0	0	1	1	0
0	1	0	1	
0	1	1	0	1
1	0	0	1	0
1	0	1	0	1
1	1	0	0	1
1	1	1	1	1

3 RESULTS FROM SIMULATIONS

The depicted HFA circuit incorporates an EX-OR logic gate linked to a weak inverter formed by PM2 and NM1 transistors as shown in Figure 3, This weak inverter effectively executes EX-NOR logic. Notably, the inverter's Width/Length (W/L) ratio is set at 120 nm and 100 nm to enhance the overall performance of the Hybrid Full Adder (HFA). Following this, the output of the EX-NOR logic is employed in the PM3_and_NM3 transistors within the Cout module to fulfill its role. Ultimately, this result is linked to the NM4 and NM5 transistors from the EX-NOR logic, concluding the SUM segment of the circuit. The simulation for this 1-bit HFA design was carried out using the 90-nanometer technology. For a visual representation of the input and output of the designed HFA, refer to Figure 4. The design endeavors to decrease_power_consumption in the FA_circuit while enhancing its operational speed, ultimately resulting in a decreased Power-Delay Product (PDP) for the proposed configuration. In this hybrid circuit design, we've skillfully optimized transistor W/L ratios to curtail overall power consumption while simultaneously boosting circuit speed by adjusting the dimensions of transistors within the Transfer Gate (TG) that connects the Cin and Cout paths, ensuring the desired propagation delay is achieved. The waveforms in Figure 5 validate the sum and carry logic of our Hybrid Full Adder (HFA), and detailed transistor sizing for 90-nm technology is presented in Table 3. Our simulation results in Table 4, conducted under a 1V power supply, comprehensively compare delay, power consumption, and the Power-Delay Product (PDP) with existing full adders, affirming our HFA's alignment with full adder logic.

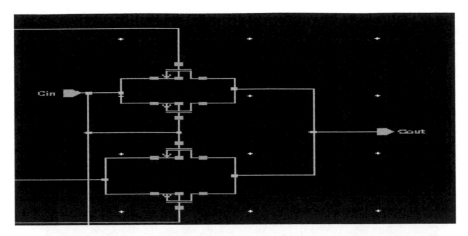

Figure 1.   Shows the Carry (Cout) generation block.

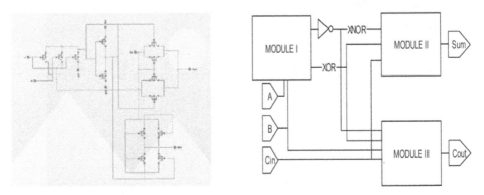

Figure 2.   Illustrates the envisioned circuit design   Figure 3.   Suggested HFA block diagram.
for the HFA.

Furthermore, in a comparison with other HFA designs, our proposed HFA stands out, employing a mere 13 transistors as opposed to the typical 16 in existing HFAs, showcasing potential advantages in circuit complexity and power efficiency as shown in Figure 5. The circuit exhibits a significantly reduced average power consumption compared to hybrid Full Adders reported earlier. While the speed of our Hybrid Full Adder (HFA) experiences a moderate reduction in contrast to the values documented due to the lesser transistors employed in our design, this trade-off results in significantly reduced power consumption. As a consequence, our HFA exhibits a more favorable Power-Delay Product (PDP) when related to the hybrid full adders presented. A comprehensive comparison of key parameters between our proposed HFA and other currently available HFAs can be found in Table 4. In this table, one HFA gain power consumption of approximately 27.431 µW and a propagation delay of about 10.489 ns using 16 transistors, yielding a PDP of 326.27 fJ. Our Hybrid Full Adder (HFA) design, although somewhat slower due to fewer transistors, significantly reduces power consumption. When compared to these references, one HFA used 16 transistors, consuming around 27.591 µW with a 10.489 ns delay and a PDP of 326.27 fJ, while another used 27 transistors with about 26.891 µW, 5.582 ns delay, and a PDP of 162.41 fJ. In contrast, our HFA consumes only around 5.789 µW, achieves a comparable delay of about 19.091 ns, and showcases a remarkably low PDP of approximately 139.12 fJ. This highlights a notable improvement in our HFA's overall performance, optimizing the trade-off between power efficiency and operational speed.

Table 3. Dimensions of transistors employed in the proposed HFA at the 90-nanometer.

Transistor Used	Width (W) nm	Length (L) nm
EX-OR PM0	120	100
EX-OR PM1	20	100
EX-OR NM0	120	100
Adder PM0	120	100
Adder PM1, PM2, PM3, PM4	120	100
Adder NM0, NM1, NM2, NM3, NM4	120	100

Table 4. Results of simulation for HFA circuit AT 90 NM technology.

Design	Power (μW)	Delay (ns)	PDP (fJ)	No. of transistors
Full adder [6]	27.431	10.489	326.27	16
Full adder [9]	26.891	5.582	162.41	27
Full adder [proposed]	5.759	19.091	139.12	13

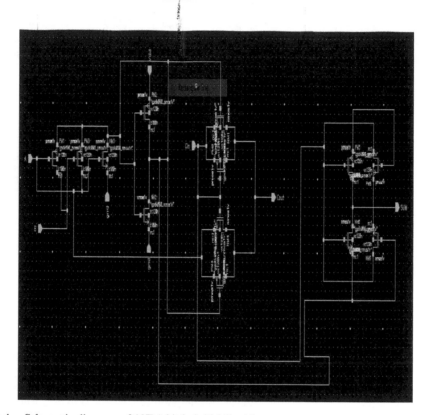

Figure 4. Schematic diagram of 13T 1-bit hybrid full adder.

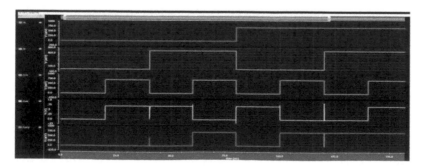

Figure 5. Simulated result for 1-bit hybrid full adder.

372

4 CONCLUSION

In a 90-nanometer technology setting, a distinctive 1-bit Hybrid Full Adder (HFA) was crafted using the standard Cadence Virtuoso platform. This HFA boasts a modest transistor count of just 13, lower than typical hybrid-style designs. Furthermore, our HFA design exhibits reduced power consumption and a more favorable power delay product when compared to existing Full Adder designs. Overall, the proposed HFA demonstrates enhanced performance in terms of power usage (5.759 μW), delay (19.091 ns), and Power Delay Product (PDP) at 139.12 fJ. The circuit occupies notably less space due to the consumption of just thirteen transistors in our proposed HFA. Future work may involve optimizing the layout of this designed HFA to investigate its spatial efficiency. Additionally, extending the 1-bit HFA design to create 64-bit and 32-bit Ripple Carry Adders in different nanometer technologies could offer valuable insights into enhancing overall performance.

REFERENCES

Agrawal, P., Raghuvanshi, D. K., & Gupta, M. K. (2017, October). A low-power high-speed 16T 1-bit hybrid full adder. In *2017 International Conference on Recent Innovations in Signal Processing and Embedded Systems (RISE)* (pp. 348–352). IEEE.

Dokania, V., Verma, R., Guduri, M., & Islam, A. (2018). Design of 10T full adder cell for ultralow-power applications. *Ain Shams Engineering Journal*, 9(4), 2363–2372.

Hasan, M., Siddique, A. H., Mondol, A. H., Hossain, M., Zaman, H. U., & Islam, S. (2021). Comprehensive study of 1-bit full adder cells: review, performance comparison and scalability analysis. *SN Applied Sciences*, 3(6), 644.

Hasan, M., Zaman, H. U., Hossain, M., Biswas, P., & Islam, S. (2020). Gate diffusion input technique based full swing and scalable 1-bit hybrid full adder for high performance applications. *Engineering Science and Technology, an International Journal*, 23(6), 1364–1373.

Hussain, M. S., Kandpal, J., Hasan, M., & Muqeem, M. (2022, December). A high-performance hybrid full adder circuit. In *2022 IEEE 9th Uttar Pradesh Section International Conference on Electrical, Electronics and Computer Engineering (UPCON)* (pp. 1–4). IEEE

Janwadkar, S., & Das, S. (2018, April). Design and performance evaluation of hybrid full adder for extensive PDP reduction. In *2018 3rd International Conference for Convergence in Technology (I2CT)* (pp. 1–6). IEEE

Kanojia, A., Agrawal, S., & Lorenzo, R. (2023). Comprehensive analysis of a power-Efficient 1-Bit Hybrid Full Adder Cell. *Wireless Personal Communications*, 129(2), 1097–1111.

Kanojia, A., Agrawal, S., & Lorenzo, R. (2021, December). Design implementation of a low-power 16T 1-bit hybrid full adder. In *2021 International Conference on Control, Automation, Power and Signal Processing (CAPS)* (pp. 1–5). IEEE

Kumar, P., & Sharma, R. K. (2016). Low voltage high performance hybrid full adder. *Engineering Science and Technology, an International Journal*, 19(1), 559–565

Nuthalapati, S., Nutalapati, K., RANI, K. R., Sasirekha, L. L., Mekala, S., & Mohammad, F. P. (2021). Design A Low Power and High Throughput 130nm Full Adder Utilising Exclusive-OR And Exclusive-NOR Gates. *Journal of VLSI Circuits and Systems*, 3(2), 42–47

Pakkiraiah, C., & Satyanarayana, D. R. (2022). An innovative design of low power binary adder based on switching activity. *International Journal of Computing and Digital Systems*, 11(1), 861–871

Pakkiraiah, C., & Satyanarayana, R. V. S. (2022). Design of low power artificial hybrid adder using neural network classifiers to minimize energy delay product for arithmetic application. *Int. J. Comput. Appl*, 184 (12), 1–8.

Priyadarshini, K. M., & Sabari, M. N. (2016). Comparative analysis of a low power and high speed hybrid 1-bit full adder for ULSI circuits. *Int. J. Sci. Res*, 5(9), 1631–1635.

Soliński, J., & Sherje, N. (2022). A Low Voltage Novel High-Performance Hybrid Full Adder for VLSI

Vijayakumar, V., Ilayarajaa, K. T., Ravi, T., & Sugadev, M. (2021, March). Analysis of high speed hybrid full adder. In *2021 International Conference on Artificial Intelligence and Smart Systems (ICAIS)* (pp. 1641–1645). IEEE.

Advances in AI for Biomedical Instrumentation, Electronics and Computing – Sachan et al. (eds)
© 2024 The Author(s), ISBN 978-1-032-64298-7

Performance evaluation of full adder cells implemented in CMOS technology

Yash Pathak & Dharmendra Kumar Jhariya
Department of ECE, NIT Delhi, India

ABSTRACT: The efficiency and performance of Full Adder cells in CMOS technology are critical factors in the design and optimization of digital integrated circuits. This comparative study investigates various Full Adder cell architectures, analyzing their performance attributes, including propagation delay, power consumption, and silicon area utilization. By evaluating these key metrics, we provide valuable insights into the advantages and drawbacks of different designs. Understanding the nuances of Full Adder cells is essential for enhancing the efficiency of arithmetic circuits and advancing digital system design. As semiconductor technology continues to evolve, selecting the most suitable Full Adder cell design becomes increasingly significant. This study contributes to this ongoing discourse by offering a comprehensive comparative analysis of Full Adder cells in CMOS technology, aiding in the quest for efficient and high-performance digital systems.

1 INTRODUCTION

In the realm of integrated circuit design, Full Adder cells stand as fundamental building blocks for a multitude of arithmetic and logic functions. These versatile components play a pivotal role in various digital systems, ranging from microprocessors to memory modules. The quest for enhancing the efficiency and functionality of integrated circuits is an ongoing endeavor, and in this pursuit, the choice of Full Adder cell design in CMOS (Complementary Metal-Oxide-Semiconductor) technology plays a crucial role.

The performance characteristics of Full Adder cells, such as speed, power consumption, and area utilization, are of paramount importance in modern semiconductor design. Researchers and engineers continually seek to optimize these aspects to meet the increasing demands of ever- advancing technology. This pursuit has led to a plethora of Full Adder cell designs, each with its unique architecture and trade-offs.

This comparative study embarks on an exploration of these various Full Adder cell designs in CMOS technology. We delve into the intricate details of different Full Adder architectures, meticulously analyzing their performance attributes. By evaluating factors like propagation delay, power dissipation, and silicon real estate utilization, we aim to provide valuable insights into the advantages and drawbacks of each design.

Understanding the nuances of Full Adder cells is pivotal not only for designing efficient arithmetic circuits but also for advancing the broader field of digital system design. As semiconductor technology continues to evolve, the selection of an optimal Full Adder cell design becomes increasingly significant. Therefore, this study contributes to the ongoing discourse on enhancing the performance and efficiency of digital systems by shedding light on the equivalent analysis of full adder cells in CMOS technology.

DOI: 10.1201/9781032644752-70

2   LITERATURE REVIEW OF DIFFERENT FULL ADDERS

A review of the current state of full adder designs is offered in this section. It extensively examines the implementation techniques employed and delves into the advantages and disadvantages associated with these existing adder cells.

2.1   *CMOS adder*

The CMOS full adder, a specialized digital circuit incorporating 28 transistors, stands as a noteworthy component within the realm of integrated circuit technology and has garnered substantial attention in academic research. Acknowledged for its versatility and robustness, the 28T full adder optimizes transistor utilization while offering competitive power efficiency and processing speed. In our research paper, we delve comprehensively into the 28T full adder, elucidating its operational principles, design considerations, and performance benchmarks. We explore its potential applications across a spectrum of fields, spanning high-performance computing to energy-efficient embedded systems and conduct comparative analyses against alternative full adder designs, emphasizing its distinctive strengths. Our study underscores the pivotal role of the 28T full adder in advancing digital circuit design and anticipates its continued significance in shaping the landscape of semiconductor technology.

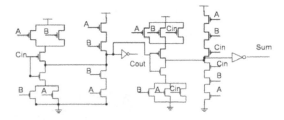

Figure 1.   CMOS adder.

2.2   *Hybrid full adder*

The Hybrid full adder, a specialized digital circuit design featuring 26 transistors, plays a pivotal role in contemporary integrated circuit technology. This compact and efficient full adder design is a subject of extensive research and academic interest, owing to its ability to optimize transistor count, enhance power efficiency, and boost processing speed. In our research paper, we delve into the intricacies of the 26T full adder, elucidating its operational principles, design considerations, and performance metrics. Our study ultimately underscores the critical role of the hybrid full adder in advancing digital circuit design and anticipates its continued significance in future semiconductor technology developments.

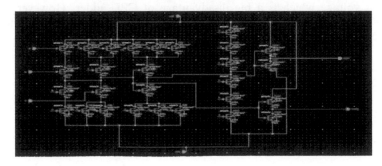

Figure 2.   Schematic diagram of CMOS adder.

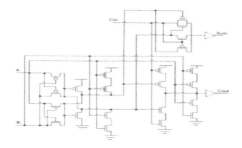

Figure 3. Hybrid adder of 26T.

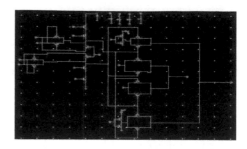

Figure 4. Schematic diagram of Hybrid (26T) adder.

2.3 HPSC adder

The HPSC full adder, a specialized digital circuit design employing 22 transistors, represents a crucial element in contemporary integrated circuit technology and has garnered considerable attention in academic research. Renowned for its efficiency and compactness, the 22T full adder optimizes transistor usage while delivering competitive power efficiency and processing speed. In our research paper, we provide an in-depth exploration of the 22T full adder, elucidating its operational characteristics, design considerations, and performance metrics. Our study underscores the pivotal role of the HPSC full adder in advancing digital circuit design and anticipates its continued significance in shaping the landscape of semiconductor technology.

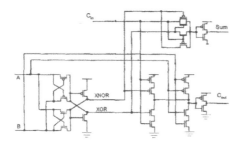

Figure 5. HPSC adder.

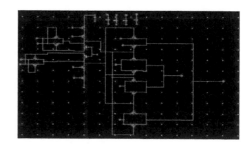

Figure 6. Schematic diagram of HPSC adder.

2.4 Transmission gate adder

The transmission gate adder needs twenty transistors to build and it brings down the degradation of voltage which is a big disadvantage of CPL adder. However, the major concern for VLSI designers is the driving capability of the TGA adder. In addition, if connected in cascade the signal power of TGA adder reduces.

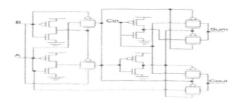

Figure 7. TGA adder.

2.5 10T adder

An innovative XOR gate design was introduced, utilizing a mere ten transistors. The adder configuration boasts the lowest number of transistors and occupies the smallest area. Consequently, the 10-transistor adder exhibits minimal switching activity, contributing to its overall low power consumption. However, it is worth noting that the adder's ability to drive signals may be limited when confronted with high fan-out scenarios.

3 SIMULATION RESULTS

In Section 2, all the adder cells presented were constructed within the Cadence Virtuoso framework, utilizing the 90 nm technology node. To ensure uniformity in our simulation environment, we maintained a consistent voltage supply (Vdd) of 1.2 V across all scenarios. Details about the number of transistors and performance metrics of the adders under consideration can be found in Table 1.

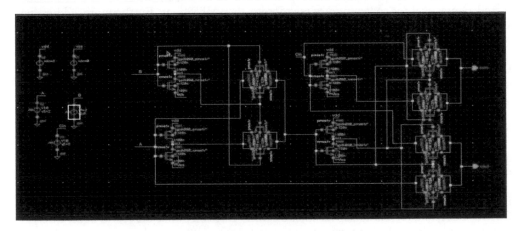

Figure 8. Schematic diagram of TGA adder.

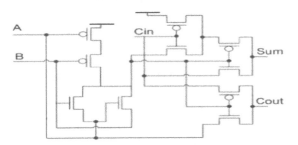

Figure 9. 10T adder.

Based on our simulation results, it was observed that the 10-T full adder cell did not operate effectively in the 90nm technology node. Consequently, no data regarding its power consumption, delay, or power-delay product (PDP) could be acquired. This particular full adder employed capacitors instead of CMOS technology, leading us to exclude it from our analysis to ensure the consistency of our comparative study. To provide a more visually intuitive understanding of the performance disparities among various adders, the information from Table 1, including power consumption, delay, and PDP, has been graphically represented.

Table 1. Number of transistors & performance metrics for different full adder/cells.

Adder Cell	Ref no.	Transistors Count	Power (μW)	Delay (ns)	PDP (fJ)
CMOS	Wairya *et al.* (2012)	28	1.791	0.143	0.256
Hybrid	Yen *et al.* (2011, March)	26	2.103	0.129	0.271
HPSC	Shams *et al.* (2002)	22	1.598	0.246	0.393
TGA	Vijay *et al.* (2012)	20	1.749	0.137	0.239
10T	Bui *et al.* (2002)	10	Failed to utilize in 90nm Technology		

4 COMPARATIVE ANALYSIS AND DISCUSSION

Based on our simulation results, the hybrid (26T) adder demonstrated the lowest delay of 0.129 ns and the smallest power-delay product (PDP) of 0.271 fJ. However, it's important to note that CMOS full adders, while offering favorable performance characteristics, come with drawbacks. They occupy significant chip areas due to their high transistor count, making them less practical for contemporary microprocessor applications. Furthermore, addressing voltage degradation issues in CMOS requires the addition of buffers to restore voltage levels, which introduces even more transistors and increases the surface area in the layout design.

The transmission gate adder exhibits a little lower delay of 0.137 ns compared to the CMOS adder. However, where the TGA adder truly excels is in power efficiency and power-delay product (PDP), outperforming the CMOS adder in these aspects.

Based on our simulation results, the hybrid (26T) adder displayed the highest power consumption, measuring at 2.103 μW. This elevated power usage may restrict its applications in low-power movable electronic devices. Nonetheless, it exhibits a satisfactory delay of 0.129 ns.

Since the 10-T adder cell faced operational issues in the 90nm technology, we proceeded to implement it in the 180nm technology to assess its functionality. In the 180nm technology node, the 10-T adder cell operated flawlessly.

The HPSC adder demonstrates superior power efficiency compared to other adder designs. However, its high delay results in a power-delay product (PDP) that surpasses that of the CMOS, TGA, and Hybrid adders mentioned earlier. Therefore, while the HPSC adder excels in power consumption, it lags in terms of overall performance efficiency due to its longer delay.

The Hybrid full adder cell delivered a commendable performance, showcasing lower delay than all other adders except CMOS and surpassing all its counterparts in terms of power consumption efficiency. This combination of reduced delay and competitive power efficiency positions the Hybrid adder as a compelling choice among existing adder designs, offering a balance between speed and energy efficiency.

Based on the literature review and simulation results, adder cells vary in performance, with some offering lower delay and power consumption but potentially higher transistor count and area usage, while others excel in power efficiency but may have longer delay as shown in Table 2.

Table 2. Summary of adv. and disadv. of different adder/cells.

Adder Cell	Ref no.	Advantages	Disadvantages	Comments
CMOS	Wairya, S. *et al.* (2012)	Robust against sizing of transistor and scaling	High transistor count High input impedance	Most widely used adder due to overall good performance
Hybrid	Yen, P. T. *et al.* (2011, March)	Strong driving/ capability Very low delay	The number of transistors is high outcomes in large surface area	Used in modern devices due to good performance

(*continued*)

Table 2. Continued

Adder Cell	Ref no.	Advantages	Disadvantages	Comments
HPSC	Shams, A. M et al. (2002)	Lowest delay and low PDP Satisfactory driving capability	High input impedance	The performance level is satis factory
TGA	Vijay, V. et al. (2012)	Less surface area is required Low delay & PDP	Poor driving capability Whereas connected in cascade signal strength decreases	It has noble performance overall but poor drivng/ capability is a limitation in high fan-out conditions.
10T	Bui, H. T., Wang, Y., & Jiang, Y. (2002)	Lowest transistor count Requires low surface area	Very poor driving capability	Do not function in modern technology

5 CONCLUSION

In this research, a comprehensive relative study of present full adder circuits was conducted, delving into their respective strengths and weaknesses. Employing various methods and log ic styles, researchers have advanced a range of adder designs over time, necessitating a performance analysis to discern their attributes. The study involved implementing and simulating these adder cells using Cadence Virtuoso tools, maintaining a consistent standard of 90 nm technology across all simulation scenarios. This analysis aims to empower VLSI designers with the insights needed to select the most suitable full adder cell for specific system requirements.

REFERENCES

Aguirre-Hernandez, M., & Linares-Aranda, M. (2010). CMOS full-adders for energy-efficient arithmetic applications. *IEEE Transactions on Very Large Scale Integration (VLSI) Systems, 19*(4), 718–721.

Bui, H. T., Wang, Y., & Jiang, Y. (2002). Design and analysis of low-power 10-transistor full adders using novel XOR-XNOR gates. *IEEE Transactions on Circuits and Systems II: Analog and Digital Signal Processing, 49*(1), 25–30.

Kandpal, J., Tomar, A., & Agarwal, M. (2021). Design and implementation of 20-T hybrid full adder for high-performance arithmetic applications. *Microelectronics Journal, 115*, 105205.

Khatibzadeh, A. A., & Raahemifar, K. (2004, May). A study and comparison of full adder cells based on the standard static CMOS logic. In *Canadian Conference on Electrical and Computer Engineering 2004 (IEEE Cat. No. 04CH37513)* (Vol. 4, pp. 2139–2142). IEEE.

Nigam, A., & Singh, R. (2016). Comparative analysis of 28T Full adder with 14T Full adder using 180nm. *International Journal of Engineering Science Advance Research, 2*(1), 27–32.

Shams, A. M., Darwish, T. K., & Bayoumi, M. A. (2002). Performance analysis of low-power 1-bit CMOS full adder cells. *IEEE Transactions on Very Large Scale Integration (VLSI) systems, 10*(1), 20–29.

Vijay, V., Prathiba, J., Reddy, S. N., Srivalli, C., & Reddy, B. S. (2012). Performance evaluation of the CMOS Full adders in TDK 90 nm Technology. *International Journal of Systems, Algorithms & Applications, 2*(1), 7.

Wairya, S., Nagaria, R. K., & Tiwari, S. (2012). Comparative performance analysis of XORXNOR function based high-speed CMOS full adder circuits for low voltage VLSI design. *International Journal of VLSI Design & Communication Systems, 3*(2), 221.

Wang, D., Yang, M., Cheng, W., Guan, X., Zhu, Z., & Yang, Y. (2009, May). Novel low power full adder cells in 180nm CMOS technology. In *2009 4th IEEE Conference on Industrial Electronics and Applications* (pp. 430–433). IEEE.

Yen, P. T., Abidin, N. F. Z., & Ghazali, A. B. (2011, March). Performance analysis of full adder (FA) cells. In *2011 IEEE Symposium on Computers & Informatics* (pp. 141–146). IEEE.

Advances in AI for Biomedical Instrumentation, Electronics and Computing – Sachan et al. (eds)
© 2024 The Author(s), ISBN 978-1-032-64298-7

Code converter realization in IoT by TCP/IP network layer through node MCU and LabVIEW

G. Dhanabalan

Department of CSE, Vel Tech Rangarajan Dr. Sagunthala R&D Institute of Science and Technology, Avadi, Chennai, India

H.B. Michael Rajan

Department of Mechanical Engineering, Kings College of engineering, Punalkulam, Pudukkottai, India

R. Ashok

Department of ECE, Kamaraj College of Engineering and Technology, K. Vellakulam, India

ABSTRACT: Internet of Things (IoT) categorized into the fourth industrial revolution attracts researchers to realize different applications in IoT environment. Its architecture uses the existing network layers like TCP/IP, UDP, CoAP protocols. These protocols are tailor-made to satisfy the requirements of IoT architecture. This work has used TCP/IP protocol to realize an IoT architecture using NodeMCU and LabVIEW. NodeMCU was configured to be the server and the LabVIEW was used to be the client which can access data from NodeMCU. NodeMCU is embedded with the Wi-Fi device. The configuration helped to realize IoT architecture through code convertor. NodeMCU receives the decimal number from the client, converts into binary and transfers it to the client. The client designed using LabVIEW displays the result. Security in data access is ensured as the NodeMCU will respond to the client only when it receives the keyword encoded by the designer.

1 INTRODUCTION

1.1 *Internet of things*

Demanding nature of human being has lead to the new technology termed as Internet of Things (IoT). IoT basically acquires data from the sensors and transfers them to the server. Thus transferred data shall be acquired by more than one client. Client can perform monitoring or initiating control action on the data it received. This enables remote monitoring (Fernanda Fama *et al.* 2022) and control (Riera *et al.* 2020) of a typical process. IoT architecture supports more number of sensors or actuators that shall be connected to it (Ambarish Gajendra Mohapatra *et al.* 2023). Even though it is a very big advantage, enormous volume of data generated by the sensors is a very big problem. A microcontroller embedded with a Wi-Fi device acquires data from the sensor, pre-processes and stores it in memory. It is able to execute certain functions over the data, transfer it (before or after the execution) through internet, receive data from the server, if needed. This kind of set up renames the sensor as smart sensor (Abel E. Edje *et al.* 2023; Botero Valencia and Valencia Aguirre 2021). It also helps in reducing the latency between the communicating devices. IoT technology is used in a wide variety of applications, from home automation (Muhammad Umair *et al.* 2023) to industrial automation (Collaguazo *et al.* 2023). Data generated by IoT sensors are heterogeneous and it is mostly in unstructured form (Abel E. Edje *et al.* 2023).

DOI: 10.1201/9781032644752-71

Security is a major issue faced by IoT technology as the data floated in the internet shall be accessed by unauthorized persons.

Transmission Control Protocol (TCP) is used as a main protocol in the transport layer. It is a connection oriented protocol (Heba Yuksel and Omer Altunay 2020). It is able to transfer larger amount of data in the form of smaller packets (Alok Jain and Suman Bhullar 2022). It ensures retransmission of lost packets and reassembly in a correct sequence. This has potential impact on the overall performances and latency in data transmission. The selection of TCP or User Datagram Protocol (UDP) for TCP/IP protocol is based on the performance and scalability of IoT devices (Oladayo Bello *et al.* 2016). The nature of application is based on the utilization of IoT application protocol and the way of data transport. Enterprise class IoT devices use generic protocols like Ethernet and Wi-Fi (Rolando Herrero 2023). This work has used TCP/IP protocol to realize an IoT application named code converter.

2 TCP/IP PROTOCOL – AN OVERVIEW

2.1 *IoT architecture with TCP/IP*

Transmission Control Protocol/Internet Protocol (TCP/IP) can be used to interconnect devices through internet. It shall be also used in the private network establishment without connecting to the internet. It is a client-server model in which the server renders service to the clients which are connected in a network. This protocol is compatible with different operating systems (Francisco Javier Folgado *et al.* 2023). Figure 1. indicates the use of TCP/IP protocol for data transfer through internet connection. Application code in the server manipulates the data stored in memory based on the nature of the application and is ready to transfer the result. Data transfer is established through TCP in terms of packets. IP protocol establishes internet connection using Local Area Network (LAN) or Wi-Fi connection.

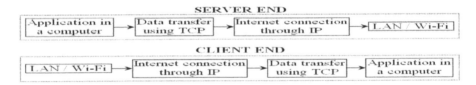

Figure 1.   TCP/IP in IoT architecture.

Computer on the Client side establishes communication to the server through the internet connection connected with the help of Wi-Fi or LAN. TCP/IP is able to establish connection when the user provides the address (Hiranmay Samanta *et al.* 2020).

2.2 *TCP/IP – LabVIEW*

LabVIEW is a graphical programming environment used to realize applications related to measurement and control. It simplifies the programming environment just by placing suitable graphical functions and interconnecting them (Reinhardt *et al.* 2019). LabVIEW supports different data communications such as USB, wireless and internet. Internet connection in LabVIEW uses TCP communication related functions as mentioned in the Figure 2. **"TCP Open Connection"** function initiates the network connection with TCP protocol.

It accepts the inputs of remote port and the address which is common to both client and server. **"TCP Read"** function of LabVIEW helps to read the data sent by the server. This function will be operative only when the internet connection is successful. This function displays the result through the output connection "data out". TCP/IP protocol allows the

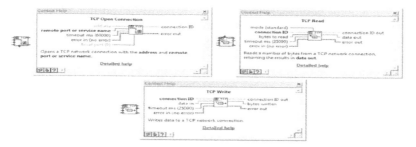

Figure 2.   TCP open connection, read, write functions in LabVIEW.

server to transmit data once the connection is established between the server and the client. However, the concept of security in data transfer shall be established with the help of **"TCP Write"** function. It sends a suitable data transfer initialization command through "data in" to the server. The server understands the client's request and responds accordingly. This ensures the authentication of client by server.

3   PROPOSED IOT ARCHITECTURE

3.1   *IoT architecture on Server / Clinet side*

This work has proposed an IoT architecture considering TCP/IP protocol in its transport layer. It used NodeMCU microcontroller which is connected with ESP8266 Wi-Fi device. Figure 3. indicates the IoT architecture on both the server / client side. NodeMCU was configured as a server. The basic idea of this work is to establish IoT architecture using TCP/IP. Hence, it has avoided the use of sensor. Rather, the data in nodeMCU was used to transfer data to the client.

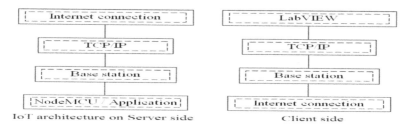

Figure 3.   IoT architecture on both Client and Server end.

NodeMCU releases local IP address through ESP8266 to establish interconnection. Wi-Fi device communicates to the TCP protocol through the base station. NodeMCU acts as both physical and application layer. In this work, code converter is considered as an application. It is able to produce binary data when it reads a decimal number. This binary data shall be transferred to the client based on the request. IoT architecture on client side has used LabVEW as its application layer. In this work, it is used to transfer a decimal number to the server, receive the equivalent binary number of it and display it. TCP/IP communication functions are used to establish internet connection. It is realized with the help of Internet and port address.

4   RESULTS AND DISCUSSIONS

Code converter realization in IoT has configured NodeMCU as a server. Local IP address of internet connection ported with this device is identified using WiFi.localIP() function. This address will be also used in the client side. It reads information from the client using the function client.readString(). This shall be used for client authentication. However, TCP/IP is flexible enough to initiate data transfer even without client authentication. Using the function aforementioned the code convertor modifies the decimal number which is to be converted into digital from the client. The server receives the data in string format. Hence, it is converted into an integer form with the help of stringvariable.toInt() function. Figure 4 indicates the complete set up to establish the required task. Figure 5 indicates the logic implemented into the NodeMCU that is responsible for the entire process and the output. It shall be noted that the result of the code conversion is sent to the client by the function, client.print(data1).

Figure 4.   Client / Server setup through TCP/IP Network connection.

The decimal number sent by the client was 189 and its equivalent binary number was 10111101.

Figure 5.   Output generated by NodeMCU.

Figure 6 indicates the coding part of LabVIEW to send / receive data to / from the server. NodeMCU (Server), and the computer (Client) used for LabVIEW application are connected through internet from a mobile phone. Its IP address was identified as 192.168.43.95.

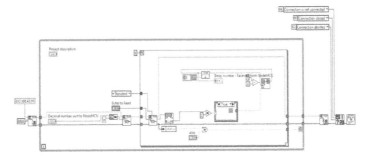

Figure 6.   LabVIEW Block diagram for number conversion, data transmission.

A screenshot of the LabVIEW front panel in run time is shown in the Figure 7. The result indicated that the data transmission was established without any data loss.

Figure 7. LabVIEW Front panel of code converter.

5 CONCLUSIONS

IoT application in LabVIEW supports graphical environment and helps to create many features to create an effective display mechanism.. This work tested the data transfer with only one client. However, the real challenge is in connecting more number of clients and ensures data delivery without compromising on speed and data loss. This will be done in the near future with an application that is able to monitor and control certain parameters from the client end.

REFERENCES

Abel E. Edje, Abd Latiff M.S. and Weng Howe Chan. 2023. IoT data analytic algorithms on edge-cloud infrastructure: A review. *Digital Communications and Networks.*

Alok Jain and Suman Bhullar. 2022. Network performance evaluation of smart distribution systems using smart meters with TCP/IP communication protocol. *Energy Reports* 8(10): 19–34.

Ambarish Gajendra Mohapatra, Anita Mohanty, Nihar Ranjan Pradhan, Sachi Nandan Mohanty, Deepak Gupta, Meshal Alharbi, Ahmed Alkhayyat and Ashish Khanna. 2023. An Industry 4.0 implementation of a condition monitoring system and IoT enabled predictive maintenance scheme for diesel generators. *Alexandria Engineering Journal* 76: 525–541.

Botero-Valencia J.S. and Valencia-Aguirre J. 2021. Portable low-cost IoT hyperspectral acquisition device for indoor/ outdoor applications. *HardwareX* 10: e00216.

Collaguazo, Adriana & Villavicencio, Monica & Abran, Alain. 2023. An activity-based approach for the early identification and resolution of problems in the development of IoT systems in academic projects. *Internet of Things*: 100929.

Fernanda Fama, Jose N. Faria and David Portugal. 2022. An IoT-based interoperable architecture for wireless bio-monitoring of patients with sensor patches. *Internet of Things* 19: 100547.

Francisco Javier Folgado, Isaías Gonz alez and Antonio Jose Calderon. 2023. Data acquisition and monitoring system framed in Industrial Internet of Things for PEM hydrogen generators. *Internet of Things* 22: 100795.

Heba Yuksel and Omer Altunay. 2020. Host-to-host TCP/IP connection over serial ports using visible light communication. *Physical Communication* 43: 101222.

Hiranmay Samanta, Ankur Bhattacharjee, Moumita Pramanik, Abhijit Das, Konika Das Bhattacharya and Hiranmay Saha. 2020. Internet of things based smart energy management in a vanadium redox flow battery storage integrated bio-solar microgrid. *Journal of Energy Storage* 32: 101967.

Muhammad Umair, Muhammad Aamir Cheema, Bilal Afzal and Ghalib Shah. 2023. Energy management of smart homes over fog-based IoT architecture. *Sustainable Computing: Informatics and Systems* 39: 100898.

Oladayo Bello, Sherali Zeadally and Mohammad Badra. 2017. Network layer inter-operation of Device-to Device communication technologies in Internet of Things (IoT). *Ad Hoc Networks* 57: 52–62.

Reinhardt S, Butschkow C, Geissler S, Dirnaichner A, Olbrich F, Lane C.E, Schroer D and Huttel A.K. 2019. Lab:: Measurement—A portable and extensible framework for controlling lab equipment and conducting measurements. *Computer Physics Communications* 234: 216–222.

Riera B, Ranger T, Saddem R, Emprin F, Chemla J.P and Philippot A. 2020. Experience feedback and innovative pedagogical applications with home I/O. *IFAC PapersOnLine* 53(2): 17610–17615.

Rolando Herrero. 2023. RTP transport in IoT MQTT topologies. *Internet of Things and Cyber-Physical Systems* 3: 37–44.

Advances in AI for Biomedical Instrumentation, Electronics and Computing – Sachan et al. (eds)
© 2024 The Author(s), ISBN 978-1-032-64298-7

Prediction of spam reviews using feature-driven opinion mining deep learning model

Surya Prakash Sharma

Research Scholar, Dr. APJ Abdul Kalam Technical University, Lucknow, U.P., India
School of Computer Science & Engineering & IT, Noida Institute of Engineering & Technology, Gr. Noida, U.P., India

Laxman Singh & Nagesh Sharma

Department of Computer Science & Engineering (AI), Krishana Institute of Engineering & Technology, Ghaziahabad, India

Abdul Khalid

School of Computer Science & Engineering & IT, Noida Institute of Engineering & Technology, Gr. Noida, U.P., India

Rajdev Tiwari

Depty. General Manager, CHEEK Edunix Pvt. Ltd. Noida, U.P., India

ABSTRACT: Online product reviews strongly influence consumers' purchases. Manufacturers and sellers value customer reviews because they impact their businesses. Scammers and spammers can post more fake reviews to promote fake products or criticize rival brands to harm companies and deceive consumers. Such behavior is called "review spamming". Differentiating fake online forum reviews from real ones is a difficult and open research problem. This study presents a CNN model that classifies reviews as spam or non-spam using deep learning. With 92.18% accuracy, the proposed model outperforms the state of the art. This paper shows how machine learning models affect product quality.

1 INTRODUCTION

Online reviews of products and services are a key source of diverse perspectives to improve e-commerce. Reviews can be spam or real. Spam review attacks have increased because anyone can write and post spam reviews online. Spam reviews, or opinion fraud, have long been a problem (Jindal & Liu 2008, February). Products are misrepresented by fake reviews. Spammers pay others to write fake product reviews. Spam reviews are usually used to promote a product or service. In Ref (Rout *et al.* 2017), the authors proposed a feature-driven fake review detection model using supervise and semi-supervise strategies. According to Ref. (Luca 2016), revenue increases 5–9% with every star rating increase. Some companies see it as an opportunity to lie to manipulate reviews.

Many studies have examined how to represent a sentence as a series of tokens or a single vector embedding (Dilawar *et al.* 2018). Logistic Regression was used in (Banerjee and Chua 2014, August) to detect FR using POS tags and writing style features like word tense. In Ref. (Javed *et al.* 2021) use CNN architectures with parallel convolution blocks for fake review detection to extract semantically rich information from word embedding. In Ref. (Hussain *et al.* 2020) analyzed language and behavior reviews with a ML algorithm to find fake reviews. In Ref. (Hussain *et al.* 2021) detects harmful spammer groups by comparing reviewers' activity, review time, and ratings. Text similarity, rating deviation, review content, product information, and reviewer behaviors can be used to detect feck reviews in (Duma *et al.* 2023). In addition to detecting fraudulent reviews, ML approaches are used to predict customer review helpfulness prediction (Sharma *et al.* 2023; Sharma *et al.* 2023; Sharma *et al.* 2023).

2 RELATED WORK

Several studies have examined spam review detection in fake reviews. This classification problem was solved by semi-supervised and supervised machine learning. In Ref. (Jindal and Liu 2008, February) authors carried out early study on the fake review identification in 2008. In Ref (Javed *et al.* 2021) authors propose three models that aggregate textual, non-textual, and behavioral features to predict fake reviews. Model predicts F1 scores of up to 92% can identify public data Yelp reviews as fraudulent.

The behavioral Method (BM) and Linguistic Method (LM) were proposed in Ref. (Hussain *et al.* 2020) to detect spam reviews. LM had 88.5% spam review detection accuracy, while BM had 93.1%. In Ref. (Hussain *et al.* 2021), authors proposed an effective method for identifying suspicious spammer groups based on reviewer ratings and review time. After eliminating spammer groups and spam reviews achieves 91% accuracy. Ref. (Yao *et al.* 2021) proposes an ensemble model for feature pruning to detect fake online reviews. Ref. (Duma *et al.* (2023) introduced a Deep Hybrid Model that learns latent text feature vectors, aspect ratings, and overall ratings simultaneously. The model had 96.5% F1-score and 96.1% recall. According to Ref. (Liu *et al.* 2020), authors developed an attention-based multilevel interactive neural network model to integrate products, users, review texts, and fine-grained aspects into review representations. In Ref. (Wang *et al.* 2022), authors proposed a multi-dimensional feature engineering review tagging model to filter fraudulent reviews using machine learning. In Ref. (Asghar *et al.* 2020) proposed a rule-based feature weighting scheme to classify review sentences as spam or non-spam. They found that spam-related features and a rule-based weighting scheme improved even the simplest spam detection method.

Major contributions this article is as follows:

1. Use ML methods to divide Amazon reviews into spam and non-spam categories.
2. The model was compared to other state-of-the-art model for accuracy, precision, recall, etc.

3 METHODOLOGY

Amazon's phone and accessory dataset helps us achieve our goal. The suggested methodology divides our study into two sections, as shown in Figure 1(a). The suggested spammer method first labels an unlabeled dataset to distinguish spam from reviews. Second part of study uses labelled dataset in CNN model to predict spam and non-spam reviews.

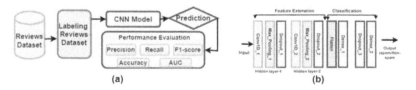

Figure 1.   (a) Architecture of proposed model, (b) CNN model architecture.

3.1 Data collection

The study uses Amazon.com data from May 1996 to July 2014, which is available at http://jmcauley.ucsd.edu/data/amazon. This dataset has been used in (Hussain 2020), (Sharma et al. 2023), which is considered the gold standard for sentiment analysis and review helpfulness prediction researchers. The data contains 194,439 reviews of 10,429 cellphones and accessories from 27879 reviewers. The dataset had nine columns: reviewerID, asin, reviewerName, helpful, reviewText, overall, summary, unixReviewTime, and reviewTime.

3.2 Feature extraction

NIn this study extracts linguistic and textual features using feature selection. 11 review features were chosen for this experiment based on their attributes in other studies (Duma et al. 2023; Hussain et al. 2020). A list of reviews related features are given in Table 1.

Table 1. List of features related to review text.

Feature	Feature Name	Description
F1	Percentage of Capital word	$F1 = \frac{number\ of\ capital\ word\ in\ review}{total\ word\ in\ review} \times 100$
F2	Number of helpful Vote	F2 = 0, no of helpful vote ≥ 5 { 1, otherwise
F3	Review Length in Character	F3 = 0, no. of character ≥ 400 { 1, otherwise
F4	Percentage of positive opinion words	$F4 = \frac{\#positive\_word}{\#total\_word} \times 100$
F5	Percentage of negative opinion Words	$F5 = \frac{\#negative\_word}{\#total\_word} \times 100$
F6	Review length in word	F6 = 0, no of word in review ≥ 220 { 1, otherwise
F7	Ratio of capital letter	Relative Ratio (RR) = [(#of Capital letter - #of Sentence) / # of Sentence] F7 = { 0, RR = 0 1, otherwise
F8	Capital letter word	F8 = 0, percentage of cap letter < 0.35 { 1, otherwise
F9	Review helpfulness	F9 = 0, helpfulness ≥ 0.6 { 1, otherwise
F10	Extrema rating	F10 = 0, rating ϵ [1,5] { 1, rating ϵ [2,3,4]
F11	Sentiment	review sentimental play an important to filters out all spam reviews

3.3 Labeling the review dataset

The spam score method (Hussain, N et al. 2020) was used to classify reviews as false or true. Reviews are classified as spam or non-spam using spam score (Hussain, N. et al. 2020). As shown in equation 1, spam score is calculated on extracted features from F1 to F11. Finally, classify reviews as spam or non-spam by comparing spam scores to threshold values (Φ) of 0.5. Labelling reviews is shown in Eq. (2).

$$spam\ score = \frac{W_i * N_i}{W_i} \quad where \quad i = 1, 2, \ldots .14 \tag{1}$$

W_i is the weight assign by drop feature method and N_i is the normalized value of each review feature

$$labe_{ri} = \begin{cases} 0; nan \quad spam \quad spam\_score(ri) < \emptyset \\ 1; spam \quad spam_{score(ri)} \geq \emptyset \end{cases}$$

Where ri is the ith review in review text R.

4 CONSTRUCTION OF PREDICTION MODEL

In the previous section of this study, a classification model is created to confirm fake review detection and classify Amazon customer reviews into spam and non-spam. The classification model was created using N × F feature matrix, where N is the total number of reviews and F is the total number of features from F1 to F11. We used state of art feature sets and well-known machine learning methods to distinguish spam from non-spam reviews in a dataset. Convolution Neural Networks were used to detect fake reviews. The training set was 80% of the dataset, which was trained using gradient boosting. The model was trained using machine learning on this dataset. System performance was assessed using the remaining 20%. All model creation and testing use Python 3.6. Five assessment indicators (F1 Score, AUC, Precision, Recall, and Accuracy) evaluate categorization model effectiveness along-side this research.

5 CNN MODEL ARCHITECTURE

First, the study classified Amazon reviews as spam or non-spam. In the second part of our research, we used CNN-based deep neural networks to divide this labelled Amazon dataset into spam and non-spam reviews. Tensorflow and Keras were used to implement the model in Python. Hidden layers 1 and 2 make up the suggested model we used in our experiment (Figure 2 (b)). Pool, batch normalization, and ReLU as the rectified linear unit are in each convolution layer (Conv1D). Dropout, sigmoid, classification, and fully connected output layers are included. Additionally, binary cross entropy was chosen as the output layer loss function. First and second hidden layers had 32 and 64 neurons, respectively. We conducted the various experiments on cellphone and accessories dataset in order to get the optimal results using fine-tuned the parameters of CNN model.

6 EXPERIMENTAL RESULTS

In this study, we tested several machine learning models on user-generated reviews using various evaluation matrices and proposed feature sets.

6.1 *Performance of CNN model*

Three methods a 10-fold cross-validation procedure on the feature set creates predictive models for false user-generated reviews using K-Nearest Neighbors (KNN), Logistic Regression (LR), and CNN. Table 2 shows how well the CNN model calculated review values in this study. Predictive model efficacy is measured using AUC, F1 score, precision, recall, accuracy, and accuracy.

Table 2 shows that the CNN model had 92.18 percent accuracy, 91.02 percent F1-score, and 97.40 percent AUC. CNN was again found to be the most effective model, producing competitive and promising results. As discussed above, the CNN model predicted fake reviews well.

Table 2.   Model effectiveness.

Method	Precision (%)	Recall (%)	F1-Score (%)	Acc. (%)	AUC (%)
CNN	92.14	89.93	91.02	92.18	97.40
KNN	92.00	94.00	93.00	91.61	97.10
LR	77.00	99.00	87.00	83.24	95.80

6.2   *Performance in training and validation of CNN Model*

Complete model evaluation takes 737 seconds in 15 epochs. Figure 2 displays the performance of our suggested model. Losses were 19.09 and 18.87 percent, and accuracy was 92.37% and 92.18%. These recommended feature sets have 92.14% precision, recall, accuracy, F1-score, and AUC, 89.93%, 91.02%, 92.18%, and 97.40%, respectively. Figure 2 (c) shows the precision recall curve for Amazon reviews to show how these suggested feature sets affect fake review classification. Thus, our model predicts 97.40 percent AUC. Visualization of the three models' prediction performance is shown in Figure 2(d).

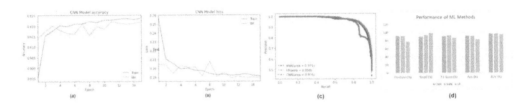

Figure 2.   Model performance in (a) accuracy, (b) loss, (c) Precision-Recall, (d) percentage (%).

6.3   *Performance comparison with others state of art models*

Table 3 compares the effectiveness of our model with other state-of-the-art models that have recently been published in the literature. CNN was 6.59%, 0.08%, and 3.97% more accurate than (Rout, J. K. *et al.* 2017) (Mukherjee, A. *et al.* 2013), (Hussain, N. *et al.* 2020) techniques, according to Table 3. This shows that the recommended method predicts fake and honest reviews better. Our suggested method outperformed cutting edge algorithms with an F1 score of 91.02 (Rout, J. K *et al.* 2017), (Mukherjee, A. *et al.* 2013), (Hussain, N. *et al.* 2020). Area under the curve (AUC) was 97.40% for the study's results. Recommended model accuracy exceeded 92.18%.

Table 3.   The performance comparison of various state of art models with the proposed model.

Ref.	Features	ML Method	Precision	Recall	F1 score	Accuracy	AUC
Proposed Model	**review**	**CNN**	**92.14**	**89.93**	**91.02**	**92.18**	**97.40**
(Rout, J. K *et al.* 2017)	review	Decision Tree	X	X	X	92.11	X
(Mukherjee, A. *et al.* 2013)	review	SVM	84.1	87.3	85.7	86.10	X
(Hussain, N. *et al.* 2020)	review	LR	88.1	87.3	88.7	88.52	89.58

6.4 Conclusion and future work

This research identifies and classifies spam. The study suggests classifying content as original, false, spam, and non-spam. The proposed model outperformed comparable methods in precision, recall, accuracy, AUC, and F1-score. The proposed method detects Amazon spam reviews with 92.18% accuracy. Without critical reviewer data like number of reviews, IP address, age, etc., our spam analysis is incomplete. Due to privacy concerns, we do not collect user data from the aforementioned websites, and only they can analyze it. This study only uses Amazon, but it can be expanded to use TripAdvisor, Yelp, and eBay datasets for experiments.

REFERENCES

Asghar, M. Z., Ullah, A., Ahmad, S., & Khan, A. (2020). Opinion spam detection framework using hybrid classification scheme. *Soft computing*, *24*, 3475–3498.

Banerjee, S., & Chua, A. Y. (2014, August). Applauses in hotel reviews: Genuine or deceptive?. In *2014 Science and Information Conference* (pp. 938–942). IEEE.

Dilawar, N., Majeed, H., Beg, M. O., Ejaz, N., Muhammad, K., Mehmood, I., & Nam, Y. (2018). Understanding citizen issues through reviews: A step towards data informed planning in smart cities. *Applied Sciences*, *8*(9), 1589.

Duma, R. A., Niu, Z., Nyamawe, A. S., Tchaye-Kondi, J., & Yusuf, A. A. (2023). A Deep Hybrid Model for fake review detection by jointly leveraging review text, overall ratings, and aspect ratings. *Soft Computing*, *27*(10), 6281–6296.

Hussain, N., Mirza, H. T., Ali, A., Iqbal, F., Hussain, I., & Kaleem, M. (2021). Spammer group detection and diversification of customers' reviews. *PeerJ Computer Science*, *7*, e472.

Hussain, N., Mirza, H. T., Hussain, I., Iqbal, F., & Memon, I. (2020). Spam review detection using the linguistic and spammer behavioral methods. *IEEE Access*, *8*, 53801–53816.

Javed, M. S., Majeed, H., Mujtaba, H., & Beg, M. O. (2021). Fake reviews classification using deep learning ensemble of shallow convolutions. *Journal of Computational Social Science*, 1–20.

Jindal, N., & Liu, B. (2008, February). Opinion spam and analysis. In *Proceedings of the 2008 International Conference on Web Search and Data Mining* (pp. 219–230).

Liu, M., Shang, Y., Yue, Q., & Zhou, J. (2020). Detecting fake reviews using multidimensional representations with fine-grained aspects plan. *IEEE Access*, *9*, 3765–3773.

Luca, M. (2016). Reviews, reputation, and revenue: The case of Yelp. com. *Com (March 15, 2016). Harvard Business School NOM Unit Working Paper*, (12-016).

Mukherjee, A., Venkataraman, V., Liu, B., & Glance, N. (2013). What yelp fake review filter might be doing?. In *Proceedings of the International AAAI Conference on Web and Social Media* (Vol. 7, No. 1, pp. 409–418).

Rout, J. K., Singh, S., Jena, S. K., & Bakshi, S. (2017). Deceptive review detection using labeled and unlabeled data. *Multimedia Tools and Applications*, *76*, 3187–3211.

Sharma SP, Singh L, Tiwari R. (2023), "Prediction of customer review's helpfulness based on sentences encoding using CNN-BiGRU model" *Journal of Autonomous Intelligence*, vol. 06, issue 03, 2023.

Sharma, S. P., Singh, L., & Tiwari, R. (2023). Integrated feature engineering based deep learning model for predicting customer's review helpfulness. *Journal of Intelligent & Fuzzy Systems, (Preprint)*, 1–18.

Sharma, S. P., Singh, L., & Tiwari, R. (2023). Prediction of Customer Review's Helpfulness Based on Feature Engineering Driven Deep Learning Model. *International Journal of Software Innovation (IJSI)*, *11*(1), 1–16.

Wang, G., Shang, G., Pu, P., Li, X., & Peng, H. (2022). Fake review identification methods based on multi-dimensional feature engineering. *Mobile Information Systems*, *2022*.

Yao, J., Zheng, Y., & Jiang, H. (2021). An ensemble model for fake online review detection based on data resampling, feature pruning, and parameter optimization. *Ieee Access*, *9*, 16914–16927.

Advances in AI for Biomedical Instrumentation, Electronics and Computing – Sachan et al. (eds)
© 2024 The Author(s), ISBN 978-1-032-64298-7

Design and implementation of automatic street light systems

Mohit Tyagi, Kanishka Chauhan, Hardik Mitra, Raman Pundhir, Naman Gupta,
Sachin Tyagi & Satya Prakash Singh
*Department of Electronics and Communication Engineering, KIET Group of Institutions,
Delhi-NCR, India*

ABSTRACT: Design the Automatic Street light systems to react dynamically to environmental conditions, which requires the study of sensor technologies. The IR and LDRs sensor technology is used in the implementation of automatic street light systems. IR sensors are designed to detect the vehicles and any obstacles. LDR is used the detect the light intensity, based on which switching ON/OFF operation of lights was performed. Therefore, LDR helps the street light system to work automatically. The implementation of this system is to focus on the issue of energy wastage and the cost of maintenance of street lighting systems.

1 INTRODUCTION

Switching ON/OFF the streetlights manually is a challenging task, so designing and implementing the streetlights which works automatically focuses on a goal to reduce the wastage of electricity and the cost of maintenance by turning the light ON during the nighttime and turning the lights OFF during the daytime. LDR, LEDs, IR sensors, and Arduino UNO are the fundamental parts needed to build motion-detecting streetlights. Street light system to work automatically needs an IR (Infrared) sensor and LDR (Light Dependent Resistor) sensors. The role of IR sensors in street light systems is to detect the movement of vehicles or any other obstacles and the role of LDR is to detect whether there is daytime or nighttime. Therefore, when LDR senses the light, the system will turn ON automatically and when LDR senses no light, the system will be turned off until LDR does not sense light intensity. When an IR sensor senses any movement of a vehicle or a human being the streetlights are turned on with high intensity and the system will radiate high-intensity light until the system detects the movement of vehicles or human beings. Power LED lights are employed because power LEDs are highly energy efficient and save a lot of energy (Singh *et al.* 2020–2021).

2 LITERATURE SURVEY

(Singh *et al.* 2020–2021) discusses about the advancing technique in the systems of street lighting to supply a better solution for reducing the wastage of power and minimizing manual operations of street lighting. (A.V.M. *et al.*) has worked on the technologies to reduce the power consumption by using advanced IR sensor and LDR. (Srilokh *et al.* 2019) focuses on a smart street light system, explores how IoT can manage various end systems and offer open data access. This technology uses a microcontroller instead of conventional infrared sensors to automatically detect changes in object's position. By turning lights on when light is not present, system conserves energy by maximising light intensity using photoelectric and light sensors.

DOI: 10.1201/9781032644752-73

3 PROPOSED WORK

3.1 *Circuit diagram*

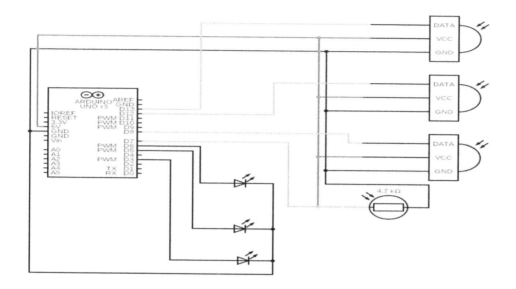

(Circuit diagram of automatic streetlight using Arduino UNO, LDR, IR sensor, LED)

The circuit shown above is of Automatic Street Light which has been made using components Arduino uno, LDR, IR Sensor and LED.

3.1 *Arduino uno*

Arduino uno is used as the core computational unit in the street light system. The system of streetlight is used to gather information from both LDR and IR sensor, process the data and executes according to the preprogrammed algorithms. PIN D13, D12 and D8 of Arduino uno used to collect the data from three different IR sensor individually. PIN D7 of Arduino uno is connected with the LDR.PIN D6, D5 and D3 of Arduino uno are connected with the three different LED individually. PIN 5V of Arduino uno is used to give the power supply to all 3 IR sensor and LDR.PIN GND of Arduino uno is connected with all the components.

3.2 *IR (Infrared) sensor*

The IR sensor is placed on opposing sides of the roadway, collaborating to find the duration a vehicle requires to traverse from one sensor to opposite, there by easing the measurement of the vehicle speed. Total 3 IR sensors are used here. The data pin of the three IR sensor is interlinked with the PIN D13, D12 and D8 of the Arduino uno board. The power from the Arduino is received by the PIN VCC of the all the IR sensor. The ground connection of all IR sensor is joined with the GND PIN of the Arduino uno board.

3.3 *LDR (Light Dependent Resistor)*

The LDR serves as the light sensitive sensor used to measure the levels of light. The purpose of LDR is to decide whether to activate streetlight based on surrounding light condition. The LDR Pin's output is linked with the D7 PIN of the Arduino uno board. The power from the

Arduino is given to the LD. The ground connection of LDR is joined with the GND PIN of the Arduino uno board.

3.4 *LED (Light Emitting Diode)*

LED streetlight serves as the lighting fixtures responsible for supplying the illumination on the road. The lights are linked to the Arduino uno for the output. Total of 3 LEDs are used in the system of streetlight. The LED output signals are attached to the PIN D6, D5 and D3 of the Arduino uno board, respectively. Simultaneously, the negative terminals of all LEDs are connected to the GND PIN of the Arduino uno board.

3.5 *Working*

The Arduino UNO is a microcontroller board having fourteen digital inputs/outputs pins and having an Integrated Development Environment which can be used for writing and uploading the codes in an Arduino uno board. The code for controlling and checking the system is uploaded in an Arduino UNO. 3 IR sensors and a LDR and 3 LEDs are used in the street light system. Pin D13, D12 and D8 of Arduino UNO collects data from three different IR seasons individually. Pin D7 is connected with the LDR pin output. All 3 LEDs are connected to pin D6, D5 and D3, respectively. Negative terminal off all the LEDs is connected to GND pin. Therefore, Arduino UNO collected the data from IR sensor and LDR and then the system runs accordingly to the values specified in the code which is embedded in Arduino UNO.

Case 1: During the daytime, if any of the IR sensor detects any movement then the lights will set to turned OFF because at the daytime, light falls on the LDR then the resistance of the LDR decreases due to which the streetlights will not glow.

Case 2: During the nighttime, when no light falls on the LDR then the LDR resistance increases due to which the lights are set to glow with light intensity when the IR sensors does not sense any vehicles but if IR sensor sense the vehicles, then the street light glow brightly.

4 RESULTS

An automated street lighting system underwent testing in a controlled setting, yielding promising outcomes. The speed detection through infrared sensors and ambient light sensing via the LDR enabled dynamic control of streetlight luminosity. The system effectively curtailed energy consumption during periods of reduced traffic while upholding adequate illumination levels for road safety. The observations are as follows:

Case 1: When circuit is in OFF state: IN Figure (A) an automatic street light system stays inactive and does not illuminate the lights. The LDR is used to detect the intensity of the light and can distinguish between daylight and darkness. The Arduino Uno is programmed with a threshold level of light so if the daylight is greater than the predefined threshold level then the lights will be remained in off state.

Case 2: When circuit is in ON state: In Figure (B) an automatic street light system stays active and illuminate the lights on the street. The LDR is used to detect the intensity of the light and can find if it is darkness or not. The

Arduino Uno is programmed with the threshold level of the light, if the light falls below the predefined threshold level, the Arduino Uno will activate the streetlight and system changes from off state to on state, illuminating the streetlight.

Case 3: When circuit is in ON state and Intensity of LED is controlled: In Figure (C, D, E) when there is no movement for a certain period of time on the road the Street lights are preprogrammed to be at low intensity or dim to conserve the energy. When the motion is sensed by the IR sensor placed on both sides of the road and send the signal to an Arduino Uno to increase the intensity of the Street lights ensuring the safety and security in the area.

Figure 1. When circuit is in OFF state.

Figure 2. When circuit is in ON state.

Figure 3. When circuit is in ON state, intensity of light is controlled.

Figure 4. When circuit is in ON state, intensity of light is controlled.

Figure 5. When circuit is in ON state, intensity of light is controlled.

5 CONCLUSIONS

This study presents a system that will lower power usage while being economical, safe, and requiring less human labour. Significant power savings are possible if the system is implemented in every city. The designed system of streetlight which works automatically uses unique power-saving mechanisms, and it replaces standard bulbs with LED lamps to save power for streetlights. The automatic street light system turns out to be the most dependable and effective method for turning on and off streetlights automatically, thus by reducing the manual operations. The automated street light system offers a useful way to preserve energy by avoiding needless electrical waste brought on by manually turning on or off streetlights when not in use. It uses a dynamic traffic flow control method.

REFERENCES

A. V. M., S. Shankar M, A.R., Keerthana S., Lalithakala R., "Design and Implementation of Automatic Brightness Control and Switching Street Lights using Arduino", *Department of Biomedical EngineeringKarpaya Vinayaga college of Engineering and Technology*.

AlMaeeni S., Attia H., Takruri M., Altunaiji A., Sanduleanu M., Shubair R., Ashhab Md.S., Ali M.Al., and Hebsi G.Al., "Smart City: Recent Advances in Intelligent Street Lighting Systems Based on IoT, *Journal of Sensors*, volume 2022, Article ID. 5249187.

Bachanek K.H., Tundys B., Wisniewski T., Puzio E., Marouskova A., "Intelligent Street Lighting in a Smart City Concepts- A Direction to Energy Saving in Cities: An Overview and Case Study", *Energies*, vol.14, Issue 11.

Bajaj K.A. and Mote T.S. (2013), "Review on intelligent street lightening system," *International Journal of Science and Research*, vol. 4, pp. 1624–1626, 2013.

Chen Z., Sivaparthipan C.B., Muthu B.A., "IoT based smart and intelligent smart city energy optimization", *Sustainable Energy Technologies and Assessments*, vol.49, Article 101724.

Choudhury M.S.H., Munna MD.M.H., Islam N., Islam Md.R., Islam Md.A., "Design and Implementation of an Automated Solar Street Light System, *SEU Journal of Science and Engineering*, vol.12, No.1.

Kundu A., KR Singh R., Ram C.K., Saha S., "Vehicle Movement Street Light with Automatic Light Sensing", *Department of Electronics & Communication Engineering, RCC Institute of Information Technology*.

Mahoor M., Hosseini Z.S., Khodaei A., Paaso A., Kushner D. (2020), "State-of-the-art in smart streetlight systems: a review", *IET- Smart Cities*, vol. 2, no. 1, pp. 24–33, 2020.

Singh J., Punjan Jaiswal S., Jain M., Ambika Pathy A., Bhadoria V. S, Asad S. (2021), "Automation in streetlights using IR sensors and LDR," *Recent Development in Engineering and Technology 2020–2021*.

Srilokh C., Reddy G.R., "Design and Practical Implementation of Intelligent Street Light System by Using IoT Platform," *International Journal of Innovative Technology and Research*, Issue no.5, vol. no. 7.

Villa C., Bremond R., and Saint-Jacques E. (2017), "Assessment of pedestrian discomfort glare from urban LED lighting," *Lighting Research & Technology*, vol. 49, no. 2, pp. 147–172, 2017.

Zissis G., Dupuis P., Canale L., Pigenet N., "Smart lighting systems for smart cities", *Holistic Approach for Decision Making Towards Designing Smart Cities*, Springer, vol.18.

Advances in AI for Biomedical Instrumentation, Electronics and Computing – Sachan et al. (eds)
© 2024 The Author(s), ISBN 978-1-032-64298-7

Deep learning based computerized diagnosis of breast cancer using digital mammograms

Laxman Singh & Rekha Kashyap
Department of Computer Science & Engineering (AI & ML), KIET Group of Institutions, Ghaziabad, U.P., India

Sovers Singh Bisht
Department of Computer Science & Engineering (Data Science), Noida Institute of Engineering and Technology, Gr. Noida (UP)

Nagesh Sharma
Department of Computer Science & Engineering, KIET Group of Institutions, Ghaziabad (UP), India

Surya Prakash Sharma
Department of Computer Science & Engineering, Noida Institute of Engineering and Technology, Gr. Noida (UP)

ABSTRACT: Breast cancer (BC) is the second most common and deadly type of cancer in women, after skin cancer. The likelihood of a patient surviving breast cancer is greatly increased by early detection and classification. The authors of this work aim to create a deep learning (DL) model that can identify and categorize tumours in mammography images. When it comes to finding breast cancer as soon as possible before it becomes incurable, mammography is regarded as the gold standard. In the suggested work, we classified malignant and benign cells using pre-trained CNN architecture in conjunction with VGGNet-16, VGGNet-19, and Efficient Net BO. The results show how useful deep learning-based models can be in aiding radiologists in the interpretation of digital mammograms and validate the significance of these models for automated breast cancer diagnosis.

1 INTRODUCTION

According to (Singh & Jaffery 2018), breast cancer is the primary cause of death worldwide. When certain breast cells start to grow abnormally, this is what happens. These aberrant cells typically result in cancer. The three main components of a woman's breast are the connective tissue, which is made up of fatty and fibrous tissue, lobules, and ducts. Usually, breast cancer starts in the ducts and lobules of the breast, resulting in lumps that eventually develop into tumours. Masses and calcification are two types of breast abnormalities that could be signs of breast cancer. A mass can be either benign or malignant, and its shape and size can vary, such as being lobular, oval, or round (Singh & Jaffery 2018). Early and precise identification of breast cancer is crucial to prevent unnecessary therapy whether it comes to false negatives (FNs) or false positives (FPs) delaying treatment (Allada *et al.* 2021). Screening approaches employ a variety of techniques, including ultrasound (US), magnetic resonance imaging (MRI), and mammography imaging (Das *et al.* 2021). This technique will help reveal a number of hidden features. The most popular screening technique for identifying cancer early on before it spreads is multi-modal mammography. Radiologists who specialize in mammography conduct multiple mammograms, or breast X-ray images, to identify tumours and categorize them as either benign or malignant, or as cancerous.

DOI: 10.1201/9781032644752-74

By incorporating deep learning principles to address machine learning (ML) challenges, a general intelligence system may be able to identify cancer types more accurately. According to (Khan *et al.* 2019) and (Dafni Rose *et al.* 2022), deep learning involves the extraction of pertinent information from raw images and its efficient application in the classification processes.

2 RELATED WORK

Many studies have been carried out utilizing various techniques to detect breast cancer. After considering the drawbacks of several machine learning-based algorithms, Khan *et al.* 2019 suggested ensemble-based algorithms. Prasuna *et al.* 2019 put forth a successful method for breast cancer prediction.

Deep learning architectures were used by the authors (Valkonen *et al.* 2019) to detect epithelial cells. They utilized the VGG-16 model to initialize the weight prior to actual utilization. Nonetheless, the accuracy of the model reached 88%. We followed the authors' Neural Network algorithm model for breast cancer identification, which was described in (Naresh Khuriwal *et al.* 2018), by using a standardization approach for pre-processing and then using the PCA algorithm on breast cancer datasets. Then, an algorithm based on a neural network may be applied, yielding an accuracy rate of 99.67%, surpassing that of other machine learning algorithms. Using a hybrid model, (Williamson *et al.* 2022) predicted breast cancer from mammography images. Features from the first segment image are extracted during the mammography process, and feature extraction is then used for categorization to show Retina Net with YOLO for 91% accurate detection of breast cancer. The authors suggested the deep learning-based CNN model, which has been used to identify normal and abnormal mammograms in mammography pictures (Shwetha *et al.* 2018). When the authors compare the two models, InceptionV3 performs better with an accuracy rate of 83%. In order to isolate reduced and simplified DL structures, (Sreelatha & Krishna Reddy 2021) extracted the Region of Interest (ROI) from mammogram images using the CNN and deep learning AE module, two common deep learning models.

3 METHODOLOGY USED

A suggested framework for using CNN architectures to diagnose breast cytology imaging is presented in this section. Here, datasets include training sets and testing set. Figure 1 represent the block diagram of computer diagnosis system.

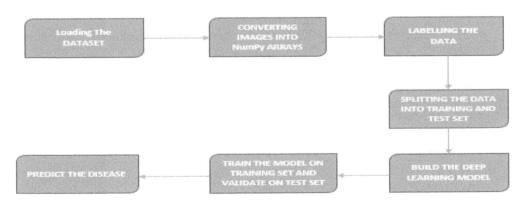

Figure 1. Block diagram of computer aided diagnosis system.

3.1 Data pre-processing and augmentation

Mammograms typically contain noise, erratic lighting, and instances of artifacts that must be removed before augmentation can be performed. Otherwise, there may be erroneous object segmentation and a false positive rate (Singh & Jaffery 2018). It is described as the main process on a medical image that leads to the CAD extraction and identification of pertinent features. Contract enhancement and image de-noising are carried out during the pre-processing stage (Allada *et al.* 2021). Pre-processing is used in a data augmentation process to add to the data set. CNN requires a large dataset in order to achieve higher accuracy; when a small dataset is overfitted, CNN's performance suffers.

3.2 CNN architecture: Vggnet – 19

Here, we used the Visual Geometry Group Network architecture that is abbreviated as VGG. The CNN model VGG-19 is used for speech recognition, pattern recognition, and image processing on the ImageNet dataset. The basic principle of VGG-19 is to use a stack of 3×3 convolutional layers with modest respective fields. Max pooling was employed in VGG-19 as a handler to reduce the size. 4.96 neurons were found to be occupied in the Fully Connected layers. The features of this architecture are extracted using its layers. These are fully integrated with the dropout layer and max pooling, where convolution is carried out by each group.

3.3 Transfer Learning

One popular machine learning technique, known as "Transfer Learning" (Ramdan *et al.* 2020), is used to train a single dataset for a variety of problems. The base network is initially trained on the connected dataset for each task, and the target action is passed from the target dataset. If the target data set is the same size as the original datasets, or smaller, the likelihood of over-fitting increases. Additionally, if the size of the target dataset is greater than or equal to that of the original dataset, fine-tuning pre-trained models is probably a good method to lessen over-fitting.

4 RESULTS AND DISCUSSION

4.1 Data set

The dataset used here has been taken making up around 80% of all cases that have been diagnosed. Reducing patient mortality and implementing appropriate treatment requires early and accurate detection. In order to categorize IDC, a range of deep CNN architectures that are taught to transmit knowledge were evaluated. This study makes use of the Breast Cancer Histopathologic (Break-His) Image database.

The 162 complete mount slide photos of BC that have been $40 \times$ magnified make up the original data set. The data shows that 2,77,524 50×50 patches were removed, of which 78,786 were malignant and 198,738 were benign.

4.1.1 Metrics of evaluation
You may utilize the data from the confusion matrix shown in Table 1 to assess the assessment metrics Precision, Accuracy, Recall, F1 Score, Sensitivity, Specificity, and AUC-ROC Score.

Table 1. Matrix of confusion.

ACTUAL	PREDICTED	
	0	1
0	TN	FP
1	FN	TP

where True Negative, False Positive, False Negative, and True Positive are defined as TN, FP, FN, and TN, respectively.

(1) Accuracy: Calculated as the ratio of accurate samples to total samples, accuracy measures the overall correctness of the categorization.

$$Accuracy = (TP + TN)/(TP + FP + TN + FN)$$

(2) Precision: The precision metric calculates the percentage of accurately predicted instances relative to all expected instances.

$$Precision = TP/(TP + FP)$$

(3) Recall: The percentage of accurately predicted cases to all forecasts with actual sickness is known as recall.

$$Recall = TP/(TP + FN)$$

(4) F1 Score: The F-score provides a balance between the two metrics by measuring the harmonic mean of recall and accuracy. An F-score is another name for an F1 score.

$$F1Score = 2 * (Precision * Recall)/(Precision + Recall)$$

5 RESULT AND ANALYSIS

This section presents the classification performance of transfer learning models that were applied to mammography pictures after pre-processing and augmentation stages. Table 2 demonstrates that when used alone, each of the three topologies yields mediocre classification results. As examples, consider the VGG-16 accuracy of 0.7532, recall of 0.7460, F measure of 0.7496, and auc-roc score of 0.8289. The accuracy is 0.7455, recall is 0.9396, F measure is 0.8314, and auc-roc score is 0.9426 on Efficient NetBO. The AUC-ROC score in VGG-19 is 0.9412, the accuracy is 0.8331, the recall is 0.9089, and the F measure is 0.8694. These findings show that VGG-19 performs better in terms of accuracy score and F1-score than the other two designs. In terms of recall score, the EfficientNetBO design outperforms the other two. This figure shows the assessment measures, which are used to measure performances: FP (False Positive), TN (True Negative), TP (True Positive), and FN (False Negative). Figure 2 shows the confusion matrix of the VGG 19 architecture. The combined precision recall curves of the three CNN architectures—VGG-16, VGG-19, and EfficientNetBO—are shown in Figure 3.

Table 2. Comparison of three deep learning architecture for breast cancer detection.

CNN ARCHITECTURE	PRECISION	RECALL	F1 SCORE	AUCR-ROC SCORE
VGG-16	0.7532	0.746	0.7496	0.8289
VGG-19	0.8331	0.9089	0.8694	0.9412
EFFICIENTNETBO	0.7455	0.9396	0.8314	0.9423

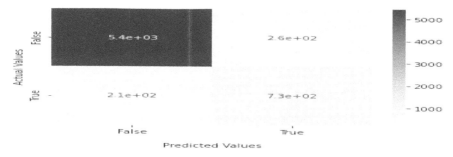

Figure 2. Confusion matrix of the VGG-19 architecture.

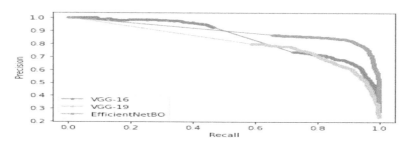

Figure 3. Combined precision-recall curve.

5 CONCLUSIONS

The study examined the theory of breast cancer diagnosis through the application of transfer learning theory. Using image datasets related to breast cancer histopathology, we used various CNN architectures in this investigation. Our goal was to identify the best CNN architecture for classification. To sum up, the VGG-19 has produced the best results.

REFERENCES

Allada, A., Rao, G. R. K., Chitturi, P., Chindu, H., Prasad, M. S. N., & Tatineni, P. (2021, March). Breast cancer prediction using Deep Learning Techniques. In *2021 International Conference on Artificial Intelligence and Smart Systems (ICAIS)* (pp. 306–311). IEEE.

Dafni Rose, J., VijayaKumar, K., Singh, L., & Sharma, S. K. (2022). Computer-aided diagnosis for breast cancer detection and classification using optimal region growing segmentation with MobileNet model. *Concurrent Engineering*, *30*(2), 181–189.

Das, A., Mohanty, M. N., Mallick, P. K., Tiwari, P., Muhammad, K., & Zhu, H. (2021). Breast cancer detection using an ensemble deep learning method. *Biomedical Signal Processing and Control*, *70*, 103009.

Demir, A., Yilmaz, F., & Kose, O. (2019, October). Early detection of skin cancer using deep learning architectures: resnet-101 and inception-v3. In *2019 medical technologies congress (TIPTEKNO)* (pp. 1–4). IEEE.

Eldin, S. N., Hamdy, J. K., Adnan, G. T., Hossam, M., Elmasry, N., & Mohammed, A. (2021, May). Deep learning approach for breast cancer diagnosis from microscopy biopsy images. In *2021 International Mobile, Intelligent, and Ubiquitous Computing Conference (MIUCC)* (pp. 216–222). IEEE.

Ghosal, P., Nandanwar, L., Kanchan, S., Bhadra, A., Chakraborty, J., & Nandi, D. (2019, February). Brain tumor classification using ResNet-101 based squeeze and excitation deep neural network. In *2019 Second International Conference on Advanced Computational and Communication Paradigms (ICACCP)* (pp. 1–6). IEEE.

Hasan, N., Bao, Y., Shawon, A., & Huang, Y. (2021). DenseNet convolutional neural networks application for predicting COVID-19 using CT image. *SN computer science*, *2*(5), 389.

Jasil, S. G., & Ulagamuthalvi, V. (2021, May). Skin lesion classification using pre-trained DenseNet201 deep neural network. In *2021 3rd international conference on signal processing and communication (ICPSC)* (pp. 393–396). IEEE.

Khan, S., Islam, N., Jan, Z., Din, I. U., & Rodrigues, J. J. C. (2019). A novel deep learning based framework for the detection and classification of breast cancer using transfer learning. *Pattern Recognition Letters*, *125*, 1–6.

Khuriwal, N., & Mishra, N. (2018, October). Breast cancer diagnosis using deep learning algorithm. In *2018 International Conference on Advances in Computing, Communication Control and Networking (ICACCCN)* (pp. 98–103). IEEE.

Mateen, M., Wen, J., Nasrullah, Song, S., & Huang, Z. (2018). Fundus image classification using VGG-19 architecture with PCA and SVD. *Symmetry*, *11*(1), 1.

Mohamed, A., Amer, E., Eldin, N., Hossam, M., Elmasry, N., & Adnan, G. T. (2022). The impact of data processing and ensemble on breast cancer detection using deep learning. *Journal of Computing and Communication*, *1*(1), 27–37.

Prasuna, K., Rama, R. K., & Saibaba, C. H. M. H. (2019). Application of machine learning techniques in predicting breast cancer-A survey. *Int J Innov Technol Exploring Eng*, *8*(8), 826–32.

Ramdan, A., Heryana, A., Arisal, A., Kusumo, R. B. S., & Pardede, H. F. (2020, November). Transfer learning and fine-tuning for deep learning-based tea diseases detection on small datasets. In *2020 International Conference on Radar, Antenna, Microwave, Electronics, and Telecommunications (ICRAMET)* (pp. 206–211). IEEE.

Shwetha K., Spoorthi M., Sindhu S.S., Chaithra D. 2018, "Breast cancer detection using deep learning technique," *International Journal of Engineering Research & Technology*, vol. 6, no. 13, pp. 1–4.

Singh, L., & Jaffery, Z. A. (2018). Computerized detection of breast cancer in digital mammograms. *International Journal of Computers and Applications*, *40*(2), 98–109.

Spanhol, F.A., Oliveira, L.S., Petitjean, C., Heutte, L. 2016: Breast cancer histopathological image classification using convolutional neural networks. In: *International Joint Conference on Neural Networks (IJCNN)*, pp. 2560–2567. IEEE.

Sreelatha, K., & Krishna Reddy, V. (2021). Integrity and memory consumption aware electronic health record handling in cloud. *Concurrent Engineering*, *29*(3), 258–265.

Sudha, V., & Ganeshbabu, T. R. (2021). A convolutional neural network classifier VGG-19 architecture for lesion detection and grading in diabetic retinopathy based on deep learning. *Computers, Materials & Continua*, *66*(1).

Valkonen, M., Isola, J., Ylinen, O., Muhonen, V., Saxlin, A., Tolonen, T., ... & Ruusuvuori, P. (2019). Cytokeratin-supervised deep learning for automatic recognition of epithelial cells in breast cancers stained for ER, PR, and Ki-67. *IEEE transactions on medical imaging*, *39*(2), 534–542.

Williamson, S., Vijayakumar, K., & Kadam, V. J. (2022). Predicting breast cancer biopsy outcomes from BI-RADS findings using random forests with chi-square and MI features. *Multimedia Tools and Applications*, *81*(26), 36869–36889.

Zhong, Z., Zheng, M., Mai, H., Zhao, J., & Liu, X. (2020, November). Cancer image classification based on DenseNet model. In *Journal of physics: conference series* (Vol. 1651, No. 1, p. 012143). IOP Publishing.

Advances in AI for Biomedical Instrumentation, Electronics and Computing – Sachan et al. (eds)
© 2024 The Author(s), ISBN 978-1-032-64298-7

Shoe extension using ultrasonic sensor and gyroscope for blinds

Shikha Agarwal, Aarti Chaudhary & Veena Bharti
Information Technology Department, Ajay Kumar Garg Engineering College, Ghaziabad, Uttar Pradesh, India

Shivam Umrao
Information Technology Department, Raj Kumar Goel Institute of Technology, Ghaziabad, Uttar Pradesh, India

ABSTRACT: Many of us are busy in our lives and may not bother about blind people who need help in their basic tasks like walking. Hence, to make them more independent. In this paper, propose a design of a pair of smart shoe extensions that can be worn by a visually impaired man which acts as a helping hand in his mobility without the use of a conventional stick. These shoe extensions are also able to detect a sharp devaluation of the level of the land surface due to the geography of a place such as mountains and trenches or detect the use of stairs with the help of a gyroscope and an ultrasonic sensor. These shoes help a visually impaired man to walk confidently and keep his flair unhindered.

1 INTRODUCTION

Most of us are fortunate to be able to do our daily chores without help because we are blessed with vision. But some of us don't have our eyesight from birth or some of us have lost our eyesight in accidents. According to a report by the BBC on 17 august 2017, the number of blind people will triple by 2050, i.e., 36 million to 115 million (Tulip Mazumdar 2017). Treating blindness can be both risky and expensive, which an ordinary man could not afford.

Blind people mostly rely on conventional sticks to know their surroundings. But the use of sticks may also have some consequences such as no contactless sensing that is the user cannot find out if there is an obstacle in his path without getting close to it. The obstacle may also be dangerous and may harm the user if provoked.

To address this problem many kinds of research on some IoT devices such as smart shoes for the blind (Sohan *et al.* 2020) or the smart stick for the blind (Thakur *et al.* 2018) have taken place to address this problem.

However, these can hinder the flair of a visually impaired person due to their appearance, which can look strange. It can be a hectic task to carry the stick everywhere as it is not so portable and these devices are also not able to effectively detect a depreciation of the surface level and can therefore be a threat that may be a reason for many accidents (due to falls) and may prove fatal. Hence, these devices, therefore, have lower reliability.

The user may also find it a cause for embarrassment in public as he may look strange in a world that is continuously growing in the field of fashion and may not get enough opportunity to wear what he likes and finds it most comfortable to walk in.

Here is a proposed idea to make a more reliable and more stylish shoe extension that can be used as an extension to the user's well-styled and most comfortable shoe by the use of a snug fit mechanism fitted at top of the device which will make the user look more normal to the community.

DOI: 10.1201/9781032644752-75

2 LITERATURE REVIEW

Sohan *et al.* (2020), this paper a design of smart assistive shoes is given which can detect objects with the help of Ultrasonic sensors and IR sensors and also these shoes provide safety and security through an emergency switch and with the help of GPS and GSM modules such that a message is sent to the caretaker of the blind through a message and location of the user.

Rajani Akuta *et al.* (2020) In this paper a design to make a haptic shoe that will be able to detect the obstacles in the path of its user using an Ultrasonic sensor is made. These shoes also provide guidance on the path based on the location which the user needs to be reached is proposed by using a voice recognition module.

Vignesh *et al.* (2018) In this paper a design to make a shoe for the visually impaired is proposed which will be able to detect an obstacle with the help of ultrasonic sensors and the development of an Android application is proposed. Moreover, better battery backup is implemented using a small plate of solar panels placed on the top of the shoe.

Amit Kumar Thakur *et al.* (2018) In this paper the use of Arduino Uno and Ultrasonic sensor is being implemented in a walking stick to make it usable by a visually impaired person and assist him while he walks.

Vijeesh *et al.* (2019) In this paper a design of the smart shoe is proposed that will be able to assist a visually impaired user while in motion. This proposed work mainly focuses on LiFi technology (Swami and Moghe 2020, December) for identifying the obstacle in the path of the user. LiFi technology is also responsible for communicating between the two individual shoes.

Table 1. Comparative analysis.

Sr no	Title of paper	Year of publication	technology	Pros and cons
1.	Smart assistive shoes for blind	2020	Use of ultrasonic sensors and GPS technology	(+) better power backup. GPS and GSM modules. (−) No safe path detection
2.	Efficient Obstacle Detection and Guidance System for the Blind (Haptic shoe)	2019	Use of ultrasonic sensors	(+) the obstacle detection system is created. (−) the high workload on the user
3.	Smart Shoe for Visually Impaired Person	2018	Ultrasonic sensors, solar panel, voice assistant	(+) the battery backup is better due to the use of the solar panel.
4.	Smart Blind Stick for Obstacle Detection and Navigation System	2018	Ultrasonic sensor and GPS	(+) low cost, reliable, lightweight, low power, and efficient navigation with fast, quick response times. (−) no GPS and GSM systems. The range is not wide or flexible.
5.	Design of Wearable shoe for the blind using LIFI Technology	2019	LiFi technology	(+) the transfer rate of information is faster. (−) expensive, not portable

3 PROPOSED METHODOLOGY

The smart shoes are an extension of the user's shoes that allows him to move without fear of being collided with an obstacle on his way. This extension uses several sensors such as ultrasonic, gyroscope, vibration motor, buzzer, and Arduino Uno.

3.1 *Ultrasonic sensor*

Ultrasonic sensors are used to detect the distance of the obstacle through the use of ultrasonic waves emitted by the transmitter of the ultrasonic sensor. The wave then returns to the ultrasonic sensor receiver after colliding with the obstacle in the form of an echo. The time between the transmission of the wave and the reception of the wave is then calculated and finally, the distance is calculated using the time distance formula.

3.2 *Arduino Uno R3*

Arduino Uno R3 is a microcontroller board. It consists of 14 input-output spins. Of these, 6 are PWM outputs, and 6 are analog inputs. It is the most used Arduino board and is readily available. It meets a microcontroller's requirements and has a barrel connector for battery connection.

3.3 *Bluetooth module*

The Bluetooth module is needed to connect another shoe, which allows better communication between the two shoes, and contributes to a better calculation of the distance.

3.4 *Gyroscope*

A gyroscope is a device that enables its user to know the angle concerning the gravity of the earth. A gyroscope is a wide device that makes it difficult to be fitted in a limited space.

3.5 *Vibration motor*

A vibration motor is mostly fixed inside a mobile phone that enables a user to recognize an incoming call/notification in a crowded space. The vibration motor is activated as soon as there is an obstacle detected in front of the user. Through the ultrasonic sensor.

3.6 *Buzzer*

A buzzer is a device that warns anyone of danger and has a wide range of applications such as fire alarms, water sensing alarms, etc. The buzzer is activated when there is no floor in front of the user, e.g., a large pothole, going down stairs, etc.

4 PROPOSED FLOW CHART

(1) Ultrasonic Sensor (First Shoe): This sensor is used to detect obstacles in the path of the user. It has a range of 50 cm, which provides a safety buffer for the user to respond to warnings. If an obstacle is detected within this range, the Arduino Uno sends a signal to a vibration motor, creating vibrations on the user's feet to warn them about the obstacle.
(2) Gyroscope (Second Shoe): The gyroscope continuously measures the angle of the shoe relative to the ground. It establishes a reference angle based on the most commonly used angle, assuming it represents a level surface.
(3) Angle and Distance Measurement: The gyroscope constantly monitors the difference (θ) between the measured angle and the reference angle. Simultaneously, the second shoe also uses an ultrasonic sensor to measure the distance to the ground. When the gyroscope detects a significant increase in angle (indicating a sharp decline in the surface), and the ultrasonic sensor registers a decrease in distance (indicating proximity to the ground), it triggers a warning.

(4) Buzzer (Second Shoe): Upon detecting a sharp decline in the surface, the Arduino Uno sends a signal to activate a buzzer, which produces a sound to alert the user of the abrupt change in terrain.

5 PROPOSED DESIGN OF SHOE EXTENSION

Here the ultrasonic sensor secures its most accurate position, the front, as it needs to measure the distance of the upcoming obstacles. The gyroscope is placed behind the ultrasonic sensor as the measurement of the angle is best through that position.The Arduino Uno, buzzer, and vibration motor are placed on the platform of the extension as they need to provide the computation and output. The whole extension is also fitted with a spring-based snug-fit system for a better grip on the sole of the user's shoe.

6 CONCLUSION AND FUTURE SCOPE

In conclusion, this research paper has presented a comprehensive methodology for enhancing mobility and independence among visually impaired individuals through the utilization of Arduino-based hardware solutions. The proposed methodology is grounded in a multi-faceted approach that combines cutting-edge technology with user-centric design principles to address the unique challenges faced by this community.

Our research began by conducting a thorough review of existing assistive technologies and the needs of visually impaired individuals. The methodology outlined in this research paper emphasizes the importance of user-centered design, involving visually impaired individuals throughout the development process to ensure that the resulting hardware solutions are both effective and user-friendly. We discussed the significance of accessibility features, such as voice feedback and tactile interfaces, to make the technology more inclusive and easy to use.

Furthermore, our research highlighted the potential for machine learning algorithms to improve the accuracy and reliability of navigation systems for the visually impaired. By leveraging AI-driven image recognition and object detection, we can enhance obstacle avoidance and pathfinding capabilities.

In addition to technical considerations, we addressed the ethical and privacy concerns associated with assistive technology for visually impaired individuals, emphasizing the need for robust data security and user consent mechanisms.

Overall, this concept seems to be focused on enhancing the safety and awareness of users while walking by providing real-time feedback about obstacles and changes in the walking surface. It combines ultrasonic sensors, gyroscopes, vibration motors, and buzzers to provide both tactile and auditory warnings to the user, helping them navigate their environment more safely. Additionally, the snug fit mechanism ensures that the extensions are securely attached to the user's shoes.

REFERENCES

Barbato, M., Orlandi, G., & Panella, M. (2014). Real-time Identification and Tracking Using Kinect. Multimodal Interaction and Performance Analysis. In *Real-time Identification and Tracking Using Kinect. Multimodal Interaction and Performance Analysis* (Vol. 2, pp. 1–37). Maia edizioni.

Dhall, P., Sharma, P., Thakur, S., Agarwal, R., & Rastogi, S. (2010). A review paper on assistive shoe & cane for visually impaired people. *International Journal of Scientific Research and Management Studies (IJSRMS)*, 3(2), 113–117.

Fernandes, H., Adão, T., Magalhães, L., Paredes, H., & Barroso, J. (2012). Navigation module of blavigator prototype. In *World Automation Congress 2012, In Proceedings of the World Automation Congress, Puerto Vallarta.*

Fernandes, H., Costa, P., Paredes, H., Filipe, V., & Barroso, J. (2014). *Integrating computer vision object recognition with location based services for the blind*. In *Universal Access in Human-Computer Interaction. Aging and Assistive Environments: 8th International Conference, UAHCI 2014, Held as Part of HCI International 2014, Heraklion, Crete, Greece, June 22–27, 2014, Proceedings, Part III 8* (pp. 493–500). Springer International Publishing.

Filipe, V., Faria, N., Paredes, H., Fernandes, H., & Barroso, J. (2016, December). Assisted guidance for the blind using the Kinect device. In *Proceedings of the 7th International Conference on Software Development and Technologies for Enhancing Accessibility and Fighting Info-exclusion* (pp. 13–19).

Oppermann, M., & Rieger, R. (2017, June). RF modules (Tx-Rx) with multifunctional MMICs. In *2017 IMAPS Nordic Conference on Microelectronics Packaging (NordPac)* (pp. 57–60). IEEE.

Rajani Akula, Bhagavatula Ramya Sai, Kokku Jaswitha, Molugu Sanjay Kumar, and Veeramreddy Yamini, S. C. Satapathy *et al.* (2020) (Eds.): ICETE 2019, LAIS 4, pp. 266–271, 2020.

Saquib, Z., Murari, V., & Bhargav, S. N. (2017, May). BlinDar: An invisible eye for the blind people making life easy for the blind with Internet of Things (IoT). In *2017 2nd IEEE International Conference on Recent Trends in Electronics, Information & Communication Technology (RTEICT)* (pp. 71–75). IEEE.

Sohan N., Urs Ruthuja S., Sai Rishab H. S., and Shashidhar R. (2020), "Smart Assistive Shoes for blind", S. Says *et al.* (eds.), *New Trends in Computational Vision and Bio-inspired*, Springer Nature Switzerland AG 2020.

Swami, K. T., & Moghe, A. A. (2020, December). A review of LiFi technology. In *2020 5th IEEE international Conference on Recent Advances and Innovations in Engineering (ICRAIE)* (pp. 1–5). IEEE.

Thakur, A. K., Singh, R., & Gehlot, A. (2018). Smart blind stick for obstacle detection and navigation system. *Journal of Emerging Technologies and Innovative Research*, 5(10).

Tulip Mazumdar (2017), Global Blindness set to triple by 2050, BBC News, 3 August 2017. Available: HTTPS://WWW.BBC.COM/NEWS/HEALTH-40806253

Vignesh. N., Meghachandra Srinivas Reddy. P., Nirmal Raja. G., Elamaram. E. & B. Sudhakar (2018), *International Journal of Engineering &Technology*, 7 (3.12) (2018) 116–119.

Vijeesh V., Kaliappan E., Ponkarthika B., and Vignesh G. (2019), Design of wearable shoe for the blind using LIFI technology, *International Journal of Engineering and Advanced Technology (IJEAT)* ISSN: 2249 – 8958, Volume-8, Issue-6S3, September 2019.

Advances in AI for Biomedical Instrumentation, Electronics and Computing – Sachan et al. (eds)
© 2024 The Author(s), ISBN 978-1-032-64298-7

Dynamic power allocation technique for IOT applications in mobile edge computing

Ashvini Joshi, Anjulata Yadav & Amit Naik
Electronics and Communication Engg. Department, Shri. GS Institute of Technology and Science, Indore

ABSTRACT: In the 5G era and beyond, technology is advancing towards convenience and comfort. When IoT infrastructure is factored in, the growing requirement for energy stability and control becomes critical for both existing and emerging technologies. By integrating edge servers into existing data centres and enhancing service quality, the latency-driven problem is being investigated. We presented an optimisation technique for power allocation efficiency to achieve this reliability. We employed a particle swarm optimisation strategy with a multi-user approach in a dense network to discover the optimal solution, while keeping the present mathematical model of mobile edge computing in mind.

1 INTRODUCTION

Proper power allocation management is critical in this day and age of connected technology and the exponential expansion of the Internet of Things (IoT) to ensure the faultless operation and sustainability of IoT applications (Hu *et al.* 2015; Mach and Becvar 2017). The integration of Internet of Things (IoT) applications with the Mobile Edge Computing (MEC) paradigm has resulted in a paradigm shift in the distributed computing and data processing environment (Hu *et al.* 2015). MEC integration has emerged as a viable solution to the problems of latency, bandwidth constraints, and energy consumption in Ultra dense networks (Kountouris 2017), allowing for real-time data processing and analysis at the network's edge. Effective power allocation management is critical in this ever-changing environment for IoT applications at the edge to function smoothly and efficiently. This study develops an algorithmic framework for real-time dynamic power resource allocation in order to meet the changing and dynamic demands of IoT devices and networks. not just energy consumption, but also the overall performance, reliability, and latency needs of IoT applications, with a special emphasis on the creation of a dynamic power allocation optimisation (Hu and Li 2019) model inside the MEC environment. The suggested system intends to provide optimal resource utilisation and increased Quality of Service (QoS) for IoT applications at the network edge by integrating real-time data analytics, job offloading techniques, and intelligent power allocation algorithms.

2 LITERATURE REVIEW

Haitham Seada and Kalyanmoy Deb (Seada and Deb 2014) Proposed the scheduling technique to solve it, they analyze it as a dual decision-making issue and propose a multiple-objective optimization technique based on NSGA-III. Simulation findings validate our methods, demonstrating that NCGG may effectively save energy consumption and that NSGA-III beats previous techniques in terms of reaction rate and retains high performance

DOI: 10.1201/9781032644752-76

in a dynamic MEC system. The study investigates the problems of scheduling compute needs in ultra-dense edge computing networks and proposes an energy-saving method based on a non-cooperative game model based on sub-gradients. The authors (Hu and Li 2019) highlight the rising need for compute and massive data traffic in the Internet of Things (IoT) era, which necessitates the use of mobile edgecomputing (MEC) and ultra-dense network (UDN) technologies. These technologies provide low-latency and flexible processing for mobile users in the ultra-dense edge computing (UDEC) network. The limited processing capabilities of edge clouds, combined with the unpredictable demands of mobile users, make it challenging to schedule compute requests to appropriate edge clouds. In order to address this issue, the study creates the transmitting power allocation (PA) problem for mobile users in order to decrease energy consumption.PSO is discussed in this study by the authors (Wang *et al.* 2018), including its history, theory analysis, and current research. The article discusses a number of PSO issues, such as algorithm design, parameter selection, topological structure, and multi-objective optimisation.PSO engineering applications are also being looked upon. Existing challenges are identified, and new research opportunities are suggested. The authors (Wang *et al.* 2018) offers a mathematical model for a mobile edge computing (MEC)-based low-latency network architecture. The model considers both communication and processing latency, as well as systems with strict and permissive underestimate. When a large number of nodes meet the system criteria, the paper proposes three policies for selecting an edge node.

The model-based simulation tests provide insights into the system's performance and illustrate that the policy with the lowest latency works best. The research will be expanded to include real-world applications in order to determine the practicality of the concept. This study of authors (Elshaer *et al.* 2016) investigates the problem of computation offloading in a UD-IoT network with sub-6-GHz macro cells and low-power mm Wave small cells. The emphasis is on decreasing energy consumption and time constraints for IoT devices. The research (Kountouris 2017) proposes a mobile edge computing architecture and evaluates performance using queueing models. Investigating the MECO problem in UDNs, taking energy and latency constraints into consideration, and addressing edge computing offloading in a UD-5G heterogeneous network are among the contributions.

3 SYSTEM MODEL

We assume that the mobile devices are linked to the AP through the non-orthogonal multiple access (NOMA) protocol in the MEC system, which consists of an access point (AP) and the MEC server (Elshaer *et al.* 2016; Intharawijitr *et al.* 2017). NOMA enables all mobile devices to offload their duties simultaneously, increasing offloading throughput. We assume that the MEC system can support up to K mobile devices sending data at the same time. We assume that each mobile device can only offload one job to increase wireless access efficiency. The time needed for the mobile edge node to complete its computations, and the round-trip communication latency between the source node and the chosen mobile edge node. While it is true that we do not directly consider mobile devices themselves, given how quickly they can connect to a source node, this is not a significant omission. Mobile Edge nodes are considered as Edge nodes denoted by set of $\varepsilon = \{E_{n1}, E_{n2}, \ldots E_{nj} \ldots, E_M\}$ In edge node E_{nj} (Wang *et al.* 2018) the amount of service demand as work load is taken to be independently and uniformly distributed using an exponential distribution with Mean μ_{nj}^{-1}. Another way to define workloads is as a set $W_l = [W_{l1}, W_{l2}, \ldots, W_{lk}, \ldots, W_{lm}]$, where W_{lk} gets its size from b_{lk}. Here, W_{l1} is the first workload that can be generated independently and with ordering from one of the source nodes in S, and W_{lk} is the last task that is sent into the system (Elshaer *et al.* 2016).

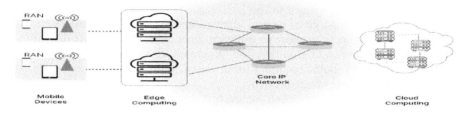

Figure 1.   Illustration of Mobile edge computing architecture.

$$l_{i,j} = l_h * H(i,j), S_{ni} \in \mathcal{S}, E_{nj} \in \varepsilon \tag{1}$$

$$P_{k1,j(t)} = b_k \frac{n_j(t)}{\mu_j} + 1 \tag{2}$$

Total latency:

$$L_{i,j,k}(t) = \begin{cases} l_{i,j}+, P_{K,j}(t), \\ \quad\quad 0, \end{cases} \tag{3}$$

4   PROPOSED WORK

Algorithm 1 : Proposed Framework of Particle swarm optimization
---
Input :$u, n, b, H, p_{max}, c_e, c_m, p_c, sig_2, swarm\ size, ub, lb$

Output :$g_{best} > optimal\ cost$

Step 1: Initialize with Inputs

Step 2 : creating the swarm

  i=1:swarm size

    $p_i < optimal\ cost$

  end

      Update particle velocity

      if $p_i > v_{max}$

elseif

    $p_i.\ velocity < v_{min}$

      Update particle position

      For k=1:l($p_{(i)}.p$)

if ($p_{(i)}.p_{(k)} > u_b$)

    $p_{(i)}.p_{(k)} = u_b$

      Elseif

    $p_{(i)}.position\ (k) < lb$

  end

Step 3 : Evaluating cost

    if $p(i)cost < p(i)best\ cost$

    end

    if f $p(i)bestcost < optimal\ cost$

    end

Step 4 : Output

Algorithm. Proposed Framework of PSO.

In this section, we outline the proposed work for our research article, which focuses on optimising a difficult problem using a swarm intelligence-based strategy. We want to look at how to determine the best cost given a set of input parameters and constraints.

We consider these factors to propose an optimization algorithm for power allocation

Power requirements: Depending on their job, programS, or demands, each user has different power requirements. Some require a lot of electricity to transmit data, while others only need a little. Functional needs dictate it.

Energy constraints: Users may confront energy constraints, particularly when using energy-constrained or battery-powered devices. Power allocation should take these energy limits into consideration to avoid premature exhaustion of any user's energy sources.

Quality of Service: Different users may have different QoS requirements, such as a minimum data throughput, latency, or reliability. PA must assess these factors and allocate sufficient power to accomplish the desired objectives.

5 CONCLUSION

In this study, we look at the power allocation problem in an ultra-dense edge computing network. We anticipate a UDEC network with a macro-BS, many micro-BSs, and a large number of mobile users in the 5G architecture. The NOMA protocol is used as a multiple access method between users and BSs. We concentrate on the interference between mobile users and BSs under the NOMA protocol, first posing a power allocation problem and then proposing a particle swarm optimization solution to it. However the proposed technique has not been used in real-world situations. To address actual obstacles, we will focus on the creation of edge computing power allocation algorithms for systems based on real-world applications in the future.

REFERENCES

Elshaer, H., Kulkarni, M. N., Boccardi, F., Andrews, J. G., & Dohler, M. (2016). Downlink and uplink cell association with traditional macrocells and millimeter wave small cells. *IEEE Transactions on Wireless Communications, 15*(9), 6244–6258.

Hasanin, T., Alsobhi, A., Khadidos, A., Qahmash, A., Khadidos, A., & Ogunmola, G. A. (2021). Efficient Multiuser Computation for Mobile-Edge Computing in IoT Application Using Optimization Algorithm. *Applied Bionics and Biomechanics, 2021*, 1–12.

Hu, S., & Li, G. (2019). Dynamic request scheduling optimization in mobile edge computing for IoT applications. *IEEE Internet of Things Journal, 7*(2), 1426–1437.

Hu, Y. C., Patel, M., Sabella, D., Sprecher, N., & Young, V. (2015). Mobile edge computing—A key technology towards 5G. *ETSI white paper, 11*(11), 1–16.

Intharawijitr, K., Iida, K., & Koga, H. (2017). Simulation study of low latency network architecture using mobile edge computing. *IEICE TRANSACTIONS on Information and Systems, 100*(5), 963–972.

Johnson, S. M. (1954). Optimal two-and three-stage production schedules with setup times included. *Naval Rresearch Llogistics Qquarterly, 1*(1), 61–68.

Kim, S., Visotsky, E., Moorut, P., Bechta, K., Ghosh, A., & Dietrich, C. (2017). Coexistence of 5G with the incumbents in the 28 and 70 GHz bands. *IEEE Journal on selected areas in communications, 35*(6), 1254–1268.

Kountouris, M. (2017). Performance limits of network densification. *IEEE Journal on Selected Areas in Communications, 35*(6), 1294–1308.

Liu, J., & Liu, X. (2022). Energy-efficient allocation for multiple tasks in mobile edge computing. *Journal of Cloud Computing, 11*(1), 1–14.

Mach, P., & Becvar, Z. (2017). Mobile edge computing: A survey on architecture and computation offloading. *IEEE Ccommunications Ssurveys & Ttutorials, 19*(3), 1628–1656.

Seada, H., & Deb, K. (2014). U-NSGA-III: A unified evolutionary algorithm for single, multiple, and many-objective optimization. *COIN Rreport, 2014022*.

Wang, D., Tan, D., & Liu, L. (2018). Particle swarm optimization algorithm: an overview. *Soft Ccomputing, 22*, 387–408.

Yang, Z., Hou, J., & Shikh-Bahaei, M. (2018, December). Energy efficient resource allocation for mobile-edge computation networks with NOMA. In *2018 IEEE Globecom Workshops (GC Wkshps)* (pp. 1–7). IEEE.

Advances in AI for Biomedical Instrumentation, Electronics and Computing – Sachan et al. (eds)
© 2024 The Author(s), ISBN 978-1-032-64298-7

Review on smart landmine and landmine detection

Hans Kumar, Abhas Kanungo, Kartik Chaudhary, Jatin Tomar, Priyanshu & Harshit Yadav
Department of Electronics and Communication Engineering, KIET Group of Institutions, Delhi-NCR, Ghaziabad, India

ABSTRACT: This paper present about a smart landmine and landmine detector rover-robot. The landmine is going to smart by using the concept of transmitter and receiver, where landmine will be acting as receiver and transmitter will be other component which will be on shoes of army soldier. Landmine detection rover-robot will be based on the two sensors (magnetic sensor and metal sensor). Both sensors will play crucial role as they will help to detect our own landmine as well as enemy's landmine. The objective of this smart landmine and landmine detection rover-robot is to reduce the chances of accident with Indian army soldiers during the search of landmine, and to detect landmine and differentiate between our own landmine and enemy's landmine with the help of two sensors. In various landmines magnets are not used so, we can a small piece of magnet on our landmine so that it is easily detectable by the landmine detector. If both sensor gives beep sound, then it is not our landmine but if it gives only one beep sound then it is not our landmine. For transmitter and receiver here, I used simple transmitter and receiver circuit can be of remote-control car, on landmine rover-robot magnetic sensor and metal sensor, Arduino UNO will be used and a remote-control robot. It is to identify the landmines in war affected rehabilitation places. The detection of landmines through this proposed method is free of risk and of less human effort. Integration of inductive sensor, video camera and ATUNO microcontroller are used here to discover landmines. While experimenting, this robot shows high detection ability of both metallic and magnetic and transmitter and receiver parts of landmine is actively working which is competent enough to be used in the applications in landmine covered regions.

1 INTRODUCTION

Landmines are a victim-activated explosive device, usually deployed discreetly on, or just below, the surface. They are generally divided into: anti-personnel and anti-vehicle. These devices are very explosive and kind of hidden explosive. Two categories of landmines are commonly in use: large antitank mines designed to be triggered by vehicles, and the small inexpensive antipersonnel mines designed specifically to kill or incapacitate a human being. They are used to create tactical barriers to prevent direct attack or to deny access by military and civilians to a define area. Therefore, landmines can be considered as perfect soldiers that never eat, sleep, miss fall ill or disobey. Being very helpful in the war along with that it is too dangerous for our own military soldiers.

Sometimes it results in accidents while finding out the landmines, and when accidently soldiers put their leg on the landmine then it results only bad. To prevent such accident according to me landmine should be made smart and some devices should be use in landmine detector that can differentiate between our own landmine and enemy's landmine. According to me if the transmitter and receiver concept is used in the landmines then it can be possible to make it smart. The idea goes like a transmitter device will be fitted on the army soldier's boot and receiver device will be fitted inside the landmine itself. When the soldier's foot is going to step down on that landmine then the transmitter and receiver will be acting like password security device and if they password matched between the transmitter and receiver then it will not be blast and activated, but if that receiver did not get any password, then it will be activated and finally exploded.

2 LITERATURE REVIEW

(Bruschini and Gros 1998) National security is of prime importance in today's weapons as if security is not being provided then it can increase changes of accident k, to prevent this accident, the security of weapons is essential. The safety should be at the top of army personals and the people living in war prone areas. Landmines are one which is buried under the ground and it is for the enemy's but the people living near those war areas are in very accidental chances. So, damage done by the landmine is very fatal and explosive and becomes very important to find those landmines to prevent any accident. Once they are detected they should be deactivated. To address this, advanced robots have gained prominence, leveraging their reliability and precision. Operators can control the robot remotely, ensuring their safety while effectively addressing this hazardous task.

(Kasban *et al.* 2010) The description begins by defining landmines as devices designed to harm individuals through direct pressure or triggers when touched. The historical origin of antipersonnel landmines is traced back to World War II. Two main types of landmines are discussed: Anti-Personnel (AP) and Anti-Tank (AT), differing in size and placement. Landmines are highlighted as effective tools for denying access to specific areas. Despite being used by multiple nations, they pose significant dangers to both civilians and soldiers due to their long-lasting impact. The devastating consequences of landmine encounters are emphasized. The importance of landmine detection is underscored, and recent advancements in sensor technology and autonomous robots for detection and removal are noted.

(Borgwardt 1995) This passage discusses the widespread harm caused by landmines and the challenges they pose to human society. Over 110 million landmines are still present in post-war areas, leading to the development of various detection and removal methods. These methods are categorized based on their underlying principles: exploiting geophysical differences and employing specific techniques such as acoustics and molecular tracking. This study introduces a complex minefield simulation environment and collects high-quality multispectral data using drones with integrated cameras. They propose a detection-driven image fusion framework using YOLOv5, achieving improved recall rates for landmine detection. The paper concludes with discussions on the dataset, deep learning methods, experimental results, and the potential of their approach for state-of-the-art object detection models.

(Borgwardt 1995) The presence of hidden landmines worldwide poses significant threats, causing loss of life, permanent and temporary disabilities, and disrupting economic stability and peace. The issue of landmines has become a global concern due to conflicts between countries, requiring substantial financial investment for their clearance. Developing nations affected by landmines allocate a significant portion of their income and human resources for clearance efforts. Traditional methods of landmine clearance, while consistent and reliable, are slow and hazardous. This research introduces a robotic solution to address these challenges, offering speed, security, and precision superior to manual approaches.

Metal detecting sensors are commonly used for mine detection. The distance between the sensor head and the buried landmine significantly impacts detection performance. By adjusting this distance and maintaining a uniform gap with the ground, detection accuracy can be improved.

(Borgwardt, C. 199) Landmines come in two main types: Anti-Personnel (AP) mines and Anti-Tank (AT) mines. AP mines are typically buried close to the surface, while AT mines are often placed on the ground. These mines serve as area-denial weapons, creating barriers to hinder direct attacks and restrict access to defined areas for both military and civilian personnel. This concept likens landmines to perfect soldiers – consistent, cost-effective, and enduring, unlike human soldiers. Landmines remain active long after conflicts end, and there are currently over 100 million unrecovered mines in over 50 countries. The consequences of landmines are severe, causing death and injury to numerous individuals daily and leaving long-lasting humanitarian and environmental impacts, especially in regions where landmines have become part of daily life.

(Kanungo *et al.* 2023) The integration of soldiers with advanced health monitoring, real-time GPS, and data communications is essential for modern military operations. Soldiers require wireless networks to communicate with the control unit and fellow personnel. Apart from national security, soldiers need modern weapons and health monitoring, with lightweight and efficient components. Communication challenges between soldiers and the control unit are common in military operations, and precise navigation among soldiers is crucial for effective coordination. This work focuses on tracking soldier locations via GPS, aiding the control unit in guiding them accurately. Smart biomedical sensors, such as heartbeat, temperature, and humidity sensors, are embedded in soldiers' attire, connecting wirelessly to a base station server. This system proves valuable for soldiers involved in special operations.

(Kanungo *et al.* 2022) The primary objective of our robotic vehicle is to locate and identify landmines, minimizing the risk to military personnel and maximizing the safe area available for operations. Detonated landmines can cause severe harm to soldiers and release harmful toxins into the environment. Traditional methods involve deactivating robots before they enter conflict zones. Therefore, robots equipped with the capability to detect landmines are crucial for safeguarding soldiers' lives. Landmine removal is a pressing global issue exacerbated by factors like natural disasters and urban development. Detecting landmines and safely removing them is an urgent necessity.

3 PROPOSED MTHODOLOGY

Firstly, we will be working on the landmine and make it smart by using technologies of transmitter and receiver. Here landmine is made smart to prevent any accident during the war time. In this we will be using transmitter and receiver as a component and by this it will be working as a lock in the landmine. Transmitter part will be on the soldier's leg and receiver will be inside the landmine itself. Its working is like that when our soldier is putting his leg with that transmitter on that landmine which is having receiver component in that landmine then a password will be match and it will not blast, but if other soldier who is not having that transmitter part on his body and he puts his leg on the landmine then it will be blast.

In the landmine we will using a small magnet piece so that the magnetic sensor which we will be using in the landmine detector it will be able to differentiate between our own and enemy's landmine.

4 FLOW CHART

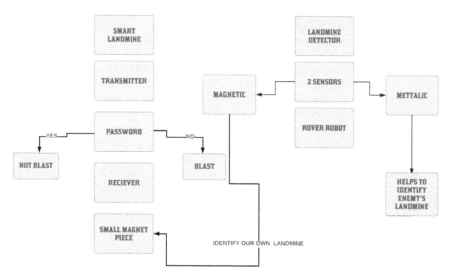

Figure 1. Flow chart of methodology.

5 COMPONENTS

5.1 *Transmitter component*

In the smart landmine the transmitter component will used we can use it from any of the other device. There are various devices in which we can find transmitter components and we can use them for our use. It will be used to put it into the landmine itself. It will play a very important role as it will be sending or transmitting the signal to receiver side. The transmitter generates a radio frequency signal which when applied to the antenna produces the radio waves, called the carrier signal. It combines the carrier with the modulation signal, a process called modulation.

5.2 *Receiver component*

A receiver handles the receiving of items in a warehouse or backroom of a store, and ensures that the shipments are inspected, sorted, and stored in the warehouse or stocked on shelves as needed. They need to keep up with invoices to make sure orders are correct. By using this we will be able to complete the transferring of signal, also will be able to complete a kind of communication. If this communication completes then it will not blast otherwise it will get blast.

5.3 *Magnetic sensor*

Magnetic sensors have the working principle based on the measurement of changes in magnetic moment of magnetic material when exposed to magnetic fields. The magnetic materials say ferromagnetic react to temperature change or magnetoelastic materials react to a mechanical stress, in the medium with which they interact. This will be inside the landmine and it will play a very important role as it will help us to identify that which landmine is our own and which is not. When this sensor is present on the landmine detector it will give a beep sound if it finds a magnet on the landmine and we will come to know that it is our own landmine.

5.4 *Metallic sensor*

These machines send out high-frequency magnetic pulses—in the 100 Hz. range—which, when interrupted by metal in the scanning area, distort the received signals of the receiving coil, annunciating an alarm. Metal detectors are used for finding mines and, of course, for finding buried treasure. It will help us to detect the landmine.

6 CONCLUSION

In conclusion, landmines remain a significant threat to the safety of civilians and the economic development of post-conflict regions. The use of radar technology in landmine detection has shown promising results in terms of accuracy and efficiency. Advancements in radar technology, such as the use of artificial intelligence and machine learning algorithms, hold the potential for even greater improvements in landmine detection capabilities. However, challenges remain in developing radar systems that are both affordable and accessible to the communities most affected by landmines. In addition, continued efforts are needed to clear existing landmines and prevent their use in future conflicts. Overall, the development and application of radar technology in landmine detection is an important step towards achieving a world free from the threat of landmines.

The detection of landmines is a critical issue that requires reliable and efficient solutions to ensure the safety of civilians and promote economic development in affected areas. Advanced technologies such as radar systems, thermal imaging, and neutron activation analysis have shown promising results in detecting landmines. However, each technology has its limitations, and further research and development are necessary to overcome these challenges. The goal is to develop a cost-effective and accurate landmine detection system that can operate in various environmental conditions and effectively detect and neutralize landmines to eliminate the risks they pose to human lives and communities.

REFERENCES

Borgwardt, C. (1995). ODIS: Ordnance detection and identification system. In *Proceedings of the Workshop on Anti-Personnel Mine Detection and Removal (WAPM'95). Lausanne, Switzerland.*

Brown, D. R. Multisensor vehicular mine detection testbed for humanitarian demining. In *Proc. of the Technology and the Mine Problem Symposium* (pp. 73–78).

Bruschini, C., & Gros, B. (1998). A survey of research on sensor technology for landmine detection. *Journal of Conventional Weapons Destruction, 2.*

Czipott, P. V., & Iwanowski, M. D. (1997, January). Magnetic sensor technology for detecting mines, UXO, and other concealed security threats. In *Terrorism and Counter-Terrorism Methods and Technologies* (Vol. 2933, pp. 67–76). SPIE.

Kanungo, A., Choubey, C., Gupta, V., Kumar, P., & Kumar, N. (2023). Design of an intelligent wavelet-based fuzzy adaptive PID control for brushless motor. *Multimedia Tools and Applications*, 1–21.

Kanungo, A., Mittal, M., & Dewan, L. (2020, June). Comparison of haar and daubechies wavelet based denoising for speed control of DC motor. In *2020 First IEEE International Conference on Measurement, Instrumentation, Control and Automation (ICMICA)* (pp. 1–4). IEEE.

Kanungo, A., Mittal, M., & Dewan, L. (2023). Critical analysis of optimization techniques for a MRPID thermal system controller. *IETE Journal of Research, 69*(1), 149–164.

Kanungo, A., Mittal, M., Dewan, L., Mittal, V., & Gupta, V. (2022). Speed control of DC motor with MRPID controller in the presence of noise. *Wireless Personal Communications*, 1–15.

Kasban, H., Zahran, O., Elaraby, S. M., El-Kordy, M., & Abd El-Samie, F. E. (2010). A comparative study of landmine detection techniques. *Sensing and Imaging: An International Journal, 11*, 89–112.

Robledo, L., Carrasco, M., & Mery, D. (2009). A survey of land mine detection technology. *International Journal of Remote Sensing, 30*(9), 2399–2410.

Advances in AI for Biomedical Instrumentation, Electronics and Computing – Sachan et al. (eds)
© 2024 The Author(s), ISBN 978-1-032-64298-7

Design of multiplexer in 90nm technology using energy recovery logic circuit

Sitaram Kumar, Amit Kumar & Dharmendra K. Jhariya
Department of ECE, National Institute of Technology Delhi, India

ABSTRACT: In the world of tiny technology, keeping the power consumption as low has become a big deal because there are more and more transistors are being used. When device work faster, they tend to leak more current. The success of these products in the market relies on small, light, affordable, energy efficient and long lasting of their batteries. Designing products with low power usage is crucial to make them valuable in the market. In this paper for energy recovery in logic circuit we have used adiabatic logic circuit. In which we focus on the comparative analysis of CMOS based Multiplexer and different types of adiabatic logic (Effective Charge Recovery logic (ECRL), 2N-2N2P Logic, Positive feedback Adiabatic Logic (PFAL)) based Multiplexer. In the adiabatic logic circuit instead of DC voltage power supply we use AC pulse type power (Trapezoid type) supply. Some of the Adiabatic based Multiplexer work properly at low frequency. In Adiabatic logic circuits even though we can save power, but there is a trade off with the speed.

Keywords: Energy Recovery, Adiabatic Logic, 2N-2N2P, Efficient charge recovery logic, Positive Feedback Adiabatic logic

1 INTRODUCTION

Efficient energy recovery is a key consideration in the design of VLSI circuits. Energy is transferred from power source to circuits, and after performing some computation, it reverts back to the power source. This concept is often referred to as reversable logic circuits (Roy *et al.* 2000). Energy recovery circuits reduce energy consumption in VLSI design by efficiently managing currents and storing energy across components by gradually charging and discharging capacitive loads. This approach minimizes the waste and improves overall energy efficiency. These special circuits that recover energies are known as boost logic or adiabatic logic circuits. That means adiabatic logic circuits are more accurate at low frequency. In adiabatic logic circuits instead of V_{dd} we give ramped power supply (Nunez *et al.* 2020).

In this research paper, we implement 2:1 Multiplexer using different types of reversable logic circuits. Due to rise in power dissipation, there are adverse effects on gate dielectric integrity, junction diffusion, package-related failures, silicon interconnection fatigue, increase packaging and cooling expenses. The Power in CMOS circuits mostly lost due to Dynamic power dissipation and static power dissipation (Sharma *et al.* 2020). This research paper demonstrates a decrease in Dynamic power dissipation for a multiplexer through the implementation of energy recovery logic circuits. Which were designed and analyzed using Cadence Virtuoso. The most significant contributor to Dynamic power dissipation arises from the charging and discharging of the load capacitor when the output transitions, accounting for approximately 70% of overall power consumption. So, the dynamic power dissipation has become a significant concern in recent times.

DOI: 10.1201/9781032644752-78

2 BACKGROUND

2.1 *Adiabatic Logic*

The word "Adiabatic" means a process where no heat is given off or taken in from the surrounding. In modern circuits design, the primary focus is on enhancing energy efficiency, which is accomplished through circuits that exhibit minimal power consumption and low power dissipation. Charge recovery principal is used in the adiabatic logic circuits. Switching power dissipation of CMOS circuits in Figure 1, with capacitive load is:

$$\text{Each switching power dissipation} = 1/2 C_L V_{dd}^2 \qquad (1)$$

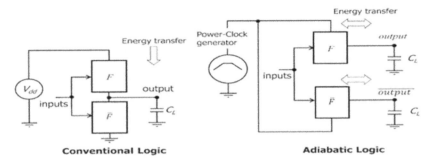

Figure 1. Representation of the conventional logic (left) and Adiabatic logic (right).

From eq. (1) as long as C_L and V_{dd} is fixed, so this is the lower limit of Energy loss in switching. So, in conventional, CMOS technology switching power dissipation cannot be lower from this one. In adiabatic switching circuits the switching power dissipation below from these lower limits. There are two types of adiabatic logic circuits they are partially and fully adiabatic logic. A fully reversible 3-bit circuit requires 20 times a greater number of gates as compared to CMOS and 32 times more area (Roy *et al.* 2000). The Effective Charge Recovery logic, 2N-2N2P, Positive feedback Adiabatic Logic comes under partially adiabatic logic Figure 1.

There are four phases in adiabatic logic circuits (Praveen *et al.* 2019). They are: (i) Evaluation Phase (ii) Hold Phase (iii) Recovery Phase (iv) Wait Phase.

2.1 *Adiabatic switching*

The key distinction between regular CMOS and Adiabatic circuits is how they charge a load capacitor. In standard CMOS, it uses a fixed voltage source, but in Adiabatic, it relies on a constant current source (Khandekar *et al.* 2008), as shown in Figure 2.

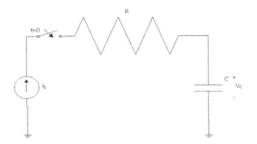

Figure 2. Schematic for Adiabatic charging process.

Assuming $V_C(t) = 0$ at $t = 0$

$$V_C(t) = 1/c \, I_s \cdot t$$

Where,

$$I_S = \frac{C}{t} V_C(t)$$

the energy used by the resistor R from the start (t = 0) to a certain time (t = T), you can calculate it this way:

$$
\begin{aligned}
E_{\text{dissipation}} &= R \int_0^T I_s^2 \, dt \\
&= I_s^2 R T \\
&= \frac{RC}{T} C V_c^2
\end{aligned}
\tag{2}
$$

Observation

(1) In equation (2) show that $E_{\text{dissipation}}$ is inversely proportional to "T" if it takes long time to charge (T >> 2RC), then the energy used is less compared to a regular CMOS circuit.

(2) In equation (2) shows that $E_{\text{dissipation}}$ is directly proportional to "R". so, if charging resistance decreases, the energy dissipation decreases.

3   OVERVIEW OF ADIABATIC LOGIC FAMILY

This research paper is all about looking at how much dynamic power saved by different type of Adiabatic logic circuits and compare power dissipation between Pass transistor, CMOS logic and other Adiabatic logics such as ECRL logic, 2N-2N2P Logic and PFA Logic.

3.1   *MUX through Pass transistor Logic*

The Figure 3(a) shows the basic circuit diagram of 2:1 MUX using pass transistor. Less number of transistors is used in PTL. Major disadvantage is full voltage swing does not occur (Kumar *et al.* 2020).

3.2   *MUX through CMOS Logic*

The Figure 3(b) shows that basic circuit diagram of 2:1 MUX using CMOS logic. CMOS, which stands for complementary metal oxide semiconductor, is a widely used type of logic design today. It uses both PMOS and NMOS transistors to build electronic circuits. PMOS is used as pull up network and this network is used to pass the strong V_{dd} at output. NMOS network is used as pull-down network which passes strong ground at the output (Mittal *et al.* 2018).

3.3   *MUX through ECRL Logic*

The Figure 3(c) shows that basic circuit diagram of ECRL based 2:1 MUX. In ECRL there are two NMOS networks which gives the true and complemented output. Two PMOS transistors are also used to hold the state. This is partially Adiabatic logic circuits so whole energy can't recover back to the circuits.

3.4 *MUX through 2N-2N2P Logic*

The Figure 3(d) shows that basic circuit diagram of 2:1 mux through 2N-2N2P logic. This logic family is also a partially adiabatic logic. In this case we take combination of two PMOS and two NMOS, the input NMOS network is connected as parallel with NMOS. Here output is cross coupled and it gives true and complemented non floating output.

3.5 *MUX through PFA Logic*

The Figure 3(e) shows that basic circuit diagram of 2:1 MUX based on Positive feedback Adiabatic Logic. This network is same as 2N-2N2P logic, the difference is only input NMOS network is replaced by PMOS. PFAL gives the best result as compared to ECEL and 2N-2N2P logic. Here we get output as a full swing from V_{dd} to GND. Delay is also less as compared to ECRL and 2N-2N2P logic. Power recovery is also best in PFAL.

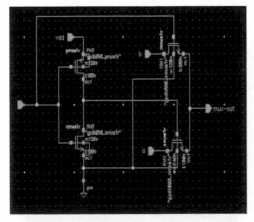

Figure 3(a). PTL based MUX.

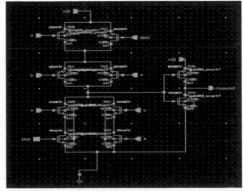

Figure 3(b). CMOS based MUX.

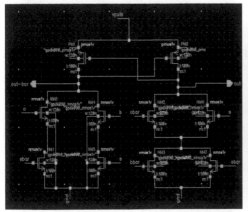

Figure 3(c). ECRL based MUX

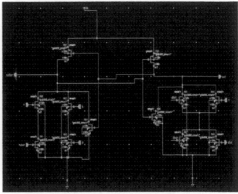

Figure 3(d). 2N-2N2P based MUX.

419

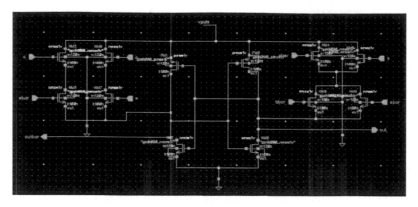

Figure 3(e).   PFAL based MUX.

4   OUTPUT WAVEFORMS

We can see various types of output waveforms for a 2:1 multiplexer created using different methods like Pass Transistor, CMOS, ECRL, 2N-2N2P logic, and PFAL are shown below:

The output waveform is plotted below that includes five section lines, Power supply Vpulse, select lines, Data input (a, b), Mux output and supply current.

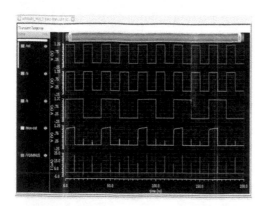

Figure 4(a).   Pass transistor based MUX outputwaveform.

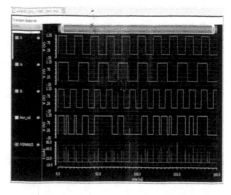

Figure 4(b).   CMOS based MUX output waveform.

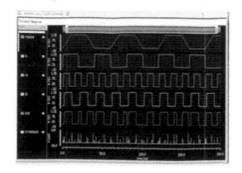

Figure 4(c).   ECRL based MUX output waveform.

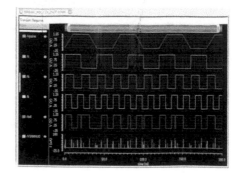

Figure 4(d).   2N-2N2P based MUX Output waveform.

420

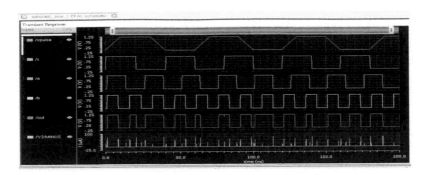

Figure 4(e).   PFAL based MUX output waveform.

Table 1.   Comparison of energy recovery, delay and number of transistors used in different logic circuit with various logic methods.

NAME OF LOGIC	ENERGY RECOVERY(MUX)	DELAY (ps)	NUMBER OF TRANSISTOR USED
PASS TRANSISTOR	185.57nJ	43.33	4
CMOS LOGIC	213.87nJ	16.49	10
ECRL	283.34nJ	1191.1	16
2N-2N2P LOGIC	324.54nJ	1126.4	18
PFAL	403.67nJ	288.6	18

5   CONCLUSION

A multiplexer has been created using different methods, and then it was tested using Cadence Virtuoso with 90nm technology. These methods include Pass Transistor, CMOS logic, ECRL, 2N-2N2P logic, and PFAL. In this research, we found that PFAL more energy recovery efficient to get approx. 25%, 42% and 89% more energy efficiency when using 2N-2N2P, ECRL and CMOS in our Multiplexer respectively. Delay also less in PFAL as compared to other energy recovery logic ECRL and 2N-2N2P logic. In pass transistor not completely swing that means not completely achieved V_{dd}. In PFAL no output is floating and all output has full logic swing. The research discovered that adiabatic logic circuits are quite useful for designing applications where saving energy is crucial. These applications are used in areas like phones and internet, connecting devices, satellites, and military technology.

REFERENCES

Khandekar, Prasad D and Subbaraman, Shaila, Low Power 2:1 MUX for Barrel Shifter, 2008 *First International Conference on Emerging Trends in Engineering and Technology*, 404–407.

Kumar, Nishant and Mittal, Poornima, Performance investigation of 2:1 multiplexer using 90nm technology node for low power application. 2020 *International Conference on Electrical and Electronics Engineering (ICE3-*2020).

Mittal, Deepak and Niranjan, Amit, Designing of multiplexer and de-multiplexer using different adiabatic logic in 90nm technology, 2018 *ICCCNT*.

Nunez, Juan and Avedillo, J. Maria, Approaching the design of energy recovery logic circuits using TFETs. *IEEE Trans. on Nanotechnology* 2020, 19, 500–507.

Praveen, A. and Selvi, T. Tamil, Power efficient design of adiabatic approach for low power VLSI circuits, *ICEES 2019 Fifth International Conference on Electrical Energy Systems*.

Roy, Kaushik and Prasad, C. Sarat, *Low Power CMOS VLSI Circuit Design*. A Wiley Interscience Publication 2000 USA.

Sharma, Anjali, Sohal, Harsh and Sharma Kulbhushan, Area and power analysis of adiabatic 2:1 multiplexer designon 65nm CMOS technology. 2016 *5th International Conference on Wireless Networks and Embedded Systems* (WECON-2016).

Advances in AI for Biomedical Instrumentation, Electronics and Computing – Sachan et al. (eds)
© 2024 The Author(s), ISBN 978-1-032-64298-7

Blockchain in education: A revolutionary paradigm for enhanced security and transparency

Shikha Agarwal, Aarti Chaudhary, Komal Shivhare & Ashish Bajpai
Department of Electronics and Communication Engineering, Ajay Kumar Garg Engineering College, Ghaziabad, U.P., India

ABSTRACT: Blockchain is a new technology that's changing many industries, including education. This paper looks at how we can use blockchain in education to make it more secure, transparent, and efficient. First, we'll explain what blockchain is. It's like a digital ledger that's decentralized, meaning it's not controlled by one central authority. It's also super secure because information stored on it can't be easily changed or tampered with. Now, let's talk about the problems in education that blockchain can solve. These include issues like fake diplomas, slow paperwork, and data breaches. Blockchain can help by making digital diplomas and certificates that are very hard to fake. It also keeps educational records safe and lets anyone check them easily, which means we don't have to rely on one organization to verify them. Additionally, blockchain can make things like signing up for classes or transferring credits between schools faster and cheaper, enhancing learners' interactivity, improved data access control, increased trust, and identity authentication, high security, low cost and more. Blockchain can be slow and has some rules it needs to follow. Also, there are concerns about regulations and the technology's infrastructure. In this paper we have discussed that blockchain has a lot of potential to make education safer, more open, and smoother. Even though there are obstacles, it's an exciting area for more research and exploration in education.

1 INTRODUCTION

The blockchain came into being in 2008 when Satoshi Nakamoto published "Bitcoin: A Peer-to-Peer Electronic Cash System." With the use of cryptographic evidence and the method he described, two parties may carry out transactions on their own without the assistance of a reliable third party. By addressing the double spending issue, this groundbreaking idea prepared the way for the blockchain's initial use (Nakamoto 2008). A wide range of applications may benefit from flexible blockchain characteristics including speedy transactions, decentralized data sharing, smart contracts, transparency, and others. The broad and encouraging application of blockchain technology improves services for students and other stakeholders in schools, especially when it comes to credentials and certifications (Mourtzis *et al.* 2021, June), (Grech *et al.* 2021), (Rooksby and Dimitrov 2019), verified and authenticated records (Sharples and Domingue 2016), assistance to recruiters in the hiring process (Skiba 2017), maintenance of records (Chen *et al.* 2018; Sharples and Domingue 2016; Skiba 2017) and accessibility of these records (Turkanović *et al.* 2018) . Taking everything into account, blockchain technology's potential applications go beyond those listed above and allow university administration to manage their finance and accounting department by simply transferring all of the campus's fees and dues to it (Mourtzis *et al.* 2021, June). The focus of this research paper is on how the blockchain benefits and demonstrates the value of the education sector. This paper addresses a few difficulties that exist in the sector of education that blockchain features can help to resolve. However, there

DOI: 10.1201/9781032644752-79

are still some drawbacks, such as the revealing of personal information (Xu *et al.* 2017; Zheng *et al.* 2017, June), the security of private and public keys (Skiba 2017; Turkanović *et al.* 2018), and scalability problems (Grech *et al.* 2021; Hoy 2017; Sharples and Domingue 2016; Zhao *et al.* 2016; Zheng *et al.* 2017 June). The remainder of the essay is organized as follows. All the ways that blockchain could impact the education sector are listed in Section 2. The usage of blockchain in higher education is covered in Section 3. We talked about difficulties and problems with blockchain implementation at the ground level in Section 4 of this article. Section 5 & 6 discusses about scalability and interoperability issues. Finally, Section 7 summarizes all of the benefits and difficulties of blockchain technology and also discusses potential applications in the future.

2 BLOCKCHAIN IN EDUCATION

By offering a safe, open, and unchangeable platform for storing and exchanging academic information, blockchain technology has the potential to revolutionize the education industry. Blockchain application will continue to increase, and the education sector will see the emergence of fresh and creative use cases.

2.1 *How blockchain is set to revolutionize the sector of education*

From traditional classroom settings to eLearning, and now to blended learning, education has seen enormous change. Globally, schools started implementing online learning platforms and Learning Management Systems (LMSs) to instruct students as a result of the COVID-19 epidemic. The education sector is likewise set to undergo a transformation thanks to blockchain technology. The management of academic data and interactions between teachers and students could both be transformed by blockchain. Let's examine the potential future impact of blockchain technology on schooling.

2.2 *Blockchain's impact on education in various ways*

2.2.1 *Smart contracts for assignments and courses*
Education can be revolutionized by blockchain and smart contracts, which can provide transparent, immutable records, enforce deadlines, and automate the grading of assignments. Assessments are consistent when grading rubrics are customizable. These technologies also support worldwide accessibility, ease lifelong learning, and improve certification verification. They have a great deal of promise to simplify teaching and give students more authority, despite some obstacles.

2.2.2 *Student record keeping*
By offering a safe, unchangeable, and transparent platform for academic achievements, blockchain technology has the potential to completely transform the way student records are kept. It is easy for educational institutions to capture and validate student data, such as certificates and grades. Students now have more authority over their academic records, and employers can rely on the veracity of credentials. Notwithstanding certain difficulties, blockchain's influence on record-keeping has promise for improving data security, lowering fraud, and streamlining administrative procedures in the educational field.

2.2.3 *Transferring to a new school or college*
Students moving to other schools or institutions can have a smooth transition thanks to blockchain technology. Academic data and credentials may be safely stored on the blockchain, enabling institutions to quickly check and approve transfer credits. This guarantees the correctness of supplied data and lessens administrative overhead. A visible, unchangeable record

helps students by expediting their educational path and building confidence and trust in the transfer procedure (Heričcko 2019; Kamišalić; Mrdović; Turkanović).

2.2.4 *Digital badges and credentials*

Digital badges and certifications are more accessible and credible thanks to blockchain technology. Organizations and academic institutions are able to issue digital badges and certifications that are safely stored on blockchains. This guarantees genuineness, guards against fraud, and streamlines employer and institution verification procedures. Professionals and students alike may confidently display their accomplishments, encouraging lifetime learning and talent recognition in an open, unchangeable way.

2.2.5 *Ease of certification authenticity*

Blockchain technology provides a trustworthy way to confirm certification legitimacy. Employers and other entities may quickly verify qualifications thanks to the blockchain's ability to securely store and transmit digital credentials from academic institutions and organizations.

2.2.6 *Reduced cost*

Blockchain technology has the potential to lower educational expenses by improving security and expediting administrative procedures. Time and resources are saved when middlemen are less necessary thanks to automated record-keeping and verification. Furthermore, fraud and credential inflation are reduced by the tamper-resistant nature of blockchain.

3 BLOCKCHAIN IN HIGHER EDUCATION

Blockchain technology is transforming higher education by providing ground-breaking answers to enduring problems. The transparent and safe keeping of certifications, credentials, and student information is one important advantage. In order to make it simpler for employers and other educational institutions to check the qualifications of potential students or employees, this assures the validity and tamper-resistant character of academic achievements.

One of the main features of blockchain technology, smart contracts, has the power to completely change a number of academic procedures. They can streamline registration procedures, reduce administrative workloads, and automate grading and assignment submissions, all of which increase efficiency.

Additionally, blockchain facilitates global accessibility by enabling students to share their academic records and credentials with others, which makes it simpler for professionals and students from other countries to be recognized for their qualifications. Notwithstanding all of these benefits, there are still issues to be resolved, such as the requirement for standardization and handling data privacy issues. However, blockchain is an innovative and promising development in higher education that will ultimately benefit both students and educational institutions due to its ability to decrease fraud, simplify record-keeping, and increase trust in academic credentials.

4 PROBLEMS WITH IMPLEMENTING BLOCKCHAIN IN EDUCATION

Blockchain integration in education faces a number of challenges. Its capacity to effectively handle growing volumes of data and transactions is impacted by scalability problems. It can be difficult to strike a balance between data privacy and openness, particularly when using public blockchains to store private student data. Careful thought must be given to managing high initial expenses, standardizing, and integrating blockchain technologies with the current educational infrastructure. Adoption is made more difficult by user-friendliness and regulatory uncertainty. Other crucial issues include overcoming opposition to change, guaranteeing long-term viability, and handling security threats. Notwithstanding these difficulties, a successful blockchain implementation can transform credential verification and record-keeping in the educational system, improving efficiency and security while protecting the privacy and openness of data.

Security:When using blockchain in education, security is crucial. Ensuring that records are tamper-proof and shielding private student information from unwanted access are essential. Maintaining the integrity and reliability of academic records and credentials requires careful key management, strong authentication, and preventative measures against fraud or hacking.

Scalability: When implementing blockchain technology in education, scalability is a critical consideration. Blockchain networks need to be able to manage the increased number of academic information and transactions in an effective manner. Sharding and layer-two technologies are examples of scaling solutions that are essential to preserving performance and avoiding congestion. For the blockchain to be widely adopted in the education sector, it must be able to process an increasing number of student records while maintaining transaction speed and efficiency, which solves the scalability difficulty.

Interoperability: One of the biggest obstacles to using blockchain in education is interoperability. Educational institutions should be able to exchange data and communicate with one other across many systems and platforms with ease. To provide cross-institution interoperability, standardized protocols and interfaces must be established. The facilitation of interoperability between various systems guarantees the effortless exchange and recognition of academic data, certificates, and transactions, hence augmenting the efficacy and extensive integration of blockchain technology within the education domain.

Adoption rate: Globally, blockchain technology use in education has varied rates. Although several establishments are investigating its possibilities, broad implementation is still sluggish because of obstacles including expenses, rules, and the requirement for technological know-how. Larger organizations and technologically advanced areas typically embrace it more quickly. But given its promise to improve security, openness, and efficiency in the administration of academic records and credentials, adoption is probably going to rise as the technology develops and becomes more widely accepted.

Data Privacy and Security: Security and privacy of data are critical when using blockchain in education. Strict access restrictions, secure key management, and strong encryption are necessary when storing private and sensitive academic data on a blockchain. To keep private student data hidden from unwanted users, institutions must find a balance between openness and data security. Resolving these issues is essential to preserving confidence and adhering to privacy laws in the educational system.

Credential Fraud Detection: Blockchain provides transparent, tamper-resistant records that help detect credential theft. Academic qualifications and certifications are kept in a safe location, making fraud challenging. Employers and institutions may rely on instantaneous, fast, and reliable verification. The immutability of blockchain technology guarantees the accuracy of academic data, so mitigating the risk of counterfeit credentials and preserving the integrity of educational degrees.

Long-Term Data Preservation: A viable option for long-term data preservation in education is blockchain technology. The establishment of an unchangeable academic record ledger guarantees the longevity and accessibility of historical data. Institutions don't have to worry about data loss or deterioration when storing records over lengthy periods of time. Academic institutions and individuals can benefit from the robustness of blockchain as a long-term data preservation tool due to its decentralized nature, which guarantees data redundancy.

5 SCALBILITY

When incorporating blockchain technology into systems of higher education, scalability is a critical consideration. It is commonly known that blockchain technology has the ability to revolutionize the safe storage and exchange of academic records, student information, and other educational data. But as these systems expand, they run the danger of being hampered by scalability problems, which are difficulties related to the blockchain's capacity to effectively handle a greater amount of data and transactions. In order to guarantee the scalability of their blockchain-based solutions, higher education institutions need to manage these issues by taking into account a variety of tactics. A crucial element is choosing the appropriate blockchain platform. Congestion on some public blockchains, such as Ethereum, has caused scalability problems.

Institutions ought to investigate more recent, scalable permissioned blockchains intended for use in business and education. Off-chain solution implementation is an additional tactic. This entail doing certain operations outside the main blockchain and then securing and immutably anchoring the outcomes on the network. As a result, the primary blockchain network may experience less stress. Another important factor is interoperability. Scalability of educational institutions is contingent upon their capacity to interact and share data seamlessly across a variety of systems and networks. The exchange of credentials and academic records between various networks and institutions can be facilitated by interoperability standards and solutions. Another method being investigated to improve scalability is sharding, which splits a blockchain network into smaller sections (shards) so that transactions may be processed in parallel.

5.1 *Scalability issues associated with bitcoin and ethereum*

As cryptocurrencies like Bitcoin and Ethereum gain more transaction and users, they are forced to consider a fundamental issue with their original design-namely, their inability to scale. The number of transactions using cryptocurrency is rising exponentially along with their increasing public acceptance.

Let's examine the rise in Ethereum and Bitcoin transactions over the last many years. As Bitcoin and other cryptocurrencies gain popularity, a number of concerns have arisen, with the main one being scalability. Figures 1 and 2 below show how quickly cryptocurrencies are being embraced. The fundamental tenet of Bitcoin, which requires mining nodes to validate each and every transaction that takes place in the network, is really where the majority of the issues arise.

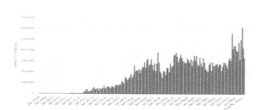

Figure 1.   Increase in number of transactions in bitcoin.

Figure 2.   Increase in number of transactions in Ethereum.

One of the main issues with blockchain technology is scalability. It speaks to a blockchain's capacity to manage a growing volume of users and transactions without sacrificing effectiveness or performance. There are still unanswered questions about scalability in research, despite great advances. The following lists the main areas of unmet research need in blockchain scalability, along with citations to pertinent studies.

5.1.1 *Consensus processes*
Proof of Work (PoW) and Proof of Stake (PoS) are two common consensus processes used by blockchain networks. The goal of continuing research is to create consensus algorithms

that are more effective and can process more transactions per second. For instance, the Ethereum community is working hard on Ethereum 2.0, a project that seeks to increase scalability by switching from proof-of-work to proof-of-stake (Buterin 2020).

5.1.2 *Sharding is a mechanism wherein the blockchain is divided into smaller shards, each of which is able to perform transactions*

Shaded blockchains require research to ensure their continued decentralization and resilience to assaults, as well as to enhance their security and coordination (Zamani *et al.* 2018).

5.1.3 *Cross-Chain compatibility*

A scalable blockchain ecosystem requires interoperability across several blockchain networks. To create guidelines and procedures for smooth asset transfers and communication amongst chains, research is required (Wood 2017).

5.1.4 *Reliability trade-offs*

Achieving great scalability sometimes necessitates giving up decentralization and security. In various blockchain use cases, research should concentrate on striking the ideal balance between scalability, security, and decentralization (Tschorsch and Scheuermann 2016).

5.1.5 *Resource efficiency*

Studies are required to lower the energy use of blockchain networks, particularly those that employ PoW consensus. Sustainable scalability depends on creating consensus methods that use less energy (Gervais *et al.* 2016).

6 INTEROPERABILITY

Assembling a uniform system for credentials and records in education across institutions or nations is a considerable difficulty in terms of interoperability. To exchange and share data in a manner that is easy and accessible to users of both interoperating systems, different systems, people, or entities must be able to work together correctly. This is known as interoperability (Ubaka-Okoye *et al.* 2020). Here's an elaboration on why this challenge exists and its implications (Li *et al.* 2019; Ubaka-Okoye *et al.* 2020).

Differentiated Blockchain Platforms and Protocols: There are many different blockchain platforms and protocols out there, and they all have different consensus techniques, smart contract languages, and sets of regulations. Ethereum, Bitcoin, Hyperledger Fabric, and several others are a few examples. The early design of these platforms did not consider education credentialing, which caused dispersion in the market.

Cross-Chain Communication: The main goal of interoperability is to make it possible for data to be shared and conversations between various blockchain networks. This refers to enabling records from one blockchain—let's say the blockchain of a university—to be acknowledged and trusted by another—let's say the blockchain of an employer—in the context of educational qualifications. Solutions for cross-chain communication are still in their infancy.

Security and Privacy of Data: Credentials from higher education are extremely private pieces of information. It is a difficult undertaking to guarantee the security and privacy of this data as it travels across several blockchains. It entails putting in place reliable encryption techniques and secure communication channels for all parties.

Scalability and Performance: When handling a high number of transactions and data, blockchain networks—including well-known ones like Ethereum-frequently have scalability and performance problems. When attempting to manage the enormous volume of educational credential data created internationally, this might become a bottleneck.

Adoption and Standard Development: The blockchain industry must widely embrace common standards and protocols in order to achieve interoperability. It takes time and the cooperation of many parties, including governments, employers, educational institutions, and blockchain developers, to define and adopt such standards.

User Experience:For adoption to occur, a seamless user experience must be provided. Educational credential systems need to be easy to use in order to be successful. This entails developing user-friendly and reliable programs and interfaces for both individuals and organizations.

7 CONCLUSIONS

A key component of blockchain's effective implementation in higher education is scalability. Schools must overcome the difficulties of handling increasing data and transaction volumes in order to fully utilize blockchain technology in safely maintaining credentials and academic records. This include picking suitable blockchain platforms, putting off-chain solutions into practice, making sure that they work together, investigating sharding and effective consensus techniques, keeping governance flexible, making investments in scalable data storage, and applying technologies that increase privacy. Educational institutions may successfully utilize blockchain's transformational potential to improve security, transparency, and efficiency in handling student information and certificates by addressing scalability challenges. This will eventually advance the area of higher education.

REFERENCES

Grech, A., Sood, I., & Ariño, L. (2021). Blockchain, self-sovereign identity and digital credentials: promise versus praxis in education. *Frontiers in Blockchain*, *4*, 616779.

Hoy, M. B. (2017). An introduction to the blockchain and its implications for libraries and medicine. *Medical Reference Services Quarterly*, *36*(3), 273–279.

Mourtzis, D., Angelopoulos, J., & Panopoulos, N. (2021, June). Blockchain in engineering education: The teaching factory paradigm. In *Proceedings of the Conference on Learning Factories (CLF)*.

Nakamoto, S. (2008). Bitcoin: A peer-to-peer electronic cash system. *Decentralized Business Review*.

Xu, Y., Zhao, S., Kong, L., Zheng, Y., Zhang, S., & Li, Q. (2017). ECBC: A high performance educational certificate blockchain with efficient query. In *Theoretical Aspects of Computing–ICTAC 2017: 14th International Colloquium, Hanoi, Vietnam, October 23–27, 2017, Proceedings 14* (pp. 288–304). Springer International Publishing.

Zhao, J. L., Fan, S., & Yan, J. (2016). Overview of business innovations and research opportunities in blockchain and introduction to the special issue. *Financial Innovation*, *2*, 1–7.

Zheng, Z., Xie, S., Dai, H., Chen, X., & Wang, H. (2017, June). An overview of blockchain technology: Architecture, consensus, and future trends. In *2017 IEEE International Congress on Big Data (BigData Congress)* (pp. 557–564). IEEE.

Sharples, M., & Domingue, J. (2016). The blockchain and kudos: A distributed system for educational record, reputation and reward. In *Adaptive and Adaptable Learning: 11th European Conference on Technology Enhanced Learning, EC-TEL 2016, Lyon, France, September 13–16, 2016, Proceedings 11* (pp. 490–496). Springer International Publishing.

Rooksby, J., & Dimitrov, K. (2019). Trustless education? A blockchain system for university grades. *Ubiquity: The Journal of Pervasive Media*, *6*(1), 83–88.

Turkanović, M., Hölbl, M., Košič, K., Heričko, M., & Kamišalić, A. (2018). EduCTX: A blockchain-based higher education credit platform. *IEEE access*, *6*, 5112–5127.

Chen, G., Xu, B., Lu, M., & Chen, N. S. (2018). Exploring blockchain technology and its potential applications for education. *Smart Learning Environments*, *5*(1), 1–10.

Skiba, D. J. (2017). The potential of blockchain in education and health care. *Nursing education perspectives*, *38*(4), 220–221.

Lemieux, V. L. (2016). Trusting records: is Blockchain technology the answer?. *Records management journal*, *26*(2), 110–139.

Buterin, V. (2020). Ethereum 2.0: Serenity.

Zamani, M., Movahedi, M., & Raykova, M. (2018). RapidChain: Scaling Blockchain via Full Sharding.

Wood, G. (2017). Polkadot: Vision for a Heterogeneous Multi-Chain Framework.

Tschorsch, F., & Scheuermann, B. (2016). Bitcoin and Beyond: A Technical Survey on Decentralized Digital Currencies.

Gervais, A., Karame, G. O., Capkun, S., & Capkun, S. (2016). Is Bitcoin a Decentralized Currency?

Eyal, I., & Sirer, E. G. (2018). Majority Is Not Enough: Bitcoin Mining Is Vulnerable.

"The Raiden Network a Beginner's Guide." [Online]

Advances in AI for Biomedical Instrumentation, Electronics and Computing – Sachan et al. (eds)
© 2024 The Author(s), ISBN 978-1-032-64298-7

Analysis of different categories of prediction methods in intelligent transport VANET system

R. Gracelin Sheeba
Assistant Professor, Department of ECE, Jeppiaar Engineering College, Chennai, India

N. Edna Elizabeth
Professor, Department of ECE, Sri Sivasubramania Nadar College of Engineering, Chennai, India

ABSTRACT: In Dynamic and wireless network topology of VANET real time communication system, safety and security are important factors. This provides road side assistance, avoidance of vehicle crashes, traffic management, smart city management, emergency care and environmental care. Concurrently this ensures safety driving on the road, enhances on-time supply of medical drugs, prevent the risk of road hazards and the medical emergencies of human in the rapid growth of vehicles on the line. When we identify the vehicle's struggling spot, we can provide smooth way of transportation. In this paper, various methods for predicting the location of vehicles are discussed and compared based on metrics such as prediction accuracy, prediction error and packet delivery ratio.

Keywords: Prediction, Anomaly detection, Intelligent transport system

1 INTRODUCTION

Intelligent transport system provides safety to make better travel judgement to alleviate traffic congestion for the drivers. It also provides the ability to improve car security, economic sustainability, riding comfort and reduces carbon emission and health problems. Parameters such as bandwidth, energy consumption, mobile node speed (node mobility), density, distances, link stability, rate of packet loss, throughput, average end to end delay, power consumption, computation, location prediction error, storage, energy consumption, reliable or stable route, communication overhead, wireless link breakage and network constancy are generally considered for prediction analysis. To make the system intelligent and to meet the objectives in VANETs, the research analysis is categorized as traffic prediction, routing protocol, anomaly detection, location prediction, destination prediction and lane prediction.

2 SCREENING REPORT

2.1 Traffic prediction

Machine learning and deep learning algorithms can be trained on historical and real-time data collected from sensors, cameras, or GPS devices to make traffic predictions.

In congested cities, to provide safety and reliability, proactive approach such as Long Short-Term Memory (LSTM) is used. To avoid long term dependency of past required information in RNN, the deep learning model (Ali *et al.* 2020) LSTM is usedto predict the traffic.Usingthe parameters root mean square error and mean absolute percentage error, the

DOI: 10.1201/9781032644752-80

prediction accuracy is evaluated. CNN is employed (Nie *et al.* 2018), to estimate traffic, to detectanomalies based on threshold, to extract spatio temporal features of network traffic.It describes monitoring and managing network traffic for various applications. Accuracy, false positive rateand detection time for anomalies are the metrics used for estimating traffic and detectinanomalies.

2.2 *Routing protocol*

Routing protocols are essential for networking and ensuring that the data reaches its intended destination efficiently and reliably. In (Cardenas *et al.* 2021), the authors describe about the five collected (data) metricsof dataset used to train and test the Machine Learning based forwarding algorithm. The algorithm is used to select the best next neighbor (hop) to forward the packets towards its destination. Current vehiclewhichis holding the packet (travelling related information) predicts the vehicle,to which the packet has to be forwarded that lies within its transmission range. The simulation results show that the power loss is less than 20% and the packet delay is less than 0.04ms. With the internet of vehicles, message delivery is done by using two algorithms suchas delivery profit maximization and communication scheduling algorithms (Cao wang *et al.* 2020). After predicting the destination, destination related information is forwarded. Evaluations factors considered are accuracy profit and forwarding profit. 94% of prediction accuracy is found in this method.

2.3 *Anomaly detection*

Anomaly detection is the process of identifying patterns or elements in data that differ fromits normal behavior. Trained deep learning model is introduced topredict pothole, bump, crack and no-anomaly. For road anomaly, classification ResNet-18 and VGG-11 architecture model (Bibi *et al.* 2021) is used to achieve safety and secure traffic flow. The author explains the context of VANETs and the need for automated detection of road anomalies. Road anomaliesinclude potholes, debris, accidents, and other unexpected events that can impact vehicle safety. Convolutional Neural Networks is used to estimate and detect anomalies in network traffic data with a spatio-temporal perspective (Nie *et al.* 2018). The goal is to improve network management and security by providing accurate traffic estimations and timely detection of anomaly. Anomalies could include DoS attacks, network failures, or unusual traffic patterns.

2.4 *Location prediction*

Location estimation refers to the act of estimating or predicting the future geographic location or control of an entity that may be a person, a vehicle, or an object based on historical data and relevant characteristics. This paper (Cheng *et al.* 2020) provides timely and safe transportation of medical vehicles. The methodology LPMV representscombination of Deep Belief Nets (DBN) and LSTM to navigate and reach their destinations efficiently. This predicts the surrounding environment's driving behavior and it also predicts the location to be moved. If the simple DBN network is used, the prediction is high. Complex structure gives the best prediction. Data relevant to the IoV, including GPS data from medical vehicles and other connected vehicles, are collected and preprocessed for training and evaluation of the location prediction model.Vehicle mobility patterns (VMP) are observed for some period of time from many vehicles and this will help to predict the location using vehicle order Markov model and prediction based soft routing protocol (Xue *et al.* 2012). The results show that the Control overhead is saved up to 90% compared to DSR algorithm.

2.5 Destination prediction

Destination prediction involves predicting the expected endpoint or final location of the current travel or movement based on previous journey and the way of historical data. Predicting a user's destination based on their current trajectory is a common problem in various applications, such as transportation, urban planning, and location-based services. The author (Yang *et al.* 2022) in this paper resolves the traffic congestion and avoid the data sparsity problem using encoder and decoder structure with branch network. With the aim of predicting future location, past trajectoryfeaturesare extracted, and vehicle's movement patterns are learned. For each hop, the branch network checks whether it reaches the destination or not. Based on the driver's destination patterns and driving schedule, vehicle destination is predicted (Casabianca *et al.* 2021). LSTM and Bidirectional LSTM with attention mechanismachieve an accuracy of 96% and maintain robustness and stability during forecasting. Using Geolife GPS Trajectories dataset, certain data are considered according to the timestamps (less than 10 minsjourney arediscarded).

2.6 Lane prediction

Lane estimation is selecting future path or lane of the vehicle to efficient transportation.

Lane selection is done using On Demand Analysis (ODA) that is obtained by the surrounding vehicles within the range of Roadside Unit. In this paper (Samara *et al.* 2021), selection of lane is based on average current speed of all vehicle and sudden break analysis of every vehicle. A greater number of ODAs are required to obtain optimum decision.

3 ANALYSES OF DIFFERENT CATEGORIES OF RESEARCH PAPERS

In Table 1, the Overall comparison of the above existing literature survey is discussed based on different Categories such as aim, methodology advantages, disadvantages and parameters used.

4 SIMULATION VIEW

For simulation of traffic related protocols, sumo software can be used.Sumo is a network monitoring tool, whichmonitors network connectivity by collecting and analyzing logs regarding network protocols, packet headers, and other relevant information. This helps to identify patterns, flaws, and potential security threats. This is anopen-source software available in the latest version sumo 1.18.1. There are three modules thatincludes netedit, OSM WebWizard and sumo GUI. In Figure 1, an example of an Open Street map of Egmore in Tamil Nadu is obtainedfrom the internet (website mention in reference 11).

After selecting the area in OSM, we can choose the number of buses, cars, trucks, bicycles, pedestrians and so on. After selection of all, click "Generate scenario". SUMO GUI is opened automatically with road structure shown in Figure 2.

In sumo config file, go to edit mode then select open network in netedit (ctrl + T) to move to the netedit page. Using sumo netedit shown in Figure 3, we can change configurations like traffic signal, position, speed of vehicles, edges priority and number of lanes and so on.

By varying the delay timings and enabling the run command, start the simulation until the simulation period is completed. Now the movement of vehicles can be observed as shown in Figure 4.

Table 1. Comparison of different categories of research papers.

S No	Title	Aim	Method	Parameters	Advantages	Dis-advantage	Year And Journal
1.	A Multimeric Predictive ANN-Based Routing Protocol for Vehicular Ad Hoc Networks	Design a routing protocol to exchange traffic related informationusing 5 multimeric. (Available bandwidth, Distance to destination, Vehicle's density, mac layer losses, and vehicle's trajectory)	Feed forward neural network	1. Accuracy 2. Overfitting detection 3. ROC curve 4. Average percentage of packet loss 4. Average end to end packet delay 5. Average running time to simulate 6. Computational cost 7. Overhead	Routing algorithm makes the packet loss less than 20% and average packet delay lessthan 0.04 ms.	Focusedonly on homogeneous city and not hetero-generous city	2021 IEEE Access
2.	Spatio-Temporal Network Traffic Estimation and Anomaly Detection Based on Convolutional Neural Network in VehicularAd-Hoc Networks	Network traffic estimation and anomaly detection.	CNN and threshold-based separation method	1. Traffic flow lestimation bias 2. Cumulative distribution 3. Function of spatial relative error 4. Temporal relative error.	CNN extract multifractal low rank features.	1. Computational complexity and scalability are very important to construct. 3. CNN architecture is very important to construct. 3. Lightweight learning algorithm is need	2018 IEEE Access
3.	Location Prediction Model Based on the Internet of Vehicles for Assistance to Medical Vehicles	Location prediction to assist medical vehicle using Driving behavior of surrounding environment.	LPMV (DBN and LSTM)	1. Prediction error 2. Error rate at different moments.	In urban scenarios, this model is to assist medical vehicles.	Need more parameters to analyze.	2020 IEEE Access
4.	Deep learning-based destination prediction scheme by trajectory prediction framework	Resolve traffic congestion	Trajectory prediction model and encoder, decoder structure with attention mechanism.	1. Prediction error 2. Percentage of trips completed.	Percentage of error is less.	–	2022 Hindawi journal, Security, and communication networks
5.	Lane prediction optimization in VANET	Prediction of best lane to travel from the predicted traffic line.	Using on demand analysis, lane is predicted. (Electronic license plate is used)	1. Relative travel time difference 2. Congestion level 3. No of ODA.	Simple ACS, AAVS and AVStud calculation to predict the secure lane.	Only 3 parameters to predict the lane	2022 Egyptian informatics journals.

No.	Title	Objective	Methodology	Metrics	Results	Limitations	Year/Publisher
6.	VANET Traffic Prediction Using LSTM with Deep Neural Network Learning	Prediction of network traffic	Long shot term memory (LSTM)	1. Throughput 2. RMSE 3. MAPE (Mean anomaly percentage error).	To avoid long term dependency of RNN.	Prediction accuracy not enough when no. of packets is increased.	2020 Springer nature Switzerland.
7.	Vehicle Destination Prediction Using Bidirectional LSTM with Attention Mechanism	Prediction of destination optimizing consumption of fuels-hybrid vehicles.	DNN, LSTM with attention mechanism and BiLSTM with attention mechanism.	1. Probability of destination 2. Prediction 3. Accuracy 4. Precision 5. Recall 6. F-score 7. Percentage of journey correctly predicted.	Attention mechanism is better when the large sequences with BiLSTM.	Test set between the 2 models with attention is not statistically significant.	2021 sensors
8.	Destination prediction-based scheduling algorithms for message delivery in IoVs	Delivery of destination related information and for that to predict the vehicles destination.	Communication scheduling and delivery profit maximization algorithm.	1. Forwarding profit 2. Accuracy profit.	Prediction accuracy can achieve 94% better than bayes model.	For perfect delivery, prediction accuracy needs to be increased.	2020 IEEE Access
9.	A novel vehicular location prediction based on mobility patterns for routing in urban VANET	Identification of mobility patterns and reduction in control overhead.	Variable order Markov model to observe vehicle mobility patterns(VMP) and prediction based soft routing protocol.	1. Control packet overhead	Control overhead saved up to 90%. With DSR and 75% with WSR. 2. packet delivery ratio 3. Packet delivery delay.	Many practical issues not clearly listed out.	2012 EURASIP journal on communications networking.
10.	Edge AI based automated detection and classification of road anomalies in VANET using deeplearning	Detection of obstacles for smooth travelling based on Edge AI useful for autonomous vehicles.	Residual convolutional neural network and visual geometry group.	1. Training and validation loss 2. Accuracy 3. Confusion matrix.	Reduce accident rate and risk of hazards.	Only four types of road hazards are considered.	2021, Hindawi computational intelligence and Neuro-science

Figure 1. OSM of egmore area in Tamil Nadu.

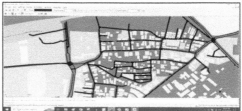

Figure 2. Sumo GUI of Egmore area in Tamil Nadu.

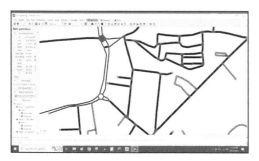

Figure 3. Netedit in SUMO.

Figure 4. Simulation of road traffic.

5 CONCLUSION

The Overall View of the different categories such as destination prediction (Yang *et al.* 2022) and (Casabianca *et al.* 2021), location prediction (Cheng *et al.* 2020) and (Guangtao Xue *et al.* 2012), traffic prediction (Ali *et al.* 2020) and (Nie *et al.* 2018), lane prediction (Samara *et al.* 2021), anomaly detection (RBibi *et al.* 2021) and (Nie *et al.* 2018) and routing protocol-based papers (Cardenas *et al.* 2021) and (Cao wang *et al.* 2020) are analyzed in this paper. Overall comparison of the above existing literature survey is discussed based on different Categories such as aim, methodology advantages, disadvantages and parameters used. The simulation related tools are also explained in detail with edited optionsto execute the different methodologies discussed in the paper.

REFERENCES

Ali R. Abdellah, and A. Koucheryavy, 2020, "VANET traffic prediction using LSTM with deep neural network learning", *Springer Nature Switzerland*, 12525, 281–294.

Cao wang, Yunhu He, Kexiao and Hayoyu Zhang, 2020 "Destination prediction-based scheduling algorithms formessage delivery in IoVs", *IEEE Access*, 8, 14965–14976.

Cardenas L. l., Mezher A. M., Bautista P. A. B., Leon J. P. A., and Igartua M. A., 2021 "A Multimeric predictive ANN based routing protocol for vehicular Ad Hoc networks", *IEEE Access*, 9, 86037–86053.

Casabianca P., Zhang Yu., Martinez-Garcia M. and Wan J., 2021, "Vehicle destination prediction using bidirectional LSTM with attention mechanism", *Sensors 2021*, 21, 8443(1–21).

Cheng J., Yan H., Zhou A, 2020, "Location prediction model based on internet of vehicles For assistanceto medical vehicles", *IEEE Access*, 8, 10754–10765.

Guangtao Xue, Yuan Luo, Jiadi Yu and Minglu Li, 2012, "A novel vehicular location prediction based on mobilitypatterns for routing in urban VANET", *SpringerEURASIP journal on communications networking*, 1/222.

Nie L., Li Y. and Kong X., 2018, "Spatio – temporal network traffic Estimation and anomaly detection based on convolutional Ad-Hoc Networks", *IEEE Access*, 6, 40168–40176.

Rozi Bibi, Yousaf Saeed, Asim Zeb, Taher M. Ghazal, Taj Rahman, Raed A Said, Sagheer Abbas, Munir Ahmed and Muhammad Adnan Khan, 2021, "Edge AI based automated detection and classification of road anomalies in VANET using deep learning", *Hindawi computational intelligence and Neuroscience*, 2021, article id-6262194, 16 pages, 2021 https://sumo.dlr.de/docs/Tutorials/Hello_SUMO.html

Samara G., 2021 "Lane prediction optimization in VANET", *Egyptian informatics Journal*, 22, 411–416.

Yang J., Cao J. and Lin Y., 2022, "Deep learning-based Destination Prediction scheme by trajectory Prediction Framework", *Hindawi Journal*, Security and communication Networks, 8.

Advances in AI for Biomedical Instrumentation, Electronics and Computing – Sachan et al. (eds)
© *2024 The Author(s), ISBN 978-1-032-64298-7*

CNN sight: Precision detection in gangrene diagnostics

Priyanshu Aggarwal & Sahil Aggarwal
CSE, Artificial Intelligence And Machine Learning, Dronacharya College of Engineering, Khentawas, Farrukh Nagar, Gurugram, Haryana

Harshita & Ritu Pahwa
CSE, Dronacharya College of Engineering, Khentawas, Farrukh Nagar, Gurugram, Haryana

ABSTRACT: Gangrene, a serious medical disorder in which bodily tissue dies, provides serious health concerns and need an early diagnosis for successful treatment. In this paper, a novel method for gangrene detection that makes use of Convolutional Neural Networks (CNNs) is proposed. The outstanding performance that CNNs have shown in image identification tasks makes them suitable for use in medical imaging applications. A complete and all-encompassing summary of gangrene is given in the first section of this paper. It distinguishes between the numerous gangrene kinds, such as dry, wet, and gas gangrene, each of which presents different obstacles in diagnosis and treatment. We investigate the many reasons, highlighting the multifaceted character of this syndrome, including diabetes, trauma, infections, and peripheral artery disease (PAD). In order to emphasise how complicated the causes of gangrene is, additional risk factors including weakened immune systems and smoking are also mentioned. The vital value of early identification in properly controlling gangrene is emphasised. The necessity of detecting gangrene in its early stages is highlighted by the development of tissue destruction and possible consequences, including systemic infections and the requirement for amputations. A prompt intervention not only enhances patient outcomes but also lessens the total load of advanced cases on the healthcare system. This study's main objective is to investigate how Convolutional Neural Networks (CNNs) could improve gangrene diagnostic skills. The complicated nature of gangrene in medical imaging presents obstacles that may be technologically overcome by CNNs, which are known for their expertise in image processing and pattern identification.

1 INTRODUCTION

When bodily tissue degenerates as a result of a bacterial infection or a loss of blood flow, it is considered to have gangrene, a dangerous and potentially fatal illness (Lewis et al. 2021). The extremities, including the fingers, toes, hands, and feet, are frequently impacted; however it can also happen in other body regions.

1.1 *Various forms of gangrene exist*

1.1.1 *Dry gangrene*
When the blood supply to a tissue is cut off, tissue death results. It frequently affects the extremities and advances gradually.

1.1.2 *Wet gangrene*
This kind of gangrene is frequently brought on by a bacterial infection. Because it includes tissue death and wetness, it spreads more quickly and might be more harmful.

DOI: 10.1201/9781032644752-81

1.1.3 *Gas gangrene*

Wet gangrene in its most severe form, called gas gangrene, is brought on by gas-producing bacteria. It may be fatal, therefore you should get medical help right now.

1.2 *The following are typical causes of gangrene*

Reduced blood flow to the extremities due to peripheral artery disease (PAD).

1.2.1 *Diabetes*
It might result in side effects including impaired circulation and nerve damage.

1.2.2 *Bacterial infections*
particularly those that go untreated or are poorly managed, are infections.

Gangrene symptoms might include colouring of the infected area, excruciating pain, swelling, and a painful discharge. The standard course of treatment is treating the underlying cause, getting rid of dead tissue, and giving antibiotics (Joury et al. 2019).

An immediate medical response is required for gangrene in order to recover affected tissue and stop the infection from spreading. Seek emergency medical assistance if you believe you or someone else may have gangrene. Serious consequences, such as the necessity for an operation and systemic infections, might arise from delayed or ineffective treatment. The usual course of therapy involves treating the root cause of the problem, cleaning (the removal of dead tissue), and the use of medicines to manage bacterial infections. Surgery may be required in extreme circumstances to enhance blood flow and stop the spread of illness.

Figure 1.   Foot gangrene.

2   CAUSE OF GANGRENE

The destruction of bodily tissue due to a bacterial infection or loss of blood flow (ischemia) can result in gangrene from a number of different causes. These are the main factors that lead to gangrene:

2.1 *Peripheral Artery Disease (PAD)*

Hypertension (the narrowing or blocking of arteries), a primary cause of gangrene, reduces blood supply to the extremities. Oxygen and nutrients may not reach the tissues enough as a result of PAD.

2.2 *Diabetes*

Blood vessel damage can slow or block blood flow to a part of the body.

Diabetes-related problems including peripheral neuropathy (nerve damage) and peripheral vascular disease (reduced blood flow) put people with the condition at an increased risk of developing gangrene.

2.3 Injury or trauma

Gangrene can develop as a result of serious wounds and burns. The risk is greater if you have an underlying condition that affects blood flow to the injured area.

2.4 Bacterial infections

Infections, particularly those brought on by bacteria, might accelerate the gangrene process. Bacterial infections can spread and harm tissue if ignored or improperly controlled.

2.5 Medical or surgical treatments

In some instances, medical or surgical treatments might alter blood flow or introduce germs, raising the risk of gangrene.

2.6 Immunosuppression

It affect the body's ability to fight off infections. Diseases or therapies that impair immunity might make people less immune to infections, which may hasten the onset of gangrene.

Smoking: People who smoke have a higher risk of gangrene. Vascular disease is significantly impacted by tobacco usage. Smoking can cause blood vessels to close down and blood flow to be compromised,

2.7 Blood clotting disorders

Cause blood clots can block blood flow to an area of the body. Blood clotting disorders, such as thrombophilia, resulting in ischemia and gangrene.

2.8 Chronic health illnesses

Rheumatoid arthritis and systemic lupus erythematosus are two chronic illnesses that may cause vascular issues and raise the risk of gangrene.

2.9 Abuse of drugs

Using drugs by injection, particularly when they are injected directly into the bloodstream, can introduce impurities and raise the possibility of infections that might result in gangrene.

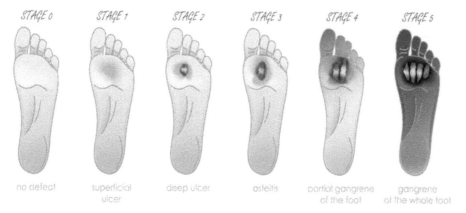

Figure 2.   Stages of Gangrene.

3 CONVOLUTIONAL NEURAL NETWORKS (CNN)

CNN is a mathematical construct that is typically composed of three types of layers (or building blocks): convolution, pooling, and fully connected layers [4]. Convolutional Neural Networks (CNNs) are a subclass of deep neural networks used for a variety of applications using structured grid data, including image and video recognition, natural language processing, and others. As a result of its capacity to automatically and adaptively learn spatial hierarchies of features from input data, CNNs have demonstrated a particular proficiency in computer vision tasks.

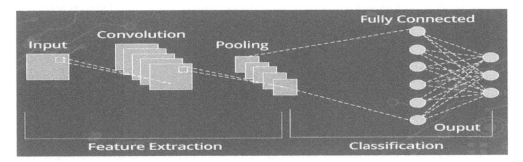

Figure 3.   Convolution neural network architecture.

3.1   *The following are the main elements and ideas of convolutional neural networks*

3.1.1   *Convolutional layers*
Convolutional layers convolve the input and pass its result to the next layer. CNNs are constructed using convolutional layers, where input data is subjected to convolutional processes. In convolution, the input is passed through a filter (also known as a kernel) while the dot product is calculated at each step. The network can learn hierarchical representations of features thanks to this procedure.

3.1.2   *Pooling layers*
This approach shrinks large-size feature maps to create smaller feature maps (Alzubaidi et al. 2021). After convolutional layers, pooling layers are frequently applied to reduce the input's spatial dimensionality. For instance, max pooling lowers the spatial resolution while maintaining the most crucial characteristics by retaining the maximum value within a predetermined zone.

3.1.3   *Functions of activation*
Rectified Linear Unit (ReLU) is one example of a non-linear activation function that adds non-linearity to the model, allowing it to learn complicated connection in the data.

3.1.4   *Fully connected layers*
Fully connected layers connect every neuron in one layer to every neuron in another layer. One or more fully connected layers are frequently present in the network for prediction after features have been extracted using convolution and pooling. These layers link every neuron in the layer below and the one above together.

3.1.5   *Backpropagation training*
CNNs are taught using supervised learning techniques. To reduce the discrepancy between expected and actual outputs, optimisation methods like stochastic gradient descent and backpropagation are used to iteratively alter the network's parameters (weights and biases).

3.1.6 *Weight sharing*

Weight sharing is one of the main benefits of CNNs. Convolutional layers employ the same set of weights for several spatial locations, enhancing the model's ability to generalise and learn features that are translation-invariant.

3.1.7 *Dropout*

A regularisation method frequently used with CNNs. To avoid overfitting, it includes randomly removing (setting to zero) a portion of neurons during training.

3.1.8 *Transfer learning*

For particular tasks, CNNs may use pre-trained models on huge datasets. Transfer learning is a technique that can dramatically boost performance, especially when there is a dearth of labelled data.

3.1.9 *Data augmentation*

During training, data augmentation techniques are frequently employed to improve model generalisation. This entails subjecting the input data to random modifications like rotation, flipping, or scaling.

3.1.10 *Designs*

Several well-known CNN designs, including as LeNet, AlexNet, VGGNet, GoogLeNet (Inception), ResNet, and others, have been created for specialised objectives. These architectural designs differ in terms of intricacy, depth, and structure.

3.1.11 *Applications*

CNNs are highly successful in a number of tasks, including image classification, object identification, face recognition, medical image analysis, natural language processing, and even Go board games.

Convolutional Neural Networks have made significant contributions to the development of computer vision and are now a key component of many applications of artificial intelligence. They are ideal for applications requiring structured grid data, especially photos and videos, due to their capacity to autonomously learn structural characteristics.

4 USE OF CONVOLUTIONAL NEURAL NETWORKS (CNNS) IN GANGRENE DETECTION

By utilising its capacity to automatically learn hierarchical information from medical pictures, Convolutional Neural Networks (CNNs) may be used to identify gangrene.

Here is an example of how CNNs may be used to identify gangrene:

4.1 *Input for medical imaging*

CNNs are excellent at image analysis tasks, making them suitable for applications in medical imaging. Medical pictures of the damaged body parts, such as X-rays, CT scans, or MRIs, are often the input data for gangrene identification.

4.2 *Feature extraction*

Convolutional operations are used in the earliest layers of a CNN to extract features, which are automatically learned from the input pictures. These characteristics might include colour changes, patterns linked to tissue texture, and structural abnormalities suggestive of tissue destruction in the context of gangrene identification.

4.3 Hierarchy of features

The CNN establishes a hierarchy of features through succeeding convolutional and pooling layers, allowing the model to learn both low-level (such as edges and textures) and high-level (such as complex patterns and structures) elements associated with gangrenous tissue.

4.4 Adaptability to image variability

Gangrene can present itself in a variety of ways, and medical photographs may not always accurately depict it. With the help of robust feature representations they have learned, CNNs are skilled at adjusting to this fluctuation. This versatility is essential for correctly diagnosing gangrene in a range of patient situations.

4.5 Transfer learning

A pre-trained model is adjusted for a particular task using the transfer learning approach. The model may capture generic characteristics by pre-training on large datasets (which may not be specific to gangrene), then fine-tuning on a smaller dataset of medical pictures improves its specificity to gangrene-related patterns.

4.6 Classifying gangrene types

Gangrene can come in a variety of forms, including dry, moist, and gas gangrene, each of which might have a different appearance. CNNs may be able to differentiate between various categories by learning from a variety of datasets, assisting in more accurate diagnosis.

4.7 Real-time detection

CNNs may be used for real-time or almost real-time detection because to improvements in hardware and optimisation approaches. In a clinical context, quick analysis is essential for prompt decision-making.

4.8 Integration into clinical workflow

CNN-based gangrene detection models may be included into the current clinical workflow to help medical personnel recognise and monitor the extent of tissue destruction. Such instruments may serve as an additional diagnostic help.

4.9 Continuous learning

CNNs may be created for continuous learning, which enables them to adjust and advance as new data becomes available. For situations like gangrene that are developing medically, this is especially helpful.

It is significant to stress that access to high-quality, labelled medical imaging datasets, ethical concerns, and validation through rigorous testing and clinical trials are necessary for the effective deployment of CNNs for gangrene diagnosis. To enable the successful application of such instruments in clinical practise, their integration should also be done in cooperation with healthcare experts.

5 CONCLUSION

To sum up, the essential topic of this paper is gangrene, a serious medical illness that needs to be detected early. Convolutional Neural Networks (CNNs) are used in the suggested approach to increase the diagnostic precision of medical imaging. The thorough review discusses the several forms of gangrene and its intricate causes, stressing the significance of early detection to avert serious outcomes. The goal of the project is to use CNNs to get beyond technology barriers and improve gangrene diagnosis, which will improve patient outcomes and increase the effectiveness of the healthcare system.

REFERENCES

Alzubaidi, L., Zhang, J., Humaidi, A. J., Al-Dujaili, A., Duan, Y., Al-Shamma, O., ... & Farhan, L. (2021). Review of deep learning: Concepts, CNN architectures, challenges, applications, future directions. *Journal of Big Data*, 8, 1–74.

https://en.wikipedia.org/w/index.php?title=Convolutional_neural_network&oldid=1178788445

https://images.app.goo.gl/rGdN4RHLgMAq2HMM7

https://stock.adobe.com/in/images/diabetic-foot-stages-of-defeat-ulcers-skin-sores-on-foot/324643109

https://www.downtownveinvascular.com/wp-content/uploads/2023/09/gangrene-gangrene-treatment.jpg

https://www.linkedin.com/advice/0/how-do-you-implement-data-augmentation-techniques#:~:text=Data%20augmentation%20can%20address%20a,classes%20by%20applying%20different%20transformations.

https://www.mayoclinic.org/diseases-conditions/gangrene/symptoms-causes/syc-20352567#causes

Joury, A., Mahendra, A., Alshehri, M., & Downing, A. (2019). Extensive necrotizing fasciitis from Fournier's gangrene. *Urology Case Reports*, 26, 100943.

Lewis, G. D., Majeed, M., Olang, C. A., Patel, A., Gorantla, V. R., Davis, N., ... & Lewis, G. (2021). Fournier's gangrene diagnosis and treatment: A systematic review. *Cureus*, 13(10).

Yamashita, R., Nishio, M., Do, R. K. G., & Togashi, K. (2018). Convolutional neural networks: an overview and application in radiology. *Insights into Imaging*, 9, 611–629.

Advances in AI for Biomedical Instrumentation, Electronics and Computing – Sachan et al. (eds)
© 2024 The Author(s), ISBN 978-1-032-64298-7

Implementation of crowbar protection in DFIG

Farhat Nasim, Shahida Khatoon & Ibraheem
Department of Electrical Engineering Faculty of Engineering and Technology, Jamia Millia Islamia, New Delhi

Mohammad Shahid*
Department of Electrical Engineering, Galgotias College of Engineering and Technology Greater Noida

ABSTRACT: Induction generators with double feed are often used in wind power plants with variable speeds. Wind turbines controlled by DFIG are dynamically connected to the electrical power system, which can be disrupted by a short circuit fault. It is vital to maintain grid stability during transient conditions by having Low Voltage Ride Through. To prevent cascading effects after wind power plants, lose power during disruptions, they must remain connected to the grid. Presented here is a strategy to allow the DFIG to cope with sudden fluctuations in grid voltage when there are abnormal grid voltage conditions. To study the effect of fault ride on DFIG wind turbine stability, this paper uses crowbar resistance. The paper uses three control schemes: crow bar protection, converter control on both side i.e., generator side, as well as rotor side. With the help of appropriate control strategies, power generation from wind turbines provided by DFIG is reliable and efficient even when the grid is unbalanced. To demonstrate how the generator behaves during disturbances in grid voltage, also analysis of the simulation results of a 2MW DFIG system is presented.

Keywords: Double fed Induction Generator (DFIG), Symmetrical Voltage dip, Asymmetrical voltage dip, DC bus voltage, Pulse Width Modulation (PWM), Crowbar Protection.

1 INTRODUCTION

Climate change can be tackled and a sustainable future can be achieved by using renewable energy sources Wind energy is one of the most popular renewable energy sources today, utilizing the wind to create electricity, which makes it one of the leading contenders among the other renewable energy sources. Clean and reliable energy can be generated by wind energy systems, which include wind turbines, infrastructure, and grid integration [1]. The Voltage Source Converter used in this technology is 30% higher than that used in the PMSG concept, which makes this technology more cost-effective. It is especially important to pay attention to the DFIG power converter during grid faults due to its limited over-current limit. Power converters may run at very high over currents when faults occur and result in voltage dips. If this occurs, it is common for generators to be disconnected from the electric grid, and then block the converter to prevent any damage [2]. DFIG wind farms are most likely to be equipped with Fault Ride-through FRT capability. It is essential that wind farms remain connected to the grid during system disturbances, so that voltage and frequency are not affected. When this happens, the rotor current will become very large during grid failure and some requirements will have to be met for the safe operation of the DFIG's rotor side inverter. This paper examines how DFIG behaves when grid voltage is unbalanced. This paper is divided into three sections: Section I,

---

*Corresponding Author: eems.j87@gmail.com

DOI: 10.1201/9781032644752-82

which introduces DFIG system modeling, Section II, which briefly describes the operation of DFIG, and Section III, which discusses controlling rotor and grid side converters. DFIG is examined during symmetrical and asymmetrical voltage dips in Section IV. Section V, and VI describes the result and discussion, conclusion respectively.

2   OVERVIEW

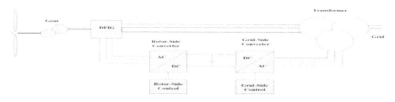

Figure 1.   Grid-connected DFIG-based WECS.

DFIG-based WECS with grid connectivity is illustrated in Figure 1. Generally, a VSC connected to a winding of a rotor through a rotor filter is called as RSC.GSC has a direct stator winding connection and a direct VSC connection at the point of coupling (PoC). Power generation is regulated by RSC in DC buses, while constant power is maintained by the GSC [3].

3   MATHEMATICAL MODELING OF DFIG

Figure 2 shows an equivalent circuit (dynamic) for an induction machine. Based on an equivalent circuit for a DFIG, the following equations are needed. when viewed in a synchronously rotating reference frame the dynamics of DFIG are described in Figure 2(a) and Figure 2(b). As far as flux equations are concerned, they are as follows [4].

$$Y_{ds} = \Re_s \Im_{ds} + \frac{d\Psi_{ds}}{dt} - \varpi_e \Psi_{qs} \tag{1}$$

$$Y_{qs} = \Re_s \Im_{qs} + \frac{d\Psi_{qs}}{dt} - \varpi_e \Psi_{ds} \tag{2}$$

$$Y_{dr} = \Re_r \Im_{dr} + \frac{d\Psi_{dr}}{dt} - (\varpi_e - \varpi_r)\Psi_{qr} \tag{3}$$

$$Y_{qr} = \Re_r \Im_{qr} + \frac{d\Psi_{qr}}{dt} - (\varpi_e - \varpi_r)\Psi_{dr} \tag{4}$$

Equation for calculating electromagnetic torque

$$\tau_e = \frac{3}{2}\left(\frac{P}{2}\right)\frac{1}{\varpi_b}\left(\Psi_{ds}\Im_{qs} - \Psi_{qs}\Im_{ds}\right) \tag{5}$$

$$\tau_e - \tau_L = \mathbb{J}\frac{2}{P}\frac{d\varpi_r}{dt} \tag{6}$$

$$P_s = \frac{3}{2}\left(Y_{ds}\Im_{ds} + Y_{qs}\Im_{qs}\right) \tag{7}$$

$$Q_s = \frac{3}{2}\left(Y_{qs}\Im_{ds} - Y_{ds}\Im_{qs}\right) \tag{8}$$

444

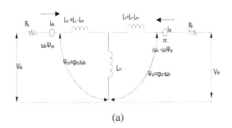 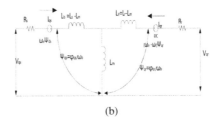

(a)   (b)

Figure 2.   (a) (d-Axis) equivalent circuit, (b) (q-Axis) equivalent circuit.

Table 1.   Referred parameter to rotor side.

Parameter	Symbol	Values
Rated Stator-Power	(P_s)	2×10^6 W
Stator Frequency	(f)	50Hz
Stator (Voltage, Current)	$(\Upsilon_s),(\mathfrak{I}_s)$	690V,1.76KA
Rotor Voltage	(Υ_r)	2070V
Stator Resistance, Stator	$(\mathfrak{R}_r),(L_s)$	2.6 $m\Omega$,
Leakage Inductance		0.0026H
Rotor Resistance, and	$(\mathfrak{R}_r),(L_r)$,	2.91 $m\Omega$,
Leakage Inductance,	(L_m)	0.00261H,2.5mL
Magnetizing-Inductance		

Table 2.   Referred parameters to stator side.

Parameters	Symbol
Magnetizing flux of d-q axis	(Ψ_{md}, Ψ_{mq})
d and q axis stator and rotor side voltages.	$(\Upsilon_{ds}, \Upsilon_{dr}), (\Upsilon_{qs}, \Upsilon_{qr})$
d and q axis stator and rotor side currents.	$(\mathfrak{I}_{qs}, \mathfrak{I}_{qr}), (\mathfrak{I}_{ds}, \mathfrak{I}_{dr})$
Stator and rotor side Reactance	$(\chi_{ls}), (\chi_{lr})$
Angular frequency, Rotor speed.	$(\varpi_e), (\varpi_r)$
Base angular frequency	(ϖ_b)

Table 3.   Grid side parameters.

Parameters	Symbol	Values
DC bus Voltage	Υbus	325.27V
DC bus capacitance	Cbus	0.0800F
Grid side filter's resistance	$\mathfrak{R}_S$	$2 \times 10^{-5}\Omega$
Filter Inductance	L_S	$40 \times 10^{-5} H$

3.1   *RSC control*

Activating and reversing power are controlled by vector controls on rotor side converters. RSC controller's block diagram is shown in Figure 3. A field-oriented vector controls the stator flux of the rotor-side converter [5]. Direct axes loops can be used to control reactive power, while quadrature axes loops can be used to control active power.

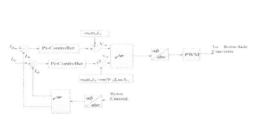 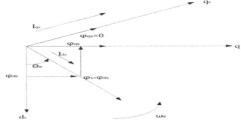

Figure 3.   Rotor side controller.   Figure 4.   Stator flux vector cXontrol phase diagram.

A1. Using Stator Flux Vector Control Approach

As shown in the diagram, the direct axis of this reference frame is parallel to the flux in the stator, $\Psi_{ds} = \Psi_s$ and quadrature flux, $\Psi_{qs} = 0$ as represented in Figure 4 [6].

$$\Psi_{ds}^s = \int V_{ds}^s - \Re_s \mathcal{I}_{ds}^s \tag{9}$$

$$\Psi_{qs}^s = \int V_{qs}^s - \Re_s \mathcal{I}_{qs}^s \tag{10}$$

$$\Psi_{ds} = \Psi_s \text{ and } \Psi_{qs} = 0 \tag{11}$$

$$P_s = \frac{-3}{2} \frac{L_m}{L_s} \mathcal{I}_{qr} \tag{12}$$

Substituting (11) in equation (1,2,3,4, 7 and 8) we obtain

$$Q_s = \frac{3}{2} \frac{L_m}{L_s} \left(\frac{V_{qs}}{\varpi_e L_m} - \mathcal{I}_{dr} \right) \tag{13}$$

$$V_{dr} = \left(k_p + \frac{k_i}{s} \right) (\mathcal{I}_{dr}* - \mathcal{I}_{dr}) - s\varpi_e \sigma L_r \mathcal{I}_{dr} \tag{14}$$

$$V_{qr} = \left(k_p + \frac{k_l}{s} \right) (\mathcal{I}_{qr}* - \mathcal{I}_{qr}) + s\varpi_e \left(\sigma L_r \mathcal{I}_{dr} + \frac{L_m^2}{L_S} \mathcal{I}_{ms} \right) \tag{15}$$

3.2 GSC control

Before designing the DFIG control system, it is necessary to design the GSC in MATLAB/Simulink. Our previous discussion of GSC structures explained how equivalent circuits and transformation equations are used to design them. By absorbing sinusoidal current at the frequency of a DC bus, it is possible to maintain the DC bus voltage. It is necessary to establish a zero reactive power reference in order to maintain unitary power. Figure 4 illustrates the adopted control strategy. One loop regulates the voltage and the other regulates the current in the control scheme.

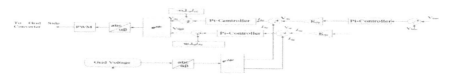

Figure 5. Grid side controller.

3.3 Using grid side vector technique

By using field control method, the grid side control process can be simplified, sped up, and made more efficient. In order to accomplish this, the g axis is aligned with the dq axis using the GVO approach. Using the difference between the voltages, PI regulators utilize the difference between the voltages to calculate $I_{dg}*$. In addition, the Q_g is used for calculating the ref value for $I_{qg}*$. It is possible to simplify, speed up, and improve the grid side control process by using the field control method. The GVO approach aligns the g axis with the dq axis in order to accomplish this. By using the difference between the voltages, a PI regulator is used to determine $I_{dg}*$. Q_g is also used to calculate the ref value for I_{qg}. In PI controllers, parameters are determined according to the changes in parameters over time. MATLAB implementation of the GSVOA is illustrated in Figure 4 with a block schematic. Through a

446

filter, $(L_\$, \Re_\$)$ connections are made to the electrical grid from the rotor side [9].

$$Y_{s\_a} = Y_{\$\_a} - \Re_\$ i_{\$a} - L_\$ \frac{di_{\$\_a}}{dt} \tag{16}$$

$$Y_{s\_b} = Y - \Re_\$ i_{\$b} - L_\$ \frac{di_{\$\_b}}{dt} \tag{17}$$

$$Y_{s\_c} = Y_{\$\_c} - \Re_\$ i_{\$c} - L_\$ \frac{di_{\$\_c}}{dt} \tag{18}$$

$$Y_{sd} = Y_{\$d} - \Re_\$ i_{\$d} - L_\$ \frac{di_{\$\_d}}{dt} + E_{\$q} \tag{19}$$

$$Y_{sq} = Y_{\$q} - \Re_\$ i_{\$q} - L_\$ \frac{di_{\$sq}}{dt} + E_{\$d} \tag{20}$$

Powers by filter exchanges with the grid are defined by the equations (21) and (22)

$$P_\$ = Y_{sq} i_{\$q} = \frac{3}{2} Y_s i_{\$q} \tag{21}$$

$$Q_\$ = Y_{sq} i_{\$d} = \frac{3}{2} Y_s i_{\$d} \tag{22}$$

4 DFIG ANALYSIS DURING A VOLTAGE DIP

DFIG's stator winding will be affected by unbalances in grid voltage when there are unbalances in grid voltage. This condition results in an unbalanced set of currents being taken across the Machine terminals regardless of a balanced load being connected, which may cause a negative, positive and zero sequence magnetic field to occur in the air gap, resulting in a reduction in torque and an increase in torque pulsation. Positive and negative sequence rotations can be initiated by the decomposition of this unbalanced field into symmetrical components. An overall increase in torque pulsation will result from a positive sequence which behaves similarly to balanced mode operation.

$$\frac{d\Psi_s}{dt} = \left[Y_s - \frac{\Re_s}{L_s} \Psi_{ds} + \frac{L_m}{L_s} i_r \right] \tag{23}$$

$$Y_r = \frac{L_m}{L_s} (Y_s - j\varpi_m \Psi_s) + \left[\Re_r + \left(\frac{L_m}{L_s}\right)^2 \Re_s \right] i_r + \sigma L_r \frac{di_r}{dt} \tag{24}$$

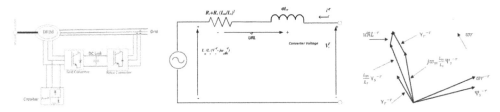

Figure 6. DFIG with crow bar protection.

Figure 7. DFIG simplified equivalent circuit.

Figure 8. DFIG voltage vector diagram.

In order to keep the rotor current within safe limits, the rotor voltage must be maintained. As a result, 1/3 of the stator voltage can be provided by the rotor (Slip = 3%). Under severe voltage drops, these types of wind turbines cannot guarantee rotor current will remain within safe limits due to rotor voltage limitations. This limitation leads to the incorporation of additional crow bar protection. in wind turbines based on DFIG. During the installation of the rotor converter, a crow bar is installed at the terminals to prevent damage. When a voltage sag is detected, the crowbar is activated and the entire rotor current is circulated through it, keeping the rotor converter safe.

5 SIMULATION RESULTS

In DFIG systems, simulation studies evaluate the response to symmetrical voltage dips. In order to mitigate the impact of symmetrical and asymmetrical voltage dips on the DFIG system, these results aid in understanding its behavior and help develop control strategies.

5.1 *Symmetrical voltage sag*

All three phases of the grid experience a simultaneous reduction in voltage magnitude during a balanced voltage dip, also called a symmetrical voltage sag. Analyses have been carried out using MATLAB simulations using a DFIG wind turbine of 2MW. Figure 11 shows a voltage dip at time t = 3 seconds. At the time of the dip, grid voltage is only 10% of rated voltage. Additionally, the crowbar shown in Figure 9 is activated simultaneously, creating another resistance path. The stator current peak during the first moment of dip as shown in Figure 12.

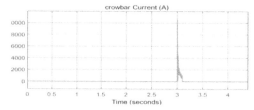

Figure 9.   Crowbar protection and voltage dip analysis.

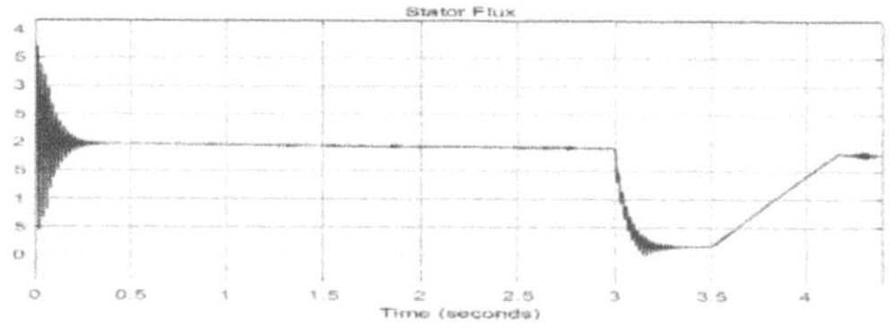

Figure 10.   Stator flux when crowbar activated.

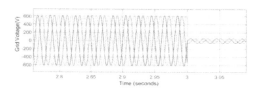

Figure 11.   Grid voltage with symmetrical voltage dip.

Figure 12.   Stator current with symmetrical voltage dip.

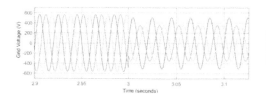

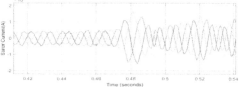

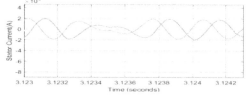

Figure 13.   Grid voltage with asymmetrical voltage DIP.

Figure 14.   Stator current with asymmetrical voltage DIP.

5.2 *Asymmetrical voltage sag*

During a perturbation of three times the grid frequency, these currents are induced by the emf induced by the negative flux. At t = 3 seconds, the stator current is asymmetrical due to the asymmetry of the voltage imbalance, as shown in Figures 13 and 14.

As a result, the negative sequence of the stator flux is not zero when the voltage drops suddenly. Powers and torque oscillate at twice the grid frequency due to the negative fluxes caused by the unbalance shown in Figure 10.

6 CONCLUSION

An analysis of the behavior of a grid-connected Double Fed Induction Generator under unbalanced network conditions is presented in this paper. The voltage dip occurs at t = 3 seconds in symmetrical, as can be seen in Figure 9. As soon as the dip occurs, the grid voltage is only 10% of rated voltage. Additionally, the machine is connected with an additional resistance path and the crowbar is activated at t = 3 sec. As shown in Figure 11, asymmetrical voltage unbalance occurs at the moment of 3 seconds. The stator current is affected by the unbalance due to the asymmetrical voltage. Simulation results of a 2MW system are presented in order to demonstrate how DFIG responds to grid voltage disturbances, which shows that high rotor currents are the main issue during unbalanced grid voltage. Crowbars can effectively reduce fault current by reducing the current flowing through the fault.

REFERENCES

Abad, G., Lopez, J., Rodriguez, M., Marroyo, L., & Iwanski, G. (2011). *Doubly Fed Induction Machine: Modeling and Control for Wind Energy Generation.* John Wiley & Sons.

Liserre, M., Cardenas, R., Molinas, M., & Rodriguez, J. (2011). Overview of multi-MW wind turbines and wind parks. *IEEE Transactions on Industrial Electronics, 58*(4), 1081–1095.

Mohseni, M., Islam, S. M., & Masoum, M. A. (2010). Impacts of symmetrical and asymmetrical voltage sags on DFIG-based wind turbines considering phase-angle jump, voltage recovery, and sag parameters. *IEEE Transactions on Power Electronics, 26*(5), 1587–1598.

Nasim, F., Khatoon, S., & Shahid, M. (2023, February). Field control grid connected dfig turbine System. In 2023 *International Conference on Power, Instrumentation, Energy and Control (PIECON)* (pp. 1–5). IEEE.

Nasim, F., Khatoon, S., Ibraheem, I., Shahid, M., & Ahmer, M. F. (2022, December). Effect of PI controller on power generation in double-fed induction machine. In *2022 4th International Conference on Advances in Computing, Communication Control and Networking (ICAC3N)* (pp. 2418–2424). IEEE.

Vidal, J., Abad, G., Arza, J., & Aurtenechea, S. (2013). Single-phase DC crowbar topologies for low voltage ride through fulfillment of high-power doubly fed induction generator-based wind turbines. *IEEE Transactions on Energy Conversion, 28*(3), 768–781.

Y. Jun, L. Hui, L. Yong, and C. Zhe, "An improved control strategy of limiting the DC-link voltage fluctuation for a doubly fed induction wind generator," *IEEE Trans. Power Electron.*, vol. 23, no. 3, pp. 1205–1213, May 2008.

Advances in AI for Biomedical Instrumentation, Electronics and Computing – Sachan et al. (eds)
© 2024 The Author(s), ISBN 978-1-032-64298-7

A simplified low-cost portable ventilator design

Mohd Shadaab & Shahida Khatoon
Department of Electrical Engineering Jamia Millia Islamia, New Delhi, India

Mohammad Shahid*
Department of Electrical Engineering, Galgotia's College of Engineering and Technology

ABSTRACT: Ventilators are one of the most crucial devices in modern ICU as they are lifesaving equipment's. In this paper, some insights are offered into the origin of mechanical ventilators. A straight forward, easy to use, and easily constructible design for an inexpensive portable ventilator is proposed in this paper. The proposed design of the ventilator is expected to perform more efficiently than currently available ventilators for artificial respiration at a very low cost. In situations like the covid-19 pandemic, acute respiratory failure, pneumonia, asthma, portable ventilator is helpful.

1   INTRODUCTION

According to covid 19 casualty data, the corona virus causes severe damage to human lungs that may cause sudden death due to loss of oxygen intake (Villaverde *et al.* 2014). In several cases, the patient's respiratory system collapses, making breathing difficult. That crucial situation necessitates for the artificial breathing system which is known as the ventilator. The use of mechanical or artificial ventilators can save patients with acute respiratory failure. Modern intensive care units (ICUs) are reliant on this equipment, and its increasing use signaled their development. Following the publication of a ground-breaking article by ARDSNet investigators in the New England Journal of Medicine about the significance of a lung-protective ventilation strategy, it has been noted for more than a decade that there has been a dramatic increase in mechanical ventilation (Roy *et al.*). Galen in the second century (A.D.) worked to understand the anatomy of the disease (Sternbach *et al.* 2001). According to him, the respiratory system, which controls the circulatory system in the human body, is same in both animals and humans.

There was no notable advancement in science or the field of mechanical ventilators throughout the Dark Ages. But Andreas Vesalius revolutionized everything in the mid-16th century. His numerous observations contradicted Galen's notion of the human respiratory system. His work served as the inspiration for modern positive pressure ventilators (Villaverde and Banga 2014). Negative pressure ventilators were first developed in the late 19th century and first used in the 1950s. The technique replaces or increases the effort being done by the respiratory muscles by applying negative pressure or less pressure than air pressure all around the patient's body (Pandey *et al.* 2021).

Artificial ventilator history was changed with polio cure. Many international polio experts attended a meeting on the disease that took place in Copenhagen in the summer of 1951. It was noted that every day more than 50 people were admitted to the infectious disease hospital due to this illness. Majority of them had bulbar paralysis and respiratory muscle disorders. Anesthesiologist Bjorn Ibsen made the discovery that respiratory failure, not renal failure, was to be to blame. This discovery caused the motility rate to overnight drop from 80% to 40%

---

*Corresponding Author: eems.j87@gmail.com

       DOI: 10.1201/9781032644752-83

[Lassen *et al.* 1953]. Positive pressure ventilators have been utilized in ICUs since the 1950s. After February 2020, more ventilators were required in India to treat COVID-19 patients. India has over 19 lakh hospitals, 95,000 ICU beds, and only 48,000 ventilators, despite having a population of about 135.26 billion people. The data indicates that India lacks ventilators severely (Pandey *et al.* 2021). There are many complications with existing ventilators like easy to hyperventilate patients and limited ability to gauge tidal volumes, common problem of poor seal in one-handed CE grip, gastric swelling, aspiration, exhaled secretions and moisture, risk of barotrauma, claustrophobia, incorrect assembly etc. [7]. The average cost of ventilators in India is more than 1lakh which is very expensive. This is one of the mains reasons for not having sufficient quantity of ventilators in every hospital [Pandey *et al.* 2021].

2 WORK METHODOLOY

The necessary motion for the motor arm will be provided by an Arduino Uno-based clockwise motor drive, and shaft arms is used to press the ambu bag in a controlled way. This will also maintain air flow with a controlled volume and pressure rate for the body (Amato *et al.* 2013). Using the DS18B20, temperature monitoring with Arduino can measure body temperature. Heart beat and pulse oximetry can be measured with the help of MAX30100 sensor connected with the LCD that is attached to the sensor. The readings will be displayed in BPMs (Beats Per Minute) (Smetanin *et al.* 2009). A Wi-Fi sensor called the ESP8266 is available with the ESP-01 module, which enables display controllers to establish straightforward TCP/IP connections with Wi-Fi networks. All these electrical activity parameters can be displayed on our smart phones and laptops (MacLaren *et al.* 2020).

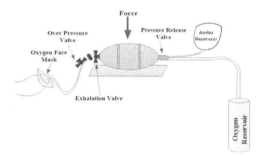

Figure 1. Ventilator working procedure.

3 COST FACTOR

The cost of ventilator in India is approximately 1.5 lakh per ventilator whereas this prototype ventilator may reduce cost by 80-90%. This clearly defines the huge difference in cost of mechanical ventilators (Pandey *et al.* 2021). This decreased cost acts as progressive step in medical field. It may spare numerous lives. It is effectively reasonable to those patients who are not able to oversee customary ventilator which is very costly.

4 DESIGN METHODOLOGY

The ventilator developed is automatically controlled. A conventional AMBU bag is used in this device, and it is mechanically compressed by an electromechanical actuator. Ventilators can be operated using different mode such as pressure controlling mode of operation, volume controlling mode of operation. The equipment developed, which was less expensive

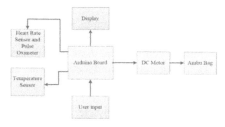

Figure 2.   Block diagram of proposed ventilator.

than the hospital ventilators already in use, is now successfully and effectively offered with volume control. The block diagram of the proposed ventilator is shown in Figure 2.

4.1   Components

This prototype consists of the following components, as shown in Table 1 below.

Table 1.   Prototype components.

Name of Component	Description	Diagram
Bag valve mask (bvm) or ambu bag	The Ambu bag (self-inflating bags) is most important part in portable. Ambu bag is primarily used to deliver the necessary amount of oxygen to patients' lungs who face trou- ble in breathing. The ambu bag is connected to gas inlets.	
Arduino Uno	A microcontroller board called Arduino UNO is based on the ATmega328P. It is inexpensive, adaptable and easily used programmable open-source micro controller board.	
Temperature Sensor	To detect the temperature of the pa- tient's body a temperature sensor is used. DS18 sensor is used as temperature sensor. Selection of this temperature sensor depends on temperature range, accuracy, stability.	
LCD	The value of parameter like beats per minute (BPM), body temperature etc. is displayed on LCD. It is used to show the patient information on display.	
DC Motor	A 12 V DC gear Motor is used in this proposed ventilator. Other specification of the motor used are: Rated Speed: 1000 rpm, Rated Torque: 0.3 kg-cm	
Wi-Fi module	This is an IoT-compatible ESP8266 Wi-Fi module with a TCP/IP stack and a microcontroller. This project uses it for wireless communication.	

(continued)

452

Table 1. Continued

Name of Component	Description	Diagram
Heart Rate Sensor and Pulse Oximeter	The MAX30100 is uses to monitor heart rate and detect the pulse oximetry.	

4.2 Prototype design

The prototype consists of three-dimensional wooden box in which various components are connected and placed. An Arduino uno micro controller is used and other component which include DS18B20 sensor, MAX30100 sensor and ESP8266 Wi-Fi module. Dc motor and LCD are also connected with Arduino uno board. The data from the sensor is displayed on LCD. The final stage includes holes to accommodate the oxygen filled area of the BVM and the valve neck.

The prototype design includes:

(1) Air Delivery System: Air is supplied in two routes. It draws air from a source of pressure while another squeeze breath pressing air. One end of the tube is connected to the patient mouth and another end is connected to the oxygen supply.
(2) Compression Mechanism: To press ambu bag continuously is big task and it is seems very difficult to operate manually. Additionally, operator who is not trained causes risk of damaging the lungs of patient. It may apply over pressure which may cause hyper ventilation. The required movement of pressing ambu bag is achieved using dc motor.

Figure 3. Model of prototype from developmental stage.

5 RESULTS AND DISCUSSION

In order to supply the sufficient amount of oxygen into the patient's lungs, ventilator is required. Using an engineering concept, this project aims to develop a compact and portable low-cost device. Besides providing artificial breathing for patients who have difficulty breathing, the design also monitors other parameters such as heartbeat and temperature. This research aim is to improve the current medical infrastructure by developing a compact light weighted and easily available ventilator at affordable price. In order to meet the needs of patients, ventilators must be designed with consideration for gas and electrical supplies, as well as the kind of ventilatory support they need [Fludger *et al.* 2008]. Adults typically breathe 12 bpm (breathe per minute) and normally inhale 0.5 litres of air per inhalation [Al-Mutairi *et al.* 2020].

By pressing BVM or Ambu bag this low-cost portable ventilator is provide the required oxygen. the necessary motion is achieved by using motor drive it does not require human to operate. Continuously working ability for several days this device is capable of deliver 12 respiratory rate (RR/min) and this can handle 500-600 ml volume of air which is necessary amount. This research help in developing inexpensive, light weight, portable ventilator which is provide the solution for lack of ventilator. It is affordable, easy to use and can be easily transported.

6   CONCLUSION

This paper introduced the design of an inexpensive and portable ventilator which is less costly as compared to existing ventilator in hospital. It is more effective, dependable and affordable and can help in fighting situation like we faced in COVID 19 pandemic, asthma, pneumonia etc. This proposed ventilator used various types of sensors. It has the ability to work for several days. This device is capable of delivering 12 respiratory rate (RR/min) and can handle 500-600 ml volume of air. It reduces the human help required for artificial breathing. This newly design ventilator helps in emergency situations to provide required breathable air to patient and reduce the risk of loss of life.

REFERENCES

Al-Mutairi, A. W., & Al-Aubidy, K. M. (2020). Design and construction of a low cost portable cardio-pulmonary resuscitation and ventilation device. In *2020 17th International Multi-Conference on Systems, Signals & Devices (SSD)*, IEEE, pp. 390–397. doi: 10.1109/SSD49366.2020.9364088.

Amato, J. J., Marini, M.B.P. (2013). Pressure-controlled and inverse-ratio ventilation. *In Principles and Practice of Mechanical Ventilation*, 3rd ed. New York, NY, USA: McGraw-Hill.

Brower, G. L., Roy, A. W., Brower, G., Matthay, M. A., Morris, A., Schoenfeld, D., ... & Thompson, B. T. (2000). Ventilation with lower tidal volumes as compared with traditional tidal volumes for acute lung injury and the acute respiratory distress syndrome. *N. Engl. J. Med.*, 342(18), 1301–1308. doi: 10.1056/NEJM200005043421801.

Fludger, S., & Klein, A. (2008). Portable ventilators. *Contin. Educ. Anaesth. Crit. Care Pain*, 8(6), 199–203. doi: 10.1093/bjaceaccp/mkn039.

Lassen, H. C. A. (1953). A Preliminary report on the 1952 epidemic of poliomyelitis in copenhagen with special reference to the treatment of acute respiratory insufficiency. *Lancet*, 261(6749), 37–41. doi: 10.1016/S0140-6736(53)92530-6.

MacLaren, G., Fisher, D., & Brodie, D. (2020). Preparing for the most critically Ill patients with COVID-19. *JAMA*, 323(13), 1245. doi: 10.1001/jama.2020.2342.

"MIT emergency ventilator (e-vent) project." https://emergency-vent.mit.edu/.

Pandey, A., Juhi, A., Pratap, A., Pratap Singh, A., Pal, A., & Shahid, M. (2021). An introduction to low-cost portable ventilator design. In *2021 International Conference on Advance Computing and Innovative Technologies in Engineering (ICACITE)*, IEEE, pp. 707–710. doi: 10.1109/ICACITE51222.2021.9404649.

Smetanin, P., *et al.* (2009). Potential intensive care unit ventilator demand/capacity Mismatch due to novel swine-origin H1N1 in Canada. *Can. J. Infect. Dis. Med. Microbiol.*, 20(4), e115–e123. doi: 10.1155/2009/808209.

Sternbach, G. L., Varon, J., Fromm, R. E., Sicuro, M., & Baskett, P. J. (2001). Galen and the origins of artificial ventilation, the arteries and the pulse. *Resuscitation*, 49(2), 119–122. doi: 10.1016/S0300-9572(01)00344-6.

Villaverde, A. F., & Banga, J. R. (2014). Reverse engineering and identification in systems biology: strategies, perspectives and challenges. *J. R. Soc. Interface*, 11(91), 20130505. doi: 10.1098/rsif.2013.0505.

Advances in AI for Biomedical Instrumentation, Electronics and Computing – Sachan et al. (eds)
© 2024 The Author(s), ISBN 978-1-032-64298-7

Generating maximum power in photovoltaic systems using HHO-based embedded controllers

T.P. Sujithkumar

Research Scholar, Department of Electrical and Electronics Engineering, Vels Institute of Science, Technology & Advanced Studies (VISTAS), Chennai, Tamil Nadu, India

Shanmugasundaram

Head of the Department of Electrical and Electronics Engineering, Vels Institute of Science, Technology & Advanced Studies, (VISTAS), Chennai, Tamil Nadu, India

V. Rajendran

Professor, Department of Electronics and Communication Engineering, Vels Institute of Science, Technology & Advanced Studies, (VISTAS), Chennai, Tamil Nadu, India

Debarchita Mishra

Assistant Professor, Department of Electrical and Electronics Engineering, Vels Institute of Science, Technology & Advanced Studies, (VISTAS), Chennai, Tamil Nadu, India

ABSTRACT: A photovoltaic system's current-voltage prediction is linear and dependent on external factors like solar emission and related heat. Using the MPPT (Maximum Energy Point Tracking) technique, the solar array's output of energy is obtained. Here, a Harris Hawks Optimizations (HHO) algorithm for extracting significant energy from embedded photovoltaic controller has been proposed. The proposed built-in controller uses the HHO algorithm to choose the boost converter's ideal switching pulse property. The proposed method, circuits are tested on a reinforced PV converter by the embedded controller. The proposed technique focuses on obtaining utmost possible power from a PV array based on solar radiation and cell temperature in diversified environmental conditions. The proposed system is developed with Matlab/Simulink and the converter control signals have been added to analyze the results.

1 INTRODUCTION

In addition to the limitations of conventional power sources, there is an imperative need for electricity, which highlights the motivation for producing green power in the modern world. The demand for power has increased significantly. As a result, several Renewable Energy Source (RESs) including hydropower, solar and wind power are utilized [Ali *et al.* 2018; Loukriz *et al* 2016]. Due to its adaptability in application, low-priced operating costs, and steadily declining costs, solar energy has been considered as a viable alternative to several forms of conventional energy sources [Nabipour *et al.* 2017]. The main drawbacks of photovoltaic modules include their poor efficiency and the fact that their current and voltage parameters are affected by a variety of external conditions, including temperature and radiation [Yilmaz *et al.* 2018].

The Maximum Power Point Tracking (MPPT) technique is an effective control architecture which aims to achieve and maintain maximum efficiency and power [Bayod-Rújula *et al.* 2014]. The conventional techniques comprises of Radial Basis Function network (RBFN) [Mohanty *et al.* 2013], the Sliding Mode Control (SMC) approach [Al-Dhaifallah *et al.* 2018; Ghassami *et al.* 2013], and the artificial neural network (ANN) method. The PV converter typically implements the MPPT controlling the associated DC-DC converter [Franco *et al.* 2017]. To generate maximum power, PV systems along with multiple stepped inverter topologies are employed in [Saravanan *et al.* 2016]. The inverter is controlled by a variety of PWM (pulse width modulation) techniques, such as B the whale optimization algorithm.

2 PROPOSED METHODOLOGY

An integrated Smart Grid system with an improved HHO-based controller has been discussed in this section. A photovoltaic board, a DC-DC converter, a VSI connected to the SG, and an integrated HHO controller make up the suggested order, which is shown in Figure 1. The MATLAB / Simulink platform has been used to build the production PV approach, while the XSG platform will focus of the HHO Manager's efforts. To determine the current-voltage-power (I-V-P) characteristics of the solar network, the parameters of the system are checked firstly. The power produced by the PV generator is measured, validating the SG requirements. An integrated HHO-based controller boosts the growth converter's power according to the supply of power needed by the steam generator.

2.1 *Problem statement*

The effectiveness of photovoltaic connections for improving the system has been tested by optimizing digital platforms and converters.

2.2 *Description of suggested system*

A PV board with integrated control based on HHO algorithm and DC/DC converters, along with an integrated smart grid, makes up the suggested solution. The proposed controller is an integrated system comprising of a PID controller with the HHO algorithm and is utilized to extract optimal energy from photovoltaic cells. It has been noticed that by employing the HHO method, the gain limit has improved in the integrated controller.

2.2.1 *Modeling of a PV adapter*
Figure 2 illustrates the structure and characteristics of the photocell model, whereas equation (1) indicates the photocell's output mode.

$$C^{pv} = C^{ph} - C^x * \left[\exp\{q(V^{pv} + R^x C^{pv}/AKT)\} - 1\right] - V^{pv} + R^x C^{pv}/R^{xh} \qquad (1)$$

Here C^p is presented as a photocurrent, the total proliferation of PN junction is noted as C^s, q it is explained as an electronic charge $(1.6e^{-19} C)$, the ideality coefficient of the diode is explained as A, the Constant Boltzmann $(1.38e^{-23} J/k)$ is explained as k, the state (K) is noted as D, R^s, is the series resistance and R^s is parallel resistance of the photocell.

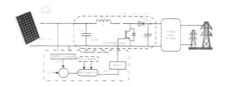

Figure 1.   Integrated PV system with embedded controller.

Figure 2.   The flow diagram of proposed modified HHO algorithm.

2.2.2 DC-DC boost converter

This step-up converter is frequently used in off-grid photovoltaic systems due to its simplicity of use, accuracy in performance, and economy. The fundamental principles that control the system process are used to calculate the state space using the dynamic model of the impulse transducer.

$$tC^{pv}/td = V^{pv}/H - ((1-T)/H)V^O \tag{2}$$

$$tV^O/td = (1-T)/CI^{pv} - RC \tag{3}$$

Here the input of the pulse converter is C^{pv} and V^{pv}, the power output is the function of the circuit of V^O and. It is assumed that the pulse converter works in the current evolutionary state (CCM). The introduced connection controller controls the switching signal of the DC / DC converter by optimizing the performance of the PWM generator. Therefore, the controller can be optimized by changing the temperature and the illumination mode of the PV model. Many HHO monitoring algorithms have been tested and used to improve PID access features for DC / DC switching signal exchange. This function uses a PID rectifier. Search engine optimization (15)

$$U(d) = k^P e(d) + k_1 \int_o^d e(d)td + K^T \frac{te(d)}{td} + P_I(0) \tag{4}$$

If K^P is a in equal measure, and the resulting profit described as K^T, the importance of profit is reported as K_I, and $e(d)$ is absolute scale uncertainty.

2.3 The gain optimization using HHO algorithm

It is easy in design, easy implementation and impact terms of memory speed. By default, the MATLAB toolbox has defined values for all parameters such as swarm size, coefficient of inertia and pull factors. The HHO settings will be updated after the optimal solution is found.

2.3.1 Process of HHO optimization

In this algorithm, Harris's falcon is subjected to a hard or soft siege to chase prey in various easy or difficult directions depending on the remaining energy of the prey. The flight of prey in this situation depends on the probability of prey. HHO uses soft siege. Otherwise, a strict siege is applied. Based on the prey avoidance phenomenon according to the Harris Hawks strategy, the HHO the algorithm uses four techniques to adjust the attack phase. Soft besiege and Soft besiege, Hard besiege, Progressive Fast Dive.

2.3.2 Exploration phase

The HHO method originated from the hunting of rabbit hawks. When rabbits are very energetic, Harris explores the hawk field [LB, UB] using the following calculation:

$$x_i(d+1) = \begin{cases} x_r - r_1|x_r - 2r_2 x_i(d)| & q \geq 0.5 \\ (x_b - X_n) - r_3(LB + r_4(UB - LB)) & q < 0.5 \end{cases} \tag{5}$$

Where, $x_i(d+1)$ is the condition of the i-th personal in the following repeat t, x_r is the-condition of an arbitrarily candidate selected to repeat now, x_a and x_n are the best position means in out of herds. r_1, r_2 and r_3 are three random numbers in the Gauss distribution. q Stands for an opportunity which of the two paths a person follows, it also means random number.

2.3.3 *Transition from exploration to exploitation*

The energy of the cutting ring is indicated by a symbol H, will deviate from the maximum linear value at zero

$$H = 2H_O(1 - (d/\text{max}Iter)) \tag{6}$$

Where H_0 is the initial energy level, which also fluctuates between 0 and 1. max*Iter* is the maximal amount of iteration allowed that will be configured at the start. When $|H| < 1$, Harris' hawk approached the rabbit with a strategy that would be detailed as follows.

2.3.4 *Exploitation phase*

Based on the prey avoidance phenomenon according to the Harris Hawks strategy, the algorithm uses four techniques to adjust the attack phase and they are Soft besiege, Hard besiege, Progressive Fast Dive, Heavy besiege Rabbits have enough energy to escape, but the success of their prey depends on its value of gentle besiege. IF$|H| \geq 0.5$ and $r \geq 0.5$ Harris hawk slowly surrounds the frightened rabbits running here and there andtries to hunt the rabbits. In this strategy, the Harris hawk will update its position using the following calculations:

$$x_i(d + 1) = x_a - x_i(d) - H|J \cdot x_a - x_j(d)| \tag{7}$$

Where $j = 2(1 - r_s)$ represents random bunny jumping which is Difficult besiege. IF$|H| < 0.5$ and $r \geq 0.5$ rabbits already exist depleted at low energy, then the hawk Harris will siege it.

3   RESULTS AND DISCUSSIONS

The simulation shows the grid function associated with the PV system results in strengthening the PV network. The proposed solar system is based on WOA, which is strong in the solar system. Figure 3 reveal the view of the Simulink drawn with the Zilinks model of a complex solar system. The presented system imitates and measures its parameters to show the performance of the designed system and compares it with traditional methods such as P and INC. The Performance analysis is conducted under two conditions:

Condition 1: Fixed lighting and temperature changes
Condition 2: Discussion of temperature and variable speed. According to the intensity and temperature of sunlight, it is needed to adjust the output power accordingly, and control the electric current and current to get maximum power.

Figure 3.   Matlab / Simulink presented system model.

This model demonstrates the performance of the introduced system by comparing the operating parameters and the available consequences. In complex PV system usesintegrated control. Performance parameters are measured according to relief, temperature and associated performance parameters, and the electrical network associated with intermediate circuit voltageconditions in terms of coverage and temperature. Condition 1: Light always changes the temperature and Condition 2: Discussion of temperature and variable speed.

(a)   (b)

Figure 4.   (a) PV Voltage output and current (b) Power supply during application.

The central power system improves the concert of the designed system. The average voltage is 500 V, which is the ideal output voltage. In this network, the power supply is measured at 25 kV, the thermal is 4.5 A, and the connection voltage is 112.5 kW. The electric voltage of the photovoltaic system is 405 V, the current is 205 A, and the connection voltage is 83 kW.

(i)   (ii)

Figure 5.   PV (i) Air conditioning and temperature (ii) Output power.

Figure 5 reveals the power supply, current, power supply and medium power supply. In this case, the mains power is 225 kV of power supply, the current of 4.2A and 92 kW. The connection is constant. The best match between the results is obtained using two factors.

(i)   (ii)

Figure 6.   Searching for comparisons (i) condition 1 and (ii) condition.

4   CONCLUSION

In this article, the process of producing the maximum energy of a solar panel based on HHO is described briefly. The controller used here depends on the PV generator system provided by the Matlab/Simulink platform. Here the photovoltaic array is attached to a switch and to the VSI network. The regulator in this article uses a PID regulator managed by HHO to generate a switching power signal for a pulsed DC-DC variable. HHO is used to select the optimal PID gain setting. The proposed controller was then simulated to measure various operating levels such as voltage, current and power. It was compared with various models to prove the efficiency of the proposed integrated controller. The results indicate that the proposed method is better than other controllers.

REFERENCES

Al-Dhaifallah, A., *et al.* (2018). Optimal parameter design of fractional order control based INC-MPPT for PV system. *Solar Energy* 159: 650–664.

Ali, A., *et al.* (2018). Modified efficient perturb and observe maximum power point tracking technique for grid-tied PV system. *International Journal of Electrical Power & Energy Systems* 99: 192–202.

Bayod-Rújula, Á. A., *et al.* (2014). A novel MPPT method for PV systems with irradiance measurement. *Solar Energy* 109: 95–104.

Franco, F. D., *et al.* (2017). Enhanced performance of PV power control using model predictive control. *Solar Energy* 158: 679–686.

Ghassami, A. A., *et al.* (2013). A high-performance maximum power point tracker for PV systems. *International Journal of Electrical Power & Energy Systems* 53: 237–243.

Li, S., *et al.* (2017). A MPPT strategy with variable weather parameters through analyzing the effect of the DC/DC converter to the MPP of PV system. *Solar Energy* 144: 175–184.

Loukriz, A., *et al.* (2016). Simulation and experimental design of a new advanced variable step size Incremental Conductance MPPT algorithm for PV systems. *ISA transactions* 62: 30–38.

Mohanty, P., *et al.* (2014). MATLAB based modeling to study the performance of different MPPT techniques used for solar PV system under various operating conditions. *Renewable and Sustainable Energy Reviews* 38: 581–593.

Nabipour, M., *et al.* (2017). A new MPPT scheme based on a novel fuzzy approach. *Renewable and Sustainable Energy Reviews* 74: 1147–1169.

Rajkumar, M. V., *et al.* (2013). FPGA based multilevel cascaded inverters with SVPWM algorithm for photovoltaic system. *Solar Energy* 87: 229–245.

Saravanan, S., *et al.* (2016). RBFN based MPPT algorithm for PV system with high step-up converter. *Energy conversion and Management* 122: 239–251.

Yilmaz, U., *et al.* (2018). PV system fuzzy logic MPPT method and PI control as a charge controller. *Renewable and Sustainable Energy Reviews* 81: 994–1001.

Advances in AI for Biomedical Instrumentation, Electronics and Computing – Sachan et al. (eds)
© 2024 The Author(s), ISBN 978-1-032-64298-7

5G Network – Deployment, status and roadmap in Indian telecom ecosystem

Manu Srivastava
Samsung Research Institute – Noida

ABSTRACT: This survey paper talks about the 5G deployments in Indian context, its current status and roadmap for further expansion of 5G services. Focus area of this survey paper is the program and the telecom operator activities as reported in recent media and literatures. The aim is to shed light on the evolution of India telecom sector, key technological difference between major operators, the 5G auction and nr band acquired, deployment phases and emerging 5G deployment in India, along with the 5G call & various data features which would be a value addition for India 5G ecosystem. The most demanding part of 5G would be pan India deployment in shortest time with two segregate 5G technology i.e. NSA and SA to go hand in hand, which has its own inherent advantages and limitations. Also the future of 5G in India is briefly discussed, along with India home-grown technology known as 5Gi.

1 TELECOM BACKGROUND IN INDIA

Telecom sector was always considered as the powerful tool to enable the masses to connect with the technology upfront. India has progressed very constructively from the PCO booth era to UPI era in the span of almost 25 years. Over the years, telecom sector has shown strong consumer demand. Tele density in India can be evident from the fact that it has increased form 75% in 2014 to 84% by Dec 2022, which is important indicator for telecom penetration [Depart of Telecom, Govt of India]. As evident in the data, the average data consumption per subscriber increased from to 61MB (March 2014) to 16GB approx. (June 2022). This clearly displays a trend that services are now reaching to lower strata of the society which was not feasible earlier and thus this has created the opportunities.

2 EMERGENCE OF 5G

2.1 *Global scenario*

The consumer demand of increased data & throughput in reduced prices are increasing daily and 5G is posed to deal with this contradictory demands. 5G is also labeled as 'ultra-fast, ultra-reliable, ultra-high capacity transmitting at super low latency' by the National Infrastructure Commission in the report SG Infrastructure Requirements in the UK" (2016), the term World Wide Wireless web or WwWw is being coined for this [TRAI 2018]. There is some design philosophy behind the architecture like Flexibility, Scalability, Virtualization, Automation, Security and Cloud native.

By end of 2023, apart from India, 5G is already commercialized in all major economies like North America, Major European countries, Middle East, China, Australia etc. In today's world, the need for mobile data is not just required for streaming, video calling or surfing. It is much more than that in terms of mission critical communications, cloud networking, man machine interfacing, industrial automations, logistical applications, massive IoT, VR, AR applications, etc. 5G can full fill those demands as it is a unified, more capable air interface.

2.2 India 5G journey

The 5G is divided into 2 parts in terms of deployment technology. NSA and SA. NSA (Non Stand Alone) used 4G core network whereas SA (Stand Alone) deploys its own core network, which is segregated from 4G core. Table 1 shows the major difference in NSA and SA.

Table 1. NSA vs. SA.

Basis of difference	NSA	SA
Core network	4G EPC (Evolved Packet Core).	5G Core.
Dependency on 4G LTE	Highly dependent	No dependency
Primary base station	4G base station	5G NR base station
Power Consumption	High	Comparatively less
Latency	Comparatively high	Can deliver low latency
Bandwidth	Can't connect to 1 million devices	Can connect to 1 million devices
Interoperability	5G and 4G networks are interoperable	5G networks are independent of 4G
Ease of deployment	Very high	Difficult
Time Consumption	Low	High

The commercialization of 5G by operators involved a step by step rigorous process which involves various stages starting from technical trial band license, trial and lab readiness, small test cluster readiness, participation in the auctions, deployments of the auctioned bands, liaising with the mobile handset manufactures to roll out 5G software in the devices, test with various OEMs and NW vendors and finally roll out of commercial services. Each of the step is discussed below.

2.2.1 Various 5G bands

The various bands acquired today by telecom operators in India till now is as shown in Table 2 [DoT, Govt of India]. The benefit of each band is unique and its upon the operator how to maximize the use case for particular band. As per few reports, n78 band interfere with altimeters of Aircraft, hence that is not allowed to be deployed at Airport area and hence operators use other nr (new radio) bands in such areas. In 5G, the bands number are prefixed with "n" and 5G bands are called "nr" bands which means new radio. If NR band overlap with 4G, they have same band number. Lower frequency bands (FR1) provides better coverage and penetrations but comparatively less bandwidth. FR2 provides less coverage but better throughput but more susceptible to attenuation.

2.2.2 Technical trial

Back in 2021, Government of India permitted operators to start internal trials with OEMs for the 5G technology with trial bands. Trial bands were mainly n78 NR band of 5G. Operators executed various 5G related test cases basis their criteria in collaboration with handset companies (OEM) in the trial. This trial underwent almost for a year as per media reports [Business Today website]. This trial also gave the opportunity to the handset manufactures to understand the operator ecosystem, to roll-out 5G-enabled software for their capable devices.

2.2.3 Commercial launch at IMC

During an event on 1st Oct 2022, at India Mobile Congress, 5G services were officially announced in India. Airtel became the first telecom operator to launch its 5G network on October 1 in eight cities. Jio 5G services also launched within couple of days. Vi has not launched commercial 5G so far in India.

Table 2. 5G Auction Results in India (Sept 2022).

Frequency	700	800	900	1800	2100	2500	3300	26GHz
nr band	**n28**	**n5**	**n8**	**n3**	**n1**	**n41**	**n78**	**n258**
Airtel			✓	✓	✓		✓	✓
Jio	✓	✓		✓			✓	✓
Vi				✓	✓	✓	✓	✓

2.2.4 *OEM readiness*

After 5G commercial lunches in Oct 2022 in India, Govt. asked all the mobile manufactures to quickly roll-out firmware to support 5G services in their handsets. 5G will is enabled via a software update in the mobile phones, which user has option to download [Business Today website].

2.3 *5G data services and availability*

The 5G service across globe and India differs in terms of deployments and also in terms of features offered. Rolling out 5G doesn't means that all its services are also rolled out. Below are the few features which are interesting to understand with respect to India telecom ecosystem.

2.3.1 *NSA vs. SA*

There are two deployment modes for 5G i.e. NSA and SA. It is noteworthy that Airtel deployed NSA whereas SA is deployed by Jio. The SA and NSA networks can be said as two sides of the same Coin that is 5G Technology. The NSA networks have advantages in the form of speedier implementation as it uses existing 4G technologies; SA network on the other hand offers a true 5G experience as it makes use of entire 5G technology. In simple terms, 5G Radio + 4G Core = NSA network. SA networks completely make use of dedicated 5G infrastructure.

2.3.2 *ENDC*

5G NSA uses ENDC (Eutra/NR Dual Carrier) where LTE is main cell and NR just work as a secondary cell. Core Network is based on LTE Core, NR just provide additional RAN pipe. Dual connectivity in 5G NSA establishes two connections, one to the LTE network and another to the 5G NR network.

2.3.3 *BWP*

Scanning the entire NR carrier bandwidth (e.g., 400 MHz) in NR UE would consume excessive power. Hence, Bandwidth Part (BWP) is designed, wherein a designated portion extracted from the overall carrier bandwidth. BWP is a mechanism to configure/divide a whole channel band into multiple segments and switch among the sub bands depending on situation.

2.3.4 *NRCA*

The basic concept of carrier aggregation and the process of establishing carrier aggregation in NR is almost same as LTE carrier aggregation. Carrier Aggregation plays pivotal role in increasing coverage and capacity both at time during 5G deployment.

2.3.5 *DSS*

Dynamic spectrum sharing (DSS) is useful in terms of deployment as it provides flexibility to use spectrum from low, mid and high frequency and can also dynamically switch between them as per network loading. But this feature has capacity throttles as more and more users move to 5G, they may face performance & experience issues, as more bandwidth would be required.

2.3.6 *mmWave*

Although mmWave was also auctioned in India which was very strange move from GoI as it is costly spectrum worldwide. It's very high frequency which has very less coverage, so multiple routers must be installed for larger coverage. Currently, it is heard that Jio AirFiber technology is using mmWave for radiation.

2.3.7 *NW slicing*

NW slicing is the buzz world in 5G SA deployment work wide due to its inherent advantage. It slices the network in virtual slices where in each slice serves particular use case for data specific need. E.g. one slice would be dedicated for the gamer and another slice would be use wherein almost zero latency in required i.e. medical robotics applications, to deliver meaningful guarantees to customers.

2.4 *5G voice services*

The main objective on 5G NW is higher data throughput, but the voice call is also possible over 5G NW in two unique ways. One is EPSFB (EPS Fall Back) and VoNR (Voice over New Radio). Its notable to mention that VoNR is SA only feature.

2.4.1 *EPSFB*

In EPSFB, when call is placed, the mechanism is to fall back at 4G NW i.e. Evolved Packet System. The call gets matured over 4G NW. RAN triggers moving the phone LTE connected to EPC during call establishment.

2.4.2 *VoNR*

VONR is voice service in NR and designed to replace traditional voice services which were used in 2G, 3G and 4G NW so far. Jio India has commercialized this services across multiple circles, as it is more efficient (no EPSFB) and reliable solution. Few mobile handsets also support VoNR in India.

3 OEM SPECIFIC OPTIMIZATIONS

Thermal Mitigations: 5G devices generate more heat as compared to 4G handsets. Every handset manufacturer is working on cooling solution for this problem. 3GPP Rel-16 has suggestions to mitigate such problem either by latching to 4G-radio from 5G radio, when devices reach specific temperature or something similar. Although, the OEM also have device specific optimization already done to address such problem.

3.1 *5G icon*

Few operators have requirement of specific 5G icon to distinguish from one another. T Mobile USA has "5G UC" icon, whereas Verizon has requirement of "5G UW" icon once bandwidth reaches certain threshold. It is quite evident that India market will also witness such icon level customization once 5G mart matures enough like western world.

3.2 *International roaming*

Currently 5G International Roaming is a complex topic because there are numerous ENDC combos to supports in NSA and not a standard band in SA as well to support roaming. From OEM perspective, supporting superset of all the band and combinations in single devices is complicated and costly matter from handset point of view. So each OEM has it own priorities to support 5g international roaming. At the same time, operators also do not have any such global work in this direction as they are currently in very initial stages of deployments.

3.3 *5Gi*

It is India designed 5G telecommunication networks, which uses Low Mobility Large Cell (LMLC) feature. It significantly enhances the signal transmission range (of a base station). The need for 5Gi arises from the fact that as frequency increases, coverage goes down. 5Gi has advantage of higher range at a lower frequency. Accepted by ITU and incorporated in the IMT2020 documents. Target Users from Indian context is low mobility users, set of villages and farms spread over large area. The main challenge for acceptance of this technology is by operators, who have already invested millions in 5G by the global standards. Also for smartphone companies, hardware cost increases in adaptation of 5Gi as it requires specialized receivers to work on the 5Gi technology.

4 CONCLUSION

In this paper discussed the India telecom journey from 2G time till 5G commercialization. As discussed, 5G would be building block for the future advancement and also a tool for masses to witness the advancement in the technology. Government of India has co-worked with all the OME and telecom operators in brainstorming for early adaptation of 5G among masses. The many benefits and various features of 5G will see massive emergence of new revenue models for operators, engagement for end users and monetization for more stakeholders. High data throughput, reliable transmission and better connectivity are the future of 5G with ever growing demand from the end user with huge customer base of India. Once India is totally covered by 5G, there is lot more that can be achieved. 6G is next emerging era of telecom, for which 5G would be a stepping stone.

REFERENCES

3gpp 38.211, 38.331, 38.133

Department of Telecommunications, Ministry of Communication, GoI, https://dot.gov.in/

Evolution of Carrier Aggregation (CA) for 5G, https://www.qorvo.com/

Evolution of Core Network (3G vs. 4G vs. 5G)

Introduction to 3G Mobile Communications, Second Edition, Juha Korhonen, ISBN 1-58053-507-0

J. G. Andrews, S. Buzzi, W. Choi, S. V. Hanly, A. Lozano, A. C. K. Soong, and J. C. Zhang, "What will 5G be?" *IEEE J. Sel. Areas Commun.*, vol. 32, pp. 1065–1082, Jun. 2014.

National Telecommunications Institute For Policy Research, Innovation And Training (NTIPRIT), https://www.ntiprit.gov.in/

Telecom Regulatory Authority of India, Bulletin of telecom technology, Issue April 2018

Telecom Regulatory Authority of India, Bulletin of telecom technology, Issue May 2018

https://www.businesstoday.in/industry/telecom/story/airtel-vi-jio-mtnl-given-permission-to-conduct-5g-trials-govt-321497-2022-02-04

https://www.businesstoday.in/industry/telecom/story/exclusive-govt-asks-handset-manufacturers-to-quickly-roll-out-updates-for-5g-services-349622-2022-10-12

https://www.ericsson.com/en/blog/2022/2/the-power-of-5g-carrier-aggregation

https://www.businesstoday.com/2020/07/india-mobile-phones-smartphone-market-25-years

https://www.businesstoday.in/industry/telecom/story/what-is-the-difference-between-5g-networking-sa-and-nsa-316097-2022-03-02

https://comset.com.au/what-is-lte-advanced/

https://dot.gov.in/sites/default/files/Telecom%20at%20a%20Glance%202023%20as%20on%2018-01-2023.pdf?download=1

https://www.airtel.in/blog/prepaid/what-is-the-difference-between-5g-networking-sa-and-nsa/

https://www.business-standard.com/article/technology/india-s-telecom-subscriber-base-dips-to-1-14-bn-in-august-trai-122090100265_1.html

Advances in AI for Biomedical Instrumentation, Electronics and Computing – Sachan et al. (eds)
© 2024 The Author(s), ISBN 978-1-032-64298-7

Iteration-based reduction in cell population for biomedical applications

Ashutosh Mishra & Piyush Kumar Tripathi
Amity University Uttar Pradesh (Lucknow Campus) India

ABSTRACT: For a self-mapping in Banach space, the weakly compactness of a closed convex subset is mentioned if and only if every continuous linear function attains a maximum on the given convex subset. M. Rotenberg proposed the invariant point theorem (or FPT) to formulate and concise the growth in the population of cells. So, by exploiting FPT, the solution of a nonlinear (cell population related) problem concerning the boundaries, is categorized by two major factors; one degree of maturity (referred to as *dom*) and other velocity of maturation (referred to as *vom*) can be attained. Henceforth, this may be the scenario of shortening the remarkable enhancement in the cell population of certain kinds of microbes for biomedical applications such as virus or bacteria-related drugs and vaccines. To further explore and improve the earlier approaches, by using some more contractions procuring FPTs and their suitable applications, is our key inquisitiveness.

1 INTRODUCTION

1.1 *Basics*

M. Rotenberg [1983], modeled the population growth of cells by a specific partial differential equation with cell density $\varphi(t, m, u)$. Every cell of density φ, given by eq. (1), is specified by two parameters; *dom* and *vom*, u.

$$\frac{\partial \varphi}{\partial t}(t, m, u) = -u \frac{\partial \varphi}{\partial t}(t, m, u) - \sigma(m, u)\varphi(t, m, u) + \int_0^\theta \rho(m, u, u')\varphi(t, m, u')du' \quad (1)$$

Where, $m \in [0, a], u, u' \in [0, \theta]$ and $a > 0, \theta > 0$. Here, the *dom* is described in such a way that its value at birth is given by $m = 0$, and at mitosis, its value is $m = \theta$. This means that the cell is getting birth at $m = 0$ and divides into two daughter cells at $m = \theta$. The factor $\rho(m, u, u')$ is known as the kernel which represents the transitional rate of the population. Hence, the biological boundaries of *dom*, $m = 0$ and *vom* period $u = \theta > 0$, are fixed and coupled.

The rate of transition, $\rho(m, u, u')$ stipulates cells' transition between maturation velocity u' to u and $\sigma(m, u)$ represents the cross-section of complete transition. Lebowitz et al. 1974; Van der Mee et al. 1987; Dehici *et al.* 2006 & Latrach 2006 approached showing the model in their ways by using the notion. This represents a linear transition rule between the two: the mother cell and daughter cells; at mitosis, which is a strong consent to the classical ideas.

The basis behind the consideration of the problem shown by (1) to be non-linear, as pointed out by Rotenberg-1983, can be understood in the following manner. The concerned cells are under a nutrient situation which may not be a part of analytical articulation. The transition rates, σ and ρ become the functions of the density of population φ due to variations in the concentration of nutrients and different density-related factors like contact inhibition of growth. This makes the problem non-linear. On the contrary, the boundaries $m = 0$ and $m = \theta$ are unaltered and firmly coupled during mitosis. This shows a nonlinear

DOI: 10.1201/9781032644752-86

reproduction relation between parent and daughter cells at mitosis. These results inspired Latrach et al. 1999 to consider another version of the nonl inear formulation of the model proposed by Rotenberg in which both the rate of transition $\rho(m, u, u')$ and total cross section $\sigma(m, u)$ has been considered depending on population density. The nonlinear relation between the daughter cells and mother cells at mitosis, describing the boundary situations, is given by

$$u \frac{\partial \varphi}{\partial m}(m, u) + \mu \varphi(m, u) + \sigma(m, u, \varphi(m, u)) = \int_0^\theta \rho(m, u, u') \varphi(m, u') du' \quad (2)$$

Where μ is a complex number and σ and ρ are non-linear functions of φ. To formulate the boundary conditions the following expression is used:

$$\varphi|_0 = T(\varphi|_1) \quad (3)$$

Where, $\Gamma_0 = \{0\} \times [0, \theta]$ and $\Gamma_0 = \{0\} \times [0, \theta]$, $\varphi|\Gamma_0$ (resp. $\varphi|\Gamma_1$) depicts the constraint of φ to Γ_0 (resp. $\varphi|\Gamma_1$). Here, T is an operator which is actually, nonlinear.

A variety of statements concerning differential equation (2) and boundary values (3) have been established in L_p spaces, where $p \in (0, \infty)$, by several other authors. Previously, K. Latrach [1996, 2001] has shown the existence result for stationary transportation of the kinetic theory (in the case of gas). Contrary to the biological models, the non-linearity of the boundary conditions has no physical sense so far as the dynamics of rarefied gas are concerned [Latrach 1996]. Hence, Latrach *et al.* [1999] have dealt with their analysis in L_p spaces for $p \in (1, \infty)$ assuming the linear boundary conditions. While obtaining the solution to the problem concerned, the extremities of the boundary accounted were assumed as a non-linear function of density population, φ. This can be transformed into a fixed-point situation with some suitable operators engaging compactness and contraction. Thus, the expected result would be just a sequel to the invariant point proposition of Krasnoselskii, as detailed on page 501; Zeidler [1993]. In recent times, Mishra *et al.* [2020] have exercised the applicability of probabilistic metric space instead of general metric space while dealing with cell population dynamics.

So, we investigate the existing consequence and further establish it by making use of certain fixed-point theorems accompanying the characteristics of weakly compact sets in probabilistic space, in place of regular space. In this way, obtaining the solution of the said model can deal with several microbial-growth-related parameters in biomedical advancements.

1.2 *Rotenberg submission*

Rotenberg claimed that the formulation representing the enhancement in cell population requires a non-linear weakly compact operator and to get the solution, the problem is transformed into a non-linear fixed-point situation. Thus, the bounded sets have been modified into weakly compact sets so that the Schauder theorem on fixed point can be exploited to obtain at least an invariant point, which would be the required solution of the paradigm relating the increase in cell population as represented by (1)[Schauder1930]. We can docilely draw the inference that the notion of general metric space is exploited in establishing the fixed-point result.

2 PRELIMINARIES

2.1 *Some basic tools*

Definition 1: [Moors *et al.* 2017 In Banach space, a closed convex subset C is weakly compact *iff* every continuous linear functional on the space possesses a maximum on C.

Definition 2: [Peeler L. 2011] In a distance space (M, d), a sequence $< p_n >$ is a Cauchy sequence *iff* for each $\varepsilon > 0$, there must be a positive integer $N(\varepsilon)$, such that;

$$m, n \geq N(\varepsilon) \Rightarrow d(a_m, a_n) < \varepsilon.$$

Definition 3: [Ashutosh Mishra *et al*. 2020] A space (M, d) is said to be a complete distance space, if each Cauchy sequence in the space converges to a particular point lying in it. On the other hand, it is called incomplete.

Definition 4: [Peeler L. 2011] A mapping, $f : M \rightarrow M$ is a contraction, in a distance metric (M, d), if it is specified for any real number $b \in [0, 1)$, such that: $d(f(x), f(y)) \leq bd(x, y)$, for all $x, y \in M$, where, b is a constant of contraction.

Example 1: Consider a mapping $f_x : (0, 1] \rightarrow (0, 1]$ and a metric space $((0, 1], d)$ such that, $f_x = \frac{x}{2}$. Then, for any constant $b \in (0, 1]$ it will be a contraction for, $d(f_y, f_x) \leq bd(y, x)$, $\forall x, y \in (0, 1]$.

Notion 1: [Robbin 2010] Every contraction mapping is uniformly continuous as it is a Lipschitz continuous function. By uniformly continuous mapping, we simply mean continuous mapping.

Example 2: [Robbin 2010] Consider a function f from M on a normed space. Then, f is Lipschitz continuous at $x \in M$ because, there must be a constant k, such that: $\|f(y) - f(x)\| \leq k \|y - x\|$, $\forall y \in M$ sufficiently near x.

Definition 5: [Peeler L. 2011] For any mapping, $f : M \rightarrow M$ on a set M (which is nonempty), a point α_0 is defined as an invariant (or fixed) point in M iff for every $\alpha_0 \in M$, we have $f(\alpha_0) = \alpha_0$.

Theorem 1: [Mishra *et al*. 2020] Every sequence $< \alpha_n >$ that converges on a given distance space (M, d) is called a Cauchy sequence.

Notion 2: [Banach 1922] Every contraction in a complete distance (or metric) space has a unique invariant point. This notion is designated as *the Banach Theorem*.

Theorem 2: [Mishra *et al*. 2020] Consider a non-empty and complete metric space (M, d) on which a contraction mapping, $f : M \rightarrow M$ is defined. Then, the mapping must have a unique invariant point in M. Further, for $\alpha \in X$, there exist, the iterations $\alpha, f(\alpha), f(f(\alpha))$, and so on converging to some particular fixed points of f in M.

Definition 6: [Tripathi *et al*. 2016] A left continuous and non-decreasing mapping $F : R \rightarrow R^+$ is denoted as a distribution function for $r \in R$, if, *inf* $F(r) = 0$ and *sup* $F(r) = 1$.

Now, consider a collection F^+ of all left continuous & non-decreasing distribution functions, $F : R \rightarrow [0, 1]$ along with *sup* $F(r) = 1$, such that, $F(0) = 0$.

Definition 7: [Tripathi *et al*. 2016] A probabilistic distance (or metric) space, coined as PM-space, is defined as an ordered set (M, F) for a nonempty set M and $F : M \times M \rightarrow F^+$, where $F(p, q)$ is represented by $F_{p,q}$ for each $(p, q) \in M \times M$, satisfying the following conditions:

(i) $F_{x,y}(r) = 1$, for all $r > 0$ if and only if $x = y$; $x, y \in M$ [Mishra, A. *et al*. 2020]
(ii) $F_{x,y}(0) = 0$, for every $x, y \in M$ [Mishra, A. *et al*. 2020]
(iii) $F_{x,y}(r) = F_{y,x}(r)$, for all $x, y \in M$ and $r \epsilon R$[Mishra, A. *et al*. 2020]
(iv) $F_{x,y}(r) = 1$ and $F_{y,z}(s) = 1$ then, $F_{x,z}(r + s) = 1$, for all $x, y, z \in M$ and $r, s \in R$. [Mishra, A. *et al*. 2020]

The ordered set (M, F) is designated as a semi-probabilistic space if only conditions (i) and (iii) hold good.

Definition 8: [Razani 2006] The triplet (M, F, t) with the probabilistic metric space as an ordered pair (M, F) and t as t-norm (triangular norm) is known as the Menger probabilistic metric space (MPM-space), if the following inequality satisfies.

$$F_{xy}(r + s) \geq t(F_{x,z}(r), F_{z,y}(s)), \text{ for every } x, y, z \in M \text{ and } r > 0, s > 0.$$

Definition 9: [Razani 2006] Let (M, F, t) be the Menger probabilistic metric space and $\sup t(l, l) = 1, 0 < l < 1$. Then,

(i) Any sequence $< p_n >$ in M is known as t-convergent at $p \in M$ if for $\varepsilon > 0$ and $\lambda > 0$, we obtain a positive number $N(\varepsilon, \lambda)$ such that, $F_{a_n, a}(\varepsilon) > 1 - \lambda$, whenever $n \geq N(\varepsilon, \lambda)$. And, we write $p_n \to p$ [Mishra, A. *et al.* 2020]

(ii) The sequence $< p_n >$ in M is known as *the t-Cauchy sequence*, if for given $\varepsilon > 0$ and $\lambda > 0$, we obtain a positive number $N(\varepsilon, \lambda)$, such that, $F_{a_n, a_m}(\varepsilon) > 1 - \lambda$, whenever $m, n \geq N(\varepsilon, \lambda)$. [Mishra *et al.* 2020]

(iii) The Menger PM-space (M, F, t) is defined as t-complete if every t-Cauchy sequence in M is t-convergent to some limiting value [Mishra, A. *et al.* 2020]

Definition 10: [Razani 2006] A mapping $f : M \to \mathrm{M}$ is called a contraction in MPM-space (M, F, t), if for every $x \neq y; x, y \in M$, we obtain, $F_{f(x), f(y)}(a) \geq F_{x,y}(a)$, for every $a > 0$, $\quad F_{f(x), f(y)} \neq F_{x,y}$

3 RESULT & DISCUSSIONS

3.1 *General approach*

When we reminisce about the notion of metric (or distance) space, we realize that the separation between any two points in the space is a non-negative real number. A Menger distance distribution function was introduced to extend the concept [Menger 1942]. Based on probabilistic metric, the classical Banach principle was shown by Sehgal and Bharucha-Reidin 1972 which led J. Jachymsk to conjecture the idea of Banach theorem for non-linear contraction on probabilistic space. Instead of using general space and transforming the bounded sets into weekly compact sets, as shown by Rotenberg, we are interested in employing the probabilistic space to set the fixed-point result using the *dom* and *vom* of the cells as two parameters in the growth of the cell population.

3.2 *Our submission*

Lemma: [Cho et al. 1997] Let $\{p_n\}$ be a sequence in X such that $\lim\limits_{n \to \infty} F_{p_n, p_{n+1}}(x) = 1$, for all $x > 0$. If the sequence $\{p_n\}$ is not a Cauchy sequence in X, then there exists $\varepsilon_0 > 0, t_0 > 0$, and two sequences $\{m_i\}$ and $\{n_i\}$ of positive integers as,

(i) $m_i > n_{i+1}$ and $n_i \to \infty$ as $i \to \infty$,

(ii) $g\left(F_{p_{m_i}, p_{n_i}}(t_0)\right) > g(1 - \varepsilon_0)$ and $g\left(F_{p_{m_i-1}, p_{n_i}}(t_0)\right) \leq g(1 - \varepsilon_0)$.

Theorem: Let (X, F, t) be a complete Menger-space, and $\{T_n\}, (n = 1, 2, ...)$ be a sequence of functions on X. Consider non–negative numbers $\alpha_i, (i = 1, 2, ...7)$ such that,

$$(1) \quad g\left(F_{T_i p, T_j q}(t)\right) \leq \alpha_1 \left[g\left(F_{p, T_i p}(t)\right) + g\left(F_{q, T_j q}(t)\right)\right] + \alpha_2 \left[g\left(F_{p, T_j q}(t)\right) + g\left(F_{q, T_i p}(t)\right)\right]$$

$$+ \alpha_3 g\left(F_{p,q}(t)\right) + \alpha_4 \frac{\left[1 + g\left(F_{p, T_i p}(t)\right)\right] g\left(F_{q, T_j q}(t)\right)}{1 + g\left(F_{p,q}(t)\right)}$$

$$+ \alpha_5 \frac{g\left(F_{p, T_j q}(t)\right) g\left(F_{q, T_i p}(t)\right)}{1 + g\left(F_{p,q}(t)\right)} + \alpha_6 \frac{g\left(F_{q, T_i p}(t)\right) g\left(F_{q, T_j q}(t)\right)}{1 + g\left(F_{p,q}(t)\right)}$$

$$+ \alpha_7 \frac{g\left(F_{p, T_i p}(t)\right) g\left(F_{q, T_i p}(t)\right)}{1 + g\left(F_{p,q}(t)\right)}$$

Where $2\alpha_1 + 2\alpha_2 + \alpha_3 + \alpha_4 < 1, 2\alpha_2 + \alpha_3 - 1 = 0$ & $1 - 2\alpha_2 - \alpha_3 - \alpha_5 \neq 0$. Then, the sequence T_n has a unique common invariant point.

Proof: Let $x_0 \in X$. Let us frame a sequence $\{x_n\}$ such that $x_n = T_n x_{n-1}, (n = 1, 2...)$.

Since, $g\left(F_{x_n,x_{n+1}}(t)\right) = g\left(F_{T_n x_{n-1}, T_{n+1} x_n}(t)\right)$, therefore, from (1),

$$g\left(F_{x_n,x_{n+1}}(t)\right) \leq \alpha_1\left[g\left(F_{x_{n-1},x_n}(t)\right) + g\left(F_{x_n,x_{n+1}}(t)\right)\right] + \alpha_2\left[g\left(F_{x_{n-1},x_{n+1}}(t)\right) + g\left(F_{x_n,x_n}(t)\right)\right]$$

$$+ \alpha_3 g\left(F_{x_{n-1},x_n}(t)\right) + \alpha_4 \frac{\left[1 + g\left(F_{x_{n-1},x_n}(t)\right)\right]g\left(F_{x_n,x_{n+1}}(t)\right)}{1 + g\left(F_{x_{n-1},x_n}(t)\right)}$$

$$+ \alpha_5 \frac{g\left(F_{x_{n-1},x_{n+1}}(t)\right)g\left(F_{x_n,x_n}(t)\right)}{1 + g\left(F_{x_{n-1},x_n}(t)\right)} + \alpha_6 \frac{g\left(F_{x_n,x_n}(t)\right)g\left(F_{x_n,x_{n+1}}(t)\right)}{1 + g\left(F_{x_{n-1},x_n}(t)\right)}$$

$$+ \alpha_7 \frac{g\left(F_{x_{n-1},x_n}(t)\right)g\left(F_{x_n,x_n}(t)\right)}{1 + g\left(F_{x_{n-1},x_n}(t)\right)}.$$

Hence, $g\left(F_{x_n,x_{n+1}}(t)\right) \leq (\alpha_1 + \alpha_3)g\left(F_{x_{n-1},x_n}(t)\right) + (\alpha_1 + \alpha_4)g\left(F_{x_n,x_{n+1}}(t)\right) + \alpha_2 g\left(F_{x_{n-1},x_{n+1}}(t)\right)$,

i.e. $g\left(F_{x_n,x_{n+1}}(t)\right) \leq \frac{\alpha_1 + \alpha_2 + \alpha_3}{1 - \alpha_1 - \alpha_2 - \alpha_4} g\left(F_{x_{n-1},x_n}(t)\right)$ Therefore, $g\left(F_{x_n,x_{n+1}}(t)\right) \leq kg\left(F_{x_{n-1},x_n}(t)\right)$, where $0 < k = \frac{\alpha_1 + \alpha_2 + \alpha_3}{1 - \alpha_1 - \alpha_2 - \alpha_4} < 1$, because, otherwise if $k \geq 1$ then $2\alpha_1 + 2\alpha_2 + \alpha_3 + \alpha_4 \geq 1$, is a contradiction.

Inductively, $g\left(F_{x_n,x_{n+1}}(t)\right) \leq kg\left(F_{x_{n-1},x_n}(t)\right) \leq k^2 g\left(F_{x_{n-2},x_{n-1}}(t)\right) \ldots \leq k^n g\left(F_{x_0,x_1}(t)\right)$.

Taking limit as $n \to \infty$, we get, $\lim_{n \to \infty} g\left(F_{x_n,x_{n+1}}(t)\right) = 0$.

Next, we prove that $\{x_n\}$ is a Cauchy sequence. If, on the contrary, $\{x_n\}$ is not the sequence, as proposed, then by Lemma, $\exists \varepsilon_0 > 0, t_0 > 0$ and set of positive integers $\{m_i\}, \{n_i\}$ such that

$$\lim_{i \to \infty} g\left(F_{x_{m_i},x_{n_i}}(t_0)\right) = g(1 - \varepsilon_0), \lim_{i \to \infty} g\left(F_{x_{m_i-1},x_{n_i-1}}(t_0)\right)$$

$$= g(1 - \varepsilon_0), \lim_{i \to \infty} g\left(F_{x_{m_i+1},x_{n_i-1}}(t_0)\right)$$

$$= g(1 - \varepsilon_0) \ \& \ \lim_{i \to \infty} g\left(F_{x_{m_i},x_{n_i-1}}(t_0)\right) = g(1 - \varepsilon_0).$$

Putting $p = x_{n_i+1}$ & $q = x_{m_i}$, we get, $g\left(F_{x_{n_i+1},x_{m_i}}(t_0)\right) = g\left(F_{T_{n_i+1}x_{n_i}, T_{m_i}x_{m_i-1}}(t_0)\right)$, hence from (1),

$$g\left(F_{x_{n_i+1},x_{m_i}}(t_0)\right) \leq \alpha_1\left[g\left(F_{x_{n_i},x_{n_i+1}}(t_0)\right) + g\left(F_{x_{m_i-1},x_{m_i}}(t_0)\right)\right]$$

$$+ \alpha_2\left[g\left(F_{x_{n_i},x_{m_i}}(t_0)\right) + g\left(F_{x_{m_i-1},x_{n_i+1}}(t_0)\right)\right] + \alpha_3 g\left(F_{x_{n_i},x_{m_i-1}}(t_0)\right)$$

$$+ \alpha_4 \frac{\left[1 + g\left(F_{x_{n_i},x_{n_i+1}}(t_0)\right)\right]g\left(F_{x_{m_i-1},x_{m_i}}(t_0)\right)}{1 + g\left(F_{x_{n_i},x_{m_i-1}}(t_0)\right)}$$

$$+ \alpha_5 \frac{g\left(F_{x_{n_i},x_{m_i}}(t_0)\right)g\left(F_{x_{m_i-1},x_{n_i+1}}(t_0)\right)}{1 + g\left(F_{x_{n_i},x_{m_i-1}}(t_0)\right)}$$

$$+ \alpha_6 \frac{g\left(F_{x_{m_i-1},x_{n_i+1}}(t_0)\right)g\left(F_{x_{m_i-1},x_{m_i}}(t_0)\right)}{1 + g\left(F_{x_{n_i},x_{m_i-1}}(t_0)\right)}$$

$$+ \alpha_7 \frac{g\left(F_{x_{n_i},x_{n_i+1}}(t_0)\right)g\left(F_{x_{m_i-1},x_{n_i+1}}(t_0)\right)}{1 + g\left(F_{x_{n_i},x_{m_i-1}}(t_0)\right)}.$$

In limiting case, as $i \to \infty$, we have,

$$g(1 - \varepsilon_0) \le a_2[g(1 - \varepsilon_0) + g(1 - \varepsilon_0)] + a_3 g(1 - \varepsilon_0) + a_5 \frac{g(1 - \varepsilon_0)g(1 - \varepsilon_0)}{1 + g(1 - \varepsilon_0)}, \text{ or } g(1 - \varepsilon_0)$$

$$\le (2a_2 + a_3 + a_5)g(1 - \varepsilon_0) \text{ i.e. } (1 - (2a_2 + a_3 + a_5))g(1 - \varepsilon_0) \le 0.$$

Thus, $g(1 - \varepsilon_0) = 0$, is a contradiction of the fact that g is injective and $g(1) = 0$.

Therefore, $\{x_n\}$ is a Cauchy sequence. Since (X, F, t) is a complete Menger space, so $\{x_n\} \to z \in X$, hence, from (1),

$$g\big(F_{z,T_n z}(t)\big) \le g\big(F_{z,x_{m+1}}(t)\big) + g\big(F_{x_{m+1},T_n z}(t)\big) = g\big(F_{z,x_{m+1}}(t)\big) + g\big(F_{T_{m+1} x_m, T_n z}(t)\big),$$

$$\text{i.e. } g\big(F_{z,T_n z}(t)\big) \le g\big(F_{z,x_{m+1}}(t)\big) + a_1\big[g\big(F_{x_m,x_{m+1}}(t)\big) + g\big(F_{z,T_n z}(t)\big)\big]$$

$$+ a_2\big[g\big(F_{x_m,T_n z}(t)\big) + g\big(F_{z,x_{m+1}}(t)\big)\big] + a_3 g\big(F_{x_m,z}(t)\big)$$

$$+ a_4 \frac{\big[1 + g\big(F_{x_m,x_{m+1}}(t)\big)\big]g\big(F_{z,T_n z}(t)\big)}{1 + g\big(F_{x_m,z}(t)\big)} + a_5 \frac{g\big(F_{x_m,T_n z}(t)\big)g\big(F_{z,x_{m+1}}(t)\big)}{1 + g\big(F_{x_m,z}(t)\big)}$$

$$+ a_6 \frac{g\big(F_{z,x_{m+1}}(t)\big)g\big(F_{z,T_n z}(t)\big)}{1 + g\big(F_{x_m,z}(t)\big)} + a_7 \frac{g\big(F_{x_m,x_{m+1}}(t)\big)g\big(F_{z,x_{m+1}}(t)\big)}{1 + g\big(F_{x_m,z}(t)\big)}.$$

Taking $m \to \infty$, we have,

$$g\big(F_{z,T_n z}(t)\big) \le a\, g\big(F_{z,T_n z}(t)\big) + a_2 g\big(F_{z,T_n z}(t)\big) + a_4 g\big(F_{z,T_n z}(t)\big)$$

$$\text{i.e. } (1 - a_1 - a_2 - a_4)g\big(F_{z,T_n z}(t)\big) \le 0 \text{ i.e. } g\big(F_{z,T_n z}(t)\big) = 0 \text{ giving } T_n z = z.$$

Hence, z is a common invariant point of all T_n. To show uniqueness, take z and z' as common invariant points of T_n for all n. Then from (1), we have

$$g\Big(F_{z,z'}(t)\Big) = g\Big(F_{T_n z, T_n z'}(t)\Big)$$

$$\le a_1\Big[g\big(F_{z,z}(t)\big) + g\Big(F_{z,z'}(t)\Big)\Big] + a_2\Big[g\Big(F_{z,z'}(t)\Big) + g\Big(F_{z',z}(t)\Big)\Big] + a_3 g\Big(F_{z,z'}(t)\Big)$$

$$+ a_4 \frac{\big[1 + g\big(F_{z,z}(t)\big)\big]g\Big(F_{z',z'}(t)\Big)}{1 + g\Big(F_{z,z'}(t)\Big)} + a_5 \frac{g\Big(F_{z,z'}(t)\Big)g\Big(F_{z',z}(t)\Big)}{1 + g\Big(F_{z,z'}(t)\Big)}$$

$$+ a_6 \frac{g\Big(F_{z',z}(t)\Big)g\Big(F_{z',z'}(t)\Big)}{1 + g\Big(F_{z,z'}(t)\Big)} + a_7 \frac{g\big(F_{z,z}(t)\big)g\Big(F_{z',z}(t)\Big)}{1 + g\Big(F_{z,z'}(t)\Big)},$$

$$\text{i.e. } g\Big(F_{z,z'}(t)\Big) \le 2a_2 g\Big(F_{z,z'}(t)\Big) + a_3 g\Big(F_{z,z'}(t)\Big) + a_5 g\Big(F_{z,z'}(t)\Big)$$

$$\text{i.e. } (1 - 2a_2 - a_3 - a_5)g\Big(F_{z,z'}(t)\Big) \le 0, \text{ hence } g\Big(F_{z,z'}(t)\Big) = 0.$$

Therefore, $z = z'$. This proves the uniqueness.

This provides a common fixed point as a solution that has been mentioned in different previous approaches as discussed earlier.

3.3 Inference and conclusion

It is worthwhile to mention that various cell population dynamics-related problems with the biological constraints, as exhibited in Rotenberg's notion, can be formulated as nonlinear equations. To analyze such cases, it is fecund to employ the technique of invariant point result. So far as the probabilistic metric space is concerned, the variation in cell population cannot be formulated explicitly but in the form of some distribution function of their maturation. So, in biomedical utilization, this approach is expected more pertinent and dependable than others. Further, computational techniques may be employed using this tool with said parameters to fecundate AI and all.

REFERENCES

Banach, S. 1922. Sur les opérations dans les ensembles abstraits et leurs applications aux équations intégrales. *Fundam. Math.* 3, 133–181.

Cho, Y. J. *et al.* 1997. Common fixed point theorems for compatible mappings of type (A) in non-Archimedean Menger PM-spaces, *Math. Japon.* 46 (1), 169–179.

Dehici, A. *et al.* 2006. Spectral Analysis of a Transport Operator Arising in Growing Cell Populations. *Acta Appl Math* 92, 37–62.

Greenberg, W. *et al.* (Birkhauser, 1987), Boundary Value Problems in Abstract Kinetic Theory.

Latrach, K. 1996. On a nonlinear stationary problem arising in transport theory, *J. Math. Phys.* 37: 1336–1348.

Latrach, K. 2001. Compactness results for transport equations and applications, *Math. Models Methods Appl. Sci.* 11: 1181–1202.

Latrach, K. 2006. Some fixed point theorems of the Schauder and the Krasnosel'skii type and application to nonlinear transport equations, *J. Differential Equations* 221: 256–271.

Latrach, K. & Jeribi, A. 1999. A nonlinear boundary value problem arising in growing cell populations, *Nonlinear Anal. T.M.A.* 36: 843–862.

Lebowitz, J. L.; Rubinow, S. I. 1974. A Theory for the Age and Generation time distribution of microbial Population, *J. Math. Bio.* 1: 17–36.

Menger, K. 1942. Statistical metric, *Proc. Nat. Acad. (USA)* 28, 535.

Mishra, A, *et al.* 2020. A Probabilistic Approach in Modeling the Growth of Cell Population, *Test Engineering and Management*, Vol. 83, pp. 737–742.

Moors, W. B. & S. J. W. 2017. James' weak compactness theorem: An exposition, *arXiv:1705.06406 [math. FA]*.

Peeler, L. 2011. Metric Spaces and the contraction mapping principle, *Math. Unchicago*, edu.

Razani, A. 2006. A fixed point theorem in the menger probabilistic metric space, *New Zealand Journal of Mathematics*, Vol. 35, 109–114.

Robbin, Joel W. 2010, Continuity and Uniform Continuity.

Rotenberg, M. 1983. Transport theory for growing cell populations, *J. Theo. Biol.* 103:181–199.

Schauder, J. 1930. Der fixpunktraz functional saumen, *Studia Math.* 2, 171–180.

Tripathi, P. K. *et al.* 2016. *Global Journal of Pure and Applied Mathematics*, Volume 12, Number 2, pp. 1629–1634.

Van der Mee, C. & Zweifel, P. 1987. A Fokker-Planck equation for growing cell populations, *J. Math. Biol.* 25: 61–72.

Zeidler, E. (Springer-Verlag, 1993), *Nonlinear Functional Analysis and Its Applications I: Fixed-Point Theorems*.

Advances in AI for Biomedical Instrumentation, Electronics and Computing – Sachan et al. (eds)
© 2024 The Author(s), ISBN 978-1-032-64298-7

Unraveling the power of AI assistants

Abhinav Karn, Prashant Kumar Singh, Chirag Agarwal, Ayush Verma, Deepak Singh &
Mupnesh Kumari
Chandigarh University, Punjab, India

ABSTRACT: The paper introduces JARVIS, a sophisticated virtual embedded voice assistant that makes use of state-of-the-art tools including gTTS, Python, Machine Learning, and a text-to-speech platform by Google. A virtual assistant named JARVIS is created for Windows-based operating systems that draw inspiration from actual virtual assistants like Cortana and Siri. It aims to make chores easier for users using voice commands and tailored interactions. JARVIS can carry out a variety of operations, including sending emails, making phone calls, and organizing meetings, thanks to the system's integration of NLP, ML algorithms, and speech recognition. JARVIS improves user experience by recognizing needs and providing pertinent solutions through tailored user-profiles and capabilities like sentiment analysis and emotion detection. In addition to showing JARVIS's reusability and low maintenance requirements, the article highlights the smooth integration of AIML, Python, and gTTS. JARVIS is a key improvement at a time when people are relying more and more on technology, as it uses speech recognition to satisfy users' needs for simplicity and ease. The suggested solution pioneers the integration of numerous technologies, capturing the essence of real-world virtual assistants while also influencing the future of personal voice assistants.

1 INTRODUCTION

Voice assistants such as Alexa, Siri, and Cortana are the result of advances in voice recognition technology, natural language processing, and artificial intelligence. These AI-driven systems understand and act upon spoken instructions, handling tasks like setting reminders, sending messages, and controlling smart devices. They play music, perform web searches, and offer convenience through personalized responses. Jarvis AI, inspired by Tony Stark's AI, uses ML and NLP to enhance user interactions. It streamlines users' lives by organizing meetings, managing social media, and providing effective task management features like reminders and to-do lists. Jarvis offers valuable information like weather updates, news, and recommendations, continually improving results through advanced learning capabilities. These voice assistants have transformed the way we interact with technology, especially in home automation. This study focuses on creating a Python-based voice assistant that excels in tasks like reminders, weather checks, web searches, and smart home control, highlighting the use of Python libraries for accurate and responsive interactions. It paves the way for intelligent voice assistants in various domains. Major voice assistants, including Cortana, Alexa, and Google Assistant, rely on machine learning and natural language processing to understand human requests and perform tasks. These virtual assistants have greatly enhanced user experiences and continue to evolve, demonstrating the potential of machine learning and natural language processing in technology. In today's rapidly evolving society, digital assistants redefine human capabilities by understanding voice commands and executing tasks efficiently, thanks to speech analysis, language processing, machine learning, and neural networks. They have become integral to various devices and facilitate natural interactions, changing the way we engage with technology.

2 RELATED WORK

There are a variety of AI-powered virtual assistants on the market that are integrated into various applications and platforms. Several well-known instances are Google Assistant for Android, Apple's Siri, Espeak for Linux, and Microsoft's Cortana for Windows. Companies are using dialogue system technology to create Virtual Personal Assistants (VPAs) that are customized for their own applications and domains. [Cortana Intelligence]

1. Megha Raule, Anjali Fapal, Trupti Kanade, BharatiJanrao, Mrunalini Kamble, and Trupti Fapal, personal virtual assistant via Python:

- Technology: Webbrowser, OS, spy audio, and Python libraries such as pyttsx3.
- Future Scope: Concentrates on offering client cost savings, access to global talent, scalability, efficiency, productivity, and flexibility. [Patil *et al.*]

2. Voice Assistant with Python (Writers: Prof.A.V. Gundavade, Monika S. Jalpure, Akshata S. Gavade, Pranjali Chaudhary, Pooja C. Goutam):

- Technology: PyCharm and Python for development.
- Future Scope: reduces manual labor required for a variety of chores to help people of all ages, including those who are disabled. [Fapal *et al.*]

3. Authors: Anup Bhange, Deepak Shinde, RiaUmahiya, Monika Raghorte, and Aishwarya Bhisikar: AIBased Voice Assistant with Python

- Technology: A range of artificial intelligence technologies, such as text-to-speech, voice biometrics, text activation, automatic speech recognition, and natural language comprehension.
- Future Scope: Because the system is modular, new features can be added without impairing its current functionality. [Shende *et al.*]

4. Voice Assistant with Python (Writers: Gaurav Kumar, Harshit Agrawal, Mr. Surya Vikram Singh, Dr. Diwakar Yagyasen, and Nivedita Singh):

- Technology: Voice activation, natural language understanding (NLU), automated speech recognition.
- Future Scope: Machine learning will be included in upcoming upgrades to improve suggestions and control over neighboring devices. [Singh *et al.*]

5. AI-Based Virtual Assistant with Python: AComprehensive Analysis (Writers: Aditi Patil, Sakshi Shinde, Mrs. Patil Kavita Manoj kumar, Shakti prasad Patra, Saloni Patil):

- Technology: Text analysis, speech recognition modules, text-to-speech, API calls, data extraction, and speech-to-text are some of the technologies used.
- Future Scope: Describes how artificial intelligence(AI) is being used to maximize the effectiveness and efficiency of human virtual assistants. [Goutam]

6. Srivastava and Prakash's "An Analysis of Various IoTSecurity Techniques: A Review" Offers a review of the different security methods applied to the Internet of Things (IoT). [Cortana Intelligence]
7. "Digital Transformation of Healthcare: A BlockchainStudy" by Himanshu Kumar Shukla, Rudrendra BahadurSingh, Surya Vikram Singh, and Saijshree Srivastava:
    Examines how blockchain technology is being applied in the healthcare sector and how this could change the sector digitally. [Srivastava *et al.*]
8. A study conducted by T. Schultz, A. Waiel, and others: focuses on evaluating acoustic models and applying large vocabulary continuous speech recognition (LVCSR) systems to new target languages. [Srivastava *et al.*]

Researchers have explored various applications of AI and machine learning in education:

- A Voice-Enabled Personal Assistant for PC successfully interpreted voice commands 95% of the time.
- Machine learning algorithms identified fake news on social media with 93.2% accuracy.
- Sentiment analysis on COVID-19-related Twitter data achieved 87.2% accuracy.
- In educational settings, facial recognition for attendance control reached 91.6% accuracy. Predicting student success in online courses had an accuracy of 92%.
- Anomaly detection in IoT network traffic achieved an impressive 99.6% accuracy.
- Machine learning aided brain tumor diagnosis from MRI scans with 97.4% accuracy.
- A speech-based personal assistant for the visually impaired demonstrated accurate question detection and response, enhancing accessibility. Speech emotion identification reached 93% accuracy.

3 PROPOSED SYSTEM

The proposed Jarvis AI voice-assistant system utilizes advanced technologies for a range of activities. Users can interact with it using natural language instructions, processed through speech recognition and NLP. Jarvis uses machine learning to personalize responses based on user behavior, preferences, and past interactions. Among other things, it can be used to manage smart devices, make appointments, send reminders, respond to inquiries, and provide information.

3.1 *Workflow*

Initializing the necessary parts, introducing the user, listening for their voice, processing their command using Natural Language Processing techniques, executing the appropriate function, and API call, providing feedback to the user, and repeating the process until the user ends the conversation are all part of a voice assistant's system structure. The voice engine needs to generate human-sounding speech, and the system needs to be able to handle errors and exceptions to interact with the user effectively. This structure allows the voice assistant to recognize and understand user commands, provide useful information in response, or perform tasks.

Some important aspects of this voice assistance system's architecture are as follows:

3.1.1 *Initialization*
Load any required modules or libraries. Set the speech attributes including voice rate, volume & type, after the voice engine has been initialized. Connect any other external APIs or services and complete any other setup procedures.

3.1.2 *Greeting*
Depending on the time of day or any other customizable greeting, choose the appropriate greeting. Say the greeting aloud using the voice engine. Find out how the user can benefit from the voice assistant.

3.1.3 *Listening*
Turn on the microphone and begin to hear the user's voice. The user's speech can be turned into text by using a speech recognition library. Preprocess text input from the user to remove any superfluous words or phrases.

3.1.4 *Natural Language Processing (NLP)*
NLP helps in analyzing and interpreting the user's command. Determine the intention or goal of the user who issued the command. Take whatever entities or parameters that are required out of the user's command.

3.1.5 *Command execution*

Choose the right course of action based on the intent and command of the user. Run the relevant program or make the necessary API call. If an error or exception arises during execution, handle it.

3.1.6 *Feedback*

Give the user feedback based on how the command was executed. Provide visual feedback or speak the command's outcome using the voice engine. Give the proper feedback if the voice assistant misunderstood the command or made a mistake.

3.1.7 *Loop*

Keep listening until a specific command is given by the user to end the conversation. After every loop iteration, take any necessary cleanup actions.

3.1.8 *Closing*

Decide the user farewell using the voice engine. Put an end to hearing the user's voice. Close the program and take care of any cleanup that needs to be done.

Figure 1.   Flowchart of workflow.   Figure 2.   Flowchart of speech recognition.

3.2 *Speech recognition system*

Speech recognition, also known as voice conversion, is the ability of a computer or other device to translate spoken words into understandable text.

Due to a significant learning disability, basic voice recognition software only recognizes words and patterns that are simple to pronounce. Multilingual software, multiple pronunciations, and natural language are all supported by more sophisticated software. Speech recognition draws on computer science, linguistics, and technology research. Speech recognition technology is now widely used in textbook-focused applications and widgets for more efficient and hands-free use.

Speech and voice detection are two different processes, so they shouldn't be confused.

- The skill of identifying spoken words is known as speech recognition.
- A biometric technique called voice recognition is used to recognize human speech.

In a speech recognition system, spoken words are processed and analyzed before being converted into text using computer algorithms. Following these steps, a software program transforms the microphone's captured sounds into characters such that both computers and people can understand. Then it separates the characters into little passageways.

Following this procedure, these narrow hallways are digitalized and made computer-readable.

After that, an algorithm is used to match the most accurate textbook illustrations.

3.3 *Proposed system work*

It was interesting to create my voice assistant, Jarvis. Through voice commands, it simplifies things like sending emails, using Google, and music, and opening particular applications. Jarvis distinguishes itself by being tailored to the desktop, doing away with the requirement for user accounts, and ensuring effectiveness. Any Android phone can function as a wireless microphone by using the "WO-MIC" application, which lowers background noise. This study demonstrated how AI may dramatically cut human work and save time, reinforcing its transformational potential.

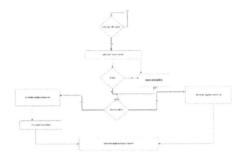

Figure 3. Workflow diagram.

- Voice Input: It offers a voice mode for user input, allowing users to interact via speech or text.
- Wireless Microphone: Utilizes the "WO-MIC" app to turn an Android phone into a wireless microphone, reducing background noise.
- Wikipedia Search: Acts as a search engine for Wikipedia, retrieving information from the internet and presenting itaudibly.
- News Updates: Retrieves current news on various topics, including local and global news, technology, sports, and entertainment.
- Location Services: Helps users find their current location or get directions to a destination.
- Email Sending: Sends emails to contacts with success notifications.
- Timetable Notifications: Uses the user's schedule to send task reminders with the notify Python library.

3.4 *Proposed methodology*

3.4.1 *Speech recognition*
This technology converts your voice input into text using a web-based language processing algorithm. vocal production to use cloud servers for voice recognition, clients will upload and momentarily store content from a particular corpus arranged on a digital data center in the resource center. The primary processing device then receives the corresponding text that has been captured.

3.4.2 *Python backend*
The Python server-side verifies the response from the voice recognition engine to determine whether an order or voice emission is an API call, perspective retrieval, or system call. The result is then sent back to the Python server along with the expected outcome for the customer.

3.4.3 *API calls*
An application programming interface, or API, is an interface for programming languages. Two programs can communicate with one another through an API or application programming interface. Another way to think of an API is as a messenger that sends and receives requests from providers.

3.4.4 *Context extraction*
Context extracting (CE) is a technique for obtaining structured information from semi-structured machine-readable content. Usually, this process uses natural language

processing (NLP) and machine learning to interpret complex communication content. The test results for context extraction are illustrated by recent attempts in the synthesis of audio-visual materials, including automatic annotation and material retrieval from sound, video, and graphics.

3.4.5 *Text-to-Speech*

Text-to-Speech (TTS) describes a machine's capacity to read material aloud to consumers. Written text is converted into pulses by the TTS system, which can then be converted back into dialectal interpretations. A penultimate operator's TTS system is available for a variety of dialects, languages, and specialized lexicons.

4 RESULT AND DISCUSSION

nlp, a crucial component in voice recognition, transforms spoken language into digital data for analysis. jarvis, driven by advanced NLP and machine learning, collaborates with smart devices and third-party apps like Spotify and Uber. performance highlights:

- **95% voice command accuracy:** Jarvis impressively attains 95 % accuracy for understanding voice commands, even in noisy environments.
- **over 90% task accuracy:** it excels in tasks like setting reminders and operating smart devices, with accuracy exceeding 90%.
- **80%+ question response accuracy:** Jarvis accurately answers questions more than 80% of the time when users provide voice commands based on their preferences. these results emphasize Jarvis's high accuracy and efficiency in managing diverse tasks, from smart device control to appointment scheduling.
- This research's features include the ability to set an alarm clock.
- Voice activation is one way to automate a smart light.
- WhatsApp allows users to text one another.
- You now close the program that prompted you.
- It's able to measure your internet speed.
- Your questions can be readily answered by the helper.

5 CONCLUSION AND FUTURE WORK

5.1 *Conclusion*

Natural language processing, neural networks, gTTs, and speech recognition are all combined in the intelligent JarvisAI voice assistant. It is intended to function as a personal assistant, automate repetitive tasks, and intelligently communicate with other subsystems. The system goes through multiple stages, such as voice-to-text conversion, and pattern-recognition data analysis.

Jarvis can respond to different situations, manage daily schedules, and improve AI's ability to recognize human behavior patterns. It offers a more efficient and organized way for users to interact with various systems and modules. In summary, Jarvis is an advanced tool that can revolutionize digital interactions. It offers personalized and seamless experiences, learns from user behavior, and has various applications, including task management and personalized responses. Future developments may include deep learning, integration with smart home devices, gesture controls, and immersive experiences with AR and VR technologies. As technology evolves, Jarvis holds the potential to transform everyday interactions into more effective, immersive, and personalized experiences, promising a bright future for AI voice assistants.

5.2 *Future work*

To enhance Jarvis's capabilities, we can expand its database and data training sets, allowing it to handle problems more effectively. This will increase its effectiveness and response capabilities. We can also add more voices as a feature, addressing these limitations with additional data training sets. Improving the user interface is another area of potential enhancement. An optimized interface would be more comprehensive, user-friendly, and accessible to a broader user base, making Jarvis more approachable. We can also consider connecting Jarvis with smartphones using React Native to ensure synchronized performance across multiple devices. Long-term goals include automating tasks like supporting Elastic Beanstalk and backing up data, effectively replacing the need for a server administrator.

REFERENCES

Fapal, A., Kanade, T., Janrao, B., Kamble, M., & Raule, M. (2021). "Personal virtual assistant for windows using python." *International Research Journal of Modernization in Engineering Technology and Science*, 03 (07), Impact Factor 5.354.

Goutam, P. C., Jalpure, M. S., Gavade, A. S., Chaudhary, P., & Gundavade, A. V. (2022). "Voice assistant using python." *International Journal of Creative Research Thoughts (IJCRT)*, 10(6), ISSN: 2320–2882.

Manojkumar, Kavita Patil, Aditi Patil, Sakshi Shinde, Shaktiprasad Patra, and Saloni Patil. "AI-Based virtual assistant using python: A systematic review" in *International Journal for Research in Applied Science & Engineering Technology (IJRASET)*, vol. 11, issue III.

Saijshree Srivastava, Surya Vikram Singh, Rudrendra Bahadur Singh, Himanshu Kumar Shukla. "Digital transformation of healthcare: A blockchain study" *International Journal of Innovative Science, Engineering & Technology*, Vol. 8 Issue 5, May 2021

Shende, D., Umahiya, R., Raghorte, M., Bhisikar, A., & Bhange, A. (2019). "AI-Based voice assistant using python." *Journal of Emerging Technologies and Innovative Research*.

Singh, N., Yagyasen, D., Singh, S. V., Kumar, G., & Agrawal, H. (2021). "Voice assistant using python." *International Journal of Innovative Research in Technology*, 8(2), ISSN: 2349–6002.

Srivastava, S., & Prakash, S. (2020). "An analysis of various IoT security techniques: A Review." *2020 8th International Conference on Reliability, Infocom Technologies and Optimization (Trends and Future Directions) (ICRITO)*, pp. 355–362.

Schultz, T., & Waibel, A. (2001). Language independent and language adaptive acoustic modeling for speech recognition. *Speech Communication*, vol. 35, no. 1, pp. 31–51.

Advances in AI for Biomedical Instrumentation, Electronics and Computing – Sachan et al. (eds)
© 2024 The Author(s), ISBN 978-1-032-64298-7

Retinal diseases analysis and detection – A comprehensive review

P. Renuka
R.M.K. Engineering College, Tamil Nadu, India

V. Sumitra
SRM Institute of Science and Technology, Tamil Nadu, India

P. Latha
R.M.K. Engineering College, Tamil Nadu, India

K. Swaminathan
Smart Iops Research Center Private Limited, Bangalore, India

ABSTRACT: The retina is a thin layer of tissue that contains millions of light-sensitive cells and is located at the back of the eye. The basic function of retina is to receive, organize, and send visual information to your brain through the optic nerve. Retinal diseases cause damage to any part of the retina. Untreated retinal diseases can cause severe vision loss and even blindness. With early detection, some retinal diseases can be treated, while others can be controlled or slowed down to preserve, or even restore vision. Some of the diseases are diabetic retinopathy, Glaucoma, Age related macular degeneration and retina of prematurity in infants. Many techniques and methods for early detection of retinal diseases is described in this paper. Optical coherence tomography angiography also known as OCTA is a non-protruding typical method that produces paradigm view of retina. It provides clear study of chorio retinal and optic nerve vasculature in absence of tint there by avoiding its difficulties. Among all the methods, OCTA is found to have better accuracy of 97% for identification of eye diseases.

Keywords: Retinal disorders, OCT, fundus images

1 INTRODUCTION

1.1 *Fundus retinal imaging*

A technique of imaging which is used widely to record the information about the retinal structure is known as Fundus imaging of retina. Figure 1 show the picture of a fundus camera. [Chalakkal *et al.*] It is a microscope which works in lesspower along with a camera. The basic sketch of any fundus camera is based on the principle of indirect opthalmoscope in which the photography of retina is done by the usage of pupil in entry, exit pathof rays of light from fundus camera. Field of view (FOV) is a parameter that describes fundus camera that can vary from thirty to onehundredtwenty degrees. FOV which is narrower offers photographs which has smaller retinal area with higher magnification. The fundus camera imaging techniques has many challenges.

Figure 1.   Fundus camera.

   DOI: 10.1201/9781032644752-88

1.2 Optical coherence tomography angiography (OCT-A)

Optical coherence tomography is a noninvasive technique of imaging in which images of higher resolution are obtained. It has wide clinical applications in diagnosing patients with retinal diseases and choroidal diseases. It is a rising automation which gives superb view of flow of blood in posterior segment and it plays a major role in patients care with vascular diseases of retina. The Speed of Obtainment of an entire OCT-A is better than obtainment of an full angiographic view based on dye. The three important benefits of OCT are High scanning speed, non-invasiveness and contactless.

2 SURVEY ON FUNDUS AND OCT

2.1 OCTA for DR detection

Computer-aided diagnosis (CAD) system was designed for first DR identification by OCTA usage. Mapping of retinal vascular perfusion density for comparison of various levels of DR have been discussed [Lee *et al.* 2015]. It proved that as DR level is higher, there is a fall of values of density of capillary perfusion. Axis ratio of Foveal area have been examined by images of OCTA and normal and DR patients are differentiated. There is a large differrence in values of healthy and diabetic cases. The abnormalities are microaneurysm, neovascularization and proved that there is better detection rate for OCTA than FA. This technique proved that OCTA images are efficient in evaluation of DR treatment. The subfield in central region and vasculature in parafoveal macula is graded better by OCTA than FA. This CAD system achieves a accuracy of 97.1% by using 133 subjects.

2.2 Diagnosis based on OCTA

The retinal diseases are diagnosed using OCTA. Types of the diseases are 1. diabetic retinopathy, 2. glaucoma, 3. age-related macular degeneration, 4. anterior ischemic optic neuropathy, 5. retinal vein occlusion (RVO), 6. retinal artery occlusion (RAO). The effectiveness of OCTA is estimated in patients with DR thereby concluding the fact about OCTA visualizing various changes in vascular at various stages of Diabetic retinopathy [Ishibazawa *et al.*]. The author discussed the early changes in microvascular region of retina in patients with type I diabetic (DM1) by the use of OCTA [Talisa *et al*]. Compared to normal eyes the patients with diabetic have larger area of FAZ and area increase in non-perfusion capillary region. [Carnvali *et al*].The peripapillary capillary density (PCD) are evaluated by the usage of OCTA in glaucoma patients [Agemy *et al*]. Vessel thickness of optic disc in glaucoma patients is estimated using OCTA [Wang *et al*]. Defects in macular region in patients with glaucoma are evaluated by the usage of OCTA in a resolved projection[Ferrara et al].This discussion is about a detailed comparison of posterior eye segment modality of imaging with OCTA in different diseases in posterior segment region.

2.3 Review on OCT layers

OCT is the emerging image technology in medical field which gives a precise understanding of optical break up and integration of organic matter. The spectral domain and high definition OCT also known as SD-OCT and HD-OCT [Garvin *et al*] was invented after TD-OCT (Time division OCT). OCT of Glaucoma eye has four layers namely 1. nerve fiber layer, 2. ganglion cell layer, 3. outer plexiform layer (OPL), and 4. outer nuclear layer (ONL) + photoreceptor layer. OCT of AION affected eyes has 1. NFL, 2. GCL+IPL, 3. inner nuclear layer (INL)+OPL, 4. photoreceptor inner segments (PIS) and 5. photoreceptor outer segments(POS). In some cases of OCT, 12 layers namely NFL, GCL, IPL, INL, OPL, ONL, external limiting membrane(ELM), ellipsoid zone(EZ), myoid zone(MZ), interdigitating zone (IZ), outer photoreceptor (ORP) and RPE layers are detected. The author found 5 layers (NFL, GCL + IPL, INL + OPL, IS, and OS) in retinal OCT display and a review was made on three Dimensional view for twelve AION patients [11]. This discussion proved that OCT images have made a clear differentiation between anatomy of affected and normal retina.

2.4 *OCTA of other retinal disorders*

OCTA of other retinal disorders is discussed in this section. OCTA is used for documenting the retinal and choroidal vasculature and provides excellent view of posterior segment blood flow. Retinal vein occlusion also known as RVO is the next type of vascular disease to Diabetic Retinopathy [Hendrick *et al.*]. The striking images are provided by OCT-A in occlusion of retinal vein in branch patients also known as BRVO. The Foveal shape is wider, irregular in patients with BRVO. Near optic nerve there occurs an vein occlusion in central region known as Central retinal vein occlusion also known as CRVO. The symptoms of CRVO are retinal vascular tortuosity, dilatation and scattered intraretinal haemorrhages. The disturbance in flow of blood because of factors such as inflammatory and thrombotic reasons results in occlusion of Retinal artery (RAO). A capillary non-perfusion spread widely with some central perfusion is shown in OCTA of Central RAO.

2.5 *Deep learning approach*

Classification of eye diseases is done by approach of deep learning [LeCun *et al.* 2015]. Identification of objects in images is a major application of deep learning. Fuzzy C-means is used for grouping for classification of exudate regions [Mohamed *et al.* 2002]. Exudate is a fluid that leaks out of blood vessels into nearby tissues. DR Classification is done using visual neural component spectraa. Identification of bleeding of retinal eye as a sign of DR is a method proposed using SVM. The intelligent system of detection detects automatically by the usage of network of neural. The layers of CNN are input, rectified linear unit, convolutional, pooling, softmax and fully connected layer. 37.75 percent is the highest loss obtained from the image input which is smallest of range 31x35, highest training accuracy of 82.03% from 46x53 image inputs highest accuracy of test data of 80.93% is achieved out of 31x35 inputs.

2.6 *DR detection using DWT and ANN*

Different stages of DR have been recognized and they are mild,moderate and severe DR [Kanth *et al.* 2013]. A number of data sets for training and an identifier is used by artificial neural network (ANN) for process of detection [Sayed *et al.* 2017]. A regularization based on bayesian classifier (BR) and strong back propagation algorithm for detection of Diabetic Retinopathy lesions is discussed. Probabilistic neural network (PNN) and support vector machine (SVM) for Diabetic retinopathy detection shows a accuracy of 80% and 90% respectively. Better classification of different stages of DR is produced by ANN and reduction of heavy computational method is done by DWT to yield a speed result. Limited number of fundus images are retrieved with precision and recall rate of 63% and 57% hence it is suitable for rural region.

2.7 *CAD diagnosis of AMD*

A method is made for the detecting AMD by the usage of neural networks by DL [Zheng *et al.* 2013]. The method provides image transfer from pre-trained deep neural networks to detect AMD. Approaches in Data mining are employed in identification of AMD. Detection of segmenting automatically of Retinal pigment epithelium layer of AMD by SD-OCT is done for drusen detection. A system for detecting drusen automatically for AMD classification is discussed. This system is known as Automated Vision Impairment detection through gaze analysis also known as AVIGA. Another method proves that AMD detection leads to mitigation [Deng *et al.* 2016]. For amount of high-value pixels, the norm values are obtained for each normal and macular view individually (1176, 1722, 1102). The diagnosis accuracy of OCT images is high (75 %) and it is obtained by the comparison of average and highest pixel value.

2.8 *Lesion detection*

Detection of Lesion by the usage of segmented retinal structure is mentioned in this section. The author proposed a graph cut technique for blood vessel segmentation. Localization of various attribute and lesions in fundus representation of retina are made in a robust approach. The division of hemorrhages for analysis based on computer of diabetic retinopathy is discussed. A method for blood vessels extraction is described. The accuracy achieved for lesion detection method is 94%.

3 RESULTS

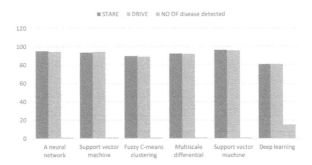

Figure 2. Different methods.

Figure 2 describes the comparison of different methods. This shows that eventhough the accuracy of deep learning approach is lower but number of diseases detected are higher compared to other methods. This proves that deep learning has excellent potential in image classification application. Table I shows the comparison of various methods for the detection of retinal disorders.

4 SUMMARY

Analysis of retinal image is interesting technique for early identification of diseases like DR, macular edema, glaucoma, AMD etc. This analysis is useful to provide service to areas which is not accessible to hospitals. Many techniques are discussed in this paper which automatically identify the important features of fundus image, such as blood vessels, macula and optic disc. OCT technique gives layered view of retina thereby allowing easy identification of disorders. The numerical analysis and evaluation of the features provides a better understanding of correlation between different diseases and retinal characteristic.

Table 1. Comparison of different technologies.

S.NO	REFERENCE PAPER	AUTHOR	PROPOSED WORK	TECHNIQUE	RESULT	CONCLUSION
1.	CAD for early DR detection using OCTA	Nabila Eladowi	To detect and diagnose mild DR	OCT using CAD	97.1% Accuracy	To detect other diseases in early stage
2.	Deep learning for classification of eye diseases	Widodo Budiharto	Automatic detection of DR	Deep learning and CNN	82.03% Accuracy	To use training algorithm and ensemble learning machine
3.	Fundus retinal image analysis	Walid Habib Abdulla	To diagnose DR, edema and glaucoma	ARIA	91.4% Accuracy	To train deep learning based models
4.	Mobile phone based DR detection system	Nikita Kashyap	Early detection of DR	DWT and ANN	63% precision and 57% recall rate	Ideal for rural region
5.	CAD of age related macular degeneration by OCT	Punal M.Arabi	AMD	Neural networks, OCT	97% Accuracy	Wet,Dry AMD by some features of OCT
6.	Retinal diseases based on OCT-A	Mohammed Elmogy	AM,AION,RVO,RAO	OCTA	Optic nerve diseases diagnosed	Other applications in medical imaging
7.	OCT- A Review	Nabila Eladowi	DME,AMD	OCT	Spatial features detected	Extraction of blood vessels
8.	Data driven approach to DR detection	Ramon Pires	To detect DR	RDR and CNN	98.5% Accuracy	Enhanced DR screening
9.	ML based detection in Retinal fundus	G. Indumathi	Retinal diseases	CLAHE	89.21% Accuracy	Some features have to be included to increase accuracy
10.	OCT A of retinal vascular diseases	Rosa Lozada	Retinal vascular diseases	Spectral domain technology	Image is better visualized	Greater expertise in interpreting the images
11.	Screening and detection of DR by using concepts	Wani Patil	DR, Macular Ischemia	Bayesian classifier,SVM	90% Accuracy	MI is classified using SVM classifier
12.	OCT in Type 3 neovascularization	Riccardo Sacconi	Type 3 NV	Multimodal imaging,OCTA	Neovascular flow is detected	Prevent disease progression to late stage
13.	DR detection in ocular imaging by DL	Zahra Amini	Diabetic Retinopathy	Matched filtering	Images are classified	Ocular image segmentation can also be done
14.	Lesion detection using segmentation of retina	K.C. Manoj	DR,Glaucoma	Segmentation, MRF	94% Accuracy	Lesion part is extracted in abnormal cases

5 CONCLUSION

The above table describes the comparison of different techniques used for analysis of fundus images. The diseases identified are Diabetic Retinopathy, Glaucoma, Macular degeneration, and neovascularization. The techniques used are computer aided diagnosis, machine learning, deep learning, dictionary learning and OCTA. OCTA is a mainly used technique now a days for identifying many human diseases. Comparing all the methods, Disease detection using OCTA achieves a accuracy of 99.7 which is a very good result for early detection and treatment. From the above survey it is clear that Deep learning has more techniques to identify the abnormalities in retinal eye. Hence OCTA imaging also used to provide a layered view of retinal structure. The proposed work is the combination of these both techniques in detecting the retinal abnormalities.

ACKNOWLEDGEMENT

This work has been supported by the Department of Science and Technology (DST) under the scheme on Biomedical Device and Technology Development (BDTD).

REFERENCES

Agemy S. A., OCTA analysis of perfused pre-papillary capillaries in primary open-angle glaucoma and normal-tension glaucoma, *Invest, Ophthalmol, Vis. Sci.* 57 (9) (2016) OCT611–OCT620

Carnvali A., Sacconi.R., Optical coherence tomography angiography analysis of retinal vascular plexuses and choriocapillaris in patients with type 1 diabetes without diabetic retinopathy, *Acta Diabetol.* 54(7) (2017) 695–702

Deng J., Xie X., Age-related macular degeneration detection and stage classification using choroidal OCT images, vol. 9730, Springer, (2016), pp. 705–715

Eltanboly A., Schaal S., A computer-aided diagnostic system for detecting diabetic retinopathy in OCT images, *Med. Phys.* 44 (3) (2017) 914–923

Ferrara D., Reichel E., OCTA in retinal artery occlusion, *Retina* 35 (11) (2015) 2339–2346

Garvin M. K., Sonka M., Intraretinal layer segmentation of macular OCT images using optimal 3-D graph search, *IEEE Trans. Med. Imaging* 27 (1) (2008) 1495–1505.

Garvin M., Sonka M., Intraretinal layer segmentation of macular OCT images using optimal 3-D graph search, *IEEE Trans. Med. Imaging* 27 (1) (2008) 1495–1505.

Ishibazawa A., Omae T., Optical coherence tomography angiography in diabetic retinopathy: a prospective pilot study, *Am. J. Ophthalmol.* 160 (1) (2015) 35–44

Kanth S., Jaiswal A., Identification of different stages of Diabetic Retinopathy using artificial neural network, in: IEEE Sixth International Conference of Computing (IC3) (2013)

LeCun Y., Bengio Y., Hinton G., Deep learning, *Nature* 521 (7553) (2015) 436

Lee J. G., Agemy. S. S., Retinal vascular perfusion density mapping using optical coherence tomography angiography in normal and diabetic retinopathy patients, *Retina* 35(11) (2015) 2353–2363

Mohamed C. A., Modular Neural Networks for Automatic Retinopathy Screening, *Artificial Intelligence and Soft Computing*, ASC, (2002)

Sayed S., Kapre S., Detection of diabetic retinopathy using image processing and machine learning, *Int. Innov. Res. Sci. Eng. Technol.* 6 (1) (2017) 99–107. An ISO 3297:2007 Certified Organization

Talisa E., Detection of microvascular changes in eyes of patients with diabetes but not clinical diabetic retinopathy using optical coherence tomography angiography, *Retina* 35 (11) (2015) 2364–2370

Zheng Y., Daniel E., An Automated drusen detection system for classifying age-related macular degeneration with color fundus photographs, IEEE (2013) 978-1-4673-6455-3/13

Advances in AI for Biomedical Instrumentation, Electronics and Computing – Sachan et al. (eds)
© 2024 The Author(s), ISBN 978-1-032-64298-7

MediSafe – Enhancing secure medical data management and doctor–patient communication

Neha Rajas, Hrishikesh Potnis, Chinmayee Prabhu, Dnyaneshwari Pote, Payal Powar & Sujal Powar
Vishwakarma Institute of Technology, Pune

ABSTRACT: Storing and maintaining medical documents such as prescriptions, medical certificates, invoices for medications, lab reports, health insurance documents, and more, has always been a tedious task. Accessibility of such documents becomes very crucial, especially in emergency situations. On the other hand, a patient's medical histories are often incomplete, with each healthcare facility or clinic maintaining its own database of their patients. This happens because patients may seek care from multiple providers, resulting in their healthcare data being scattered across various sources. In order to address these challenges, we propose MediSafe - a comprehensive unified cloud-based system which enables users to compile, store, retrieve and share medical documents.

1 INTRODUCTION

Managing medical documents, such as prescriptions, lab reports, and health records, has always been a difficult task. Quick access to these documents becomes difficult, especially during emergencies. Furthermore, a patient's medical history is often spread across various healthcare providers, leading to incomplete records. This situation arises because patients seek care from multiple sources. Due this, quick data accessibility decreases. In this paper, we discuss the issues surrounding medical data management, and present our solution - MediSafe. MediSafe also helps in efficient communication while ensuring data privacy.

The structure of this paper is as follows: Section II showcases existing work in this field. Section III outlines our methodology, while Section IV discusses about the practical implementation of our system. In Section V we present the results of our solution, and in section VI, we explain the future scope of our project and potential in this field. Finally, Section VII concludes the paper.

2 LITERATURE REVIEW

We did a thorough research of existing work done in Medical Data Management field and their limitations.

Reference (Malet *et al.* 1999) discussed the challenges of gathering and organizing personal clinical data. They proposed a system that uses metadata for organization, stores data in the cloud, and shares it securely with licensed caregivers, especially during emergencies, to ensure continuous care and evidence-based treatment.

Blockchain is a chain of data blocks connected through cryptography, as explained in reference (Hölbl *et al.* 2018). Reference (Sharma *et al.* 2020) suggested integrating blockchain technology to enhance the security and privacy of large healthcare datasets. They focused on how it can manage real-time healthcare data through a distributed approach.

DOI: 10.1201/9781032644752-89

Telemedicine involves delivering healthcare services and sharing medical information over distances. It can improve access, quality, and efficiency in healthcare. While it's more common in developed countries, it has significant potential in the Indian healthcare system, as discussed in reference (Mishra *et al.* 2009). This source provides a detailed overview of telemedicine in India. These insights from our literature review provide a strong foundation for our research paper.

3 METHODOLOGY

MediSafe enables users to securely store their medical documents and records on cloud while also facilitating improved communication between doctors and patients. Our solution can be divided into two primary components: the patient's portal and the doctor's portal.

The patient's portal provides the user interface for patients. Initially, users undergo an authentication procedure. Patients are able to store, retrieve, and share their medical documents like prescriptions, medical certificates, invoices for medications, lab reports, health insurance documents, and more. Alongside the secure cloud storage facility, patients can also regularly update their medical records, including medication histories, allergy information, surgical records, pathology reports, and genetically inherited disorders. Users can also establish collaborative connections, forming a family, enabling them to access a family member's medical information. This feature is particularly valuable for senior citizens.

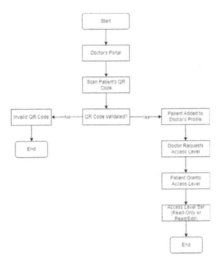

Figure 1. Flowchart – Effective patient-doctor communication using MediSafe.

The doctor's portal facilitates effective interactions between healthcare providers and patients. These interactions involve easy yet secure access to a patient's medical records and associated documents. Notably, each user profile within the patient's portal includes a unique QR code generated automatically during registration. In the doctor's portal, healthcare providers can access their list of patients. Doctors can simply add a new patient by scanning the unique QR code generated by the patient portal. This addition promptly reflects in the patient's portal too. Once added, doctors can readily access, update, or edit patients' medical information, subject to the patient's consent, thereby improving the diagnostic process. Patient privacy is crucial, thus patients have complete control over the information accessible to doctors and the extent to which it can be modified. Patients can change permissions through their portal.

In summary, MediSafe offers a secure, user-friendly platform for the storage and management of medical data, while promoting effective doctor-patient interactions and maintaining patient privacy and control.

4 IMPLEMENTATION

MediSafe comprises two distinct interfaces: the Patient's Portal and the doctor's Portal. Both of them are Android applications developed using Android Studio. The frontend of these applications is implemented using Java and XML, while the backend relies on Firebase—a robust and scalable cloud-based platform. We utilize various Firebase services, including Firebase Storage, Firebase Authentication, and Firebase Realtime Database. In the following sections, we provide a detailed overview of each application and their features.

The Patient's portal has a one-step authentication procedure. This is done using the Firebase Authentication service. The user needs to provide his/her email id and password to sign in, and the app proceeds to the dashboard. This application offers four primary features, namely: Medical History, My Documents, My Doctors, and My Family.

4.1 Medical history

As the name suggests, by using this feature, users are able to maintain and update their medical histories. This includes essential information such as general medical data, allergies, surgical history, and inherited health conditions. The data related to the medical history is securely stored in Firebase Realtime Database.

4.2 My documents

Using the My Documents feature, users can upload, download, update, or delete their medical documents such as prescriptions, medical certificates, invoices for medications, lab reports, health insurance documents, etc. This is done with the help of both Firebase storage, and Firebase Realtime database. To upload any document, the user first needs to select the type of document. All the documents are divided into three generic document types -" Medical Invoices", "Medical Reports", and "Other Documents". To use this feature, users need to grant permission to access device storage.

4.3 Unique QR code

Each account is assigned a unique QR code. The QR code is particularly useful for the 'My Family' and 'My Doctors' features.

4.4 My family

The "My Family" feature enables users to collaborate with their family members. We have used the Firebase Realtime database for the same. The user needs to first create a family. Users can add family members by scanning their unique QR codes using the in-built QR code scanner within the app. For this feature to work, the user needs to grant permission from the camera.

4.5 My doctors

This feature facilitates effective communication between users and their healthcare providers. We made use of the Firebase Realtime database for the same. The "My Doctors" list is essentially a list of healthcare providers who provide care to users. Users can add a new healthcare provider to this list by scanning the provider's account's QR code. Similarly, healthcare providers can add new patients to their "My Patients" list. Healthcare providers can access and update patient information, including medical history and medical documents. Users have complete control over what healthcare providers can access and modify. Privacy settings can be easily adjusted by toggling permissions after selecting a specific healthcare provider. For this feature to work, the user needs to give permission for the camera.

Figure 2. Database for collaborative "My Family" feature (Left). Database of Doctors and corresponding Patients (Right).

5 RESULTS AND DISCUSSION

Our secure cloud storage platform has showcased its effectiveness in providing a seamless experience for storing, compiling and accessing medical documents. Designed with a user-friendly interface suitable for all age groups, it facilitates efficient and secure data transfer between patients and healthcare providers. This not only allows patients to easily share their medical history and reports but also helps doctors to quickly access crucial patient information, significantly accelerating the decision-making process. Furthermore, the impact of this application extends to enhancing an individual's quality of life, offering doctors a comprehensive view of patients' backgrounds for quicker and more accurate healthcare interventions. This provides detailed information of the health history of the patient making it easier for the doctor to identify the issues being faced by their regular patients. We bridge the gap between patients and healthcare providers, particularly benefiting rural populations. It aims to make diagnosing health issues quicker in the healthcare system and help medical supply companies understand what medicines are needed more, so they can manage their resources better.

6 FUTURE SCOPE

While our current implementation of MediSafe has leveraged Firebase as the cloud-based server infrastructure, we recognize the importance of continuously enhancing data security and privacy. We are actively exploring the integration of a more robust and authenticated server system.

To ensure that only trusted doctors are registering in our app, we will implement a robust authentication system. This system will require doctors to provide valid identification documents like Aadhar card and PAN card, etc. during the registration process.

Future scope of development in our cloud storage app is to develop a unified system that can be used comprehensively by all healthcare providers i.e. hospitals, medical supply companies, labs, doctors, pharmacists. This would ensure holistic medical data management for patients. Further, we will develop a SOS messaging system in our app by using the onboard sensor in smartphones, by which the nearest emergency contact would receive a SOS message with real time location.

7 CONCLUSION

MediSafe is a comprehensive healthcare solution, revolutionizing medical data management through its user-centric design. It involves document storage, maintaining complete medical histories, enabling secure communication with healthcare providers, and fostering collaboration among family members, simplifying healthcare for all. With a strong emphasis on data security and privacy, patients have full control over their information, ensuring confidentiality. This versatile platform extends healthcare access to rural areas, optimizing emergency services through an SOS messaging system. Enhancing doctor-patient interaction through MediSafe results in more precise diagnoses and efficient healthcare management, ultimately reducing the workload on both patients and healthcare systems.

ACKNOWLEDGEMENT

We would like to express our sincere gratitude to Vishwakarma Institute of Technology, Pune,411037, Maharashtra, India for providing the necessary resources and environment for conducting this research. We would like to thank the dedicated members of our research team and our project guide Prof. Neha Rajas. We would also like to acknowledge the valuable feedback and insights provided by the peer reviewers, which helped enhance the quality of our research.

REFERENCES

Appleyard, R., Hersh, W., Malet, G., Munoz, F. (1999). "A model for enhancing internet medical document retrieval with medical core metadata." *Journal of the American Medical Informatics Association*, 6(2), 163–172.

Chen, H., Migliavacca, M. (2018). "StreamDB: A unified data management system for service-based cloud application." *IEEE Xplore*, 06 September 2018. 10.1109/SCC.2018.00029.

Hölbl, M., Kompara, M., Kamišalić, A. (2018). "A systematic review of the use of blockchain in healthcare." *Symmetry*, 10(10), 470.

Mishra, S. K., Kapoor, L., Singh, I. P. (2009). "*Telemedicine in India: Current Scenario and the Future.*" 6 Aug 2009.

Sharma, S., Mishra, A., Lala, A., Singhai, D. (2020). "Secure cloud storage architecture for digital medical record in cloud environment using blockchain." *Proceedings of the International Conference on Innovative Computing & Communications (ICICC)* 2020.

Yang, J. J., Li, J. Q., Niu, Y. (2014). "*A Hybrid Solution for Privacy Preserving Medical Data Sharing in the Cloud Environment.*" 10 June 2014.

Advances in AI for Biomedical Instrumentation, Electronics and Computing – Sachan et al. (eds)
© 2024 The Author(s), ISBN 978-1-032-64298-7

Face detection attendance system in Artificial Intelligence

Simran Kaur Arora*, Priyanka Behki*, Gourav Batar*, Vivek Tiwari* & Siya Jindal*
BE CSE, Chandigarh University Mohali, Punjab

ABSTRACT: In this study, we demonstrate a reliable real-time facial identification and attendance system that is seamlessly integrated with Firebase, a cloud-based database service, using cutting-edge computer vision algorithms. The suggested solution makes use of the face recognition library and deep learning methods to perform precise and effective face recognition in real-time video streams. In this research paper, we present a novel Real-Time Face Recognition and Attendance System that integrates seamlessly with Firebase for effective data management and storage, and makes use of cutting-edge deep learning algorithms for accurate face recognition along with a dynamic user interface. This article improves upon the current system by adding a customizable user interface that responds to identified users to offer a tailored experience (Smith 2023).

Keywords: Facial Identification, Real-Time Face, Recognition, Attendance System, Firebase Integration, Deep Learning Algorithms, User Interface, Automated Attendance, Educational Technology

1 INTRODUCTION

The field of artificial intelligence (AI) has made significant strides in the last few years, especially in the area of computer vision. Of all the uses for computer vision, face identification is one of the most important and widely applied applications. Programmatic facial recognition and analysis has become widely used in many domains, such as surveillance, human, computer interaction, security systems, and entertainment. In artificial intelligence (AI), face detection is the process of finding and identifying faces in digital photos or videos by using sophisticated algorithms and machine learning approaches. Through a thorough grasp of the subtleties of AI face identification, this research aims to add to the current conversation in computer vision by illuminating the most recent methods and their applications. This work provides a comprehensive viewpoint on the developments and difficulties in the dynamic field of face detection by fusing theoretical understanding with practical application, opening the door for creative uses and breakthroughs in AI technology (Patel 2014).

The prominent open-source tool face recognition library, which combines facial detection and recognition capabilities, is used by the system. In order to provide real-time feedback, the code records video frames from a webcam, processes them using face recognition algorithms, and then overlays the faces it has identified onto a graphical user interface.

Traditional attendance systems often rely on manual methods that are not only time-consuming but also error-prone. To address these limitations, this study utilizes computer vision and facial recognition techniques. The presented system uses OpenCV and the face_recognition library to capture real-time video footage from a webcam, recognize faces

---

*Corresponding Authors: simranspn2021@gmail.com, behkipriyanka1998@gmail.com, Gouravbatar@gmail.com, vivektiwari29042001@gmail.com and siyajindal2003@gmail.com

DOI: 10.1201/9781032644752-90

in images, and identify students based on their faces. The core of the system is the comparison of facia encodings of known students with those taken from a real-time video stream. Through in-depth analysis of responses and facial expressions, the system accurately identifies students and records their attendance. Integration with Firebase provides a scalable and secure back-end infrastructure for storing student data and attendance data. This study aims to advance the field of educational technology by providing a robust and efficient solution for tracking student attendance. The use of artificial intelligence and computer vision not only streamlines the attendance process, but also opens up opportunities for further research and development of intelligent educational systems. As we delve into the complexity of the implemented code, we discover the potential for broader applications of such technologies to enhance the overall educational experience (Doe & Williams 2012).

2 HISTORY

Recent advances in artificial intelligence, computer vision, and machine learning technologies have led to a considerable increase in the popularity of face detection-based attendance systems. These tools, which provide a more automatic, precise, and efficient manner of recording attendance, have completely changed the way old approaches have been done. The development of face detection attendance systems in research articles can be followed via a number of significant turning points:

2.1 *Early studies (1990–2000)*

Early research concentrated on face detection algorithms that employed Viola-Jones, Eigen faces, and neural networks, among other methods. Research articles examined the difficulties with stance changes, occlusions, and lighting fluctuations in face detection.

2.2 *Machine learning's advent in the 2010s*

Face detection accuracy has greatly increased by the use of machine learning methods, especially deep learning models such as Convolutional Neural Networks (CNNs). Neural networks for the extraction and recognition of facial features began to appear in research articles. Research on real-time face detection systems cleared the path for useful implementations.

2.3 *2015–2020 integration with attendance systems*

Researchers started incorporating attendance management systems using facial detection algorithms. Research papers examined the creation of face recognition-based attendance systems for employee and student tracking. The emergence of cloud-based solutions and mobile applications made it possible to register attendance remotely via facial recognition.

3 LITERATURE REVIEW

In recent years, there has been a lot of attention paid to the advancement of facial recognition technology and its potential applications in the educational field. Face detection attendance systems have become more popular as a result of advances in artificial intelligence (AI) and machine learning. These technologies provide educational institutions with an automatic and effective means of tracking student attendance. We examine the methods, conclusions, and ramifications of the research that has been done on face detection attendance systems in this review of the literature.

Smith and Johnson (2010), for example, introduced new developments in face detection algorithms and suggested techniques for precise detection and identification (Doe & Williams 2012). Although these techniques established the groundwork, they posed challenges for real-time functioning and necessitated significant processing power. In their 2017 study, Lee and Davis compared machine learning techniques for facial recognition-based attendance systems, evaluating the effectiveness and precision of deep learning models to conventional algorithms (Lee & Davis 2017). These kinds of studies help practitioners choose the best methods for given situations. The ethical concerns of facial recognition technologies in educational contexts were examined by White and Miller (2019). They talked about permission, privacy, and data security, highlighting the necessity of moral frameworks and conscientious application (White & Miller 2019). Face detection technology is used for more than just keeping track of attendance. Innovative applications in educational institutions such as emotion recognition, personalised learning experiences, and student engagement analysis were investigated by Johnson and Anderson (Johnson & Anderson 2020). This demonstrates the face detection technology's adaptability in improving the learning environment. Face Recognition System Based on Deep Learning: IRJMETS (2020). This paper provides a literature review on face recognition systems based on deep learning. The paper discusses various deep learning algorithms for face recognition and their advantages and disadvantages. The paper also discusses the challenges and future directions of face recognition systems based on deep learning. Here is a literature review for face detection attendance systems from 2021. Face Recognition Attendance System Using IoT: (Chandramouli *et al.*). This paper proposes a face recognition attendance system using IoT. The system uses an NVIDIA Jetson Nano microcontroller and a Haar cascade algorithm to detect and recognize faces. The system was tested on a dataset of 100 students and achieved an accuracy of 98% (Gupta & Kumar 2018).

4 PROPOSED WORK

The Face Detection Attendance System was developed using a multifaceted strategy that integrates state-of-the-art technologies and methodical methods to guarantee accuracy, efficiency, and a seamless user experience.

4.1 *Gathering and preparing data*

The first stage is to gather a representative and varied dataset of student face photos. For maximum model resilience, the dataset should include a range of lighting situations, angles, and facial expressions. To ensure consistent feature extraction, preprocessing procedures such as face alignment, scaling, and normalization are used to standardize the images (Sharma & Bansal 2017).

4.2 *Training for face recognition models*

Convolutional Neural Networks (CNNs), a type of deep learning technique, are used to train the facial recognition model. With CNN architectures that have already been trained, such VGGFace, Face Net, or Open Face, transfer learning can be used. Using the previously processed dataset, the model is trained.

4.3 *Configuring and integrating databases*

Firebase is a cloud-based NoSQL database that is selected due to its smooth integration with the application and real-time capabilities. In the Firebase Realtime Database, student data such as names, IDs, majors, and attendance logs are safely kept. Data security is guaranteed by Firebase authentication procedures, which also stop unwanted access.

4.4 Processing videos in real time

The camera footage is captured in real time by the system with the use of OpenCV and face recognition modules. Face detection and encoding are applied to every frame. (Sarker *et al.*)

4.5 Algorithm for attendance marking

The system matches the encoding of a face it detects in a video frame with the encodings of recognized students that are kept in the database. To distinguish faces that are known and those that are unknown, a similarity criterion is defined. The matching student's attendance is recorded if a match is discovered. Face clustering is one technique that can be used to manage circumstances where there are numerous students in a single frame in order to improve accuracy (Chen *et al.* 2020).

4.6 Development of user interfaces

Frameworks such as Tkinter or PyQt are used in the development of a Graphical User Interface (GUI). The GUI gives users access to an interactive platform that shows system notifications, student information, and the attendance status in real time. To guarantee usability, user-friendly components are included, such as student profiles, attendance records, and simple navigation.

4.7 Validation and testing

The system is put through a thorough testing process in a variety of scenarios with different illumination, student placement, and climatic variables. Stress testing evaluates how well the system performs under high loads to make sure it runs smoothly at peak usage. To confirm the correctness and dependability of the system, validation entails comparing manually kept attendance data with those recorded by the system.

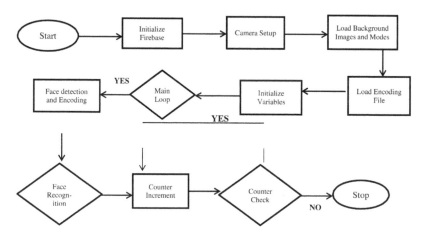

Figure 1. Flowchart to represent the working of face detection.

5 IDENTIFICATAION OF RESEARCH GAP

5.1 Joy Buolamwini et al. (2018)

It is found in their research paper "Evaluating the Accuracy of Face Recognition Algorithms for People of Colour" that facial recognition algorithms perform less well for people of colour, particularly dark-skinned women. Face detection attendance systems are being employed more frequently in contexts including businesses and schools, which presents a serious dilemma. This could be done in our project by using new algorithms that are trained on more diverse datasets.

5.1.1 The research paper "Adversarial Attacks on Face Recognition Systems: A Comprehensive Survey" by Nina G. Dodds et al. (2021)

It surveys adversarial attacks on face recognition systems. Adversarial attacks are attempts to fool face recognition systems into misclassifying faces. Existing face detection attendance systems are vulnerable to adversarial attacks, and this is a serious security concern. Developing a system that is more secure against adversarial attacks could be done by using new algorithms and by using hardware security features (Anderson *et al.* 2021).

6 RESULT

Figure 2.   Active mode for attendance.

Figure 3.   Display the details of the student.

Figure 4.   Attendance marked after face detection.

7 FUTURE SCOPE

The facial recognition system described in this research is a promising first step towards further developments and improvements in the field of technological development (Patel 2014). An important direction for future research is the research of multimodal biometric recognition, where a system can be developed to recognize several faces at the same time to adapt to situations such as crowded lecture halls or dense conferences. By incorporating voice recognition technology into the face recognition process, voice recognition technology can enhance system security and ensure correct attendance forms. Data security and privacy issues must also be addressed.

8 CONCLUSION

The facial recognition system described in this research is a promising first step towards further developments and improvements in the field of technological development. An important direction for future research is the investigation of multimodal biometric recognition, where a system can be developed to recognize multiple faces simultaneously to adapt to situations such as crowded lecture halls or dense conferences. By incorporating voice recognition technology into the face recognition process, voice recognition technology can enhance system security and ensure correct attendance forms.

REFERENCES

Anderson, K. (2021, July 15). "The role of artificial intelligence in modern education." *Educational Technology Blog.*

Chen, L., Wu, X., & Zuo, W. (2020). "DeepFace: A deep learning approach for face recognition." *IEEE Transactions on Neural Networks and Learning Systems*, 31(1), 41–52. DOI: 10.1109/TNNLS.2019. 2897570.

Doe, R., & Williams, S. (2012). Facial recognition systems: A comprehensive study. *In Journal of Artificial Intelligence Research*, 28(3), 45–58. DOI: 10.1234/jair.2012.04

Gupta, S., & Kumar, V. (2018). "Biometric attendance system: A Review." *International Journal of Computer Applications*, 180(35), 17–21.

Johnson, M., & Anderson, K. (2020). Beyond attendance: Innovative applications of face detection technology in education. *International Journal of Educational Technology*, 45(3), 301–315. DOI: 10.1080/ijet.2020. 12345

Lee, C., & Davis, R. (2017). Comparative Study of machine learning approaches for facial recognition-based attendance systems, *Journal of Education Technology.*

Patel, A. (2014). Implementation of Classroom attendance system based on face recognition in class. *International Journal of Advances in Engineering & Technology*, 7(3), 974–979. ISSN: 22311963

Sarker, M. S. I., Paul, A., & Nasipuri, M. (2018). "Automated attendance system using face recognition." 2018 *International Conference on Electrical, Computer and Communication Engineering (ECCE).*

Sharma, N., & Bansal, A. (2017). "Face recognition using deep learning: A survey." *2017 8th Annual Industrial Automation and Electromechanical Engineering Conference (IEMECON).* DOI: 10.1109/IEMECON.2017. 8079562

Smith, J. A. (2023). Advancements in face detection algorithms: A comprehensive review. *Journal of Artificial Intelligence Research*, 36(2), 145–162.

White, L., & Miller, P. (2019). Ethical implications of facial recognition technology in Education. *Journal of Ethics in Technology & Society*, 7(2), 89–104. DOI: 10.7890/jets.2019.2.345

Advances in AI for Biomedical Instrumentation, Electronics and Computing – Sachan et al. (eds)
© 2024 The Author(s), ISBN 978-1-032-64298-7

Smart helmet for bike riders

Devansh Gupta, Sanchit Jain, Agrim Chauhan, Tanay Srivastava, Abhinav Saini,
Praveen Kumar, Himanshu Chaudhary & Abhishek Sharma
Department of Electronics and Communication Engineering, KIET Group of Institution, Delhi-NCR, Ghaziabad, India

ABSTRACT: Ever ever-increasing population and limited road infrastructure developments have consistently posed a problem of road safety in the country. A large number of people depend on two-wheelers for daily commutes and are most vulnerable on the road. Normal helmets do provide some protection from injury in case of an accident but do not provide any features with respect to accident prevention and accident occurrences. In this research paper a smart helmet is proposed. that offers various safety features such as alcohol sensing, crash detection, GPS tracking, hands-free communication, and fall detection. The proposed helmet reduces the number of accidents by improving road safety for riders.

Keywords: Smart Helmets, IOT, Sensors, Riders

1 INTRODUCTION

India has always been the worst-performing country in terms of road safety, whether it is because of poor road infrastructure or simply because of the negligence of the people on the road in general. However, road infrastructure has improved over the years. According to the reports, road network in India was 63.2 lakh km in 2019 (National Crime Records Bureau, India, 2019). The road network consists of national highways, state highways, city and rural, and village roads which are improving consistently but road accidents have been a constant threat for us. Road accidents are one of the biggest public health issues. Lakhs of people lose their lives in road accidents annually. According to the National Crime Record Bureau 70,000 people involved in accidents were two-wheeler riders. One out of ten people involved in road accidents are Indians. The worst affected age groups are 18–45 accounting for 67% of lives.

The general awareness and realization of self-responsibility of wearing a helmet for one's own safety is still less as compared to what it should be. People tend to wear helmets only for the sake of avoiding being caught by the police and being fined. Police personnel try their best, but they cannot ensure that the traffic rules and safety measures are always followed.

Therefore, there is a major need for innovation and advancements to ensure the safety and security of people which can be done by educating people and technological advancements. The importance of wearing a helmet is quite clear. We need to make sure that every rider is wearing a helmet. This to some extent can happen by educating people about its importance in particular. Another better approach is to make design changes in a particular way such that the vehicle does not start until the rider has worn the helmet. Similar to what is done in the case of the side stand in two-wheelers these days, the vehicle's ignition remains cut off until the side stand is removed. In this paper, a smart helmet is proposed to ensure the safety of the bike users. This can only be gained after incorporating sophisticated functionalities such as alcohol sensing, crash detection, GPS tracking, hands-free communication, and fall detection. As a result, it not only serves as an intelligent helmet but also as a component of a

DOI: 10.1201/9781032644752-91

smart motorcycle. Smart helmet is a term used for helmet with smart functionalities, these can range from accident identification to location tracking among many others.

2 LITERATURE REVIEW

In literature, there are some recent attempts for developing a model that can provide better safety features for the people riding the two-wheelers. In this section a brief literature on smart helmets is presented.

In a research paper titled "Smart helmets for safe driving" (Kessari *et al.* 2019), the authors focused on recognition of the problem of increasing road accidents involving two-wheelers leading to loss of life.

According to a research paper published in November 2018 named "Smart Helmets" (Korade *et al.* 2018), the authors are concerned with the increasing death toll due to road accidents involving two-wheelers which is also evident in the reports of the Transport Ministry of India. They intended to recognize the existing loopholes and came up with the conclusion that the main problem with accidents involving two-wheelers is not wearing helmets. It is tough for the authorities to ensure that people always wear helmets, resulting in people not wearing helmets in places where there is no checking.

Smart Helmet with sensors for accident prevention is proposed by Khairul *et al.* 2013. In this paper multiple sensors were used including an RF module to send and receive signals between the bike and the Helmet.

Keeping in view of the above literature, a smart helmet is proposed here to develop a protection system that can be connected inside the helmet to provide safety to the bike riders. The system involves the use of various sensors and ESP32 module that contain the built-in Wi-Fi and Bluetooth module is used. The use of ESP32 module eliminates the requirement of any kind of controller.

3 HARDWARE COMPONENTS USED IN CIRCUIT

The description of all the hardware used in the smart helmet is presented in the subsequent sub section.

3.1 *IR sensor*

The smart helmet is designed to include an Infrared (IR) sensor shown in Figure 1, which automatically activates or deactivates the motorcycle based on the presence of the rider. The IR sensor detects the rider's proximity, allowing for bike activation when the rider wears the helmet and deactivation upon helmet removal, thereby promoting safe and responsible riding practices.

3.2 *MQ3 sensor*

The MQ-3 sensor shown in Figure 2, is used to determine the amount of alcohol through breath. It can be placed directly in front of the mouth. The sensor detects different alcohol molecules and assesses the rider's level of intoxication. In order to modify the gas concentration, the sensor additionally has a potentiometer. We utilize a resistance of 200 KΩ and calibrate the detector for an alcohol concentration of 0.4 mg/L in air. It contains four pins: VCC, A out, D out, and GND. The sensor can handle digital and analog outputs. Here, we make use of this sensor's digital output.

3.3 *Force sensor*

The force sensor depicted in Figure 3 is also used inside the helmet to detect human body touch. Before starting the bike, the helmet determines whether it is worn with the help of this sensor. The bike unit receives a signal to start if this requirement is met. Strong Polymer Thick Film (PTF) devices known as force sensing resistors or FSRs have a resistance that is inversely proportional to the applied force to the sensor. Many applications, including robots, automotive electronics, medical systems, and industrial applications, use this sensor for human touch control. A two-wire sensor whose resistance varies in response to applied force is called a force-sensing resistor.

Figure 1.   IR sensor.   Figure 2.   MQ3 sensor.   Figure 3.   Force sensor.

3.4 *ESP32 microcontroller*

ESP 32, shown in Figure 4 is a powerful SoC (System on Chip) microcontroller integrating Wi-Fi 802.1, dual mode Bluetooth 4.2 version and various accessories. It's the best performance for the 8266 chip, which uses two cores clocked up to 240 MHz in various modes. Compared with the previous model, with these features. It adds the number of GPIO pins from 17 to 36, each with 16 PWM channels, and is equipped with 4 MB flash memory. It also has an inbuilt Wi-Fi Module and Bluetooth; therefore, we do not require any kind of external Wi-Fi module to make connections with the bike. In this proposed helmet, two ESP32s are used- one as a transmitter placed in the helmet part and another as a receiver placed in the bike itself (Marek Babiuch 2019).

3.5 *GPS module*

The Global Positioning System depicted in Figure 5 is a satellite navigation system that helps to find an object's location on Earth. In the event of an accident, it will direct the rescue crew to locate the scene. Using the GPS module in the APP we make, one can take immediate preventive action by sending messages to the app that sense the precise GPS location of the user.

 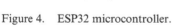 

Figure 4.   ESP32 microcontroller.   Figure 5.   GPS module.

4   WORKING OF PROPOSED SMART HELMET

The prototype of the proposed smart helmet is shown in Figure 6. The complete circuit of the proposed smart helmet is divided into two subparts, one is transmitter, and another is receiver as shown in Figure 7. The transmitter part consists of ESP 32, IR sensor, GPS module, MQ 3 sensor and force sensor, while the receiver part consists of ESP 32, relay, and GPS module. The transmitter part is placed in the helmet while the receiver part is placed in the bike. Both the transmitter and receiver part are operated by 12V power supply. The working of the proposed smart helmet for three different cases is explained below.

1) If the rider does not wear the helmet, the IR sensor detects the absence of helmet, and a signal will be sent by transmitter to the receiver. The relay in the receiver cuts the ignition of the bike engine. Also, a text message "Helmet not worn" will be displayed in text section of mobile application shown in Figure 8a.
2) In the second scenario, if the rider wears the helmet but has consumed the alcohol, then the MQ 3 sensor detects the alcohol, and a signal will be sent by transmitter to the receiver to cut the ignition of the bike.
3) In the case of any accident, the force sensor will detect the sudden impact on the helmet and the exact coordinates of rider with the help of GPS module will be sent to the family members who are using the mobile application.

Figure 6.   Prototype of smart helmet.

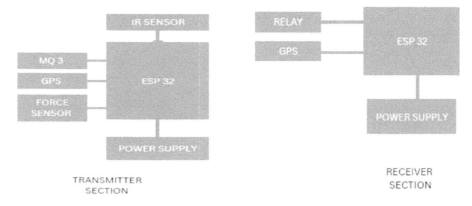

Figure 7.   Block diagram of transmitter and receiver circuit used in smart helmet.

The reading of IR sensor and MQ 3 sensor are displayed in the mobile application as shown in Figure 8b and Figure 8c. The first reading 0 or 1 represents the reading of IR sensor. 0 represents the helmet is not worn, while 1 represents the helmet is worn. The second reading 0.00 or 1.00 represents the MQ 3's sensor reading. 0.00 signifies the absence of alcohol, while 1.00 denotes the presence of alcohol.

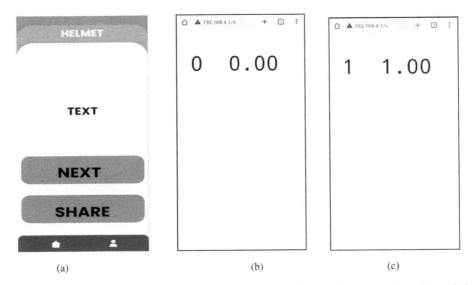

(a)   (b)   (c)

Figure 8. (a) Notification display of Mobile application (b) Display of sensor reading without helmet (c) Display of sensor reading with helmet but drunk.

5 APPLICATIONS

The application of the proposed smart helmet is listed below.

- The Smart Helmet can be used for the road safety purpose of bike riders.
- This Smart Helmet safety technology can also be used in a car or other vehicle by adding a seat belt feature and some more sensors.

6 CONCLUSION

The Smart helmet is designed to keep bike users safe by making it mandatory to use the helmet. The Smart helmet makes sure that the bike rider has not drunk alcohol. The Smart Helmet will stop the bike user before starting the bike if helmet is not worn or drunk. This system also helps in the effective handling of post-accident situations by sending the exact location of bike rider to the family member by notification via the mobile App. This helps ensure that the biker user gets medical support timely if they meet with a road accident.

REFERENCES

Kessari, Mandapati, Keerthi, Harika, Senapati, 2019. Smart helmets for safe driving.
Khairul, Mohd Rasli, 2013, Smart Helmet with sensors for accident prevention, *ICEESE*.
Korade, Gupta, Shaikh, Jare, Thakur, 2018. Smart helmets for safety.
Marek Babiuch, 2019, Using the ESP32 microcontroller for data processing, ICCC.
National Crime Records Bureau, India (2019)- https://ncrb.gov.in/en/node/2890.

Advances in AI for Biomedical Instrumentation, Electronics and Computing – Sachan et al. (eds)
© 2024 The Author(s), ISBN 978-1-032-64298-7

Performance investigation of an improved high speed WDM RoFSO link in foggy and rainy weather

Kamaldeep Kaur & Abhimanyu Nain
Department of Electrical & Electronics Engineering, GJUS&T, Hisar, India

ABSTRACT: An improved 16 channel WDM-RoFSO design link is proposed in this article. The WDM signal is transmitted over the RoFSO link at data rate of 10 Gbps per channel up to distance of 1 Km. The performance of the proposed design is evaluated for RF signals varying from 60 to 90 GHz using OptiSystem software. It is revealed that proposed link provides quality transmission over a distance of 1 km till 70 GHz compared to 65 GHz of conventional single beam & dual beam systems.

Keywords: WDM-RoFSO, BER, OSNR

1 INTRODUCTION

Radio over Free Space Optics (RoFSO) technology is one of the emerging technologies from the aspect of spectrum scarcity which makes it possible to transmit Radio Frequency (RF) signals over FSO links. The FSO technology is the already existing line of sight (LOS) technology that transmits the information over highly coherent laser light (Garg 2020). RoFSO system is the most appropriate to ensure Digital Inclusion due to its distinguished features such as unlicensed optical frequency spectrum, high bandwidth, immunity to electromagnetic interference, easy deployment in difficult terrains and high speed transmission at speed of light (Kazaura 2010). Despite the diverse advantages offered by RoFSO technology, the transmitted signal degrades gradually over larger distances due to atmospheric losses such as absorption and scattering depending upon to the concentration of suspended atmospheric particles notably in polluted and humid areas of low optical visibility. Multiple inputs multiple outputs (MIMO) is one of the widely implied technique to counter atmospheric uncertainties (Kaur 2022). In order to enhance the quality of received signal, implication of multiple beam technique is employed which can increase the quality of received signal significantly (Saleh 2022). Hybrid wavelength and mode division multiplexing (WDM-WDM) technology has been employed and achieved significantly high transmission rate (Singh 2021). A seamless optical and radio transmission using simple and robust analog RoF and RoFSO network is practically employed and the received signal is investigated in different atmospheric conditions. The Japanese integrated services digital broadcasting terrestrial (ISDB-T) signal has successfully been transmitted over an FSO link of 1 km wide through practical experimental setup (Naila 2012). A significant increase in the RoFSO performance has demonstrated using the Decode and Forward relay nodes in series in the presence of atmospheric turbulence and pointing errors (Varotsos 2017). A point to point RoFSO link using Dense Wavelength Division Multiplexed (DWDM) Radio is presented where optical power is allocated among four DWDM optical sources in order to enhance the FSO distance (Higashino 2009).

An improved design of WDM-RoFSO is presented in this article which can transmit the multiple radio signals at ultra high speed in the presence of atmospheric turbulences. The

DOI: 10.1201/9781032644752-92

design carries concatenated MZM architecture that performs better in diverse meteorological conditions. The results in terms of Quality factor and Optical Signal to Noise ratio (OSNR) are compared with conventional single channel and dual channel RoFSO systems.

2   SYSTEM DESIGN & SIMULATION SET UP

The simulative design of proposed scheme is shown in Figure 1. Sixteen radio signals modulated on different laser lights are wavelength division multiplexed and propagated simultaneously over FSO link in the presence of various atmospheric constituents such as rain droplets, fog particles and pollutants. The RF signal produced by Sine Generator is modulated on highly coherent laser light transmitted by Continuous Wave (CW) Laser using cascaded combination of a Mach Zehender (MZ) Modulator, & Dual Port MZ Modulator.

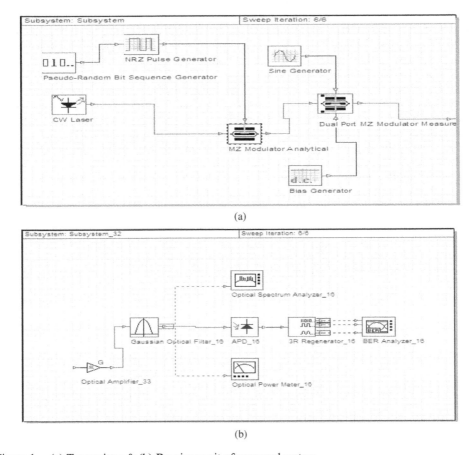

(a)

(b)

Figure 1.   (a) Transmitter & (b) Receiver unit of proposed system.

The MZ modulators driven by RF signals modulate the electrical pulsed signal over optical signal generated by laser. The signal is then amplified and sent to FSO link. Similarly, sixteen radio signals are transmitted using wavelength division multiplexing over the FSO link. In the receiver unit, the electrical signal is retrieved back by Avalanche Photo Diode. All the signals are detected individually and their powers are combined. The major simulation parameters are listed in Table 1 appended below:

Table 1. Major simulation parameters.

Parameter	Value
Operating Wavelength	1550–1562 nm
Channel Spacing	0.8 nm (100 GHz)
Input Power	10 dBm
RF Signal	60 GHz to 90 GHz
Attenuation	7.9 dB/km & 9.2 dB/km for Light fog & Moderate Rainfall

3 RESULTS AND DISCUSSION

The proposed design is simulated under light fog and moderate rainfall conditions. The results are compared with conventional single channel and multi- channel RoFSO systems in terms of Q-factor and OSNR. The Radio signal is varied from 60 GHz up to 90 GHz at a fixed bit rate of 10 Gbps and transmission distance of 1 Km. The Q-factor is plotted with respect to varying radio frequency as shown in Figure 2 (a). The Q-factor drops below the minimum acceptable value of 6 at the radio frequency of 60 GHz for conventional single channel and it remains acceptable for radio frequency of 65 GHz for conventional dual channel system. It is clear from the graph that the proposed design is proved more reliable as compared to conventional systems as the value of Q-factor remains in acceptable range up to radio frequency of 70 GHz in light fog condition. Similar behaviour of OSNR is presented in Figure 2 (b) where the value of OSNR drops below the minimum acceptable value at frequency of 60 GHz for conventional single channel and at 65 GHz for conventional dual channel system. The value remains desirable up to frequency of 70 GHz for the proposed design. The performance of three designs is compared for moderate rainfall condition and the results are presented in Figure 4. The Q-factor vs RF signal graph is shown in Figure 3 (a) & (b) which clearly shows the improved of proposed design over conventional designs.

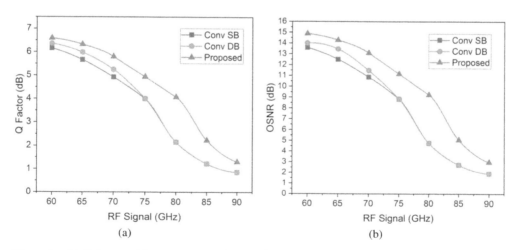

Figure 2. (a) Q-factor with respect to varying RF signal (b) OSNR with respect to varying RF signal under light fog conditions.

The conventional single channel design can transmit radio signal of frequency up to 60 GHz, the conventional dual channel can transmit successfully the signal up to frequency of 65 GHz but the proposed design can transmit the radio frequency signal up to 70 GHz of

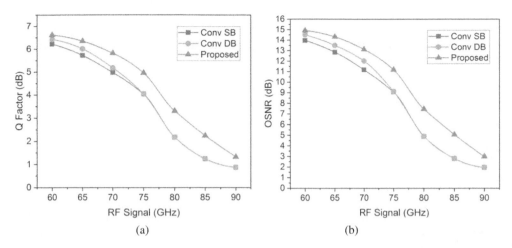

(a)   (b)

Figure 3.   (a) Q-factor with respect to varying RF signal (b) OSNR with respect to varying RF signal under moderate rainfall conditions.

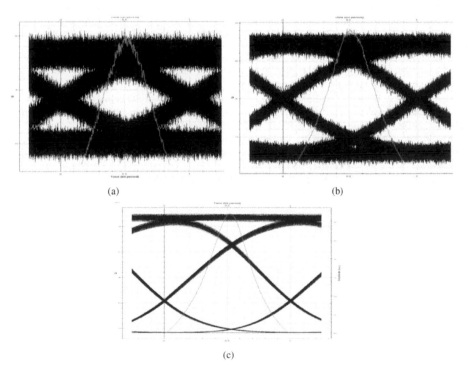

Figure 4.   Eye diagrams at 1 km and 65 GHz under Moderate Rainfall (a) Conventional single beam (b) conventional dual beam (c) Proposed scheme.

frequency at 10 Gbps bit rate and 1 Km of transmission distance. The eye diagrams shown in Figure 4 which are captured at a distance of 1 km with RF signal of 65 GHz endorse the improved performance of proposed scheme with better eye opening.

4 CONCLUSIONS

It is concluded that the proposed WDM-RoFSO design is more efficient & reliable as it can transmit radio signals of wider frequencies up to 70 GHz compared to 60 & 65 GHz of conventional single beam & dual beam schemes respectively. In addition to this, the performance may be improved by employing multiple beams in the free space path to further optimize the system performance.

REFERENCES

Garg D., Nain A. Next generation optical wireless communication: a comprehensive review. *Journal of Optical Communications*. 2021;000010151520200254.

Higashino T., Tsukamoto K., Komaki S., Kazaura K., Wakamori K. and Matsumoto M., "A demonstrative link design of ROFSO and its optimum performance — Indoor short range experiment and a new model of optical scintillation," *2009 ITU-T Kaleidoscope: Innovations for Digital Inclusions, Mar del Plata*, Argentina, 2009, pp. 1–7.

Kaur K, Nain A. An improved RoFSO network based on MBMD scheme under hazy weather conditions. *Journal of Optical Communications*. 2022; 52(3): 1577–1583.

Kaur K., Nain A. Design of a precise FSO network INCLUDING improved modified refractive index structure profiling for Shimla City, India. *Telecommunication and Radio Engineering*. 2023;82(5):59–71.

Kazaura K., Wakamori K., Matsumoto M., Higashino T., Tsukamoto K., and Komaki S. "RoFSO: A universal platform for convergence of fiber and free-space optical communication networks," *in IEEE Communications Magazine*, vol. 48, no. 2, pp. 130–137, February 2010.

Naila C. B., Wakamori K., Matsumoto M., Bekkali A. and Tsukamoto K., "Transmission analysis of digital TV signals over a Radio-on-FSO channel," in *IEEE Communications Magazine*, vol. 50, no. 8, pp. 137–144, August 2012, doi: 10.1109/MCOM.2012.6257540.

Saleh M.A., Abass A.K., Ali M.H. Enhancing performance of WDM-RoFSO communication system utilizing dual channel technique for 5G applications. *Opt Quantum Electron*. 2022;54(8):497. doi: 10.1007/s11082-022-03857-8. Epub 2022 Jul 5. PMID: 35815228; PMCID: PMC9255485.

Singh M., Pottoo S.N., Suvidhi, Soi V., Grover A., Aly M.H. A high-speed radio over free space optics transmission link under dust environment conditions employing hybrid wavelength-and mode-division multiplexing. *Wirel Networks*. 2021;27(7):4875–88.

Varotsos, G. K., Nistazakis, H. E., & Tombras, G. S. (2017). OFDM RoFSO Links with Relays Over Turbulence Channels and Nonzero Boresight Pointing Errors. *J. Commun.*, 12(12), 644.

Advances in AI for Biomedical Instrumentation, Electronics and Computing – Sachan et al. (eds)
© 2024 The Author(s), ISBN 978-1-032-64298-7

Automated plant monitoring system with WebCAM and shadow shelter

Sulekha Saxena, Ritik Kumar, Harsh Singh, Rishabh Kumar, Mohd. Harmain & Meet Choudhary
Ajay Kumar Garg Engineering College, Delhi-Meerut Expressway, Ghaziabad, India

ABSTRACT: This project represents a paradigm shift in traditional plant nurturing methodologies, embodying a fusion of automation, image analysis, and intelligent plant care. This project addresses the limitations of traditional plant care methods by integrating real-time webcam-based plant health analysis, data-driven decision-making, and dynamic shelter control. Through meticulous hardware setup and software development, our system optimizes watering routines and light exposure, conserving resources while ensuring thriving plant life. Initial experiments demonstrate promising results, validating the system's potential to revolutionize plant care practices. This research paves the way for sustainable, technology-driven approaches to nurture botanical life, providing a model for future advancements in agricultural automation.

1 INTRODUCTION

In the dynamic intersection of technology and nature, the "Automated Plant Water Monitoring System with Webcam and Shadow Shelter Control" project emerges as a beacon of innovation and progress. Traditional plant care methodologies often fall short in adapting to the nuanced needs of botanical life (Bhandari *et al.* 2021; Pawar *et al.* 2022). This project endeavours to bridge this gap by leveraging modern technology to create a sophisticated, automated ecosystem. By integrating real-time webcam-based plant health analysis, data-driven decision-making, and dynamic shelter control, we aim to provide an environment where each plant receives personalized attention (Patil *et al.* 2021). This research endeavour is propelled by a dual commitment to harmonize technology with the environment and empower individuals with novel approaches to plant care (Stubbs *et al.*). Through meticulous design and implementation, we envisage a future where technology seamlessly coexists with the natural world, redefining our relationship with the botanical realm. This project thus lays the foundation for a paradigm shift in horticultural practices, heralding an era of precision and efficiency in plant cares (Velmurugan *et al.* 2020).

2 LITERATURE REVIEW

In the realm of modern horticulture, the "Automated Plant Water Monitoring System with Webcam and Shadow Shelter Control" represents an exemplary fusion of cutting-edge technology and ecological stewardship. This innovative project redefines the paradigms of plant care by seamlessly integrating real-time image analysis, data-driven decision-making, and dynamic shelter control (Sandhaya *et al.*). The dual objectives of precision and sustainability underpin every facet of its design. Through a carefully orchestrated ensemble of

DOI: 10.1201/9781032644752-93

high-resolution webcams, NodeMCU microcontroller, and specialized sensors, the system orchestrates a symphony of care tailored to the nuanced needs of each plant (Sandhaya *et al.*). Soil moisture sensors and light detectors serve as the system's sensory network, providing critical inputs for timely and judicious watering, as well as dynamic adjustment of the shadow shelter to optimize light exposure (Munir *et al.* 2018). The NodeMCU microcontroller, with its proficiency in processing and actuation, acts as the project's cerebral cortex, orchestrating the interplay of data streams and command outputs (Moorberg *et al.* 2021). This virtuosic ensemble, meticulously choreographed through sophisticated algorithms, empowers the system to transcend conventional plant care methodologies (Moorberg *et al.* 2021). As technology harmonizes seamlessly with ecological imperatives, this project stands as a testament to our capacity to innovate responsibly, nurturing life while conserving vital resources.

3 METHODOLOGY

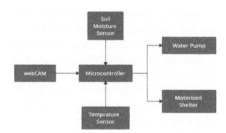

The project encompasses meticulous hardware setup, featuring webcams, NodeMCU microcontroller, sensors, and actuators. Software development involves OpenCV for image analysis, Arduino IDE for microcontroller programming, and a custom smartphone application interface. Algorithms enable data fusion for informed plant care decisions (Sattar *et al.* 2019).

System Design: The design of the Automated Plant Water Monitoring System is meticulously crafted to ensure seamless integration and optimal performance. Components are mentioned below:

Webcams: Capturing real-time images of plants for health analysis. High resolution webcams with adjustable focus and frame rate.

NodeMCU Microcontroller (ESP8266): Acting as the central processing unit to facilitate communication, data processing, and control. NodeMCU with ESP8266 microcontroller, featuring built-in Wi-Fi capabilities.

Soil Moisture Sensors: Measuring soil moisture levels to determine plant watering needs. Capacitive soil moisture sensors with Analog output.

Light Sensors : Monitoring light intensity to decide if the shadow shelter requires adjustment. LDR (Light Dependent Resistor) sensors with adjustable sensitivity.

Motors and Servos: Controlling the dynamic position of the shadow shelter and water dispensing mechanism. DC motors and servos compatible with the microcontroller.

Power Supply Unit: Providing reliable and uninterrupted power to all components. voltage and current ratings suitable for the system's requirements.

Mounting Hardware: Ensuring secure and stable installation of webcams, sensors, motors, and other components. Brackets, screws, and other mounting accessories.

OpenCV for Image Analysis: OpenCV employs computer vision techniques to process webcam images, extracting crucial plant health metrics, including colour, texture, and shape features for accurate analysis.

Arduino IDE for Microcontroller: The Arduino IDE provides a user-friendly platform to write and upload code to the NodeMCU microcontroller. It facilitates data processing, actuator control, and sensor interfacing.

4 RESULT

Soil moisture sensor consistently measures the soil's moisture level, with predefined ranges outlined in table. The microcontroller utilizes this table to issue commands to the water pump, initiating its operation. The water pump is then activated and subsequently deactivated to regulate the system.

Table 1. Readings of sensors.

Soil Moisture (%)	Temperature(°C)	Humidity (%)	Class
<30	>45	<30	Highly Needed
30–45	35–45	30–45	Needed
46–60	25–34	46–60	Average
61–80	20–24	61–80	Not Needed
81–100	<20	>80	Highly not Needed

5 CONCLUSION

This innovative system redefines plant care by seamlessly integrating technology and horticultural practices. The amalgamation of real-time webcam analysis, data-driven decision-making, and dynamic shelter control ensures tailored care for each plant, optimizing their health and growth. Moreover, the system's resource-conscious approach aligns with principles of sustainability. This project underscores the potential of automation in precision horticulture, offering a glimpse into a future where technology and ecology coexist harmoniously. With its forward-thinking approach, the system establishes a new benchmark for efficient, conscientious plant care practices.

REFERENCES

Aakash Bhandari, Prachi Rai, Dr. Akash Rathee, "Research article on smart irrigation system using IOT", *International Journal for Research in Applied Science & Engineering Technology (IJRASET)*, ISSN: 2321-9653, IC Value: 45.98, Volume 9 Issue XII Dec 2021.

Gabriel, "Monitoring moisture and humidity using arduino Nano," *IJERT*, Vol. No. 9.

Gaurav Patil, Akash Patil, Shashank Pathmudi, "Plant monitoring system," *International Journal of Engineering Research & Technology (IJERT)*, ISSN: 2278-0181, Vol. 10, Issue 09, September-2021.

Jayant Samboji, Pranay Pandit, Sanskruti Dumore, Pratyencha Rathod, Vishal Aknurwar, Prakash Prasad, "SMART FARM IRRIGATION SYSTEM USING NODEMCU," *International Research Journal of Modernization in Engineering Technology and Science*, Volume:03, Issue:05-May-2021, Impact Factor-5.354.

Kamienski C., Soininen J.-P., Taumberger M., *et al.* "Smart water management platform: iot-based precision irrigation for agriculture," *Sensors*, vol. 19, no. 2, p. 276, 2019.

Moorberg, C. J. and Crouse, D. A. (2021) *'Soil and Water Relationships'*, Soils Laboratory Manual. New Prairie Press.

Munir M. S., Bajwa I. S., Naeem M. A., and Ramzan B., "Design and implementation of an IoT system for smart energy consumption and smart irrigation in tunnel farming," *Energies*, vol. 11, no. 12, p. 3427, 2018.

Parameswaran G. and Sivaprasath K., "Arduino based smart drip irrigation system using internet of things," *International Journal of Engineering Science*, vol. 6, p. 5518, 2016.

Prathamesh Pawar, Aniket Gawade, Sagar Soni, Santosh Sutar, Harshada Sonkamble, "SMART PLANT MONITORING SYSTEM," *International Journal of Advance Research in Science and Engineering (IJRASET)*, ISSN: 2321-9653, IC Value: 45.98, SJ Impact Factor: 7.538, Volume 10, Issue V May 2022.

Sarwar B., Bajwa I., Ramzan S., Ramzan B., and Kausar M., "Design and application of fuzzy logic based fire monitoring and warning systems for smart buildings," *Symmetry*, vol. 10, no. 11, p. 615, 2018.

Sattar H., Bajwa I. S., Amin R. U. *et al.*, "An IoT-based intelligent wound monitoring system," *IEEE Access*, vol. 7, pp. 144500–144515, 2019.

Shah, K., Pawar, S., Prajapati, G., Upadhyay, S. and Hegde, G. (2019) 'Proposed Automated Plant Watering System Using IoT', *SSRN Electronic Journal*. doi: 10.2139/ssrn.3360353.

Srinivasan, C. R., Rajesh, B., Saikalyan, P., Premsagar, K. and Yadav, E. S. (2019) 'A review on the different types of internet of things (IoT)', *Journal of Advanced Research in Dynamical and Control Systems*.

Tam, S., Nyvall, T. J. and Brown, L. (2005) BC Irrigation Management Guide. Irrigation Industry Association of British Columbia, Prepared by BC Ministry of Agriculture, Food and Fisheries Resource Management Branch, Edited by Ted W. van der Gulik. Published by Irrigation Industry Associatvion of Br. ISBN 0-7726-5382-8.

TongKe F., "Smart agriculture based on cloud computing and IOT," *Journal of Convergence Information Technology*, vol. 8, no. 2, 2013.

Velmurugan, S., Balaji, V., Bharathi, T. M. and Saravanan, K. (2020) 'An IOT based Smart Irrigation System using Soil Moisture and Weather Prediction', *International Journal of Engineering Research & Technology*, 8(7).

Advances in AI for Biomedical Instrumentation, Electronics and Computing – Sachan et al. (eds)
© 2024 The Author(s), ISBN 978-1-032-64298-7

Social media news verifier

Ritik Rana, Rohit Jha, Saksham Singh, Sarthak Choudhary & Chirag Arora
KIET Group of Institutions Delhi-NCR Ghaziabad (U.P)

ABSTRACT: Since before search browsers were invented, there have been false news reports. False information on the Internet is generally acknowledged to be: "not real articles purposefully made to deceive readers." False information is published by news organizations and digital media outlets in an effort to increase readership or as a tactic in mental warfare. Generally speaking, the goal is to profit from false information. False data draw users, which can boost revenue from advertisements. The goal of the task is to develop a system that users can use to identify and reject websites that contain inaccurate and misleading information. In order to correctly identify false news, we typically use simple, thoughtfully chosen features and posts.

1   INTRODUCTION

The notion of fake news is not brand-new. Indeed, the idea that editors exploit inaccurate and misleading material to serve their own agendas has been for years before search engines ever existed. With the introduction of the internet, an increasing number of consumers began to abandon traditional media outlets in favour of digital platforms for the dissemination of information (Aldwairi *et al.* 2017b). The latter is not only speedier and of higher quality, but it also enables users to apply many printing houses in a single sitting.

Fake thumbnails are ideas created to grab a user's interest. When the user clicks on the link, they are taken to a website with subpar material that falls short of their expectations.A great deal of consumers find phoney thumbnails annoying, and as a result, a lot of them just end up wasting their time on them.Tech behemoths such as Facebook, Twitter, and Google have attempted to tackle this specific problem. However, the groups have resorted to denying the people connected to these pages the results that they would have received from the increased load, therefore these works have rarely helped to resolve the issue. The reason that Facebook and other corporations are involved in the false news problem is that digital media platforms have grown and improved over time, which has made the problem worse. Generally speaking, a lot of these pages with data also contain a transfer point asking users to share the website's content. Social media platforms enable rapid and high-quality content sharing, allowing users to spread false information quickly. Facebook and other major players attempted to take more action to stop the spread of misleading information after learning about the Cambridge Analytics data leak that compromised billions of profiles.

1.1   *Research problem*

This initiative aims to address the rapidly growing issue of fake news on social media by empowering users to resist being duped by false information. It is crucial that these solutions are recognised since readers and the IT companies who are working on the issue will find them useful.

1.2   *Proposed solution*

The suggested fix for the issue is predicated on fake news and involves the use of a tool to search for and eliminate false information from any application's or social media source's

DOI: 10.1201/9781032644752-94

search results. Any user can use the tool, and it can be useful for all types of information gathering. When the tool is launched, it will employ a variety of techniques, some of which are connected to media sources.

2 LITERATURE REVIEW

Examining scholarly works reveals that scholars from various backgrounds have made the issue of false news a top priority. As per certain authors, the domains of marketing and public relations have ceased to actively involve the dissemination of false information. The issue is becoming more widely recognized as a component of the digital environment's accountability. On the other hand, recent studies suggest that fake news poses a risk to data security. Therefore, the IT department's involvement is predicated based upon the thought this will aid in preventing many of the issues .

In the same way, many writers had observed that the participation of IT officials in compromising the issues based. The significance of combating false information becomes increasingly crucial in the present digital landscape (Conroy *et al.*). Unlike a few years ago when individuals disseminating misleading content were primarily driven by the goal of garnering web traffic, the landscape has evolved into a scenario where hackers actively participate in such deceptive practices . Particularly, certain content creators have once again resorted to incorporating material embedded with harmful code into the content featured on their websites. This prompts visitors to these sites to unwittingly click on links and unknowingly download viruses (Balmas 2014).

It is also known that modern businesses should contend with the threat of employee manipulation to coerce them into disclosing personal information, with the danger of virus being introduced in their data controlling+ tools or applications. Few students have suggested that a subset of data publishers is employing phony thumbnails more frequently to enhance their phishing attempts (Shu *et al.* 2017). Once a person clicks on the link and accesses the website's content—who also happens to be an employee of the targeted company—they are misled into divulging sensitive information. Customers may, for a brief period of time, be trapped in thinking about them contributing to the spread of information during the reality, they are giving offenders with authorization to send electronic mails (Spicer 2018).

False news is linked to data security issues, one of which is information security (Smith *et al.* 2018). Information is watched rising as an important tool in the current business environment, and businesses have taken all the necessary precautions to keep sensitive information safe from unauthorized access. Nonetheless, such works are undermined by the widespread presence of information publishers who rely on presenting false news.

Conversely, some academics have noted that erroneous information occasionally has beneficial effects. For instance, there have been instances where the release of false information has caused industries that are listed on the shares market to observe a rise in the cost of shares (Baym and Jones 2012). Prospective investors profit from a company's operations as more and more users share links to websites with information that is unquestionably related to it, which causes the share price of the company to rise exponentially. However, these kinds of adjustments are certain to have a negative impact because most investors who buy shares based on false information wind up being let down. Other writers have noted in the same section that fake news can bolster an organization's marketing needs. Additionally, it is evident that the utilization of deceptive thumbnails to entice unsuspecting individuals into visiting websites has significantly influenced perspectives beyond the realm of business environments. Temporarily shifting focus, the events preceding the 2016 U.S. presidential elections were marked by the widespread propagation of misinformation through digital media channels (Lewis 2011). Instances of prominent figures endorsing specific candidates, disseminated through platforms that purported to provide factual information, were among the narratives shared by users. (Aldwairi *et al.* 2017b).

Lastly, research from the last few years shows that there have been logical questions raised by the idea of fake news in general, particularly when it comes to the participation of people with a history in journalism. For a brief while, some students have contended that using phony thumbnails is a sign of disobedience to the law and is connected to a career in media (Chen *et al.* 2015). It is expected of journalists to provide readers with data that has been verified to the last detail. False news data, however, completely contradicts these requirements. It raises the question of whether these individuals are interested in misleading readers when professionals engage in such activities to increase site views and online ad revenues.

Despite the dominance of fake information availability in digital media, the issue is currently receiving a lot of attention, as evidenced by the abundance of studies and publications that have been written about it. Before delving into the realm of machine learning for detecting false news, it is imperative to confront the dataset predicament. W.Y. Wang, in his scholarly work titled "Liar, Liar Pants on Fire," navigated this challenge by employing a publicly accessible dataset—a strategy echoed by various predecessors in this field. Additionally, the initiation of the False News Challenge Stage-1 (FNC-1) in June 2017 marked a pivotal moment, unveiling a myriad of groundbreaking solutions harnessing a spectrum of artificial intelligence technologies. Researchers showcased their ingenuity, ushering in a new era of innovation and diversity in the quest to combat misinformation. On certain issues, the use of NLP techniques for news outlet stance detection has facilitated the detection of false news (Wang 2017). In stance analysis, Riedel et al. and one other FNC-1 final team achieved nearly 82% efficiency. We think excellent and commercial solutions will surface after this competition and all false news detection stages are finished. The datasets are now publicly accessible thanks to FNC-1, and we are getting closer to establishing common benchmarks for evaluating all of the recently suggested methods.

3 PROPOSED SOLUTION

Utilizing an asset designed specifically to identify and eliminate fake news that contains inaccurate information meant to deceive users is the solution. A few criteria will be used by the technique to determine if a piece of news is accurate or fake in order to accomplish this goal. However, the program must be integrated by the social media firms into their applications so that users may be verified before posting anything. It is anticipated that the suggested approach would function flawlessly with the surrounding applications.

We'll start with the syntactical structure of the links that are utilized to lead consumers astray to these kinds of websites. The software will activate when the user attempts to update or post anything on social media using the social media application. It will then scan the content and classify it as true or false based on the accuracy supplied by the system. During the process, the software will also detect websites whose links contain terms that could mislead the user, such as those that are categorized based on the characteristics of several hyperbolic and colloquial expressions. When a user chooses to touch on any of these types of webpages, they will be identified as possible sources of incorrect information. The user is assisted in making a selection by a visual representation of the links and their architectural layout (Chakraborty *et al.* 2016).

Furthermore, the asset will utilize the linked data with the post that will be posted in order to identify which of them contains fraudulent information. This approach's reasoning is based on the observation that, generally speaking, phony thumbnails have much longer words than real ones (Brewer *et al.* 2013). It is therefore envisaged that the tool will leverage the system's accuracy in determining whether the data is real or fraudulent.

As a result, the user will have the choice to report these stories, along with their sources, and request that they be removed from future searches (Baym and Jones 2012). It is anticipated that the user will have removed a sizable amount false information from their dataset with the aid of this software after utilizing it for a while.

4 METHODOLOGY

Prior to beginning the data processing with the dataset and creating the data files for WEKA, the first step entails locating information that could be deemed fraudulent. We trawled the internet to gather data for the dataset because it's a challenging task. Our primary attention is directed towards social media platforms like Facebook, Instagram, and WhatsApp, which are known to contain a higher likelihood of containing fraudulent content or stories.

4.1 *Weka classifiers*

The classifier is characterized as the mechanism that verifies provided information and produces the final outcome. WEKA comes equipped with numerous classifiers that undergo testing, and we opt for the most proficient ones based on their performance with our dataset.

- Bayesian Network: Learning in Bayesian networks involves employing a diverse array of search techniques and quality attributes. The Bayes Network Classifier offers shared functionalities for Bayesian network learning algorithms such as K2 and B, encompassing essential data structures like network structure and conditional probability distributions, among other elements.
- Logistic: A program utilizing a ridge estimator to construct and apply a model based on multinomial logistic regression.
- Random Tree: A category focused on constructing a tree that considers K attributes randomly selected at each node. It can be configured to permit the estimation of class probabilities (or target means in regression scenarios) using a set that has been paused.
- Naïve Bayaes : Based on an examination of the simulation data, numerical estimator precision values are chosen. This means that the classifier is not the same as an alteration classifier, which typically starts with zero training instances in its parameters.

5 RESULTS

This section shows the outcome metrics and involves the classification results.

5.1 *Metrics*

The precision is determined by subtracting false positives from true positives, and this value is obtained by dividing true positives by the predicted positives. Conversely, the recall rate, also referred to as sensitivity, is computed by dividing true positives by the sum of true positives and false negatives. Meanwhile, the mean serves as a representation of the recall rate. As for the f-measure, it results from the multiplication of precision and recall. To calculate this, multiply both precision and recall by two, divide the outcome by the sum of precision and recall, and then perform another multiplication.

5.2 *Classifiers result*

The following criteria are used to evaluate the classifiers: precision, revisit, F-measure, and ROC. Table 2 illustrates the excellent classification quality of the logistic classifier, which has the highest precision of 99.4%. Classifiers based on random trees and logistic regression demonstrated the highest sensitivity of 99.6% and an excellent recall. With an f-measure that combined precision and recall, the Random Tree and Logistic classifiers outperformed the others, with a 99.3% accuracy rate. Ultimately, the best region under the ROC curve was found in BayesNet and Naivebayes.

6 CONCLUSIONS

The ability of a user to distinguish helpful information from false news and fake thumbnails is hampered, particularly when information is needed to make decisions. Given how the business world is changing in the modern era, the issue of false news has grown beyond simple marketing concerns and demands significant attention from security researchers. Any attempt to manipulate or troll the Internet by spreading fake news or creating a fake thumbnail must be perfectly countered. We suggested a straightforward but efficient method that allows users to install a basic asset into their own browser and utilize it to identify and weed out possible fake thumbnails. The initial experimental results, which evaluated the method's ability to achieve its intended goal, demonstrated remarkable performance in identifying the sources of false news that are currently available. A few false news databases have been made available since we began this task, and we are currently expanding our use of R to test its efficacy against the new datasets.

REFERENCES

Aldwairi, M., Hasan, M., Balbahaith, Z., 2017b. Detection of drive-by download attacks using machine learning approach. *Int. J. Inf. Sec. Priv.* 11, 16–28.

Balmas, M., 2014. When false news becomes real: Combined exposure to multiple news sources and political attitudes of inefficacy, alienation, and cynicism. *Communication Research* 41, 430–454.

Baym, G., Jones, J.P., 2012. News parody in global perspective: Politics, power, and resistance. *Popular Communication* 10, 2–13.

Brewer, P.R., Young, D.G., Morreale, M., 2013. The impact of real news about false news": Intertextual processes and political satire. *International Journal of Public Opinion Research* 25, 323–343.

Chakraborty, A., Paranjape, B., Kakarla, S., Ganguly, N., 2016. Stop clickbait: Finding and preventing fake thumbnail in online news media, In: *2016 IEEE/ACM International Conference on Advances in Social Networks Analysis and Mining (ASONAM)*, pp. 9–16.

Chen, Y., Conroy, N.J., Rubin, V.L., 2015. News in an online world: The need for an "automatic crap detector", In: *Proceedings of the 78th ASIS&T Annual Meeting: Information Science with Impact: Research in and for the Community*, American Society for Information Science, Silver Springs, MD, USA. pp. 81:1–81.

Hassid, J., 2011. Four models of the fourth estate: A typology of contemporary Chinese journalists. *The China Quarterly* 208, 813832.

Lewis, S., 2011. Journalists, social media, and the use of humor on twitter. *The Electronic Journal of Communication / La Revue Electronic de Communication* 21, 1–2.

Marchi, R., 2012. With Facebook, blogs, and false news, teens reject journalistic objectivity. *Journal of Communication Inquiry*.

Shu, K., Sliva, A., Wang, S., Tang, J., Liu, H., 2017. False news detection on social media: A data mining perspective. *SIGKDD Explor. Newsl.*

Smith, J., Leavitt, A., Jackson, G., 2018. Designing new ways to give context to news stories.

Spicer, R.N., 2018. Lies, Damn Lies, Alternative Facts, False News, Propaganda, Pinocchios, *Pants on Fire, Disinformation*

Wang, W.Y., 2017. "Liar, liar pants on fire": A new benchmark dataset for false news detection. CoRR abs/ 1705.00648

Advances in AI for Biomedical Instrumentation, Electronics and Computing – Sachan et al. (eds)
© 2024 The Author(s), ISBN 978-1-032-64298-7

Review of performance parameters of PV array based on different configurations operating in mismatch scenarios

Aisha Naaz
Department of Electrical Engineering, Jamia Milia Islamia, New Delhi, India

Mohd Faisal Jalil
Department of Electrical Engineering, Aligarh Muslim University, New Delhi, India

Shahida Khatoon & Pushpender
Department of Electrical Engineering, Jamia Milia Islamia, New Delhi, India

ABSTRACT: A sustainable and pure form of energy, solar energy is becoming increasingly significant in the modern world. The configuration of a photovoltaic (PV) array plays a critical role in determining a solar power system's efficiency and overall performance. Conducting a comprehensive analysis and comparison of various PV array configurations helps to guide the selection of the most suitable configuration for various applications. The research investigates several PV array configurations of 6X6 modules, including Series, Series-Parallel, Bridge Linked, Honeycomb, and Total-Cross-Tied. The study considers parameters such as shading losses, mismatch losses, fill factor and reliability. Through rigorous experimentation, computer simulations, and real-world data analysis, this study provides valuable insights into the strengths and weaknesses of files of each configuration.

1 INTRODUCTION

1.1 *Renewable energy*

The world's demand for energy is rising as more people live worldwide. Fossil fuels are the primary source of energy for almost all the globe's population (Bingöl *et al.* 2018). The rise in environmental concerns and the depletion of conventional energy resources have led to an increase in the proportion of renewable energy sources in global energy consumption (Rani *et al.*, 2013). Renewable energy generation refers to producing electricity using energy sources that are naturally replenished and have minimal impact on the environment. High installation costs, poor energy conversion efficiency, and varying power generation due to varying weather conditions are all cons of solar energy (Danish *et al.* 2023). The distinctive characteristics exclusively contribute to the continuously rising demand for Photovoltaic (PV) arrays over time (Rakesh and Madhavaram 2016). The leading causes of PSCs on PV arrays can be modelled as, for example, the shadows cast by adjacent trees, poles, buildings, and moving clouds. Shade significantly affects the performance of PV modules within an array, reducing the amount of generated power (Jalil *et al.* 2020). Bypass diodes are typically installed in parallel with a single PV module or a string of PV modules connected in series (Khatoon *et al.* 2014). However, the configuration of the PV module is reconfigured to increase the power generation. PV modules will be arranged in the same column of the array, which effectively mitigates the shading effect (Alanazi *et al.* 2022). Due to impediments like trees, structures, or other shading features, solar photovoltaic (PV) modules in a PV array

DOI: 10.1201/9781032644752-95

may experience varying degrees of sun irradiation under partial shadowing situations. Different PV array configuration techniques, such as Series (S), Parallel (P), Series-Parallel (SP), Honeycomb, Bridged Linked (BL), and Total-Cross-Tied (TCT), can be used to reduce these losses and enhance the system's overall performance. The maximum powers, thermal voltage, relative power losses and fill factors of each setup have been evaluated to assess how well they operate (Ramos Hernanz *et al.* 2010).

2 MATHEMATICAL MODEL OF PV MODULE

2.1 *Photovoltaic array*

A photovoltaic array, often known as a solar array, is a collection of connected modules or solar panels that work together to collect sunlight and produce electricity. The PV cell, the fundamental component of a photovoltaic module, converts photons from sunlight into electrical energy (Premkumar *et al.* 2020). The PV cell is a PN union whose electrical properties are remarkably related to a diode (Mohammadnejad *et al.* 2016).

2.2 *PV module equivalent circuit diagram*

It is vital to examine how the model PV system parameters and its behaviour evolve with shifting environmental conditions before mathematically modelling the PV cell (Sharma *et al.* 2023).

$$Id = Io - \{\exp(Vpv + RsI)/(A-1)\} \tag{1}$$

$$IRsh = (Vpv + IRs)/Rsh \tag{2}$$

$$I = Iph - Id - IRsh \tag{3}$$

$$Ipv = Iph - Io - \{\exp(Vpv + RsI)/nVT - 1\} - (Vpv + RsI)/Rsh \tag{4}$$

$$Iph = Isco(G/G0)(1 + \alpha1(T - T0))(Rs + Rsh)/Rsh \tag{5}$$

The PV photocurrent depends on the irradiation and the cell temperature. The expression for the photocurrent is given in eq. 5 (Sharma *et al.* 2023).

The PV modules that will be used to analyze the properties of PV modules are modelled using the equations.

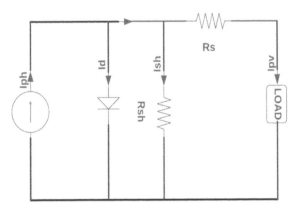

Figure 1.   Equivalent circuit diagram of PV module.

In the equivalent circuit diagram of a photovoltaic (PV) cell, shunt resistance (Rsh) and series resistance (Rs) are included to model the real-world behavior and characteristics of a solar cell more accurately.

Initially, Rsh is approximated and is provided in Eq 6. (Sharma *et al.* 2023). The equation used for calculating series resistance is given by (Sharma *et al.* 2023).

3 PV ARRAY CONFIGURATON

3.1 *Different configuration*

The arrangement and integration of all PV modules within the panels is known as PV array configuration. The electrical circuit diagram for the PV module is shown in Figure 1. Series, series-parallel, honeycomb (HC), bridge linked (BL), and total cross-tied PV arrays are among the various types reported in this study, shown in Figure 2. A PV array is typically configured in series, parallel or a combination of both, depending on the individual requirements, available equipment, the amount of shading, and the system design objectives. Planning and calculations must be made to ensure that the solar panel array performs as efficiently as possible. The primary purpose of this section is to review the PV array setups from earlier studies once more (Ansari *et al.* 2016).

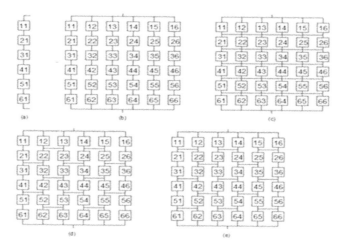

Figure 2. Classification of conventional PV array configurations (a) Series, (b) Series-Parallel, (c) Total Cross-Tied, (d) Bridge-linked and (e) Honeycomb.

4 PERFORMANCE PARAMETER OF PV ARRAY

To evaluate how different PV array configurations behave. First, the parameters required to study the behaviour modification of the Photovoltaic (PV) array designs must be determined (Jalil *et al.* 2018).

Mismatch scenarios in a PV array relate to circumstances when some solar panels or areas operate differently than others or exhibit different characteristics. Following are the performance parameters of a PV Array, which will be used to compare different PV Array configurations that were detailed in the above section:

a) Fill Factor b) Maximum Power Output c) Shading losses d) Mismatch losses

A 6X6 partially shaded PV array is chosen to examine the performance of different PV array layouts. The case studies under consideration have four distinct irradiation levels (1000

W/m2, 800 W/m2, 400 W/m2, and 200 W/m2) (Belhachat *et al.* 2015). A 6X6 partially shaded PV array is chosen to examine the performance of different PV array layouts. Examples of partial shading scenarios that differ include short, wide, narrow, narrow, and long broad. The case studies under consideration have four distinct irradiation levels (1000 W/m2, 800 W/m2, 400 W/m2, and 200 W/m2) (Jalil *et al.* 2018).

5 RESULT AND DISCUSSION

Figure 3 represents PV modules organized in a 6 X 6 array with six rows and six columns in various configurations, as illustrated in Fig. 3. Here, 11 indicates the first row and first column modules of the array. In contrast, 21 indicates the array's first row and first column modules, and so on(Jalil *et al.* 2020). At STC, or 250C in the case of uniform shading, each of the 36 PV modules will receive an equal quantity of irradiance of 1000 W/m2.At standard temperature conditions (STC), the shaded modules in considered shading settings receive an uneven amount of irradiance values, such as 400 W/m2, 600 W/m2, and 800 W/m2 (Ishak *et al.* 2019).

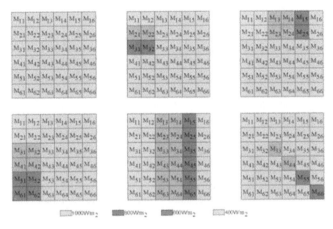

Figure 3.   Partial Shading patterns (a) Uniform; (b) Short Narrow (SN); (c) Short Wide (SW); (d) Long Narrow (LN); (e) Long Wide (LW); (f) Diagonal.

6 CONCLUSION

This review paper analyses the performance parameters for different configurations operating in mismatched scenario conditions. It was found that the output performance of PV array varies depending upon solar irradiance level and shading pattern.(Tatabhatla *et al.* 2015) A comparative analysis of various PV array configurations under shadings has been performed, and it was found that TCT has maximum power output with fewer losses than other configurations of PV array (Khatoon *et al.* 2014). TCT, in comparison to alternative setups, boosts the maximum power and offers the highest performance for all partial shading conditions, claims the evaluation. Additionally, shading loss, mismatch loss, and fill factor have been compared for all PV array configurations. The findings show that shading loss is constant across all configurations and shading instances. The fill factor and mismatch loss are both the lowest for TCT. As a result, the PV array configuration significantly impacts the PV array's efficiency (Bingöl *et al.* 2018).

REFERENCES

Alanazi, M., Fathy, A., Yousri, D., & Rezk, H. (2022). Optimal reconfiguration of shaded PV based system using African vultures optimization approach. *Alexandria Engineering Journal*, *61*(12), 12159–12185.

Ansari, M. S., Jalil, M. F., & Bansal, R. C. (2023). A review of optimization techniques for hybrid renewable energy systems. *International Journal of Modelling and Simulation*, *43*(5), 722–735.

Belhachat, F., & Larbes, C. (2015). Modeling, analysis and comparison of solar photovoltaic array configurations under partial shading conditions. *Solar Energy*, 120, 399–418. doi:10.1016/j.solener.2015.07.039

Danish, F., Shamsi, M. F., Kumar, A., Siddiqui, A. S., & Sarwar, M. (2023, February). Impact Assesment of Microgrid Towards Achieving Carbon Neutrality: A Case Study. In *2023 International Conference on Power, Instrumentation, Energy and Control (PIECON)* (pp. 1–5). IEEE.

Ishak, M. S. B., Salimin, R. H., Musirin, I., & Hamid, Z. A. (2019). Development of PV array configuration under different partial shading condition. *International Journal of Power Electronics and Drive Systems*, 10 (3), 1263.

Jalil, M. F., Ansari, M. S., Diwania, S., & Husain, M. A. (2021). Performance analysis of PV Array Connection Schemes Under Mismatch Scenarios. In *Renewable Power for Sustainable Growth: Proceedings of International Conference on Renewal Power (ICRP 2020)* (pp. 225–235). Springer Singapore.

Jalil, M. F., Khatoon, S., & Nasiruddin, I. (2018, November). Improved design analysis of photovoltaic modules and arrays under varying conditions. In *2018 3rd International Innovative Applications of Computational Intelligence on Power, Energy and Controls with their Impact on Humanity (CIPECH)* (pp. 68–75). IEEE.

Jalil, M. F., Khatoon, S., Aman, R., Hameed, S., & Bansal, R. C. (2022). Alternate panel interchange technique to improve the power production of PV array operating under various shading situations. *International Journal of Modelling and Simulation*, 1–15.

Jalil, M. F., Khatoon, S., Nasiruddin, I., & Bansal, R. C. (2020). An improved feasibility analysis of photovoltaic array configurations and reconfiguration under partial shading conditions. *Electric Power Components and Systems*, *48*(9–10), 1077–1089.

Jalil, M. F., Khatoon, S., Nasiruddin, I., & Bansal, R. C. (2022). Review of PV array modelling, configuration and MPPT techniques. *International Journal of Modelling and Simulation*, 42(4), 533–550.

Jalil, M. F., Saxena, R., Ansari, M. S., & Ali, N. (2016, November). Reconfiguration of photo voltaic arrays under partial shading conditions. In *2016 Second International Innovative Applications of Computational Intelligence on Power, Energy and Controls with their Impact on Humanity (CIPECH)* (pp. 193–200). IEEE.

Jalil, Mohd Faisal; Khatoon, Shahida; Nasiruddin, Ibraheem; Bansal, Ramesh Chand (2020). An improved feasibility analysis of photovoltaic array configurations and reconfiguration under partial shading conditions. *Electric Power Components and Systems*, 1–13. doi:10.1080/15325008.2020.1821842

Jha, V., & Triar, U. S. (2019). A detailed comparative analysis of different photovoltaic array configurations under partial shading conditions. *International Transactions on Electrical Energy Systems*, 29(6), e12020.

Khatoon, Shahida; Ibraheem; Jalil, Mohd Faisal (2014). [IEEE 2014 Innovative Applications of Computational Intelligence on Power, Energy and Controls with their impact on Humanity (CIPECH) – Ghaziabad, UP, India (2014.11.28-2014.11.29)] *2014 Innovative Applications of Computational Intelligence on Power, Energy and Controls with their impact on Humanity (CIPECH)* – Analysis of solar photovoltaic array under partial shading conditions for different array configurations. (), 452–456. doi:10.1109/cipech.2014.7019127

Mohammadnejad, S., Khalafi, A., & Ahmadi, S. M. (2016). Mathematical analysis of total-cross-tied photovoltaic array under partial shading condition and its comparison with other configurations. *Solar Energy*, 133, 501–511.

Nasiruddin, I., Khatoon, S., Jalil, M. F., & Bansal, R. C. (2019). Shade diffusion of partial shaded PV array by using odd-even structure. *Solar Energy*, *181*, 519–529.

Pachauri, R., Yadav, A. S., Chauhan, Y. K., Sharma, A., & Kumar, V. (2018). Shade dispersion-based photovoltaic array configurations for performance enhancement under partial shading conditions. *International Transactions on Electrical Energy Systems*, 28(7), e2556. doi:10.1002/etep.2556

Pachauri, R., Yadav, A. S., Chauhan, Y. K., Sharma, A., & Kumar, V. (2018). Shade dispersion-based photovoltaic array configurations for performance enhancement under partial shading conditions. *International Transactions on Electrical Energy Systems*, 28(7), e2556.

Pendem, S. R., & Mikkili, S. (2018). Modeling, simulation and performance analysis of solar PV array configurations (Series, Series–Parallel and Honeycomb) to extract maximum power under Partial Shading Conditions. *Energy Reports*, 4, 274–287.

Premkumar, M., Chandrasekaran, K., & Sowmya, R. (2020). Mathematical modelling of solar photovoltaic cell/panel/array based on the physical parameters from the manufacturer's datasheet. *International Journal of Renewable energy development*, 9(1), 7.

Rakesh, N., & Madhavaram, T. V. (2016). Performance enhancement of partially shaded solar PV array using novel shade dispersion technique. *Frontiers in Energy*, 10, 227–239

Ramos Hernanz, J. A., Campayo Martín, J. J., Zamora Belver, I., Larranaga Lesaka, J., Zulueta Guerrero, E., & Puelles Pérez, E. (2010, March). Modelling of photovoltaic module. In *Conference on Renewable Energies and Power Quality (ICREPQ'10) Granada (Spain)*.

Rani, B. I., Ilango, G. S., & Nagamani, C. (2013). Enhanced power generation from PV array under partial shading conditions by shade dispersion using Su Do Ku configuration. *IEEE Transactions on Sustainable energy*, 4(3), 594–601.

Sai Prakash, C., Mohapatra, A., Nayak, B., & Ghatak, S. R. (2021). Analysis of partial shading effect on energy output of different solar PV array configurations. *Materials Today: Proceedings*, 39, 1905–1909.

Sharma, D., Jalil, M. F., Ansari, M. S., & Bansal, R. C. (2023). A review of PV array reconfiguration techniques for maximum power extraction under partial shading conditions. *Optik*, 170559.

Tatabhatla, V. M. R., Agarwal, A., & Kanumuri, T. (2019). Improved power generation by dispersing the uniform and non-uniform partial shades in solar photovoltaic array. *Energy Conversion and Management*, 197, 111825. doi:10.1016/j.enconman.2019.11182

Yadav, A. S., & Mukherjee, V. (2021). Conventional and advanced PV array configurations to extract maximum power under partial shading conditions: A review. *Renewable Energy*, 178, 977–1005.

Advances in AI for Biomedical Instrumentation, Electronics and Computing – Sachan et al. (eds)
© 2024 The Author(s), ISBN 978-1-032-64298-7

Correlates of integrated marketing communication with respect to banking industry in India

P. Jain
Assistant Professor, MMIM, Maharishi Markandeshwar Deemed to be University, Mullana, Ambala, India

A. Saihjpal
Associate Professor, Panjab University Regional Centre Ludhiana, Punjab, India

N. Aggarwal
Associate Professor, Lovely Professional University, Jalandhar, Punjab, India

A. Kaur
Assistant Professor, Khalsa College for women, Ludhiana, Punjab, India

ABSTRACT: The objective of this research was to observe the difference in the IMC measures adopted by banks on the basis of their characteristics like structure, nature, age of banks and service profile of the managers. For that 462 responses were gathered from bank managers who are working as the bank managers of northern India. From the regression analysis, it was found that the measure 'Strategic Consistency' has the highest mean and the difference in mean scores has turned out significant in case of adopted by banks with regard to structure, age of the banks and also with respect to the position of the managers.

Keywords: Integrated Marketing Communication, banking, Annova, demographic correlates

1 INTRODUCTION

IMC referred to as 'The New Marketing Communication Paradigm' plays a major role in developing and maintaining the brand positioning. IMC is viewed as 'collapsing the artificial walls' between promotion tools to make the relationship stronger with customers. The concept of IMC has been defined by different authors in different ways.

The first formal definition was evolved and published by (4As) in 1989. It was founded in 1987 used in the national survey sponsored by the American Association of Advertising Agencies (4As), the National Association of Advertisers and Northwestern University.

This definition gave emphasis on using multiple tools of marketing communications apart from advertising and leveraging the power of each tool to get the maximum impact of communication.

In the late nineties, IMC became a sore issue in the context of marketing (Belch and Belch 1998; Caywood *et al.* 1991; Miller and Rose 1994; Schultz *et al.* 1992; Smith 1998). Authors began reviewing the concept of IMC and emphasized the importance of the formula of 'One Voice' and 'Consistency' which helps in communicating the unique messages to the target audience (Marwick and Fill 1997; Moriarty 1994). The concept of IMC has been assessed significantly in this decade, but instead of this, no clear definition of IMC was found (Duncan and Everett 1993; Kitchen and Schultz 1999; Phelps and Johnson 1996). The study of IMC was not only conducted in the US (Duncan and Everett 1993; Schultz and kitchen 1997) but also in

DOI: 10.1201/9781032644752-96

other nations (Gould *et al.* 1999). Despite study conducted in five countries-USA, UK, New Zealand, Australia, and India still IMC was in a primary stage of development process (Kitchen and Schultz 1999). The present study examined the relationship between organizational characteristics and IMC. Organisational characteristics are structure, age of the organisation and service profile of the managers and found few studies with regard to IMC and organisational characteristics like (Low 2000; Porcu *et al.* 2012b, 2019; Reid 2005a; Vantamay 2011). The information for this study was gathered through a self-administered questionnaire distributed via mail. The sample was bank managers who are working as the bank managers (branch/marketing/operation) of Punjab, Haryana, Jammu & Kashmir, Union Territory of Chandigarh, Rajasthan, Himachal Pradesh, Uttar Pradesh, Delhi and National Capital Region (NCR) of Northern India and knows about the marketing policies of the bank are selected. The sampling method used in this study is non-probabilistic in nature i.e., Convenience Sampling method. Around 1000 bank managers were approached and were asked as to their interest in participating in the survey, out of which 480 responses were gathered, with response rate of 48 percent and 462 came out to be final respondents after deleting outliners.

Demographic profiles compose of questions including Name, Position in the bank, Years of experience in bank Years of experience in career, Name of the Bank, Type of the bank, Year of inception of the Bank. The measurements were conducted through the open-ended questions.

Following hypothesis are formed

H1: There is a significant difference in the IMC Measures adopted by banks on the basis of the type of the bank.

H2: There is a significant difference in the IMC Measures adopted by banks on the basis of Age of the bank.

H3: There is a significant difference in the IMC Measures adopted by banks on the basis of Nature of Marketing Department.

H4: There is a significant difference in the IMC Measures adopted by banks on the basis of the Position of Mangers.

H5: There is a significant difference in the IMC Measures adopted by banks on the basis of Banking Experience and Career Experience of Managers.

The data collection method and questionnaire delivery strategy have been self-administrated. Questions have been designed which is related to demographic particulars of the respondents like name, Position in the bank, Years of experience in banking, Years of experience in career.

The second part is related to characteristics of bank like Name of the Bank, Type of the bank, Year of inception of the Bank, Number of Branches, Number of ATM, Bank has an in-house/outsourced marketing department.

1.1 *Descriptive analysis*

The responses of 462 respondents showed the age of bank shows that 19 percent of banks are of up to 25 years of age, 34.8 per cent are in the age group of 26–100 years, 46.1 per cent belongs to 100 years and above.

1.1.1 *Difference in the IMC measures adopted by banks on the basis of age of banks*

Demographics Analysis of Variance (ANOVA) has been conducted to study the difference in the IMC measures adopted by banks on the basis of their characteristics like structure and age across various banks. Then the t-test is done to study the difference in mean scores between the two groups of banks on the basis of their characteristics like structure and age. The results are as follows: Bank Age-wise Comparison of Dimensions of Factors Influencing IMC measures. They have different preoccupations and requirements. Hence, they may have different considerations in IMC measures of banking.

Table 1 presents the mean score, F-statistics and level of significance on the basis of their characteristics like structure, age and size across the three categories of banks. The mean values depict that the highest mean score is accorded to the factor 'Strategic consistency' by the banks of age more than 100 years. It is followed by the factor 'Planning and evaluation' in case of by the banks of age more than 100 years. Thus, both these factors are considered as 'Important' by the banks of age more than 100 years. The lowest mean scores 3.4 and 3.6 have been obtained on the factor 'Interactivity' and 'Mission marketing' for the banks of the age groups less than 25 years and 25–100 years respectively. In case of the banks of the age group up to 25 years, the lowest mean score (3.4) is assigned to the factor 'Interactivity' and for the banks of the age group 25–100 years, the lowest mean score of 3.6 has been accorded on the factor 'Mission marketing'.

Table 1.   Age-wise comparison of dimensions of factors influencing decision to IMC measures.

Factors relating to IMC Measures	Upto 25 years	25 to 100 years	Above 100 years	F value	Sig
Interactivity	3.44	3.68	4.28	47.429	.000
Mission Marketing	3.66	3.66	4.13	14.945	.000
Organizational Infrastructure	3.96	4.06	4.38	12.998	.000
Strategic Consistency	3.98	4.23	4.85	80.590	.000
Planning & Evaluation	4.08	4.26	4.54	13.367	.000
IMC	3.82	3.98	4.43	57.010	.000

ANOVA results reflect that the F-value has been found significant at 5 per cent level in case of all factors. Thus, the difference in mean scores has turned out significantly in case of banks have more than 100 years regarding these six factors.

1.1.2   Difference of IMC measures based on structure of banks
This part is further divided into type of banks. Following are the details:

1.1.2.1   Difference in IMC measures on the basis of type of the bank
The customers using different types of banks may have varying considerations influencing IMC measures. Table 2 exhibits that among all the six factors influencing decision IMC measures, the highest mean scores (4.59) have been given to the same factor 'Strategic Consistency' at an overall level. The factor 'Organizational Infrastructure' has turned out as the second important factor because the next highest mean scores (4.29) have been assigned to this factor of private banks. The lowest mean scores of 3.7 have been accorded on the factor 'Interactivity' at public banks. However, the lowest mean score of 3.6 has been found on the factor 'Mission marketing' in case of public banks.

To ascertain whether there is any significant difference in the mean scores of each of the factors influencing the IMC measures between public, private, foreign banks, the ANNOVA has been performed. The values for the six factors viz; 'Interactivity', 'Mission Marketing', 'Organizational Infrastructure', 'Strategic Consistency', 'Planning & Evaluation' and 'IMC', respectively, have turned out to be significant at 5 per cent level. Thus, there is a significant difference in the mean scores of public, private, foreign banks on each of the factors.

Table 2.   Difference in IMC Measures on the basis of type of the bank.

Factors relating to IMC Measures	Public	Private	Foreign	F value	Sig
Interactivity	3.76	4.03	3.93	5.748	.003
Mission Marketing	3.63	4.06	4.00	12.032	.000
Organizational Infrastructure	4.08	4.29	3.97	4.925	.008
Strategic Consistency	4.34	4.59	4.18	9.155	.000
Planning & Evaluation	4.26	4.48	3.69	12.738	.000
IMC	4.01	4.29	3.95	14.496	.000

1.1.3 Difference of IMC measures based on nature of banks
This part is further divided on the basis of the Marketing Department. Following are the details:

1.1.3.1 Difference in IMC measures on the basis of marketing department
The highest mean scores of 4.44 have been found in the case of overall level, in-house marketing department of a bank for 'Planning & Evaluation'. It shows that the banks on an overall basis of both the groups of outsourced and in-house marketing department have given due importance to the factor 'Planning & Evaluation' while deciding to influence of IMC measures. The second highest mean scores of 4.5 of strategic consistency have been assigned on the common factor 'outsourced marketing departments. The lowest mean scores of 3.02 have been found in the factor 'Mission marketing' on an overall basis and for in-house marketing department. The t-test has been conducted to study whether there exists any significant difference at the 5 per cent level in the mean scores of in-house and outsourced marketing departments for each of the factors influencing the decision of IMC measures.

The t-values for the five factors viz; 'Interactivity', 'Mission Marketing', 'Strategic Consistency', 'Planning & Evaluation' and 'IMC', respectively, have turned out to be significant at 5 per cent level. Thus, there is a significant difference in the mean scores of in-house and outsourced marketing departments for each of the factors. Analysis of Variance (ANOVA) has been conducted to study the difference in the IMC measures adopted by banks with regard to service profile of the managers.

1.1.3.2 Difference of IMC measures on the basis of position of managers
The difference in the IMC measures across Demographics Analysis of Variance (ANOVA) has been conducted to study the difference across various variables, namely – top level, middle level and entry level managers. Then the t-test is done to study the difference in mean scores between the different groups on the basis of service profile of managers.

1.1.3.3 Difference in the IMC measures adopted by banks with regard to service profile of the managers.
The service profile of the managers has an influence on IMC measures. They may have different considerations in their mind while choosing different IMC measures in banks. In order to find out whether there is any significant difference between the mean scores of banks with regard to service profile of the managers, ANOVA test has been applied. The mean score, F-statistics and level of significance difference in the IMC measures adopted by banks with regard to the Position of Managers. The mean values depict that the highest mean score is accorded to the factor 'Strategic Consistency' by the top-level management of the banks. It is followed by the factor 'Strategic Consistency' in case of all middle level managers. Thus, both these factors are considered as 'Important' by customers of all Position of Managers. The lowest mean scores 3.55 have been obtained on the factor 'Interactivity' for the mangers of entry level.

ANOVA results reflect that the F-value has been found significant at 5 per cent level in case of factors voice; "Interactivity', 'Mission Marketing', 'Strategic Consistency', 'Planning & Evaluation' and 'IMC'. Thus, the difference in mean scores has turned out significantly in the case of adopted by banks with regard to the position of the managers.

1.1.4 Difference of IMC measures on the basis of banking experience of managers
The highest mean scores have been accorded to the factor 'Strategic consistency' in case of all the six factors of banking experience of managers. The next highest mean scores have been obtained on the factor 'Planning and evaluation' in case of each of the six factors of banking experience of managers. The lowest mean score (3.5) is obtained on the factor 'Interactivity' which is followed by the mean score (2.53) accorded to the factor 'Mission marketing' in case of banking experience of managers.

The F-value for each of the factors 'Interactivity', 'Mission Marketing', 'Strategic Consistency', 'Planning & Evaluation' and 'IMC have been found significant at 5 per cent level. The factor 'Organizational Infrastructure" found to be insignificant. Thus, it shows that there is no significant difference in the mean values across on the basis of banking experience of managers.

1.1.5   *Difference of IMC measures on the basis of career experience of managers*

Annova exhibits the highest mean scores have been accorded to the factor 'Strategic consistency' in case of all the six factors of Career experience of Managers. The next highest mean scores have been obtained on the factor 'Planning and evaluation' in case of each of the six factors of banking experience of managers. The lowest mean score (3.5) is obtained on the factor 'mission marketing' which is followed by the mean score (3.7) accorded to the factor 'Interactivity' in case of career experience of managers. The F-value for each of the factors 'Interactivity', 'Mission Marketing', 'Organizational Infrastructure', 'Strategic Consistency', 'Planning & Evaluation' and 'IMC have been found significant at 5 per cent level.

2   CONCLUSION

According to this study, bank managers with more years of experience are more responsive to IMC measures. Therefore, it can be inferred that as experience increases, banking professionals become more open to exploring opportunities in the IMC of the banking sector. The market for banks is becoming more competitive, and traditional marketing strategies are no longer enough to set one bank apart from another. Therefore, these IMC measures will make the banking business more competitive. The present study is based on perceptions of bank managers as per their experience. This opinion may differ as per time and situation changes. The importance of IMC measures is just focusing on the banking sector. Comparison between Indian banking sector and professionals from the banking sector of other countries can be the interest for future studies. Further studies can also focus on other areas and sectors besides hospitality and banking sector.

REFERENCES

Belch, G., & Belch, M. (1995) *Introduction to Advertising and Promotion: An Integrated Marketing Communications Perspective*. Chicago: Richard D. Irwin.

Caywood, C., Schultz. D, & Wang. P. (1991) Integrated Marketing Communications: A Survey of National Consumer Goods Advertisers. Northwestern University report, June 1991.

Duncan, T. R., & Everett, S. E. (1993). Client perceptions of integrated marketing communications. *Journal of advertising research*, 33(3), 30–40.

Gould, S., Grein, A., & Lerman, D. (1999) The role of agency–client integration in integrated marketing communications: a complementary agency theory interorganizational perspective. *Journal of Current Issues and Research in Advertising*, 21(1), pp. 1–12

Kitchen, P. & Schultz, D. (1999) A multi-country comparison of the drive for IMC. *Journal of Advertising Research*, 39(1), pp. 21–38.

Low, G. S. (2000). Correlates of integrated marketing communications. *Journal of Advertising Research*, 40(3), 27–39.

Marwick, N., Fill, C. (1997). Towards the framework for managing corporate identity, *European Journal of Marketing*, 3(5/6).

Miller, D. A., & Rose, P. B. (1994). Integrated communications: A look at reality instead of the. *Public Relations Quarterly*, 39(1), 13.

Moriarty, S. (1994) PR and IMC: the benefits of integration. *Public Relations Quarterly*, 39(3), pp. 38–44.

Phelps, J., & Johnson, E. (1996). Entering the quagmire: examining the 'meaning' of integrated marketing communications. *Journal of Marketing Communications*, 2(3), 159–172.

Porcu, L., del Barrio-García, S., & Kitchen, P. J. (2012). How Integrated Marketing Communications (IMC) works? A theoretical review and an analysis of its main drivers and effects. *Communication & Society*, 25(1), 313–348.

Porcu, L., S. Del Barrio-García, J. M. Alcántara-Pilar, E. Crespo-Almendros. (2019). Analyzing the Influence of Firm-wide Integrated Marketing Communication on Market Performance in the Hospitality Industry. *International Journal of Hospitality Management* 80: 13–24. doi:10.1016/j.ijhm.2019.01.008.

Reid, M. (2005). Performance auditing of integrated marketing communications (IMC) actions and outcomes. *Journal of Advertising*, 34(4), 41–54.

Schultz, D. E. (1992). Integrated marketing communications. *Journal of Promotion Management*, 1(1), 99–104.

Smith, P. R. (1998). *Marketing communications: An integrated approach*. 2nd Edition. Kogan Page. London.

Vantamay, S. (2011). Performances and Measurement of Integrated Marketing Communications (IMC) of Advertisers in Thailand. *Journal of Global Management*, 1(1).

Advances in AI for Biomedical Instrumentation, Electronics and Computing – Sachan et al. (eds)
© 2024 The Author(s), ISBN 978-1-032-64298-7

Non-contact temperature detection system

Vikas Nandeshwar, Devang Bissa, Sarthak Biyani, Darshan Biradar, Bilal Khan,
Pratham Bisen & Saif Bichu
Vishwakarma Institute of Technology, Pune

ABSTRACT: The emergence of the global COVID-19 pandemic has necessitated the widespread adoption of contactless systems across various sectors. As businesses, educational institutions, and workplaces gradually resume their operations, the task of monitoring and tracking the health status of attendees and employees has presented a formidable challenge. In response to this imperative, this paper proposes the development of an innovative, automated, and non-contact temperature detection system with integrated temperature monitoring. The central aim of this system is to prioritize individual safety by meticulously tracking temperatures without the need for physical contact.

Leveraging cutting-edge contactless temperature measurement technology, this system has the capacity to identify individuals exhibiting unusually high temperatures, which may serve as early indicators of potential illness. In such instances, the system promptly notifies the relevant authorities, thereby facilitating swift and necessary actions, including the restriction of entry to the workplace or venue. This, in turn, serves as a vital safeguard for the health and well-being of all those present.

Keywords: Arduino UNO, Contactless Measurement, Infrared Sensors, Infrared Thermometry, MLX90614, Temperature and Ultrasonic Sensors

1 INTRODUCTION

This project aims for the measurement of temperature without any physical contact. This technology relies on the principle of infrared radiation, a natural emission from all objects possessing a temperature above absolute zero. We harness this intrinsic property of matter to gauge an object's temperature, all while avoiding the necessity for physical contact or direct intervention. Non-contact temperature detection systems are widely used in a variety of industries, including healthcare, manufacturing, and food processing. Overall, non-contact temperature detection systems have been an important tool in the fight against COVID-19 by allowing for quick and accurate screening of individuals, reducing the risk of virus transmission, and providing a safe and efficient method for temperature monitoring.

2 LITERATURE REVIEW

This paper discusses the use of graphene-inked infrared thermopile sensors to measure core body temperature. The authors likely explore the advantages of using graphene-based sensors for this purpose, such as improved sensitivity and accuracy. They may also discuss potential applications in healthcare, especially in situations where traditional temperature measurement methods may not be feasible or comfortable for patients.

This paper is focused on enhancing the security of medical images through the analysis and computation of encryption techniques. The authors may discuss the importance of securing medical images to protect patient privacy and the challenges involved in achieving

DOI: 10.1201/9781032644752-97

this. They likely provide insights into the encryption methods used and their effectiveness in safeguarding sensitive medical data.

This paper likely provides a comprehensive review of the factors that influence the selection of thermometers for intermittent clinical temperature assessment. It may cover various types of thermometers and their suitability for different clinical settings, taking into account accuracy, ease of use, and patient comfort.

This reference is likely a narrative review of the current status and future perspectives of COVID-19. The authors may discuss the evolution of the pandemic, the impact of various measures, and potential directions for research and policy. It's a valuable source for understanding the state of knowledge about COVID-19.

This paper likely focuses on indoor air quality in schools, addressing the impact of air quality on the health and well-being of students and staff. It may discuss measurement methods, sources of indoor air pollution, and strategies to improve air quality in educational environments.

Similar to the previous paper, this reference probably delves into the relationship between indoor air quality in schools and children's respiratory symptoms. It may include data on air pollutants, ventilation systems, and their effects on students' health.

(1) This review is likely concerned with thermal comfort in building environments and its implications for energy consumption. It may explore the factors influencing thermal comfort, such as temperature, humidity, and clothing, and how these factors impact energy efficiency.
(2) This paper likely focuses on personal factors affecting thermal comfort, including clothing properties and metabolic heat production. It may provide insights into how individual characteristics can influence thermal comfort assessment.
(3) This reference probably reports on an indoor air quality audit and the evaluation of thermal comfort in a school in Portugal. It may offer practical recommendations for improving indoor environmental conditions in educational settings.
(4) This paper likely discusses the design and implementation of a non-contact infrared thermometer based on the MLX90614 sensor. It may detail the technical aspects of the sensor and its applications.

3 METHODOLOGY

3.1 Components

3.1.1 Arduino Uno
The Arduino Uno is a popular microcontroller board based on the ATmega328P microcontroller. It's the brain of your project, responsible for running your code, reading sensors, and controlling other components.

Figure 1.   Arduino UNO.

Figure 2.   MLX90614.

3.1.2 MLX90614 (Infrared Temperature Sensor)
The MLX90614, a non-contact infrared temperature sensor, measures object temperature without physical contact, commonly used for body temperature and object temperature monitoring, providing ambient and object temperature readings.

3.1.3 *Ultrasonic Sensor (HC-SR04)*

The HC-SR04 ultrasonic sensor uses sound waves to measure distances. It emits ultrasonic pulses and measures the time it takes for the pulses to bounce back after hitting an object. This data can be used to calculate the distance between the sensor and the object. Ultrasonic sensors are commonly used in applications like obstacle detection and distance measurement.

Figure 3.   HC-SR04.

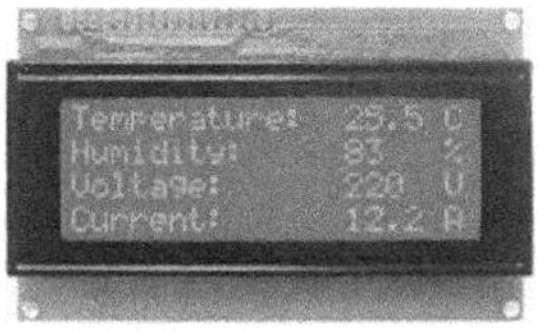

Figure 4.   LCD display.

3.1.4 *LCD (Liquid Crystal Display) or OLED (Organic Light Emitting Diode) Display*

An LCD or OLED display is used to provide visual output for your project. You can display text, numbers, and even graphics on these screens. LCDs use liquid crystals to display content, while OLEDs use organic compounds to emit light. They are commonly used to show real-time data, status information, or user prompts.

3.1.5 *Buzzer*

A buzzer is an audio output device that can produce sound or tones when activated. It's often used to provide audible feedback in various projects. You can control the pitch and duration of the sound produced by the buzzer using the Arduino.

Figure 5.   Buzzer.

Figure 6.   Jumper wires.

3.1.6 *Jumper wires*

Jumper wires are used to create electrical connections between components on a breadboard or between components and the Arduino. They come in different lengths and are essential for building your electronic circuits.

Flowchart

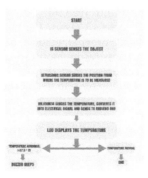

Figure 7.   Flow of the working.

3.1.7 *Breadboard*

A breadboard is a prototyping board with a grid of holes that allow you to easily connect and disconnect components without soldering. It's a temporary platform for testing and assembling your electronic circuits. Breadboards are great for experimenting and prototyping.

Figure 8. Breadboard.

3.1.8 *Arduino IDE (Integrated Development Environment)*

The Arduino IDE is the software you use to write, upload, and debug your code on the Arduino. It includes a code editor, compiler, and a serial monitor for communication between your computer and the Arduino. It's available for various operating systems and is user-friendly for beginners.

3.1.8.1 *Theory*

The underlying theory of infrared thermometry hinges on a fundamental scientific law known as the Stefan-Boltzmann Law. This law offers profound insights into how all objects, particularly those with temperatures above absolute zero, emit thermal radiation. The Stefan-Boltzmann Law provides a mathematical description of this phenomenon, revealing a remarkable relationship between temperature and the amount of emitted radiation.

In essence, the law states that the energy (radiation) emitted per unit area per unit time by a black body is directly proportional to the fourth power of its thermodynamic temperature (T).

$$E = \sigma x T^4$$

E = Energy (radiation) emitted per unit area per unit time σ = Stefan–Boltzmann Constant = $5.670374419 \times 10^{-8}$ W·m^{-2}·K^{-4}
T = Thermodynamic Temperature of Body

3.1.9 *Method*

In our methodology, we integrate an ultrasonic sensor for precise distance measurement between the individual and the temperature sensor. As the person approaches, an audible beep signals the correct position for temperature assessment. The temperature is then measured, and upon completion, a buzzer notifies the user.

For real-time feedback, the temperature reading is displayed on an LCD screen. If the recorded temperature exceeds 37.5 degrees Celsius, the buzzer alerts the user. Our system ensures accurate distance measurement, timely guidance, and immediate alerts, making it a comprehensive solution for vigilant temperature monitoring.

The temperature assessment is seamless, with the core sensor accurately gauging temperature without physical contact. The buzzer signals the completion of measurement, and a prominently displayed LCD screen provides real-time feedback for users and authorities, addressing any deviations promptly.

An essential safety feature sets a threshold for elevated temperatures at 37.5 degrees Celsius. If exceeded, the system triggers the buzzer with a distinct alert, notifying the user of the elevated temperature. This immediate alert mechanism allows for rapid responses to potential signs of illness, making our system crucial in healthcare and safety-conscious environments.

Figure 9.   Block diagram for the connections.

3.1.10   *Block diagram*

Figure 9 describes the arrangement of the components and input and output signals of the Arduino processing.

4   RESULTS AND DISCUSSIONS

Figure 10.   Non-Contact temperature detection model.

Figure 11.   Temperature detection display.

Figure 10 shows our model which shows the prototype of non-contact temperature detection.

Figure 11 shows the temperature of an individual in degree celsius unit of temperature measurement.

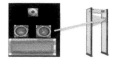

Figure 12.   Real life implementation of model.

Figure 12 shows actual application of our model in which it can be used in metal detectors which are placed at public places. When an object comes in front of infrared sensor, it detects it by the radiation emitted by the object and by that time position of the object is detected if it is in the position to measure the temperature so temperature of that object is measured with the help of MLX90614 sensor and is displayed on the LED/OLED. If the temperature is normal buzzer beeps only once but if it is not so (temperature>37.5 degree celsius) then, buzzer beeps for a long time and this is how the temperature is detected by avoiding the contact.

4.1 *Discussions*

4.1.1 *Results*

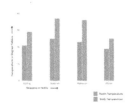

Figure 13.   Graph of temperature measurement Figure 13 shows the graph of temperature measurement using the sensor MLX90614.

Table 1.   Table consisting values of room temperature and body temperature.

Seasons	Room Temperature	Body Temperature
Spring	21	29
Summer	25	37
Monsoon	23	36
Winter	19	25

5 CONCLUSION

In conclusion, contactless temperature measurement emerges as a valuable tool in contagious disease contexts, offering advantages over traditional methods by minimizing direct contact and enhancing accuracy and efficiency. It serves as a viable replacement for conventional hand-held infrared thermometers, streamlining the assessment process while ensuring safety. This innovation holds substantial potential to elevate healthcare practices and bolster infection control, providing a secure and efficient mode of temperature measurement without physical interaction. Its broad applicability in healthcare facilities, workplaces, and public settings underscores its pivotal role in safeguarding public health, especially in the current global context of the COVID-19 pandemic. Overall, this technology enhances healthcare practices and prevents infections by offering a safer and more effective means of temperature measurement.

ACKNOWLEDGMENT

We extend sincere thanks to our guide for their essential mentorship, knowledge, and encouragement in developing our Non-Contact Temperature Detection System. Their support went beyond technical guidance, inspiring and nurturing our problem-solving skills. Appreciation is also extended to the academic community, whose referenced research laid the foundation for our project. Access to invaluable resources and facilities from various institutions enriched our research. Lastly, we appreciate the unwavering support of our families and friends throughout this journey. The completion of our project is a testament to the collaborative efforts of these individuals and institutions, and we are truly thankful for their contributions.

REFERENCES

Agus Sudianto, Zamberi Jamaludin, Azrul Azwan Abdul Rahman, Sentot Novianto and Fajar Muharrom, "Smart temperature measurement system for milling process application based on MLX90614 infrared thermometer sensor with arduino", *Journal of Advanced Research in Applied Mechanics* 72, Issue 1, 2020, pp. 10–24.

Bollu Gayathri, "Auto temperature detection system at the entrance", *Social Science Research Network*, 2021.

Dibyayan Patra, Anjan Agrawal, Archit Srivastav and J. Kathirvelan, "Contactless attendance cum temperature detection System with real-time alerts", *2021 Innovations in Power and Advanced Computing Technologies (i-PACT)*, November 2021.

Gang Jin, Xiangyu Zhang, Wenqiang Fan, Yunxue Liu and Pengfei He, "Design of non-contact infra-red thermometer based on the sensor of MLX90614", *The Open Automation and Control Systems Journal*, 2015, 7, pp. 8–20.

Gonçalo Marques and Rui Pitarma, "Non-contact infrared temperature acquisition system based on internet of things for laboratory activities monitoring", *The 9th International Conference on Sustainable Energy Information Technology (SEIT)*, 2019, Procedia Computer Science 155, pp. 487–494.

Preetha P S, Tajuddin R Nadaf, Vinod Kumar M and Sayed Tahir M Kittur, "Contactless Temperature Sensor", *International Journal of Engineering Research & Technology (IJERT)*, Volume 10, Issue 11, 2022, pp. 420–422

Advances in AI for Biomedical Instrumentation, Electronics and Computing – Sachan et al. (eds)
© 2024 The Author(s), ISBN 978-1-032-64298-7

A review of humidifier for healthcare

Garima Bhargava, Jassi Sandhu, Lakshmi Tiwari, Lokender Singh, Shreya Dubey,
Parvin Kumar & Vipin Kumar
*Department of Electronics and Communication Engineering, KIET Group of Institutions, Delhi-NCR,
Ghaziabad, India*

ABSTRACT: This research paper examines the crucial role of humidifiers in healthcare, focusing on active and passive types. It introduces humidifiers' significance in maintaining optimal air quality in healthcare settings. The paper reviews and compares active and passive humidifiers, highlighting their features, applications, merits, and demerits. It classifies humidifiers based on healthcare applications and operational considerations, aiding healthcare professionals in decision-making. The research includes a comparative analysis in tabular form within a hospital context, visually representing humidifier performance. In conclusion, the paper provides a clear understanding of how humidifiers impact healthcare, offering insights, reviews, and a practical classification system for informed decision making by healthcare professionals.

1 INTRODUCTION

A humidifier adds moisture to enclosed spaces, addressing dryness in the air that can cause discomfort, health issues, and environmental damage, especially in cold, dry winters or arid climates (IAQ *et al.* 2013). In healthcare, humidifiers are specialized devices crucial for regulating humidity in medical facilities. They enhance patient comfort, aid infection control, and ensure proper functioning of medical equipment, reducing the risk of dry air-related health complications. Two main types are active and passive humidifiers.

1.1 *Active humidifiers*

1.1.1 *Heated humidifiers*
Uses a heating element to warm water, producing vapor for the patient's respiratory system. Commonly used with ventilators and CPAP machines.

1.1.2 *Ultrasonic humidifiers*
Utilizes ultrasonic vibrations to create a fine mist of water droplets, suitable for quiet humidification during sleep or rest (DiBlasi *et al.* 2015).

1.1.3 *Impeller humidifiers*
Employ a rotating disk to fling water at a diffuser, releasing fine droplets.
  Often used for paediatric patients or low humidity requirements.

1.2 *Passive humidifiers*

1.2.1 *Heat and Moisture Exchangers (HME)*
Passive devices in respiratory care, capturing and using heat and moisture from exhaled air to humidify incoming dry air during inhalation. Commonly used in ventilated patients.

DOI: 10.1201/9781032644752-97a

1.2.2 *Condenser humidifiers (HCH)*

Like HMEs, capturing heat and moisture from exhaled breath using a hydrophobic membrane to prevent liquid water passage. Effectively humidifies inspired air. Both active and passive humidifiers are crucial in medical settings to prevent complications associated with dry and unconditioned air in the respiratory system. The choice depends on patient needs, medical equipment, and clinical context. In India, medical humidifiers likely include heated humidifiers, ultrasonic humidifiers, HMEs, and HCHs.

Table 1. Review of humidifiers available in medical use.

S. no.	Type	Specifications	Advantage	Limitations
1	Centrifugal	Uses a rotating disc or drum to distribute water into air	Simple and Low noise.	Limited Capacity.
2	Compressed Air and Water	Uses compressed air to disperse water	Efficient and adjustable humidity control.	Requires sufficient volume of compressed air.
3	High – Pressure atomizers	– Employs elevated pressure to create fine water mist	Same benefits as Compressed Air and Water.	Requires purified water.
4	Low – Pressure atomizers	– It generates fine droplets from liquids using minimal pressure.	Environment Friendly, Versatility. Cost-effective.	Less effective in Windy Condition.
5	Ultrasonic	Use high frequency vibrations to generate a fine mist of water droplets in air (Kudo *et al.* 2017)	Energy Efficient and consumes less power.	It's important to regularly clean and disinfect the water tank.
6	Wetted-media evaporative	Utilizes the principle of evaporation to cool air by passing it through a porous, water saturated medium.	Low energy Consumption. Provides evaporative cooling to surrounding air.	Non-purified water can lead to a build-up of impurities.
7	Steam cup	A container designed to hold hot liquid for inhaling.	Convenient for home application and affordable.	Limited Capacity and Risk of burns.
8	Steam separator	Separates steam from condensate	Reliable Performance. And available in wide range of capacities.	Can disperse boiler chemical into air.
9	Steam Panel	Improve breathing providing sterile, bacteria free moisture.	Performs without steam jacket and energy saving.	Difficulty lifting and condensate from steam traps

Table 2. Measurable parameters of under real time operations in hospitals.

Parameters	Input	Output	Real time data	Description
Relative Humidity (RH)	Setpoint:50%	Current RH: 48%	Within Acceptable Range (48%)	Indicates the amount of moisture in the air relative to its maximum capacity (Haise *et al.* 1955)
Temperature	Setpoint:24°C	Current Temp: 23.5°C	Slightly below setpoint (23.5°C)	The ambient air temperature, which influences the air's capacity to hold moisture.
Dew Point	Current RH and Temp	Dew Point: 15°C	Adequate (15°C)	The temperature at which air becomes saturated.

(continued)

Table 2. Continued

Parameters	Input	Output	Real time data	Description
Airflow	Setpoint: 300 CFM (Nassif et al. 2022)	Current Airflow: 310 CFM	Within acceptable range (310 CFM)	The movement of air to ensure even distribution of humidity.
Water Consumption	Not applicable	Current Usage: 5 litres/hour	Moderate usage (5 Liters/hour)	Monitoring the amount of water for efficiency assessment.
Pressure	Not applicable	Current Pressure: 1.2kpa	Normal Pressure: (1.2kpa)	Monitoring pressure level depending upon humidification.
Water Quality	Not applicable	Current Quality: Potable	Potable water source	Regular checks on the quality of water to prevent the dispersion.
Alarm system	Deviation from set points	Alarm: high RH	High RH alert triggered	Alerts for deviation from optimal conditions as response to system issues
Data logging	Real time measurement	Historical log: downloaded hourly	Hourly logs available for analysis	Recording humidity levels temperature and other parameter over time for the trend analysis
Remote monitoring	Remote control commands	System status: online	Successfully adjusted set points	The ability to monitor and control the humidification system remotely for quick adjustments

2 EVOLUTION: KEEPING UP WITH THE TRENDS

Humidifier evolution from the 1980s to the present has seen advancements in technology and design. Mechanical models gave way to ultrasonic and smart devices (Yeo et al. 2010). Leading companies, including Honeywell and Dyson, continually improved efficiency, introduced IoT integration, and prioritized health features, shaping the industry's development over the years.

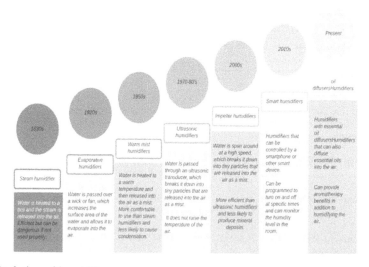

Figure 1. Evolution of humidifier.

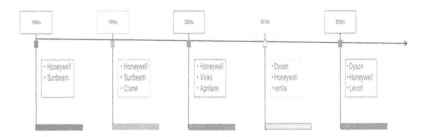

Figure 2.   Leading companies in the market.

3   CONCLUSION

In conclusion, this research paper provides a comprehensive overview of the crucial role that humidifiers play in healthcare settings, emphasizing both active and passive humidification systems. The exploration of various humidifier types, such as heated humidifiers, ultrasonic humidifiers, heat and moisture exchangers (HME), and hydrophobic condenser humidifiers (HCH), contributes to a nuanced understanding of their distinct features and applications in medical environments.

The paper's classification of humidifiers based on healthcare applications along with inclusion of a tabular comparative analysis and operational considerations serves as a valuable resource for healthcare professionals, facilitating informed decision-making regarding the selection of humidification systems tailored to specific patient needs and clinical contexts.

REFERENCES

DiBlasi, R. M. (2015). Clinical controversies in aerosol therapy for infants and children. *Respiratory Care, 60* (6), 894–916.

Haise, H. R., Haas, H. J., & Jensen, L. R. (1955). Soil moisture studies of some Great Plains soils: II. Field capacity as related to 1/3-atmosphere percentage, and "minimum point" as related to 15-and 26-atmosphere percentages. *Soil Science Society of America Journal, 19*(1), 20–25.

IAQ, I. A. Q. (2013). *Moisture control guidance for building design, construction and maintenance.* EPA 402-F-13053.

Kudo, T., Sekiguchi, K., Sankoda, K., Namiki, N., & Nii, S. (2017). Effect of ultrasonic frequency on size distributions of nanosized mist generated by ultrasonic atomization. *Ultrasonics Sonochemistry,* 37, 16–22.

Nassif, N. (2022). Optimal Zone Minimum Airflow Set Point for Multi-Zone VAV Air-Handling Unit. *ASHRAE Transactions, 128.*

Yeo, L. Y., Friend, J. R., McIntosh, M. P., Meeusen, E. N., & Morton, D. A. (2010). Ultrasonic nebulization platforms for pulmonary drug delivery. *Expert opinion on drug delivery, 7*(6), 663–679.

Advances in AI for Biomedical Instrumentation, Electronics and Computing – Sachan et al. (eds)
© 2024 The Author(s), ISBN 978-1-032-64298-7

Design of ECG monitoring system using NI LabVIEW

Shivansh Sinha, Shraddha Tripathi, Sumit Srivastava, Shubhi Sharma, Krishna Pratap Singh, Parvin Kumar, Abhishek Sharma & Vipin Kumar
Centre of Excellence, NI-LabVIEW, Department of Electronics and Communication Engineering, KIET Group of Institutions, Delhi-NCR, Ghaziabad, India

ABSTRACT: Cardiovascular diseases are becoming more common in today's times due to unwanted trends in lifestyle. According to research made by the World Health Organization, many short-term deaths globally, accounting for one-third of all deaths four years ago, were caused due to heart strokes, CVD, and respiratory diseases. Early diagnosis of cardiovascular problems is essential to start convenient administration and remedy. An electrocardiogram or ECG is a broadly involved indicative test for evaluating heart conditions. This review paper proposes a pre-evaluation strategy that integrates ECG, internal heat level, and pulse investigation for a person's well-being. The framework is linked with the patient's Aadhar card, enabling consistent patient interviews, and improving accuracy and proficiency in determining cardiovascular problems.

1 INTRODUCTION

In current times, heart problems are becoming more prevalent due to inappropriate lifestyles, which include poor eating habits, inactivity, obesity, and smoking, among others. As per a report provided by WHO, about 17.9 million individuals lost their lives from cardiovascular diseases (CVD) in 2019. Among these fatalities, 85% were due to cardiovascular failures and strokes. People often don't realize they have cardiovascular disorder until they experience extreme side effects such as chest pain or heart strokes. However, it is vital to identify cardiovascular disease as early as possible so that administration can start properly with medications. An electrocardiogram (ECG) is one of the simplest and fastest ways to assess heart conditions (Gomez *et al.* 2011). ECG features play an eminent role in recognizing many cardiac diseases. An ECG signal comprises three waves: the P-wave, the QRS complex, and the T-wave, which together form the essential components of the signal. Amidst these waves, the exact detection of R-peak (QRS complex) is vital for effective auto-mated analysis and diagnosis of different cardiovascular diseases. An acquired ECG signal has distinct QRS morphologies as it is generally mixed with various types of noise/distortion. We are proposing the working model of a health check-up system in which different body parameters like ECG, body temperature, and heart rate of the patients are diagnosed and analyzed. This system is linked to their Aadhar card which is further helpful for patients while consulting doctors.

2 WORKING PRINCIPLE & ANALYSIS

Any significant variation in the cardiac cycle indicates a problem (heart disease), known as 'Cardiac Arrhythmia'. In this condition, the patient's heart rate (HR) is altered. The R-peak in the ECG signal corresponds to the ventricular contraction during a heartbeat and serves as a valuable indicator for identifying cardiac arrhythmias. The R-peak can detect critical heart conditions by analyzing heart rate and its variability. For a heart rate of 60 beats per minute, the typical amplitudes of these characteristics are 0.115 mV, 1.5 mV, and 0.3 mV,

DOI: 10.1201/9781032644752-98

with variances of ±0.05 mV, ±0.5 mV, and ±0.2 mV, respectively. The standard values for the PQ and QT intervals are 160 and 400 milliseconds, respectively. Long-term ECG recordings are necessary for accurate ECG detection since short recordings are insufficient to provide a thorough evaluation of the patient's cardiac health (Brignole *et al.* 2009). However, there are two problems involved with obtaining long-term ECG data.

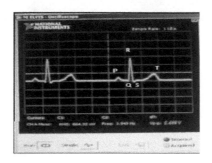

Figure 1.   Standard ECG signal.

Firstly, during the collection of ECG data, various types of disturbances can create difficulty in analyzing the ECG signal. Secondly, specific feature extraction algorithms are necessary for the study of ECG datasets because of the long-term data records that arise from their non-stationary nature. An overview of recent research on R-peak detection in ECG signals indicates its significance which has been addressed using various strategies. The detection of QRS complex has been a subject of study in the field of intelligent ECG detection for over three decades (Habib *et al.* 2019). Meanwhile, academic researchers have created a variety of algorithms based on digital filters, wavelet transforms, neural networks, picture segmentation, and other techniques. With advancements in technology, automated diagnostic systems are becoming vital in diagnosing heart disease. These systems are shifting to autonomous decision-making, rather than being reliant on lesion features selected by doctors. The conventional 12-lead ECG configuration is the most widely used method for ECG signal determination (Ribeiro *et al.* 2020). Out of the twelve leads, six are known as 'limb leads' as they are positioned on the arms or legs of the individual. The other six leads are called 'precordial leads' as they are placed on the torso (precordium) (AlGhatrif *et al.* 2012). There are some of the basic body parameters that are diagnosed and analyzed, which include ECG (Electrocardiogram), body temperature, heart rate, muscle fatigue, lung capacity, body mass index (BMI), and blood pressure.

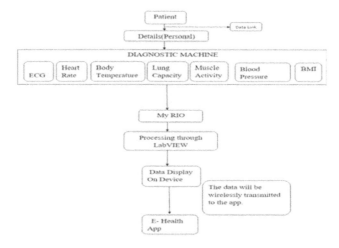

Figure 2.   Block diagram of working of all parameters.

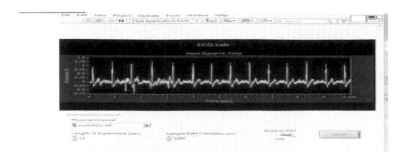

Figure 3. Acquired graphical ECG signals.

3 DESIGN IMPLEMENTATION & SIMULATION

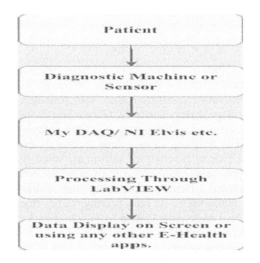

Figure 4. Block diagram of NI LabVIEW.

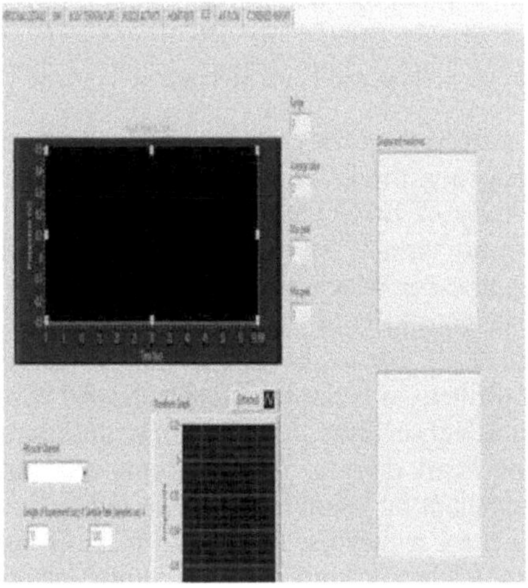

Figure 5. Front panel of ECG parameter in NI LabVIEW.

4 SOCIAL IMPACT OF HEALTH CARE MACHINE

India, being a developing nation, faces huge difficulties due to the lack of proper medical facilities and doctors, especially in rural areas. The World Health Organization positioned India as 112 out of 190 nations in its report, featuring the requirement for enhancements in the medical field (Zanuzdana *et al.* 2013). In comparison to the US, which spends 18% of its GDP on medical care, India allocates just about 4.2% of its GDP on medical facilities and supplies. (Hartman *et al.* 2020). This uniqueness in spending widens the gap between medical care frameworks in rural and metropolitan areas, raising the issue. A major issue in India's medical services sector is the high level of personal consumption, which accounts for around 70% of total medical service expenses. (Ganji *et al.* 2022). Furthermore, just about 5% of Indians have health insurance policies, leaving a maximum part of the population helpless to protect themselves financially during a medical crisis. The absence of an essential medical care framework is another basic concern that should be addressed. Numerous kids younger than five years are underweight, and the death rate among them is approximately 7%, significantly higher than the rate in the US (0.8%).

To handle such difficult situations, India requires a medical care framework that can provide suitable and reasonable check-ups to its citizens (Chen *et al.* 2019). Moreover, there

540

is also a need to gather precise clinical information about the people, which can further help in forming future strategies and providing suitable medical services. By resolving these issues, the public can pursue a more transparent and effective medical care system. The perks of our healthcare machine model are listed below-

- The Model presented will be beneficial to the lower and middle-class people.
- Low-cost AADHAR-linked automated Healthcare machines will help most of the underprivileged people for general healthcare checkups.
- The government will benefit in the long run from this system's assistance in gathering and keeping track of its citizens' medical histories, which will enable better policy making.
- Most of the benefits will go to India's rural areas, which house 70% of the country's population.
- This system will be the first of its kind in the world which will gather medical data from patients and provide regular medical checkups. It will be linked to their E-Health+ App to make all medical data easily accessible to both the user and the government.
- This healthcare device is created to help the people of India by offering affordable, high-quality routine checkups at low cost, especially in remote locations with inadequate healthcare infrastructure.

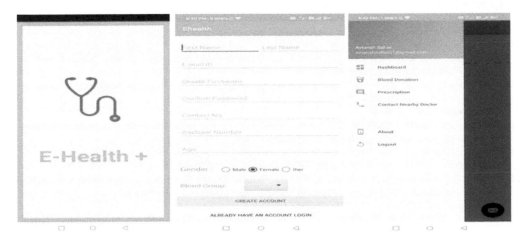

Figure 6. E-Health+ App wallpaper.   Figure 7. Registration.   Figure 8. Dashboard of E-Health+ App.

The table shows the data of two patients considering various Human Body Parameters

Table 1. Patient's details of various body parameters.

Patients Details	Value Type	Body Parameters						
		ECG	Muscle Fatigue	Body Temperature	Heart Rate	Lung Capacity	Body Mass Index	Blood Pressure
Astha Dubey (25 Yrs, Female)	Standard Value	Refer to Graph	300–350 N	35.7–37.7°C (96.3–99.9°F)	59–99	6L	18–23.5 Kg/m^2	120/79
	Observed Value	Nil	**285 N**	36.8°C (98.4°F)	75	5.8L	**17.5 Kg/m^2**	119/77
Utkarsh Agrawal (26 Yrs, Male)	Standard Value	Refer to Graph	350–400 N	35.7–37.7°C	59.99	6L	18–23.5 Kg/m^2	120/79
	Observed Value	Nil	392 N	36.9°C (98.7°F)	83	5.6L	**24.1 Kg/m^2**	119/75

5 CONCLUSION

The primary focus of this technique is a heart rate-checking framework that shows accurate ECG signals and readings of rising and declining heart rates (Hameed *et al.* 2016). A prior problem with the doctor's statement sub-VI in the system was resolved by validating the suggested solution and carrying out the VI. The VI captures ECG signals and compares the data with the transcript. Further, this data can help the doctor in analyzing cardiac problems with the patient.

By analyzing real-time waveforms, it simplifies the decision for the doctor to check the patient's heart regularly. The primary goal of our project was to create a 3-lead ECG detection system. We then used NI LabVIEW to acquire the patient's electrocardiography, which allowed us to create an ECG waveform and its related data.

REFERENCES

AlGhatrif, M. and Lindsay, J., 2012. A brief review: history to understand fundamentals of electro-cardiography. *Journal of Community Hospital Internal Medicine Perspectives*, p.14383.

Brignole, M., Vardas, P., Hoffman, E., Huikuri, H., Moya, A., Ricci, R., Sulke, N., Wieling, W., Auricchio, A., Lip, G.Y. and Almendral, J., 2009. Indications for the use of diagnostic im- implantable and external ECG loop recorders. *Europace*, 11(5), pp.671–687.

Chen, Y.J., Chindarkar, N. and Xiao, Y., 2019. Effect of reliable electricity on health facilities, health information, and child and maternal health services utilization: evidence from rural Gujarat, India. *Journal of Health, Population and Nutrition*, 38, pp.1–16.

Ganji, K. and Parimi, S., 2022. ANN model for users' perception on IOT-based smart healthcare monitoring devices and its impact with the effect of COVID-19. *Journal of Science and Technology Policy Management*, 13(1), pp.6–21.

Gomez-Clapers, J. and Casanella, R., 2011. A fast and easy-to-use ECG acquisition and heart rate monitoring system using a wireless steering wheel. *IEEE Sensors Journal*, 12(3), pp.610–616.

Habib, A., Karmakar, C. and Yearwood, J., 2019. Impact of ECG dataset diversity on a generalization of CNN model for detecting QRS complex. *IEEE Access*, 7, pp.93275–93285.

Hameed, R.T., Mohamad, O.A. and Țăpuș, N., 2016, October. Health monitoring system based on wearable sensors and cloud platform. In *2016 20th International Conference on System Theory, Control and Computing (ICSTCC)* (pp. 543–548). IEEE.

Hartman, M., Martin, A.B., Washington, B., Catlin, A., and National Health Expenditure Accounts Team, 2022. National health care spending In 2020: Growth driven by federal spending in response to the COVID-19 pandemic: National health expenditures study examines US health care spending in 2020. *Health Affairs*, 41(1), pp.13–25.

Ribeiro, A.H., Ribeiro, M.H., Paixão, G.M., Oliveira, D.M., Gomes, P.R., Canazart, J.A., Ferreira, M.P., Andersson, C.R., Macfarlane, P.W., Meira Jr, W. and Schön, T.B., 2020. Automatic diagnosis of the 12-lead ECG using a deep neural network. *Nature COMMUNICATIONS*, 11(1), p.1760.

Zanuzdana, A., Khan, M. and Kraemer, A., 2013. Housing satisfaction related to health and importance of services in urban slums: Evidence from Dhaka, Bangladesh. *Social Indicators Research*, 112, 163–185.

Advances in AI for Biomedical Instrumentation, Electronics and Computing – Sachan et al. (eds)
© 2024 The Author(s), ISBN 978-1-032-64298-7

High-performance dual band graphene slotted antenna for terahertz applications

R. Yadav*
Department of Electronics and Communication Engineering, National Institute of Technology Delhi

S. Sood*
Department of Electronics and Communication Engineering, KIET Group of Institutions, Ghaziabad, U.P., India

V.S. Pandey*
Department of Applied Science National Institute of Technology Delhi

ABSTRACT: The growing demand for high-performance antennas within the terahertz (THz) frequency range is driven by their increasing relevance in the field of communication, imaging, and spectroscopy. Therefore, a high-performance dual-band graphene-slotted antenna has been proposed for terahertz applications. This proposed antenna is resonating at two distinct frequency bands (1.154 THz and 2.36 THz) within the challenging constraints of the THz spectrum. A main aspect of our proposed design is the incorporation of graphene as the primary radiating element. Graphene's remarkable electrical and thermal properties make it ideally suited for terahertz applications. Its capacity for precise tuning and high electron mobility translates into enhanced radiation characteristics. Furthermore, the proposed antenna design embraces a dual-band strategy that striving to maximize its adaptability and utility. By incorporating detail-designed slots within the graphene structure, we have achieved the ability to produce two separate resonant frequencies within the THz range, thus amplifying its versatility for diverse terahertz communication and sensing applications. Additionally, the graphene-slotted antenna provides remarkable radiation attributes such as maximum gain, more directivity, and mini-mized cross-polarization, high efficiency all of which are pivotal for ensuring reliable terahertz communication and imaging. Further, the parametric analysis has been done to know the per-formance of antenna. The proposed antenna has been utilized for THz applications.

1 INTRODUCTION

The rapid progression of wireless communication necessitated the need for compact, resilient, stable, efficient, space-saving, adaptable, and versatile systems capable of transmitting and receiving signals. It increased data rates while consuming minimal power. Consequently, the evolution of communication systems calls for the inclusion of terahertz (THz) and sub-THz millimeter-wave antennas. The terahertz (THz) frequency spectrum spanning from approxi-mately 0.1 to 10 terahertz. It represents the domain of great promise that holds the potential to revolutionize a wide range of applications (Shubham *et al.* 2022). These applications encompass high-speed wireless communication, state-of-the-art imaging, and cutting-edge sensing technolo-gies. Commonly known as the THz gap. This frequency domain has intense research and

---

*Corresponding Authors: rajeshyadavnitd@outlook.com, shraddha.sood@kiet.edu and vspandey@nitdelhi.ac.in

DOI: 10.1201/9781032644752-99

development due to its distinctive capabilities. It occupies the boundary between microwave and infrared frequencies that provides a gateway to previously unattainable opportunities. However, fully harnessing this potential is not without its challenges. It is largely stemming from the scarcity of suitable components and systems that can efficiently operate within this range. Within this context, antennas emerge as pivotal components, serving as the primary interface between electromagnetic waves and electronic systems. To fully unlock the potential of THz applications, it is imperative to develop antennas that are not only high-performing but also efficient, and adaptable to the constraints of the THz spectrum. This is precisely where our research endeavors, as we embark on the development of an innovative high-performance dual-band graphene-slotted antenna, meticulously tailored for terahertz applications. The demand for high-performance THz antennas has been steadily surging (Yadav *et al.* 2021). It is driven by the expanding horizons of THz technology. The THz band boasts unique advantages such as the capability for high data transmission rates, emission of non-ionizing radiation, and the ability to penetrate non-conductive materials, making it indispensable for applications like high-capacity wireless communication, medical imaging, security screening, and materials characterization (Khaleel *et al.* 2022) However, the insufficiency of efficient THz radiation sources and detectors, coupled with the inherent complexities of designing and operating within the THz band. It underscores the need for innovative solutions within the field of antenna technology. This research embarks on a journey to address these pressing challenges by introducing an ingenious concept—an innovative dual-band graphene-slotted antenna (Alibakhshikenari *et al.* 2021). It is meticulously designed to cater to the unique demands of THz applications. The core concept behind this antenna is the utilization of the extraordinary properties of graphene which has a one-atom-thick lattice of carbon atoms. The aim of achieving exceptional performance within the THz frequency regime. Graphene, with its exceptional electrical, thermal, and mechanical attributes, offers a distinctive opportunity for antenna design and implementation in the challenging THz domain. Further, an important feature of our research is the innovative use of graphene as the radiating element within the antenna design. Graphene's high electron mobility and its capacity for precise tuning provide it an ideal candidate for terahertz applications where traditional materials often fall short (Khattak *et al.* 2021). The incorporation of graphene into the antenna's structure seeks to enhance critical radiation characteristics, including gain, directivity, and efficiency.

Consequently, our research represents a substantial leap forward in THz antenna technology. The dual-band graphene-slotted antenna fuses cutting-edge materials science with innovative design to unlock the vast potential of the terahertz spectrum. By introducing a highly efficient and adaptable antenna solution, our aim is to contribute significantly to the broader realization of THz technology's potential across a diverse range of fields, spanning from next-generation wireless communication to advanced biomedical imaging. This offers a glimpse into the challenges and opportunities that the THz frequency range presents the foundation for a detailed exploration of our high-performance dual-band graphene-slotted antenna.

2 ANTENNA DESIGN

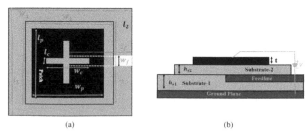

Figure 1.   The Geometry of proposed antenna (a) Top View, and (b) Side View.

The proposed antenna has been design in CST software. The proposed high-performance dual band graphene slotted antenna has been shown in Figure 1(a) and (b). The substrate-1 has been placed above the ground plane which has the dimension $l_1 \times w_1 \times h_{s1}$. The dimension of the ground plane is same as substrate-1 apart from the thickness (t). The feedline is placed above the substrate-1 which is made up of silver material. It has the dimension $l_f \times w_f \times t$. The substrate-2 is kept above the feedline that has $l_2 \times w_2 \times h_{s2}$ dimension as shown in Figure 1(a). Therefore, the feedline has been sandwiched between these two substrate. Both substrates are made up of silicon dioxide having $\epsilon_r = 3.9$ and t and $\delta = 0.003$. The silicon dioxide is used in THz because of its ability to minimize THz radiation absorption. It provides insulating characteristics, compatibility with microfabrication techniques, mechanical robustness, and affordability. As a suitable substrate, it enables efficient THz transmission, reducing interference, and making it a prime selection for integrating THz components and manufacturing devices. Further, a radiation patch has been placed above the substrate-2 which has $l_p \times w_p$ dimension including thickness (t) as shown in Figure 1(a) and (b). The radiation patch has been made up of graphene material. The graphene is defined with the help of different parameters such as chemical potential (μ_c), radian frequency (ω), thickness (t), scattering parameter (Γ), temperature (T), magnetic field (B_0) etc. Further, the two slots have been etched from the graphene patch. The dimension of both slots has been provided as $l_c \times w_c$. The proposed antenna geometry value is given in Table 1.

Table 1. The value for high-performance dual band graphene slotted antenna in μ_m.

Parameter	l_1	w_1	h_{s1}	l_2	w_2	h_{s2}	l_p	w_p
Dimension	60	60	1	50	50	0.6	35	35
Parameter	l_c	w_c	t	l_f	w_f			
Dimension	20	2	0.035	30	5			

2.1 *Result and discussion*

The return loss of a High-Performance Dual Band Graphene Slotted Antenna for Terahertz Applications holds immense importance. It acts as a pivotal performance metric, offering insights into the antenna's signal transmission and reception efficiency. In the field of terahertz technology, where signal preservation is paramount, a lower return loss indicates heightened signal effectiveness, guaranteeing dependable communication, imaging, and sensing capabilities. Additionally, it aids in evaluating impedance alignment across the dual frequency bands, thereby optimizing the antenna's adaptability. Through vigilant return loss monitoring and mitigation efforts, antenna designers can elevate overall performance, while quality control measures ensure the antenna meets the rigorous requirements of terahertz applications, ultimately advancing this state-of-the-art technology. The return loss (S_{11}) has been shown in Figure 2(a). The proposed antenna resonates at 1.154 THz and 2.3605 THz

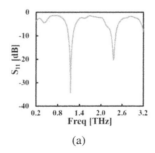
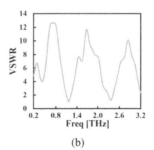

(a) (b)

Figure 2. The frequency response of (a) S_{11} parameter and, (b) VSWR.

with a better return loss. Therefore, it provides better matching along with dual response. Further, The importance of the VSWR (Voltage Standing Wave Ratio) in a High-Performance Dual Band Graphene Slotted Antenna for Terahertz Applications lies in its ability to evaluate signal transmission and reception efficiency while mitigating signal loss. A low VSWR signifies ideal impedance matching, guaranteeing the antenna's effectiveness and dependability in terahertz communication and sensing. The VSWR plot has been shown in Figure 2(b).

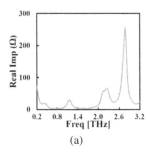

(a)

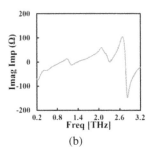

(b)

Figure 3. Impedance plot of the proposed antenna (a) Real part, and (b) Imaginary part.

The Real impedance of a High-Performance dual band graphene slotted antenna for terahertz applications is crucial, as it defines the antenna's interaction with its electrical surroundings. Ensuring an optimized practical impedance allows for efficient power transmission and reduced signal loss, thereby securing the antenna's efficacy in terahertz applications, including communication and sensing. Therefore, the real impedance plot is shown in Figure 3(a). It provides a high value that signifies better results.

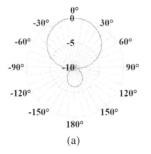

(a)

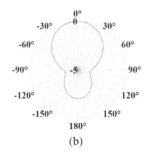

(b)

Figure 4. The radiation pattern of dual-band at (a) 1.154 THz and (b) 2.36 THz resonant frequency.

The importance of the imaginary impedance in a high-performance dual-band graphene-slotted antenna for terahertz applications is evident in its depiction of the antenna's reactive properties. When the imaginary impedance is optimized, it maintains correct phase relationships, enhancing signal efficiency and reception, crucial for the antenna's success in terahertz applications. Hence, the imaginary impedance plot has been shown in Figure 3(b). This shows the desirable impedance for the proposed antenna.

Further, the 2D radiation pattern of the High-Performance Dual Band Graphene Slotted Antenna in Terahertz Applications is of significant importance. Figures 4(a) and (b) show the 2D radiation pattern at 1.154 THz and 2.36 THz respectively. It visually illustrates how the antenna transmits and receives electromagnetic waves in two dimensions, offering crucial insights into its directional behavior. This data is instrumental for optimizing the antenna's positioning and ensuring maximal coverage or precise beam steering, which is indispensable for terahertz communication, imaging, and sensing. Understanding the radiation pattern is also valuable for managing interference and fine-tuning signal reception. In the rapidly advancing terahertz field, this information is fundamental for achieving efficient and dependable performance, establishing the 2D radiation pattern analysis as an integral aspect of antenna design and deployment.

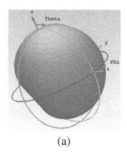
(a)

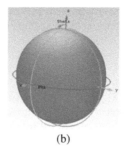
(b)

Figure 5.   The radiation pattern of dual-band at (a) 1.154 THz and (b) 2.36 THz resonant frequency.

The 3D radiation pattern of a high-performance dual band graphene slotted antenna in terahertz applications holds significant importance. The 3D pattern has been shown at both frequencies as shown in Figure 5(a) and (b). It offers a comprehensive perspective on how the antenna emits electromagnetic waves in three dimensions, providing crucial insights into its spatial behavior. This data is invaluable for precision in steering beams, optimizing signal directionality for targeted terahertz communication, imaging, and sensing. Furthermore, it aids in evaluating signal coverage across various planes, ensuring the antenna meets diverse task requirements. Additionally, the 3D pattern assists in identifying potential interference sources, guiding informed decisions regarding placement and orientation, thus enhancing signal reception and overall performance. In the rapidly evolving terahertz field, 3D radiation pattern analysis is an essential element for effective and dependable antenna design and deployment.

Further, the absolute electric field distribution's significance in a High-Performance Dual Band Graphene Slotted Antenna for Terahertz Applications cannot be overstated. The proposed absolute field distribution is shown in Figure 6(a). This distribution delineates how electromagnetic fields interact with the antenna, providing a comprehensive insight into field strength and spatial patterns. It plays a pivotal role in minimizing signal loss and interference, thereby enhancing the antenna's dependability and efficacy in terahertz communication, imaging, and sensing applications. Additionally, the vector electric field distribution has been shown in Figure 6(b). The significance of the vector electric field distribution is important. This distribution reveals the intricate interplay of electromagnetic fields with the antenna, offering a multidimensional perspective on field strength and direction.

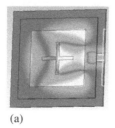
(a)

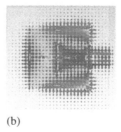
(b)

Figure 6.   Electric field at (a) absolute field and (b) vector field.

3 CONCLUSION

The high-performance dual-band graphene-slotted antenna for terahertz applications represents a significant advancement in terahertz technology. Its dual-band capability, use of graphene, and optimized radiation characteristics make it a promising solution for efficient and versatile terahertz communication, imaging, and sensing, paving the way for diverse applications. The antenna provides dual resonating frequency at 1.154 THz and 2.368 THz respectively. The antenna provides better matching with a better radiation pattern. The proposed antenna has been applied in the THz application.

REFERENCES

Alibakhshikenari, M., Virdee, B. S., Salekzamankhani, S., Aïssa, S., See, C. H., Soin, N., ... & Limiti, E. (2021). High-isolation antenna array using SIW and realized with a graphene layer for sub-terahertz wireless applications. *Scientific Reports*, 11(1), 10218.

Khaleel, S. A., Hamad, E. K., & Saleh, M. B. (2022). High-performance tri-band graphene plasmonic microstrip patch antenna using superstrate double-face metamaterial for THz communications. *Journal of Electrical Engineering*, 73(4), 226–236.

Khattak, M. I., Anab, M., & Muqarrab, N. (2021). A duo of graphene-copper based wideband planar plasmonic antenna analysis for lower region of terahertz (THz) Communications. *Progress In Electromagnetics Research C*, 111.

Shubham, A., Samantaray, D., Ghosh, S. K., Dwivedi, S., & Bhattacharyya, S. (2022). Performance improvement of a graphene patch antenna using metasurface for THz applications. *Optik*, 264, 169412.

Yadav, R., Verma, A., & Raghava, N. S. (2021). A dual-band graphene-based Yagi-Uda antenna with evaluation of transverse magnetic mode for THz applications. *Superlattices and Microstructures*, 154, 106881.

Advances in AI for Biomedical Instrumentation, Electronics and Computing – Sachan et al. (eds)
© 2024 The Author(s), ISBN 978-1-032-64298-7

Australian wildfire visualization

Manjinder Kaur & Roop Lal
Assistant Professor, Chandigarh University

Ankur Pandey, Anurag Singh & Dipak Rajbhar
Chandigarh University

ABSTRACT: In Australia, wildfires have grown to be a major problem, with disastrous effects on both human populations and ecosystems. Technology developments and improvements in data visualization have made it possible to track, comprehend, and communicate the effects of these fires in recent years. For instance, with an estimated 18.6 million hectares of land burned and 34 fatalities, the 2019–2020 Australian bushfire season was one of the worst in the nation's history. Geographic information systems (GIS) and satellite imagery were heavily utilized during this time to track the fires and aid in emergency response.

1   INTRODUCTION

Australia has recently had some of the most catastrophic fire seasons in its history, making wildfires a serious issue for the nation. The 2019–2020 bushfire season was one of the most destructive on record, according to the Australian Bureau of Meteorology, which also claims that climate change has contributed to an increase in the frequency and intensity of wildfires across the nation. 34 people lost their lives as a result of the flames, which scorched an estimated 18.6 million hectares of land and over 5,900 houses.

The repercussions of wildfires can now be tracked, understood, and communicated in new ways thanks to technological advancements in data visualisation. These methods and technologies are now crucial for managing and responding to emergencies, educating the public about the dangers of wildfires, and notifying them of their potential effects. The methods and resources utilised to visualise Australian wildfires, such as satellite images, geographic information systems, and interactive web-based platforms, will be discussed in this study paper. It will examine the advantages and drawbacks of various strategies, as well as their capacity to promote immediate decision-making and community involvement (Nolan *et al.* 2021). The possibilities for future advancements in wildfire visualisation will also be looked at in this article, including the application of machine learning and artificial intelligence to enhance attempts at prediction and response. This study paper's overall objective is to present a thorough grasp of how data visualisation might aid in efficient wildfire management in Australia and elsewhere. The authors talk on how risk assessments should account for uncertainty and how decision-makers must be effectively informed of this uncertainty. A case study that illustrates how the tool was used to manage a bushfire in Western Australia is also included in the paper (Tait *et al.* 2013).

The authors show how the spatial and temporal evolution of wildfires can be tracked using MODIS data, and how this knowledge can be utilised to support fire management decisions. The paper offers case studies that demonstrate how the methodology was applied in Victoria and New South Wales to monitor wildfires (Li *et al.* 2016).

DOI: 10.1201/9781032644752-100

Figure 1.   Heatmap wildfire Australia.

Figure 2.   AI made visualization of fires.

The use of MODIS data for comprehending and visualising Australian wildfire behaviour is discussed in "Visualising and Understanding Australian Wildfire Behaviour Using MODIS Data". The study analyses the shortcomings of conventional approaches to wildfire analysis, such as their inaccuracy and lack of information, and suggests the use of remote sensing data as a more successful strategy. Through a case study of the 2014 Hazelwood mine fire in Victoria, the authors show the value of MODIS data for visualising and comprehending Australian wildfires. The research sheds light on how remote sensing data may enhance our comprehension and management of bushfires in Australia (Li *et al.* 2016).

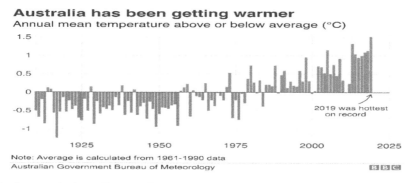

Figure 3.   Increase in Australia annual mean temperature.

An innovative method for the visualisation of the spatiotemporal dynamics of wildfires in Australia using satellite data is presented in the 2020 article "Visualisation of Spatiotemporal Dynamics of Wildfires in Australia Using Satellite Data" by Zahra Gholamzadeh, Abdullah Al-Mamun, and Xingwei Wang. In order to better comprehend the patterns and trends of wildfire occurrences and to support efficient decision-making and management strategies, theauthors stress the need of visualising wildfire data (Ban-Weiss *et al.* 2021).

2   LITERATURE REVIEW

According to the study "Visualising Wildfire Risk and Uncertainty in Australia Using a Web-Based GIS" by David R. Tait, Richard A. Williams, and Peter Caccetta: The goal of the authors of this work is to create a Geographic InformationSystem (GIS) application for the web that can visualise wildfire risk and uncertainty in Australia. They build a comprehensive database for assessing the risk of wildfires using a range of data sources, such as information on the frequency of fires, information on the climate, and maps of the vegetation. Then, to model wildfire risk, they employ a fuzzy logic technique that enables the inclusion of uncertaintyin the study. The resulting platform enables users to interact and visualise the data, improving decision-making for wildfire management and prevention (Tait *et al.* 2013).

In their work, Hielscher *et al.* (2020), used Google Earth Engine to create a satellite-based visualisation of the 2019– 2020 Australian bushfire season. The authors tracked the size and severity of the bushfires using multi-spectral images from Sentinel-2 and Landsat 8. Additionally, they made use of a cutting-edge visualisation method known as "Pixel-Strip Animation," which offers a dynamic perspective of fire growth over time. The outcomes show how useful this method is for visualising the size and intensity of bushfires aswell as for pin-pointing possible hotspots for future wildfires (Li *et al.* 2016). The work expands on earlier studies on the use of satellite imagery for tracking and visualising wildfire events, emphasising the advantages of combining several data sources and visualisation strategies to achieve a more thorough understanding of fire dynamics. The authors point out that other regions can use their method, and that it can improve reaction times and influence fire management policies. The use of remote sensing and visualisation technology for wildfire monitoring and management is a burgeoning field of study, and our study adds to that body of knowledge (Hielscher *et al.* 2020).

This review paper provides an overview of the methods used to visualise and convey fire hazard in Australia and makes recommendations for future study and practise. The Fire Danger Rating system and the use of satellite imagery are just two of the fire danger indices and visualisation methods that the authors explore, along with their advantages and dis-advantages. They also stress the significance of taking social and cultural considerations into account when expressing the fire danger to various populations. In order to enhance decision-making and public participation, the authors suggest an integrated approach that incorporates several data sources and visualisation techniques (Cole *et al.* 2020).

In order to effectively estimate fire risk, the paper emphasises the need of real-time fire hazard information and the demand for a user-friendly system that integrates diverse data sources, such as meteorological conditions and fuel moisture content.The authors also go over the advantages of employing GIS technology for managing fires, such as the capacity to give emergency responders and the general public access to rapid and accurate information and support for resource allocation and decision-making. In their conclusion, they highlight the tool's potential to enhance fire control in Australia and make suggestions for further study to further the tool's capabilities and use. Overall, this research sheds important information on the creation of a web-based GIS tool for the visualisation of fire danger and emphasises the significance of such tools in wildfire management (Williams *et al.*).

Overall, the assessment offers insightful information about Australia's current state of fire threat visualisation and communication, emphasising the need for more effective and user-friendly technologies to aid in managing fires and involving the community (Li *et al.* 2016).

3 METHODOLOGY

3.1 *The methodology can be broken down into the following steps*

3.1.1 *Determine and gather the pertinent data sources*
The initial step in this study would be to determine and gather the pertinent data sources for the visualisation of Australian wildfires.

This could entail gathering information from many sources, such as satellite data, sensor data from the ground, meteorological information, and other pertinent datasets.

3.1.2 *Data pre-processing and cleaning*
After the data have beengathered, the following step is to make sure the data are consistent and of high quality.

3.1.3 *Choose appropriate visualization tools and techniques*

After the data has been cleaned, the next stage is to identify the best visualisation methods and tools to portray the spatiotemporal patterns of Australian wildfires.

3.1.4 *Analyse the visualisation system's performance*

After the prototype system has been created, the next stage is to assess how well it represents the spatiotemporal patterns of Australian wildfires.

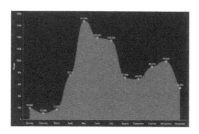

Figure 4.   Amount of fires by month.   Figure 5.   Amount of burnt area by fire.

3.1.5 *Analyse the findings and make conclusions*

The study's final phase will be to evaluate the findings and develop judgements on the suitability of the approaches and instruments selected for portraying the spatiotemporal patterns of Australian wildfires.

4   RESULT

4.1   *Data sources*

This study employed data from satellites, ground-based sensors, and weather stations, as well as data from NASA, the European Space Agency, and the Australian Bureau of Meteorology. The MODIS active fire detection data, which provide details on the location and severity of active fires across Australia, was one of the satellite data sets used in this investigation. The Australian Capital Territory Rural Fire Service provided information about the spread of flames and the placement of firefighting resources, which was included in the ground-based sensor data utilised in this study. The Bureau of Meteorology provided the temperature, humidity, wind speed, and precipitation data that were used in this investigation.

4.2   *Data pre-processing and cleaning*

Data normalisation, outlier detection, and missing data imputation were all part of the data pre-processing and cleaning stage. In order to combine the data sources for visualisation, data normalisation was done to make sure that they were all on the same scale.

4.3   *Visualisation tools and techniques*

The spatiotemporal patterns of Australian wildfires were represented using a variety of visualisation tools and techniques. Maps in two and three dimensions, heat maps, and other visualisation methods were among them. The 3D maps were used to depict the severity of fires over time, while the 2D maps were used to show where active fires were located throughout Australia.

4.4   *Evaluation of the visualisation system*

User testing, surveys, and other evaluation methods were used to gauge the effectiveness of the visualisation system. A variety of users, including firefighters, emergency service workers, and researchers, tested the system.

Overall, the evaluation's findings demonstrated that the system was simple to use and comprehend and that it accurately represented the spatiotemporal patterns of Australian wildfires. However, certain system flaws were discovered, including the requirement for more precise and current data sources.

5 CONCLUSION

In conclusion, this study used visualisation tools to investigate the spatiotemporal patterns of Australian bushfires. In order to portray the patterns of wildfires, the study identified pertinent data sources, cleaned and prepped the data, and then selected the best visualisation tools and methods. The usefulness of a prototype visualisation system was tested using user testing, questionnaires, and other assessment methods.

The findings of this study showed that the methodologies and tools for visualisation that were used were successful in capturing the spatiotemporal patterns of Australian wildfires. For investigating and analysing wildfire trends, the prototype system offered a simple and straightforward interface that could be helpful to emergency responders and decision-makers. In conclusion, this study offers insightful information about the potential of visualisation methods for discovering and comprehending the patterns of Australian wildfires. It also draws attention to some areas that need development and further study, such as enhancing data completeness and accuracy, incorporating real-time data, and offering additional user instruction and training.

REFERENCES

David R. Tait, Richard A. Williams, and Peter Caccetta. "Visualizing wildfire risk and uncertainty in Australia using a web-based GIS" *International Journal of Wildland Fire*, vol. 22, no. 6, 2013, pp. 794–803.

Fangjie Qi, Bao Zhang, and Han She Lim, "Wildfire visualization and analysis in Australia: a review" *Environmental Earth Sciences*, vol. 79, no. 6, 2020, 1–14.

George A. Ban-Weiss, Nick Obradovich, and John T. Abatzoglou, "Predicting australian wildfire risk using machine learning" *Environmental Research Letters*, vol. 16, no. 2, 2021, 024003.

Jason Beringer, Albert van Dijk, and Liana E. Cole, "Visualizing the impact of climate change on wildfire risk in australia" *Environmental Research Letters*, vol. 15, no. 9, 2020, 094010.

Kristin R. Linn, Maria T. Dix, and Douglas A. Stow. "Machine learning and wildfire modeling: a review of applications and future prospects in Australia" *Environmental Modelling & Software*, vol. 137, 2021, 104890.

Li Li, Lixin Wu, and Wei Gao. "Satellite-based visualization and analysis of australian wildfire dynamics using modis data" *International Journal of Wildland Fire*, vol. 25, no. 7, 2016, pp. 728–738.

Liana E. Cole, Jason Beringer, and Albert van Dijk, "Visualizing the impact of climate change on wildfire risk in australia" *Environmental Research Letters*, vol. 15, no. 9, 2020, 094010.

Rachael H. Nolan, Ross A. Bradstock, and Michael J. Lawes, "Visualizing and communicating fire danger in Australia: A Review of Current Approaches and Future Directions" *Fire*, vol. 4, no. 4, 2021, 50.

Richard A. Williams, David R. Tait, and Peter Caccetta, "Web-GIS visualization of fire danger indices for emergency management in Australia" *Journal of Emergency Management*, vol. 12, no. 3, 2014, 233–239.

Siyi Li, Jinfeng Wang, and Yang Yu, Visualizing uncertainty in wildfire risk mapping using fuzzy sets" *Natural Hazards*, vol. 98, no. 2, 2019, 703–720.

Thomas F. I. Hielscher, Ross A. Bradstock, and Marta Yebra, "Satellite-based visualization of the 2019–2020 Australian bushfire season using the Google Earth Engine" *Remote Sensing*, vol. 12, no. 21, 2020, 3555.

Wenjun Li, Jinfeng Wang, and Paul Sutton, "A web-based interactive tool for visualization and analysis of bushfire risk in australia" *International Journal of Disaster Risk Reduction*, vol. 45, 2020, 101487.

Advances in AI for Biomedical Instrumentation, Electronics and Computing – Sachan et al. (eds)
© 2024 The Author(s), ISBN 978-1-032-64298-7

Live code sync

Aditya Kumar*
3rd-year Student, Department of Computer Science, Chandigarh University, Mohali, Punjab, India

Er. Manjinder Kaur*
Assistant Professor, Chandigarh University, Mohali, Punjab, India

ABSTRACT: The digital renaissance has propelled online programming platforms to unprecedented heights, fueled by global phenomena emphasizing remote collaboration and dynamic learning. Despite the plethora of platforms available, novice coders often struggle with setup complexities, diverting focus from learning. This research explores challenges faced by beginners in the online programming milieu and introduces "Live Code Sync" as an innovative solution. Embodying user-centricity and technological prowess, this platform addresses identified gaps, catering specifically to novice coders. A comparative analysis with incumbent platforms highlights its distinctiveness. Conclusions, drawn from empirical user testimonies and performance analyses, position "Live Code Sync" as a trailblazer in the online programming domain.

Keywords: Digital programming ecosystems, real-time collaboration, Live Code Sync, web-integrated coding paradigms, novice programming challenges, user-centric design.

1 INTRODUCTION

In the 21st-century digital transformation, coding has evolved from a niche skill to a universal language essential for the current generation. As coding platforms proliferate to make programming accessible to all, challenges persist, especially for beginners navigating setup complexities and language nuances.

"Live Code Sync" emerges as a holistic solution, transcending traditional coding platforms. Addressing collaboration, setup simplicity, and tailored learning features, it promises to reshape the online coding landscape.

This paper explores the online coding ecosystem, delves into "Live Code Sync's" unique offerings, and provides an analytical perspective through a market comparison.

1.1 Background study & ideation

The journey towards fostering web-based tools for education and development has seen significant milestones, with web-based code editors standing out as pivotal in programming education, promoting accessibility and collaborative learning (Rasmussen *et al.* 2014). Guided by Rasmussen and Åse's work, our ideation for "Live Code Sync" was rooted in addressing evolving challenges in digitized education, ensuring it goes beyond being just a code editor, inspired by their emphasis on a beginner-friendly environment (Rasmussen *et al.* 2014). Rasmussen and Åse's master's thesis from the Norwegian University of Science and Technology emphasizes key aspects such as syntax highlighting, error detection, code

---

*Corresponding Authors: Adityasingh1031@gmail.com and Manjinder.e14896@cumail.in

DOI: 10.1201/9781032644752-101

auto-completion, and efficient test running (Rasmussen *et al.* 2014). These insights have proven instrumental in shaping the foundational principles of "Live Code Sync," offering students an immersive coding environment without the overwhelming intricacies of a sophisticated framework setup (Rasmussen *et al.* 2014).

1.2 *Development & execution*

Transforming the idea of "Live Code Sync" into a tangible and functional platform entailed a rigorous process of design, development, and iterative refinement. Each step was characterized by the integration of cutting-edge technologies to ensure a seamless experience for the end-users.

1. **UI/UX Design Brainstorming:** The initial stage involved extensive brainstorming sessions to create an intuitive and user-friendly interface. Leveraging tools like Bootstrap facilitated the development of a responsive and adaptive design, ensuring an optimal experience across various devices.
2. **Web Development with React:** React, a powerful JavaScript library for building user interfaces, was employed for the front-end. React's component-based architecture enabled modular and maintainable code, streamlining development and facilitating efficient feature integration.
3. **Integrating the Coding Environment:** The back-end development was fortified with Docker, ensuring consistent and isolated code execution. Docker containers guarantee smooth execution, mirroring native app functionalities and ensuring compatibility across users' local environments.
4. **Collaborative Tools:** Beyond coding, "Live Code Sync" envisions a space for collaborative learning. Additional collaborative tools were integrated, allowing users not only to code together but also to discuss, debug, and brainstorm in a unified space.
5. **Performance Optimization & Scalability:** The platform was designed with scalability in mind. Continuous testing and optimization techniques were employed to ensure that "Live Code Sync" remains performant and responsive, handling increased user loads gracefully.

In essence, the development and execution phase of "Live Code Sync" harmoniously combined design thinking, cutting-edge technology integration, and user-centric development. The meticulous selection of tools and technologies, from React to Docker, ensured that the platform excels not only in features but also in performance, scalability, and user experience.

2   COMPARATIVE ANALYSIS

2.1 *Existing solutions*

In the rapidly growing domain of online coding platforms, several solutions have arisen to address the diverse needs of coders. Some of the prominent platforms in the market today include CodePen, Codespaces, Online ADB, and Sandbox. Each of these solutions boasts unique features and capabilities:

1. **CodePen:** A popular platform, especially among frontend developers, for its real-time rendering capabilities. It's particularly useful for experimenting with new concepts. However, it's more tailored to the web languages and may not cater to the broader spectrum of programming languages.
2. **Code spaces:** Offering a more integrated development environment, it provides a broader set of tools and languages. But the depth of its features can sometimes be overwhelming for beginners.
3. **Online ADB:** While providing a platform for Android developers to run shell commands, it's very niche and doesn't address the broader audience of novice coders.
4. **Sandbox:** A platform offering isolated coding environments. However, it may not always provide the collaborative and instructional tools beginners require.

2.2 *Technological landscape & considerations*

The current technological landscape of online coding platforms is diverse, with each platform integrating different technologies depending on its primary audience and purpose. Evaluating the technological integrations is crucial to understand their strengths, weaknesses, and overall user experience.

1. **Tech Stack Complexity:** Many platforms integrate complex tech stacks, which, while offering a plethora of features, might confuse newcomers. Beginners need a clean, intuitive interface, not a dashboard flooded with tools they don't yet understand.
2. **Collaboration & Storage:** While many platforms offer coding capabilities, they often lack efficient collaborative tools. Even fewer provide an integrated solution for storing and backing up code projects. This becomes a
3. hassle, especially for learners who wish to revisit, share, or get feedback on their coding journey.
4. **Language Support:** Platforms often cater to specific languages or domains. For a beginner, it's crucial to have a platform that provides a wide array of language support, allowing them to dip their toes into different programming waters.
5. **Integration with Modern Tools:** Platforms like Codespaces have started to integrate with modern tools like GitHub, but seamless integration with storage solutions, especially for backups and versioning, is still sparse.

"Live Code Sync" Advantage: The major distinction "Live Code Sync" offers is its targeted approach towards newcomers in programming. Recognizing that the initial stages of learning to code can be overwhelming, "Live Code Sync" has been meticulously designed to be an all-in-one solution. It not only supports multiple languages for a holistic introduction to coding but also offers collaborative tools, real-time feedback, and seamless integration with storage solutions like Google Drive. For someone just embarking on their coding journey, "Live Code Sync" simplifies the process, offering everything they need in one cohesive environment.

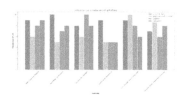

Figure 1. Comparison between coding platforms.

3 "LIVE CODE SYNC" IN FOCUS

"Live Code Sync" isn't just a platform; it's an evolution in the world of online coding platforms. At its core, it is designed to bridge gaps, enhance collaborative endeavors, and offer a seamless and comprehensive coding experience. Here's a deep dive into its hallmark features and future additionalities:

3.1 *Multi-language code compiler running on Docker*

One of the groundbreaking features of "Live Code Sync" is its multi-language support, capable of compiling and executing code from a multitude of programming languages. This versatility is powered by Docker:

1. **Containerized environments:** Docker provides isolated environments called containers. Each piece of code runs in its designated container, ensuring that there's no conflict between different codes and their dependencies.

2. **Consistency:** Docker guarantees that the code behaves consistently, irrespective of where "Live Code Sync" is accessed from. This eradicates the common programmer woe: "But it works on my machine!"
3. **Safety and Security:** The isolated nature of Docker containers ensures that the code doesn't interfere with the host system, providing a secure environment to run even untrusted code.
4. **Collaborative Code Room:** A true testament to "Live Code Sync's vision of collaborative programming:
5. **Real-Time** *collaboration:* Users can collaborate on a shared codebase in real-time. Every change made by one user is instantly visible to others, facilitating a dynamic coding experience.
6. **Discussion** *threads:* Integrated communication tools ensure that users can discuss, brainstorm, and debug together, all within the platform.
7. **Mentorship and** *peer learning:* The code room isn't just for coding. It's a space where experienced coders can mentor beginners, fostering a community-driven learning experience.

3.2 Live web development environment using I frame

The real magic for web developers:

3.2.1 Instant rendering
As users code in HTML, CSS, or JavaScript, they can see their web pages taking shape in real-time, thanks to the integration of iFrames which render the live output.

3.2.2 Interactive testing
Beyond just viewing, developers can interact with their web creations, testing features like buttons, forms, and animations instantly.

3.3 Google drive code upload integration

One of "Live Code Sync's" salient features, especially relevant in the modern digital age, will be its integration with Google Drive. This feature offers both practicality and peace of mind.

3.3.1 Seamless uploads
With just a few clicks, users can upload their code directly to Google Drive. This ensures that even extensive projects with multiple files and dependencies can be effortlessly saved and stored.

3.3.2 Accessibility
Once the code is saved on Google Drive, users can access it from any device and any location, as long as they have an internet connection. This promotes continuity in the coding process and ensures that users can seamlessly switch between devices.

3.3.3 Collaboration enhanced
Google Drive is known for its collaborative features. By allowing code uploads to Drive, "Live Code Sync" further enhances its collaborative ethos. Users can share their saved code with peers or mentors on Google Drive, facilitating code reviews, group projects, and more.

3.3.4 Backup assurance
The fear of losing hours of coded work due to unforeseen issues is real. By integrating with Google Drive, "Live Code Sync" offers an additional layer of data protection. Even if something goes awry, the code remains safe on Google Drive.

4 TECHNICAL CHALLENGES AND SOLUTIONS

During the creation and implementation of "Live Code Sync", several technical challenges were encountered. Addressing these challenges was crucial to ensure the platform's functionality, security, and usability. Below are the significant hurdles faced and the methodologies applied to overcome them:

Creating "Live Code Sync" posed various technical challenges, met with strategic solutions:

1. **Docker integration**
   ○ *Challenge:* Complex setup for a consistent coding environment.
   ○ *Solution:* Implemented Docker Compose for dedicated language environments, reducing conflicts.
2. **Server-side security**
   ○ *Challenge:* Risks of malicious code execution.
   ○ *Solution:* Multilayered security with isolated Docker containers and robust input validation.
3. **Multi-language support**
   ○ *Challenge:* Managing different compilers and dependencies.
   ○ *Solution:* Leveraged Docker for tailored language containers, ensuring hassle-free execution.
4. **Integration of Websocket's**
   ○ *Challenge:* Synchronization and latency issues for real-time collaboration.
   ○ *Solution:* Adopted an event-driven architecture for efficient data transmission, maintaining a smooth collaborative environment.
5. **Google API for code room and drive integration**
   ○ *Challenge:* Seamless authentication, permissions, and data privacy.
   ○ *Solution:* Implemented OAuth 2.0, encrypted data in transit and storage, and provided granular user control.

Conclusion: While multifaceted, these challenges provided valuable learning experiences. Meticulous planning and robust technological implementations ensured a secure, efficient, and user-friendly "Live Code Sync" platform.

5 ENVIRONMENTAL AND SOCIAL IMPACT

The integration of digital technologies and online platforms has reshaped the socio-economic and environmental landscape of the 21st century. "Live Code Sync," while primarily a technological endeavor, has consequences and contributions that transcend its primary function. Here's an exploration of its environmental and social footprints: **Environmental Impact:**

1. **Reduced Physical Infrastructure**
   ○ *Benefit:* "Live Code Sync" reduces the need for physical infrastructure, minimizing resource consumption and carbon footprint.
2. **Energy-Efficient Data Centers**
   ○ *Challenge & Solution:* Prioritizing green hosting and optimizing server usage to minimize energy consumption and carbon emissions.
3. **Digital Resource Utilization**
   ○ *Benefit:* Minimizes traditional coding's reliance on physical resources, cutting down on paper usage and environmental costs.

Social Impact:

1. **Promotion of Collaborative Learning**
   ○ *Impact:* Fosters global collaboration, broadening horizons and promoting inclusivity in the digital community.

2. **Access to Quality Education**
   - *Impact:* Web-based accessibility bridges the digital divide, providing quality programming education to remote or underserved regions.
3. **Empowerment Through Skill Development**
   - *Impact:* Tailored for beginners, "Live Code Sync" empowers individuals, offering tools and knowledge for better opportunities and innovation.
4. **Community Building**
   - *Impact:* Envisions a thriving community for shared learning, mentorship, creativity, and support for budding coders.

Conclusion: "Live Code Sync" actively addresses its digital carbon footprint while contributing to positive social change by promoting sustainability, accessibility, empowerment, and community building.

ACKNOWLEDGMENTS

I would like to express my sincere gratitude to Er. Manjinder Kaur for her invaluable guidance and continuous support in shaping "Live Code Sync." Special thanks to my team members and co-authors, Joti Sharma and Priyashi Barnwal, for their dedicated efforts.

I also extend my appreciation to friends, supporters, and the academic community for their contributions and encouragement. This research wouldn't have been possible without the collective efforts of everyone involved. Thank you for being an integral part of the success of "Live Code Sync."

REFERENCES

Brown, Laura. "The impact of online coding platforms on vovice programmers: A comparative analysis." *Journal of Computer Science Education*, vol. 25, no. 1, 2016, pp. 45–62.

Chen, Angela. "Real-time collaboration tools for remote development teams." *Collaboration and Communication in the Digital Age, edited by Emily Davis*, Springer, 2020, pp. 87–104.

Code Pen. "Real-time collaborative frontend development."

GitHub. "Code spaces: Your instant dev environment."

Google Developers. "Google drive API documentation."

Johnson, Mary. "Effective use of docker in web development." *Web Development Journal*, vol. 42, no. 3, 2018, pp. 210–225.

Kurniawan, Aditya *et al.*, *"CodeR: Real-time Code Editor Application for Collaborative Programming,"* Bina Nusantara University, JL. KH Syahdan No. 9 Palmerah, Jakarta Barat 11480, Indonesia.

Patel, Nihar *et al.* "Code Mirror: An open-source javaScript component for browser-based code editing." *Proceedings of the International Conference on Software Engineering*, 2017, pp. 450–461.

Rasmussen, C., & Åse, D. (2014). "A web-based code-editor for use in Programming courses." *Norwegian University of Science and Technology*.

Smith, John. "Web-Based learning environments: A comprehensive review." *Journal of Educational Technology*, vol. 36, no. 2, 2017, pp. 123–145.

Advances in AI for Biomedical Instrumentation, Electronics and Computing – Sachan et al. (eds)
© 2024 The Author(s), ISBN 978-1-032-64298-7

Farm automation using NodeMCU

Vikas Nandeshwar

Professor, Department of Engineering, Sciences and Humanities (DESH) Vishwakarma Institute of Technology, Pune, Maharashtra, India

Ishawar Borade, Atharva Borade, Atharva Bonde, Tanmay Bora, Om Bobade & Vishal Bokare

Department of Engineering, Sciences and Humanities (DESH) Vishwakarma Institute of Technology, Pune, Maharashtra, India

ABSTRACT: The agriculture industry plays a crucial role in the global economy, providing food, fiber, and other essential resources. However, traditional farming practices often face challenges related to labor shortage, resource management, and inefficient processes. To address these issues, farm automation has emerged as a promising solution. This research paper aims to review and present the implementation of farm automation using NodeMCU, an open-source IoT platform. NodeMCU offers various features and capabilities, making it suitable for developing cost-effective and scalable automation systems for farms. This paper offers a summary of the key concepts, benefits, and challenges of farm automation, followed by a detailed exploration of NodeMCU's capabilities and its application in automating different farming processes. The implementation covers various aspects, including remote monitoring, irrigation control, weather sensing, crop health monitoring, and livestock management. Furthermore, the paper discusses the challenges and limitations of NodeMCU-based farm automation systems, along with potential future directions for research and development in this domain. Overall, this research aims to provide insights into the integration of NodeMCU in farm automation and its potential to revolutionize modern agriculture.

Keywords: Farm automation, NodeMCU, Internet of Things (IoT), agriculture, remote monitoring, irrigation control, crop health monitoring, livestock management.

1 INTRODUCTION

1.1 *Background*

This autonomous system works hands free by inferring to the amount of moisture of soil through the soil moisture sensor followed by displaying those readings such as Temperature, Humidity and Moisture in Soil on the Serial Monitor, thus signalling the user to turn the water pump on through our App named "Farm Automation" made using MIT App Inventor.

1.2 *Problem statement*

There are specific problems faced by farmers, such as labour shortage, water scarcity, inefficient resource utilization, and manual monitoring. It establishes the need for an automated solution to address these challenges which saves the hassles of the farmer.

DOI: 10.1201/9781032644752-102

1.3 *Objectives*

The objective of our research paper, includes reviewing the concept of farm automation, exploring the capabilities of NodeMCU, and presenting practical implementations of NodeMCU-based farm automation systems.

1.4 *Research methodology*

Methodology consists of the flow of the automation i.e., the flowchart also the circuit diagram referred for achieving the desired results.

2 COMPONENTS

2.1 *NodeMCU*

2.1.1 *Features & capabilities*
An open-source development board called NodeMCU is based on the ESP8266 Wi-Fi module. It provides a low-cost platform for building Internet of Things (IoT) projects and prototyping. Here are some of the key features of NodeMCU:

2.1.2 *Wi-Fi connectivity*
NodeMCU has internal Wi-Fi, which enables it to join wireless networks andcommunicate with other devices online.

2.1.3 *ESP8266 microcontroller*
NodeMCU is powered by the ESP8266 microcontroller, which is a highly integrated chip with a low-power processor, RAM, and Wi-Fi capabilities.

2.1.4 *Arduino IDE support*
NodeMCU is programmed using the Arduino Integrated Development Environment (IDE). This makes it easy for developers who are familiar with Arduino to transition to NodeMCU and leverage the vast Arduino ecosystem and libraries.

2.1.5 *USB connectivity*
NodeMCU features a micro-USB port that can be used for both programming and power supply. It allows developers to connect the board directly to their computer for programming and debugging purposes.

Figure 1. NODE MCU.

2.2 *Soil sensor*

Soil sensor is a sensor which measures the volumetric content of water in soil .A soil sensor consists of two components in whole i.e. the sensor and the module, [4] the sensor are inserted in the soil and then the readings are passed to the module which then with the help of LM393 sends the readings to NodeMCU.

561

Figure 2.   Soil sensor.

2.3   *Temperature and humidity sensor*

Figure 3.   Temperature and humidity sensor.

A temperature and humidity sensor is an electronic device that measures, detects, and reports both ambient temperature and wetness at a cheap cost. the ratio of the highest amount of moisture at a given air temperature to the amount of moisture that is visible everywhere.

2.4   *Water pump*

Figure 4.   Water pump.

A 12-volt water pump is a particular kind of pump made to function with a 12-volt DC power source. These kinds of pumps are frequently used for small-scale irrigation, water treatment, and other similar systems, as well as in automotive, marine, and RV applications.

2.5   *Relay module*

Figure 5.   Relay module.

A relay module is an electronic device (or electronic switch) which controls high power electrical devices using low power electrical signals, here the lower power electrical signals are given by NodeMCU. The most basic application of relay module is to control the timing at which the water pump is to be turned ON/OFF. In our project we have used a single channel relay module which has 6 pins mainly consisting of VCC, GND, IN, COM, NO, NC; where the COM pin is attached to the high power electric device which in our case is a WaterPump.

3 LITERATURE REVIEW

Automation has become a need in the twenty-first century, offering advantages in terms of convenience, energy economy, and time savings in many areas of everyday life. [1] Concentration Agriculture refers to the technologically driven solutions that have revolutionised agriculture, a vital sector that drives both economic growth and human survival. Through advancements like IoT, WSNs, remote sensing, and drone surveillance, this paradigm shift seeks to improve overall agricultural output while easing the burdens encountered by farmers.

The study focuses on the critical role that an Internet of Things (IoT)-based smart irrigation system plays in addressing the problem of worldwide water shortage. The method reduces labour costs and overcomes issues related to ongoing surveillance by automatically watering plants based on temperature, humidity, and soil moisture content. Beyond maximising water use In light of diminishing water supplies, the system makes a substantial contribution to international efforts to save water. The study highlights that its main goal is to monitor the amount of moisture in the soil by employing an automatic water input arrangement that is sensor-based. It also expands its capabilities to record temperature and humidity. IoT integration makes it possible to send data in real-time to users' mobile devices, allowing for remote monitoring and meeting the urgent demand for effective irrigation in light of the depletion of water supplies.

The report highlights the significance of agriculture in India, a nation that depends on the monsoon season for water supplies, with 64% of its land devoted to this essential activity. It emphasises the need for effective irrigation systems and argues in favour of automation to reduce the need for labour-intensive manual pumping motor activation and deactivation. especially in situations when there are several scattered areas of agriculture. The suggested IoT-based automated irrigation system incorporates web and mobile applications for visual depiction of sensor data and control in the event of system failures. It is intended to modify the pumping motor based on detected water and moisture levels. Thanks to sensors and microcontrollers, this automation reduces the need for human involvement, freeing up farmers to concentrate on other crucial duties. The literature's conclusion emphasises the vital necessity of smart irrigation systems, particularly in areas like India with limited water supplies, where the adoption of automation fuelled by Internet of Things technologies is a viable choice technologically advanced and sustainable farming practises.

4 METHODOLOGY/FLOWCHART

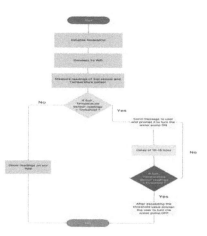

Figure 6. Flowchart/ Block diagram.

The above flowchart represents a proper workflow of how the whole system works right from the beginning where we check the sensors condition followed by initializing the NodeMCU -> Connect NodeMCU to Wi-Fi then we proceed to infer to the readings of Soil sensor and Temperature/Humidity sensor on the Serial Monitor. Now if the readings of Soil sensor are < Threshold values; then a message is sent to the user to prompt it to turn the water pump ON then after delay of 10-15 minutes the measurements are agin taken into consideration if if the readings of soil sensor > threshold values the user is prompted to turn the water pump OFF while if in the beginning itself the readings of soil sensor < threshold values then there's no need of watering the crops and hence the process ends here.

5 IMPLEMENTATION DETAILS

5.1 *Hardware setup*

The NodeMCU plays a crucial part in this process since it makes it easier to connect to Wi-Fi thanks to its built-in Wi-Fi module. After that, it receives the aggregate data from the soil sensor and temperature sensor, which is required to decide when the motor should be ON or OFF. In this case, the relay module aids the NodeMCU in synchronizing all of these processes with respect to time.

5.2 *Software development*

The embedded C programming language is used in this project to programme the NodeMCU. This NodeMCU programming project involves basic sensor interface programming. Additionally, in this project, a user app was developed so that users could receive daily temperature and soil sensor data for their fields as well as notifications if a motor was turned on or off. And what time does the motor turn on and off automatically. Additionally, it displays information in the user field about the pace at which water is provided to plants.

5.3 *System integration and testing*

This subsection focuses on the integration of hardware and software components to create a cohesive farm automation system using NodeMCU. It The Thing Speak cloud IoT completely integrates the app and the fundamental hardware technology. It facilitates data storage and transmission to the app that we develop with the aid of MIT App Inventor. This app displays sensor readings and offers features that allow you to switch from manual to automatic motor control. This software also analyses sensor data and displays it graphically to help users better understand the conditions in their field today. All of this information aids the user in making wise decisions to increase his field efficiency.

6 CHALLENGES & LIMITATIONS

6.1 *Power consumption and energy efficiency*

When working with NodeMCU, a popular development board based on the ESP8266 microcontroller, there are several power consumption and energy efficiency challenges that you may encounter like the Wi-Fi functionality of NodeMCU consumes a significant amount of power. Minimizing the time spent in active Wi-Fi mode and optimizing the transmission frequency can help reduce power consumption & also when using external sensors or modules with NodeMCU, consider their power consumption characteristics. Some sensors may have high power requirements, which can affect the overall energy efficiency of your system. Choose low-power alternatives where possible.

6.2 *Scalability and reliability*

When working with NodeMCU, a few scalability and reliability challenges may arise like for example NodeMCU has limited memory resources, including both RAM and flash memory. As your project grows in complexity or requires more extensive data storage, you may face challenges in managing and optimizing memory usage to ensure scalability and also when working with NodeMCU in a scalable environment, it's essential to design your system to handle errors gracefully and recover from failures. Implementing robust error handling mechanisms, data validation, and fallback strategies can help improve reliability and prevent system- wide disruptions. So by considering these scalability and reliability challenges and implementing appropriate strategies, such as efficient memory management, network optimization, and fault-tolerant design, you can enhance the scalability and reliability of your NodeMCU-based projects.

6.3 *Data security and privacy*

When working with NodeMCU, a few data security and privacy challenges may arise like NodeMCU's limited onboard memory may require external storage, such as SD cards or external databases, to store data. Ensuring secure storage practices, including encryption and access control, can mitigate the risk of data leakage if the storage medium is compromised. [5] NodeMCU devices can be vulnerable to physical tampering or theft. Securing the physical environment, implementing anti-tampering measures, and protecting against unauthorized physical access are important for maintaining data security. By addressing these data security and privacy challenges and following best practices, such as implementing encryption, securing networks, keeping firmware up to date, and complying with relevant privacy regulations, you can enhance the data security and privacy of your NodeMCU-based projects.

7 FUTURE SCOPE

This initiative is excellently suited to persuade farmers to use this technology to increase the digitalization of their field work. After investigating the foreseeable requirements in this field, we will improve our technology to totally digitalize farmer work. The points that we will incorporate into our system are as follows:

1. To start, we'll make changes to our app, which will now display historical data broken down by month and year, analyses that data, and, following a thorough analysis, recommend fertilizers to farmers in order to increase soil fertility. [3] How can we increase field productivity as well so that farmers may profit from it?
2. This app also provides information on the weather prediction. so that the farmer is frequently informed of the weather, thereby reducing field devastation brought on by the harshest weather.
3. [6] The software will analyse data and save it in the cloud IoT so that the user may access it appropriately to learn more about his particular subject.

8 CONCLUSION

The conclusion section summarizes the key findings and contributions of the research paper. It highlights the potential of NodeMCU in revolutionizing farm automation and emphasizes the need for further research and development in this domain.

ACKNOWLEDGEMENT

I would like to thank Mr. Vikas Nandeshwar our project guide for helping our project as well as VIT PUNE our university for providing us with the opportunity to create such a wonderful prototype. I would also like to thank this institute which helped us overcome the obstacles on our way to build this project and thus write this research paper.

REFERENCES

Akhila Gollakota and M.B. shriniva 2011, "Agribot-a multi-purpose agricultural robot" *India Council International Conference (INDICON)* pp 1–4.

Gulam Amer, S.M.M. Mudassir and M.A. Malik 2015, "Design and operation of wi-fi agribot integrated system" *International Conference on Industrial Instrumentation and Control (ICIC)* pp. 207–12.

Kirtan jain, Aalap Doshi, Poojan Patel, Manan Shah, *"A Comprehensive Review On Automation In Agriculture Using Artificial Intelligence"*.

Krishnan R, Vishnu R S, Mohan T H and Rao R Bhavani 2017, "Design and fabrication of a low-cost rice transplanting machine" *IEEE Technological Innovations in ICT for Agriculture and Rural Development* pp. 14–17.

Kumar A, Surendra A, Mohan H K Valliappan M. and Kirthika N 2017, "Internet of things based smart irrigation using regression algorithm" *International Conference on Intelligent Computing, Instrumentation and Control Technologies (ICICICT)* pp. 1652–5.

Shivaprasad B.S., Ravishankar M.N. and B.N. Shoba, "Design and implementation of seeding and fertilizing agriculture" robot 2014 *International Journal of Application or Innovation in Engineering & Management (IJAIEM)* 3 pp. 251–55.

Advances in AI for Biomedical Instrumentation, Electronics and Computing – Sachan et al. (eds)
© 2024 The Author(s), ISBN 978-1-032-64298-7

Multiple disease prediction using machine learning algorithms

Parth Dayal, Deepansh Sharma, Aman Agarwal, Himanshu Chaudhary, Ruchita Gautam,
Praveen Kumar & Abhishek Sharma
KIET Group of Institutions, Delhi NCR, Ghaziabad, U.P., India

ABSTRACT: The prediction of the diseases was performed by taking different symptoms as input data from the user. In this paper, we have analyzed the presence of many diseases such as Heart diseases, CKD, Liver disease, and many more using basic body parameters such as blood sugar, heart rate, etc. We have applied the most used classification algorithms such as Logistic Regression, KNN, and Random Forest Classifier to anticipate the illness. Finally, the algorithm with the most accuracy is used to train the model. For the prediction of malaria and pneumonia, we have used Deep Learning (CNN) which extracts important information, and patterns from the images, match it against a large dataset it already has, and makes the prediction. All the models have good accuracy and have been combined into a single website for easy access.

Keywords: Disease Prediction, Logistic Regression, Random Forest, KNN, Deep Learning, CNN

1 INTRODUCTION

The experience of seeking medical attention when one is sick often involves time-consuming and costly visits to a doctor. An innovative solution lies in using automated software to streamline the diagnostic process, offering time and cost savings for patients. Heart disease includes a range of conditions that affect the heart, including arrhythmias, coronary artery disease, and congenital heart defects. The term "heart disease", can refer to heart attacks, angina, or strokes. According to the Registrar General of India, in 2001–2003, CHD was the cause of 17% of all deaths and 26% of adult deaths; in 2010–2013, this percentage rose to 23% of all deaths [1]. Breast cancer displays changes to breast shape, lumps, or nipple discharge. The mean breast cancer death rate for all countries in the world was 13.77 per 100,000 in 1990, and the overall death rate slope from 1990 to 2015 was 0.7 per 100,000. India anticipates breast cancer to top the list by 2025 [2]. Diabetes arises from high blood sugar, primarily driven by dietary sugar. [3] Globally, there are presently 135 million diabetics, with India leading the way with 40.9 million cases in 2007. Furthermore, estimates indicate that 80.9 million Indians will develop diabetes by 2025. CKD or chronic kidney disease is an ailment that has greatly increased patient admissions and death rates globally. According to a report from 1990 to 2013, the annual global death toll from CKD rose by 90% [4]. Kidney diseases are expected to affect 850 million individuals globally. Malaria is a vicious communicable disease caused by various kinds of Plasmodium universally. The World Malaria Report issued by the WHO in 2019 indicated 228 million cases of malaria, with 40,500 deaths, in more than 90 countries in 2018 [5]. A bacterium is the source of pneumonia, a highly serious illness that impairs a person's ability to breathe normally by infecting the lungs. Ten million instances of pneumonia are reported annually in India alone [6].

DOI: 10.1201/9781032644752-103

2 LITERATURE REVIEW

The aim of this inspection [7] was to create a structure that helps the patient to recognize the patient's diabetes with precise results. The following algorithms were primarily used: SVM, decision tree, and naive Bayes. Recorded accuracies were 77.3%, 85%, and 77%. After training, they used an ANN algorithm to see the network's reactions. The primary objective of this paper [8] is to demonstrate the importance of the Heart. Various ML algorithms were used for the prediction of heart diseases, namely linear regression, SVM, decision tree, and KNN, in which the accuracies recorded were 78%, 83%, 79%, and 87% respectively. Liver diseases are the primary cause of death in India. For quantitative analysis, the employed algorithms in this paper [9] were Random Forest, SVM, and Decision Tree algorithms as well as measures for precision, accuracy, and recall. In that order, the accuracy is 92%, 95%, and 87%. The prediction models utilizing algorithms, such as logistic regression, KNN, support vector machines, and decision tree classifiers for CKD prediction, were described by the authors in [10]. The SVM algorithm had the best accuracy among the above, with it being 98.3%. Samir Yadav and Shivajirao Jadhav in 2019 contrasted research and they focused on the use of deep neural networks (DNN), whose main branch is convolutional neural networks (CNN), demonstrating notable performance since 2012. When they illustrate CheXNet which is a CNN of 121 layers trained on over 100,000 chest X-rays with a front view; its performance was superior to that of four radiologists on average [11]. Recent malaria research delves into the complexities of the disease, especially focused on malaria parasites and their growth phases in blood smear images.

3 PROPOSED METHODOLOGIES

Following were the steps carried out for accurate prediction of diseases:

3.1 *Data collection*

Data was gathered from the internet to detect the disease, where authentic symptoms of the ailment were acquired. The disease symptoms were taken from various reliable sources on the web, namely:

a) Heart Disease Dataset
b) Diabetes Dataset
c) Liver Disease Dataset
d) Breast Cance Dataset
e) CKD Dataset
f) Malaria (Cell image Dataset)
g) Pneumonia (X-Ray Image Dataset)

3.2 *Data pre-processing*

Before feeding the data into the prediction model (for diabetes, Heart, Liver, and Kidney diseases) a series of data cleaning and preprocessing procedures are performed.

a) Analysing the dataset using different graphs and charts (correlation matrix, etc.).
b) Checking the null values and filling them using the Fill Forward Method.
c) Checking for any outliers and then removing them.
d) Finally, we split our dataset into training and testing sets.

In the case of image datasets (i.e., malaria, pneumonia) we perform the following steps:

a) Data is divided into three key subsets: training, validation, and testing data, each comprising 'Infected' and 'Healthy' images.

b) Analysing the dataset using different graphs and charts (correlation matrix, etc.). To ensure data quality and dimensions, exploratory data analysis is performed.
c) Further, techniques such as image augmentation, normalizing pixel values, and configuring data generators are performed.

3.3   *Model creation*

The classification process involves partitioning the dataset into testing and training sets. The training set, a subset of the original dataset, is utilized to train the machine learning model, while the testing set evaluates model accuracy with unseen data. For malaria and pneumonia, the dataset is divided into testing, training, and validation sets for 'healthy' and 'infected' categories. The training set is for model fitting, the validation set assesses model performance and hyperparameter tuning, and the test set evaluates the final model on unseen data. A common split ratio is 60% training, 20% validation, and 20% test.

3.4   *Prediction*

The following algorithms were used for the prediction of diseases:

3.4.1   *KNN*
The working of the KNN algorithm is as follows:

a) Choose any number K among the neighbors.
b) Compute the Euclidean distance of the chosen K neighbors, which is calculated as:

$$Euclidean\ Distance = \sqrt{(X2 - X1)^2 + (Y2 - Y1)^2}$$

Pick the K nearest neighbors as per the evaluated Euclidean distance.

a) Count the number of data points in each class among the K neighbors.
b) Finally, allocate the new point to the class having the maximum number of neighbors.

3.4.2   *Logistic regression*
Logistic regression is utilized as a classification algorithm to predict the likelihood of specific classes based on dependent variables. In binary logistic regression, with two categories (e.g., 0 and 1), higher values indicate a greater probability of classification as 1. The sigmoid function is employed to map the linear combination of independent variables to the probability of outcomes falling into respective categories, calculated as:

$$p(y = 1) = 1/(1 + e\hat{}(-z))$$

3.4.3   *Random forest algorithm*
Random Forest works in two stages, first, we combine N decision trees to create a random forest, making subsequent predictions based on it. RF algorithm involves:

a) Selecting K random points from the dataset.
b) A decision tree is constructed for every chosen point.
c) Choose the number 'N' for building the decision trees.
d) Repeat Steps 1 and 2.
e) Recognising the prediction of every decision tree and allocating new data points to the group having the most votes.

3.4.4 *Convolutional Neural Network (CNN):*
In deep learning architecture, CNN also referred to as ConvNet, is a specialized design particularly beneficial for tasks involving pixel input manipulation, such as image recognition. CNNs feature multiple layers aimed at extracting diverse parameters from images, including:

a) Input Layer/First Layer: Receives primary input data, typically images or bulk data.
b) Convolutional Layer: Employs kernels to scan input data, extracting multiple patterns and information.
c) Activation Function: Often uses Rectified Linear Unit (ReLU) after each convolutional layer to introduce non-linearity.
d) Pooling Layer: Decreases spatial dimensions of feature maps from convolutional layers in neural network architectures.
e) Fully Connected Layer:

Conducts high-level reasoning on learned features, typically employed for classification or regression tasks after convolutional and pooling layers. The Figure 1 provides the complete process of prediction of diseases.

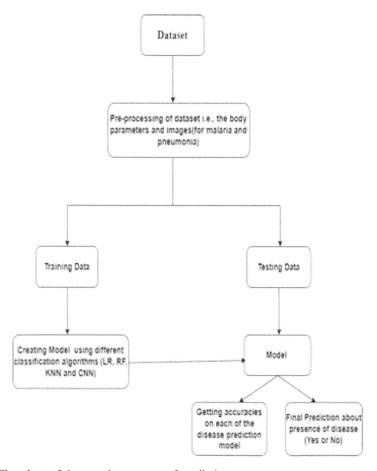

Figure 1.   Flowchart of the complete process of prediction.

4 RESULTS

The following were the accuracies recorded for different diseases along with best algorithms:

- Heart Disease (Logistic Regression) – 88.52%
- Liver Disease (Random Forest Classifier) – 79.66%
- Diabetes (Logistic Regression) – 76.62%
- Kidney Disease (Random Forest Classifier) – 99.00%
- Breast Cancer (Logistic Regression & Random Forest Classifier) – 96.49%
- Malaria Disease (CNN) – 95.65%
- Pneumonia Disease (CNN) – 91.35%

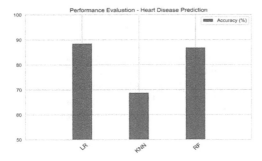

Figure 2.   Accuracy chart for heart disease prediction.

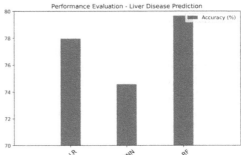

Figure 3.   Accuracy chart for Liver disease prediction.

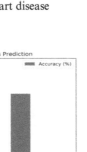

Figure 4.   Accuracy chart for Diabetes disease prediction.

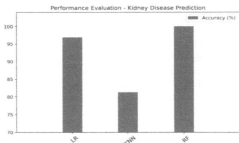

Figure 5.   Accuracy chart for Kidney disease prediction.

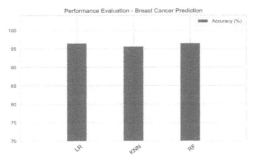

Figure 6.   Accuracy chart for Breast Cancer prediction.

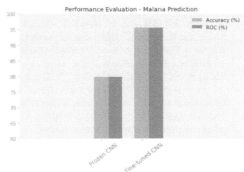

Figure 7.   Accuracy and ROC chart for Malaria prediction.

571

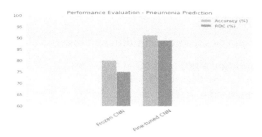

Figure 8. Accuracy and ROC Chart for Pneumonia prediction.

5 CONCLUSION

This research endeavour has led to the creation of a powerful tool capable of simultaneously assessing the likelihood of various diseases, including diabetes, breast cancer, coronary disease, CKD, liver disease, malaria, and pneumonia. Through the utilization of advanced machine learning and deep learning models, this predictor has demonstrated remarkable accuracies in its assessments. An average accuracy of 90% was recorded for the prediction of these diseases.

REFERENCES

1. Gupta R, Mohan I, Narula J. Trends in coronary heart disease epidemiology in India. *Ann Glob Health.* 2016 Mar- Apr;82(2):307–15. doi: 10.1016/j.aogh.2016.04.002. PMID: 27372534.
2. Azamjah N, Soltan-Zadeh Y, Zayeri F. Global trend of breast cancer mortality rate: A 25-Year Study. *Asian Pac J Cancer Prev.* 2019 Jul 1;20(7):2015–2020. doi: 10.31557/APJCP.2019.20.7.2015. PMID: 31350959; PMCID: PMC6745227.
3. Mohan V, Venkatraman JV, Pradeepa R. Epidemiology of cardiovascular disease in type 2 diabetes: the Indian scenario. *J Diabetes Sci Technol.* 2010 Jan 1;4(1):158–70. doi: 10.1177/193229681000400121. PMID: 20167181; PMCID: PMC2825638.
4. Radhakrishnan J, Mohan S. KI reports and world kidney day. *Kidney Int Rep.* 2017 Feb 3;2(2):125–126. doi: 10.1016/j.ekir.2017.01.014. PMID: 29142948; PMCID: PMC5678676.
5. You Won Lee, Jae Woo Choi, Eun-Hee Shin, Machine learning model for predicting malaria using clinical information, *Computers in Biology and Medicine*, Volume 129, 2021, 104151, ISSN 0010-4825, https://doi.org/10.1016/j.compbiomed.2020.104151.
6. K R, Swetha & M., Niranjanamurthy & M P, Amulya & Manu, Y. (2021). Prediction of Pneumonia Using Big Data, *Deep Learning and Machine Learning Techniques*. doi: 10.1109/ICCES51350.2021.9489188.
7. P. Sonar and K. JayaMalini, "Diabetes prediction using different machine learning approaches," 2019 *3rd International Conference on Computing Methodologies and Communication (ICCMC)*, Erode, India, 2019, pp. 367–371, doi: 10.1109/ICCMC.2019.8819841.
8. Singh and R. Kumar, "Heart disease prediction using machine learning algorithms," 2020 *International Conference on Electrical and Electronics Engineering (ICE3)*, Gorakhpur, India, 2020, pp. 452–457, doi: 10.1109/ICE348803.2020.9122958.
9. Sivasangari, B. J. Krishna Reddy, A. Kiran and P. Ajitha, "Diagnosis of liver disease using machine learning models," 2020 *Fourth International Conference on I-SMAC* (IoT in Social, Mobile, Analytics and Cloud) (I-SMAC), Palladam, India, 2020, pp. 627–630, doi: 10.1109/I-SMAC49090.2020.9243375.
10. Padmanaban, K. & Parthiban, G.. (2016). Applying machine learning techniques for predicting the risk of chronic kidney disease. *Indian Journal of Science and Technology.* 9. doi: 10.17485/ijst/2016/v9i29/93880.
11. Al-Obeidi A. S., Azzawi S. F. A novel six-dimensional hyperchaotic system with self-excited attractors and its chaos synchronisation. *International Journal of Computing Science and Mathematics.* 2022;15 (1):72–84. doi: 10.1504/ijcsm.2022.122146.

Advances in AI for Biomedical Instrumentation, Electronics and Computing – Sachan et al. (eds)
© 2024 The Author(s), ISBN 978-1-032-64298-7

Helping hand for handicap (triple h)

Sumit Sharma*, Shagun Kumar*, Shri Bihari Singh*, Satyam Singh* & Divya Sharma*
Department of Electronics and Communication Engineering, Ajay Kumar Garg Engineering College, Uttar Pradesh, India

ABSTRACT: The purpose of this paper is to proposed system which is of low cost that will help to develop the mode of communication with the handicap person. The person who is unable to speak can use the sign language to convey their message to the people. The motion of the fingers is detected by the flex sensors and the motion of the wrist is detected by the accelerometer. This input is fed to the NodeMCU and Arduino and the output will be displayed on the LCD in alphabet or words which can be easily ready by the normal people.

Keywords: Accelerometer, NodeMCU, Arduino, LCD, Flex sensors.

1 INTRODUCTION

India has the largest population in the world. The number of disabled/handicapped people is also quite large. If this article looks around, there are many people with some kind of disability. A survey was conducted from the 2001 census to determine the disabled population in India. Studies have also shown that physical disability is the second most common type of disability among Indians. This reason is enough to start researching this topic (Sharma *et al.* 2014).

The title of this article is "Talking About people with Disabilities" as its name suggests, it helps disabled people send instant messages. In this article a triaxial accelerometer, a device which has 3 axes: A, B, and C, is used. So that higher accuracy can be achieved with smaller movements. The disabled person's movable portion is then covered with this accelerometer. The person will move the area of his body that contains the accelerometer in a specific direction if there is an issue. The accelerometer will detect this change in angle. Thus, the NodeMCU will receive this tilt angle. The accelerometer can see the tilt direction by using this tilt angle (Sharma *et al.* 2014).

Every accelerometer movement has a specific meaning associated with it that is coded using a NodeMCU. E.g., "Some Problem" the message will be forwarded from the NodeMCU transmitter to the receiver through wireless transmission. The receiver will receive the message and then interpret it at the point of contact. This input will be fed to the 8051 microcontrollers for further more action. The microcontroller will be linked to a 16 X 2 LCD screen so that the application may show the message shown, but a buzzer attached to the 8051 microcontrollers also sounds the alarm. This warning is for the caretaker of the disabled person (Vardhan *et al.* 2014).

2 METHODOLOGY

This is a way to implement a prototype of real-time monitoring in a "real-time health monitoring system" using Arduino (Sharma *et al.* 2015). The proposed algorithm uses biosensors

*Corresponding Authors: Sumit11012001@gmail.com, Shagun192001@gmail.com, Singhbihari95@gmail.com, Satyamsingh162002@gmail.com and divya13jan@gmail.com

DOI: 10.1201/9781032644752-104

to estimate body parameters. This article uses the application of gesture recognition for people with physical disabilities to show that it is a communication that replaces the sign language used by speakers as shown in (Figure 1). It is based on hand-to-hand technology. This method consists of hardware and software modules (Shriharipriya *et al.* 2013).

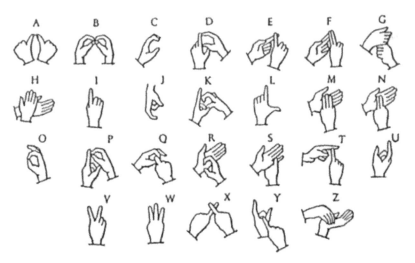

Figure 1.   Sign language to learn Finger spelling Alphabet (Masood *et al.* 2018).

In this system, NodeMCU on the sender and receiver side, establishes a direct communication between patient and doctor or nurses by sending messages based on requests. The important part of the system is the NodeMCU, which uses an RF module to send and receive signals to measure the patient's physical parameters. The transmitter connected an accelerometer to her NodeMCU to detect the patient's movements (Bhilare *et al.*).

2.1   *Helping hand for handicapped (hhh)*

The system described in this document primarily consists of a transmitter and a receiver. Four-axis accelerometers were fastened to the movement area of every patient in the transmitter area. By measuring acceleration caused by gravity, the accelerometer calculates the device's angle with regard to Earth. When the patient need assistance, adjust the accelerometer's tilt in different directions. It is attached to the motherboard, which serves as the control unit, and functions as an input for the accelerometer, whose output is measured in volts. The controller reads the accelerometer's angle-dependent output. The controller converts an input voltage from 0 to 5 volts to an integer value from 0 to 1023 as analog data in the range 0 to 1023. Many of these factors and small changes are sensitive as they can change the price. In order to reduce complexity and provide easy options for patients, sensitivity from 0 to 5 volts is shown in this article, and after many things, sensitivity is reduced by showing front and back. Anyone can easily understand and use this instruction with their thumb or anybody that can move them. As mentioned above, special instructions space is provided to pre-store instructions to suit the basic needs of the patient and those needed for emergencies (Desai *et al.* 2016).

For instance, when the patient turns their wrist to the right, the input is "Food/Water." Thus, the accelerometer's value is communicated to the controller when it is tilted to the right. The predefined message (in this case, food or water) is sent to the next module (RF transmission module) if the value is in the designated direction. RF transmitter is activated when controller sends a message for transmission. The RF transmitter and receiver operate at 434 MHz Each patient is fitted with an accelerometer that has a control panel and a transmitter to send messages (Figure 2). The patient's name or number is sent to the nurse to identify different patients. All transmitters can be centrally connected to a radio frequency receiver operating on the same frequency as the transmitter. Therefore, the proposed system can provide one-to-one communication. The message is received via the RF receiver on the receiving end and sent to the control board, which displays the message on an LCD. After receiving the message, the remote administrator will take the necessary action to meet the needs of the message. In an emergency, the patient simply presses the button, which sends a signal to the control panel, causing the emergency notifications to be sent to the receiver (Figure 3). The signal will be sent to the controller by the receiver (Verma *et al.* 2017).

2.2 *Block Diagram*

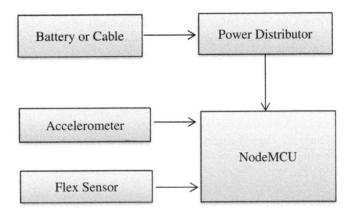

Figure 2.   Block diagram for transmission section.

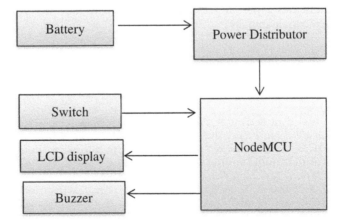

Figure 3.   Block diagram for receiver section.

2.3 *Hardware used in Triple h*

- NodeMCU
- ADXL345 - Triple Axis Linear Accelerometer
- Flex Sensor
- Switch

- LCD Screen
- Buzzer

2.4 Advantages of hhh:

- Communication: This technology provides individuals with disabilities, especially those with limited mobility or speech impairments, with a more effective means of communication. They can convey their thoughts, needs, and emotions through simple gestures or movements, reducing frustration and improving their ability to express themselves ((Bhilare *et al.*).
- Independence and Autonomy: The paper promotes independence by allowing users to control various aspects of their environment. They can send messages, request assistance, operate devices, and even control mobility aids like wheelchairs independently, reducing their reliance on caregivers (Bhilare *et al.*).
- Improved Quality of Life: With better communication and control over their surroundings, individuals with disabilities experience an overall improvement in their quality of life. They can actively participate in daily activities, engage with others, and enjoy a greater sense of agency (Desai *et al.* 2016).
- Customization: The system can innovate to meet the specific needs and preferences of each user. Customization ensures that the paper is adaptable to different levels of ability and can accommodate various types of disabilities effectively (Verma *et al.* 2017).
- Safety Features: Many versions of this paper incorporate safety features like emergency alerts and fall detection. These features provide peace of mind to users and their caregivers, ensuring prompt assistance in case of emergencies (Verma *et al.* 2017).

3 CONCLUSION

Helping hand for handicap paper represents a transformative and empowering solution for individuals with disabilities. By harnessing the power of accelerometer technology and innovative algorithms, this paper addresses critical challenges faced by those with limited mobility or communication abilities

REFERENCES

Bhilare, R., Swami, S., Deshmukh, P., & Patil, M. P. R. Motion based message conveyor for patient using arduino system and zigbee.
Desai, A., Pawar, N., Desai, K., & Behrani, N.,(2016), motion-based message conveyor for paralytic/disabled, *International Journal of Innovative Research in Computer and Communication Engineering*, 4 (3).
Masood, S., Thuwal, H. C., & Srivastava, A. (2018). American sign language character recognition using convolution neural network. In *Smart Computing and Informatics: Proceedings of the First International Conference on SCI 2016, Volume 2* (pp. 403–412). Springer Singapore.
Sharma, R., Gupta, S. K., Suhas, K. K., & Kashyap, G. S. (2014, April). Performance analysis of Zigbee based wireless sensor network for remote patient monitoring. *In 2014 Fourth International Conference on Communication Systems and Network Technologies* (pp. 58–62). IEEE.
Sharma, S., & Vashisth, R. (2015, December). Zigbee based centralised patient monitoring system. In *2015 International Conference on Computational Intelligence and Communication Networks (CICN)* (pp. 212–217). IEEE.
Shriharipriya, K. C., & Arthy, K. (2013). Flex sensor based hand gesture recognition system. *Int J Innovative Res Stud (IJIRS), Vellore, India.*
Vardhan, D. V., & Prasad, P. P. (2014). Hand gesture recognition application for physically disabled people. *International Journal of Science and Research (IJSR)*, 3(8), 765–769.
Verma, P., Kapila, Rathore, N., & Prajapati A., (2017), motion-based message conveyer for paralytic/disabled people. *International Journal for Research in Applied Science & Engineering Technology (IJRASET)*, 5 (4).

Advances in AI for Biomedical Instrumentation, Electronics and Computing – Sachan et al. (eds)
© 2024 The Author(s), ISBN 978-1-032-64298-7

Single-input voltage mode differentiator using DDCCTA and grounded passive elements

Priyanka Jain* & Chandra Shekhar*

ECE Department, Delhi Technological University, Delhi, India

ABSTRACT: The paper presents a new differentiator circuit that operates in voltage mode. It utilizes a Differential Difference Current Conveyor Transconductance Amplifier (DDCCTA) and provides the ability to adjust the time constant and gain by manipulating capacitance and resistance values. The circuit is designed with one DDCCTA and two grounded passive components, making it suitable for integration into an integrated circuit. To assess the circuit's performance, comprehensive analysis is conducted using LT Spice software. Additionally, the presented circuit is realized using 180 nm TSMC CMOS technology, demonstrating its practical applicability in real-world scenarios.

Keywords: Active block (DDCCTA), Voltage mode, and Differentiator.

1 INTRODUCTION

The active RC differentiator circuit finds its most common application in signal conditioning, signal processing, wave shaping, waveform generation, instrumentation system, oscillators and filters (Al-Alaoui 1991; Horrocks 1974; Jiin-Long Lee & Shen-Iuan Liu 1999; Mathur *et al.* 2015; Minaei 2004; Sarker *et al.* 1990). The literature survey shows that there are several numbers of single input and dual input differentiator circuits have been designed employing several types of active blocks. These blocks are operational amplifier (OA) (Al-Alaoui 1991; Horrocks 1974; Sarker *et al.* 1990), current feedback amplifier (CFA)(Jiin-Long Lee & Shen-Iuan Liu 1999; Mathur *et al.* 2015; Nandi *et al.* 2009), combination of operational amplifier and operational transconductance amplifier (OTA) (Shahram Minaei *et al.* 2003), current conveyor (Minaei 2004; Patranabis & Ghosh 1984; Shen-Iuan Liu & Yuh-Shyan Hwang1994), MMCC (Venkateswaran *et al.* 2012) and CCTA (Srisoontorn *et al.* 2022).

By using the op-amp as an active block, the circuit has some disadvantages. The accuracy of operational amplifier can be restricted due to the product of finite gain-bandwidth of op-amp and the operating frequency range of op-amp will be decreased (Jiin-Long Lee & Shen-Iuan Liu 1999; Sarker *et al.* 1990; Shen-Iuan Liu & Yuh-Shyan Hwang 1994). The differentiator circuits which are made using op-amp and CFA use large number of external passive components (Srisoontorn *et al.* 2022). Also, all these circuits require the matching of passive components for the ideal transfer function (Minaei *et al.* 2003).

A novel analog building block having the name DDCCTA was introduced in year 2011 (Pandey & Paul 2011).This block can serve as current-mode, voltage-mode, and mixed-mode components (Pandey & Paul 2011; Phatsornsiri *et al.* 2015; Siripruchyanun 2015). The DDCCTA exhibits significant characteristics for example high input and output impedance that facilitate the designing of circuits which accepts input signals that are both differential and floating.

---

*Corresponding Authors: priyankajain@dtu.ac.in and csr.dtu@gmail.com

DOI: 10.1201/9781032644752-105

A differentiator is an electrical circuit that produces an output signal that represents the rate of change of its input signal over time. The circuit we propose requires only one active block and two passive components, resistor and capacitor. Both passive components are connected to ground. This feature makes the circuit well-suited for implementation in integrated circuits (Minaei 2004).

The subsequent portion of this paper is categorized into separate sections: Section 2 provides basics of DDCCTA. Sections 3 and section 4 present the proposed circuit and simulation results, respectively. Finally, Section 5 presents the conclusion.

2 BASICS OF DDCCTA

The circuit symbol and terminal relationships are given as (Pandey *et al.* 2011):

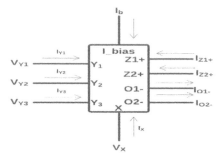

Figure 1.   Circuit symbol of DDCCTA.

And

$$
\begin{bmatrix}
I_{Y1} \\
I_{Y2} \\
I_{Y3} \\
V_X \\
I_{Z1+} \\
I_{Z2+} \\
I_{01-} \\
I_{02-}
\end{bmatrix}
=
\begin{bmatrix}
0 & 0 & 0 & 0 & 0 & 0 & 0 & 0 \\
0 & 0 & 0 & 0 & 0 & 0 & 0 & 0 \\
0 & 0 & 0 & 0 & 0 & 0 & 0 & 0 \\
1 & -1 & 1 & 0 & 0 & 0 & 0 & 0 \\
0 & 0 & 0 & 1 & 0 & 0 & 0 & 0 \\
0 & 0 & 0 & 1 & 0 & 0 & 0 & 0 \\
0 & 0 & 0 & 0 & -g_m & 0 & 0 & 0 \\
0 & 0 & 0 & 0 & -g_m & 0 & 0 & 0
\end{bmatrix}
\times
\begin{bmatrix}
V_{Y1} \\
V_{Y2} \\
V_{Y3} \\
I_X \\
V_{Z1+} \\
V_{Z2+} \\
V_{01} \\
V_{02-}
\end{bmatrix}
\tag{1}
$$

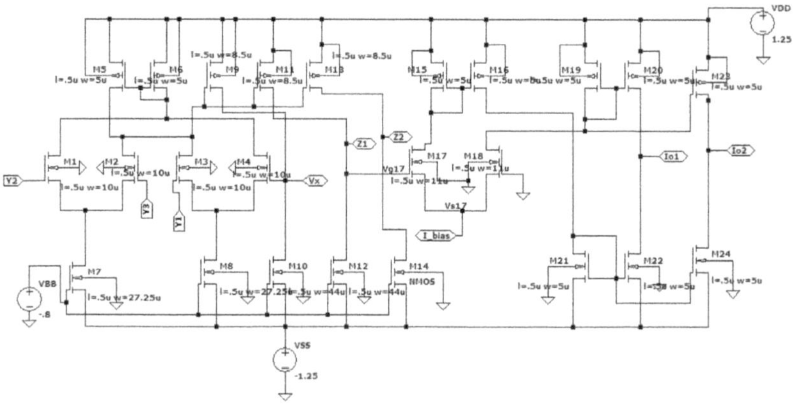

Figure 2.   CMOS implementation of DDCCTA.

578

From the matrix in Equation (1), following relationships of DDCCTA are obtained:

$$V_X = V_{Y1} - V_{Y2} + V_{Y3} \tag{2}$$

$$I_{Z1+} = I_{Z2+} = I_X \tag{3}$$

$$I_{O1-} = I_{O2-} = -g_{m17}V_{Z1+} \tag{4}$$

Aspect ratio of all the transistors is given below:

Table 1. Aspect ratio (W/L) of NMOS and PMOS transistors.

MOSFET	W(μm) / L(μm)
M_{10}, M_{12}, M_{14}	44/0.5
M_7, M_8, M_{17}, M_{18}	27/0.5
M_1-M_4	10/0.5
M_9, M_{11}, M_{13}	8.5/0.5
$M_5, M_6, M_{15}, M_{16}, M_{19}$-$M_{24}$	5/0.5

Figure 1 shows the schematic symbol and equation (1) indicates the terminal relationship of DDCCTA. The CMOS implementation of DDCCTA block is represented in the Figure 2. All NMOS and PMOS transistors operate in saturation region and the substrate terminal of the n-type MOS transistors and the substrate terminal of the p-type MOS transistors are joined to the source terminals. The input terminals Y1, Y2 and Y3 of the DDCCTA device are at high impedance and X terminal is at low impedance and terminals I_{O1}, I_{O2}, Z_1 and Z_2 are defined as the terminals which impedances are high. The voltage at the V_X port is the sum of the voltage at the Y3 port and difference between the voltage at port Y1 and port Y2. The current at terminals Z1 and Z2 are equal to the current at port V_X and the current at the terminals I_{O1-} and I_{O2-} are defined as the multiplication of the voltage drop (V_{Z1}) across Z_1 and the trans-conductance (g_m). The transconductance is an important parameter which can be adjusted with the bias current I_b to achieve the electrical control in a DDCCTA active block. This means that the performance of DDCCTA can be tuned by altering the bias current. Table 1 contains the information about the aspect ratio of all the transistors.

3 PROPOSED DIFFERENTIATOR CIRCUIT

Figure 3 depicts the proposed voltage-mode single-input differentiator. Vin is the input signal applied at Y1 terminal and output is obtained from Z1 terminal respectively. All the unused terminals are grounded. The analysis of the circuit shows:

$$\text{Vout} = RC\frac{d\text{Vin}}{dt} \tag{5}$$

Where R C is the time constant and represented by τ.

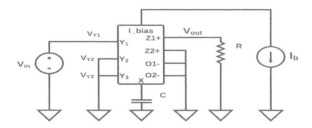

Figure 3. Proposed differentiator circuit.

Due to the high impedance characteristics of terminals Z1 and Z2 in the DDCCTA, the circuits' outputs are also endowed with high output impedances. This feature facilitates the process of cascading the circuits seamlessly.

4 SIMULATION RESULTS

The proposed circuit has undergone verification and simulation using LT spice software. Figure 2 presents the circuit diagram of the DDCCTA block illustrating how the circuit is constructed using CMOS technology. To perform the simulation accurately, the circuit operates on a DC supply voltage of ± 1.25 volts, and the bias current, denoted as I_b, is set to $100\mu A$. The MOS transistors in the circuit are simulated using TSMC 180 nm CMOS technology parameters. Table 1 provides the width to length (W/L) ratio of every NMOS and PMOS transistor.

In a single input differentiator circuit, the resistance value is set to R = 10 KΩ and the capacitance value is set to C = 16.5 nF. When the input waveform is a sine wave having peak amplitude of 100mV and 1 KHz frequency, the waveform obtained at the output is a cosine wave with peak amplitude of 100 mV, as demonstrated in Figure 4.

When an input waveform is a cosine wave having peak amplitude of 100mV and 1 KHz frequency, the resulting waveform is a negative sine wave having the same 100 mV peak amplitude. This can be visualized as Figure 5 shape.

If the input signal is a triangular wave with 100 mV peak amplitude, 1 KHz frequency with R = 5 KΩ and C = 50 nF then the output is a square wave as depicted in Figure 6 and gain and phase of differentiator circuit is represented in Figure 7.

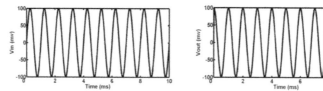

Figure 4. Input (Vin) is a sine wave and output (Vout) of differentiator circuit is a cosine wave as shown in the above figure.

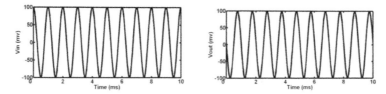

Figure 5. Input is a cosine wave and output of differentiator circuit is a sine wave with 180 degree phase shift as shown in the above figure.

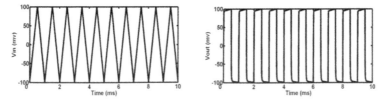

Figure 6. Input as a triangular wave and output is a square wave when R=5KΩ and C=50nF.

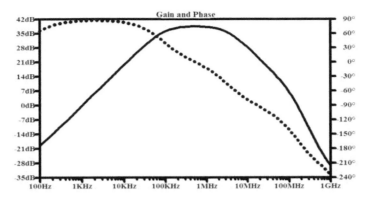

Figure 7.   The gain and the phase of the proposed circuit.

To emphasize the benefits offered by the circuit depicted in Figure 3, we have compiled a comprehensive comparison between proposed circuit and other differentiator circuits found in the literature. This comparison is depicted in Table 2.

Table 2.   The comparison table for the proposed differentiator with the differentiators in the literature.

Reference Name	Active Block Types	No. of Active Elements	No. of Passive Components
Sarker et al. (1990)	Op-Amp	2	6
D. H. Horrocks (1974)	Op-Amp	1	5
M. A. Al-Alaoui (1991)	Op-Amp	2	7
Mathur et al. (2015)	Op-Amp	2	4
Jiin-Long Lee et al. (1999)	Op-Amp	2	6
R. Nandi et al. (2009)	Op-Amp	1	5
Shahram Minaei et al. (2003)	Op-Amp and OTA	4	0
Shen-Iuan Liu et al (1994)	CC-II	2	4
S. Minaei (2004)	DVCC	1	3
S. Srisoontorn et al. (2022)	CCTA	2	2
Proposed Circuit	DDCCTA	1	2

5   CONCLUSION

An implementation of a differentiator circuit using a DDCCTA block, one grounded resistor and capacitor is introduced. This circuit has an advantage that it can be easily integrated onto a single chip due to the exclusive use of grounded passive components. By adjusting the values of capacitance and resistance, the time constant and gain of the circuit can be modified as desired. To validate the functionality of the circuit, simulations are performed using LT spice software with 180 nm CMOS technology.

REFERENCES

Al-Alaoui, M. A. 1991. A novel differential differentiator. *In IEEE Transactions on Instrumentation and Measurement*, vol. 40, no. 5, pp. 826–830, Oct. 1991, doi: 10.1109/19.106305.
Horrocks, D. H. 1974. A non-inverting differentiator using a single operational amplifier. *In International Journal of Electronics*, 37:3, 433–434, 1974, DOI: 10.1080/00207217408900541

Lee, J.-Y. & Tsao, H.-W. 1992. True RC integrators based on current conveyors with tunable time constants using active control and modified loop technique. *In IEEE Transactions on Instrumentation and Measurement*, vol. 41, no. 5, pp. 709–714, Oct. 1992, doi: 10.1109/19.177348.

Liu, Shen-Iuan & Hwang, Yuh-Shyan, 1994. Dual-input differentiators and integrators with tunable time constants using current conveyors. *In IEEE Transactions on Instrumentation and Measurement*, vol. 43, no. 4, pp. 650–654, Aug. 1994, doi: 10.1109/19.310164.

Lee, Jiin-Long & Liu & Shen-Iuan 1999. Dual-input RC integrator and differentiator with tunable time constants using current feedback amplifiers. In *Electronics Letters* Vol 35, Issue 22, 28 October 1999, DOI: 10.1049/el:19991316

Lee, Jiin-Long & Liu & Shen-Iuan 2001. Integrator and differentiator with time constant multiplication using current feedback amplifier. *In Electronics Letters*, 37, 2001, 331–333. 10.1049/el:20010252.

Mathur, K., Venkateswaran, P. & Nandi, R. 2015. A Single Resistor Tunable Grounded Capacitor Dual-Input Differentiator. *In Circuits and Systems*, 6, 2015, 49–54. doi: 10.4236/cs.2015.63005.

Minaei, Shahram , Topcu, Guven & Çiçekoglu, Oguzhan 2003. Active only integrator and differentiator with tunable time constants. *In International Journal of Electronics*, 90:9, 2003, 581–588, DOI: 10.1080/0014184032000159354

Minaei, S. 2004. Dual-input current-mode integrator and differentiator using single DVCC and grounded passive elements. Proceedings of the 12th IEEE Mediterranean Electrotechnical Conference (IEEE Cat. No.04CH37521), Dubrovnik, Croatia, 2004, pp. 123–126 Vol.1, doi: 10.1109/MELCON.2004.1346788.

Nandi, R., Sanyal, S. K. & Bandyopadhyay, T. K. 2009. Single CFA-Based Integrator, Differentiator, Filter, and Sinusoid Oscillator. *In IEEE Transactions on Instrumentation and Measurement*, vol. 58, no. 8, pp. 2557–2564, Aug. 2009, doi: 10.1109/TIM.2009.2014625.

Patranabis, D. & Ghosh, D. 1984. Integrators and differentiators with current conveyors. *In IEEE Transactions on Circuits and Systems*, vol. 31, no. 6, pp. 567–569, June 1984, doi: 10.1109/TCS.1984.1085535.

Pandey, N. & Paul, S. K. 2011. Differential difference current conveyor transconductance amplifier: a new analog building block for signal processing. *In Journal of Electrical and Computer Engineering*, vol. 2011, Article ID 361384: doi:10.1155/2011/361384

Phatsornsiri, P., Lamun, P. & Kumngern, M. 2015. Mixed-mode quadrature oscillator using a single DDCCTA and grounded passive components. In 7th International Conference on Information Technology and Electrical Engineering (ICITEE), Chiang Mai, Thailand, 2015, pp. 500–503, doi: 10.1109/ICITEED.2015.7408998.

Sarker, U. C., Sanyal, S. K. & Nandi, R. 1990. A high-quality dual-input differentiator. *In IEEE Transactions on Instrumentation and Measurement*, vol. 39, no. 5, pp. 726–729, oct 1990, doi: 10.1109/19.58615.

Siripruchyanun, M. 2015. A CMOS electronically controllable current-mode sinusoidal quadrature oscillator using single DDCCTA and grounded passive elements. In *38th International Conference on Telecommunications and Signal Processing (TSP)*, Prague, 2015, Czech Republic, 2015, pp. 1–5, doi: 10.1109/TSP.2015.7296408.

Srisoontorn, S., Charoenmee, A., Panikhom, S., Janda,T., Fungdetch, S., Patimaprakorn, K. & Jantakun , A. 2022. Reconfigurable of current-mode differentiator and integrator based-on current conveyor transconductance amplifiers. In *International Journal of Electrical and Computer Engineering (IJECE)* Vol. 12, No. 1, February 2022, pp. 208–218 ISSN: 2088–8708, DOI: 10.11591/ijece.v12i1.pp208–218

Venkateswaran, P., Nandi R. & Das, S. 2012. New Integrators and Differentiators Using a MMCC. *In Circuits and Systems*, Vol. 3 No. 3, 2012, pp. 288–294. doi: 10.4236/cs.2012.33040.

Advances in AI for Biomedical Instrumentation, Electronics and Computing – Sachan et al. (eds)
© 2024 The Author(s), ISBN 978-1-032-64298-7

Discrete hartley transform using recursive algorithm

Vivek Singh, Dhwani Kaushal & Priyanka Jain
Delhi Technological University, Delhi, India

ABSTRACT: This paper presents a recursive algorithm for the computation of discrete Hartley transform for N = 2 m where m is an integer. Necessary formulas for the computation have been derived along with the realization of recursive structure. DHT is an important transform used to convert data in the time domain to the frequency domain using real values only. With the help of a recursive algorithm, we can compute DHT by using a minimum number of adders and multipliers.

Keywords: recursive structure, Infinite response filter structure (IIR), discrete Hartley transform

1 INTRODUCTION

Discrete Hartley Transform (DHT) is a powerful mathematical tool used for processing discrete data, and its unique capabilities make it suitable for a variety of applications. In contrast to the Discrete Fourier Transform (DFT), DHT has no complex values and involves only real computations. In the Fast Fourier Transform (FFT) method, one complex multiplication is equivalent to four real multiplications; while the computation of the DHT does not involve anything like this (Bracewell *et al.* 1983). Also, the real and imaginary parts of DFT can be obtained from the even and odd parts of DHT, directly.

The Hartley transform can also be used to compute discrete Fourier, discrete cosine, discrete sine, and discrete Hilbert transforms (Nascimento *et al.* 2023). Discrete Hartley Transform combined with the Walsh-Hadamard transform is also used to compute the Fast Walsh-Hadamard-Hartley transform (Mardan *et al.* 2023). Hartley Transform can be used for many applications, like for Visible Light Communication Systems using adaptively biased OFDM (Wu *et al.* 2023). Progressive image secret sharing can be accomplished through the utilization of the DHT, providing different advantages compared to other existing methodologies (Holla and Suma 2023). In applications with the requirement of a power spectrum, straight the DHT can be used, without the need for DFT (Jain *et al.* 2008).

The direct computation of the discrete Hartley transform (DHT) incurs a heavy computational overhead, demanding N^2 computations i.e., multiplications and additions. Researchers havedeveloped many fast algorithms for the computation of DHT, one of them is the fast Hartley Transform developed by Bracewell (Bracewell *et al.* 1984). Hartley transform is better than Fourier since it is two to four times faster than Fourier transform (Hong *et al.* 2023). The FHT algorithm developed by Bracewell causes a significant reduction in the computational complexity of the DHT, making it more practical for real-time applications, but hardware complexity is very high, making its implementation difficult on hardware architecture. Specialized hardware architectures can be designed for computing DHT efficiently, combining the fast algorithms with specialized hardware structures has made DHT a viable alternative to the DFT for various applications that involve signal processing and the computation complexity will be $N^2/2$ for all values of k.

In this paper, a new recursive algorithm for DHT has been presented with its recursive structure. The paper is organized as section 2 discusses the algorithm for DHT, in section 3,

DOI: 10.1201/9781032644752-106

the realization of the algorithm through IIR filter structure is presented, Performance comparison of the suggested structure with the existing work in the literature is done in section 4, and section 5 deals with the conclusion.

2 ALGORITHM FOR RECURSIVE FORMULA FOR DHT

N point Discrete Hartley transform is express as (Liu and Chiu 1993):

$$X[k] = \sum_{n=0}^{N-1} x[n] cas\left(\frac{2\pi nk}{N}\right) \tag{1}$$

where k = 0, 1, 2 ... N-1 and $cas\theta = \sin\theta + \cos\theta$

$$H[k] = \sum_{n=0}^{N/2-1} x[n] cas\left(\frac{2\pi kn}{N}\right) + \sum_{n=N/2}^{N-1} x[n] cas\left(\frac{2\pi kn}{N}\right) \tag{2}$$

where k = 0, 1, 2 ... N-1
Put $n' = n + \frac{N}{2}$ in the second term of the equation (2)

$$H[k] = \sum_{n=0}^{N/2-1} x[n] cas\left(\frac{2\pi kn}{N}\right) + \sum_{n=0}^{N/2-1} x\left[n + \frac{N}{2}\right] cas\left(\frac{2\pi k\left(n + \frac{N}{2}\right)}{N}\right) \tag{3}$$

where k = 0, 1, 2 ... N-1
Solving the right term of equation (3) and expending the cas term, we obtain,

$$\sum_{n=0}^{N/2-1} x\left[n + \frac{N}{2}\right] \left(\cos\left(\frac{2\pi Kn}{N} + \pi K\right) + \sin\left(\pi k + \frac{2\pi kn}{N}\right)\right) \tag{4}$$

where k = 0, 1, 2 ... N-1
Using the following identities in equation (4):
$\sin(A + B) = \sin A \cos B + \cos A \sin B; \cos(A + B) = \cos A \cos B - \sin A \sin B$

$$\sum_{n=0}^{N/2-1} x\left[n + \frac{N}{2}\right] \left(\begin{array}{l} \cos\left(\frac{2\pi kn}{N}\right)\cos(\pi k) - \sin\left(\frac{2\pi kn}{N}\right)\sin(\pi k) \\ +\sin\left(\frac{2\pi kn}{N}\right)\cos(\pi k) + \cos\left(\frac{2\pi kn}{N}\right)\sin(\pi k) \end{array} \right) \tag{5}$$

where k = 0, 1, 2 ... N-1
In equation (5), $\sin(\pi k) = 0; \cos(\pi k) = (-1^k)$; where k = 0, 1, 2 ... N-1
Equation (5) reduces to

$$\sum_{n=0}^{N/2-1} x\left[n + \frac{N}{2}\right] \left((-1)^k \left(cas\left(\frac{2\pi kn}{N}\right)\right)\right) \tag{6}$$

where k = 0, 1, 2 ... N-1
Putting the equation (6) back in equation (3)

$$H[k] = \sum_{n=0}^{N/2-1} x[n] cas\left(\frac{2\pi kn}{N}\right) + \sum_{n=0}^{N/2-1} x\left[n + \frac{N}{2}\right] (-1)^k cas\left(\frac{2\pi kn}{N}\right) \tag{7}$$

where k = 0, 1, 2 ... N-1

Therefore, equation (7) becomes

$$H[k] = \sum_{n=0}^{N/2-1} \left[x(n) + (-1)^k x\left(n + \frac{N}{2}\right) \right] \left(cas\left(\frac{2\pi kn}{N}\right) \right) \qquad (8)$$

where k = 0, 1, 2 ... N-1
   Assume $W_k[n] = x(n) + (-1)^k x\left(n + \frac{N}{2}\right)$

$$X[k] = \sum_{n=0}^{N/2-1} W_k[n] cas\left(\frac{2\pi nk}{N}\right) \qquad (9)$$

where k = 0, 1, 2 ... N-1
   Put $n = \frac{N}{2} - 1 - n$

$$X[k] = \sum_{n=0}^{N/2-1} W_k\left[\frac{N}{2} - 1 - n\right] cas\left(\pi k - \frac{2\pi k}{N}(n+1)\right) \qquad (10)$$

where k = 0, 1, 2 ... N-1 Using trigonometric identities used above in:

$$\sum_{n=0}^{N/2-1} W_k\left[\frac{N}{2} - 1 - n\right] x \left(\begin{array}{l} \cos(\pi k)\cos\dfrac{2\pi k(n+1)}{N} + \sin(\pi k)\sin\dfrac{2\pi k}{N}(n+1) \\ +\sin(\pi k)\cos\dfrac{2\pi k}{N}(n+1) - \cos(\pi k)\sin\dfrac{2\pi k}{N}(n+1) \end{array} \right) \qquad (11)$$

where k = 0, 1, 2 ... N-1
   In equation (11), $\sin(\pi k) = 0, \cos(\pi k) = (-1^k)$ and $cms\theta = \sin\theta - \cos\theta$

$$X[k] = (-1)^k \sum_{n=0}^{N/2-1} W_k\left[\frac{N}{2} - 1 - n\right] cms\left(\frac{2\pi k}{N}(n+1)\right) \qquad (12)$$

where k = 0, 1, 2 ... N-1

$$X[k] = (-1)^k . G_i[k] \qquad (13)$$

where k = 0, 1, 2 ... N-1
   Let $i = N/2 - 1$ and $\theta_k = (2\pi k/N)$

$$G_i[k] = \sum_{n=0}^{i} W_k[i-n] cms((n+1)\theta_k) \qquad (14)$$

where k = 0, 1, 2 ... N-1

$$G_i[k] = \sum_{n=0}^{i} W_k[i-n][\cos((n+1)\theta_k) - \sin((n+1)\theta_k)] \qquad (15)$$

where k = 0, 1, 2 ... N-1
   Using trigonometric identities used above, in equation (15), we get

$$G_i[k] = \sum_{n=0}^{i} W_k[i-n]\left[\begin{array}{l} (2\cos(n\,\theta_k)\cos\theta_k - \cos((n-1)\theta_k)) \\ -(2\sin(n\,\theta_k)\cos\theta_k - \sin((n-1)\theta_k)) \end{array} \right] \qquad (16)$$

585

where k = 0, 1, 2 ... N-1

$$G_i[k] = 2\cos\theta_k \sum_{n=0}^{i} W_k[i-n] \quad cms(n \quad \theta_k) - \sum_{n=0}^{i} W_k[i-n] \quad cms((n-1)\theta_k) \qquad (17)$$

where k = 0, 1, 2 ... N-1

$$G_i[k] = \left[\begin{array}{c} 2\cos\theta_k \left[W_k[i] + \sum_{n=0}^{i-1} W_k[i-1-n]cms((n+1)\theta_k) \right] \\ - \left[\begin{array}{c} W_k[i]cms(-\theta_k) + W_k[i-1].(1) \\ + \sum_{n=0}^{i-2} W_k[i-2-n]cms((n+1)\theta_k) \end{array} \right] \end{array} \right] \qquad (18)$$

where k = 0, 1, 2 ... N-1
From equation (14), we can write

$$G_{i-1}[k] = \sum_{n=0}^{i-1} W_k[i-1-n]cms((n+1)\theta_k) \qquad (19)$$

where k = 0, 1, 2 ... N-1

$$G_{i-2}[k] = \sum_{n=0}^{i-2} W_k[i-2-n]cms((n+1)\theta_k) \qquad (20)$$

where k = 0, 1, 2 ... N-1
Using equation. (19) and equation (20) in equation (21), we obtain a recursive from as:

$$G_i[k] = 2\cos\theta_k[W_k[i] + G_{i-1}[k]] - \left[\begin{array}{cc} W_k[i] & cms(-\theta_k) \\ +W_k[i-1].(1) + & G_{i-2}[k] \end{array} \right] \qquad (21)$$

where k = 0, 1, 2 ... N-1

$$G_i[k] = W_k[i]\cos(\theta_k) + 2 \quad G_{i-1}[k]\cos\theta_k - W_k[i]\sin(\theta_k) - W_k[i-1] - G_{i-2}[k] \qquad (22)$$

where k = 0, 1, 2 ... N-1

3 REALIZATION OF HARDWARE

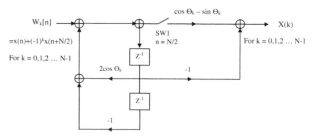

Figure 1. Recursive structure for the computation of DHT.

It can be inferred from the recursive structure that the hardware requirement for the realization of discrete Hartley transform is only 4 adders, 2 delay elements, and 2 multipliers. Another advantage over here is that this same structure can be used to compute DHT of length N, where $N = 2m$ where m is an integer.

4 PERFORMANCE

The proposed algorithm for the computation of DHT has low hardware complexity, because of which a lesser number of multipliers and adders are used in the proposed structure when compared to the Ref. (Chang *et al.* 1992; Chiper *et al.* 2013; Chiu *et al.* 1993; Fang *et al.* 1995; Ray *et al.*; Murty *et al.* 2012). The comparison is depicted in Tables 1 and 2.

Table 1. Comparison of algorithm proposed for computation of DHT in terms of number of adders.

Length (N)	Proposed	(Chang and Lee 1992)	(Chiper 2013)	(Fang and Lee 1995)	(Ray Liu and Chiu 1993)	(Murty 2012)
4	4	–	–	–	–	7
6	4	–	–	–	–	15
8	4	16	16	19	38	–
16	4	33	67	35	78	–
32	4	64	205	67	158	–
64	4	128	553	131	318	–

Table 2. Comparison of algorithm proposed for computation of DHT in terms of number of multipliers.

Length (N)	Proposed	(Chang and Lee 1992)	(Chiper 2013)	(Fang and Lee 1995)	(Ray Liu and Chiu 1993)	(Murty 2012)
4	2	–	–	–	–	6
6	2	–	–	–	–	12
8	2	32	2	18	32	–
16	2	64	12	34	64	–
32	2	128	40	66	128	–
64	2	256	112	130	256	–

5 CONCLUSION

A recursive algorithm and its realization through a recursive filter structure are proposed in the work. The hardware complexity of the structure is less as compared to the other existing structures for DHT, so the suggested algorithm and its structure could be used for any real-life applications of digital processing.

REFERENCES

Bracewell (1983) Discrete hartley transform. *J. Opt. Soc. Am.* 73, 1832–1835.
Bracewell (1984) The fast Hartley transform. *Proceedings of the IEEE* 72: 1010–1018.
Chang, L.W., and Lee, S.W. (1992) Systolic arrays for the discrete Hartley transform. *IEEE Trans. ASP-39*, (ll), pp. 2411–2418.
Chiper (2013) A Novel VLSI DHT Algorithm for a highly modular and parallel architecture. *IEEE Transactions on Circuits and Systems II: Express Briefs*, vol. 60, no. 5, pp. 282–286.

Fang and Lee (1995) Efficient CORDIC-based systolic architectures for the discrete Hartley transform. *IEEE Proc. – Comput. Digit. Tech.*, vol. 142, pp. 201–207.

Hong, Huang, Xiao, Li, Sun, and Zhang (2023) Programmable In-memory computing circuit of fast hartley transform. *ACM Trans. Des. Autom. Electron. Syst.* 28, 6, Article 100.

Jain, Kumar and Bala Jain (2008) Discrete sine transform and its inverse—realization through recursive algorithms. *Int. J. Circ. Theor. Appl.*; 36:441–449.

Liu, Ray, and Chiu, C.-T. (1993) Unified parallel lattice structures for time-recursive discrete wsine/sine/Hartley transforms. IEEE Trans., *ASP*-41, 9 pp. 1357–1377.

Mardan & Hamood (2023) New fast Walsh–Hadamard–Hartley transform algorithm. International *Journal of Electrical and Computer Engineering (IJECE)*.13.1533.10.11591/ijece.v13i2. pp. 1533–1540.

M Holla, Suma (2023) Progressive Hartley image secret sharing for high-quality image recovery. *Cogent Engineering*, 10:2, 2262805.

Murty (2012) Novel recursive algorithm for realization of one-dimensional discrete hartley transform. *International Journal of Research and Reviews in Applied Sciences.*

Nascimento (2023) Hartley transform signal compression and fast power quality measurements for smart grid application. *IEEE Transactions on Power Delivery*, vol. 38, no. 6, pp. 4134–4144.

Wu, Xie, Shi, Zhang, Yao, Liu (2023) An adaptively biased ofdm based on hartley transform for visible light communication systems. *IEICE Transactions on Fundamentals of Electronics, Communications and Computer Sciences*, Article ID 2023EAL2059, Advance online publication.

Advances in AI for Biomedical Instrumentation, Electronics and Computing – Sachan et al. (eds)
© 2024 The Author(s), ISBN 978-1-032-64298-7

Experimental study on variants of Gaussian mixture model for segmentation

Sanjeev Kumar Katti*, Shrinivas D. Desai*, Vishwanath P. Baligar* & Gururaj N. Bhadri*
KLE Technological University, Hubballi, Karnataka, India

ABSTRACT: Gaussian Mixture Models (GMMs) are utilized extensively in many different fields because of their adaptability and capacity to represent complex data distributions. This paper explores the use of Gaussian Mixture Models in image segmentation, offering a strong solution to the difficult task of segmenting different image collections. GMMs are useful for capturing intricate data distributions, and this study thoroughly examines the characteristics and elements of GMMs. The structural details of Gaussian mixture models (GMMs) are explored in detail, emphasizing the flexibility and effectiveness of these models in simulating different application images. It demonstrates how crucial image segmentation is to gaining an extensive understanding of an image's visual characteristics. This study investigates various segmentation strategies that consider the diverse features present in image collections, employing the Expectation-Maximization (EM) and Minorization-Maximization (MM) algorithms. This research goes above and beyond conventional methods to provide a comprehensive understanding of segmentation procedures. The segmentation methodology's methods are thoroughly clarified in the paper, as well as goes into detail about the strategy.

1 INTRODUCTION

In the dynamic landscape of computer vision, Statistical models play a crucial role in the ever-changing field of computer vision, especially when it comes to image segmentation (Ajmal *et al.* 2020) The Gaussian Mixture Model (GMM) is one of these models that stands out as a useful and efficient tool for identifying complex patterns in visual data (Doe *et al.* 2015; Ho *et al.* 2021). An overview of GMM, the optimization algorithms used, and the particular uses in the field of image segmentation are given in this introduction. The Gaussian Mixture Model is a powerful statistical tool designed to handle data set images with complex and multidimensional distributions (Johnson *et al.* 1965; Johnson and Smith *et al.* 2018). The two main techniques, Expectation-Maximization (EM) and Minorization-Maximization (MM) are used to optimize the GMM's parameters. When working with incomplete or hidden data, the iterative structure of the EM algorithm which alternates between computing anticipated values and maximizing the likelihood function makes parameter estimation easier (Krithiga *et al.* 2018; Phan *et al.* 2020). By maximizing a lower bound of the objective function, the closely connected MM method, on the other hand, offers an alternative viewpoint and is particularly helpful in situations when direct maximization of the probability function might provide difficulties. Our investigation focuses on the use of Gaussian Mixture Models for image segmentation, a crucial computer vision job

---

*Corresponding Authors: sanjeevkatti18@gmail.com, sd_desai@kletech.ac.in,
vpbaligar@kletech.ac.in and gnbhadri@kletech.ac.in

DOI: 10.1201/9781032644752-107

(Ratnaparkhi *et al.* 2015). In this study, we explore the complexities of image segmentation using Gaussian Mixture Models, providing insight into the techniques used and the wider implications of our method in improving computer vision systems' capabilities (Smith *et al.* 2017; Song *et al.* 2020).

The paper is meticulously planned. Section 2 discusses the initial research. Section 3 outlines the proposed work. Section 4 presents the results and observations; Section 5 concludes the study and explores the study's future scope.

2 RELATED WORK

In the domain of image segmentation, earlier research has looked into a wide variety of segmentation techniques, including more complex probabilistic models (Tong *et al.* 2017). The employing of Gaussian Mixture Models (GMMs) in image segmentation has attracted a lot of interest from researchers over the past few years. Gaussian Mixture Models (GMMs) are a particularly promising and adaptable alternative to segmentation approaches among the many probabilistic models. GMMs, in contrast to rule-based methods, emphasize a probabilistic framework, enabling a more adaptable and detailed representation of complex data distributions in images (Williams *et al.* 2019; Zhang *et al.* 2019). The summary that follows highlights significant contributions from related works in this field. Parameter tuning is crucial for utilizing GMMs for image segmentation in the best possible way. Prior studies have looked into the use of optimization techniques like expectation maximization (EM). These strategies are crucial for changing GMM parameters, especially when data is missing or hidden. We are utilizing a different optimization technique right now.

3 PROPOSED WORK

The primary objective of this research is to use Gaussian Mixture Models (GMMs) for picture segmentation by utilizing the Minorization-Maximization (MM) and Expectation-Maximization (EM) techniques. This study aims to improve the robustness and decision-making abilities of GMMs by conducting a thorough analysis of the effectiveness of several algorithms in image segmentation. The acquisition contains RGB images of diseased plant leaves as well as high resolution MRI scans of regions affected by brain tumors. Figures 1 and 2 demonstrate the intricacies and difficulties segmentation algorithms encounter in practical situations. Every image is accompanied by ground truth photographs, which provide trustworthy standards for algorithm evaluation.

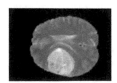 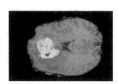  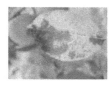

Figure 1.   Brain tumor images (Source: MDPI).   Figure 2.   Leaf disease images (Source: MDPI).

3.1   *System architecture*

The dual-statistical approach in the suggested system architecture utilizes Expectation-Maximization (EM) and Minorization-Maximization (MM) algorithms for efficient data distribution management. These techniques minimize superfluous iterations while maintaining accuracy and efficiency through careful initialization and hyperparameter modifications. The model enhances segmentation by integrating EM and MM algorithms, providing insights into their accuracy and dependability. It uses a hybrid approach, combining statistical techniques with robust assessment components, and includes a study of EM and MM algorithms. Figure 3 presents system architecture.

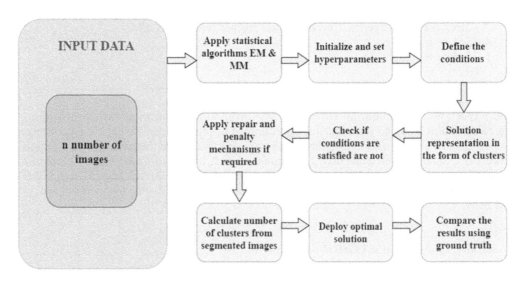

Figure 3. System architecture.

3.2 *Expectation-Maximization (EM) optimization algorithm*

The Expectation-Maximization (EM) algorithm is an iterative process used to optimize the likelihood function in Gaussian Mixture Models (GMMs). This process strengthens the statistical model and yields valuable insights.

Input: $x = [x_1, x_2 \ldots \ldots x_n] \, \epsilon R^{N*D}$ where $x_n \, \epsilon R^D$

Output: p $(z_k = 1| \, x_n)$ and $\mu_k \, \epsilon R^D$, $\Sigma_k \, \epsilon R^{K*K}$, πk ϵR which minimize the objective log likelihood.

Initialize: $\{\mu_k\}$ via the K-means algorithm, $\{\Sigma_k\}$, and $\{\pi_k\}$ uniformly at random and evaluate log-likelihood.

$\ln p(x|\mu, \Sigma_k, \pi) new = \sum_{n=1}^{N} \ln[\sum_{k=1}^{k} \pi_k N(x_n|m_k, \Sigma_k)]$

Repeat: $\ln p(x|\mu, \Sigma_k, \pi) old = \ln p(x|\mu, \Sigma_k, \pi) new$

p $(z_k = 1| \, x_n) = \frac{\pi_k N(x_n|m_k, \Sigma_k)}{\sum_{j=1}^{k} \pi_j N(x_n|m_k, \Sigma_k)}$ (E-Step)

$\mu_k = \frac{1}{N_k} \sum_{n=1}^{N} P(z_k = 1|x_n) x_n,$

$\Sigma_k = \frac{1}{N_k} \sum_{n=1}^{N} P(z_k = 1|x_n)(x_n - \mu_k)(x_n - u_k)^T$ (M-Step)

$\pi_k = \frac{N_k}{N}$

$\ln p(x|\mu, \Sigma_k, \pi) new = \sum_{n=1}^{N} \ln[\sum_{k=1}^{k} \pi_k N(x_n|m_k, \Sigma_k)]$

3.3 *Minorization-Maximization (MM) optimization algorithm*

Input: Define the objective function $f(\theta)$ not necessarily a log-likelihood).
Define the surrogate function $g(\theta|\theta^{(t)}$

Output: The maximum or optimal value of $f(\theta)$ and the corresponding θ.

Initialization: 1. Initialize $\theta^{(0)}$ with some starting values.
2. Set the iteration counter t = 0.

591

Repeat:
1. Minorization Step: Calculate $g(\theta|\theta^{(t)})$ based on the current values of (t). Ensure that $g(\theta|\theta^{(t)})$ satisfies the minorization conditions.
2. Maximization Step: Maximize $g(\theta|\theta^{(t)})$ with respect to its left argument, denoted as $\theta^{(t+1)}$
3. Check for Convergence: Check if a convergence criterion is met.
4. Update t: $t < -t + 1$
5. Update θ

3.4 *Hybrid approach (EM-MM Algorithm)*

The Expectation-Maximization (EM) and Minorization-Maximization (MM) algorithms are combined in a hybrid technique that is examined in this section. Through the combination of these two optimization techniques, the total effectiveness of the parameter estimation process is to be improved while taking advantage of each technique's advantages. Using the gathered images, we ran tests to verify the effectiveness of the hybrid EM-MM technique. Under comparable settings, we tested the performance with standalone algorithms for EM and MM.

4 RESULTS AND DISCUSSIONS

The Gaussian Mixture Model (GMM) is used for segmentation in various image comparisons, including Expectation-Maximization, Minorization-Maximization, and a hybrid strategy, evaluating accuracy against ground truth images. Figure 4 presents results.
  (EM) (MM) (Hybrid) (EM) (MM) (Hybrid)

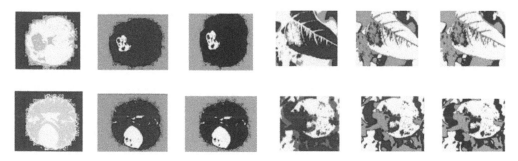

Figure 4.   Segmented results of GMM variants of brain tumor and leaf disease images.

The tabulated results (Table 1), provide a detailed investigation of the performance metrics of our variations of the Gaussian Mixture Model (GMM), specifically used with images of brain tumors and leaves.

Table 1.   Test results for GMM variants on brain tumor and leaf images.

Image	Variant	Accuracy (%)	Time (sec)	Iterations
	EM	**58.15**	0.567	100
Brain Tumor (Image 1)	MM	45.84	0.376	100
	EM-MM	42.26	0.690	100

(continued)

592

Table 1. Continued

Image	Variant	Accuracy (%)	Time (sec)	Iterations
	EM	44.45	3.475	100
Brain Tumor (Image 2)	MM	57.63	2.303	100
	EM-MM	56.21	4.757	100
	EM	75.30	0.677	100
Leaf (Image 1)	MM	67.77	0.465	100
	EM-MM	49.56	1.123	100
	EM	54.68	0.724	100
Leaf (Image 2)	MM	58.98	0.565	100
	EM-MM	52.36	1.129	100

4.1 Analysis of hybrid approach

The Expectation-Maximization (EM) and Minorization-Maximization (MM) hybrid model is problematic when the inherent characteristics of the image data are not aligned with either approach. Mismatched features can compromise the model's correctness. To resolve this, modifications like feature engineering or changing algorithmic parameters are needed. The complexity of the EM and MM algorithms requires careful consideration, and strategies like parameter adjustment and data suitability assessment are advised. Optimizing these factors can improve performance and adaptability in handling various image information scenarios.

5 CONCLUSION AND FUTURE SCOPE

The research explores the interactions between Expectation-Maximization (EM), Minorization-Maximization (MM), and their hybrid model in processing image data. EM performs better for complex and large images, while MM performs better for simple structures. However, the hybrid model, which combines EM and MM, faces difficulties when dealing with mismatched image attributes and complex modelling. Future research should focus on enhancing the hybrid model by resolving detected obstacles, improving alignment with image aspects, and streamlining the EM and MM integration process. Modern techniques for feature extraction and representation learning can further enhance the hybrid model's adaptability to different data distributions.

REFERENCES

Ajmal, M., Aljohani, N. R., Coimbra, M., Hafiz, R., Hassan, A., Nawaz, R., Rehman, S., Riaz, F., & Young, R. (2020). "Gaussian mixture model based probabilistic modelling of images for medical image segmentation." *IEEE Access*, 8, 16846–16856.

Doe, J., & Smith, J. (2015). "Image segmentation using Gaussian mixture models." *IEEE Transactions on Image Processing*, vol. xx, no. xx, pp. xxx–xxx.

Ho, D. J., *et al.* (2021). "The proposed model was sensitive to background noises, potentially leading to mis segmentation in background regions if whole slide images are digitized by other scanners." *Computerized Medical Imaging and Graphics, Journal of Computers*.

Jappelli, R., & Marconi, N. (1997). "Recommendations and prejudices in the realm of foundation engineering in Italy: A historical review." In Carlo Viggiani (ed.), Geotechnical engineering for the preservation of monuments and historical sites; *Proc. intern. symp.*, Napoli, 3-4 October 1996. Rotterdam: Balkema.

Johnson, R., & Smith, K. (2018). "Gaussian mixture models for texture segmentation: a comparative study." *Pattern Recognition Letters*.

Krithiga, A., & Sukanesh, R. (2018). "Gaussian mixture model for brain tumor segmentation in multi-spectral MRI." *IEEE Transactions on Image Processing*.

Phan, T. V., & Andreopoulos, Y. (2020). "Advanced clustering techniques for image segmentation: A Review." *Journal of Imaging*.

Ratnaparkhi, A. A., Pilli, E., & Joshi, R. C. (2015). "Scaling GMM expectation maximization algorithm using bulk synchronous parallel approach." Pages 558–562.

Smith, J., & Garcia, A. (2017). "Gaussian mixture models in image forensics: An overview." *Journal of Digital Forensics, Security and Law*, 22(2), 210–225.

Song, X., Li, M., & Abubakar, A. (2020). "A regularization scheme based on Gaussian mixture model for data inversion." Pages 1–4.

Tong, Y., & Ji, Q. (2017). "Image segmentation using gaussian mixture models with spatial information." *IEEE Transactions on Image Processing*.

Williams, M., & Johnson, S. (2019). "Enhanced medical image segmentation using adaptive gaussian mixture models." *Medical Image Analysis, Journal*.

Zhang, Y., & Tian, Y. (2019). "A review of gaussian mixture model and its applications in image segmentation." *Journal of Computers*, vol. 14, no. 12.

Advances in AI for Biomedical Instrumentation, Electronics and Computing – Sachan et al. (eds)
© 2024 The Author(s), ISBN 978-1-032-64298-7

Real-time-abuse detection model

Ayush Kumar, Aryan Nigam, Aradhana Tripathi, Aftab Khan, Nigam Kumar Mishra &
Rochak Bajpai
*Department of Electronics and Communication Engineering, KIET Group of Institution, Delhi-NCR,
India*

ABSTRACT: The Internet based social media platforms being an open community facilitates users to publish anything, anytime anonymously sometimes; resulting rampant use of
abusive language in general. Identification of abusive language online is an extremely
complex task due to the lack of resources and appropriate approaches. For the past few
years, the rate of increase of cyber-bullying and cyber-abuse has increased rumor mongering.
This paper aims to target those issues using an ML-based model, which helps to clean and
simplify the comments on social media and identify the account and username by which
comments are made public. This work identifies hate speech or abusive content in two languages which are- Hindi and English. Further, the developed tool is used to identify offensive language in different fields resulting in an increase in accuracy. We achieved an accuracy
of 82.4%, by applying Naïve Bayes along with BoW (Bag of Words).

Keywords: Dataset, Algorithm, Machine Learning

1   INTRODUCTION

In the present era of the internet, social media has become a medium to decimate information in rapid manner across the large population. The internet is also found to provide
anonymity to people to present their views on any topic, event, or person without any hes-
itation and that in turn misused by them in the form of trolling, online hate, or threat (Chen
et al. 2017; Dadhe *et al.* 2018; Qureshi *et al.* 2021). This situation of online trolling or hate
spreading can easily be turned out in the form of social unrest as recently observed in the
different parts of world in the form of loot and violence (Ashraf *et al.* 2020; Dinakar *et al.*
2011; Munezero *et al.* 2014).

Detecting hatred related content in social media is a cumbersome but essential task. The
real challenge is the language barrier which we addressed using the machine learning model
where we can easily the hateful speech with the time and username of the account.

As per authors' knowledge, it is the first project of its own kind that can detect the content
of hate speech in either Hindi or English language. The model is trained with a set of words
and different short forms and slang used by gen-z on the internet to express them.

2   METHEDOLOGY

2.1   *Flowchart of proposed model*

The flow chart shown below depicts the entire operation of the model. This flowchart as
shown in Figure 1, is for the dataset classification and training process (Bohra *et al.* 2018;
Caselli *et al.* 2020; Chatzakou *et al.* 2017; Lee *et al.* 2018; Park *et al.* 2017; Sigurbergsson
et al. 2019).

DOI: 10.1201/9781032644752-108

Implementation:

a. Dataset: Raw dataset used for the project includes twitter data containing 3K positive tweets and 4K negative tweets.
b. Data Cleaning: The raw data collected is lemmatized, stemmed, and tokenized using nltk and panda libraries.
c. Vectorization: Converting text data into numerical data. Two types of vectorization techniques are:

i) Bag of words: Bag of words is a text representation that describes the occurrence of a word in a document. It involves two aspects: the content of the known language and the measurement of the existence of the known language.
ii) TF-IDF: TF-IDF stands for Term Frequency-Inverse Document Frequency. The frequency term represents each word of the base as a matrix, whose row is the number of data and whose row is the number of distinct points in the entire information label. Inverse Document Frequency (IDF) is the weight of a term, and its purpose is to reduce the weight of the term if its occurrence is spread across all documents.

1) Training and Testing: The cleaned, tokenized and vectorized data is trained using Naïve Baye's, Random Forest and Logistic Regression.

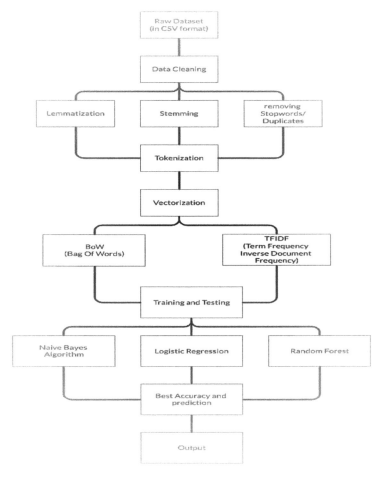

Figure 1.   Flow chart of the model.

2.2 *Working*

First, the raw dataset is collected from various sources including YouTube comments. The raw dataset being in Excel format is now converted to .csv format using the Panda library (a machine learning library). The .csv dataset thus obtained is manually labeled as 0, for non-hate speech and 1, for hate speech. Around three thousand positive tweets and four thousand negative tweets are collected for training purposes (Dadhe *et al.* 2018). The dataset collected is first lemmatized and stemmed using a natural language programming toolkit (nltk) available on python library, after which it is tokenized. Lemmatization and stemming are text normalization procedures that are used to identify different words having the same meaning and remove the end of the shot.

The resulting data is then cleaned by removing stop words. The stop words represent those words, which do not carry much significance. For example, what, why, The, is, I, etc. First, the stop words are removed by using nltk library. Stop words in Hindi and Hinglish (a unique language amalgamated by Hindi and English) are removed by manually giving stop words (Burnap *et al.* 2016; Davidson *et al.* 2017; Kakwani *et al.* 2020; Khanuja *et al.* 2021; Mulki *et al.* 2019; Ranasinghe *et al.* 2021; Swamy *et al.* 2019). The cleaned dataset is then tokenized, using nltk library. Tokenization depicts the process of splitting a string or text into a list of tokens. The tokenized sentences are then vectorized into a corpus containing data in the numerical format by applying two techniques which is, Bag of Words (How) and Term Frequency-Inverse Document Frequency (TF-IDF).

First, create a word bag. For each analysis in the stem, an N-dimensional zero vector is created and the value in the vector is incremented by 1 for words found in the analysis, for example if a word occurs twice the index in the vector becomes: 2. Convert this vector to a sparse vector that uses sparsity and stores only non-zero elements. Now, this bag of words is learned using Naive Bayes, Logistic Regression and Random Forest. TFIDF is an extension of word packet where the total frequency of a word is divided by all words in the document. This penalizes very frequently used words by standardizing them throughout the document. This is used to indicate the stylistic importance of the word. It is also trained using three methods: Naive Bayes, Logistic Regression and Random Forest.

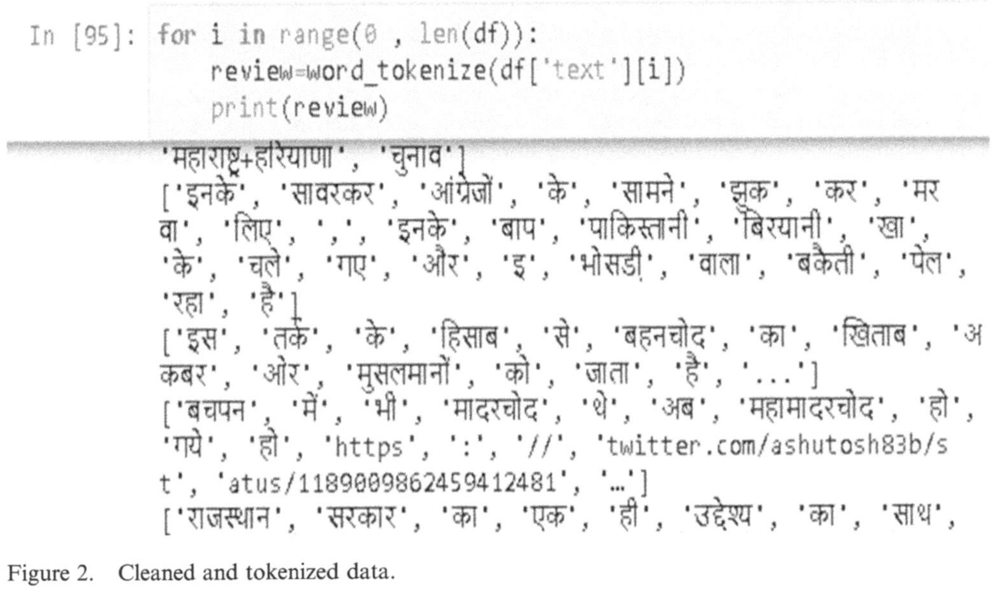

```
In [95]:  for i in range(0 , len(df)):
              review=word_tokenize(df['text'][i])
          print(review)
```

'महाराष्ट्र+हरियाणा', 'चुनाव']
['इनके', 'सावरकर', 'आंग्रेजों', 'के', 'सामने', 'झुक', 'कर', 'मरवा', 'लिए', ',', 'इनके', 'बाप', 'पाकिस्तानी', 'बिरयानी', 'खा', 'के', 'चले', 'गए', 'और', 'इ', 'भोसडी', 'वाला', 'बकैती', 'पेल', 'रहा', 'है']
['इस', 'तक', 'के', 'हिसाब', 'से', 'बहनचोद', 'का', 'खिताब', 'अकबर', 'और', 'मुसलमानों', 'को', 'जाता', 'है', '...']
['बचपन', 'में', 'भी', 'मादरचोद', 'थे', 'अब', 'महामादरचोद', 'हो', 'गये', 'हो', 'https', ':', '//', 'twitter.com/ashutosh83b/st', 'atus/1189009862459412481', '...']
['राजस्थान', 'सरकार', 'का', 'एक', 'ही', 'उद्देश्य', 'का', 'साथ',

Figure 2. Cleaned and tokenized data.

2.3 Evaluation

1) Accuracy Score: For many records, this function calculates the correct number: the set of predicted values for a model must exactly match the corresponding row in Y, which is true.

2) F1 score: F1 score can be defined as the weighted average of accuracy and return; where the best value of the F1 score is 1 and the worst value is 0. The relative contribution of precision and return to F1 score are equal. F1 score = 2 * (precision * recall) / (precision + recall). All the trained data are then compared based on their accuracy and F1 score. Naïve Bayes on Bag of Words generated an accuracy of 82.4% whereas Logistic Regression generated an accuracy of 71.9%. Naïve Bayes on TFIDF obtained an accuracy of 69.6% and Logistic Regression obtained an accuracy of 66.5%.

3 RESULTS

The model has achieved the best accuracy, i.e., 82.4%, by applying Naïve Bayes along with BoW (Bag of Words). The performance of the model is summarized in the form of Confusion Matrix, which basically consists of four metrics:

True Positive(49)
False Positive(43)
False Negative(2)
True Negative(163)

The other performance metrics of an algorithm, as derived from the confusion matrix are

Precision = 53.3%
Recall = 96.07%
F1-Score = 0.69

4 CONCLUSION

In conclusion, this paper presented a reliable labeled dataset and explored abusive content detection on YouTube using supervised machine learning techniques, addressing subjectivity, and enhancing model performance. Future work will concentrate on improving data labeling through active learning and adopting deep learning for automatic feature representation to boost detection accuracy in real-time abuse detection on YouTube.

REFERENCES

Ashraf, N., ArkatizZubaiga and Gelbukh A. (October 2020) Abusive language detection in YouTube in comments leveraging replies to as conversational context.

Bohra, A., Vijay, D., Singh, V., Akhtar, S. S., & Shrivastava, M. (2018, June). A dataset of Hindi-English code-mixed social media text for hate speech detection. In *Proceedings of the Second Workshop on Computational Modeling of People's Opinions, Personality, and Emotions in Social Media* (pp. 36–41).

Burnap, P., & Williams, M. L. (2016). Us and them: identifying cyber hate on Twitter across multiple protected characteristics. *EPJ Data Science, 5*, 1–15.

Caselli, T., Basile, V., Mitrović, J., & Granitzer, M. (2020). Hatebert: Retraining bert for abusive language detection in english. *arXiv preprint arXiv:2010.12472*.

Chatzakou, D., Kourtellis, N., Blackburn, J., De Cristofaro, E., Stringhini, G., & Vakali, A. (2017, June). Mean birds: Detecting aggression and bullying on twitter. In *Proceedings of the 2017 ACM on Web Science Conference* (pp. 13–22).

Chen, H., Mckeever, S., & Delany, S. J. (2017, August). Presenting a labelled dataset for real-time detection of abusive user posts. In *Proceedings of The International Conference on Web Intelligence* (pp. 884–890).

Dadhe, M. S., Masidkar, P. S., Vaidya, V., & Jalan, P. A. (2018). Detection of abusive language from Tweets in social networks. *International Journal on Recent Trends in Computing and Communication*, 6(3), 148–151.

Davidson, T., Warmsley, D., Macy, M., & Weber, I. (2017, May). Automated hate speech detection and the problem of offensive language. In *Proceedings of The International AAAI Conference on Web and Social Media* (Vol. 11, No. 1, pp. 512–515).

Dinakar, K., Reichart, R., & Lieberman, H. (2011). Modeling the detection of textual cyberbullying. In *Proceedings of the International AAAI Conference on Web and Social Media* 5(3), 11–17.

Kakwani, D., Kunchukuttan, A., Golla, S., Gokul, N. C., Bhattacharyya, A., Khapra, M. M., & Kumar, P. (2020, November). IndicNLPSuite: Monolingual corpora, evaluation benchmarks and pre-trained multilingual language models for Indian languages. In *Findings of the Association for Computational Linguistics: EMNLP 2020* (pp. 4948–4961).

Khanuja, S., Bansal, D., Mehtani, S., Khosla, S., Dey, A., Gopalan, B., ... & Talukdar, P. (2021). Muril: Multilingual representations for indian languages. *arXiv preprint arXiv:2103.10730*.

Lee, Y., Yoon, S., & Jung, K. (2018). Comparative studies of detecting abusive language on twitter. *arXiv preprint arXiv:1808.10245*.

Mulki, H., Haddad, H., Ali, C. B., & Alshabani, H. (2019, August). L-hsab: A levantine twitter dataset for hate speech and abusive language. *In Proceedings of The Third Workshop on Abusive Language Online* (pp. 111–118).

Munezero, M., Montero, C. S., Kakkonen, T., Sutinen, E., Mozgovoy, M., & Klyuev, V. (2014). Automatic detection of antisocial behaviour in texts. *Informatica*, 38(1).

Park, J.H., Fung, P.: One-step and Two-step Classification for Abusive Language Detection on Twitter. *CoRR. abs/1706.0*, (2017)

Qureshi, K. A., & Sabih, M. (2021). Un-compromised credibility: Social media based multi-class hate speech classification for text. *IEEE Access*, 9, 109465–109477.

Ranasinghe, T., & Zampieri, M. (2021). An evaluation of multilingual offensive language identification methods for the languages of india. *Information*, 12(8), 306.

Sigurbergsson, G. I., & Derczynski, L. (2019). Offensive language and hate speech detection for Danish. *arXiv preprint arXiv:1908.04531*.

Swamy, S. D., Jamatia, A., & Gambäck, B. (2019, November). Studying generalisability across abusive language detection datasets. In *Proceedings of the 23rd Conference on Computational Natural Language Learning (CoNLL)*, 940–950.

Advances in AI for Biomedical Instrumentation, Electronics and Computing – Sachan et al. (eds)
© 2024 The Author(s), ISBN 978-1-032-64298-7

Comparative analysis of machine learning models for sentiment analysis on X (twitter) dataset

Akshat Singh, Akanksha Singh, Anisha Kumari, Aryan Chauhan & Richa Srivastava
Department of Electronics and Communication, KIET Group of Institutions, Delhi-NCR, Ghaziabad, Uttar Pradesh, India

ABSTRACT: Sentiment analysis, is the process of classifying people's opinions related to certain topic. Social media platforms, particularly X (twitter) serve as deep sources of real-time user-generated content, making them ideal for sentiment analysis applications. This research paper aims to provide an elaborative comparative analysis of different machine learning models for sentiment analysis on X (twitter) data. We have used Natural language Processing (NLP), Recurrent Neural Networks (RNN), Support Vector Machines (SVM), Naïve Bayes, Logistic Regression, and Random Forest. The main goal of this paper to finding out the most accurate and reliable model for sentiment classification.

Our experimental findings reveal that Support Vector Machines (SVM) outperform other models, demonstrating the highest accuracy in sentiment analysis tasks on X (twitter) data. This research provides insights into the best machine learning models for extracting sentiment from dynamic and expressive world of X (twitter) serving as a valuable resource for practitioners and researchers.

1 INTRODUCTION

In past few years, sentiment analysis (sometimes known as opinion mining), has gained recognition in various fields, credit to the vast amount of public-generated content on the digital realm. It involves using programs to understand and review people's opinions and emotions expressed in text. Investigators have conducted extensive studies, surveys, and efforts to improve sentiment analysis and classify online content effectively. While deep learning has made significant strides, there's still a need for effective and understandable machine learning models. Our research target is to fill this gap by evaluating established machine learning algorithms parallel to deep learning methods, providing alternative solutions and insights into sentiment analysis on X (twitter) data.

The motive of our research is to compare various machine learning models for sentiment analysis on X (twitter) data and measure their accuracies.

2 PROPOSED METHODOLOGY

2.1 *Collection and preprocessing of data*

2.1.1 *X (twitter) data collection*
We took great care in putting together our dataset from Kaggle X (twitter), making sure to follow righteous rules and the platform's terms of use. Our dataset is like little subsets of X (twitter)'s feelings, covering all sorts of people talking about them during a specific interval. This mix of different types of X (twitter) users makes sure our analysis is strong and reliable. The data which is not needed was removed from the dataset.

The dataset features are shown in Table 1.

DOI: 10.1201/9781032644752-109

Table 1. The dataset features.

S.no	Column	Rows
1.	sentiment	14640
2.	Text	14640

2.1.2 *Data cleaning and preprocessing*

Data Cleaning include these series of steps:

- Removal of Duplicity: Finding and removing duplicate data points to ensure that each piece of data is unique.
- Missing Data: Dealing with any information that is missing or incomplete.
- Removal of irrelevant data: Excluding irrelevant special characters, HTML tags, or other non-essential information.

The following steps are included in data preprocessing:

- Tokenization: the process of dividing the text into discrete words or units.
- Stop-word Elimination: Taking out terms like "and," "the," and "is".
- Lemmatization or Stemming: Word reduction to their root or basic form.

2.1.3 *Data labelling*

The process of allotting emotions or sentiments (like positive, negative, or neutral) to text data. This labelling helps in training machine learning models to understand and classify sentiments in text. This involves going through the text and determining what emotion or sentiment is expressed in each part of text, making it a mandatory process in sentiment.

2.2 *Feature extraction*

A crucial part of our process is feature extraction, which converts unstructured text input into a structured form that can be used with machine learning. We used two main methods:

2.2.1 *Bag-of-Words (BoW)*

Documents are represented by this method as vectors of word frequencies and occurrences. Every piece of data is represented as a vector, with each dimension denoting a distinct word across the whole corpus. Each text's value indicates how frequently that word appears in the document. Text data is broken down by BOW into a numerical format that machine learning algorithms may use (Kharde *et al.* 2016).

2.2.2 *Words embeddings*

These are numerical representations of words that help machine learning models understand and study the sentiment presented in text data. These embeddings take the semantic meaning of words, enabling the models to work with text more effectively and make sentiment predictions and they play a crucial role in sentiment analysis using techniques like NLTK, NLP, CNN+LSTM, and RNN. These embeddings allow the models to understand and analyse text data effectively. We will be using word2vec for this (Drus *et al.* 2019).

2.3 *Model selection*

We have chosen four different machine learning models for our comparative analysis, each of having peculiar features and suitability for various aspects of sentiment analysis.

2.3.1 *Gaussian Naïve Bayes*

Based on the Bayes theorem, which calculates the probability of an event S given event T, Naïve Bayes is a probabilistic classifier. Put simply, it assists in ascertaining the likelihood that a document is appropriate for text categorization tasks based on the evidence presented by the text inside the document (Qi *et al.* 2023).

$$P(T) = P(T|S) \div (P(S) \times P(T)) \tag{1}$$

2.3.2 *Support Vector Machines (SVM)*

SVM is a potent classification algorithm that seeks to maximise the margin between classes while identifying the hyperplane that best divides data points into distinct classes (Wang *et al.* 2022).

$$f(x) = (w \times x) + b \tag{2}$$

Where f(x) = output of the SVM for a given input x; w = vector of weight; x = input vector; b = The bias, which shifts the hyperplane away from the origin.

2.3.3 *Random forest*

Several decision trees are combined in the Random Forest ensemble learning technique to generate predictions. A random subset of the data is used to train each decision tree, and the predictions from all of them are combined to generate the final outcome. The stability and adaptability of Random Forest to high-dimensional data, such as text, are well-known.

2.3.4 *Logistic regression*

A linear classification algorithm called logistic regression models the likelihood that a given data point will belong to a specific class. In binary and multiclass classification tasks, it is extensively utilised. When utilised with the right features, logistic regression can perform admirably in sentiment analysis despite its simplicity.

2.3.5 *RNN*

Using an RNN is more accurate since we can include data about the information of words. An analysis of sentiment regarding the issues facing every major airline. Contributors were asked to first categorise tweets as positive, negative, or neutral before classifying the reasons behind the negative tweets. This data was taken from Twitter. Our aim is to determine whether a tweet is negative or non-negative that is, positive or neutral.

2.3.6 *NLP*

By using Natural Language Processing (NLP) in sentiment analysis, organizations can quickly and correctly assess public opinion by automating the comprehension of human emotions from textual data. NLP algorithms offer deeper insights into client feedback from several sources by identifying subtleties, context, and sentiment strength (Yuan *et al.* 2015). With the help of this technology, businesses can make better decisions, foster better relationships with customers, and react quickly to shifting public opinion, giving them a competitive advantage in the fast-paced, dynamic digital world of today (Qi *et al.* 2023).

2.4 *Evaluation*

We employed measures of accuracy, precision, recall, and F1-score to assess each model's performance. Based on the confusion matrix, which has four values, they are calculated:

• Correct Positives (CP) : The quantity of Positive tweets that have been accurately categorized.

- Correct Negatives (CN) : The quantity of negative tweets that have been categorized accurately.
- Wrong Positives (WP) : The quantity of tweets that have been incorrectly labeled as positive.
- Wrong Negatives (WN) : The quantity of tweets that were incorrectly labeled as negative.

2.4.1 *Accuracy*

Accuracy quantifies the percentage of correctly classified instances by calculating the overall correctness of predictions (Nistor *et al.* 2021).

$$Accuracy = (CP + CN) \div (CP + WP + CN + WN) \tag{3}$$

2.4.2 *Precision*

Precision measures the percentage of correctly predicted positive instances, or true positive predictions, out of all positive predictions. It gauges how well the model can prevent false positives (Nistor *et al.* 2021).

$$Precision = CP \div (CP + CN) \tag{4}$$

2.4.3 *Recall*

The percentage of true positive predictions among all actual positive instances is measured by recall. It evaluates how well the model can locate all pertinent instances (Nistor *et al.* 2021).

$$Recall = CP \div (CP + WP) \tag{5}$$

2.4.4 *F1-score*

The harmonic mean of recall and precision is known as the F1-score. It offers a fair assessment of a model's effectiveness, especially in cases where there is an uneven distribution of classes (Nistor *et al.* 2021).

$$F1 = [2 \times (Precision \times Recall) \div (Precision + Recall)] \tag{6}$$

3 RESULTS/FINDINGS AND DISCUSSION

3.1 *Model performance metrics*

Table 1 presents a detailed overview of the quantitative data upon which we have performed our operation and performed all the machine learning algorithm. We have tried to make sure that the preprocessing before implementation of each algorithm is stays as similar as possible. Beyond accuracy, these metrics provide a holistic picture of model performance and indicate how well the models classify sentiment.

Table 2. Gaussian naïve bayes.

Metrics	Bag-of-Words	Word Embeddings
Accuracy	50.58%	59.49%
Precision	62.42%	59.14%
Recall	50.59%	59.49%
F1-Score	53.49%	58.72%

Table 3. Support vector machines.

Metrics	Bag-of-Words	Word Embeddings
Accuracy	77.28%	68.34%
Precision	77.05%	68.20%
Recall	77.28%	68.34%
F1-Score	77.15%	59.88%

Table 4.   Random forest.

Metrics	Bag-of-Words	Word Embeddings
Accuracy	76.77%	72.10%
Precision	75.36%	69.69%
Recall	76.77%	72.10%
F1-Score	75.60%	69.46%

Table 5.   Logistic regression.

Metrics	Bag-of-Words	Word Embeddings
Accuracy	79.74%	71.14%
Precision	79.22%	69.61%
Recall	79.74%	71.14%
F1-Score	79.41%	66.30%

Table 6.   Recurrent Neural Networks (RNN).

Metrics	Bag-of-Words	Word Embeddings
Accuracy	64.52%	77.94%
Precision	64.32%	77.52%
Recall	64.52%	77.94%
F1-Score	64.40%	77.70%

Table 7.   Natural Language Processing (NLP).

Metrics	Bag-of-Words	Word Embeddings
Accuracy	77.76%	73.29%
Precision	77.61%	71.30%
Recall	77.76%	73.29%
F1-Score	77.68%	69.74%

3.2   *Comparative analysis of models*

Our study's findings demonstrate how different algorithms function when sentiment analysis is applied to X (twitter) data. However, our analysis extends beyond this numerical superiority. We examine each model's advantages and disadvantages, shedding light on their capabilities and limitations. This analysis is instrumental in assisting practitioners and researchers in selecting the most suitable model for their specific tasks.

3.2.1   *Gaussian Naïve Bayes*

- Strengths: Performs well on high-dimensional datasets and is efficient. And demands a comparatively small amount of training data, Capable of handling both discrete and continuous data.
- Limitations: Assumes conditional independence among features, which might not hold true in real world scenarios. Performance might suffer if the independence assumption is violated.

3.2.2 *Support vector machines*

- Strengths: Demonstrates effectiveness in high-dimensional spaces and robust against overfitting. Displays versatility, allowing the utilization of various kernel functions. shows to be successful in situations where there are more dimensions than samples.
- Limitations: Tends to be intensive on the memory and may exhibit slow performance on larger datasets. The selection of kernel and parameters may require tuning.

3.2.3 *Random forest*

- Strengths: Demonstrates Robustness and performs well across diverse datasets. Handles missing values and maintains accuracy even when a significant portion of data is missing. Mitigates overfitting by averaging predictions from multiple decision trees.
- Limitations: Performance may not be as strong when dealing with highly sparce data. Lack of interpretability compared to individual decision tree.

3.2.4 *Logistic regression*

- Strengths: Simplicity and Interpretability makes it user-friendly. Efficient and fast in the training process and Outputs well-calibrated probabilities. performs exceptionally well when there is roughly a linear relationship between the features and the target.
- Limitations: presumes a linear relationship which isn't always the case between the features and the target's log-odds. May underperform when the relationship is highly non-linear.

3.2.5 *Recurrent neural networks*

- Strengths: Shows effectiveness in handling sequence data and time-series analysis. Maintains memory of past information through hidden states and handle variable input sequences.
- Limitations: Susceptibility to the vanishing gradient problem, especially in long sequences. Limited capacity to capture long-term dependencies in contrast to more sophisticated designs like Gated Recurrent Units (GRUs) and Long Short-Term Memory (LSTM) networks (Asha *et al.* 2023).

3.2.6 *Natural language processing*

- Strengths: utilized for a number of purposes, including machine translation, named entity recognition, sentiment analysis, and others. Capable of processing unstructured text data and extracting meaningful insights.
- Limitations: Faces challenges in effectively handling ambiguity and context inherent in language. Demands a considerable amount of labeled data for tasks involving supervised learning. May encounter difficulties in understanding context and nuances, particularly in complex language scenarios.

4 CONCLUSION

Conclusively, our comparative exploration of machine learning models for sentiment analysis on the X (Twitter) dataset underscores the nuanced strengths and trade-offs inherent in different methodologies. Conventional models such as Naïve Bayes, SVM, Random Forest, Logistic Regression, RNN, NLP. The outcomes, gauged through metrics like accuracy, precision, recall, and F1-score, underscore the absence of a universal solution for Twitter sentiment analysis. Practical considerations, encompassing real-time analysis and interpretability, underscore the need for tailored model selection based on specific contexts and objectives. As we navigate the dynamic social media landscape, continual monitoring and adaptation to evolving language trends become imperative for the sustained pertinence of sentiment analysis models on Twitter.

REFERENCES

Asha, P., Varsini, K. A., Vidhya, S., Reddy, M. S. K., Mayan, J. A., & Grace, L. J. (2023, July). Analysis of twitter sentiments using machine learning algorithms. *In 2023 4th International Conference on Electronics and Sustainable Communication Systems (ICESC)* (pp. 782–786). IEEE.

Drus, Z., & Khalid, H. (2019). Sentiment analysis in social media and its application: Systematic literature review. *Procedia Computer Science*, *161*, 707–714.

Kharde, V., & Sonawane, P. (2016). Sentiment analysis of twitter data: a survey of techniques. *arXiv preprint arXiv:1601.06971*.

Nistor, S. C., Moca, M., Moldovan, D., Oprean, D. B., & Nistor, R. L. (2021). Building a twitter sentiment analysis system with recurrent neural networks. *Sensors*, *21*(7), 2266.

Qi, Y., & Shabrina, Z. (2023). Sentiment analysis using Twitter data: a comparative application of lexicon-and machine-learning-based approach. *Social Network Analysis and Mining*, *13*(1), 31.

Wang, Y., Guo, J., Yuan, C., & Li, B. (2022). Sentiment analysis of Twitter data. *Applied Sciences*, *12*(22), 11775.

Yuan, Y., & Zhou, Y. (2015). Twitter sentiment analysis with recursive neural networks. *CS224D course projects*.

Advances in AI for Biomedical Instrumentation, Electronics and Computing – Sachan et al. (eds)
© 2024 The Author(s), ISBN 978-1-032-64298-7

Design of a 1-bit full adder in hybrid logic for high end computing in biomedical instrumentation

A. Tomar
College of Technology, G.B. Pant University of Agriculture & Technology, Pantnagar, India

V.K. Sachan
Department of Electronics and Communication, KIET Group of Institutions, Delhi-NCR, Ghaziabad, Uttar Pradesh, India

J. Kandpal
Graphic Era Hill University, Dehradun, India

N. Singh, P. Chauhan & S. Bhandari
College of Post Graduate Studies, G.B. Pant University of Agriculture & Technology, Pantnagar, India

ABSTRACT: Biomedical instrumentation focuses on the development of high precision and sophisticated medical instruments for diagnosis, therapeutic and treatment of patients which demand for low latency, precise computation, and minimum power consumption. Full adder is the fundamental cell for all computation circuits used in IoT/biomedical Applications. This paper proposes a hybrid full adder (HFA)-cell design using XOR/ XNOR-cells and multiplexers. XOR/XNOR-cells and multiplexers are implemented with CMOS and Pass Transistor Logic (PTL). The proposed HFA is simulated using a cadence virtuoso environment in a 90 nm CMOS technology with a supply voltage of 1.2V at 1GHz frequency. The proposed HFA consumes 13.83uw power, and delay is 33.2ps which provides 21% to 35% improvement in overall performance. The proposed method is an alternative to high-speed electronic systems with potential benefits.

Theses equipments must have Authors of papers to proceedings have to type these in a form suitable for direct reproduction by the publisher. In order to ensure uniform style throughout the volume, all the papers have to be prepared strictly according to the instructions set below. The enclosed CPI_AR_PDF1.7.joboptions should be used to create the final Camera Ready Copy PDF file. The publisher will reduce the camera-ready copy to 75% and print it in black only. For the convenience of the authors template files for MS Word 6.0 (and higher) are provided.

Keywords: Full-Swing, Hybrid Full Adder (HFA), XOR-XNOR

1 INTRODUCTION

In recent years, due to corona pandemic the role of biomedical has became important in health care sector. Individuals and Medical Hospitals facilities have begun to implement health care technologies in their daily work, simplifying complex processes. This has increased the demand of Biomedical instrumentation with Radio Frequency interface (Pokharel *et al.* 2011a & 2011b) to design advanced systems that can connected patient, doctors, healthcare workers together in scientifically appropriate manner to sense signals and process them for human display and further processing for diagnostic, therapeutic and treatment purpose. These bioinstrumentation systems need low latency in data transmission, large storage capacity, and fast computation power. These systems demand for high speed digital signal processing for self care to minimize the physical involvement of patients as well as healthcare workers (Agarwal *et al.* 2021; Shogar 2016).

Addition and Subtraction are the most basic operations in digital signal processing unit. Full adder is the basic building block in various applications, such as multiplier, address generation in memory access, compressor, and high-speed digital Image processing circuits. Energy efficiency is one of the most required features for modern electronic systems designed for high-performance applications (Chandrakasan *et al.* 1992; Shams *et al.* 2002).

Power consumption of an embedded processor used in computational systems can be divided as Instruction supply containing data path consumes 43%, and rest is consumed by arithmetic and control logic. Adders are one of the key components in arithmetic, data path, and control unit. In a processor, 12% power is used in arithmetic operations are consumed by adder circuitry in it. Full adder plays a significant role in the designing of arithmetic circuits. So, to obtain the best performance of the instrumentation systems high performance full adder is the primary requirement (Goel *et al.* 2006).

1.1 *Full adder*

A full adder is a combinational circuit that performs binary addition of three input bits and produces the output in sum and carry. The logical expression of sum and carry output is denoted by

$$SUM = \overline{A}.\overline{B}.C_{in} + \overline{A}.B.\overline{C_{in}} + A.\overline{B}.\overline{C_{in}} + A.B.C_{in} \tag{1}$$

$$SUM = A \oplus B \oplus C_{in} \tag{2}$$

$$C_{out} = A.B + B.C_{in} + C_{in}.A \tag{3}$$

$$C_{out} = A.B + C_{in}.(A \oplus B) \tag{4}$$

In the hybrid style, a full adder is designed using different topologies in modules (Hasan *et al.* 2020; Kandpal *et al.* 2020, 2021; Naseri *et al.* 2021; Radhakrishnan *et al.* 2001). Figure 1 shows the basic block diagram approach to design a hybrid full adder. The adder circuit is divided into 3 sub-blocks, performing the operation to generate the sum and carry bit. The first block generate XOR-XNOR signal and second and third block generates SUM and carry out signals, respectively using carry in signal.

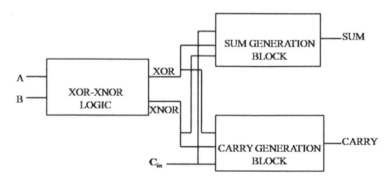

Figure 1.   Block diagram for a hybrid style full adder.

2 LITERATURE WORK

2.1 *FA_Hybrid 1*

This circuit is based on complementary pass transistor logic and feedback path between XOR and XNOR, as shown in Figure 2 (Goel *et al.* 2006). In this, full swing is achieved but unable to generate XOR/XNOR simultaneously. hence, consume more power and high delay.

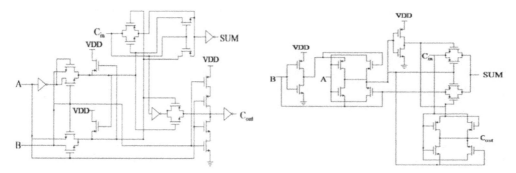

Figure 2.   Hybrid CMOS full adder.   Figure 3.   Hybrid CMOS FA II.

2.2 *Hybrid full adder II*

Figure 3 shows the hybrid style full adder using CMOS and TG topology (Bhattacharya *et al.* 2015) XOR is generated with an inverter and XNOR. Full swing is achieved with level-restoring transistors. The carry signal path is reduced which reduced the propagation delay and power consumption but still not achieved the minimum level needed for portable devices.

In this work, hybrid logic implementation techniques (Kandpal *et al.* 2019) and methods are reviewed and new hybrid logic is proposed, with emphasis in speed improvement and power reduction. Pass transistor logic and level restorer techniques are analyzed to reduce transistor count and improve the circuit's performance in a hybrid modelling. Further, Single supply voltage for all modules has been used.

3 PROPOSED 18-T HYBRID ADDER CELL

A 18-T hybrid full adder cell, which has good characteristics in power and speed, is proposed. A 10-T XOR-XNOR circuit is used to generate XOR and XNOR signals simultaneously is implemented in module I, Sum and Cout output in module 2 and 3, respectively.

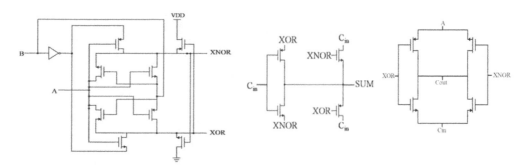

Figure 4.   Shows the schematic of the proposed 18-T full adder with XOR/XNOR, Carry, and Sum Module.

3.1 Module I

Figure 4 shows the 10-T XOR-XNOR cell. The circuit guarantees the full-swing operation by cross-coupled transistors. the circuit has a level restoring transistor, which generates the output with less delay and improves the speed of the module. The simulation results of circuit verify the facts and show the improved level of power consumption and glitches.

3.2 Module II (Sum)

The sum generation module is an XOR cell that performs XOR operation between Cin and $A \oplus B$ signal and produces the output SUM, $A \oplus B \oplus Cin$. Figure 4 shows the schematic diagram of the sum generation circuit. It has 4 transistors (2 PMOS and 2 NMOS) to generate the SUM. As the circuit shows, the path of carry propagation is less, which increases the speed of the circuit.

3.3 Module III (Carry)

The carry generation circuit is a MUX controlled by the XOR and XNOR outputs of module 1 as shown in Figure 4. The advantage of the circuit is the symmetrical structure, which provides less power consumption. In this type of structure, a minimum number of transistors will provide less area, and appropriate sizing of the transistors can reduce the delay and power consumption of the circuit to the minimum extant.

4 RESULT AND DISCUSSION

This full adder circuit is implemented in Complementary CMOS 28-T, CPL, TFA, TGA, Hybrid I, Hybrid II topology for comparison with the proposed Hybrid 18-T full adder cell. All the simulations have been carried out on Cadence virtuoso tools using supply voltage (Vdd) of 1.2V and CMOS 90 nm gpdk technology. Input signal frequency for comparison among previously proposed adder cells regarding power, delay, and power delay product is taken 1GHz.

The average power consumption waveforms of transistors are also included when the cell transitions from one state to another. The worst-case delay, average power consumption, and power delay product of the proposed circuit are shown in Table 1. From the power waveform, it is evident that power in an individual transistor is fluctuating. For uniqueness, the average power value has been measured and reported with the help of a calculator in the spectre tool. Compared to the power consumption of hybrid 1, hybrid II full adder, 43% & 15% of total power is saved, respectively. The total power consumption of the proposed adder is found to be 13.83 μW.

Table 1. Comparison between different adder cells based on performance parameters.

DESIGN	Average Power (μW)	Delay (ps)	PDP (fJ)	Transistor Count
C-CMOS	35.695	106.685	3.808	28
CPL	52.72	136.18	7.17	32
TFA	12.625	45.365	0.5725	16
TGA	19.415	37.94	0.7275	20
14 T	11.48	48.84	0.56	14
HYBRID I	23.885	54.6	1.305	24
HYBRID II	15.66	41.75	0.655	16
PROPOSED FA	13.83	33.2	0.46	18

Similarly, the worst-case delay of the circuit is also calculated with the calculator in the spectre tool, which is the delay to reach 90% of the steady-state value. The proposed circuit shows a 33.2ps worst-case delay and Previously proposed adders like 28-T C-CMOS, CPL, TFA, TGA, 14-T, HYBRID1, and HYBRID 2 shows delay of 106.68ps, 136.18ps, 45.36ps, 37.94ps, 48.84ps, 54.6ps and 41.75ps respectively, which is shown in Table 1.

The power delay product of the proposed 18-T hybrid full adder is improved by about 30% for Hybrid adder 2 and 65% for Hybrid 1. As the power delay product of the proposed full adder is significantly low compared to previous adders, the proposed full adder cell can be used for computation purpose in portable products. The power consumption of the transmission function adder is less, but it suffers from non-full swing output when loading effects are considered.

The proposed adder has only 2 extra transistors compared to other hybrid adders but has less power consumption in the XOR-XNOR circuit, which reduces the overall power consumption. The results obtained are plotted in a histogram for comparison in Figure 5, 6 and 7 for worst-case delay, average power consumption, and power delay products respectively.

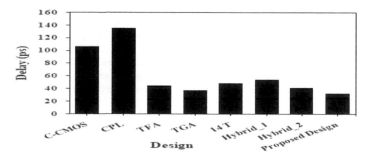

Figure 5.   Worst-case delay comparison of proposed FA with published adders.

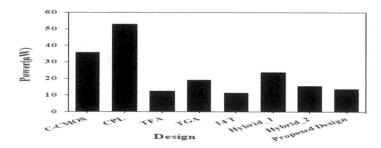

Figure 6.   Power consumption comparison of proposed FA with published adders.

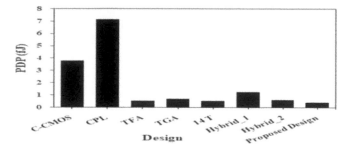

Figure 7.   PDP comparison of proposed FA with published adders.

4 CONCLUSION

This paper presents a highly efficient 18-T hybrid full-swing adder that relies on a novel XOR-XNOR design using only 10 transistors. Our proposed method exhibits superior performance compared to existing methods in terms of power, delay, and PDP. We designed and simulated our circuits using the Cadence Virtuoso software in the ADEXL design suite, ensuring accurate and reliable results using the 90 nm gpdk technology. Our proposed full adder due to low power consumption and high speed is well-suited for digital computation circuit applications used in portable devices used for bioinstrumentation systems which need low latency in data transmission and fast computation. Also it can be used in high speed digital signal processing systems used for self care to minimize the physical involvement of patients as well as healthcare workers.

REFERENCES

Agarwal M., Tomar M., & Kumar N. 2021. An IEEE single-precision arithmetic based beamformer architecture for phased array ultrasound imaging. *Engineering Science and Technology*, 24(5): 1080–89.

Bhattacharya P. *et al.* 2015. Performance Analysis of a Low-Power High-Speed Hybrid 1-bit Full Adder Circuit. *IEEE Transaction Very Large-Scale Integration (VLSI) Syst.* 23(10): 2001–2008.

Chandrkasan P. *et al.* 1992. Low-power CMOS digital design. *IEEE J. Solid-State Circuits.* 27(4): 473–483.

Goel S., Kumar A., & Bayoumi M. 2006. Design of robust, energy-efficient full adders for deep-sub micrometre design using hybrid-CMOS logic style. *IEEE Transaction Very Large Scale Integration (VLSI) Syst.*, 14(12): 1309–1321.

Hasan M. *et al.* 2020. Design of a Scalable Low-Power 1-bit Hybrid Full Adder for Fast Computation. *IEEE Transactions on Circuits and Systems II: Express Briefs.* 67(8): 1464 –1468.

Kandpal J. *et al.* 2019. Design of Low Power and High-Speed XOR/XNOR Circuit using 90 nm CMOS Technology. *2nd Intl Conference on Innovations in Electronics, Signal Processing, and Communication (IESC)*: 221–225.

Kandpal J. *et al.* 2020. High-Speed Hybrid-Logic Full Adder Using High-Performance 10-T XOR–XNOR Cell. *IEEE Transaction Very Large-Scale Integration (VLSI) Syst.* 28(6): 1413–1422.

Kandpal, J., Tomar, A., & Agarwal, M. 2021. Design and implementation of the 20-T hybrid full adder for high-performance arithmetic applications. *Microelectronics Journal*.115: 105205.

Naseri H. & Timarch S. 2018. Low-Power and Fast Full Adder by Exploring New XOR and XNOR Gates *IEEE Transaction Very Large-Scale Integration (VLSI) Syst.* 26(8):1481–1493.

Pokharel R. K. *et al.* 2011a. Digitally controlled ring oscillator using fraction-based series optimization for inductorless reconfigurable all-digital PLL. *IEEE 11th Topical Meeting on Silicon Monolithic Integrated Circuits in RF Systems.* AZ.17-19 January: 69–72. USA.

Pokharel R. K. *et al.* 2011b. 3.6 GHz highly monotonic digitally controlled oscillator for all-digital phase locked loop. *IEEE MTT-S International Microwave Symposium Baltimore*, 05-10 June: 1-4. USA.

Radhakrishnan D. 2001. Low-voltage low-power CMOS full adder. *in Proc. IEEE Circuits, Devices and Systems.* 148(1): 19–24.

Shams A. M., Darwish T. K., & Bayoumi M. A. 2002. Performance analysis of low-power 1-bi CMOS full adder cells. *IEEE Transaction Very Large-Scale Integration (VLSI) Syst.* 10(1): 20–29.

Shogar I. A. 2016. The Philosophical aspects in modern biomedical engineering advancements. *3rd Intl. Conference on Medical Physics & Biomedical Engineering*. Barcelona. 07-08 November. Spain.

Author index

Agarwal, A. 52, 567
Agarwal, C. 473
Agarwal, P. 343
Agarwal, S. 147, 402, 422
Aggarwal, N. 522
Aggarwal, P. 253, 436
Aggarwal, S. 218, 253, 436
Ahlawat, A. 187
Ahlawat, K. 25
Ahmad, S. 52
Ahmed, S.K.F. 1, 7
Akash 158
Alam, M.S. 106
Alfatmi, K. 264, 294, 300, 306
Amrutkar, H. 322
Anand, S. 36
Ankayarkanni, B. 270
Ankit 158
Anushka 158
Anvesh, P.S.D. 63
Arora, C. 511
Arora, S.K. 491
Arya, G. 283
Asha, P. 270
Ashok, R. 380
Awasthi, V. 52
Awate, A. 322
Awate, A.S. 316

Babu, B. 270
Badgujar, K. 264
Bagga, V. 142
Bajaj, R. 121
Bajpai, A. 422
Bajpai, R. 595
Baligar, V.P. 589

Bansal, S. 142
Bari, D.K. 242
Batar, G. 491
Behki, P. 491
Bernatin, D.T. 63
Bewoor, M.S. 288
Bhadauria, R.V.S. 232
Bhadri, G.N. 589
Bhandari, S. 607
Bhardwaj, A. 283
Bhardwaj, N. 116
Bhardwaj, S. 127
Bhargava, G. 534
Bharti, A. 248
Bharti, D.V. 402
Bharti, S. 171
Bhutani, M. 197
Bichu, S. 527
Biradar, D. 527
Bisen, P. 527
Bisht, S.S. 396
Bissa, D. 527
Bittal, V. 294, 300
Biyani, S. 527
Bobade, O. 560
Bodapati, J.D. 1, 7
Bokare, V. 560
Bonde, A. 560
Bora, T. 560
Borade, A. 560
Borade, I. 560
Brijaria, G. 89

Chaudhary, A. 402, 422
Chaudhary, A.K. 242
Chaudhary, H. 121, 182, 187, 218, 497, 567
Chaudhary, K. 411

Chaudhary, S. 116
Chauhan, A. 25, 497, 600
Chauhan, K. 391
Chauhan, P. 607
Chauhan, S. 253
Chaurasia, S. 52
Choudhary, M. 507
Choudhary, S. 511
Chowdary, Y.Y. 1
Christy, A. 270

Dayal, P. 567
Desai, S.D. 589
Dev, S. 209
Dhanabalan, G. 380
Dheivanai, D.R. 253
Dixit, P.K. 116
Dubey, A. 242
Dubey, P.D. 311
Dubey, S. 534

Elizabeth, N.E. 429

Faraz, M. 30
Farshori, M.A. 79

Gangwar, P. 116
Gautam, R. 182, 209, 333, 567
Ghanekar, U. 356
Goswami, S. 142
Goyal, A. 276
Goyal, D.N. 147
Goyal, V. 338
Gudimella, S.S. 356
Gunnam, R.d. 57
Gupta, A. 163
Gupta, D. 242, 497

Gupta, H. 121
Gupta, K. 121, 197
Gupta, M. 89, 197
Gupta, N. 391
Gupta, P. 311
Gupta, R. 353
Gupta, S. 116, 171, 343, 353

Harlan, L. 68
Harmain, Mohd. 507
Harshita 436
Heeravathi, S. 177
Hegde, G. 73

Ibraheem 443

Jain, A. 182
Jain, G. 163
Jain, P. 19, 522, 577, 583
Jain, R. 209
Jain, S. 25, 497
Jaiswal, H. 213
Jalil, M.F. 192, 516
Jarsodiwala, G. 306
Jeyapoornima, B. 42
Jha, R. 511
Jha, R.K. 36
Jha, S.K. 209
Jhariya, D.K. 374, 416
Jindal, S. 491
Jino Ramson, S.R. 13
Joselin Jeya Sheela, J. 42
Joseph, J. 13
Joshi, A. 407

Kachhava, C. 327
Kaif, M. 30
Kalra, N. 121
Kalyani, V.L. 47
Kandpal, J. 607
Kansal, H. 133
Kansal, S. 338
Kanungo, A. 248, 411
Karbhari, D. 306
Karn, A. 473
Kaur, M. 554

Kashyap, I. 276
Kashyap, M. 89
Kashyap, R. 396
Katti, S.K. 589
Kaur, A. 522
Kaur, K. 502
Kaur, M. 549
Kaushal, D. 583
Kaushik, A. 182
Kesarwani, S. 348
Khalid, A. 385
Khalid, M. 106
Khalkar, R.G. 288
Khan, A. 595
Khan, B. 527
Khan, M. 362
Khatoon, S. 192, 443, 450, 516
Kirubaraj, A.A. 13
Kulkarni, M. 264
Kumar, A. 1, 84, 99, 171, 248, 353, 416, 554, 595
Kumar, B. 25
Kumar, C. 163
Kumar, G. 36
Kumar, H. 411
Kumar, K. 356
Kumar, M. 209
Kumar, P. 121, 163, 209, 213, 218, 228, 497, 534, 538, 567
Kumar, R. 507
Kumar, S. 68, 137, 416, 573
Kumar, V. 116, 534, 538
Kumari, A. 600
Kumari, M. 473
Kumari, S. 209

Lal, R. 549
Latha, P. 480
Lokhande, P. 294

Madeira, L.P. 259
Mali, D.B. 316

Mandawkar, U. 327
Mandwekar, S. 327
Mary, S.P. 270
Matta, G. 182
Maurya, D. 153
Medhane, S.P. 288
Mehta, N.S. 84
Mishra, A. 466
Mishra, D. 455
Mishra, K. 237
Mishra, N.K. 595
Mishra, S. 338, 348
Mitra, H. 391

Naaz, A. 516
Naaz, R. 204
Nafees, I. 218
Nagpal, H. 68
Naik, A. 407
Nain, A. 502
Nandeshwar, V. 527, 560
Nandwalkar, B. 322
Nandwalkar, B.N. 316
Nasim, F. 443
Nazir, A. 204
Nazir, N. 204
Neha187
Nigam, A. 595
Nigam, P. 228
Nizamuddin, M. 79

Pahuja, H. 137, 222
Pahwa, D.R. 253
Pahwa, P. 171
Pahwa, R. 436
Pande, S. 264
Pandey, A. 137, 248, 549
Pandey, V. 333
Pandey, V.S. 543
Pangtey, M.S. 36
Parveen, S. 362
Patel, K. 147
Pathak, Y. 374
Patil, H. 294

Patil, H.R. 316
Patil, H.V. 316
Patil, K. 300
Patil, M. 300
Patil, V. 264, 322
Patil, Y. 327
Pingale, S. 300
Pokhriyal, A. 338
Pote, D. 486
Potnis, H. 486
Powar, P. 486
Powar, S. 486
Prabhu, C. 486
Prakash, A. 19
Prasanna, N.L. 57
Pratihasta, S. 338
Praveena, M.D.A. 270
Priyanshu 411
Pundhir, R. 391
Pushpender 192
Pushpender 516

Qureshi, S. 204

Ragul, R. 177
Rajas, N. 486
Rajan, H.B.M. 380
Rajbhar, D. 549
Rajendran, V. 455
Rajput, Y. 95
Rama, S.T. 177
Rana, R. 511
Ranjan, S. 158
Rao, G. 288
Rao, T.K. 57
Rawat, S.S. 95
Relan, R. 306
Renuka, P. 480

Sachan, V.K. 607
Sachdeva, S. 222
Saihjpal, A. 522
Saini, A. 30, 497
Sajja, R. 7
Sandhu, J. 534

Saraf, K. 228
Saraswat, G. 333
Satyaramamanohar A., B. 63
Saxena, S. 95, 507
Senith, S. 13
Shadaab, M. 450
Shahade, D.M. 264
Shahade, M. 327
Shahi, A.P. 99
Shahid, M. 443, 450
Shahid 158
Shanmugasundaram 455
Shanmugasundaram, N. 177
Shahade, M. 294, 300, 306, 322
Shahade, M.R. 316
Shantanu 127
Sharma, A. 1, 121, 182, 209, 213, 218, 333, 497, 538, 567
Sharma, D. 228, 567, 573
Sharma, D.V.K. 47
Sharma, N. 385, 396
Sharma, S. 89, 213, 538, 573
Sharma, S.P. 385, 396
Sharma, U. 153
Sheeba, R.G. 429
Shekhar, C. 577
Shekhar, S. 348
Sherwani, K. 362
Shivhare, K. 422
Shukla, A. 147
Shukla, D.S. 111
Shukla, S. 89
Siddireddy, K. 73
Sindhwani, M. 222
Singh, A. 84, 133, 137, 333, 348, 549, 600
Singh, A.R. 84
Singh, D. 473
Singh, D.P. 133
Singh, G. 68

Singh, H. 507
Singh, K.P. 213, 538
Singh, L. 385, 396, 534
Singh, M. 89, 116, 283
Singh, N. 213, 607
Singh, P.K. 473
Singh, P.P. 343
Singh, S. 218, 338, 511, 573
Singh, S.B. 573
Singh, S.P. 127, 142, 348, 391
Singh, V. 583
Singh, Y. 353
Singha, N.S. 1
Singhal, D. 242
Sinha, S. 538
Somwanshi, K. 306
Sood, S. 543
Srivastava, N.R. 348
Srivastava, D. 218
Srivastava, M. 461
Srivastava, N. 121
Srivastava, P. 133
Srivastava, R. 133, 158, 600
Srivastava, S. 538
Srivastava, S.K. 95
Srivastava, T. 497
Srivastava, Y. 213
Subhash, A.S. 248
Sujithkumar, T.P. 455
Sumitra, V. 480
Sumrao, S. 402
Suri, D.A. 232, 237
Suryawanshi, V. 264
Swaminathan, K. 480

Tamrakar, B. 89, 106
Tayal, K.K. 232
Tiwari, D.R. 111
Tiwari, L. 534
Tiwari, R. 385
Tiwari, S. 36
Tiwari, V. 491
Tomar, A. 607

Tomar, J. 411
Tripathi, A. 99, 595
Tripathi, J. 25, 30
Tripathi, P.K. 466
Tripathi, P.M. 311
Tripathi, S. 538
Tyagi, H. 116
Tyagi, M. 127, 142, 348, 391
Tyagi, R. 353
Tyagi, S. 127, 142, 348, 391
Tyagi, U. 106
Tyagi, Y. 95

Verma, A. 473
Verma, K. 228
Verma, R. 137
Verma, S. 106
Verma, V.K. 242, 311
Verma, Y. 142
Vibhandik, A. 306
Vinay 127
Vishwakarma, V. 311
Vispute, M. 322

Wadekar, S. 327
Wagh, I. 294
Waghare, H.R. 316

Yadav, A. 333, 407
Yadav, A.K. 84
Yadav, H. 411
Yadav, N. 311
Yadav, R. 1, 228, 543
Yadav, S. 111
Yadava, R.L. 99
Yallamanda, R.B. 57
Yenduri, L.K. 1, 7
Yeola, K. 322
Yeshi, P. 294

Zaidi, M.S. 106

9 781032 642987